Studies in Logic
Mathematical Logic and Foundations
Volume 75

Theory of
Effective Propositional
Paraconsistent Logics

Volume 65
Elementary Logic with Applications. A Procedural Perspective for Computer Scientists
D. M. Gabbay and O. T. Rodrigues

Volume 66
Logical Consequences. Theory and Applications: An Introduction.
Luis M. Augusto

Volume 67
Many-Valued Logics: A Mathematical and Computational Introduction
Luis M. Augusto

Volume 68
Argument Technologies: Theory, Analysis, and Applications
Floris Bex, Floriana Grasso, Nancy Green, Fabio Paglieri and Chris Reed, eds

Volume 69
Logic and Conditional Probability. A Synthesis
Philip Calabrese

Volume 70
Proceedings of the International Conference. Philosophy, Mathematics, Linguistics: Aspects of Interaction, 2012 (PhML-2012)
Oleg Prosorov, ed.

Volume 71
Fathoming Formal Logic: Volume I. Theory and Decision Procedures for Propositional Logic
Odysseus Makridis

Volume 72
Fathoming Formal Logic: Volume II. Semantics and Proof Theory for Predicate Logic
Odysseus Makridis

Volume 73
Measuring Inconsistency in Information
John Grant and Maria Vanina Mrtinez, eds.

Volume 74
Dictionary of Argumentation. An Introduction to Argumentation Studies
Christian Plantin. With a Foreword by J. Anthony Blair

Volume 75
Theory of Effective Propositional Paraconsistent Logics
Arnon Avron, Ofer Arieli and Anna Zamansky

Studies in Logic Series Editor
Dov Gabbay   dov.gabbay@kcl.ac.uk

Theory of Effective Propositional Paraconsistent Logics

Arnon Avron
Ofer Arieli
and
Anna Zamansky

© Individual author and College Publications, 2018
All rights reserved.

ISBN 978-1-84890-270-1

College Publications
Scientific Director: Dov Gabbay
Managing Director: Jane Spurr

http://www.collegepublications.co.uk

Printed by Lightning Source, Milton Keynes, UK

All rights reserved. No part of this publication may be reproduced, stored in a retrieval system or transmitted in any form, or by any means, electronic, mechanical, photocopying, recording or otherwise without prior permission, in writing, from the publisher.

To Carolina Blasio, who was consistently fascinated by paraconsistency.

Contents

ReadMe: An Essential Preface   xv

I   Reasoning With Inconsistency   1

1   Propositional Logics: Preliminaries   3
- 1.1   General Background 3
  - 1.1.1   What is a Propositional Logic? 3
  - 1.1.2   Some Basic Notions 6
  - 1.1.3   Basic Connectives 9
- 1.2   Semantic Approaches to Defining Logics 13
  - 1.2.1   Satisfiability Relations 13
  - 1.2.2   Classical Logic 14
  - 1.2.3   Intuitionistic Logic 20
- 1.3   Syntactic Approaches to Defining Logics 23
  - 1.3.1   Proof Systems 23
  - 1.3.2   Hilbert-type Systems 25
  - 1.3.3   Ordinary Gentzen-type Systems 31
  - 1.3.4   Calculi of Hypersequents 44
- 1.4   Bibliographical Notes and Further Reading 46

2   Negation and Paraconsistency   47
- 2.1   General Considerations 47
  - 2.1.1   What is a Negation? 47
  - 2.1.2   What is a Paraconsistent Logic? 49
- 2.2   Maximality Properties 53
- 2.3   Proof Theory of Paraconsistent Logics 57
- 2.4   Summary – Desiderata List 59
- 2.5   Bibliographical Notes and Further Reading 60

II Multi-Valued Truth-Functionality   61

3 Paraconsistent Matrices   63
3.1 Preliminaries 63
3.2 Paraconsistent Matrices and Their Logics 72
    3.2.1 Matrices with Negation 72
    3.2.2 Paraconsistent Matrices 75
    3.2.3 Maximality of Paraconsistent Many-Valued Logics . . 77
3.3 Bibliographical Notes and Further Reading 82

4 Three-Valued Matrices   83
4.1 An Introduction 83
4.2 Paraconsistent Three-Valued Matrices 87
4.3 Properties of Paraconsistent 3-Valued Logics 94
4.4 Central Paraconsistent Three-Valued Logics 98
    4.4.1 The Logics $\mathbf{P_1}$, $\mathbf{P_1^K}$, and $\mathbf{P_1^S}$ 99
    4.4.2 The Logic $\mathbf{SRM_{\rightarrow}}$ 102
    4.4.3 The Logic $\mathbf{LP}$ and Its Main Monotonic Extensions . . 105
    4.4.4 The Logics $\mathbf{PAC}$ ($\mathbf{RM_3}$) and Its Main Extensions . . 110
    4.4.5 The Logic $\mathbf{PAC_{\supset}}$ 114
    4.4.6 The Logic $\mathbf{PE3}$ 116
4.5 Proof Systems 121
    4.5.1 Gentzen-type Systems 121
    4.5.2 Hilbert-type Systems 128
4.6 Bibliographical Notes and Further Reading 133

5 Paradefinite Matrices   135
5.1 The Four-Valued Paradefinite Framework 135
5.2 Dunn-Belnap's Matrix $\mathcal{FOUR}$ 138
    5.2.1 Motivation 138
    5.2.2 Other Useful Connectives 140
    5.2.3 Properties of $\mathbf{L}_{\mathcal{FOUR}}$ 142
5.3 Important Extensions of $\mathcal{FOUR}$ 146
    5.3.1 A Maximal Extension 147
    5.3.2 A Maximal Monotonic Extension 149
    5.3.3 A Maximal Classical Paraconsistent Extension 151
    5.3.4 A Maximal Non-Exploding Extension 153
    5.3.5 A Maximal Flexible Extension 154
    5.3.6 The Classical Extension 156
5.4 Proof Systems 158
    5.4.1 Gentzen-type Systems 158
    5.4.2 Hilbert-type Systems 161

	5.5	Bilattices and Their Logics 164
		5.5.1 Bilattice-Valued Matrices 164
		5.5.2 Characterization in $\mathcal{FOUR}$ 167
	5.6	Bibliographical Notes and Further Reading 172

III Non-Determinism   173

6 Non-Deterministic Matrices   175
  6.1 Motivation 175
  6.2 Basic Theory of Nmatrices 176
  6.3 Operations on Nmatrices 181
  6.4 Paraconsistent Nmatrices 184
  6.5 Maximality and Pre-Maximality 185
  6.6 Bibliographical Notes and Further Reading 186

7 Using the Basic Four Truth-Values   187
  7.1 ¬-Fundamental Nmatrices 187
  7.2 Bivalence 192
  7.3 Maximality and Pre-maximality 196
  7.4 Minimality and Modularity 202
  7.5 Processing Information from a Set of Sources 219
      7.5.1 The logic **EIP** 220
      7.5.2 A Gentzen-type System for **EIP** 226
  7.6 Bibliographical Notes and Further Reading 232

8 Logics of Formal (In)consistency   233
  8.1 Introduction 233
  8.2 The Basic C-systems 239
  8.3 Propagation of Consistency 249
  8.4 da Costa's Consistency Operator(s) 263
      8.4.1 The Strict Case 263
      8.4.2 The Non-Strict Case 285
  8.5 Bibliographical Notes and Further Reading 290

IV Possible Worlds   291

9 Negation as a Modal Operator   293
  9.1 Replacement in Paraconsistent Logics 293
  9.2 The Modal Logic **B** as a Paraconsistent Logic 294
      9.2.1 The Logic **NB** and Its Proof Systems 295
      9.2.2 Semantics of **NB** 298

	9.2.3	Basic Properties of **NB** 304
	9.2.4	**NB** as an LFI 305
	9.2.5	**NB** as a Version of the Modal Logic **B** 309
9.3	**S5** as a Paraconsistent Logic 310
	9.3.1	The Logic **NS5** 310
	9.3.2	A Cut-free Hypersequential System for **NS5** 314
9.4	A Fully Maximal Extension of **NS5** 319
9.5	Using Other Standard Modal Logics 326
	9.5.1	**K** as a Paraconsistent Logic 327
	9.5.2	**T** as a Paraconsistent Logic 328
	9.5.3	**S4** as a Paraconsistent Logic 329
9.6	Bibliographical Notes and Further Reading 330

10 Constructivity and Intuitionism   331
10.1 The Semantic Framework 333
10.2 Analyticity, Conservativity, and Decidability 341
10.3 **N4** and Other Important Systems 347
10.4 Gentzen-type Systems 352
    10.4.1 The N_4 family 352
    10.4.2 The Constructive Logics with Excluded Middle 358
10.5 Bibliographical Notes and Further Reading 360

V   Relevance   361

11 What is a Relevant Logic?   363
11.1 General Considerations 364
    11.1.1 The Basic Relevance Criterion 364
    11.1.2 A Plausible Semantic Criterion 366
11.2 Relevant Entailments 368
11.3 Basic Relevant Connectives 371
    11.3.1 Weak Relevant Implication 371
    11.3.2 (Strong) Relevant Implication 376
    11.3.3 Relevant Conjunction 382
    11.3.4 Relevant Negation 385
11.4 Relevant Logics 390
11.5 Bibliographical Notes and Further Reading 394

12 The Minimal Relevant Logic   395
12.1 The Logic $\mathbf{R}_{\to}$ and Its Proof Systems 395
12.2 Minimality and Other Properties of $\mathbf{R}_{\to}$ 399
12.3 Advantages and Drawbacks of $\mathbf{R}_{\to}$ 405

| CONTENTS | xi |

12.4 Bibliographical Notes and Further Reading 406

13 Two Maximal Relevant Logics — 407
13.1 A Maximal Strong Relevant Logic 407
    13.1.1 The Logic $\mathbf{RMI}_\to^\neg$ and Its Proof Systems 407
    13.1.2 Weakly Characteristic Matrix for $\mathbf{RMI}_\to^\neg$ 410
    13.1.3 Maximal Strong Relevance of $\mathbf{RMI}_\to^\neg$ 415
    13.1.4 Axiomatic Simple Extensions of $\mathbf{RMI}_\to^\neg$ 418
    13.1.5 Expressive Power of the Language of $\mathbf{RMI}_\to^\neg$ 420
    13.1.6 Strongly Characteristic Matrix for $\mathbf{RMI}_\to^\neg$ 424
    13.1.7 More General Useful Semantics for $\mathbf{RMI}_\to^\neg$ 428
13.2 A Maximal Normal Relevant Logic 432
    13.2.1 The Normal Simple Extensions of $\mathbf{RMI}_\to^\neg$ 432
    13.2.2 Properties of $\mathbf{L}_{\mathcal{A}_\omega}$ 434
    13.2.3 Gentzen-type System for $\mathbf{L}_{\mathcal{A}_\omega}$ 439
13.3 Bibliographical Notes and Further Reading 446

14 Relevance in Richer Languages — 447
14.1 Adding Propositional Constants 447
14.2 Anderson and Belnap's $\mathbf{R}$ 451
    14.2.1 The Construction of $\mathbf{R}$ 451
    14.2.2 Some Useful Facts about $\mathbf{R}$ 454
    14.2.3 Main Properties of $\mathbf{R}$ 456
14.3 $\mathbf{RMI}$ — A Purely Relevant Logic 458
    14.3.1 Relevant Conjunction and $\mathbf{RMI}_{min}$ 458
    14.3.2 The Logic $\mathbf{RMI}$ and Its Semantics 464
    14.3.3 Main Properties of $\mathbf{RMI}$ 469
    14.3.4 Cut-free Gentzen-type System for $\mathbf{RMI}$ 474
14.4 Bibliographical Notes and Further Reading 476

15 A Maximal Semi-Relevant Logic — 477
15.1 The Logic $\mathbf{RM}$ and Its Semantics 477
15.2 The Nice Properties of $\mathbf{RM}$ 486
15.3 Cut-free Gentzen-type System for $\mathbf{RM}$ 493
15.4 $\mathbf{RM}$ with Propositional Constants 504
15.5 $\mathbf{RM}$ as a Fuzzy Logic 505
15.6 Bibliographical Notes and Further Reading 511

Index — 537

List of Figures

1.1 The proof systems $HIL_\supset$, HIL^+, HIL, $HCL_\supset$, HCL^+ and HCL ... 27
1.2 The multiple-conclusion Gentzen-type system LK^+ ... 41
1.3 The multiple-conclusion proof system LK ... 41
1.4 A derivation tree in LK ... 42
1.5 The single-conclusion Gentzen-type system LJ^+ ... 43
1.6 The proof system GLC ... 45

2.1 A quasi-canonical system for Łukasiewicz three-valued logic ... 59

4.1 $\mathcal{THREE}_\top$ ... 106
4.2 Relative strength of some logics with Kleene's negation ... 115
4.3 The proof system $G_{\mathbf{P_1}}$ ... 123
4.4 The proof system $G_{\mathbf{J_3^B}}$ ... 124
4.5 The proof system $G_{\mathbf{SRM}_{\vec{\neg}}}$ ... 125
4.6 The Gentzen-type system $G_{\mathbf{PE3}}$ ... 126
4.7 The proof systems $H_{\mathbf{PAC}}$ and $H_{\mathbf{J_3}}$ ... 129
4.8 The proof system $H^*_{\mathbf{P_1}}$ ... 133

5.1 $\mathcal{FOUR}$ ... 141
5.2 Inference rules for $-$, N, $\otimes$ and $\oplus$ ... 159
5.3 Axioms for $\oplus$, $\otimes$, $-$, N, B ... 162
5.4 $\mathcal{DEFAULT}$ (left) and $\mathcal{NINE}$(right) ... 165
5.5 $\mathcal{FIVE}$ ... 167

7.1 Gentzen rules induced by the negation axioms of A_{PAC} ... 213
7.2 Gentzen rules induced by other axioms of A_{PAC} ... 214
7.3 The system $LK^+(\{[\mathsf{t}], [\mathsf{c}]\})$ for $\mathbf{C}_{min}$ ... 215
7.4 The Gentzen-type system $G_{\mathbf{EIP}}$ ... 227

8.1 Propagation axioms for the connectives of $\mathcal{L}_C$ ... 250
8.2 Gentzen rules induced by the a-axioms ... 254
8.3 Gentzen rules induced by the o-axioms ... 255

9.1 The proof system $GNS5^h$ 315

10.1 $LJ^+(\mathsf{N_4})$: A proof system for $\mathbf{N4}$ 354

11.1 The proof system $HLL_\rightarrow$ 372

12.1 The proof system $HR_{\overrightarrow{\neg}}$ 396

13.1 The lower-semilattice $\langle A_\omega, \preceq^1_\otimes \rangle$ 412
13.2 An example of an $RMI_{\overrightarrow{\neg}}$-tree 429
13.3 The proof system $GSRMI_{\overrightarrow{\neg}}$ 439

14.1 The proof system HR 453

15.1 The proof system GRM 494

ReadMe: An Essential Preface

Purpose of The Book

Perhaps the most counterintuitive property of classical logic (as well as of its most famous rival, intuitionistic logic) is the fact that it allows the inference of any proposition from a single pair of contradicting statements. This principle (known as the principle of explosion, 'ex contradictione sequitur quodlibet') has repeatedly been attacked on philosophical ground[1], as well as because of practical reasons: in its presence every inconsistent theory or knowledge-base is totally trivial, and so useless — something that seems to contradict our daily experience, and also the history of science (including mathematics). As a result, a lot of work and efforts have been devoted over the years to develop alternatives to classical logic that do not have this drawback. Those alternatives are nowadays called 'paraconsistent systems', and the corresponding research area — *paraconsistent reasoning*.

As indicated by the extensive list of references at the end of this book (which is still rather partial!) paraconsistent reasoning is by now a very active area of research, with a lot of interesting results. However, the *mathematical* depth and breadth of it are not well known outside the community of those working on the subject (and frequently not even by researchers inside the community). One of the reasons for this unfortunate state of affairs is the lack of an advanced methodological and comprehensive textbook that is dedicated to the mathematical side of all the main approaches to the subject.[2] The purpose of this book is to (partially) fill this gap by

---

[1] See e.g. [247] for a description and further references.

[2] When we started to write this book there were only several collections of papers and surveys (like [84, 94, 112, 280], and later [1, 95, 281, 182]). There were also books (like [4, 5]) which were exclusively devoted to relevant logics. (The latter is one of the most important approaches to the subject.) Recently, and well after we started to work on this book, two additional important textbooks appeared: [192] on constructive paraconsistent logics, and [111], which mainly describes the work of the Brazilian's school of paraconsistent logics. Nevertheless, there is still no comprehensive mathematical textbook that covers all (or at least most of) the main approaches. (The only previous textbook we know which is devoted to several approaches to paraconsistent reasoning is [108], but this is just an introductory book.)

presenting a very significant part of the rich mathematical theory which exists concerning paraconsistent reasoning.

Our original goal when we started to write this book was rather ambitious: we planned to provide a textbook that would fully describe the abovementioned rich theory. However, at a certain point we realized that by now this cannot be adequately done in just one volume. Accordingly, the scope of the present book has considerably been limited, as explained below. We hope that future textbooks (written either by us or by other authors) will continue the work and describe the many parts of the theory of paraconsistent reasoning that are not covered here.

What is Included in The Book (and What is Not)

First of all, let us note one important side of the subject that is *not* treated in this book: the philosophical one. As we have explained above, our goal here is to describe the mathematical theory. Accordingly, philosophical discussions are reduced here to a minimum. Our intention is just to provide the relevant *facts*. As for their philosophical significance — we leave it to the readers to decide, using other sources.[3]

Turning to what *can* be found in the book, we note that its actual content is summarized in its title. What follows are detailed clarifications and explanations.

1. The first thing to note is that this book is exclusively devoted to the *propositional level* of paraconsistent reasoning. This is not because we think that the first-order level is not important. On the contrary: we strongly believe that only first-order paraconsistent logics may really be useful. However, the main problems involved in developing such logics (like the 'Lewis argument' discussed in Chapter 2) arise already on the propositional level, and all the various ideas about how to tackle them are connected with this level as well. To put it in other words: the propositional fragment is the heart of every paraconsistent logic ever studied. Extending it to the full first-order language is usually a technical matter that does not involve essentially new ideas (even though the technicalities might be complex). Therefore, we believe that it is better to present first the various approaches to paraconsistency in the propositional context (as we do here).

2. Even on the propositional level we are not trying here to describe all forms of paraconsistent reasoning, but confine ourselves to the most

---

[3] Again, [247] is a good starting point.

basic one (on which any more advanced approach is based): that of paraconsistent propositional *logics*. Now, the term 'propositional logic' is very overloaded, so already at this point we should better clarify what *we* mean by that[4]. So, in this book a propositional logic is a propositional language equipped with a structural and non-trivial Tarskian consequence relation. Thus, no form of non-monotonic reasoning is described in this book. Again, this is not because we think that non-monotonic approaches are not important (all of us have actually done a lot of research on them), but because of the limited scope of this volume. An example of a very important class of paraconsistent non-monotonic systems that are out of this scope is Batens' class of 'adaptive logics' (see, e.g., [79, 80, 82, 274]), which are not logics in the sense used in this book (except for the most fundamental one, which is indeed studied here). Again, we hope and believe that future books will do justice to non-monotonic paraconsistent reasoning in general, and to the adaptive approach in particular.

3. The next concept that appears in the title of the book and is in a strong need of clarification is that of *paraconsistent* propositional logic. We have already vaguely characterized it as a logic in which not every pair of contradicting statements implies everything. However, once we leave the realm of classical logic, it is not clear what the term 'contradicting statements' exactly means. Usually it is taken in the literature on paraconsistent reasoning to be any pair of the form $\{\neg\varphi, \varphi\}$, where φ is the "negation" of the logic. (We call the corresponding notion 'pre-$\neg$-paraconsistency' in what follows.) However, this raises the question whether we wish to treat any unary connective as a potential negation. The answer one gets from studying the literature is obviously "no". This forces us to define in exact terms what we mean by 'a negation for a logic **L**'. After a careful study of the literature, we decided to take a certain notion of *coherence with classical logic* (precisely defined in Chapter 2) as our criterion for being a negation, and we define the notion of paraconsistent logic (more accurately: $\neg$-paraconsistent logic) accordingly. The reader should therefore keep in mind that the notion we use of 'paraconsistent logic' is in principle less general than the (imprecise) one found in the literature. However, this does not involve much loss of generality, because almost all the main 'paraconsistent logics' ever studied are indeed paraconsistent in the sense of this book.[5] The scope of the book is actually

---

[4]An exact definition is given in Chapter 1.
[5]Actually, our focus is almost exclusively on logics which we call *strongly* paraconsistent. See Chapter 2.

further restricted by a stronger constraint: with very few exceptions in Chapters 4 and 5 (that are briefly included there for the sake of completeness), all the logics studied in this book are $\neg$-*contained in classical logic*, not just coherent with it. What this means is again defined in a precise way in Chapter 2. Here it suffices to say that we mainly concentrate on logics which include a sufficiently rich class of standard classical connectives, each (including $\neg$) having a standard classical (two-valued) interpretation. The reason for this constraint is again our wish to devote this book to the most basic type of paraconsistent systems, on which all other such systems depend — and again almost every paraconsistent logic ever studied is an extension of one that satisfies this constraint.

4. Finally, our attitude towards paraconsistent reasoning is pragmatic. Accordingly, only *effective* propositional logics are studied in this book, since in our opinion these are the only propositional logics that have *some* chance of being used in practice. Now, by 'an effective logic' we mean here, first of all, a logic that has a corresponding *analytic proof system* (that can be implemented and used). Being on the propositional level, this implies that we seriously study here only *decidable* logics. So, although we do mention and provide some information on certain undecidable logics that are of particular importance (like the relevant logic **R**), we do not *prove* here any result about them.

Having an analytic proof system is for us only a minimal requirement. To be taken as truly effective and useful, we demand here a logic to have in addition a concrete semantics that can also be used for a decision procedure. Almost all the logics studied in this book indeed satisfy this criterion too.[6]

Content and Organization of the Book

The first part of the book describes and defines in precise terms all the basic notions connected with paraconsistency, and reviews all necessary preliminaries. It ends with a desiderata list (Section 2.4) that guides us from that point. All the other parts of the book are devoted to the study of thousands(!) of particular paraconsistent logics. To be able to do that in a reasonable way, we have divided these logics into four big families. Each such family is connected in our presentation (which does not always follow

---

[6] The sole exception is the relevant logic $\mathbf{R}_{\to}$, which does not have at present useful semantics. However, it has a crucial minimality property that partially compensates for this lack of effective semantics.

the historical development and motivations) with one central notion or idea:

Truth-Functionality The logics in this family remain faithful to the main classical semantic principles: their semantics is based on assigning *truth values* to formulas, and the value assigned to a complex formula is a *function* of the values assigned to its immediate subformulas. The main difference from the classical case is the use of extra truth-values besides the two classical ones. These ideas lead to the use of algebraic structures called *(logical) matrices*. Part II of the book is mainly devoted to logics which are induced by the simplest and best motivated types of matrices: the three-valued and the four-valued matrices. The most basic logic among the latter is Dunn-Belnap logic, which is also the logic which is induced by *logical bilattices*. Therefore also this important class of matrices is briefly investigated in Part II.[7]

Non-Determinism One of the most famous approaches to paraconsistency is that of da Costa's Brazilian School, which has developed the large family of paraconsistent logics known now as Logics of Formal Inconsistency (LFIs). It has turned out that the principle of truth-functionality is not always adequate for this family. However, it is possible to provide in a modular way simple and useful semantics for these logics by using a natural generalization of the class of standard matrices, known as *non-deterministic matrices* (Nmatrices). The semantics of Nmatrices is still multi-valued, but the principle of truth-functionality is relaxed in it by allowing the truth-value of a formula to be chosen non-deterministically from some nonempty set of options. This non-deterministic approach is applied in Part III to many useful paraconsistent logics, from Batens' basic logic **CLuN** (which requires just two truth values) to the LFIs (some of which need infinitely many).

Possible Worlds There are two main approaches to providing effective semantics for propositional logics. The previous two parts of the book dealt with one of them: the use of multiple-valued structures. Part IV of the book is devoted to the applications to paraconsistent logics of the other: that of Kripke-style possible-worlds semantics, which is the main type of semantics which is used in the study of modal logics and intuitionistic logic (as well as intermediate logics). This part is divided into two chapters, each corresponding to one of the two main versions of possible-worlds semantics.

---

[7]Various matrices (including infinite-valued ones) are used also in other parts of the book. However, they serve there only as an instrument for investigating logics which are motivated by other ideas, and have originally been introduced using other methods.

- Chapter 9 uses the Kripke-style semantics for modal logics. In fact, every logic which is studied in this chapter is actually equivalent to some well-known modal logic (like **KTB** or **S5**). However, they are all presented and studied as paraconsistent logics in the standard language of classical logic. Unlike almost all the logics which are investigated in previous chapters, those studied in Chapter 9 enjoy the important *replacement property*.

- All the various logics considered in the book up to Chapter 9 (inclusive) are essentially based on positive classical logic by being conservative extensions of it. In contrast, the paraconsistent logics studied in Chapter 10 are *constructive* in the sense of being based on positive *intuitionistic* logic. Accordingly, the semantics of those systems is based on various versions (many of them are again non-deterministic) of Kripke frames for positive intuitionistic logic. Since a special book ([192]), devoted to constructive paraconsistent logics is now available, this part of the book is less comprehensive than the others, and except for the most fundamental logics in this class (like Nelson's paraconsistent logic), it mainly includes material not covered in [192].

Relevance The last part of the book (Part V) is devoted to one of the oldest and most well-studied approaches to paraconsistency. It is based on the idea that the principle of explosion should be rejected because it allows to infer a proposition from a set of assumptions that might be totally irrelevant to it. Obviously, the demand of *relevance* between premises and conclusion is not connected just with negation. Therefore, its implementation requires more radical changes in accepted logical principles than the previous approaches. In particular: one cannot use anymore as a basis the classical or intuitionistic positive logics, but has to develop appropriate alternatives. This is exactly what has been done by the Anderson and Belnap's school of relevance logics. Unfortunately, the main systems of Anderson and Belnap (like **R**) are not effective (or even decidable). Therefore, although we do explain in Part V the approach of Anderson and Belnap and describe the logic **R**, most of this part is devoted to closely related logics which *are* effective, like $\mathbf{R}_{\to}$ (the implication-negation fragment of **R**), the semi-relevant logic **RM**, and some purely relevant relatives of the latter.

After explaining the organization of the material and the classification of the logics, we turn to the way we treat each particular logic. Here we put particular emphasis on the following three major issues:

Proof Systems We provide such systems for *every* logic we describe. For this we use two types of proof systems (each of which is fully explained in the book, so no prior knowledge of them is required):

Gentzen-type Systems: We believe that analytic Gentzen-type systems are the most important type of proof systems[8]. Therefore, we provide such a system for every logic we seriously investigate. In almost all cases the systems we provide are not only analytic, but actually admit cut-elimination, while all their rules other than cut respect some useful version of the subformula property.

Hilbert-type Systems: This is the simplest type of proof systems, and finding one is frequently quite valuable for understanding a new logic and its relations with other logics. In fact, most of the logics investigated in this book have historically been first introduced using Hilbert-type systems. Therefore, we provide such a system for every logic that we describe, provided that it has an appropriate implication connective (which satisfies some reasonable version of the deduction theorem).[9]

Semantics As noted above (see Footnote 6), with one exception we provide effective semantics for every logic we study.

Maximality and Minimality A special attention is payed to the question whether a given logic is minimal or maximal with respect to some important property, like being a maximal paraconsistent logic (as all three-valued paraconsistent logics turn to be) or the minimal paraconsistent relevant logic (which is the characteristic property of $R_\to$). We believe that such properties have great importance for understanding the nature of a logic, as well as for deciding what paraconsistent logic to use for a given goal.

In addition to the semantics and proof theory of a given logic, and to its maximality/minimality properties, we also present of course other interesting properties that it might have. Thus, for example, we extensively study the expressive power of the language of many of the logics, especially those which are dealt with in Chapters 4 and 5.

---

[8] Thus both tableaux and their dual are essentially Gentzen-type systems in disguise.

[9] Although it is not (or at least should not be) a part of the definition of a Hilbert-type system, a respectable Hilbert-type system is usually expected to be based mainly on axiom schemes, with very few (preferably just one!) rules of inference. Thus, it does not make much sense in our opinion to try to develop 'Hilbert-type systems' for logics that lack an appropriate implication connective (even though this is done, e.g., in [251]).

A Few More Important Comments

We end this preface with some more comments about the material in this book and its presentation.

- The class of intended readers of the book is rather diverse, and includes computer scientists, mathematical logicians, and philosophers. The book is in principle completely self-contained, and every prerequisite material is reviewed in the first chapter. Still, we do expect its potential readers to have basic logical background, and to be familiar with the material taught in standard introductory courses on logic.

- Proofs of known results that are not directly related to paraconsistent reasoning are omitted in this book, and appropriate references are given instead. On the other hand, we do provide proofs for almost all other non-trivial theorems and propositions. In the rare cases in which we have chosen not to do so, we provide an explanation why.

- As we have already said, the area of paraconsistent logics is expanding rapidly, with new good works published continuously. In order to be able to publish the book in our life time, we had at a certain point no choice but to ignore works that were published after a certain date. The main exceptions here are works done by ourselves, especially for the sake of the completeness of this book (according to its main subjects and themes), and in order to fill in important gaps that had existed in the related knowledge before that. In fact, a very significant part of our publications on paraconsistent logics since 2011 is due to our work on this book and to the research that this work has led to.

- Our limited knowledge, as well as space constraints, force us to omit from the book also a lot of worthy relevant work that was made prior to the date mentioned in the previous item. As a partial remedy, the last section of each chapter includes historical remarks, notes about related works, and suggestions for further reading. This section also provides references for the results of that chapter, and sometimes some other final comments that we find relevant to the chapter's topic.

Finally, we would like to declare that we have done our best to keep the presentation in the book *coherent* and *consistent* (using classical logic in all our proofs). Any failure in doing so is by no means an attempt to challenge the skills of our readers in drawing conclusions in the presence of inconsistencies. Notes on such cases (and any other comment that might help improving the text in the future) are greatly welcome.

Acknowledgments

We are very grateful to João Marcos for carefully reading great parts of the book, and providing helpful comments and suggestions. We also thank Dov Gabbay and Jane Spurr for their valuable help with the preparation of this book. The work on the book was supported by the Israel Science Foundation (grant No. 817/15).

Part I

Reasoning With Inconsistency

Chapter 1

Propositional Logics: Preliminaries

In this chapter we review all the pre-requested knowledge about propositional logics which will be needed in the rest of the book. Although the chapter includes some new definitions and propositions, most of it is well-known, and so we omit most of the proofs. The only exceptions are proofs of new propositions, or "folklore" that is difficult to find in standard textbooks.

1.1 General Background

1.1.1 What is a Propositional Logic?

In order to describe paraconsistent (propositional) logics, we first have to clarify what we mean by a (propositional) *logic*. This is what we do in this section. In addition, we shall define some related notions.

In this section $\mathcal{L}$ denotes some *propositional language*, i.e. a structure consisting of a (countably infinite) set of primitives, called the *propositional variables* (or just variables) of $\mathcal{L}$, and a (finite) set $\mathcal{C}(\mathcal{L})$ of *logical connectives*, each having a specific natural number as its arity. (0-ary connectives are called *propositional constants*, or just constants.) Given $\mathcal{L}$, we define:

- A *formula* of $\mathcal{L}$ is inductively built up from other formulas as follows:

    - Every propositional variable and every propositional constant is a formula. Such formulas are called *atomic*.

    - If $\psi_1, \ldots, \psi_n$ are formulas, and $\diamond$ is an n-ary logical connective of $\mathcal{L}$, then $\diamond(\psi_1, \ldots, \psi_n)$ is a formula. If $n > 0$ then such a formula is called *complex*.

    In the sequel we shall usually follow standard conventions and abbreviations in writing formulas. Thus, for instance, in case of unary operators we shall frequently write $\diamond\psi$ instead of $\diamond(\psi)$, and when $\diamond$ is a binary connective we shall usually write $\psi \diamond \varphi$ for $\diamond(\psi, \varphi)$. Other conventions concerning parentheses omissions will be used (perhaps for specific languages) whenever this is the common practice.

We shall assume that all propositional languages share the same set $\mathsf{Var} = \{P_1, P_2, \ldots\}$ of variables, and use p, q and r to vary over this set. The set of formulas of $\mathcal{L}$ is denoted by $\mathcal{W}(\mathcal{L})$, and $\varphi, \psi, \phi, \sigma, \tau$ (and sometimes A, B, and C) will vary over its elements.

- Because of our assumption above about Var, we shall identify a propositional language $\mathcal{L}$ with $\mathcal{C}(\mathcal{L})$ (and frequently write just $\mathcal{L}$ instead of $\mathcal{C}(\mathcal{L})$). Accordingly, we say that $\mathcal{L}_1$ is a *sub-language* of $\mathcal{L}_2$ (and denote this by $\mathcal{L}_1 \subseteq \mathcal{L}_2$) if $\mathcal{C}(\mathcal{L}_1) \subseteq \mathcal{C}(\mathcal{L}_2)$.

- We denote by $2^{\mathcal{W}(\mathcal{L})}$ the power set of $\mathcal{W}(\mathcal{L})$, the elements of which are called *theories*.[1] We shall use the symbols $\mathcal{T}$, $\mathcal{S}$ to vary over theories, and Γ, Δ to vary over *finite* theories. We denote by $\mathsf{Var}(\varphi)$ ($\mathsf{Var}(\mathcal{T})$) the set of variables that appear in φ (in the formulas of $\mathcal{T}$) and by $\mathsf{SF}(\varphi)$ ($\mathsf{SF}(\mathcal{T})$) the set of the subformulas of φ (the set of the subformulas of the formulas in $\mathcal{T}$). Following the usual conventions, we shall abbreviate $\mathcal{T} \cup \{\psi\}$ by $\mathcal{T}, \psi$ (omitting the '$\mathcal{T},$' in case $\mathcal{T}$ is empty) and $\mathcal{T} \cup \mathcal{S}$ by $\mathcal{T}, \mathcal{S}$. The notation $\emptyset$ for the empty set will frequently be omitted. (Thus we write below $\vdash \psi$ instead of $\emptyset \vdash \psi$.)

- An $\mathcal{L}$-*substitution* θ (or just a substitution, when the language is clear from the context) is a finite set of pairs $\{(\psi_1, p_1), \ldots, (\psi_n, p_n)\}$, where $\psi_1, \ldots, \psi_n$ are formulas of $\mathcal{L}$, and $p_1, \ldots, p_n$ are n distinct propositional variables. Given a substitution $\theta = \{(\psi_1, p_1), \ldots, (\psi_n, p_n)\}$ and a formula σ, we denote by $\theta(\sigma)$ or by $\sigma[\psi_1/p_1, \ldots, \psi_n/p_n]$ the formula which is obtained from σ by replacing each occurrence of p_i in it by ψ_i ($i = 1, \ldots, n$). Given a theory $\mathcal{T}$ and a substitution θ, we shall denote by $\theta(\mathcal{T})$ the set $\{\theta(\sigma) \mid \sigma \in \mathcal{T}\}$.

In [282], Tarski introduced the following basic notion:

Definition 1.1 (tcr) A (Tarskian) *consequence relation* (tcr) for a language $\mathcal{L}$ is a binary relation $\vdash$ between theories in $2^{\mathcal{W}(\mathcal{L})}$ and formulas in $\mathcal{W}(\mathcal{L})$, satisfying the following three conditions:

[R]   Reflexivity: $\{\psi\} \vdash \psi$.
[M]   Monotonicity: if $\mathcal{T} \vdash \psi$ and $\mathcal{T} \subseteq \mathcal{T}'$, then $\mathcal{T}' \vdash \psi$.
[C]   Cut (Transitivity):   if $\mathcal{T} \vdash \psi$ and $\mathcal{T}', \psi \vdash \varphi$ then $\mathcal{T}, \mathcal{T}' \vdash \varphi$.

---
[1] Note that in many textbooks the word "theory" has a different meaning, which depends also on the underlying logic (unlike our use of the term, which depends only on the underlying language).

1.1 General Background

Definition 1.2 (additional properties of a tcr) Let $\vdash$ be a Tarskian consequence relation for $\mathcal{L}$.

- $\vdash$ is *structural* if for every $\mathcal{L}$-substitution θ and every $\mathcal{T}$ and ψ: if $\mathcal{T} \vdash \psi$ then $\theta(\mathcal{T}) \vdash \theta(\psi)$.

- $\vdash$ is *non-trivial* (or *consistent*) if there exist some non-empty theory $\mathcal{T}$ and some formula ψ such that $\mathcal{T} \not\vdash \psi$.[2]

- $\vdash$ is *finitary* if for every theory $\mathcal{T}$ and every formula ψ such that $\mathcal{T} \vdash \psi$, there is a *finite* theory $\Gamma \subseteq \mathcal{T}$ such that $\Gamma \vdash \psi$.

Definition 1.3 (propositional logic)

- A (propositional) *logic* is a pair $\mathbf{L} = \langle \mathcal{L}, \vdash_{\mathbf{L}} \rangle$, where $\mathcal{L}$ is a propositional language, and $\vdash_{\mathbf{L}}$ is a structural and non-trivial tcr for $\mathcal{L}$.[3]

- A logic $\langle \mathcal{L}, \vdash_{\mathbf{L}} \rangle$ is *finitary* if so is $\vdash_{\mathbf{L}}$.

Note 1.4 Structurality is what distinguishes between theories and *logics*, and what gives logics the generality and content-independence which characterize them. Consistency is convenient for excluding trivial logics (those in which every formula follows from every theory, or every formula follows from every non-empty theory). Finitariness is essential for practical reasoning, where a conclusion is always derived from a finite set of premises. In particular, every logic that has a decent proof system is finitary. Therefore *in this book we are interested only in finitary logics*.[4]

Definition 1.5 (tautologies) Let $\mathbf{L} = \langle \mathcal{L}, \vdash_{\mathbf{L}} \rangle$ be a logic. A formula φ *follows* in $\mathbf{L}$ from a theory $\mathcal{T}$ if $\mathcal{T} \vdash_{\mathbf{L}} \varphi$. A formula φ is called a *valid* formula (or a *theorem*, or a *tautology*) of $\mathbf{L}$ iff it follows in $\mathbf{L}$ from the empty set (i.e. iff $\vdash_{\mathbf{L}} \varphi$)[5]. We denote by $Th(\mathbf{L})$ the set of tautologies of $\mathbf{L}$.

Tarski's notion of a consequence relation was generalized to the multiple-conclusion case by Scott in [263], and by Shoesmith and Smiley in [267].

---

[2] When $\vdash$ is structural, the above definition of non-triviality is equivalent to the requirement that $p \not\vdash q$ for distinct variables $p, q \in \mathsf{Var}$. The definition above is adequate also for non-structural consequence relations.

[3] In the literature, sometimes weaker conditions are required from a 'logic'. Thus, 'non-monotonic logics' refers to relations in which monotonicity (and sometime transitivity or even reflexivity) is violated. The condition of non-triviality is also not always demanded. To keep our terminology coherent, we use the term 'reasoning' for the general process of drawing conclusions from a given theory (e.g. 'non-monotonic reasoning').

[4] Actually, already the formulation of condition [C] in our definition of a tcr (Definition 1.1) is really adequate only for finitary consequence relations. The general case requires a more complicated condition. See [267] or [304].

[5] Recall that $\vdash_{\mathbf{L}} \varphi$ is an abbreviation for $\emptyset \vdash_{\mathbf{L}} \varphi$.

Definition 1.6 (Scott consequence relation, scr) A *consequence relation* in the sense of Scott (an scr) for a language $\mathcal{L}$ is a binary relation $\vdash$ between theories in $2^{\mathcal{W}(\mathcal{L})}$, satisfying the following three conditions:

[R]   Reflexivity: $\{\psi\} \vdash \{\psi\}$.
[M]   Monotonicity: if $\mathcal{T} \vdash \mathcal{S}$ and $\mathcal{T} \subseteq \mathcal{T}'$, $\mathcal{S} \subseteq \mathcal{S}'$, then $\mathcal{T}' \vdash \mathcal{S}'$.
[C]   Cut (Transitivity):   if $\mathcal{T} \vdash \psi, \mathcal{S}$ and $\mathcal{T}', \psi \vdash \mathcal{S}'$ then $\mathcal{T}, \mathcal{T}' \vdash \mathcal{S}, \mathcal{S}'$.

Structural, non-trivial, and finitary Scott consequence relations are defined by obvious generalizations of the definitions in the single-conclusion case.[6]

The most common ways of defining propositional logics may be divided into semantic (model-theoretical) approaches and syntactic (proof-theoretical) approaches. In Section 1.2 we briefly describe the first approach, and in Section 1.3 — the other one.

1.1.2 Some Basic Notions

Next, we review some basic properties of logics and relations between logics.

Definition 1.7 (fragment of a logic) Let $\mathbf{L}_1 = \langle \mathcal{L}_1, \vdash_{\mathbf{L}_1} \rangle$, and suppose that $\mathcal{L}_2 \subseteq \mathcal{L}_1$. The $\mathcal{L}_2$-*fragment* of $\mathbf{L}_1$ is the logic $\langle \mathcal{L}_2, \vdash_2 \rangle$, where $\vdash_2$ is the restriction of $\vdash_{\mathbf{L}_1}$ to $\mathcal{L}_2$ (i.e. for $\mathcal{T}$ and φ in $\mathcal{L}_2$, $\mathcal{T} \vdash_2 \varphi$ iff $\mathcal{T} \vdash_{\mathbf{L}_1} \varphi$).

Notation 1.8 Let $\mathbf{L} = \langle \mathcal{L}, \vdash_{\mathbf{L}} \rangle$ be a propositional logic, and suppose that $\mathcal{T}' \cup \{\psi'\}$ is a set of sentences in a propositional language $\mathcal{L}'$ such that $\mathcal{L} \subseteq \mathcal{L}'$. We say that $\mathcal{T}' \vdash_{\mathbf{L}} \psi'$ iff there exists a theory $\mathcal{T}$ and a formula ψ in $\mathcal{L}$, and an $\mathcal{L}'$-substitution θ such that $\mathcal{T}' = \theta(\mathcal{T})$, $\psi' = \theta(\psi)$, and $\mathcal{T} \vdash_{\mathbf{L}} \psi$.

Definition 1.9 (extensions) Let $\mathbf{L}_1 = \langle \mathcal{L}_1, \vdash_{\mathbf{L}_1} \rangle$ and $\mathbf{L}_2 = \langle \mathcal{L}_2, \vdash_{\mathbf{L}_2} \rangle$ be propositional logics.

- $\mathbf{L}_1$ is an *extension* of $\mathbf{L}_2$ if $\mathcal{L}_2 \subseteq \mathcal{L}_1$ and $\vdash_{\mathbf{L}_2} \subseteq \vdash_{\mathbf{L}_1}$.

- $\mathbf{L}_1$ is a *simple extension* of $\mathbf{L}_2$ if $\mathcal{L}_2 = \mathcal{L}_1$, and $\vdash_{\mathbf{L}_2} \subseteq \vdash_{\mathbf{L}_1}$.

- $\mathbf{L}_1$ is a *proper extension* of $\mathbf{L}_2$ if $\mathcal{L}_2 \subseteq \mathcal{L}_1$ and $\vdash_{\mathbf{L}_2} \subsetneq \vdash_{\mathbf{L}_1}$.

- $\mathbf{L}_1$ is a *strongly proper* extension of $\mathbf{L}_2$ if $\mathcal{L}_2 \subseteq \mathcal{L}_1$ and there is a sentence φ of $\mathcal{L}_2$ such that $\vdash_{\mathbf{L}_1} \varphi$ but $\not\vdash_{\mathbf{L}_2} \varphi$.

---

[6]The content of footnote 4 is relevant here too. Actually, the generalization proposed by Scott in [263] only dealt with the case involving *finite* sets of formulas. Our Definition 1.6 is a combination of his approach and the one taken by Shoesmith and Smiley in [267].

1.1 General Background

- $\mathbf{L}_1$ is a *conservative extension* of $\mathbf{L}_2$ if it is an extension of $\mathbf{L}_2$, and $\mathcal{T} \vdash_{\mathbf{L}_2} \psi$ whenever $\mathcal{T} \cup \{\psi\} \in 2^{\mathcal{W}(\mathcal{L}_2)}$ and $\mathcal{T} \vdash_{\mathbf{L}_1} \psi$.

- $\mathbf{L}_1$ is a *weakly conservative extension* of $\mathbf{L}_2$ if it is an extension of $\mathbf{L}_2$, and $\vdash_{\mathbf{L}_2} \psi$ whenever $\psi \in \mathcal{W}(\mathcal{L}_2)$ and $\vdash_{\mathbf{L}_1} \psi$.

- Suppose that $\mathcal{L}_2 \subseteq \mathcal{L}_1$.
  - Let $\mathcal{S} \subseteq \mathcal{W}(\mathcal{L}_1)$. $\mathbf{L}_1$ is the *axiomatic extension* of $\mathbf{L}_2$ by $\mathcal{S}$ if $\vdash_{\mathbf{L}_1}$ is the minimal tcr $\vdash$ on $\mathcal{L}_1$ for which $\vdash_{\mathbf{L}_2} \subseteq \vdash$, and $\vdash \varphi$ for every $\varphi \in \mathcal{S}$.
  - $\mathbf{L}_1$ is an *axiomatic extension* of $\mathbf{L}_2$ if and there is a set $\mathcal{S} \subseteq \mathcal{W}(\mathcal{L}_1)$ such that $\mathbf{L}_1$ is the axiomatic extension of $\mathbf{L}_2$ by $\mathcal{S}$.

Note 1.10 It is easy to see that if $\mathbf{L}_2 = \langle \mathcal{L}_2, \vdash_{\mathbf{L}_2} \rangle$ is finitary, $\mathcal{L}_2 \subseteq \mathcal{L}_1$, and $\mathcal{S} \subseteq \mathcal{W}(\mathcal{L}_1)$, then $\mathbf{L}_1 = \langle \mathcal{L}_1, \vdash_{\mathbf{L}_1} \rangle$ is the axiomatic extension of $\mathbf{L}_2$ by $\mathcal{S}$ iff for every $\mathcal{T} \subseteq \mathcal{W}(\mathcal{L}_1)$ and $\psi \in \mathcal{W}(\mathcal{L}_1)$, $\mathcal{T} \vdash_{\mathbf{L}_1} \psi$ iff $\mathcal{T} \cup Cl_{\mathcal{L}_1}(\mathcal{S}) \vdash_{\mathbf{L}_2} \psi$, where $Cl_{\mathcal{L}_1}(\mathcal{S}) = \bigcup_\theta \theta(\mathcal{S})$. (Here θ varies over the set of all $\mathcal{L}_1$-substitutions.)

Definition 1.11 (rule, application) Let $\mathcal{L}$ be a propositional language.

- A *rule* in $\mathcal{L}$ is a pair $\langle \Gamma, \psi \rangle$, where $\Gamma \cup \{\psi\}$ is a finite set of formulas in $\mathcal{L}$. The formulas in Γ are called the *premises* of the rule and ψ is the *conclusion* of the rule. We shall henceforth denote such a rule by Γ / ψ. If $\Gamma = \emptyset$ the rule is called an *axiom schema*, or just an axiom (usually we shall simply refer to the axiom ψ rather than the rule $\emptyset / \psi$). Rules with a non-empty set of premises are called *inference rules*.

- An *application* of the rule Γ/ψ is a pair $\langle \theta(\Gamma), \theta(\psi) \rangle$, where θ is an $\mathcal{L}$-substitution. $\theta(\psi)$ is called the *conclusion* of such an application, and the formulas in $\theta(\Gamma)$ are its *premises*. We also say that $\theta(\psi)$ is *inferable* or *derivable* or *deducible* from $\theta(\Gamma)$ by the rule Γ/ψ.

Notation 1.12 In what follows we shall usually follow standard conventions, and use variables for formulas (rather than actual formulas) and a horizontal line (rather than '/') in formulations of rules. Thus, we shall formulate the well-known rule [MP] as $\frac{\psi \quad \psi \supset \varphi}{\varphi}$ instead of our official form: $\{P_1, P_1 \supset P_2\}/P_2$.

Definition 1.13 (admissibility of rules) A rule Γ/ψ is *admissible* in a logic $\mathbf{L} = \langle \mathcal{L}, \vdash_{\mathbf{L}} \rangle$ if for every $\mathcal{L}$-substitution θ it holds that $\vdash_{\mathbf{L}} \theta(\psi)$ in case $\vdash_{\mathbf{L}} \theta(\varphi)$ for every $\varphi \in \Gamma$.

Definition 1.14 (extension by rules) Let $\mathbf{L} = \langle \mathcal{L}, \vdash_{\mathbf{L}} \rangle$ be a finitary logic, and let S be a set of rules in a language $\mathcal{L}'$ such that $\mathcal{L} \subseteq \mathcal{L}'$.

- $C_\mathbf{L}(S)$, the **L**-*closure of* S,[7] is inductively defined as follows:
  - $\langle \theta(\Gamma), \theta(\psi) \rangle \in C_\mathbf{L}(S)$, whenever θ is an $\mathcal{L}'$-substitution, Γ is a *finite* theory in $2^{\mathcal{W}(\mathcal{L}')}$, and either $\Gamma \vdash_\mathbf{L} \psi$ or $\Gamma/\psi \in S$.
  - If the pairs $\langle \Gamma_1, \varphi \rangle$ and $\langle \Gamma_2 \cup \{\varphi\}, \psi \rangle$ are both in $C_\mathbf{L}(S)$, then so is the pair $\langle \Gamma_1 \cup \Gamma_2, \psi \rangle$.

- The *extension of* **L** *by* S is the pair $\mathbf{L}^* = \langle \mathcal{L}', \vdash_{\mathbf{L}^*} \rangle$, where $\vdash_{\mathbf{L}^*}$ is the binary relation between theories in $2^{\mathcal{W}(\mathcal{L}')}$ and formulas in $\mathcal{W}(\mathcal{L}')$, defined by: $\mathcal{T} \vdash_{\mathbf{L}^*} \psi$ if there is a finite $\Gamma \subseteq \mathcal{T}$ such that $\langle \Gamma, \psi \rangle \in C_\mathbf{L}(S)$.

Note 1.15 It is easy to see that extending a finitary logic **L** by a set of axioms S (Definition 1.9 and Note 1.10) amounts to extending it by the set of rules $\{\emptyset/\varphi \mid \varphi \in S\}$. Axiomatic extensions of **L** will therefore henceforth taken as a special case of extensions by rules.

The following proposition is easily verified:

Proposition 1.16 *Let* **L**, $\mathcal{L}'$ *and* S *be like in Definition 1.14, and let* $\mathbf{L}^*$ *be the extension of* **L** *by* S.

1. *Unless* $\langle \{p\}, q \rangle \in C_\mathbf{L}(S)$ *for some distinct variables p and q,*[8] $\mathbf{L}^*$ *is a finitary propositional logic. Moreover,* $\mathbf{L}^*$ *is in this case the minimal extension of* **L** *in which φ follows from Γ whenever* $\Gamma/\varphi \in S$.

2. *For every finitary extension* $\mathbf{L}^*$ *of* **L** *there is a set S of rules such that* $\mathbf{L}^*$ *is the extension of* **L** *by* S.

The following property is essential for the *usefulness* of a logic:

Definition 1.17 (decidability of a logic) A logic $\langle \mathcal{L}, \vdash \rangle$ is called *decidable* if both membership and non-membership in $\vdash$ can effectively be determined for every finite theory and every formula in $\mathcal{L}$. (That is: there is an algorithm that for every finite theory Γ and formula ψ determines whether $\Gamma \vdash \psi$ or $\Gamma \nvdash \psi$.) $\langle \mathcal{L}, \vdash \rangle$ is *semi-decidable* if there is an algorithm that given a finite theory Γ and a formula ψ terminates iff $\Gamma \vdash \psi$. Similarly, $\langle \mathcal{L}, \vdash \rangle$ is *co-semi-decidable* if there is an algorithm that given a finite theory Γ and a formula ψ terminates iff $\Gamma \nvdash \psi$.[9]

Another useful property that a logic *may* have is the following:

---

[7] Actually, we should have denoted it by $C_\mathbf{L}^{\mathcal{L}'}(S)$, because the identity of $C_\mathbf{L}(S)$ depends also on the identity of $\mathcal{L}'$. However, this will always be clear from the context.

[8] This condition can easily be verified in all cases considered in the sequel.

[9] Note that a logic is decidable iff it is both semi-decidable and co-semi-decidable.

1.1 General Background

Definition 1.18 (replacement property) Let $\mathbf{L} = \langle \mathcal{L}, \vdash_\mathbf{L} \rangle$ be a logic.

- Formulas $\psi, \varphi \in \mathcal{W}(\mathcal{L})$ are *equivalent* in $\mathbf{L}$, denoted by $\psi \dashv\vdash_\mathbf{L} \varphi$, if $\psi \vdash_\mathbf{L} \varphi$ and $\varphi \vdash_\mathbf{L} \psi$.

- Formulas $\psi, \varphi \in \mathcal{W}(\mathcal{L})$ are *congruent* (or *indistinguishable*) in $\mathbf{L}$, denoted by $\psi \equiv_\mathbf{L} \varphi$, if for every formula σ and variable p it holds that $\sigma[\psi/p] \dashv\vdash_\mathbf{L} \sigma[\varphi/p]$.

- $\mathbf{L}$ has the *replacement property*, or is *self-extensional* ([304]), if any two formulas which are equivalent in $\mathbf{L}$ are congruent in it.

Note 1.19 Some authors consider also the *substitution property*, which $\mathbf{L}$ enjoys if $\mathcal{T}, \sigma[\psi/p] \vdash_\mathbf{L} \sigma[\varphi/p]$ whenever both $\mathcal{T}, \psi \vdash_\mathbf{L} \varphi$ and $\mathcal{T}, \varphi \vdash_\mathbf{L} \psi$. However, this notion is practically useless for paraconsistent logics. Thus if we take $\mathcal{T} = \{\varphi, \psi\}$ we get that if $\mathbf{L}$ has the substitution property then $\varphi, \psi, \neg\psi \vdash_\mathbf{L} \neg\varphi$. Since, of course, $\neg\varphi, \psi, \neg\psi \vdash_\mathbf{L} \neg\varphi$ as well, it follows that if $\neg$ has in $\mathbf{L}$ the very desirable property (at least for paraconsistent logics) of being complete for $\mathbf{L}$ (see Definition 2.5), then $\psi, \neg\psi \vdash_\mathbf{L} \neg\varphi$ for every φ, ψ, and so $\mathbf{L}$ is not strongly paraconsistent (see Definition 2.7). However, the latter is a particularly natural property which we demand in this book from every logic we study. (See Note 2.10 for an explanation.)

1.1.3 Basic Connectives

The usefulness of a logic $\mathbf{L} = \langle \mathcal{L}, \vdash_\mathbf{L} \rangle$ strongly depends on the question of what kind of connectives are available in it. Here one should not consider only the *primitive* connectives of $\mathcal{L}$ (i.e., the elements of $\mathcal{C}(\mathcal{L})$), but also the connectives that are *definable* in $\mathbf{L}$. Next is a definition of the latter notion.

Definition 1.20 (defined connectives) Let $\mathbf{L} = \langle \mathcal{L}, \vdash_\mathbf{L} \rangle$ be a logic.

- $\mathbf{L}' = \langle \mathcal{L}', \vdash_{\mathbf{L}'} \rangle$ is a *basic extension by definitions* of $\mathbf{L}$ iff there are $\diamond \in \mathcal{C}(\mathcal{L}')$ of arity m, and $\varphi \in \mathcal{W}(\mathcal{L})$ such that:

    - $\diamond \notin \mathcal{C}(\mathcal{L})$ and $\mathcal{C}(\mathcal{L}') = \mathcal{C}(\mathcal{L}) \cup \{\diamond\}$.
    - $\mathsf{Var}(\varphi) \subseteq \{P_1\}$ if $m = 0$; $\mathsf{Var}(\varphi) \subseteq \{P_1, \ldots, P_m\}$ otherwise.[10]
    - $\mathbf{L}'$ is the minimal extension of $\mathbf{L}$ in which $\diamond(P_1, \ldots, P_n)$ and φ are congruent.[11]

---

[10] The reason for the special treatment of the case $m = 0$ is to allow for 0-ary connectives to be defined even in languages which have no such connectives.

[11] It is not difficult to see that $\mathbf{L}'$ is the extension of $\mathbf{L}$ by all rules of the form: $\dfrac{\psi[\varphi/q]}{\psi[\diamond(P_1,\ldots,P_n)/q]}$ and $\dfrac{\psi[\diamond(P_1,\ldots,P_n)/q]}{\psi[\varphi/q]}$, where $\psi \in \mathcal{W}(\mathcal{L})$, and q is a variable.

The formula φ is called *the definition of* $\diamond$ *in* **L**.

- A logic $\mathbf{L}' = \langle \mathcal{L}', \vdash_{\mathbf{L}'} \rangle$ is an *extension by definitions* of **L** iff there are logics $\mathbf{L}_1, \ldots, \mathbf{L}_k$ ($k \geq 1$) such that $\mathbf{L}_1 = \mathbf{L}$, $\mathbf{L}_k = \mathbf{L}'$, and $\mathbf{L}_{i+1}$ is a basic extension by definitions of $\mathbf{L}_i$ for each $1 \leq i \leq k-1$.

- If $\diamond \in \mathcal{C}(\mathcal{L}')$ for some extension by definitions $\langle \mathcal{L}', \vdash_{\mathbf{L}'} \rangle$ of **L**, then $\diamond$ is called a *defined connective* of **L**, or just a *connective of* **L**.[12]

The following proposition is not difficult to show:

Proposition 1.21 *Let* $\mathbf{L}' = \langle \mathcal{L}', \vdash_{\mathbf{L}'} \rangle$ *be an extension by definitions of the logic* $\mathbf{L} = \langle \mathcal{L}, \vdash_{\mathbf{L}} \rangle$.

1. *There is a function* $tr : \mathcal{W}(\mathcal{L}') \to \mathcal{W}(\mathcal{L})$ *such that:*
   - $tr(\psi)$ *is congruent to* ψ *in* $\mathbf{L}'$ *for any* $\psi \in \mathcal{W}(\mathcal{L}')$.
   - $\mathcal{T} \vdash_{\mathbf{L}'} \psi$ *iff* $\{tr(\varphi) \mid \varphi \in \mathcal{T}\} \vdash_{\mathbf{L}} tr(\psi)$.
   - $tr(\psi) = \psi$ *for any* $\psi \in \mathcal{W}(\mathcal{L})$.

2. $\mathbf{L}'$ *is a conservative extension of* **L**.

Note 1.22 Because of the last proposition, an extension by definitions of a logic $\mathbf{L} = \langle \mathcal{L}, \vdash_{\mathbf{L}} \rangle$ can be viewed as practically identical with **L**. Accordingly, from now on we shall simply write: '$\mathcal{T} \vdash_{\mathbf{L}} \psi$' instead of '$\mathcal{T} \vdash_{\mathbf{L}'} \psi$' (or '$\{tr(\varphi) \mid \varphi \in \mathcal{T}\} \vdash_{\mathbf{L}} tr(\psi)$') in case $\mathcal{T} \cup \{\psi\}$ is in $2^{\mathcal{W}(\mathcal{L}')}$ for some extension by definition $\mathbf{L}' = \langle \mathcal{L}', \vdash_{\mathbf{L}'} \rangle$ of **L**. Similarly, when we talk about a 'connective of **L**' it usually does not matter whether it is a primitive connective of **L** or a defined one. Therefore (unless otherwise stated), we shall not distinguish from now on between these two types of connectives.

Next, we introduce three particularly important types of connectives.

Definition 1.23 (basic connectives for a logic) Let $\mathbf{L} = \langle \mathcal{L}, \vdash_{\mathbf{L}} \rangle$ be a propositional logic.

- A binary connective $\supset$ of $\mathcal{L}$ is called an *implication for* **L** if the classical deduction theorem holds for $\supset$ and $\vdash_{\mathbf{L}}$:

$$\mathcal{T}, \varphi \vdash_{\mathbf{L}} \psi \text{ iff } \mathcal{T} \vdash_{\mathbf{L}} \varphi \supset \psi.$$

- A binary connective $\wedge$ of $\mathcal{L}$ is called a *conjunction for* **L** if it satisfies the following condition:

$$\mathcal{T} \vdash_{\mathbf{L}} \psi \wedge \varphi \text{ iff } \mathcal{T} \vdash_{\mathbf{L}} \psi \text{ and } \mathcal{T} \vdash_{\mathbf{L}} \varphi.$$

---
[12] Note that every primitive connective of $\mathcal{L}$ is a (defined) connective of **L**.

1.1 General Background

- A binary connective $\vee$ of $\mathcal{L}$ is called a *disjunction for* **L** if it satisfies the following condition:

$$\mathcal{T}, \psi \vee \varphi \vdash_{\mathbf{L}} \sigma \text{ iff } \mathcal{T}, \psi \vdash_{\mathbf{L}} \sigma \text{ and } \mathcal{T}, \varphi \vdash_{\mathbf{L}} \sigma.$$

Note 1.24 The other standard connective of classical logic, negation, will be considered in later sections. See Notes 1.60, 1.74, and Section 2.1.1.

Note 1.25

- An implication connective for a finitary logic **L** reflects the underlying consequence relation of **L**. Its availability makes it possible to directly reduce all inferences from finite sets of premises to theoremhood: $\psi_1, \ldots, \psi_n \vdash_{\mathbf{L}} \varphi$ iff $\vdash_{\mathbf{L}} \psi_1 \supset (\psi_2 \supset (\ldots \supset (\psi_n \supset \varphi) \ldots))$. It follows that if $\mathbf{L}_1$ and $\mathbf{L}_2$ are two finitary logics in the same language which have a common implication connective, then $\mathbf{L}_1$ and $\mathbf{L}_2$ are identical iff they have the same tautologies.

- It is easy to verify that $\wedge$ is a conjunction for $\mathbf{L} = \langle \mathcal{L}, \vdash_{\mathbf{L}} \rangle$ iff the following three conditions hold for every $\psi, \varphi \in \mathcal{W}(\mathcal{L})$:

  1. $\psi \wedge \varphi \vdash_{\mathbf{L}} \psi$
  2. $\psi \wedge \varphi \vdash_{\mathbf{L}} \varphi$
  3. $\psi, \varphi \vdash_{\mathbf{L}} \psi \wedge \varphi$

  On the basis of this it is not difficult to show that one may equivalently define a conjunction for **L** as a connective $\wedge$ such that

  $$\mathcal{T}, \psi \wedge \varphi \vdash_{\mathbf{L}} \sigma \text{ iff } \mathcal{T}, \psi, \varphi \vdash_{\mathbf{L}} \sigma$$

  It follows that a conjunction connective for **L** assures that for every $\psi_1, \ldots, \psi_n, \varphi \in \mathcal{W}(\mathcal{L})$: $\psi_1, \ldots, \psi_n \vdash_{\mathbf{L}} \varphi$ iff $\psi_1 \wedge \ldots \wedge \psi_n \vdash_{\mathbf{L}} \varphi$.[13]

- It is easy show that in the case of Scott consequence relations one may equivalently define disjunction using a condition in which $\vee$ appears on the right-hand side (like in the definition of conjunction):

  $$\mathcal{T} \vdash_{\mathbf{L}} \psi \vee \varphi \text{ iff } \mathcal{T} \vdash_{\mathbf{L}} \psi, \varphi.$$

For most parts of this book, the definition above of an implication is very useful. However, this is not the case with respect to the last part, in which relevance logics are studied. In such logics it is reasonable to demand the derivability of $\varphi \supset \psi$ from $\mathcal{T}$ only in case φ is absolutely necessary for the derivability of ψ from $\mathcal{T} \cup \{\varphi\}$. Hence for these logics we shall use a somewhat weaker notion of implication.

---

[13] Here $\psi_1 \wedge \ldots \wedge \psi_n$ may be taken to stand, e.g., for $(\ldots((\psi_1 \wedge \psi_2) \wedge \psi_3) \ldots) \wedge \psi_n$.

Definition 1.26 (RDP) Let $\mathbf{L} = \langle \mathcal{L}, \vdash_\mathbf{L} \rangle$ be a logic, and let $\supset$ be a (primitive or defined) connective of $\mathcal{L}$. $\supset$ has in $\mathbf{L}$ (or $\mathbf{L}$ has with respect to $\supset$) the *relevant deduction property* (RDP) if it satisfies the following condition:

$$\mathcal{T}, \varphi \vdash_\mathbf{L} \psi \text{ iff either } \mathcal{T} \vdash_\mathbf{L} \psi \text{ or } \mathcal{T} \vdash_\mathbf{L} \varphi \supset \psi.$$

Definition 1.27 (semi-implication) Let $\mathbf{L} = \langle \mathcal{L}, \vdash_\mathbf{L} \rangle$ be a propositional logic. A binary connective $\supset$ of $\mathcal{L}$ is called a *semi-implication for* $\mathbf{L}$ if $\supset$ has in $\mathbf{L}$ the RDP, and in addition there are formulas φ and ψ such that $\vdash_\mathbf{L} \varphi \supset \psi$ but $\not\vdash_\mathbf{L} \psi \supset \varphi$.[14]

Note 1.28 Like implication, a semi-implication for $\mathbf{L}$ (or even just a connective which has in $\mathbf{L}$ the RDP) reflects the consequence relation of $\mathbf{L}$, but it does it in a more complicated way: $\psi_1, \ldots, \psi_n \vdash_\mathbf{L} \varphi$ iff there is a subset $\{\varphi_1, \ldots, \varphi_k\}$ of $\{\psi_1, \ldots, \psi_n\}$ such that $\vdash_\mathbf{L} \varphi_1 \supset (\varphi_2 \supset (\ldots \supset (\varphi_k \supset \varphi) \ldots))$.

Proposition 1.29 *Any implication for* $\mathbf{L}$ *is also a semi-implication for* $\mathbf{L}$.

Proof. Let $\supset$ be an implication for $\mathbf{L}$. Obviously, $\supset$ has in $\mathbf{L}$ the RDP. To show that it satisfies also the second condition in the definition of a semi-implication, suppose for contradiction that $\vdash_\mathbf{L} \psi \supset \varphi$ whenever $\vdash_\mathbf{L} \varphi \supset \psi$. Since $\vdash_\mathbf{L} p \supset (q \supset q)$ (because $p, q \vdash_\mathbf{L} q$), this implies that $\vdash_\mathbf{L} (q \supset q) \supset p$. But $\vdash_\mathbf{L} q \supset q$ as well (because $q \vdash_\mathbf{L} q$). It follows that $\vdash_\mathbf{L} p$, contradicting the fact that $\mathbf{L}$ is a logic. □

Semi-implications and their basic properties are investigated in Section 11.3.2 of Chapter 11. The next proposition presents one of these properties, which is important already at this stage.

Proposition 1.30 *If* $\mathbf{L}$ *has a conjunction or a disjunction, and* $\supset$ *is a semi-implication for* $\mathbf{L}$, *then* $\supset$ *is also an implication for* $\mathbf{L}$.

Proof. See Proposition 11.52. □

Proposition 1.31

1. *If* $\mathbf{L_1}$ *is an extension of* $\mathbf{L_2}$, *then* $\wedge$ *is a conjunction for* $\mathbf{L_2}$ *iff it is a conjunction for* $\mathbf{L_1}$.

2. *Suppose* $\mathbf{L_1}$ *is a finitary axiomatic extension of* $\mathbf{L_2}$.

    (a) *If* $\supset$ *is a (semi-) implication for* $\mathbf{L_2}$, *then it is also a (semi-) implication for* $\mathbf{L_1}$.

---

[14]The second condition is needed in order to distinguish a semi-implication from an equivalence connective, because the later sometimes has the RDP. This happens, e.g., in the case of the pure equivalence fragment of classical logic. (See [276] or [183, p.1132].)

(b) If $\vee$ is a disjunction for $\mathbf{L_2}$, then it is also a disjunction for $\mathbf{L_1}$.

We believe that a decent logic should provide at least some connectives of the types defined above. Accordingly, *every logic studied in this book is at least semi-normal* according to the definition below.

Definition 1.32 ((semi-)normal logics)

- A logic is called *semi-normal* if it has either a semi-implication, or a conjunction, or a disjunction.

- A logic is called *normal* if it has all the basic connectives (conjunction, disjunction, implication).[15]

Note 1.33 From Proposition 1.30 it follows that a logic is normal iff it has a semi-implication, a conjunction, and a disjunction.

The next definition and the proposition that follows it provide an important property possessed by each of the above basic connectives.

Definition 1.34 (positive connective) Let $\mathbf{L} = \langle \mathcal{L}, \vdash_\mathbf{L} \rangle$ be a propositional logic. An n-ary connective $\diamond$ of $\mathcal{L}$ is called *positive* (with respect to $\mathbf{L}$) if $p_1, \ldots p_n \vdash_\mathbf{L} \diamond(p_1, \ldots, p_n)$ for every set of variables $\{p_1, \ldots p_n\}$.[16] A logic $\mathbf{L} = \langle \mathcal{L}, \vdash_\mathbf{L} \rangle$ is called positive if every connective of $\mathbf{L}$ is positive.

Note 1.35 It is easy to see that $\mathbf{L}$ is positive iff all its primitive connectives are positive.

The proof of the following proposition is straightforward:

Proposition 1.36 *If $\diamond$ is either an implication for $\mathbf{L}$, or a conjunction for $\mathbf{L}$, or a disjunction for $\mathbf{L}$, then $\diamond$ is positive with respect to $\mathbf{L}$.*[17]

1.2 Semantic Approaches to Defining Logics

1.2.1 Satisfiability Relations

The semantic (model-theoretical) way of defining a consequence relation is based on the notion of a *satisfaction relation* for $\mathcal{L}$, as defined below:

---

[15] Our notion of normality should not be confused with the notion of normality used in modal logics, or the notion of normal theory used in [4].

[16] This is a generalization (from the classical case to arbitrary logic) of Post's notion (in [242]) of '1-preserving functions'.

[17] As we shall see in the last part of this book, this is not necessarily true in case $\diamond$ is a semi-implication for $\mathbf{L}$.

Definition 1.37 (denotational semantics)

- A *(denotational) semantics* for a language $\mathcal{L}$ is a pair $\mathsf{S} = \langle S, \models_S \rangle$, where S is a nonempty set, and $\models_S$ (the *satisfaction* relation of S) is a relation from S to $\mathcal{W}(\mathcal{L})$ for which the following holds: for every $p, q \in \mathsf{Var}(\mathcal{L})$ such that $p \neq q$, there is a $\nu \in S$ for which $\nu \models_S p$ and $\nu \not\models_S q$.

- Let ν be an element of S and ψ a formula in $\mathcal{W}(\mathcal{L})$. If $\nu \models_S \psi$ then ν is called an S-*model* of ψ (alternatively, we say that ν *satisfies* ψ).

- Let $\mathsf{S} = \langle S, \models_S \rangle$ be a denotational semantics for $\mathcal{L}$. Given a theory $\mathcal{T}$, $\nu \in S$ is an S-*model* of $\mathcal{T}$ if it is an S-model of every $\psi \in \mathcal{T}$.

Definition 1.38 ($\vdash_\mathsf{S}$) The relation $\vdash_\mathsf{S}$ that is induced by a denotational semantic S is defined as follows:

$$\mathcal{T} \vdash_\mathsf{S} \psi \text{ if every } \mathsf{S}\text{-model of } \mathcal{T} \text{ is also an } \mathsf{S}\text{-model of } \psi. \tag{1.1}$$

The following proposition is straightforward.

Proposition 1.39 *Let* $\mathsf{S} = \langle S, \models_S \rangle$ *be a denotational semantics for* $\mathcal{L}$. *Then the relation* $\vdash_\mathsf{S}$ *defined in (1.1) is a non-trivial tcr for* $\mathcal{L}$.

Note 1.40 One may consider a multiple-conclusion counterpart of (1.1), where $\mathcal{T} \vdash_\mathsf{S} \mathcal{S}$ if every S-model of $\mathcal{T}$ is also an S-model of some formula in $\mathcal{S}$. This relation on $2^{\mathcal{W}(\mathcal{L})} \times 2^{\mathcal{W}(\mathcal{L})}$ is obviously a non-trivial scr for $\mathcal{L}$.

Note 1.41 Despite Proposition 1.39, the pair $\langle \mathcal{L}, \vdash_\mathsf{S} \rangle$ is not always a logic. While the condition on $\models_S$ in Definition 1.37 guarantees non-triviality, the condition of structurality may not be satisfied. In this book we consider only *logical* semantics, i.e. denotational semantics S for which $\langle \mathcal{L}, \vdash_\mathsf{S} \rangle$ is a propositional logic.

In the rest of this section we present two fundamental examples of the use of the semantic approach.

1.2.2 Classical Logic

The most important and useful logic is classical logic. In this section we introduce this logic and review the main facts about it that are needed for the rest of the book.

From the semantic point of view, classical logic is based on the use of two *truth values*: t (for truth) and f (for falsity), as well as *truth-functional* interpretations of the connectives.

1.2 Semantic Approaches to Defining Logics 15

Definition 1.42 (bivalent interpretation) Let $\mathcal{L}$ be propositional language.

- A *two-valued truth table* for an n-ary connective $\diamond$ of $\mathcal{L}$ is an n-ary function $\tilde{\diamond} : \{t, f\}^n \to \{t, f\}$. (Note that if $n = 0$ then $\tilde{\diamond} \in \{t, f\}$.) $\tilde{\diamond}$ is called an *interpretation* of $\diamond$.

- A *bivalent interpretation* for $\mathcal{L}$ is a function $\mathbf{F}$ that assigns a two-valued truth table to each primitive connective of $\mathcal{L}$.

- Let $\mathbf{F}$ be a bivalent interpretation for $\mathcal{L}$, and let $\tilde{\diamond}$ abbreviate $\mathbf{F}(\diamond)$. An $\mathbf{F}$-*valuation* is a function $\nu : \mathcal{W}(\mathcal{L}) \to \{t, f\}$ such that for every n-ary primitive connective $\diamond$ of $\mathcal{L}$ and every $\psi_1, \ldots, \psi_n \in \mathcal{W}(\mathcal{L})$,

$$\nu(\diamond(\psi_1, \ldots, \psi_n)) = \tilde{\diamond}(\nu(\psi_1), \ldots, \nu(\psi_n)).$$

We denote by $\Lambda_{2\mathbf{F}}$ the set of $\mathbf{F}$-valuations for $\mathcal{L}$.

- A *classical (two-valued) semantics* for $\mathcal{L}$ is a pair $\langle \Lambda_{2\mathbf{F}}, \models_\mathbf{F} \rangle$ in which $\mathbf{F}$ is a bivalent interpretation of $\mathcal{L}$, and the satisfaction relation $\models_\mathbf{F}$ is defined by $\nu \models_\mathbf{F} \psi$ if $\nu(\psi) = t$. We denote by $\vdash_{2\mathbf{F}}$ the consequence relation that is induced by the denotational semantics $\langle \Lambda_{2\mathbf{F}}, \models_\mathbf{F} \rangle$.

Note 1.43 Obviously, if $\mathbf{F}$ is a bivalent interpretation of $\mathcal{L}$, then $\langle \Lambda_{2\mathbf{F}}, \models_\mathbf{F} \rangle$ is a denotational semantics for $\mathcal{L}$, $\langle \mathcal{L}, \vdash_{2\mathbf{F}} \rangle$ is a decidable logic, and

$$\mathcal{T} \vdash_{2\mathbf{F}} \psi \text{ iff } \nu(\psi) = t \text{ for every } \nu \in \Lambda_{2\mathbf{F}} \text{ such that } \forall \varphi \in \mathcal{T}\ \nu(\varphi) = t. \quad (1.2)$$

Theorem 1.44 (classical compactness, version I) *If $\mathbf{F}$ is a bivalent interpretation of $\mathcal{L}$ then $\langle \mathcal{L}, \vdash_{2\mathbf{F}} \rangle$ is finitary.*[18]

In this book we officially take as "classical logic" a certain specific logic **CL** which is defined below. However, it should be emphasized that it is just one version of what people call "classical logic", because its language is just one possible choice of 'the language of classical logic'. Thus, some texts use the symbol "&" for conjunction while we use $\wedge$. What is more, some texts use only $\neg$, $\vee$ and $\wedge$ as primitive connectives, while still others (like [212]) prefer e.g. to officially use just $\neg$ and $\supset$. Most of the time the exact choices do not matter. However, they might become significant when we try to compare other logics with "classical logic". In particular, being 'contained in classical logic' is in what follows a crucial property that a logic may or may not have. Since we do not want this important property to depend on our particular choice of 'the language of classical logic' (or on any other choice), we next provide a language-independent definition of this notion.

---
[18] This is a particular case of Theorem 3.17 below.

Definition 1.45 (containment in classical logic) Let $\mathbf{L} = \langle \mathcal{L}, \vdash_\mathbf{L} \rangle$ be a propositional logic.

- Let $\mathbf{F}$ be a bivalent interpretation of $\mathcal{L}$. We say that $\mathbf{L}$ is $\mathbf{F}$-*contained in classical logic* if $\varphi_1, \ldots, \varphi_n \vdash_{2_\mathbf{F}} \psi$ for every $\varphi_1, \ldots, \varphi_n, \psi \in \mathcal{W}(\mathcal{L})$ such that $\varphi_1, \ldots \varphi_n \vdash_\mathbf{L} \psi$.

- $\mathbf{L}$ is *contained in classical logic* if it is $\mathbf{F}$-contained in it for some bivalent interpretation $\mathbf{F}$ for $\mathcal{L}$.

Example 1.46 Almost every logic which is studied in the literature in general, and in this book in particular (beginning with those dealt with in this chapter: classical logic and intuitionistic logic), is contained in classical logic. The only exceptions in this book are the logics in Chapters 4 and 5 whose language includes propositional constants for non-classical truth-values. Another well-known example of a family of logics which are not contained in classical logic is given by the family of Post's n-valued logics for $n > 2$. (See, e.g., [176].)

Note 1.47 If $\mathbf{F}$ is a bivalent interpretation of $\mathcal{L}$, and $\diamond$ is a *defined* connective of $\mathbf{L}$, then $\mathbf{F}$ determines in an obvious way a two-valued truth table (denoted by $\mathbf{F}(\diamond)$) for $\diamond$, which all $\mathbf{F}$-valuations respect. Hence, if $\mathbf{L}$ is ($\mathbf{F}$)-contained in classical logic then so is every extension by definitions of $\mathbf{L}$.

Next, we introduce our official version of classical logic:

Definition 1.48 ($\mathcal{L}_{CL}$, CL)

- $\mathcal{L}_{CL}$ is the language of $\{\supset, \wedge, \vee, \neg\}$, where $\neg$ is unary, while the other three connectives are binary.

- $\mathbf{F}_{CL}$ is the bivalent interpretation of $\mathcal{L}_{CL}$ defined by $\mathbf{F}_{CL}(\diamond) = \tilde{\diamond}$, where for every $a, b \in \{t, f\}$:

$$a \mathbin{\tilde{\supset}} b = \begin{cases} f & \text{if } a = t \text{ and } b = f, \\ t & \text{otherwise.} \end{cases}$$

$$a \mathbin{\tilde{\wedge}} b = \begin{cases} t & \text{if } a = t \text{ and } b = t, \\ f & \text{otherwise.} \end{cases}$$

$$a \mathbin{\tilde{\vee}} b = \begin{cases} t & \text{if } a = t \text{ or } b = t, \\ f & \text{otherwise.} \end{cases}$$

$$\tilde{\neg} a = \begin{cases} f & \text{if } a = t, \\ t & \text{if } a = f. \end{cases}$$

1.2 Semantic Approaches to Defining Logics   17

- **CL** is the logic $\langle \mathcal{L}_{CL}, \vdash_{CL} \rangle$, where $\vdash_{CL}$ is the relation that is obtained from $\mathbf{F}_{CL}$ by (1.2) in Note 1.43.

Proposition 1.49 *CL is decidable and self-extensional.*

One reason for our choice of **CL** is its following well-known property:

Proposition 1.50 *With $\mathbf{F}_{CL}$ interpreting its connectives, $\mathcal{L}_{CL}$ is functionally complete: for every $g : \{t,f\}^n \to \{t,f\}$ there is a connective $\diamond$ of **CL** such that $\mathbf{F}_{CL}(\diamond) = g$ if $n > 0$, and $\mathbf{F}_{CL}(\diamond) = \lambda x \in \{t,f\}.g$ otherwise.*

Note 1.51 The last proposition holds for each sublanguage of $\mathcal{L}_{CL}$ which includes $\neg$ and at least one of the other primitive connectives of $\mathcal{L}_{CL}$.

Corollary 1.52 *Let **L** be a logic that is **F**-contained in classical logic for some bivalent interpretation **F**, and let $\diamond$ be a connective of **L**. Then there exists a connective $\diamond_{CL}$ of **CL** such that $\mathbf{F}_{CL}(\diamond_{CL}) = \mathbf{F}(\diamond)$.*

The next proposition provides the *main* reason for our choice of **CL**.

Proposition 1.53 *$\supset$ is an implication for **CL**, $\vee$ is a disjunction for **CL**, and $\wedge$ is a conjunction for **CL**. Hence **CL** is normal.*

Next, we show that the interpretations given by $\mathbf{F}_{CL}$ to the binary primitive connectives of **CL** are the only ones that make Proposition 1.53 true.

Proposition 1.54 *Let $\mathbf{L} = \langle \mathcal{L}, \vdash_{\mathbf{L}} \rangle$ be a logic, and let $\mathbf{F}$ be a bivalent interpretation for $\mathcal{L}$ such that **L** is **F**-contained in classical logic.*

1. *If $\diamond$ is an implication for **L**, or even just a semi-implication for **L**, then $\mathbf{F}(\diamond) = \mathbf{F}_{CL}(\supset)$.*

2. *If $\diamond$ is a conjunction for **L**, then $\mathbf{F}(\diamond) = \mathbf{F}_{CL}(\wedge)$.*

3. *If $\diamond$ is a disjunction for **L**, then $\mathbf{F}(\diamond) = \mathbf{F}_{CL}(\vee)$.*

Proof.

1. The more general case is proved in Corollary 11.48. Here we prove only the case where $\diamond$ is an implication for **L**. So, assume this, and let $\mathbf{F}(\diamond) = \diamond_{\mathbf{F}}$. Since $p \vdash_{\mathbf{L}} p$, also $\vdash_{\mathbf{L}} p \diamond p$. Hence, $\vdash_{2_{\mathbf{F}}} p \diamond p$, and so necessarily $t \diamond_{\mathbf{F}} t = f \diamond_{\mathbf{F}} f = t$. Next, $p \vdash_{\mathbf{L}} q \diamond q$, and since $\diamond$ is an implication for **L**, $\vdash_{\mathbf{L}} p \diamond (q \diamond q)$. Hence, also $\vdash_{2_{\mathbf{F}}} p \diamond (q \diamond q)$. Since $t \diamond_{\mathbf{F}} t = f \diamond_{\mathbf{F}} f = t$, this implies that $f \diamond_{\mathbf{F}} t = t$. Finally, $p \diamond q \vdash_{\mathbf{L}} p \diamond q$, and since $\diamond$ is an implication for **L**, $p \diamond q, p \vdash_{\mathbf{L}} q$. Hence, also $p \diamond q, p \vdash_{2_{\mathbf{F}}} q$, and so $t \diamond_{\mathbf{F}} f = f$. (Otherwise $\nu(p) = t, \nu(q) = f$ would have been a counter-example.)

2. Suppose that $\diamond$ is a conjunction for **L**, and let $\mathbf{F}(\diamond) = \diamond_\mathbf{F}$. Since $p \diamond q \vdash_\mathbf{L} p$ (see Note 1.25), also $p \diamond q \vdash_{2\mathbf{F}} p$, implying that $f \diamond_\mathbf{F} t = f$ and $f \diamond_\mathbf{F} f = f$. Similarly, since $p \diamond q \vdash_\mathbf{L} q$ also $p \diamond q \vdash_{2\mathbf{F}} q$, and so $t \diamond_\mathbf{F} f = f$. Finally, since $p, q \vdash_\mathbf{L} p \diamond q$ also $p, q \vdash_{2\mathbf{F}} p \diamond q$, and so $t \diamond_\mathbf{F} t = t$. (Otherwise, $\nu(p) = \nu(q) = t$ would have been a counter-example.)

3. Suppose that $\diamond$ is a disjunction for **L**, and let $\mathbf{F}(\diamond) = \diamond_\mathbf{F}$. Since $p \diamond q \vdash_\mathbf{L} p \diamond q$, and $\diamond$ is a disjunction for **L**, $p \vdash_\mathbf{L} p \diamond q$, and $q \vdash_\mathbf{L} p \diamond q$. Hence, also $p \vdash_{2\mathbf{F}} p \diamond q$, and $q \vdash_{2\mathbf{F}} p \diamond q$, implying that $t \diamond_\mathbf{F} t = t \diamond_\mathbf{F} f = f \diamond_\mathbf{F} t = t$. Finally, since $p \vdash_\mathbf{L} p$ and $p \vdash_\mathbf{L} p$, the assumption that $\diamond$ is a disjunction for **L** implies that $p \diamond p \vdash_\mathbf{L} p$, and so $p \diamond p \vdash_{2\mathbf{F}} p$. Hence $f \diamond_\mathbf{F} f = f$. (Otherwise, $\nu(p) = f$ would have been a counter-example.) □

Next, we describe a particularly important fragment of **CL**.

Definition 1.55 ($\mathcal{L}_{CL}^+$ **and CL**$^+$) **CL**$^+$ is the $\mathcal{L}_{CL}^+$-fragment of **CL**, where $\mathcal{L}_{CL}^+ = \{\supset, \wedge, \vee\}$.

Proposition 1.56

1. **CL**$^+$ is a positive logic. (In particular: it is a logic.)

2. Let $\diamond$ be an n-ary connective of **CL**$^+$, and let $\mathbf{F}_{CL}(\diamond) = \tilde{\diamond}$. Then $\tilde{\diamond}(t, t, \ldots, t) = t$. Conversely, if g is a function from $\{t, f\}^n$ to $\{t, f\}$ such that $n > 0$ and $g(t, t, \ldots, t) = t$, then there exists an n-ary connective $\diamond$ of **CL**$^+$ such that $\mathbf{F}_{CL}(\diamond) = g$.

Note 1.57 Since $\varphi \vee \psi$ is equivalent in **CL**$^+$ to $(\varphi \supset \psi) \supset \psi$, the last proposition holds already for the $\{\supset, \wedge\}$-fragment of **CL**$^+$.

Corollary 1.58 Let $\diamond$ be a positive connective of a logic **L** which is **F**-contained in classical logic for some bivalent interpretation **F**. Then there exists a connective $\diamond_{cl^+}$ of **CL**$^+$ such that $\mathbf{F}_{CL}(\diamond_{CL^+}) = \mathbf{F}(\diamond)$.

Proof. Since $\diamond$ is a positive connective of **L**, $p_1, \ldots, p_n \vdash_\mathbf{L} \diamond(p_1, \ldots, p_n)$, and so $p_1, \ldots, p_n \vdash_{2\mathbf{F}} \diamond(p_1, \ldots, p_n)$, because **L** is **F**-contained in classical logic (see Notes 1.22 and 1.47). This immediately implies that $\mathbf{F}(\diamond)(t, t, \ldots, t) = t$. Hence, Proposition 1.56 entails the existence of the required $\diamond_{cl^+}$. □

Note 1.59 In less precise words, Proposition 1.56 and Corollary 1.58 mean that a classical connective is positive iff it is definable in **CL**$^+$, and that a logic which is **F**-contained in classical logic is positive iff it is actually **F**-contained in **CL**$^+$. Accordingly, **CL**$^+$ is called the *positive fragment of classical logic*, or just *positive classical logic*.

1.2 Semantic Approaches to Defining Logics

Note 1.60 By the second part of Proposition 1.56, $\mathcal{L}_{CL}^+$ (with the interpretations of its connectives as provided by $\mathbf{F}_{CL}$) is not functionally complete for $\{t, f\}$. It can be shown that by adding to it any new n-ary connective $\diamond$ whose interpretation is not positive (i.e., $\tilde{\diamond}(t, \ldots, t) = f$), we get a language which *is* functionally complete for $\{t, f\}$. The simplest choices are:

- The classical negation $\neg$, as defined above.

- The propositional constant F, which is assigned f by any bivalent valuation. ($\neg \varphi$ is then defined as $\varphi \supset \mathsf{F}$).

Of these two choices, the more natural for classical logic is the one we have made above: $\neg$. However, the other one is useful too. (Thus it is more natural to use it instead of $\neg$ in the context of intuitionistic logic, which is described later in this chapter). Hence, we define:

Definition 1.61 ($\mathcal{L}_{CL}^{\mathsf{F}}$, $\mathcal{L}_{CL}^{\mathsf{F}\neg}$, $\mathbf{CL}^{\mathsf{F}}$, $\mathbf{CL}^{\mathsf{F}\neg}$)

- $\mathcal{L}_{CL}^{\mathsf{F}}$ is the language of $\{\wedge, \vee, \supset, \mathsf{F}\}$. $\mathbf{CL}^{\mathsf{F}}$ is classical logic in $\mathcal{L}_{CL}^{\mathsf{F}}$.

- $\mathcal{L}_{CL}^{\mathsf{F}\neg}$ is the language of $\{\wedge, \vee, \supset, \neg, \mathsf{F}\}$. $\mathbf{CL}^{\mathsf{F}\neg}$ is classical logic in $\mathcal{L}_{CL}^{\mathsf{F}\neg}$.

Next, we list the most basic properties that $\neg$ has with respect to $\vdash_{CL}$.

Definition 1.62 (consistent and inconsistent theories) A theory $\mathcal{T}$ in $\mathcal{L}_{CL}$ is *(classically) inconsistent* if there exists a formula ψ such that both $\mathcal{T} \vdash_{CL} \psi$ and $\mathcal{T} \vdash_{CL} \neg\psi$. $\mathcal{T}$ is *(classically) consistent* if it is not inconsistent.

Proposition 1.63 *Let $\mathcal{T}$ be a theory and ψ, φ be formulas in $\mathcal{L}_{CL}$.*

1. *If $\mathcal{T}, \psi \vdash_{CL} \varphi$ and $\mathcal{T}, \neg\psi \vdash_{CL} \varphi$ then $\mathcal{T} \vdash_{CL} \varphi$.*

2. *If $\mathcal{T}$ is inconsistent then $\mathcal{T} \vdash_{CL} \varphi$. Equivalently: $\psi, \neg\psi \vdash_{CL} \varphi$.*

3. $\psi \vdash_{CL} \neg\neg\psi$.

4. $\neg\neg\psi \vdash_{CL} \psi$.

5. *If $\mathcal{T}, \psi \vdash_{CL} \varphi$ then $\mathcal{T}, \neg\varphi \vdash_{CL} \neg\psi$.*

6. *If $\mathcal{T}, \psi \vdash_{CL} \varphi$, and $\mathcal{T}, \psi \vdash_{CL} \neg\varphi$, then $\mathcal{T} \vdash_{CL} \neg\psi$.*

Theorem 1.64 (classical compactness, version II) *A theory is classically consistent iff every finite subtheory of it is classically consistent.*

The next proposition describes a very important property of classical logic and its positive fragment.

Proposition 1.65 (maximality of $\mathbf{CL}^+$ and $\mathbf{CL}$) *No logic properly extends $\mathbf{CL}$ in its language. The same is true for $\mathbf{CL}^+$, $\mathbf{CL}^{\mathsf{F}}$, and $\mathbf{CL}^{\mathsf{F}\neg}$.*

1.2.3 Intuitionistic Logic

Another important propositional logic is intuitionistic logic — the central logic in the family of constructive logics ([85, 145, 284]). The main idea behind these logics is that a statement can be taken as true only if it has a *constructive proof*. However, in this section we introduce propositional intuitionistic logic using a *semantic* characterization of it (that will be useful in some chapters of the book, especially Chapter 10).

Definition 1.66 (Kripke frames) A *Kripke frame* for $\mathcal{L}_{CL}^+$ is a triple $\langle W, \leq, \nu \rangle$[19] in which $\langle W, \leq \rangle$ is a nonempty finite[20] partially ordered set (of "worlds"), and $\nu : W \times \mathcal{W}(\mathcal{L}_{CL}^+) \to \{t, f\}$ satisfies the following conditions:

- If $w' \geq w$ and $\nu(w, \psi) = t$ then $\nu(w', \psi) = t$.[21]
- $\nu(w, \psi \wedge \varphi) = t$ iff $\nu(w, \psi) = t$ and $\nu(w, \varphi) = t$,
  $\nu(w, \psi \vee \varphi) = t$ iff $\nu(w, \psi) = t$ or $\nu(w, \varphi) = t$,
  $\nu(w, \psi \supset \varphi) = t$ iff $\nu(w', \varphi) = t$ whenever $w' \geq w$ and $\nu(w', \psi) = t$.

Definition 1.67 ($\mathbf{IL}^+$)

- $\mathbf{IL}^+$, the Kripke denotational semantics for $\mathcal{L}_{CL}^+$, is the pair $\langle S, \models_S \rangle$ in which S is the class[22] of Kripke frames for $\mathcal{L}_{CL}^+$, and the satisfiability relation $\models_S$ is defined by:

  $\langle W, \leq, \nu \rangle \models_S \psi$ if $\nu(x, \psi) = t$ for every $x \in W$.

- Let $\mathbf{IL}^+ = \langle \mathcal{L}_{CL}^+, \vdash_{IL^+} \rangle$, where $\vdash_{IL^+}$ is the consequence relation that is induced by IL^+. $\mathbf{IL}^+$ is called *positive intuitionistic logic*.

Note 1.68 Obviously, if $S = \langle W, \leq, \nu \rangle$ is a frame, then for every $w \in W$ the function $\lambda \psi. \nu(w, \psi)$ behaves like an ordinary classical valuation with respect to $\wedge$ and $\vee$, but not with respect to $\supset$.

Proposition 1.69 (properties of $\mathbf{IL}^+$)

1. $\mathbf{IL}^+$ *is a positive logic (in particular: it is a logic).*

---

[19] In the literature one usually means by a "frame" just the pair $\langle W, \leq \rangle$. Here we find it convenient to follow e.g. [225], and use this technical term a little bit differently, so that the valuation ν is an integral part of it.

[20] In the propositional case we may restrict ourselves to finite frames. This ensures the decidability of the induced logic. In the first-order case this restriction should be dropped.

[21] Demanding this just for the case in which ψ is a propositional variable suffices here for having it for every ψ.

[22] In order to make S a set rather than a class (as demanded in Definition 1.37), we may assume without loss of generality that "worlds" are just natural numbers.

1.2 Semantic Approaches to Defining Logics 21

2. $\mathbf{IL}^+$ is decidable.

3. $\mathbf{IL}^+$ is self-extensional.

4. $\mathbf{IL}^+$ is $\mathbf{F}_{CL}$-contained in classical logic (and actually in the positive fragment of classical logic).

5. $\mathbf{IL}^+$ is normal: $\supset$ is an implication for IL^+, $\vee$ is a disjunction for it, and $\wedge$ is a conjunction for it.

The next proposition is a sort of converse to the last item of Proposition 1.69, and describes a very important property of $\mathbf{IL}^+$.

Proposition 1.70 (minimality of $\mathbf{IL}^+$) Let $\mathbf{L}$ be a finitary logic.

1. $\mathbf{L}$ has an implication $\supset$ iff it is an axiomatic extension of the $\{\supset\}$-fragment of $\mathbf{IL}^+$.

2. $\mathbf{L}$ is a normal logic, with an implication $\supset$, a conjunction $\wedge$, and a disjunction $\vee$, iff $\mathbf{L}$ is an axiomatic extension of $\mathbf{IL}^+$.

Corollary 1.71 $\mathbf{CL}^+$ and $\mathbf{CL}$ are axiomatic extensions of $\mathbf{IL}^+$.

Note 1.72 $\mathbf{IL}^+$ serves as the basis of all the constructive paraconsistent logics studied in this book. (See Chapter 10.)

Definition 1.73 ($\mathbf{IL}$)

- A *Kripke frame* for $\mathcal{L}_{CL}^{\mathsf{F}}$ is a triple $\langle W, \leq, \nu \rangle$, in which $\langle W, \leq \rangle$ is again a nonempty partially ordered set, $\nu : W \times \mathcal{W}(\mathcal{L}_{CL}^{\mathsf{F}}) \to \{t, f\}$, and ν satisfies all the conditions listed in Definition 1.66, as well as the following one:

    - For every $w \in W$, $\nu(w, \mathsf{F}) = f$.

- IL, the Kripke-style denotational semantics for $\mathcal{L}_{CL}^{\mathsf{F}}$, is the pair $\langle S, \models_S \rangle$, in which S is the class (see Footnote 22) of Kripke frames for $\mathcal{L}_{CL}^{\mathsf{F}}$, and $\models_S$ is defined like in Definition 1.67.

- Let $\mathbf{IL} = \langle \mathcal{L}_{CL}^{\mathsf{F}}, \vdash_{IL} \rangle$, where $\vdash_{IL}$ is the consequence relation that is induced by IL. $\mathbf{IL}$ is called (propositional) *intuitionistic logic*.

Note 1.74 The connective that serves in intuitionistic logic as its official negation is defined in $\mathcal{L}_{CL}^{\mathsf{F}}$ as follows:

$$\neg \varphi =_{Df} \varphi \supset \mathsf{F}.$$

It is easy to see that if $\langle W, \leq, \nu \rangle$ is a Kripke frame for $\mathcal{L}_{CL}^{\mathsf{F}}$, then

- $\nu(w, \neg\psi) = t$ iff $\nu(w', \psi) = f$ for every $w' \geq w$.

Since F is in turn definable in terms of the intuitionistic implication and negation, it is possible also to take $\mathcal{L}_{CL}$ as the language of **IL**, and we shall do so whenever it would be convenient. (The last item in the next proposition is a case in point.)

Proposition 1.75 (properties of IL)

1. *IL is a normal logic (with $\supset$, $\vee$, and $\wedge$ having their usual roles).*
2. *IL is decidable.*
3. *IL is self-extensional.*
4. *IL is $\mathbf{F}_{CL}$-contained in classical logic.*

Now we list the most important properties of the intuitionistic negation.

Proposition 1.76 (properties of the intuitionistic negation)

1. *If $\mathcal{T} \vdash_{IL} \psi$ and $\mathcal{T} \vdash_{IL} \neg\psi$ then $\mathcal{T} \vdash_{IL} \varphi$. Equivalently: $\psi, \neg\psi \vdash_{IL} \varphi$.*
2. *$\psi \vdash_{IL} \neg\neg\psi$.*
3. *If $\mathcal{T}, \psi \vdash_{IL} \varphi$ then $\mathcal{T}, \neg\varphi \vdash_{IL} \neg\psi$.*
4. *If $\mathcal{T}, \psi \vdash_{IL} \varphi$, and $\mathcal{T}, \psi \vdash_{IL} \neg\varphi$, then $\mathcal{T} \vdash_{IL} \neg\psi$.*

Corollary 1.77 $\mathbf{F}_{CL}$ *is the only bivalent interpretation $\mathbf{F}$ of $\mathcal{L}_{CL}$ such that IL is $\mathbf{F}$-contained in classical logic.*

Proof. $\mathbf{F}_{CL}$ has this property by the last item of Proposition 1.75. For the converse, assume that **IL** is $\mathbf{F}$-contained in classical logic. Proposition 1.54 and Item 1 of Proposition 1.75 imply that $\mathbf{F}(\diamond) = \mathbf{F}_{CL}(\diamond)$ for $\diamond \in \mathcal{L}_{CL}^+$. It remains to prove that $\neg$ has this property too. Let $\mathbf{F}(\neg) = \neg_\mathbf{F}$. Since $p, \neg p \vdash_{IL} q$ (first item of Proposition 1.76), necessarily $\neg_\mathbf{F}(t) = f$ (otherwise $\nu(p) = t, \nu(q) = f$ would have been a counter-example). It follows from this and from $p \vdash_{IL} \neg\neg p$ (second item of Proposition 1.76) that $\neg_\mathbf{F}(f) = t$ (otherwise $\nu(p) = t$ would have been a counter-example). □

In the next proposition we list the main classical tautologies which are not intuitionistically valid:

Proposition 1.78

1. $\not\vdash_{IL} \neg p \vee p$.

2. $\not\vdash_{IL} \neg\neg p \supset p$.

3. $\not\vdash_{IL} ((p \supset q) \supset p) \supset p$.

4. $\not\vdash_{IL} (p \supset q) \vee (q \supset p)$.

Note 1.79 *Johansson's minimal logic* [187] is defined like **IL**, but without the condition about ν and F given in Definition 1.73. In this logic $\neg$ has the properties listed in Proposition 1.76, with the exception of the first one.

1.3 Syntactic Approaches to Defining Logics

1.3.1 Proof Systems

The syntactic (or proof-theoretic) approach to defining logics is based on the notion of a *proof* in some formal calculus. In practice, a formal proof of a formula ψ in a language $\mathcal{L}$ is a finite sequence (or tree) of finite syntactical expressions which are based on $\mathcal{L}$. Since this explanation is somewhat vague, we next provide an abstract definition of what is meant in this book by a "proof system".[23]

Definition 1.80 (proof system) Let $\mathcal{L}$ be a propositional language.

- A *proof system* for $\mathcal{L}$ is a ternary relation P which has the following properties:

  1. *Effective enumerability of proofs:* There is an effectively enumerable set Pr (i.e. there is an effective function from the set of natural numbers *onto* Pr) such that all elements of P are triples of the form $\langle \mathcal{T}, \psi, d \rangle$ in which $\mathcal{T} \subseteq \mathcal{W}(\mathcal{L})$, $\psi \in \mathcal{W}(\mathcal{L})$, and $d \in Pr$.
  2. *Reflexivity:* For every $\psi \in \mathcal{W}(\mathcal{L})$ there is some $d \in Pr$ such that $\langle \{\psi\}, \psi, d \rangle \in$ P.
  3. *Monotonicity:* if $\langle \mathcal{T}, \psi, d \rangle \in$ P and $\mathcal{T} \subseteq \mathcal{T}'$, then $\langle \mathcal{T}', \psi, d \rangle \in$ P.
  4. *Transitivity:* there is an effective binary operation *append* on Pr such that if $\langle \mathcal{T}_1, \psi, d_1 \rangle \in$ P and $\langle \mathcal{T}_2 \cup \{\psi\}, \varphi, d_2 \rangle \in$ P, then $\langle \mathcal{T}_1 \cup \mathcal{T}_2, \varphi, append(d_1, d_2) \rangle \in$ P.
  5. *Structurality:* There is an effective procedure for extending any $\mathcal{L}$-substitution θ to Pr so that for all $\mathcal{T}, \psi, d, \theta$: if $\langle \mathcal{T}, \psi, d \rangle \in$ P, then also $\langle \theta(\mathcal{T}), \theta(\psi), \theta(d) \rangle \in$ P.

---

[23] While the following definition covers most of the known proof systems in the literature, it is not exhaustive. For instance, proof systems for non-monotonic formalisms like adaptive logics, conditional logics, or argumentation-based logics, are not monotonic (see, e.g., [22, 230, 274]).

6. *Non-triviality:* If p and q are distinct variables, then there is no $d \in Pr$ such that $\langle \{p\}, q, d \rangle \in \mathsf{P}$.

7. *Finitariness:* if $\langle \mathcal{T}, \psi, d \rangle \in \mathsf{P}$, then there is a *finite subset* $\Gamma \subseteq \mathcal{T}$ such that $\langle \Gamma, \psi, d \rangle \in \mathsf{P}$.

8. *Effectiveness:* there is an effective procedure that for every finite $\Gamma \subseteq \mathcal{W}(\mathcal{L})$, $\psi \in \mathcal{W}(\mathcal{L})$ and $d \in Pr$, determines whether $\langle \Gamma, \psi, d \rangle \in \mathsf{P}$ or not.

- Let P be a proof system for $\mathcal{L}$. We define the relation $\vdash_\mathsf{P}$ between theories and formulas of $\mathcal{L}$ by: $\mathcal{T} \vdash_\mathsf{P} \psi$ if $\langle \mathcal{T}, \psi, d \rangle \in \mathsf{P}$ for some $d \in Pr$. In this case d is called a *proof* (or a *derivation*) in P of ψ from $\mathcal{T}$, and we say that ψ is *derivable* in P from $\mathcal{T}$. Formulas that are derivable in P from the empty set are called *theorems* of P.

- Let P be a proof system for $\mathcal{L}$. Define $\mathbf{L_P} = \langle \mathcal{L}, \vdash_\mathsf{P} \rangle$.

The proof of the following Proposition is straightforward.

Proposition 1.81 *If P is a proof system for $\mathcal{L}$ then $\mathbf{L_P}$ is a propositional logic. This logic is finitary and semi-decidable (see Definition 1.17).*

Definition 1.82 (soundness and completeness) Let $\mathbf{L} = \langle \mathcal{L}, \vdash_\mathbf{L} \rangle$ be a propositional logic and let P be a proof system for $\mathcal{L}$.

- P is *sound* for $\mathbf{L}$ if $\mathbf{L}$ is an extension of $\mathbf{L_P}$.

- P is *weakly sound* for $\mathbf{L}$ if $\vdash_\mathbf{L} \psi$ whenever $\vdash_\mathsf{P} \psi$.

- P is *complete* for $\mathbf{L}$ if $\mathbf{L_P}$ is an extension of $\mathbf{L}$.

- P is *weakly complete* for $\mathbf{L}$ if $\vdash_\mathsf{P} \psi$ whenever $\vdash_\mathbf{L} \psi$.

In what follows, when saying that a proof system P is sound/complete for $\vdash$, we shall mean that it is sound/complete for its logic, i.e. for $\langle \mathcal{L}, \vdash \rangle$.

There are a number of different types of proof systems that are widely used in the literature, like Hilbert-type systems, natural deduction systems, sequent calculi (also known as Gentzen-type systems[24]), and analytic tableaux. In this book we shall use Hilbert-type systems and sequent calculi. Below we briefly present both, together with some useful basic examples.

---

[24]Both natural deduction systems and Gentzen-type systems are due to Gerhard Gentzen [170].

1.3 Syntactic Approaches to Defining Logics

1.3.2 Hilbert-type Systems

Hilbert-type proof systems (also known as Frege-type proof systems) form the simplest and most widely used class of proof systems.

Definition 1.83 (Hilbert-type proof system) Let $\mathcal{L}$ be a propositional language.

- A *Hilbert-type proof system* is a proof system (Definition 1.80) in which the set Pr of proofs is the set of finite sequences of formulas of $\mathcal{L}$.

- Let S be a finite set of rules in $\mathcal{L}$ (Definition 1.11). The *Hilbert-type proof system for $\mathcal{L}$ which is based on S*, denoted H_S, is the set of all triples $\langle \mathcal{T}, \psi, d \rangle$ such that:
  - d is a finite sequence $\varphi_1, \ldots, \varphi_n$ of formulas in $\mathcal{L}$.
  - $\varphi_n = \psi$.
  - Each φ_i ($1 \leq i \leq n$) is either an element of $\mathcal{T}$, or is the conclusion of an application of some rule in S whose premises are included in $\{\varphi_1, \ldots, \varphi_{i-1}\}$.[25]

Note 1.84 Let $\mathbf{L}_0 = \langle \mathcal{L}, \vdash_0 \rangle$ be the trivial propositional logic on $\mathcal{L}$ for which $\mathcal{T} \vdash_0 \psi$ iff $\psi \in \mathcal{T}$. It is easy to see that in the notations of Definitions 1.80 and 1.14, $\mathcal{T} \vdash_{\mathsf{H}_S} \psi$ iff there is a finite $\Gamma \subseteq \mathcal{T}$ such that $\langle \Gamma, \psi \rangle \in C_{\mathbf{L}_0}(S)$.

Note 1.85 It is easy to see that if S is a finite set of rules in $\mathcal{L}$ then H_S satisfies all the conditions listed in Definition 1.80, with one possible exception: the condition of non-triviality. It follows that $\mathbf{L}_{\mathsf{H}_S}$ is a logic iff there is no proof in it of q from $\{p\}$ in case p and q are two distinct variables. This condition is easily verified (using soundness with respect to some obvious two-valued semantics) for every set of rules we use in the sequel. *Therefore, we shall always (implicitly) leave this task to the reader.*

Definition 1.86 (axiomatic extension of a system) Suppose that H_S is a Hilbert-type system in the language $\mathcal{L}$. A Hilbert-type system $\mathsf{H}_{S'}$ in the language $\mathcal{L}'$ is an *axiomatic extension* of H_S if $\mathcal{L} \subseteq \mathcal{L}'$, $S \subseteq S'$, and $S' - S$ contains only axioms (i.e. rules of the form $\emptyset/\psi$ — See Definition 1.11).

Proposition 1.87 *If a Hilbert-type system $\mathsf{H}_{S'}$ in the language $\mathcal{L}'$ is an axiomatic extension of a Hilbert-type system H_S in the language $\mathcal{L}$ then $\mathbf{L}_{\mathsf{H}_{S'}}$ is an axiomatic extension of $\mathbf{L}_{\mathsf{H}_S}$.*

---

[25] That is, there is a rule $\Gamma/\varphi \in S$, an $\mathcal{L}$-substitution θ, and a set $\Gamma' \subseteq \{\varphi_1, \ldots, \varphi_{i-1}\}$ such that $\theta(\Gamma) = \Gamma'$ and $\theta(\varphi) = \varphi_i$.

Note 1.88 In practice, most of the rules in a set S which is used for constructing a Hilbert-type system $\mathbf{H}_S$ are axiom schemas, and the general tendency is to use as few inference rules (see Definition 1.11) as possible. In fact, most of the Hilbert-type systems presented in this book employ just one inference rule, and all their other rules are axioms. The systems for classical logic and for intuitionistic logic presented next are good examples.

Hilbert-type systems for the basic logics

Definition 1.89 ($HIL_\supset$, HIL^+, HIL, $HCL_\supset$, HCL^+, HCL) In Figure 1.1 there are Hilbert-type proof systems for the four basic logics which were introduced in the previous section: HCL and HCL^+ (respectively) for classical propositional logic and its purely positive fragment; HIL and HIL^+ (respectively) for intuitionistic propositional logic and its purely positive fragment. In addition, it contains Hilbert-type proof systems for the pure implicational fragments of these logics (called $HCL_\supset$ and $HIL_\supset$). As this figure shows, all these Hilbert-type systems share the same single inference rule [MP], but they differ with respect to their set of axioms. (Note that in the formulation of these axioms in Figure 1.1 the association of nested implications is taken to the right.)

Theorem 1.90 (soundness and completeness)

1. $HIL_\supset$ is sound and complete for the $\{\supset\}$-fragment of $\mathbf{IL}$.

2. HIL^+ is sound and complete for $\mathbf{IL}^+$.

3. HIL is sound and complete for $\mathbf{IL}$.

4. $HCL_\supset$ is sound and complete for the $\{\supset\}$-fragment of $\mathbf{CL}$.

5. HCL^+ is sound and complete for $\mathbf{CL}^+$.

6. HCL is sound and complete for $\mathbf{CL}$.

Note 1.91 (variations of HCL) The system HCL is used, e.g., in [153]. Its main advantage for the purposes of this book is that its two axioms for $\neg$ exactly reflect (and also dictate) the truth table of $\tilde{\neg}$: [t] implies that $\tilde{\neg}f = t$, while $[\neg\supset]$ implies that $\tilde{\neg}t = f$. Moreover: the use of HCL allows us to exactly separate between the paraconsistent part of classical logic (HCL without $[\neg\supset]$) and the direct source of the fact that a single contradiction implies in it every formula ($[\neg\supset]$). However, in the literature there are many other Hilbert-type representations of $\mathbf{CL}$. We review some below.

1.3 Syntactic Approaches to Defining Logics   27

Inference Rule: [MP] $\dfrac{\psi \quad \psi \supset \varphi}{\varphi}$

Axioms of $HIL_\supset$:

  [$\supset 1$] $\psi \supset (\varphi \supset \psi)$
  [$\supset 2$] $(\psi \supset (\varphi \supset \tau)) \supset ((\psi \supset \varphi) \supset (\psi \supset \tau))$

Axioms of HIL^+: The axioms of $HIL_\supset$ and:

  [$\wedge \supset$] $\psi \wedge \varphi \supset \psi,\ \psi \wedge \varphi \supset \varphi$
  [$\supset \wedge$] $\psi \supset (\varphi \supset \psi \wedge \varphi)$
  [$\supset \vee$] $\psi \supset \psi \vee \varphi,\ \varphi \supset \psi \vee \varphi$
  [$\vee \supset$] $(\psi \supset \tau) \supset ((\varphi \supset \tau) \supset (\psi \vee \varphi \supset \tau))$

Axioms of HIL: The axioms of HIL^+ and:

  [$F \supset$] $F \supset \psi$

Axioms of $HCL_\supset$: The axioms of $HIL_\supset$ and:

  [$\supset 3$] $((\psi \supset \varphi) \supset \psi) \supset \psi$

Axioms of HCL^+: The axioms of HIL^+ and [$\supset 3$]

Axioms of HCL: The axioms of HIL^+ and:

  [t] $\neg \psi \vee \psi$
  [$\neg \supset$] $\neg \psi \supset (\psi \supset \varphi)$

Figure 1.1: The proof systems $HIL_\supset$, HIL^+, HIL, $HCL_\supset$, HCL^+ and HCL

- Let HCL_1 be the system which is obtained from HCL by replacing axiom [t] with either $(\neg\varphi \supset \varphi) \supset \varphi$ or [26]

$$[\supset \neg] \quad (\psi \supset \varphi) \supset (\neg\psi \supset \varphi) \supset \varphi.$$

Then, unlike HCL, HCL_1 is *well-axiomatized* with respect to $\{\neg, \supset\}$. This means that each of its fragments that contains $\supset$ and $\neg$ is sound and complete with respect to the corresponding fragment of **CL**. For example: the Hilbert-type system which has MP as its inference rule and $[\supset 1]$, $[\supset 2]$, $[\supset \neg]$, and $[\neg \supset]$ as axioms, is sound and complete with respect to the $\{\neg, \supset\}$-fragment of classical logic.

- Let HCL_2 be the system which is obtained from HCL^+ (not just HIL^+!) By the addition of $[\neg \supset]$ and the following axiom:

$$[\supset \neg]_I \quad (\psi \supset \varphi) \supset (\psi \supset \neg\varphi) \supset \neg\psi.$$

Then HCL_2 is well-axiomatized with respect to $\{\supset\}$ (or just *well-axiomatized*): each of its fragments that contains $\supset$ is sound and complete with respect to the corresponding fragment of **CL** (note that HCL_1 does not have this property, and that $[\supset 3]$ is derivable in HCL and HCL_1, but is needed here).

- Let HCL_3 be the system which is obtained from HCL by replacing its two axioms for negation (i.e. [t] and $[\neg \supset]$) with $[\supset \neg]_I$ and the following double-negation axiom:

$$[c] \quad \neg\neg\psi \supset \psi.$$

HCL_3 too is sound and complete for **CL**. Its main advantage is that its axioms exactly mirror the usual natural deduction rules of **CL**.

- Finally, let HCL^F be the Hilbert-type system in the language $\mathcal{L}_{CL}^F$ (see Definition 1.73 and Note 1.60) which is obtained from HCL^+ by adding to it the axiom $[F \supset]$ of HIL. Then HCL^F is sound and complete for $\mathbf{CL}^F$. (Still another Hilbert-type system for $\mathbf{CL}^F$ is obtained by adding to HIL^+ the axiom $((\psi \supset F) \supset F) \supset \psi$.)

Note 1.92 (a variation of *HIL*) A Hilbert-type system in $\mathcal{L}_{CL}$ which is sound and complete for **IL** is obtained by deleting $[\supset 3]$ from HCL_2 (i.e. by adding to HIL^+ the axioms $[\supset \neg]_I$ and $[\neg \supset]$).

---

[26]$[\supset \neg]$ is equivalent to [t] over HIL^+, and to the schema $(\neg\varphi \supset \varphi) \supset \varphi$ over $HCL_\supset$.

1.3 Syntactic Approaches to Defining Logics 29

Note 1.93 (Hilbert system for minimal logic) A Hilbert-type system for Johansson's minimal logic (see Note 1.79) is obtained by extending the system HIL^+ to the language $\mathcal{L}_{CL}^F$ (i.e., retaining the same axioms and rules, but applications are made in the richer language). A Hilbert-type system for this logic in the language $\mathcal{L}_{CL}$ is obtained by adding to HIL^+ the axiom $[\supset \neg]_I$. (By Note 1.92, intuitionistic logic is obtained by adding also $[\neg \supset]$). An alternative system for minimal logic is obtained by adding to HIL^+ the axioms $(\varphi \supset \neg \varphi) \supset \neg \varphi$ and $\neg \psi \supset (\psi \supset \neg \varphi)$.

We end this section with some useful properties of the specific Hilbert-type systems introduced in it, as well as of all their axiomatic extensions.

Theorem 1.94 (deduction theorem) *Suppose $\mathcal{L}$ is a language which includes $\supset$. Let H be an axiomatic extension (Definition 1.86) of $HIL_\supset$ in $\mathcal{L}$.[27] Then $\supset$ is an implication for $\mathbf{L}_H$: $\mathcal{T} \vdash_H \varphi \supset \psi$ iff $\mathcal{T}, \varphi \vdash_H \psi$.*

The following is an easy corollary of Theorem 1.94 and the choice of axioms in HIL^+:

Theorem 1.95 *Let $\mathcal{L}$ be a language which includes $\supset$, $\vee$, and $\wedge$, and let H be an axiomatic extension of HIL^+ in $\mathcal{L}$. Then $\supset$ is an implication for $\mathbf{L}_H$, $\vee$ is a disjunction for it, and $\wedge$ is a conjunction for it.*

Note 1.96 It follows from Theorems 1.95 and 1.90 that $\supset$ is an implication for $\mathbf{CL}$, $\mathbf{CL}^+$, $\mathbf{IL}$, $\mathbf{IL}^+$, $\mathbf{CL}^F$, and $\mathbf{CL}^{F\neg}$; $\vee$ is a disjunction for these logics; and $\wedge$ is a conjunction for them. (These facts follow directly also from the semantic definitions of these logics, and so have already been noted above.)

Theorem 1.97 *Let H be an axiomatic extension of $HIL_\supset$ in the language $\mathcal{L}$. If $\mathcal{T} \nvdash_H \psi$ then there exists a theory $\mathcal{T}^*$ in $\mathcal{L}$ with the following properties:*

1. *$\mathcal{T} \subseteq \mathcal{T}^*$, and for every φ: $\varphi \in \mathcal{T}^*$ iff $\mathcal{T}^* \vdash_H \varphi$.*

2. *$\mathcal{T}^* \nvdash_H \psi$.*

3. *For every φ, either $\mathcal{T}^* \vdash_H \varphi$ or $\mathcal{T}^* \vdash_H \varphi \supset \psi$.*

4. *For every φ and τ, if $\mathcal{T}^* \vdash_H \varphi \supset \tau$ then either $\mathcal{T}^* \nvdash_H \varphi$ or $\mathcal{T}^* \vdash_H \tau$, and if $\mathcal{T}^* \vdash_H \tau$ then $\mathcal{T}^* \vdash_H \varphi \supset \tau$.*

5. *If the axioms $[\wedge \supset]$ and $[\supset \wedge]$ of HIL^+ are derivable in H, then for every φ and τ, $\mathcal{T}^* \vdash_H \varphi \wedge \tau$ iff both $\mathcal{T}^* \vdash_H \varphi$ and $\mathcal{T}^* \vdash_H \tau$.*

---
[27] Here, e.g., this means that H is a Hilbert-type system for $\mathcal{L}$ which has [MP] for $\supset$ as its sole rule of inference, and $\vdash_H \varphi$ for every $\mathcal{L}$-instance φ of $[\supset 1]$ and $[\supset 2]$.

6. If the axioms [∨⊃] and [⊃∨] of HIL^+ are derivable in H, then for every φ and τ, $\mathcal{T}^* \vdash_H \varphi \vee \tau$ iff either $\mathcal{T}^* \vdash_H \varphi$ or $\mathcal{T}^* \vdash_H \tau$.

7. If the axiom [⊃¬] of HCL_1 (see first item of Note 1.91) is derivable in H, then for every φ, either $\mathcal{T}^* \vdash_H \varphi$ or $\mathcal{T}^* \vdash_H \neg\varphi$.[28]

8. If the axiom [¬⊃] of HCL is derivable in H, then for every φ, either $\mathcal{T}^* \nvdash_H \varphi$ or $\mathcal{T}^* \nvdash_H \neg\varphi$.

Proof. Let $\mathcal{T}^*$ be a maximal extension of $\mathcal{T}$ in $\mathcal{L}$ such that $\mathcal{T}^* \nvdash_H \psi$.[29] Obviously, $\mathcal{T}^*$ has Properties 1 and 2. Now assume that $\mathcal{T}^* \nvdash_H \varphi$. Then $\varphi \notin \mathcal{T}^*$, and so the maximality of $\mathcal{T}^*$ implies that $\mathcal{T}^*, \varphi \vdash_H \psi$. Hence $\mathcal{T}^* \vdash_H \varphi \supset \psi$ by the deduction theorem (Theorem 1.94). This proves Property 3. As for Property 4, its first part follows from the fact that MP is a rule of H, while the second is proved using MP and axiom [⊃1]. It is easy to see that $\mathcal{T}^*$ has Property 5. Next, we show that it has Property 6. Obviously, the axiom [⊃∨] imply that $\mathcal{T}^* \vdash_H \varphi \vee \tau$ if either $\mathcal{T}^* \vdash_H \varphi$ or $\mathcal{T}^* \vdash_H \tau$. To show the converse, assume that $\mathcal{T}^* \nvdash_H \varphi$ and $\mathcal{T}^* \nvdash_H \tau$. Then $\mathcal{T}^* \vdash_H \varphi \supset \psi$ and $\mathcal{T}^* \vdash_H \tau \supset \psi$ by Property 3. Using axiom [∨⊃] this implies that $\mathcal{T}^* \vdash_H \varphi \vee \tau \supset \psi$, and so $\mathcal{T}^* \nvdash_H \varphi \vee \tau$ by Property 2. To prove Property 7, assume that [⊃¬] is derivable in H, but that there is a sentence φ such that $\mathcal{T}^* \nvdash_H \varphi$ and $\mathcal{T}^* \nvdash_H \neg\varphi$. Then by Property 3, $\mathcal{T}^* \vdash_H \varphi \supset \psi$ and $\mathcal{T}^* \vdash_H \neg\varphi \supset \psi$. Using [⊃¬] we get that $\mathcal{T}^* \vdash_H \psi$, in contradiction to Property 2. We leave the proof of Property 8 to the reader. □

Theorem 1.98 *Let H be an axiomatic extension of $HCL_\supset$ in the language $\mathcal{L}$. Assume that $\mathcal{T} \nvdash_H \psi$. Then there exists a theory $\mathcal{T}^*$ in $\mathcal{L}$ that has all the properties listed in Theorem 1.97, as well as the following one:*

- *For every φ and τ, if $\mathcal{T}^* \nvdash_H \varphi$ then $\mathcal{T}^* \vdash_H \varphi \supset \tau$.*[30]

Proof. The proof follows that of Theorem 1.97. We only need to add a proof of the additional property. Now, Property 3 from Theorem 1.97 implies that for every sentence τ, if $\mathcal{T}^* \nvdash_H \psi \supset \tau$ then $\mathcal{T}^* \vdash_H (\psi \supset \tau) \supset \psi$. By [⊃3] this means that $\mathcal{T}^* \vdash_H \psi$, which is a contradiction to Property 2 of $\mathcal{T}^*$. It follows that $\mathcal{T}^* \vdash_H \psi \supset \tau$ for every τ. Now suppose that $\mathcal{T}^* \nvdash_H \varphi$. Then $\mathcal{T}^* \vdash_H \varphi \supset \psi$ by Property 3. Since by the deduction theorem it follows that $\varphi \supset \psi, \psi \supset \tau \vdash_H \varphi \supset \tau$ for every τ, we also have $\mathcal{T}^* \vdash_H \varphi \supset \tau$ for every τ. □

---

[28] Note that this applies in particular to the case in which one can derive in H the axioms [t] and [∨⊃] of HCL, since [⊃¬] is easily derived from them in $HIL_\supset$.

[29] There are standard mathematical methods to show that such a maximal extension of $\mathcal{T}$ exists. (The strongest one uses Zorn's lemma.) We omit the details.

[30] Note that this implies that $\mathcal{T}^* \vdash_H \varphi \supset \tau$ iff either $\mathcal{T}^* \nvdash_H \varphi$ or $\mathcal{T}^* \vdash_H \tau$.

1.3 Syntactic Approaches to Defining Logics

Note 1.99 Theorem 1.98 can be used for an easy proof of the completeness of $HCL_\supset$, HCL^+, and HCL for their classical two-valued semantics (Theorem 1.90): Let H be one of these systems, and suppose that $\mathcal{T} \nvdash_H \psi$. Let $\mathcal{T}^*$ be like in Theorem 1.98. Define an assignment ν in $\{t, f\}$ by letting $\nu(\varphi) = t$ iff $\varphi \in \mathcal{T}^*$. Then Theorem 1.98 ensures that ν is indeed a valuation, and that it is a model of $\mathcal{T}$ but not of ψ.

1.3.3 Ordinary Gentzen-type Systems

Gentzen-type proof systems form another particular family of proof systems. These proof systems are based on *sequent calculi* [170]. Given a language $\mathcal{L}$, a sequent calculus in $\mathcal{L}$ is practically a Hilbert-type system that instead of formulas of $\mathcal{L}$ handles higher-level constructs, called *sequents* of $\mathcal{L}$. There are two main different sorts of sequents:

single-conclusion sequents: These have the form $\Gamma \Rightarrow \varphi$, where φ is a formula of $\mathcal{L}$, and Γ is either a finite set, or a finite multiset, or a finite sequence, of formulas of $\mathcal{L}$.

multiple-conclusion sequents: These have the form $\Gamma \Rightarrow \Delta$, where both Γ and Δ are either finite sets, or finite multisets, or finite sequences, of formulas of $\mathcal{L}$.[31]

Here $\Rightarrow$ should be a new symbol, not used in $\mathcal{L}$. Given a sequent, what is written in it to the left of $\Rightarrow$ is called the left-hand side (l.h.s.), or *antecedent*, of the sequent, and what is written in it to the right of $\Rightarrow$ is called the right-hand side (r.h.s.), or *succedent*, of the sequent.

Note 1.100 *With the exception of Part V, we shall officially use in this book sequent calculi which employ finite sets of formulas rather than sequences or multisets (in most cases this choice is only a matter of convenience, see Note 1.111 below).*

Variables and Notations. We shall use s as a variable for sequents. Depending on the sort of sequent calculus under discussion, we shall use the Greek letters Γ and Δ as meta-variables for finite sets, or finite multisets, or finite sequences, of formulas. We shall follow all standard conventions. Thus, usually we omit brackets from both sides of $\Rightarrow$, and write, e.g., $\varphi_1, \ldots, \varphi_n \Rightarrow \psi_1, \ldots, \psi_k$ instead of $\{\varphi_1, \ldots, \varphi_n\} \Rightarrow \{\psi_1, \ldots, \psi_k\}$. Similarly,

---

[31] In the literature one may find "single-conclusion sequent calculi" which in addition to sequents of the form $\Gamma \Rightarrow \varphi$ allow also sequents in which the r.h.s. may be empty (i.e. have the form $\Gamma \Rightarrow$), and multiple-conclusion sequent calculi in which *all* sequents have an empty l.h.s. (i.e. have the form $\Rightarrow \Delta$).

we write $\Gamma \Rightarrow$ instead of $\Gamma \Rightarrow \emptyset$, and $\Rightarrow \varphi$ (or $\Rightarrow \Gamma$) in case the l.h.s. of a sequent is empty. If Γ and Δ denote finite sets then usually we shall write Γ, Δ instead of $\Gamma \cup \Delta$, and Γ, φ (or φ, Γ) instead of $\Gamma \cup \{\varphi\}$ (or $\{\varphi\} \cup \Gamma$). Similar conventions are used in case Γ and Δ denote finite multisets. If Γ and Δ denote finite sequences then Γ, Δ denotes the result of appending Γ and Δ, while Γ, φ denotes the result of appending φ to Γ. The notation φ, Γ should be understood in a similar way.

Terminology and Conventions. Exactly like a Hilbert-type system, a sequent calculus G for a language $\mathcal{L}$ is determined by a finite set of *Gentzen-type rules*. The usual form of such a rule is exemplified in Figures 1.2 and 1.5 below: Each such rule r has the form $\frac{s_1,\ldots,s_n}{s_c}$, where $s_1, \ldots, s_n$ and s_c are *meta-sequents*, that use variables for formulas instead of propositional variables, and include also variables for sets/multisets/sequences of formulas (depending on the type of the sequents in G). The *context* of r consists of the latter. In the single-conclusion case, if the right-hand side of s_c is a variable formula τ, then this τ belongs to the context of r as well. The *active part* of r consists of everything in r which does not belong to its context. *Applications* (or *instances*) of r are again denoted by $\frac{s_1^*,\ldots,s_n^*}{s_c^*}$, where this time $s_1^*, \ldots, s_n^*$ and s_c^* are sequents of $\mathcal{L}$, obtained by simultaneously substituting in $s_1, \ldots, s_n$ and s_c (respectively) formulas of $\mathcal{L}$ for formula variables, and sets/multisets/sequences of formulas for the other variables.[32] In both cases the (meta-)sequents above the "fraction line" are called the *premises*, and the (meta-)sequent below it - the *conclusion*. There may be some constraints on what may be substituted for sets/multisets/sequences variables, but not on formula variables. Thus, the set of applications of a Gentzen-type rule is always closed under $\mathcal{L}$-substitutions.

The formulas in the conclusion of (an application of) a rule that belong to (are obtained by substitutions from) its active part are called the *principal* formulas of that rule (application), while the formulas in its premises which belong to (are obtained from) its active part are its *side* formulas.

The notion of a proof (or derivation) of a sequent in a sequent calculus is defined in a way which is very similar to that of a proof (or derivation) of a formula in a Hilbert-type system (cf. Definition 1.83):

Definition 1.101 (derivation, provability) Let G be a sequent calculus.

---

[32]It is customary to omit the "fraction line" in case $n = 0$. Such rules are called *axiom schemas*, or sometimes just *axioms*. (Actually, the term "axiom" should be reserved for $\mathcal{L}$-instances of axiom schemas, which are real sequents of $\mathcal{L}$. Nevertheless, we shall usually follow standard terminological confusion, and use the term "axioms" in both cases, relying on the context to determine what interpretation of the word we have in mind.) Again, examples can be found in Figures 1.2 and 1.5.

1.3 Syntactic Approaches to Defining Logics

- A *proof* (or *derivation*) in G of a sequent s from a set S of sequents is a finite sequence of sequents which ends with s, and every element in it either belongs to S, or is an axiom of G, or is obtained from previous elements of the sequence by one of the rules of G.

- We say that s *follows* from S in G (notation: $S \vdash_G s$) if there is a proof in G of s from S.

- A sequent s is *provable* in G (notation: $\vdash_G s$) if it follows in G from the empty set of sequents.

A proof in a sequent calculus is usually represented by a finite *tree* rather than by a sequence of sequents (see, e.g., Figure 1.4 below).

Next, we define two important properties of rules in sequent calculi:

Definition 1.102 (admissibility of rules) A Gentzen-type rule R is *admissible* in a sequent calculus G if the conclusion of any application of R is provable in G whenever all its premises are provable in G.

Definition 1.103 (invertibility of rules in G) A Gentzen-type rule R is *invertible* in a sequent calculus G if $C \vdash_G C_i$ for every $1 \leq i \leq k$ whenever $\frac{C_1,\ldots,C_k}{C}$ is an application of R.

Our next goal is to define how sequent calculi are used as proof systems in the sense of Definition 1.80, and so as a tool for deriving formulas from theories. For this we need a few more definitions.

Definition 1.104 (identity axioms, cut) Let G be a sequent calculus.

- Instances of the axiom schema $\psi \Rightarrow \psi$ are called *identity axioms* (and the schema itself — the identity axiom schema).

- The *cut* rule is the following Gentzen-type rule:

  - Single-conclusion case:

    $$\text{cut:} \quad \frac{\Gamma_1 \Rightarrow \psi \quad \Gamma_2, \psi \Rightarrow \varphi}{\Gamma_1, \Gamma_2 \Rightarrow \varphi}$$

  - Multiple-conclusion case:

    $$\text{cut:} \quad \frac{\Gamma_1 \Rightarrow \Delta_1, \psi \quad \Gamma_2, \psi \Rightarrow \Delta_2}{\Gamma_1, \Gamma_2 \Rightarrow \Delta_1, \Delta_2}$$

Definition 1.105 (Gentzen-type systems) Let G be a sequent calculus. G is *logical*, or a *Gentzen-type system*[33], if the following conditions hold:

- G is based on a finite set of Gentzen-type rules which includes the identity axiom schema and the cut rule, and all its elements are *effective* (i.e. for each rule there is an effective procedure that given a finite set $S \cup \{s\}$ of sequents determines whether or not s is derivable from S using that rule).

- $\not\vdash_\mathsf{G} p \Rightarrow q$ in case p and q are distinct variables.

Note 1.106 Given a description of a Gentzen-type system G, it is usually straightforward to check that it satisfies the first condition in Definition 1.105. That also the second condition (of non-triviality) is satisfied by G is a less trivial matter. However, in this book this will always be an immediate corollary of results about G that *will* be proved. More specifically: the non-triviality of G will always follow either from the fact that it admits cut-elimination (see Definition 1.120 below) or from its soundness with respect to some semantics. (Usually it will follow from both.) Accordingly, *whenever we present some sequent calculus in this book, we leave to the reader the easy task of verifying that it is indeed logical.*

Definition 1.107 ($\vdash_\mathsf{G}$, $\vdash_\mathsf{G}^m$) Let G be a Gentzen-type system.

- The tcr $\vdash_\mathsf{G}$ induced by G is defined by: $\mathcal{T} \vdash_\mathsf{G} \varphi$ if there exists a finite Γ such that $\vdash_\mathsf{G} \Gamma \Rightarrow \varphi$, and Γ consists only of elements of $\mathcal{T}$.[34]

- If G is multiple-conclusion then the scr $\vdash_\mathsf{G}^m$ induced by G is defined by: $\mathcal{T} \vdash_\mathsf{G}^m \mathcal{S}$ if there exist finite Γ and Δ such that $\vdash_\mathsf{G} \Gamma \Rightarrow \Delta$, where Γ consists only of elements of $\mathcal{T}$, and Δ consists only of elements of $\mathcal{S}$.

Note 1.108 It should be noted that while $\vdash_\mathsf{G}^m$ is defined only for multiple-conclusion Gentzen-type systems, $\vdash_\mathsf{G}$ is defined for *all* Gentzen-type systems. Moreover: the definition applies as is to all sorts of these systems (single-conclusion or multiple-conclusion; systems which employ sets of formulas, or multisets, or sequences).

Note 1.109 Definitions 1.105 and 1.107 mean that a Gentzen-type system G can be seen as a proof system in the sense of Definition 1.80, in which

---

[33] The reader should be aware that in the literature there is usually no difference between the notions of "sequent calculus" and "Gentzen-type system", and that the distinction made in our Definition 1.105 is peculiar to this book. The motivation for our definition is given by Note 1.109 and Proposition 1.110 below.

[34] The notation $\vdash_\mathsf{G}$ is now overloaded: it is used in Definition 1.101 for a relation between sequents, and for a relation between formulas here. This should not be a problem.

1.3 Syntactic Approaches to Defining Logics

the set Pr of proofs consists of derivations in G (see Definition 1.101) from the empty set of sequents. A triple $\langle \mathcal{T}, \psi, d \rangle$ is in G if and only if d is a derivation in G of a sequent of the form $\Gamma \Rightarrow \psi$, where Γ consists only of elements of $\mathcal{T}$. Accordingly, soundness and completeness of a Gentzen-type system G are defined just as in Definition 1.82.

The following proposition easily follows from our definition of a Gentzen-type system (Definition 1.105):

Proposition 1.110 *Let G be a Gentzen-type system for a language $\mathcal{L}$.*

1. $\mathbf{L}_G = \langle \mathcal{L}, \vdash_G \rangle$ *is a finitary logic.*

2. *If G is multiple-conclusion then $\vdash_G^m$ is a finitary scr which is structural and non-trivial.*

Note 1.111 As we have said in Note 1.100, until Part V we officially use in this book only Gentzen-type systems which employ finite sets of formulas. This is only a matter of convenience. We could e.g. follow Gentzen's original formulation, and use sequences instead. This would not change $\vdash_G$ or $\vdash_G^m$, provided we add to our systems the following *basic structural rules*:

single-conclusion case:

permutation: $\dfrac{\Gamma_1, \psi, \varphi, \Gamma_2 \Rightarrow \tau}{\Gamma_1, \varphi, \psi, \Gamma_2 \Rightarrow \tau}$

contraction: $\dfrac{\Gamma_1, \psi, \psi, \Gamma_2 \Rightarrow \tau}{\Gamma_1, \psi, \Gamma_2 \Rightarrow \tau}$

expansion: $\dfrac{\Gamma_1, \psi, \Gamma_2 \Rightarrow \tau}{\Gamma_1, \psi, \psi, \Gamma_2 \Rightarrow \tau}$

multiple-conclusion case:

permutation: $\dfrac{\Gamma_1, \psi, \varphi, \Gamma_2 \Rightarrow \Delta}{\Gamma_1, \varphi, \psi, \Gamma_2 \Rightarrow \Delta}$ $\qquad$ $\dfrac{\Gamma \Rightarrow \Delta_1, \psi, \varphi, \Delta_2}{\Gamma \Rightarrow \Delta_1, \varphi, \psi, \Delta_2}$

contraction: $\dfrac{\Gamma_1, \psi, \psi, \Gamma_2 \Rightarrow \Delta}{\Gamma_1, \psi, \Gamma_2 \Rightarrow \Delta}$ $\qquad$ $\dfrac{\Gamma \Rightarrow \Delta_1, \psi, \psi, \Delta_2}{\Gamma \Rightarrow \Delta_1, \psi, \Delta_2}$

expansion: $\dfrac{\Gamma_1, \psi, \Gamma_2 \Rightarrow \Delta}{\Gamma_1, \psi, \psi, \Gamma_2 \Rightarrow \Delta}$ $\qquad$ $\dfrac{\Gamma \Rightarrow \Delta_1, \psi, \Delta_2}{\Gamma \Rightarrow \Delta_1, \psi, \psi, \Delta_2}$

Note 1.112 It should be noted that in case Γ is a finite set of formulas, the fact that $\Gamma \vdash_G \psi$ does *not* imply (according to Definition 1.107) that the sequent $\Gamma \Rightarrow \psi$ is provable in G. Similarly, if Γ and Δ are finite sets of formulas, then the fact that $\Gamma \vdash_G^m \Delta$ does *not* imply that the sequent $\Gamma \Rightarrow \Delta$ is provable in G. However, this somewhat strange state of affairs is avoided in the following family of Gentzen-type systems:

Definition 1.113 (standard system) A Gentzen-type system G is called *standard* if its set of rules includes the relevant basic structural rules[35], as well as the following structural rule of *weakening*:

- Single-conclusion case:

$$\text{weakening } [W]: \quad \frac{\Gamma \Rightarrow \varphi}{\Gamma, \Gamma' \Rightarrow \varphi}$$

- Multiple-conclusion case:

$$\text{weakening } [W]: \quad \frac{\Gamma \Rightarrow \Delta}{\Gamma, \Gamma' \Rightarrow \Delta, \Delta'}$$

The proof of the following proposition is straightforward:

Proposition 1.114 *Let G be a standard Gentzen-type system.*

1. *For every finite Γ and formula ψ we have that $\Gamma \vdash_G \psi$ iff $\vdash_G \Gamma \Rightarrow \psi$.*

2. *If G is multiple-conclusion, then the following obtains for every finite Γ and Δ: $\Gamma \vdash_G^m \Delta$ iff $\vdash_G \Gamma \Rightarrow \Delta$.*

Note 1.115 (types of rules) The rules of a Gentzen-type system are usually divided into two groups:

Structural rules which include the identity axiom schema, cut, weakening (if the system is standard), and the relevant basic structural rules in case sequences or multisets are used for constructing sequents.

Logical rules which treat the connectives of the language.

There are also two particularly useful *forms* that rules (of both groups) may take: the *additive* form, and the *multiplicative* form.[36] In an additive form of a rule the contexts are the same in all of the sequents (the premises

---

[35]I.e. contraction in case G is based on multisets, contraction and permutation in case it is based on sequences. Note that there is no need to explicitly include the expansion rule, since it is a special case of the weakening rule defined below.

[36]This terminology is due to [174].

1.3 Syntactic Approaches to Defining Logics

and the conclusion). In a multiplicative form of a rule, on the other hand, the premises of the rule may have different contexts, while the context of the conclusion is obtained by combining the contexts of the premises. The logical rules in Figures 1.2 and 1.5 have an additive form. In contrast, the cut rule is presented in Definition 1.104 in its multiplicative form. All these rules have also useful forms of the other type. Thus, the additive form of the cut rule is in the single-conclusion case the following:

$$\frac{\Gamma \Rightarrow \psi \quad \Gamma, \psi \Rightarrow \varphi}{\Gamma \Rightarrow \varphi}.$$

Similarly, the multiplicative form of $[\supset\Rightarrow]$ (from Figure 1.2) is:

$$\frac{\Gamma_1 \Rightarrow \psi, \Delta_1 \quad \Gamma_2, \varphi \Rightarrow \Delta_2}{\Gamma_1, \Gamma_2, \psi \supset \varphi \Rightarrow \Delta_1, \Delta_2}$$

In standard Gentzen-type systems the additive and the multiplicative form of a rule are equivalent: each can easily be derived from the other using the structural rules.[37] However, this is not the case in non-standard Gentzen-type systems (like those which are used in Part V of the book).

Note 1.116 The relation $\vdash_G$ that is induced by G is not the only natural tcr that may be associated with G. Another important tcr that is frequently associated with an arbitrary sequent calculus G is defined by: $\mathcal{T} \Vdash_G \psi$ if the sequent $\Rightarrow \psi$ follows in G from the set of sequents $\{\Rightarrow \psi_i \mid \psi_i \in \mathcal{T}\}$. Here a formula ψ is practically identified with the sequent $\Rightarrow \psi$. It is easy to see that if G is non-trivial (in the sense that $\{\Rightarrow p\} \not\vdash_G \Rightarrow q$) and closed under substitutions, then $\langle \mathcal{L}, \Vdash_G \rangle$ is a propositional logic (even if G is a sequent calculus which is not a Gentzen-type system). Obviously, if G is a Gentzen-type system then $\vdash_G \subseteq \Vdash_G$ (i.e. if $\psi_1, \ldots, \psi_n \vdash_G \psi$ then $\psi_1, \ldots, \psi_n \Vdash_G \psi$, by n consecutive applications of the cut rule on $\psi_1, \ldots, \psi_n \Rightarrow \psi$ and $\Rightarrow \psi_i$). The converse is easily seen to hold for what are called in [34] *pure* standard Gentzen-type systems. Hence, it holds for all the Gentzen-type systems which are studied in the first nine chapters of this book except LJ_m and LJ_m^+, since all of these systems are of this type.[38]

Next, we introduce a crucial notion concerning Gentzen-type systems:

Definition 1.117 (analytic proofs, subformula property) Let G be a Gentzen-type system in a language $\mathcal{L}$.

---

[37] In this book we have usually chosen to present the additive forms of the logical rules of standard Gentzen-type systems.

[38] Chapter 9 is actually the only chapter in which impure standard Gentzen-type systems are employed, and indeed $\Vdash_G \not\subseteq \vdash_G$ for them.

- Let $\mathcal{F}$ be a set of formulas in $\mathcal{L}$. A proof in G is called $\mathcal{F}$-*analytic* if every formula which occurs in it belongs to $\mathcal{F}$.

- Let $S \cup s$ be a set of sequents in $\mathcal{L}$. A proof in G of s from S is called *analytic* if it is $\mathcal{F}$-analytic, where $\mathcal{F}$ is the set of subformulas of formulas in $S \cup s$.

- G has the *(strong) subformula property* if whenever $\vdash_G s$ ($S \vdash_G s$), there is an analytic proof of s (from S).

Note 1.118 In the literature, the subformula property is usually attributed to a Gentzen-type system G if every provable sequent of it has an analytic proof. Definition 1.117 generalizes this to derivability *from assumptions*.

Note 1.119 The subformula property entails that if a sequent is provable at all, then it is provable "without detours". This is usually the key for developing proof searching methods for (automatically) constructing proofs in the underlying calculus. On the propositional level the subformula property also almost always implies the decidability of the corresponding logic.

Next comes what is usually the key in showing the subformula property.

Definition 1.120 (cut elimination) Let G be a Gentzen-type system.

- We say that G admits *cut-elimination*, or that it is *cut-free*, if every sequent that is provable in G has in it a cut-free proof (i.e., a proof in which the cut rule is not used).

- G admits *strong* cut-elimination if for every set S of sequents and every sequent s, if $S \vdash_G s$ then s has a proof in G from S in which all cuts are on formulas which occur in some sequent of S (in particular, the case $S = \emptyset$ is just the standard cut-elimination defined above).

- G admits *analytic* cut-elimination if every sequent s that is provable in G has in it a proof in which all cuts are on subformulas of formulas that occur in s.

- G admits *strong analytic* cut-elimination if for every set S of sequents and every sequent s, if $S \vdash_G s$ then s has a proof in G from S in which all cuts are on subformulas of formulas which occur in either s or in some sequent of S.

Note 1.121 The importance of (analytic) cut elimination is due to the fact that in most of the useful Gentzen-type systems (including all the Gentzen-type systems mentioned in this book) the cut rule is the only non-analytic

1.3 Syntactic Approaches to Defining Logics 39

rule. In other words, it is the only rule that the premises of an instance of it may include formulas which are not related in any reasonable way to those in its conclusion. Thus, a system which admits (analytic) cut elimination almost always enjoys some effective version of the subformula property.

Theorem 1.122 *Let G be a standard multiple-conclusion Gentzen-type system in which all the logical rules of G are either additive or multiplicative.*

1. *If G admits cut-elimination then it also admits strong cut-elimination.*

2. *If G admits analytic cut-elimination then it also admits strong analytic cut-elimination.*

Proof.

1. It suffices to prove that if S is finite, and $S \vdash_G \Gamma \Rightarrow \Delta$, then $\Gamma \Rightarrow \Delta$ has a proof in G from S in which all cuts are on formulas which occur in some sequent of S. So assume that $S = \{\Gamma_i \Rightarrow \Delta_i \mid 1 \leq i \leq n\}$. We prove the claim by induction on n. The case where $n = 0$ follows from the assumption that G admits cut-elimination. So suppose that $n > 0$, and let $S' = S - \{\Gamma_n \Rightarrow \Delta_n\}$. Suppose that $\varphi \in \Delta_n$. Since all rules of G are either additive or multiplicative, by adding φ to the l.h.s. of all the sequents in a given proof in G of s from S, we get a proof in G of $\varphi, \Gamma \Rightarrow \Delta$ from $\{\varphi, \Gamma_i \Rightarrow \Delta_i \mid 1 \leq i \leq n\}$. Since $\varphi \in \Delta_n$, this proof can easily be transformed (using the identity axioms and the weakening rule) into a proof of $\varphi, \Gamma \Rightarrow \Delta$ from S'. It follows that $S' \vdash_G \varphi, \Gamma \Rightarrow \Delta$ for every $\varphi \in \Delta_n$. Similarly, $S' \vdash_G \Gamma \Rightarrow \Delta, \psi$ for every $\psi \in \Gamma_n$. By induction hypothesis on S', there exist proofs of all these sequents from S' in which all cuts are on formulas which occur in some sequent of S'. Now, from $\{\varphi, \Gamma \Rightarrow \Delta \mid \varphi \in \Delta_n\} \cup \{\Gamma \Rightarrow \Delta, \psi \mid \psi \in \Gamma_n\} \cup \{\Gamma_n \Rightarrow \Delta_n\}$ one can easily derive $\Gamma \Rightarrow \Delta$ using only cuts on formulas of $\Gamma_n \Rightarrow \Delta_n$. Altogether we get a proof as desired of $\Gamma \Rightarrow \Delta$ from S.

2. The proof is similar to that of the first part. □

Next, we discuss the semantics of Gentzen-type systems. More accurately: we describe the usual connections between a denotational semantics for a given logic and a Gentzen-type system for that logic.

Definition 1.123 (semantics of sequents) Let $S = \langle S, \models_S \rangle$ be a denotational semantics for $\mathcal{L}$.

- Let ν be an element of S. We say that ν is an *S-model* of a sequent $\Gamma \Rightarrow \Delta$, or that ν *satisfies* $\Gamma \Rightarrow \Delta$ (notation: $\nu \models_S \Gamma \Rightarrow \Delta$) if $\nu \not\models_S \varphi$ for some φ in Γ, or $\nu \models_S \psi$ for some ψ in Δ. ν is an *S-model* of a sequent $\Gamma \Rightarrow \psi$ (notation: $\nu \models_S \Gamma \Rightarrow \psi$) if $\nu \models_S \Gamma \Rightarrow \{\psi\}$.

- We say that a sequent s *follows* in S from a set A of sequents (notation: $A \vdash_\mathsf{S} s$) if every S-model of A is also an S-model of s.

- A sequent s is S-*valid* (notation: $\vdash_\mathsf{S} s$) if $\nu \models_S s$ for every $\nu \in S$.

The following proposition is immediate from the definitions:

Proposition 1.124 *Suppose that* S *is a denotational semantics for* $\mathcal{L}$. *Then* $\vdash_\mathsf{S} \Gamma \Rightarrow \varphi$ *iff* $\Gamma \vdash_\mathsf{S} \varphi$.

Definition 1.125 (soundness, completeness) Let G be a Gentzen-type system for $\mathcal{L}$, and let $\mathsf{S} = \langle S, \models_S \rangle$ be a denotational semantics for $\mathcal{L}$.

- G is *sound* for S if for every sequent s of $\mathcal{L}$, if $\vdash_\mathsf{G} s$ then $\vdash_\mathsf{S} s$.

- G is *strongly sound* for S if for every set $A \cup \{s\}$ of sequents of $\mathcal{L}$, if $A \vdash_\mathsf{G} s$ then $A \vdash_\mathsf{S} s$.

- G is *complete* for S if for every sequent s of $\mathcal{L}$, if $\vdash_\mathsf{S} s$ then $\vdash_\mathsf{G} s$.

- G is *strongly complete* for S if for every set $A \cup \{s\}$ of sequents of $\mathcal{L}$, if $A \vdash_\mathsf{S} s$ then $A \vdash_\mathsf{G} s$.

Proposition 1.126 *Let* G *be a Gentzen-type system for a language* $\mathcal{L}$, *and let* $\mathsf{S} = \langle S, \models_S \rangle$ *be a denotational semantics for* $\mathcal{L}$.

1. *If* G *is sound for* S, *and* $\mathcal{T} \vdash_\mathsf{G} \psi$, *then* $\mathcal{T} \vdash_\mathsf{S} \psi$.

2. *If* G *is sound and complete for* S, *and* $\vdash_\mathsf{S}$ *is finitary, then for every* $\mathcal{T}$ *and* ψ *it holds that* $\mathcal{T} \vdash_\mathsf{G} \psi$ *iff* $\mathcal{T} \vdash_\mathsf{S} \psi$.

3. *If* G *is strongly sound and strongly complete for* S, *then* $\vdash_\mathsf{S}$ *is finitary. (And so, by part 2,* $\mathcal{T} \vdash_\mathsf{G} \psi$ *iff* $\mathcal{T} \vdash_\mathsf{S} \psi$ *for every* $\mathcal{T}$ *and* ψ).

Proof. All the parts easily follow from the definitions and Proposition 1.124. We only note that a crucial fact for the last item is that by definitions, $\mathcal{T} \vdash_\mathsf{S} \psi$ iff $\{\Rightarrow \varphi \mid \varphi \in \mathcal{T}\} \vdash_\mathsf{S} \Rightarrow \psi$. □

Note 1.127 The proof of soundness or strong soundness of a given Gentzen-type system G is usually done by a straightforward induction on the structure of proofs. Since the standard structural rules (identity, cut, weakening, and the basic structural rules) always preserve satisfaction, the proof boils down to showing that so do the logical rules.

Note 1.128 Using an induction similar to that used in the proof of Theorem 1.122, it is possible to show that if G is complete for S, $A \cup \{s\}$ is a finite set of sequents, and $A \vdash_\mathsf{S} s$, then $A \vdash_\mathsf{G} s$. It follows that if $\vdash_\mathsf{S}$ is finitary as a relation between sequents then completeness of G implies strong completeness of G.

1.3 Syntactic Approaches to Defining Logics

Gentzen-type systems for the basic logics

Definition 1.129 (LK^+) LK^+ denotes the multiple-conclusion standard Gentzen-type system for $\mathcal{L}_{CL}^+$ that is given in Figure 1.2. (Recall that until Part V, Γ and Δ vary through finite sets of formulas).

Axioms: $\varphi \Rightarrow \varphi$

Rules of LK^+: cut, weakening, and the following logical rules:

$$[\wedge \Rightarrow] \quad \frac{\Gamma, \varphi, \psi \Rightarrow \Delta}{\Gamma, \varphi \wedge \psi \Rightarrow \Delta} \qquad [\Rightarrow \wedge] \quad \frac{\Gamma \Rightarrow \Delta, \varphi \quad \Gamma \Rightarrow \Delta, \psi}{\Gamma \Rightarrow \Delta, \varphi \wedge \psi}$$

$$[\vee \Rightarrow] \quad \frac{\Gamma, \varphi \Rightarrow \Delta \quad \Gamma, \psi \Rightarrow \Delta}{\Gamma, \varphi \vee \psi \Rightarrow \Delta} \qquad [\Rightarrow \vee] \quad \frac{\Gamma \Rightarrow \Delta, \varphi, \psi}{\Gamma \Rightarrow \Delta, \varphi \vee \psi}$$

$$[\supset \Rightarrow] \quad \frac{\Gamma \Rightarrow \varphi, \Delta \quad \Gamma, \psi \Rightarrow \Delta}{\Gamma, \varphi \supset \psi \Rightarrow \Delta} \qquad [\Rightarrow \supset] \quad \frac{\Gamma, \varphi \Rightarrow \psi, \Delta}{\Gamma \Rightarrow \varphi \supset \psi, \Delta}$$

Figure 1.2: The multiple-conclusion Gentzen-type system LK^+

Definition 1.130 (LK) LK is the multiple-conclusion standard Gentzen-type system for $\mathcal{L}_{CL}$ that is obtained from LK^+ by adding to it the two classical rules for negation given in Figure 1.3.

Axioms: $\varphi \Rightarrow \varphi$

Rules of LK: The rules of LK^+ and the following two rules:

$$[\neg \Rightarrow] \quad \frac{\Gamma \Rightarrow \Delta, \varphi}{\Gamma, \neg \varphi \Rightarrow \Delta} \qquad [\Rightarrow \neg] \quad \frac{\Gamma, \varphi \Rightarrow \Delta}{\Gamma \Rightarrow \Delta, \neg \varphi}$$

Figure 1.3: The multiple-conclusion proof system LK

Note 1.131

- As was indicated in Note 1.115, the rules of LK are presented in the additive form. (The only exception is cut, that is presented in Definition 1.104 in its multiplicative form, but in standard Gentzen-type systems may be taken also in the additive form.)

- In Gentzen's original formulation [170] the rules $[\wedge\Rightarrow]$ and $[\Rightarrow\vee]$ were split into two rules, each with one side formula only:

$$\frac{\Gamma,\psi\Rightarrow\Delta}{\Gamma,\psi\wedge\varphi\Rightarrow\Delta} \quad \frac{\Gamma,\varphi\Rightarrow\Delta}{\Gamma,\psi\wedge\varphi\Rightarrow\Delta} \quad \frac{\Gamma\Rightarrow\Delta,\psi}{\Gamma\Rightarrow\Delta,\psi\vee\varphi} \quad \frac{\Gamma\Rightarrow\Delta,\varphi}{\Gamma\Rightarrow\Delta,\psi\vee\varphi}$$

- In the literature one can find formulations of LK in which the axioms have the form $p \Rightarrow p$ (or $\Gamma, p \Rightarrow \Delta, p$), where p is a variable. Such axioms are not legitimate Gentzen-type rules according to our use of the term (see the discussion before Definition 1.101). However, it is easy to see that by using the other rules of LK^+ it is possible to derive all the identity axioms from this restricted set of axioms.

- If the language of classical logic is taken to be $\mathcal{L}_{CL}^{\mathsf{F}}$ ($\mathcal{L}_{CL}^{\mathsf{F}\neg}$), then LK^{F} ($LK^{\mathsf{F}\neg}$), the corresponding version of LK, is obtained from LK^+ (LK) by adding to its rules the following axiom: $\mathsf{F}\Rightarrow$

Example 1.132 Figure 1.4 shows an example of a schematic derivation tree, proving in LK one side of one of De Morgan's laws. As usual, each leaf (i.e. a most-upper line) of the tree contains an instance of the axiom schema of LK, its root (i.e. the bottom line) contains the proven sequent, and transitions from one node of the tree to another are justified by applications of the inference rules (recall that $[W]$ denotes the weakening rule).

$$\cfrac{\cfrac{\cfrac{\cfrac{\cfrac{\varphi\Rightarrow\varphi}{\varphi\Rightarrow\neg\psi,\varphi}[W]}{\Rightarrow\neg\psi,\neg\varphi,\varphi}[\Rightarrow\neg]}{\Rightarrow\neg\psi\vee\neg\varphi,\varphi}[\Rightarrow\vee] \quad \cfrac{\cfrac{\cfrac{\psi\Rightarrow\psi}{\psi\Rightarrow\neg\varphi,\psi}[W]}{\Rightarrow\neg\psi,\neg\varphi,\psi}[\Rightarrow\neg]}{\Rightarrow\neg\psi\vee\neg\varphi,\psi}[\Rightarrow\vee]}{\Rightarrow\neg\psi\vee\neg\varphi,\psi\wedge\varphi}[\Rightarrow\wedge]}{\cfrac{\neg(\psi\wedge\varphi)\Rightarrow\neg\psi\vee\neg\varphi}{\Rightarrow\neg(\psi\wedge\varphi)\supset\neg\psi\vee\neg\varphi}[\Rightarrow\supset]}[\neg\Rightarrow]$$

Figure 1.4: A derivation tree in LK

The most important facts about the classical Gentzen-type systems are:

Theorem 1.133 (cut elimination) LK^+ *admits cut elimination. The same is true for LK, LK^{F}, and $LK^{\mathsf{F}\neg}$.*

Theorem 1.134 (soundness+completeness) LK^+ *is sound and complete for* $\mathbf{CL}^+$; LK *is sound and complete for* $\mathbf{CL}$; LK^{F} *is sound and complete for* $\mathbf{CL}^{\mathsf{F}}$; *and* $LK^{\mathsf{F}\neg}$ *is sound and complete for* $\mathbf{CL}^{\mathsf{F}\neg}$.

1.3 Syntactic Approaches to Defining Logics

Note 1.135 By Proposition 1.126 and Theorems 1.90 and 1.134, we have that $\mathcal{T} \vdash_{HCL+} \psi$ iff $\mathcal{T} \vdash_{LK+} \psi$ and $\mathcal{T} \vdash_{HCL} \psi$ iff $\mathcal{T} \vdash_{LK} \psi$.

Another important property of the classical Gentzen-type systems (that can easily be verified using the cut rule) is given in the next proposition.

Proposition 1.136 *All the logical inference rules in our presentation of LK^+ are invertible in it. The same is true for LK, LK^{F}, and $LK^{\mathsf{F}\neg}$.*[39]

Next, we turn to Gentzen-type systems for intuitionistic logic. Here we present two calculi. One is (a version of) Gentzen's original single-conclusion system for **IL**. The other is Maehara's multiple-conclusion system ([279]).

Definition 1.137 (LJ^+, LJ)

- LJ^+ is the single-conclusion standard Gentzen-type system for $\mathcal{L}_{CL}^+$ that is given in Figure 1.5.

- LJ is the single-conclusion standard Gentzen-type system in $\mathcal{L}_{CL}^{\mathsf{F}}$ that is obtained from LJ^+ by adding to its rules the axiom: $\mathsf{F} \Rightarrow \varphi$

Axioms: $\varphi \Rightarrow \varphi$

Rules of LJ^+: cut, weakening, and the following logical rules:

$$[\wedge \Rightarrow] \quad \frac{\Gamma, \varphi, \psi \Rightarrow \tau}{\Gamma, \varphi \wedge \psi \Rightarrow \tau} \qquad [\Rightarrow \wedge] \quad \frac{\Gamma \Rightarrow \varphi \quad \Gamma \Rightarrow \psi}{\Gamma \Rightarrow \varphi \wedge \psi}$$

$$[\vee \Rightarrow] \quad \frac{\Gamma, \varphi \Rightarrow \tau \quad \Gamma, \psi \Rightarrow \tau}{\Gamma, \varphi \vee \psi \Rightarrow \tau} \qquad [\Rightarrow \vee] \quad \frac{\Gamma \Rightarrow \varphi}{\Gamma \Rightarrow \varphi \vee \psi} \quad \frac{\Gamma \Rightarrow \psi}{\Gamma \Rightarrow \varphi \vee \psi}$$

$$[\supset \Rightarrow] \quad \frac{\Gamma \Rightarrow \varphi \quad \Gamma, \psi \Rightarrow \tau}{\Gamma, \varphi \supset \psi \Rightarrow \tau} \qquad [\Rightarrow \supset] \quad \frac{\Gamma, \varphi \Rightarrow \psi}{\Gamma \Rightarrow \varphi \supset \psi}$$

Figure 1.5: The single-conclusion Gentzen-type system LJ^+

Definition 1.138 (LJ_m^+, LJ_m)

- LJ_m^+ is the multiple-conclusion standard Gentzen-type system that is obtained from LK^+ by replacing $[\Rightarrow \supset]$ by the following weaker rule:

$$\frac{\Gamma, \psi \Rightarrow \varphi}{\Gamma \Rightarrow \psi \supset \varphi}.$$

---
[39] Note that Proposition 1.136 would not have been true had we chosen to use the multiplicative form of the logical rules.

- LJ_m is the multiple-conclusion standard Gentzen-type system in the language $\mathcal{L}_{CL}^{\mathsf{F}}$ (Definition 1.73) that is obtained from LJ_m^+ by adding to its rules the axiom: $\mathsf{F} \Rightarrow$

Theorem 1.139 (cut elimination) LJ_m^+, LJ_m, LJ^+, and LJ admit cut elimination.

Theorem 1.140 (soundness+completeness) *Both of LJ and LJ_m are sound and complete for* **IL**, *while both of LJ^+ and LJ_m^+ are sound and complete for* **IL**$^+$.

Note 1.141 Again, by Proposition 1.126 and Theorems 1.90 and 1.140, we also have that $\mathcal{T} \vdash_{HIL^+} \psi$ iff $\mathcal{T} \vdash_{LJ^+} \psi$ iff $\mathcal{T} \vdash_{LJ_m^+} \psi$, and $\mathcal{T} \vdash_{HIL} \psi$ iff $\mathcal{T} \vdash_{LJ} \psi$ iff $\mathcal{T} \vdash_{LJ^m} \psi$.

Note 1.142 Gentzen-type systems for Johansson's minimal logic (Notes 1.79 and 1.93) can be obtained from LJ_m^+ and LJ^+ by extending them to the language $\mathcal{L}_{CL}^{\mathsf{F}}$, without any change in the rules (though the sets of instances of these rules *are* extended, as they are now applied to a richer language). The resulting systems are again sound, complete, and cut-free.

1.3.4 Calculi of Hypersequents

Calculi of ordinary sequents of the type described in the previous section are not sufficient for constructing cut-free (or at least analytic) proof systems for every logic studied in this book. Therefore, on few occasions we shall use Gentzen-type systems which employ a richer data structure: hypersequents.

Definition 1.143 A *hypersequent* is a finite multiset of ordinary sequents. The elements of this multiset are called its *components*. A hypersequent is called *single-conclusion* if all its components are single-conclusion, and *multiple-conclusion* otherwise. We denote by $s_1 \mid \cdots \mid s_n$ the hypersequent whose components are $s_1, \ldots, s_n$, and use G, H as metavariables for (possibly empty) hypersequents.[40]

Logical rules in hypersequential calculi are usually obtained from standard logical rules of ordinary sequential calculi by allowing also side components in applications of the rules (in addition to the presence of side formulas). Again, we have here a choice between treating side components in a multiplicative manner or in an additive one. On the other hand, the fact that the use of hypersequent adds an extra layer of structure opens the door to more complicated logical rules, as well as to new structural rules.

---

[40]Semantically, the interpretation of '$\mid$' is usually taken to be disjunctive.

1.3 Syntactic Approaches to Defining Logics

Thus, in addition to the old *internal* structural rule of weakening (called [IW] below), which allows to add a formula on a side of a component of a hypersequent, we usually have now also an *external* rule of weakening ([EW]), which allows to add a new component to a hypersequent. What is more, the use of hypersequents makes it possible to introduce also completely new types of structural rules. This is demonstrated by the Splitting rule [Sp] and the Communication rule [Com] in the example below.

Example 1.144 (Gödel-Dummett logic $\mathbf{G}_\infty$) Let HG_∞ be the extension of HIL by the linearity axiom $(\varphi \supset \psi) \vee (\psi \supset \varphi)$, and let $\mathbf{G}_\infty$ be the logic in $\mathcal{IL}$ which is induced by HG_∞.[41] Figure 1.6 presents the hypersequential calculus GLC for $\mathbf{G}_\infty$ that was given in [31]. It admits cut-elimination ([213, 48]), and is sound and complete for $\mathbf{G}_\infty$ in the sense that $\Gamma \vdash_{\mathbf{G}_\infty} \varphi$ iff $\vdash_{GLC} \Gamma \Rightarrow \varphi$. (Actually, $\vdash_{GLC} \Gamma_1 \Rightarrow \varphi_1 \mid \cdots \mid \Gamma_n \Rightarrow \varphi_n$ iff $\vdash_{\mathbf{G}_\infty} (\wedge \Gamma_1 \supset \varphi_1) \vee \cdots \vee (\wedge \Gamma_n \supset \varphi_n)$, where $\wedge \{\varphi_1, \ldots, \varphi_n\} = \varphi_1 \wedge \cdots \wedge \varphi_n$.)

Axioms: $\varphi \Rightarrow \varphi \qquad \mathsf{F} \Rightarrow \varphi$

Logical rules:

$$[\wedge \Rightarrow] \quad \frac{G|\Gamma, \varphi, \psi \Rightarrow \tau}{G|\Gamma, \varphi \wedge \psi \Rightarrow \tau} \qquad [\Rightarrow \wedge] \quad \frac{G|\Gamma \Rightarrow \varphi \quad G|\Gamma \Rightarrow \psi}{G|\Gamma \Rightarrow \varphi \wedge \psi}$$

$$[\vee \Rightarrow] \quad \frac{G|\Gamma, \varphi \Rightarrow \tau \quad G|\Gamma, \psi \Rightarrow \tau}{G|\Gamma, \varphi \vee \psi \Rightarrow \tau} \qquad [\Rightarrow \vee] \quad \frac{G|\Gamma \Rightarrow \varphi \quad G|\Gamma \Rightarrow \psi}{G|\Gamma \Rightarrow \varphi \vee \psi \quad G|\Gamma \Rightarrow \varphi \vee \psi}$$

$$[\supset \Rightarrow] \quad \frac{G|\Gamma \Rightarrow \varphi \quad G|\Gamma, \psi \Rightarrow \tau}{G|\Gamma, \varphi \supset \psi \Rightarrow \tau} \qquad [\Rightarrow \supset] \quad \frac{G|\Gamma, \varphi \Rightarrow \psi}{G|\Gamma \Rightarrow \varphi \supset \psi}$$

Structural rules:

$$[\text{Cut}] \quad \frac{G \mid \Gamma_1 \Rightarrow \varphi \quad G \mid \varphi, \Gamma_2 \Rightarrow \tau}{G \mid \Gamma_1, \Gamma_2 \Rightarrow \tau} \qquad [\text{IW}] \quad \frac{G \mid \Gamma \Rightarrow \tau}{G \mid \varphi, \Gamma \Rightarrow \tau}$$

$$[\text{EC}] \quad \frac{G \mid s \mid s}{G \mid s} \qquad [\text{EW}] \quad \frac{G}{G \mid s}$$

$$[\text{Sp}] \quad \frac{G \mid \Gamma_1, \Gamma_2 \Rightarrow \varphi}{G \mid \Gamma_1 \Rightarrow \varphi \mid \Gamma_2 \Rightarrow \varphi} \qquad [\text{Com}] \quad \frac{G \mid \Gamma_1 \Rightarrow \varphi_1 \quad G \mid \Gamma_2 \Rightarrow \varphi_2}{G \mid \Gamma_1 \Rightarrow \varphi_2 \mid \Gamma_2 \Rightarrow \varphi_1}$$

Figure 1.6: The proof system GLC

---

[41]$\mathbf{G}_\infty$ is the most famous *intermediate* logic (that is: a logic between intuitionistic logic and classical logic). It is also one of the three most basic standard *fuzzy* logics ([120]). More information about it is given in Chapter 15. (See in particular Note 15.89.)

Here is an example of a cut-free proof in GLC of the linearity axiom. Note that the applications of $[\Rightarrow\supset]$ and $[\Rightarrow \vee]$ are done in it in parallel:

$$\dfrac{\dfrac{\dfrac{\dfrac{\varphi \Rightarrow \varphi \qquad \psi \Rightarrow \psi}{\varphi \Rightarrow \psi \mid \psi \Rightarrow \varphi}\,[Com]}{\Rightarrow \varphi \supset \psi \mid \Rightarrow \psi \supset \varphi}\,[\Rightarrow\supset]}{\Rightarrow (\varphi \supset \psi) \vee (\psi \supset \varphi) \mid \Rightarrow (\varphi \supset \psi) \vee (\psi \supset \varphi)}\,[\Rightarrow \vee]}{\Rightarrow (\varphi \supset \psi) \vee (\psi \supset \varphi)}\,[EC]$$

Note 1.145 For all the hypersequential calculi used in this book we could have taken hypersequents as *sets* of sequents rather than multisets. However, since external contraction ([EC]) is the main source of problems for hypersequential calculi (both for proving cut-elimination and for producing efficient proof-search procedures), we have chosen to make its use explicit.

1.4 Bibliographical Notes and Further Reading

All the theorems and propositions to which proofs have been given in it are new. The only exceptions are Theorem 1.97 and Theorem 1.98, which are "folklore". Also new here (to the best of our knowledge) are Definitions 1.34 and 1.45, as well as the material in Sections 1.2.1 and 1.3.1.

The concept of semi-implication (Definition 1.23) was introduced and investigated in [53] (from where the related results quoted here are taken), but it is essentially due already to Church [118]. The notion of a normal logic (Definition 1.32) was first introduced in [58].

A good introduction to intuitionistic logic is Chapter 5 of [291]. The main textbooks on the subject are [85, 145, 284].

Gentzen original paper [170] (in which Gentzen-type systems were first introduced and the cut-elimination theorems for LK and LJ were first proved) and the books [279] and [283] are good sources for Gentzen-type proof systems. References [162] and [272] are good textbooks about analytic tableau and other proof systems, while [256] is a comprehensive source of information about proof systems in general.

Classical logic is presented and studied in any introductory book on mathematical logic, such as [152] and [212].

Hypersequents were independently introduced by Mints in [215]), Pottinger in [243], and Avron in [27] and [36]. Among other applications (including several in this book), they now provide the main framework for the proof theory of fuzzy logics ([213]).

Chapter 2

Negation and Paraconsistency

Traditional logics, most notably classical logic and intuitionistic logic, accept the following principle of explosion, 'ex contradictione sequitur quodlibet', which states that "If one claims something is both true and not true, one can logically derive any conclusion":

$$\psi, \neg\psi \vdash \varphi. \tag{2.1}$$

In contrast, paraconsistent logics reject (2.1). This rejection does not come without a price. For example, it is very easy to see that any logic which reject this rule and has a disjunction $\vee$ in its language must reject also at least one of the following two intuitively valid principles:[1]

- *The introduction of disjunction:* from ψ infer $\psi \vee \varphi$,
- *The disjunctive syllogism:* From $\neg\psi$ and $\psi \vee \varphi$ infer φ.

In this chapter we define in precise terms what paraconsistency means in this book, and what properties a useful paraconsistent logic might be expected to have. Then, in the proceeding chapters we investigate different methods for introducing such logics. Obviously, each one of these methods (explicitly or implicitly) challenges at least one of the two principles mentioned above, or avoids the use of disjunction in its language.

2.1 General Considerations

2.1.1 What is a Negation?

By (2.1), it is evident that the notion of paraconsistency depends on a corresponding notion of a 'negation'. So, for defining paraconsistency we

---
[1] This observation, originally due to Pseudo-Scotus in the 13th century [248], is known in the literature as the 'Lewis independent argument' for the validity of (2.1) [197, 198].

first need to understand the latter notion. Unfortunately, while standard abstract definitions are available for all the basic positive connectives (see Definition 1.23), "negation" lacks such a definition. What is more, there is no general agreement about the properties that it should have. Below we give a *minimal* condition that we believe any "negation" connective should satisfy, and is indeed satisfied by almost any connective that officially serves as a "negation" in some paraconsistent logic that is investigated as such in the literature.[2] Intuitively, it states that a logic that has such a connective should not admit entailments in the classical language that do not hold in classical logic. Next, we formulate this in precise terms.

Definition 2.1 (bivalent ¬-interpretation) Let $\mathcal{L}$ be a language with a unary connective ¬. A *bivalent ¬-interpretation* for $\mathcal{L}$ is a bivalent interpretation **F** for $\mathcal{L}$ (Definition 1.42) such that **F**(¬) is the classical truth table for negation (see Section 1.2.2).

Definition 2.2 (¬-containment in CL, ¬-coherence with CL) Let $\mathcal{L}$ be a propositional language with a unary connective ¬, and let $\mathbf{L} = \langle \mathcal{L}, \vdash_\mathbf{L} \rangle$ be a logic for $\mathcal{L}$.

- **L** is ¬-*contained in classical logic* if it is **F**-contained in it (Definition 1.45) for some bivalent ¬-interpretation **F**.

- **L** is ¬-*coherent with classical logic* if it has a semi-normal fragment (Definition 1.32) which is ¬-contained in classical logic.

Definition 2.3 (negation) Let $\mathbf{L} = \langle \mathcal{L}, \vdash_\mathbf{L} \rangle$ be a propositional logic for a language $\mathcal{L}$. A unary connective ¬ of $\mathcal{L}$ is called a *negation* of **L** if **L** is ¬-coherent with classical logic.

Proposition 2.4 *Let ¬ be a negation of* $\mathbf{L} = \langle \mathcal{L}, \vdash_\mathbf{L} \rangle$. *Then for every variable p it holds that* $p \nvdash_\mathbf{L} \neg p$ *and* $\neg p \nvdash_\mathbf{L} p$.[3]

Proof. Immediate from the fact that $p \nvdash_{CL} \neg p$ and $\neg p \nvdash_{CL} p$. □

As demonstrated by Proposition 2.4, ¬-coherence with classical logic is useful for assuring *negative* properties of a negation. It would be useful to have also a list of *positive* properties that negation has in classical logic, and might be desirable for a 'negation'. We use for this Proposition 1.63, but (for reasons that are explained in Note 2.10 and Footnote 4) we replace the last two properties listed there by a weaker one, which usually suffices for ensuring the desirable replacement property (Definition 1.18).

---

[2]The only exception we are aware of are Wansing's paraconsistent connexive logics ([298, 299]).

[3]This result can be made a bit stronger, to cover all the minimal conditions that are called "verification" and "falsification" in [209].

2.1 General Considerations

Definition 2.5 (common properties of negations) Let $\mathbf{L} = \langle \mathcal{L}, \vdash_\mathbf{L} \rangle$ be a propositional logic for a language $\mathcal{L}$ with a unary connective $\neg$.

- $\neg$ is *explosive* (in $\mathbf{L}$) if $\neg\varphi, \varphi \vdash_\mathbf{L} \psi$ for every φ and ψ (equivalently: when φ and ψ are variables).

- $\neg$ is *complete* (in $\mathbf{L}$) if it satisfies the following version of the *law of excluded middle* for every φ, ψ, and $\mathcal{T}$:

  (LEM) $\quad \mathcal{T} \vdash_\mathbf{L} \varphi$ whenever $\mathcal{T}, \psi \vdash_\mathbf{L} \varphi$ and $\mathcal{T}, \neg\psi \vdash_\mathbf{L} \varphi$.

- $\neg$ is *right-involutive* (in $\mathbf{L}$) if $\varphi \vdash_\mathbf{L} \neg\neg\varphi$ for every formula φ (equivalently: when φ is a variable), and is *left-involutive* (in $\mathbf{L}$) if $\neg\neg\varphi \vdash_\mathbf{L} \varphi$ for every formula φ (equivalently: when φ is a variable). $\neg$ is *involutive* (in $\mathbf{L}$) if it is both right- and left-involutive (in $\mathbf{L}$).

- $\neg$ is *contrapositive* (in $\mathbf{L}$) if $\neg\varphi \vdash_\mathbf{L} \neg\psi$ whenever $\psi \vdash_\mathbf{L} \varphi$.

Note 2.6 It is easy to verify that:

- If $\mathbf{L}$ has an implication $\supset$, then
  - $\neg$ is explosive in $\mathbf{L}$ iff $\vdash_\mathbf{L} \neg\varphi \supset (\varphi \supset \psi)$ for every φ and ψ.
  - $\neg$ is right-involutive in $\mathbf{L}$ iff $\vdash_\mathbf{L} \varphi \supset \neg\neg\varphi$ for every φ.
  - $\neg$ is left-involutive in $\mathbf{L}$ iff $\vdash_\mathbf{L} \neg\neg\varphi \supset \varphi$ for every φ.
  - $\neg$ is contrapositive in $\mathbf{L}$ iff $\vdash_\mathbf{L} \neg\varphi \supset \neg\psi$ whenever $\vdash_\mathbf{L} \psi \supset \varphi$.

- If $\mathbf{L}$ has a disjunction $\vee$ then $\neg$ is complete in $\mathbf{L}$ iff $\vdash_\mathbf{L} \neg\varphi \vee \varphi$ for every φ (equivalently: if $\vdash_\mathbf{L} \varphi \vee \neg\varphi$ for every φ).

- **CL** is the minimal normal logic which has a negation which is both complete and explosive.

2.1.2 What is a Paraconsistent Logic?

The notion of paraconsistency, informally considered in the introduction, may now be defined in exact terms:

Definition 2.7 ((strong) pre-$\neg$-paraconsistency) Let $\mathcal{L}$ be a language with a unary connective $\neg$, and let $\mathbf{L} = \langle \mathcal{L}, \vdash_\mathbf{L} \rangle$ be a logic for $\mathcal{L}$.

- $\mathbf{L}$ is *pre-$\neg$-paraconsistent* if $\neg$ is not explosive in it.

- $\mathbf{L}$ is called *strongly pre-$\neg$-paraconsistent* if there are variables p, q such that $p, \neg p \not\vdash_\mathbf{L} \neg q$. (Equivalently: if $p, \neg p \not\vdash_\mathbf{L} \neg q$ whenever p and q are distinct variables.)

Note 2.8 Since **L** is a *logic* (Definition 1.3), pre-paraconsistency (Definition 2.7) can easily be seen to be equivalent to da-Costa's notion of paraconsistency [127] (as formalized in [116]), which requires that there would be a theory $\mathcal{T}$ and formulas ψ, φ such that $\mathcal{T} \vdash_\mathbf{L} \psi$, $\mathcal{T} \vdash_\mathbf{L} \neg\psi$, but $\mathcal{T} \not\vdash_\mathbf{L} \varphi$.

Definition 2.7 intends to capture the idea that a contradictory set of premises needs not entail every formula. However, talking about 'contradictory set' makes sense only if the underlying $\neg$ somehow represents a 'negation' operation. For assuring this, one needs stronger notions.

Definition 2.9 ((strong) $\neg$-paraconsistency) Let **L** be a logic for a language with a unary connective $\neg$. **L** is *(strongly) $\neg$-paraconsistent* if it is (strongly) pre-$\neg$-paraconsistent, and $\neg$ is a negation of **L**.

Note 2.10 It should again be emphasized that our notion of paraconsistency has *two* components. In addition to the usual demand that a sentence ψ and its "negation" $\neg\psi$ do not imply everything, we also demand that the connective $\neg$ can indeed be taken to be a sort of *negation*. Our notion of *strong* $\neg$-paraconsistency goes one more step in this direction: while $\neg$-paraconsistency intuitively means that a single $\neg$-contradiction does not entail that every formula is true, strong $\neg$-paraconsistency intuitively means that a single $\neg$-contradiction does not entail that every formula is *false*. It seems to us that the notion of strong $\neg$-paraconsistency better reflects the intuitions here. Indeed, all of the logics investigated in this book are actually strongly $\neg$-paraconsistent.[4]

Convention: Henceforth we shall frequently omit the $\neg$ sign (if it is clear from the context), and simply refer to *(strongly) (pre-) paraconsistent logics*.

Our definitions immediately imply the following result:

Proposition 2.11 *Every conservative extension of a (strongly) paraconsistent logic is also (strongly) paraconsistent.*

Note 2.12 Note that in order to ensure the validity of Proposition 2.11, our definitions do *not* imply that a paraconsistent logic should be semi-normal itself, but only that it has an appropriate semi-normal fragment. Indeed, a connective may, e.g., be an implication for a fragment of a logic, but not for the logic as a whole.[5] Note also that an extension of a paraconsistent

---

[4] It can easily be seen that a logic in which $\neg$ has either of the last two properties of classical negation listed in Proposition 1.63 is not strongly $\neg$-paraconsistent. This is the reason why in Definition 2.5 we did not retain these two properties as they are.

[5] For example, let $HS4$ be some standard Hilbert-type system for the modal logic **S4** (see Chapter 9) which employs necessitation as a primitive rule. Then $\supset$ is an implication for the $\{\neg, \supset\}$-fragment of the logic induced by $HS4$, bur not for the whole logic, since $\varphi \vdash_{HS4} \Box\varphi$, but $\not\vdash_{HS4} \varphi \supset \Box\varphi$.

2.1 General Considerations

logic which is not conservative may not be paraconsistent even if it is pre-paraconsistent, because it might not have a semi-normal fragment anymore.

Example 2.13

- By Propositions 1.63 and 1.76, **CL** and **IL** are *not* (pre-) paraconsistent with respect to their official negations.

- Johansson's minimal logic (see Note 1.79) is paraconsistent, but not strongly paraconsistent.

- Define the logic $\mathbf{L_s}$ in $\mathcal{L}_{CL}$ as follows: $\mathcal{T} \vdash_{\mathbf{L_s}} \psi$ if either $\psi \in \mathcal{T}$, or ψ is a classical $\mathcal{L}_{CL}$-tautology. It is easy to verify that $\mathbf{L_s}$ is a finitary logic. This logic is obviously $\neg$-contained in classical logic, and even has the same logically valid formulas as **CL**. It is also strongly pre-$\neg$-paraconsistent (since $p, \neg p \not\vdash_{\mathbf{L_s}} \neg q$). However, we do not take $\mathbf{L_s}$ to be a paraconsistent logic, because *it has no semi-normal fragment*.

The next proposition shows that in normal strongly $\neg$-paraconsistent logics the connective $\neg$ cannot have all the properties listed in Definition 2.5.

Proposition 2.14 *Let $\neg$ be a negation of $\mathbf{L}$. Then $\mathbf{L}$ cannot be strongly $\neg$-paraconsistent if it has a conjunction, and its negation $\neg$ is complete, right-involutive, and contrapositive.*

Proof. Suppose $\mathbf{L}$ is a logic with a conjunction $\wedge$ in which $\neg$ has the three properties. Since $p \wedge \neg p \vdash_{\mathbf{L}} p$, the contrapositivity of $\neg$ implies that $\neg p \vdash_{\mathbf{L}} \neg(p \wedge \neg p)$, and so also $p \wedge \neg p \vdash_{\mathbf{L}} \neg(p \wedge \neg p)$. Obviously, $\neg(p \wedge \neg p) \vdash_{\mathbf{L}} \neg(p \wedge \neg p)$ as well. Hence the completeness of $\neg$ in $\mathbf{L}$ implies that $\vdash_{\mathbf{L}} \neg(p \wedge \neg p)$, and so $q \vdash_{\mathbf{L}} \neg(p \wedge \neg p)$. It follows by contrapositivity that $\neg\neg(p \wedge \neg p) \vdash_{\mathbf{L}} \neg q$. Since $\neg$ is right-involutive, this implies that $p \wedge \neg p \vdash_{\mathbf{L}} \neg q$, and so $p, \neg p \vdash_{\mathbf{L}} \neg q$. □

Note 2.15 Strongly paraconsistent logics reject one particular basic principle of explosion: $p, \neg p \vdash \neg q$. Our next definitions introduce other notions of explosion and paraconsistency that have been considered in the literature.

Definition 2.16 (controllably/partially $\neg$-explosive logics) Let $\mathbf{L} = \langle \mathcal{L}, \vdash_{\mathbf{L}} \rangle$ be a logic for a language $\mathcal{L}$ with negation $\neg$, and let $\sigma \in \mathcal{W}(\mathcal{L})$.

- $\mathbf{L}$ is $\neg$-*explosive* if $\neg$ is explosive in it.

- $\mathbf{L}$ is *controllably $\neg$-explosive in contact with* σ if for some (equivalently: for every) variable q such that $q \notin \mathsf{Var}(\sigma)$: $\sigma, \neg\sigma \vdash_{\mathbf{L}} q$, while neither $\sigma \vdash_{\mathbf{L}} q$, nor $\neg\sigma \vdash_{\mathbf{L}} q$. $\mathbf{L}$ is *controllably $\neg$-explosive* if there is a formula σ such that $\mathbf{L}$ is *controllably $\neg$-explosive in contact with* σ.

- **L** is *partially ¬-explosive with respect to* σ (or σ-*partially ¬-explosive*) if for some (or for every) variable q such that $q \notin \mathsf{Var}(\sigma)$: $q, \neg q \vdash_\mathbf{L} \sigma$ while $\not\vdash_\mathbf{L} \sigma$. **L** is *partially ¬-explosive* if there is a formula σ such that **L** is *partially ¬-explosive with respect to* σ.

Definition 2.17 (non-exploding logics) A logic $\langle \mathcal{L}, \vdash \rangle$ is *non-exploding* if $\mathcal{T} \not\vdash q$ for every theory $\mathcal{T}$ such that $\mathsf{Var}(\mathcal{T}) \neq \mathsf{Var}$ and every variable q such that $q \notin \mathsf{Var}(\mathcal{T})$.[6] $\langle \mathcal{L}, \vdash \rangle$ is *strongly non-exploding* if $\mathcal{T} \not\vdash \psi$ for every ψ such that $\mathsf{Var}(\mathcal{T}) \cap \mathsf{Var}(\psi) = \emptyset$ and $\not\vdash \psi$.

Example 2.18

- Classical logic and intuitionistic logics are ¬-explosive.

- Johansson's minimal logic (Note 1.79) is partially ¬-explosive with respect to $\neg p$ whenever p is a variable.

- Sette's three-valued logic $\mathbf{P}_1$ (see Chapter 4) is controllably ¬-explosive in contact with $\neg p$ whenever p is a variable.

- Obviously, if a logic is ¬-explosive then it is exploding (i.e., it is not non-exploding). *The converse fails* for many paraconsistent logics (e.g., the LFIs defined in Chapter 8).

- Relevance logics provide good examples of strongly non-exploding logics. This is assured by the *basic relevance criterion*. (See Chapter 11.)

Definition 2.19 (boldly paraconsistent logic) A logic $\mathbf{L} = \langle \mathcal{L}, \vdash_\mathbf{L} \rangle$ for a language $\mathcal{L}$ with a negation ¬ is *boldly ¬-paraconsistent* if it is not partially ¬-explosive, that is: if $q, \neg q \vdash_\mathbf{L} \sigma$, then either $\vdash_\mathbf{L} \sigma$ or $q \in \mathsf{Var}(\sigma)$.

Proposition 2.20

1. Let **L** be strongly non-exploding logic with negation ¬. Then **L** is boldly ¬-paraconsistent, and not controllably ¬-explosive.

2. Every boldly ¬-paraconsistent logic is strongly ¬-paraconsistent.

Proof. The first part is obvious. The second part follows from the fact that by definition, a logic **L** with negation ¬ is strongly ¬-paraconsistent iff for every (or some) variable p, **L** is not partially ¬-explosive with respect to $\neg p$. (Note that $\not\vdash_\mathbf{L} \neg p$ in case **L** is ¬-coherent with classical logic.) □

---

[6] An equivalent condition is that for every theory $\mathcal{T}$ such that $\mathsf{Var}(\mathcal{T}) \neq \mathsf{Var}$ there is a formula ψ such that $\mathcal{T} \not\vdash \psi$.

Note 2.21 We have introduced our general notion of paraconsistency in order to allow the language of a paraconsistent logic to include connectives that have no classical counterparts. However, most of this book is devoted to logics which provide paraconsistent alternatives to classical logic in its own language (more precisely: in some natural language for classical logic). Hence, we mainly concentrate in what follows on paraconsistent logics which are actually *contained* in classical logic, not just coherent with it. (Exceptions can be found in Chapters 4 and 5.) One natural class of such logics is provided by the next definition. It includes almost all the logics which are studied in the next two parts of the book.

Definition 2.22 (classical $\neg$-paraconsistent logic) Let $\mathcal{L}_{CL} \subseteq \mathcal{L}$, and let $\mathbf{L} = \langle \mathcal{L}, \vdash_{\mathbf{L}} \rangle$ be a logic. We say that $\mathbf{L}$ is a *classical $\neg$-paraconsistent logic* if it is an axiomatic extension of $\mathbf{CL}^+$ which is pre-$\neg$-paraconsistent and $\neg$-contained in classical logic.

Proposition 2.23 *Let $\mathcal{L}_{CL} \subseteq \mathcal{L}$, and let $\mathbf{L} = \langle \mathcal{L}, \vdash_{\mathbf{L}} \rangle$ be a logic for $\mathcal{L}$. Then $\mathbf{L}$ is a classical $\neg$-paraconsistent logic iff it is finitary; pre-$\neg$-paraconsistent; $\neg$-contained in classical logic; normal, with the connectives $\wedge$, $\vee$ and $\supset$ of $\mathcal{L}_{CL}$ serving as conjunction, disjunction and implication (respectively) for it; and its $\{\wedge, \vee, \supset\}$-fragment is $\mathbf{CL}^+$.*

Proof. This follows from Propositions 1.70, 1.65, and Corollary 1.71. □

Proposition 2.24 *If $\mathbf{L}$ is a classical $\neg$-paraconsistent logic then it is $\neg$-paraconsistent, and its $\mathcal{L}_{CL}$-fragment is contained in classical logic.*

Proof. This follows from the definitions and Proposition 1.54. □

2.2 Maximality Properties

A property of paraconsistent logics which is frequently considered desirable is *maximality*. In his seminal paper [127], Newton da-Costa implicitly indicated that paraconsistency by itself is not sufficient, and that a paraconsistent logic should be maximal in the sense that it should retain as much of classical logic as possible, while still allowing non-trivial inconsistent theories. Below we formalize two possible explications of this intuition.

The first notion of maximality that we consider is absolute in the sense that it is not defined with respect to any particular logic.

Definition 2.25 (strongly maximal paraconsistent logics) Let $\mathbf{L} = \langle \mathcal{L}, \vdash_{\mathbf{L}} \rangle$ be a finitary paraconsistent logic.[7]

---
[7]See Note 1.4 for an explanation why we concentrate on finitary logics here and later.

- **L** is *maximal* if every simple finitary extension of **L** (in the sense of Definition 1.9), whose set of theorems *properly includes* that of **L**, is not pre-paraconsistent.

- **L** is *strongly maximal* if every proper simple finitary extension of **L** is not pre-paraconsistent.

Note that maximal paraconsistency is based only on extending the underlying set of *theorems*, while the strong sense of maximal paraconsistency is based on extending the underlying *consequence relation*. Clearly, strong maximality implies maximality. In Chapter 4 we shall show that the converse does not hold (see example in Note 4.43).

The second notion of maximality that we consider is relative in the sense that it is defined with respect to classical logic.

Definition 2.26 (F-maximality relative to classical logic) Let **F** be a bivalent ¬-interpretation of a language $\mathcal{L}$ with a unary connective ¬.

- An $\mathcal{L}$-formula ψ is a *classical* **F**-*tautology* if $\vdash_{2_\mathbf{F}} \psi$ (i.e., ψ is satisfied by every two-valued valuation which respects all the truth-tables that **F** assigns to the connectives of $\mathcal{L}$).

- A logic $\mathbf{L} = \langle \mathcal{L}, \vdash_\mathbf{L} \rangle$ is **F**-*complete* if $\vdash_\mathbf{L} \psi$ whenever $\vdash_{2_\mathbf{F}} \psi$ (i.e., its set of theorems consists of all the classical **F**-tautologies).

- **L** is **F**-*maximal relative to classical logic* if the following hold:
  - **L** is **F**-contained in classical logic.
  - If ψ is a classical **F**-tautology not provable in **L**, then by adding ψ to **L** as a new axiom schema we obtain an **F**-complete logic.

Definition 2.27 (maximality relative to classical logic) A logic **L** in a language with a unary connective ¬ is *maximally* ¬-*paraconsistent relative to classical logic* if it is semi-normal, pre-¬-paraconsistent, and **F**-maximal relative to classical logic for some bivalent ¬-interpretation **F**. (Note that by definitions, such a logic is necessarily ¬-paraconsistent.)[8]

A seemingly natural strengthening of the notion of maximal paraconsistency relative to classical logic (which unfortunately turns out to be too strong, as Proposition 2.31 below indicates) would be obtained if again we consider the *consequence relation* of the underlying logic, rather than just its set of theorems (analogously to the above notion of strong maximality):

---

[8]Note 4.43 includes an example of a paraconsistent logic which is maximally paraconsistent relative to classical logic, but not strongly maximal. Proposition 4.50 includes examples of logics which are strongly maximal, but not contained in classical logic.

2.2 Maximality Properties

Definition 2.28 (strong relative F-maximality) Let **F** be a bivalent $\neg$-interpretation of a language $\mathcal{L}$ with $\neg$. A logic $\mathbf{L} = \langle \mathcal{L}, \vdash_\mathbf{L} \rangle$ is *strongly* **F**-*maximal relative to classical logic* if the following conditions hold:

- **L** is **F**-contained in classical logic.

- Let Γ be a finite set of $\mathcal{L}$-formulas and ψ an $\mathcal{L}$-formula such that $\Gamma \nvdash_\mathbf{L} \psi$, but $\Gamma \vdash_{2_\mathbf{F}} \psi$. Then the extension of **L** by the rule Γ/ψ (Definition 1.11) is the logic $\mathbf{L_F} = \langle \mathcal{L}, \vdash_{2_\mathbf{F}} \rangle$.

Definition 2.29 (strong maximal paraconsistency relative to CL) A semi-normal logic **L** in a language with a unary connective $\neg$ is *maximally $\neg$-paraconsistent in the strong sense relative to classical logic* if it is pre-$\neg$-paraconsistent, and strongly **F**-maximal relative to classical logic for some bivalent $\neg$-interpretation **F**.

A somewhat weaker demand, which also has turned out to be too strong, is given in the next lemma:

Lemma 2.30 *Suppose that* $\mathbf{L} = \langle \mathcal{L}, \vdash_\mathbf{L} \rangle$ *is a logic which is maximally $\neg$-paraconsistent in the strong sense relative to classical logic. Then there exists a bivalent $\neg$-interpretation* **F** *such that:*

1. **L** *is* **F**-*contained in classical logic.*

2. *The extension of* **L** *by the rule* $\{p, \neg p\}/q$ *(where p and q are distinct variables) is the logic* $\mathbf{L_F} = \langle \mathcal{L}, \vdash_{2_\mathbf{F}} \rangle$.

Proof. Immediate from Definition 2.29, and the fact that $p, \neg p \vdash_{2_\mathbf{F}} q$ for every bivalent $\neg$-interpretation **F**, while $p, \neg p \nvdash_\mathbf{L} q$. □

Proposition 2.31 *No semi-normal finitary logic is maximally paraconsistent in the strong sense relative to classical logic.*

Proof. By Lemma 2.30, it suffices to prove that if $\mathbf{L} = \langle \mathcal{L}, \vdash_\mathbf{L} \rangle$ is a semi-normal, pre-$\neg$-paraconsistent finitary logic, then there cannot exist a bivalent $\neg$-interpretation **F** that satisfies the two conditions given in Lemma 2.30. So, suppose for contradiction that there is such **F**. Define $\mathcal{T} \vdash_{\mathbf{L}+} \varphi$ if either $\mathcal{T} \vdash_\mathbf{L} \varphi$, or $\mathcal{T}$ is $\mathbf{L_F}$-inconsistent. We show that $\langle \mathcal{L}, \vdash_{\mathbf{L}+} \rangle$ is a finitary logic. Reflexivity and non-triviality of $\vdash_{\mathbf{L}+}$ follow from the reflexivity and non-triviality of $\vdash_\mathbf{L}$ (and the fact that $\{p\}$ is $\mathbf{L_F}$-consistent for a variable p). Structurality of $\vdash_{\mathbf{L}+}$ follows from the structurality of $\vdash_\mathbf{L}$ and of $\vdash_{2_\mathbf{F}}$. For monotonicity, let $\mathcal{T} \subseteq \mathcal{T}'$ and $\mathcal{T} \vdash_{\mathbf{L}+} \psi$. If $\mathcal{T} \vdash_\mathbf{L} \psi$, then $\mathcal{T}' \vdash_{\mathbf{L}+} \psi$ by the monotonicity of $\vdash_\mathbf{L}$. Otherwise, $\mathcal{T}$ is $\mathbf{L_F}$-inconsistent. Hence so is

$\mathcal{T}'$, and so again $\mathcal{T}' \vdash_{L+} \psi$. That $\vdash_{L+}$ is finitary follows from the finitariness of **L** and Theorem 1.64 (which easily entails a similar theorem for $\mathbf{L_F}$). Finally, for transitivity assume that $\mathcal{T} \vdash_{L+} \varphi$ and $\mathcal{T}', \varphi \vdash_{L+} \psi$. We show that $\mathcal{T}, \mathcal{T}' \vdash_{L+} \psi$ by considering the possible cases:

1. Suppose $\mathcal{T} \vdash_\mathbf{L} \varphi$ and $\mathcal{T}', \varphi \vdash_\mathbf{L} \psi$. Then $\mathcal{T}, \mathcal{T}' \vdash_{L+} \psi$ by the transitivity of $\vdash_\mathbf{L}$.

2. Suppose $\mathcal{T} \vdash_\mathbf{L} \varphi$ and $\mathcal{T}' \cup \{\varphi\}$ is $\mathbf{L_F}$-inconsistent. Since $\vdash_\mathbf{L} \subset \vdash_{2\mathbf{F}}$, we have that $\mathcal{T} \vdash_{2\mathbf{F}} \varphi$. This implies that $\mathcal{T} \cup \mathcal{T}'$ is $\mathbf{L_F}$-inconsistent, and so $\mathcal{T} \cup \mathcal{T}' \vdash_{L+} \psi$.

3. Suppose $\mathcal{T}$ is $\mathbf{L_F}$-inconsistent. Then so is $\mathcal{T} \cup \mathcal{T}'$, and so $\mathcal{T} \cup \mathcal{T}' \vdash_{L+} \psi$.

Now, $\mathbf{L_F}$ is obviously an extension of $\langle \mathcal{L}, \vdash_{L+} \rangle$. On the other hand, since $\{p, \neg p\}$ is $\mathbf{L_F}$-inconsistent, $p, \neg p \vdash_{L+} q$, and so $\langle \mathcal{L}, \vdash_{L+} \rangle$ is an extension of the extension of **L** by the rule $\{p, \neg p\}/q$. Hence, our assumptions about **F** imply that $\vdash_{L+} = \vdash_{2\mathbf{F}}$. We end the proof by showing that this is impossible.

- Suppose that **L** has a semi-implication $\supset$. Then $\neg p \not\vdash_\mathbf{L} p \supset q$, because $\neg p, p \not\vdash_\mathbf{L} q$. Therefore, it follows from the $\mathbf{L_F}$-consistency of $\{\neg p\}$ that $\neg p \not\vdash_{L+} p \supset q$. On the other hand, $\neg p \vdash_{2\mathbf{F}} p \supset q$ (by Proposition 1.54). This is a contradiction.

- Suppose that **L** has a disjunction $\vee$. Since $\neg p, p \not\vdash_\mathbf{L} q$, the fact that $\vee$ is a disjunction for **L** implies that $\neg p, p \vee q \not\vdash_\mathbf{L} q$. This in turn implies that $\neg p, p \vee q \not\vdash_{L+} q$. (This is due to Proposition 1.54 and the fact that $\{\neg p, p \vee q\}$ is $\mathbf{L_F}$-consistent.) On the other hand, $\neg p, p \vee q \vdash_{2\mathbf{F}} q$ (by Proposition 1.54 again). A contradiction.

- Suppose that **L** has a conjunction $\wedge$. By Proposition 1.54, if $\Gamma \cup \{\psi\}$ is a finite set of sentences in the language $\{\neg, \wedge\}$, and $\Gamma \vdash_{CL} \psi$ then $\Gamma \vdash_{2\mathbf{F}} \psi$, and so our assumption implies that $\Gamma \vdash_{L+} \psi$. By definition of $\vdash_{L+}$, this in turn implies that $\Gamma \vdash_\mathbf{L} \psi$ whenever $\Gamma \cup \{\psi\}$ is a finite set of sentences in the language $\{\neg, \wedge\}$, $\Gamma \vdash_{CL} \psi$, and Γ is classically consistent. It follows in particular that $\vdash_\mathbf{L} \neg((\neg p \wedge p) \wedge \neg q)$ and $p, \neg(p \wedge \neg q) \vdash_\mathbf{L} q$. Since $\vdash_\mathbf{L}$ is structural, the latter fact implies that $\neg p \wedge p, \neg((\neg p \wedge p) \wedge \neg q) \vdash_\mathbf{L} q$. Together with $\vdash_\mathbf{L} \neg((\neg p \wedge p) \wedge \neg q)$, this implies (by transitivity of $\vdash_\mathbf{L}$) that $\neg p \wedge p \vdash_\mathbf{L} q$. Since $\wedge$ is a conjunction for **L**, also $\neg p, p \vdash_\mathbf{L} \neg p \wedge p$. Hence we get that $\neg p, p \vdash_\mathbf{L} q$, contradicting the pre-paraconsistency of **L**.

Since **L** is semi-normal, we got a contradiction in all cases. $\square$

Because of Proposition 2.31, in the case of maximality relative to classical logic we shall be satisfied with the weaker notion that is based on extending the *set of theorems* of the underlying logic. The maximality of its full *consequence relation* will still be demanded in the case of absolute maximality. These considerations lead to our following precise (and optimal) counterpart of da-Costa's intuitive notion of "maximality":

Definition 2.32 (fully maximal logic) Let **L** be a finitary logic. **L** is a *fully maximal* paraconsistent logic if it is both maximally paraconsistent relative to classical logic and strongly maximal.[9]

Note 2.33 It immediately follows from the definitions that if **L** is fully maximal paraconsistent logic, then it is indeed ¬-paraconsistent, as well as ¬-contained in classical logic.

2.3 Proof Theory of Paraconsistent Logics

All the Gentzen-type systems which were presented in Chapter 1 are *canonical* ([67, 69]). This means that each of their logical (i.e., non-structural) rules has the following properties:[10]

1. Each (application of a) rule has exactly one principal formula, which is introduced on one fixed side of its conclusion.

2. There is a fixed n-ary connective $\diamond$ of the language such that the principal formula in applications of the rule is always of the form $\diamond(\psi_1, \ldots, \psi_n)$. (Such a rule is called *canonical $\diamond$-rule*.)

3. Let $\diamond(\psi_1, \ldots, \psi_n)$ be the principal formula in some application of the rule. Then all the side formulas of that application belong to $\{\psi_1, \ldots, \psi_n\}$.

4. There are no restrictions on the contexts in applications of the rule (i.e. every context is legitimate).

Now, it is easy to see that the only canonical ¬-rules which can be used in a Gentzen-type system G such that $\mathbf{L_G}$ (Proposition 1.110) is ¬-contained in classical logic are the classical two rules for ¬:

$$[\neg \Rightarrow] \quad \frac{\Gamma \Rightarrow \Delta, \varphi}{\Gamma, \neg\varphi \Rightarrow \Delta} \qquad [\Rightarrow \neg] \quad \frac{\Gamma, \varphi \Rightarrow \Delta}{\Gamma \Rightarrow \Delta, \neg\varphi}$$

---

[9] What is called here 'fully maximal and normal' was called *ideal* in [18, 58]. In this book we have chosen to use a more neutral terminology (see also [15]).
[10] See Section 1.3.3 for the terminology used below.

However, if G is a standard Gentzen-type system which has $[\neg\Rightarrow]$ as one of its rules, then $\mathbf{L_G}$ is obviously not pre-$\neg$-paraconsistent. Hence, $[\Rightarrow\neg]$ is the only canonical $\neg$-rule which can be used in standard Gentzen-type systems for paraconsistent logics. This is not sufficient for deriving most properties that $\neg$ might have in such a logic. For providing useful standard Gentzen-type systems for paraconsistent logics we need therefore a more extensive class of systems, in which $\neg$ is allowed to play an active role. The simplest such class is that of $\neg$-*quasi-canonical* systems (henceforth just *quasi-canonical*), in which every logical rule has the following properties:

1. Each (application of a) rule has exactly one principal formula, which is introduced on one fixed side of its conclusion.

2. There is a fixed n-ary connective $\diamond$ of the language such that the principal formula in applications of the rule is always of the form $\diamond(\psi_1,\ldots,\psi_n)$, or always of the form $\neg\diamond(\psi_1,\ldots,\psi_n)$. (Such a rule is called *quasi-canonical $\diamond$-rule* in the first case, *quasi-canonical $\neg\diamond$-rule* in the second.)

3. Let $\diamond(\psi_1,\ldots,\psi_n)$ or $\neg\diamond(\psi_1,\ldots,\psi_n)$ be the principal formula in some application of the rule. Then all the side formulas of that application belong to $\{\psi_1,\ldots,\psi_n,\neg\psi_1,\ldots,\neg\psi_n\}$.

4. There are no restrictions on the contexts in applications of the rule (i.e. every context is legitimate).

Example 2.34 Figure 2.1 presents the cut-free quasi-canonical system for Łukasiewicz three-valued logic (which is $\neg$-contained in classical logic, but *not* paraconsistent) that has been given in [42].[11]

We are now able to characterize the class of Gentzen-type systems to which most of the Gentzen-type systems used in this book belong. These are *quasi-canonical standard Gentzen-type systems which either have no canonical $\neg$-rules at all, or the classical $[\Rightarrow\neg]$ is their sole canonical $\neg$-rule.* (Note that such systems may still have $\neg\neg$-rules.)

Example 2.35 As noted above, the quasi-canonical system presented in Example 2.34 is not paraconsistent. However, by changing its $[\neg\Rightarrow]$ rule to the classical $[\Rightarrow\neg]$ rule, or by simply deleting $[\neg\Rightarrow]$ from it, we do get two paraconsistent systems of the type just described.

---

[11]To shorten the presentation of the system without losing its illustrative power, we take here the language of Łukasiewicz three-valued logic to be $\{\neg,\wedge,\to\}$. The connective $\vee$ can be defined there in terms of $\neg$ and $\wedge$ in the usual way, and so its rules are just dual to those of $\wedge$. (Note that $\wedge$ too can actually be defined here in terms of $\neg$ and $\to$.)

Axioms: $\psi \Rightarrow \psi$

Rules: cut, weakening, and the following logical rules:

$$[\neg \Rightarrow] \quad \frac{\Gamma \Rightarrow \Delta, \psi}{\Gamma, \neg\psi \Rightarrow \Delta}$$

$$[\neg\neg \Rightarrow] \quad \frac{\Gamma, \varphi \Rightarrow \Delta}{\Gamma, \neg\neg\varphi \Rightarrow \Delta} \qquad [\Rightarrow \neg\neg] \quad \frac{\Gamma \Rightarrow \Delta, \varphi}{\Gamma \Rightarrow \Delta, \neg\neg\varphi}$$

$$[\wedge \Rightarrow] \quad \frac{\Gamma, \varphi, \psi \Rightarrow \Delta}{\Gamma, \varphi \wedge \psi \Rightarrow \Delta} \qquad [\Rightarrow \wedge] \quad \frac{\Gamma \Rightarrow \Delta, \varphi \quad \Gamma \Rightarrow \Delta, \psi}{\Gamma \Rightarrow \Delta, \varphi \wedge \psi}$$

$$[\neg\wedge \Rightarrow] \quad \frac{\Gamma, \neg\varphi \Rightarrow \Delta \quad \Gamma, \neg\psi \Rightarrow \Delta}{\Gamma, \neg(\varphi \wedge \psi) \Rightarrow \Delta} \qquad [\Rightarrow \neg\wedge] \quad \frac{\Gamma \Rightarrow \Delta, \neg\varphi, \neg\psi}{\Gamma \Rightarrow \Delta, \neg(\varphi \wedge \psi)}$$

$$[\to \Rightarrow] \quad \frac{\Gamma, \neg\varphi \Rightarrow \Delta \quad \Gamma, \psi \Rightarrow \Delta \quad \Gamma \Rightarrow \varphi, \neg\psi, \Delta}{\Gamma, \varphi \to \psi \Rightarrow \Delta}$$

$$[\Rightarrow \to] \quad \frac{\Gamma, \varphi \Rightarrow \psi, \Delta \quad \Gamma, \neg\psi \Rightarrow \neg\varphi, \Delta}{\Gamma \Rightarrow \varphi \to \psi, \Delta}$$

$$[\neg\to \Rightarrow] \quad \frac{\Gamma, \varphi, \neg\psi \Rightarrow \Delta}{\Gamma, \neg(\varphi \to \psi) \Rightarrow \Delta} \qquad [\Rightarrow \neg\to] \quad \frac{\Gamma \Rightarrow \varphi, \Delta \quad \Gamma \Rightarrow \neg\psi, \Delta}{\Gamma \Rightarrow \neg(\varphi \to \psi), \Delta}$$

Figure 2.1: A quasi-canonical system for Łukasiewicz three-valued logic

2.4 Summary – Desiderata List

We end this chapter with a list of properties that we expect from a logic intended to be useful for handling inconsistencies involving $\neg$:

1. It should be strongly $\neg$-paraconsistent.

2. It should be normal (or at least semi-normal. See Note 2.12).

3. Its negation should be complete, and either involutive, or contrapositive and left-involutive. (Note that the need for a choice here is forced by Proposition 2.14.)

4. It should have a useful denotational semantics and a standard Gentzen-type proof system (preferably quasi-canonical one) with nice proof-theoretical properties (like cut-elimination).

5. It should have nice maximality properties. Being fully maximal is the best here, unless this demand is in conflict with properties that are

taken to be more important for the logic, like being a constructive logic (Part IV) or a relevant logic (Part V). In such a case other appropriate maximality properties will be sought.

In the proceeding chapters we shall investigate the main approaches for defining logics with the above properties.

2.5 Bibliographical Notes and Further Reading

Paraconsistent thinking may be traced back already to Aristotelian logic. The foundations of paraconsistent reasoning in modern times where laid at the beginning of the twentieth century by the Russian logician Vasiliev and the Polish philosopher Lukasiewicz, who offered new interpretations to formal systems in which contradictions could make sense. (See [203] for a synopsis of Lukasiewicz's 1910 paper, in which he criticizes Aristotle's law of non-contradiction.) Other paraconsistent systems where later introduced independently by the pioneering works of Jaśkowski [186], Nelson [224], Anderson and Belnap [3], and da-Costa [127]. Since then the investigation of paraconsistent systems grew rapidly. Nowadays there is enormous amount of works covering different aspects of this field. As said in the preface, Priest's survey [247] is a good starting point to study the subject in general, including its philosophy, history, and other subjects not treated in this book. Other general sources for further study are given in Footnote 2 of the preface.

Investigations of the formal properties of a negation are well-studied in the literature (In fact, the collection [167] is devoted to this issue.) We refer, e.g., to [41] and [209], where negation operators are investigated in the context of paraconsistent reasoning.

A systematic study of the desirable properties of paraconsistent logics can be found in [18] and [19]. Definitions 2.16 and 2.19 are essentially taken from [113, 116] (with some adaptations and simplifications due to the notion of a *paraconsistent logic* as used in this book). The notions of strong maximality are considered in [17, 19, 58]. The other desirable properties listed in Section 2.4 are investigated in [18, 20].[12] This includes the properties of containment in classical logic and of maximality relative to classical logic, both of which have been widely used (though not always precisely defined) in the literature (see, e.g., [113, 141, 193, 208]).

For a different approach to the methodology of paraconsistent logics in general, and to the subject of maximality in particular, see [301].

---

[12]Logics that satisfy the desiderata list are called in these papers *ideal*. See footnote 9.

Part II

Multi-Valued Truth-Functionality

Chapter 3

Paraconsistent Matrices

As we have shown in the previous chapters, classical logic and the standard two-valued matrix are not adequate for paraconsistent reasoning (see also Corollary 3.57 below). To overcome this shortcoming of two-valued logics, we consider in this part of the book and in its next part one of the most common approaches to paraconsistent reasoning, based on *multiple-valued semantics*. In this approach, the set of truth values is enriched with new elements other than the two classical ones t and f. The additional truth values are often intuitively understood as representing different degrees of inconsistency or incompleteness, and this allows to define different consequence relations that are useful for reasoning with uncertainty.

In this part of the book we concentrate on *algebraic approaches* to multi-valued reasoning, using *matrices*.

3.1 Preliminaries

In this section we review basic notions and facts about matrices. We omit proofs of propositions that appear elsewhere in the literature. (See Section 3.3.)

Definition 3.1 (matrices) A (multi-valued) *matrix* for a language $\mathcal{L}$ is a triple $\mathcal{M} = \langle \mathcal{V}, \mathcal{D}, \mathcal{O} \rangle$, where

- $\mathcal{V}$ is a non-empty set of truth values;

- $\mathcal{D}$ is a non-empty proper subset of $\mathcal{V}$ (whose elements are called the *designated* elements of $\mathcal{V}$);

- $\mathcal{O}$ is a function that associates an n-ary function $\widetilde{\diamond}_{\mathcal{M}} : \mathcal{V}^n \to \mathcal{V}$ with every n-ary connective $\diamond$ of $\mathcal{L}$.

Note 3.2 The pair $\langle \mathcal{V}, \mathcal{O} \rangle$ in Definition 3.1 is an *algebraic structure*.

Notation 3.3 $\overline{\mathcal{D}} = \mathcal{V} \setminus \mathcal{D}$. (Note that if $\langle \mathcal{V}, \mathcal{D}, \mathcal{O} \rangle$ is a matrix, then $\overline{\mathcal{D}}$ too is a non-empty proper subset of $\mathcal{V}$.)

Definition 3.4 (submatrix) Let $\mathcal{M} = \langle \mathcal{V}, \mathcal{D}, \mathcal{O} \rangle$ and $\mathcal{M}_s = \langle \mathcal{V}_s, \mathcal{D}_s, \mathcal{O}_s \rangle$ be matrices for $\mathcal{L}$. $\mathcal{M}_s$ is a *submatrix* of $\mathcal{M}$ if $\mathcal{V}_s$ is a subset of $\mathcal{V}$ which is closed under the operations in $\mathcal{O}$, $\mathcal{D}_s = \mathcal{D} \cap \mathcal{V}_s$, and for every $\diamond \in \mathcal{C}(\mathcal{L})$, $\tilde{\diamond}_{\mathcal{M}_s}$ is the restriction of $\tilde{\diamond}_{\mathcal{M}}$ to $\mathcal{V}_s$.

Definition 3.5 (extension/reduction of a matrix) Let $\mathcal{M} = \langle \mathcal{V}, \mathcal{D}, \mathcal{O} \rangle$ be a matrix for $\mathcal{L}$, and let $\mathcal{M}_e = \langle \mathcal{V}_e, \mathcal{D}_e, \mathcal{O}_e \rangle$ be a matrix for $\mathcal{L}_e$. $\mathcal{M}_e$ is an *extension* of $\mathcal{M}$ to $\mathcal{L}_e$, and $\mathcal{M}$ is a *reduction* of $\mathcal{M}_e$ to $\mathcal{L}$, if $\mathcal{L} \subseteq \mathcal{L}_e$, $\mathcal{V} = \mathcal{V}_e$, $\mathcal{D} = \mathcal{D}_e$, and $\mathcal{O}_e(\diamond) = \mathcal{O}(\diamond)$ for every connective $\diamond$ of $\mathcal{L}$.[1]

Definition 3.6 (finite and effective matrices) Let $\mathcal{M} = \langle \mathcal{V}, \mathcal{D}, \mathcal{O} \rangle$ be a matrix for $\mathcal{L}$.

- We say that $\mathcal{M}$ is *(in)finite*, if so is $\mathcal{V}$.

- We call $\mathcal{M}$ *effective*, if it satisfies the following conditions:

    a) $\mathcal{V}$ can effectively be enumerated (and so is at most countable).

    b) For each connective $\diamond$ of $\mathcal{L}$, the interpretation function $\mathcal{O}(\diamond)$ is computable.

    c) The set $\mathcal{D}$ of the designated truth values is decidable.

The set $\mathcal{D}$ of the designated elements in $\mathcal{V}$ is used for defining satisfiability and validity:

Definition 3.7 (valuations, satisfiability) Let $\mathcal{M} = \langle \mathcal{V}, \mathcal{D}, \mathcal{O} \rangle$ be a matrix for a propositional language $\mathcal{L}$.

- A *partial $\mathcal{M}$-valuation* for $\mathcal{L}$ is a function $\nu : \mathcal{W} \to \mathcal{V}$ for some subset $\mathcal{W}$ of $\mathcal{W}(\mathcal{L})$ that satisfies the following conditions:

    – The set $\mathcal{W}$ is closed under subformulas.

    – For every n-ary connective $\diamond$ of $\mathcal{L}$ and every $\psi_1, \ldots, \psi_n$ such that $\diamond(\psi_1, \ldots, \psi_n) \in \mathcal{W}$,
    $$\nu(\diamond(\psi_1, \ldots, \psi_n)) = \tilde{\diamond}_{\mathcal{M}}(\nu(\psi_1), \ldots, \nu(\psi_n)).$$

- A partial $\mathcal{M}$-valuation is a (full) *$\mathcal{M}$-valuation* if its domain is $\mathcal{W}(\mathcal{L})$. We denote the set of all the $\mathcal{M}$-valuations by $\Lambda_{\mathcal{M}}$.

---
[1] Note the difference between extensions of matrices and of logics (Definition 1.9).

3.1 Preliminaries

- A valuation $\nu \in \Lambda_{\mathcal{M}}$ is an $\mathcal{M}$-model of a formula ψ, or ν $\mathcal{M}$-satisfies ψ (notation: $\nu \models_{\mathcal{M}} \psi$), if $\nu(\psi) \in \mathcal{D}$. We say that ν is an $\mathcal{M}$-model of a theory $\mathcal{T}$ (notation: $\nu \models_{\mathcal{M}} \mathcal{T}$) if it is an $\mathcal{M}$-model of every element of $\mathcal{T}$. We denote the set of $\mathcal{M}$-models of ψ by $mod_{\mathcal{M}}(\psi)$, and the set of $\mathcal{M}$-models of $\mathcal{T}$ by $mod_{\mathcal{M}}(\mathcal{T})$. (So $mod_{\mathcal{M}}(\psi) = \{\nu \in \Lambda_{\mathcal{M}} \mid \nu(\psi) \in \mathcal{D}\}$, and $mod_{\mathcal{M}}(\mathcal{T}) = \cap_{\psi \in \mathcal{T}} mod_{\mathcal{M}}(\psi)$.)

- A formula ψ is $\mathcal{M}$-satisfiable if $mod_{\mathcal{M}}(\psi) \neq \emptyset$; It is an $\mathcal{M}$-tautology if $mod_{\mathcal{M}}(\psi) = \Lambda_{\mathcal{M}}$. A theory $\mathcal{T}$ is $\mathcal{M}$-satisfiable if $mod_{\mathcal{M}}(\mathcal{T}) \neq \emptyset$.

Note 3.8 Obviously, for every matrix $\mathcal{M}$ the pair $\langle \Lambda_{\mathcal{M}}, \models_{\mathcal{M}} \rangle$ is a denotational semantics in the sense of Definition 1.37.

Proposition 3.9 Let $\mathcal{M}$ be a matrix for $\mathcal{L}$. The semantics induced by $\mathcal{M}$ has the following properties:

TRUTH-FUNCTIONALITY. *The truth value that an $\mathcal{M}$-valuation assigns to a complex formula is uniquely determined by the truth values it assigns to its subformulas, and so by the truth value it assigns to its variables.*

ANALYTICITY. *Any partial $\mathcal{M}$-valuation can be extended to a full valuation.*

SUBSTITUTION-CLOSURE. *If $\nu \in \Lambda_{\mathcal{M}}$ and θ is an $\mathcal{L}$-substitution, then the composition $\nu \circ \theta$ of ν and θ is in $\Lambda_{\mathcal{M}}$ as well.*

Definition 3.10 (representability, functional completeness) Let $\mathcal{M}$ be a matrix for a language $\mathcal{L}$, where $\mathcal{M} = \langle \mathcal{V}, \mathcal{D}, \mathcal{O} \rangle$.

- Let ψ be a formula in $\mathcal{L}$ such that $\mathsf{Var}(\psi) \subseteq \{P_1, \ldots, P_n\}$. $F_\psi^n : \mathcal{V} \to \mathcal{V}$ is defined by: $F_\psi^n(a_1, \ldots, a_n) = \nu(\psi)$, where $\nu \in \Lambda_{\mathcal{M}}$ is any $\mathcal{M}$-valuation such that $\nu(P_i) = a_i$ for $1 \leq i \leq n$.

- For $n > 0$, a function $g : \mathcal{V}^n \to \mathcal{V}$ is *represented in* $\mathcal{M}$ by a formula ψ such that $\mathsf{Var}(\psi) \subseteq \{P_1, \ldots, P_n\}$ if $g = F_\psi^n$. We say that $g : \mathcal{V}^0 \to \mathcal{V}$ (i.e., an element g of $\mathcal{V}$) is represented in $\mathcal{M}$ by a formula ψ such that $\mathsf{Var}(\psi) \subseteq \{P_1\}$ if $F_\psi^1(a) = g$ for every $a \in \mathcal{V}$.

- We say that $\mathcal{L}$ *includes in* $\mathcal{M}$ a function g, if g can be represented in $\mathcal{M}$ by a formula in $\mathcal{W}(\mathcal{L})$.

- $\mathcal{L}$ is *functionally complete in* $\mathcal{M}$ if for every $n \geq 0$, every function from $\mathcal{V}^n$ to $\mathcal{V}$ is representable in $\mathcal{M}$ by some formula in $\mathcal{W}(\mathcal{L})$.

Convention When $\mathcal{M}$ is clear from the context, we might omit the "in $\mathcal{M}$" or the prefix '$\mathcal{M}$-' from the notions above, or say "in $\mathcal{L}$" instead of "in $\mathcal{M}$".

The following is a special case of schema (1.1) in Definition 1.38:

Definition 3.11 ($\vdash_\mathcal{M}$, $\mathbf{L}_\mathcal{M}$, $Th(\mathcal{M})$) Let $\mathcal{M}$ be a matrix for $\mathcal{L}$. $\vdash_\mathcal{M}$, the consequence relation that is induced by (or associated with) $\mathcal{M}$, is defined by: $\mathcal{T} \vdash_\mathcal{M} \psi$ if $mod_\mathcal{M}(\mathcal{T}) \subseteq mod_\mathcal{M}(\psi)$.[2] We shall denote by $\mathbf{L}_\mathcal{M}$ the logic $\langle \mathcal{L}, \vdash_\mathcal{M} \rangle$ which is induced by $\mathcal{M}$, and by $Th(\mathcal{M})$ the set $Th(\mathbf{L}_\mathcal{M})$.

Example 3.12 Let $\mathcal{M}_{CL}$ be the two-valued matrix $\langle \{t, f\}, \{t\}, \mathbf{F}_{CL} \rangle$ for $\mathcal{L}_{CL}$. Then $\mathbf{CL} = \mathbf{L}_{\mathcal{M}_{CL}}$. (See Definition 1.48.) Similarly, $\mathbf{CL}^+ = \mathbf{L}_{\mathcal{M}_{CL}^+}$, where $\mathcal{M}_{CL}^+$ is the reduction of $\mathcal{M}_{CL}$ to $\mathcal{L}_{CL}^+$.

Proposition 3.13 *For every propositional language $\mathcal{L}$ and a matrix $\mathcal{M}$ for $\mathcal{L}$, $\mathbf{L}_\mathcal{M}$ is a propositional logic.*

Proof. The proof is easy. We only note that the non-triviality of $\vdash_\mathcal{M}$ is due to the fact that we demand $\mathcal{D}$ (where $\mathcal{M} = \langle \mathcal{V}, \mathcal{D}, \mathcal{O} \rangle$) to be a *nonempty proper* subset of $\mathcal{V}$. Therefore if p and q are distinct variables then there is a valuation ν such that $\nu(p) \in \mathcal{D}$, while $\nu(q) \notin \mathcal{D}$. Hence $p \nvdash_\mathcal{M} q$. □

Proposition 3.14 *Suppose that $\mathcal{M} = \langle \mathcal{V}, \mathcal{D}, \mathcal{O} \rangle$ is a matrix for $\mathcal{L}$, and $\mathcal{M}' = \langle \mathcal{V}', \mathcal{D}', \mathcal{O}' \rangle$ is a matrix for $\mathcal{L}'$.*

1. *If $\mathcal{L} \subseteq \mathcal{L}'$, and $\mathcal{M}'$ is an extension of $\mathcal{M}$, then $\vdash_{\mathcal{M}'}$ is a conservative extension of $\vdash_\mathcal{M}$, i.e. if $\mathcal{T}$ and φ are in $\mathcal{L}$, then $\mathcal{T} \vdash_{\mathcal{M}'} \varphi$ iff $\mathcal{T} \vdash_\mathcal{M} \varphi$.*

2. *If $\mathcal{L} = \mathcal{L}'$, and $\mathcal{M}'$ is a submatrix of $\mathcal{M}$, then $\vdash_\mathcal{M} \subseteq \vdash_{\mathcal{M}'}$, i.e., if $\mathcal{T}$ and φ are in $\mathcal{L}$, and $\mathcal{T} \vdash_\mathcal{M} \varphi$, then $\mathcal{T} \vdash_{\mathcal{M}'} \varphi$.*

Proposition 3.15 *If $\mathcal{M} = \langle \mathcal{V}, \mathcal{D}, \mathcal{O} \rangle$ is a matrix for $\mathcal{L}$, and each $a \in \mathcal{V}$ is represented in $\mathcal{M}$, then $\mathbf{L}_\mathcal{M}$ has no proper extension in its language.*

Note 3.16 We say that a matrix $\mathcal{M} = \langle \mathcal{V}, \mathcal{D}, \mathcal{O} \rangle$ has n elements (where n is a natural number or even an infinite cardinal) if $\mathcal{V}$ has n elements. In the literature a logic $\mathbf{L}$ is usually called *n-valued* if $\mathbf{L} = \mathbf{L}_\mathcal{M}$ for some matrix $\mathcal{M}$ that has n elements. We too shall use sometimes this terminology in this part of the book. However, the meaning of "n-valued logic" will considerably be extended in Part III of the book. (See Note 6.14.)[3]

The following is an important property of finite matrices:

Theorem 3.17 (compactness) $\mathbf{L}_\mathcal{M}$ *is a finitary propositional logic for every propositional language $\mathcal{L}$ and a finite matrix $\mathcal{M}$ for $\mathcal{L}$.*

---

[2] A Scott (multiple-conclusion) consequence relation is defined similarly by $\mathcal{T} \vdash_\mathcal{M} \mathcal{S}$ if for every $\nu \in mod_\mathcal{M}(\mathcal{T})$ there is some $\psi \in \mathcal{S}$ such that $\nu \in mod_\mathcal{M}(\psi)$.

[3] For still other meanings of "n-valued logic" see Section 3.3.

3.1 Preliminaries

Definition 3.11 and Proposition 3.13 can be generalized as follows:

Definition 3.18 ($\vdash_{\mathcal{F}}$) Let $\mathcal{F}$ be a class of matrices for $\mathcal{L}$. The relation $\vdash_{\mathcal{F}}$ that is induced by $\mathcal{F}$ is defined by: $\mathcal{T} \vdash_{\mathcal{F}} \psi$ if $\mathcal{T} \vdash_{\mathcal{M}} \psi$ for every $\mathcal{M} \in \mathcal{F}$.[4]

Proposition 3.19 $\langle \mathcal{L}, \vdash_{\mathcal{F}} \rangle$ *is a propositional logic for any propositional language $\mathcal{L}$ and any set $\mathcal{F}$ of matrices for $\mathcal{L}$.*

Example 3.20 The semantics for intuitionistic logic described in the first chapter (Section 1.2.3) is isomorphic to the semantics induced by the family of matrices of the form $\langle \mathcal{V}, \mathcal{D}, \mathcal{O} \rangle$, where $\mathcal{V}$ is the set of upward-closed nonempty subsets of some partially ordered set $\langle W, \leq \rangle$, $\mathcal{D}$ is $\{W\}$, and the operations in $\mathcal{O}$ are determined by using $\leq$. As the next theorem shows, this observation is just an instance of the general principle that matrices-based semantics can replace any other type of denotational semantics.

Theorem 3.21 *For every propositional logic there exists some set of matrices which induce it.*

Next we characterize the logics which are induced by a *single* matrix.

Definition 3.22 (soundness and completeness) Let $\mathbf{L} = \langle \mathcal{L}, \vdash_{\mathbf{L}} \rangle$ be a propositional logic, and let $\mathcal{M}$ be a matrix for $\mathcal{L}$.

- If $\mathbf{L}_{\mathcal{M}}$ is an extension of $\mathbf{L}$ we say that $\mathbf{L}$ is *sound* for $\mathcal{M}$.

- If $\mathbf{L}$ is an extension of $\mathbf{L}_{\mathcal{M}}$ we say that $\mathbf{L}$ is *complete* for $\mathcal{M}$.

- $\mathcal{M}$ is a *characteristic* matrix for $\mathbf{L}$ if $\mathbf{L} = \mathbf{L}_{\mathcal{M}}$ (i.e. if $\mathbf{L}$ is both sound and complete for $\mathbf{L}_{\mathcal{M}}$).

- As in Definition 1.82, one may consider weaker versions of the above notions as follows: $\mathbf{L}_{\mathcal{M}}$ is *weakly sound* for $\mathbf{L}$ if for every $\psi \in \mathcal{W}(\mathcal{L})$, $\vdash_{\mathcal{M}} \psi$ implies that $\vdash_{\mathbf{L}} \psi$, and $\mathbf{L}_{\mathcal{M}}$ is *weakly complete* for $\mathbf{L}$ if $\vdash_{\mathbf{L}} \psi$ implies that $\vdash_{\mathcal{M}} \psi$.

- $\mathcal{M}$ is a *weakly characteristic* matrix for $\mathbf{L}$ if $\mathbf{L}$ is both weakly sound and weakly complete for $\mathbf{L}_{\mathcal{M}}$ (i.e. $\vdash_{\mathbf{L}_{\mathcal{M}}} \psi$ iff $\vdash_{\mathbf{L}} \psi$).

Definition 3.23 (uniformity) Let $\mathbf{L} = \langle \mathcal{L}, \vdash_{\mathbf{L}} \rangle$ be a logic.

- A theory $\mathcal{T}$ is called $\vdash_{\mathbf{L}}$-*non trivial*, if there exists a formula $\psi \in \mathcal{W}(\mathcal{L})$ such that $\mathcal{T} \not\vdash_{\mathbf{L}} \psi$.

---

[4] Again, a corresponding multiple-conclusion consequence relation may be defined in the obvious way.

- **L** is *uniform* if for every two theories T_1, T_2 and formula $\psi \in \mathcal{W}(\mathcal{L})$, we have that $T_1 \vdash_\mathbf{L} \psi$ in case $T_1, T_2 \vdash_\mathbf{L} \psi$, and T_2 is a $\vdash_\mathbf{L}$-non trivial theory that has no variables in common with $T_1 \cup \{\psi\}$.

Theorem 3.24 (Łoś-Suszko) *A finitary propositional logic $\langle \mathcal{L}, \vdash_\mathbf{L} \rangle$ has a single characteristic matrix iff it is uniform.*

Corollary 3.25 *The logic $\mathbf{L}_\mathcal{M}$ induced by a matrix $\mathcal{M}$ is non-exploding (Definition 2.17) iff it satisfies the following condition: for every two theories T_1, T_2 and formula $\psi \in \mathcal{W}(\mathcal{L})$, $T_1 \vdash_{\mathbf{L}_\mathcal{M}} \psi$ in case $T_1, T_2 \vdash_{\mathbf{L}_\mathcal{M}} \psi$ and T_2 is a theory that has no variables in common with $T_1 \cup \{\psi\}$.*

Note 3.26

1. Under a more complicated definition of uniformity (see [290]), Theorem 3.24 can be rephrased also for propositional logics that are not necessarily finitary.

2. The condition given in Corollary 3.25 is called in Chapter 11 "the basic relevance criterion", and is (as its name suggests) the most basic condition that every relevant logic should satisfy.

Below are some facts about entailment checking in matrix-based logics.

Theorem 3.27 (decidability) *Let Γ be a finite theory, ψ a formula, and $\mathcal{M}$ an effective matrix for $\mathcal{L}$.*

1. *The question whether $\Gamma \vdash_\mathcal{M} \psi$ is co-semi-decidable.*

2. *If $\mathcal{M}$ is finite, then the question whether $\Gamma \vdash_\mathcal{M} \psi$ is decidable.*

3. *If $\vdash_\mathcal{M}$ has a sound and complete effective proof system, then $\vdash_\mathcal{M}$ is decidable even for infinite $\mathcal{M}$.*

Proof. Part 1 is by an exhaustive search for a partial valuation which is a model of Γ, falsifies ψ, and whose domain is $\mathsf{SF}(\Gamma \cup \{\psi\})$. By analyticity, we have that $\Gamma \nvdash_\mathcal{M} \psi$ iff such a counter-model exists, and so once such a valuation is found the search terminates. Part 2 is by a similar algorithm, which always terminates, since due to the finiteness of Γ and $\mathcal{M}$, the search space is also finite. Part 3 follows from Part 1 and the fact that provability with respect to an effective proof system is semi-decidable. □

Now we turn to study the basic connectives (Definition 1.23) in the context of matrices-based logics.

3.1 Preliminaries

Definition 3.28 Let $\mathcal{M} = \langle \mathcal{V}, \mathcal{D}, \mathcal{O} \rangle$ be a matrix for $\mathcal{L}$, and let $\mathcal{A} \subseteq \mathcal{V}$.

- An n-ary connective $\diamond$ of $\mathcal{L}$ is called $\mathcal{A}$-*closed*, if $\tilde{\diamond}(a_1, \ldots, a_n) \in \mathcal{A}$ for every $a_1, \ldots, a_n \in \mathcal{A}$.

- An n-ary connective $\diamond$ of $\mathcal{L}$ is called $\mathcal{A}$-*limited*, if the following two conditions hold for some $X \subseteq \{1, \ldots, n\}$:
  - The value of $\tilde{\diamond}(a_1, \ldots, a_n)$ depends only on $\{a_i \mid i \in X\}$. In other words: $\tilde{\diamond}(a_1, \ldots, a_n) = \tilde{\diamond}(b_1, \ldots, b_n)$ whenever $a_i = b_i$ for every $i \in X$.
  - $\tilde{\diamond}(a_1, \ldots, a_n) \not\in \mathcal{A}$ whenever $a_i \not\in \mathcal{A}$ for some $i \in X$.

Definition 3.29 (basic $\mathcal{M}$-connectives) Let $\mathcal{M} = \langle \mathcal{V}, \mathcal{D}, \mathcal{O} \rangle$ be a matrix for a language $\mathcal{L}$.

- A connective $\wedge$ in $\mathcal{L}$ is called an $\mathcal{M}$-*conjunction* if it is $\mathcal{D}$-closed and $\mathcal{D}$-limited, i.e. if for every $a, b \in \mathcal{V}$, $a \tilde{\wedge} b \in \mathcal{D}$ iff both $a \in \mathcal{D}$ and $b \in \mathcal{D}$.

- A connective $\vee$ in $\mathcal{L}$ is called an $\mathcal{M}$-*disjunction* if it is $\overline{\mathcal{D}}$-closed and $\overline{\mathcal{D}}$-limited, i.e. if for every $a, b \in \mathcal{V}$, $a \tilde{\vee} b \in \mathcal{D}$ iff $a \in \mathcal{D}$ or $b \in \mathcal{D}$.

- A connective $\supset$ in $\mathcal{L}$ is called an $\mathcal{M}$-*implication* if for every $a, b \in \mathcal{V}$, $a \tilde{\supset} b \in \mathcal{D}$ iff either $a \not\in \mathcal{D}$ or $b \in \mathcal{D}$.

Note 3.30 It is easy to verify that if $\supset$ is an $\mathcal{M}$-implication then the connective $\vee$ defined by $\varphi \vee \psi = (\varphi \supset \psi) \supset \psi$ is an $\mathcal{M}$-disjunction.

The next two propositions, relating the concepts in Definitions 1.23, 3.4, 3.5, and Definition 3.29, are easily verified:

Proposition 3.31 *Let $\diamond \in \mathcal{L}$, and let $\mathcal{M}$ be a matrix for $\mathcal{L}$. If $\diamond$ is an $\mathcal{M}$-conjunction ($\mathcal{M}$-disjunction, $\mathcal{M}$-implication) then it is also an $\mathcal{M}'$-conjunction ($\mathcal{M}'$-disjunction, $\mathcal{M}'$-implication) for any matrix $\mathcal{M}'$ which is either a submatrix of $\mathcal{M}$, or an extension of $\mathcal{M}$ to some richer language.*

Proposition 3.32 *Let $\mathcal{M} = \langle \mathcal{V}, \mathcal{D}, \mathcal{O} \rangle$ be a matrix for a language $\mathcal{L}$, and let $\diamond$ be a connective of $\mathcal{L}$.*

1. *$\diamond$ is an $\mathcal{M}$-conjunction iff it is a conjunction for $\mathbf{L}_\mathcal{M}$.*

2. *If $\diamond$ is an $\mathcal{M}$-disjunction then it is also a disjunction for $\mathbf{L}_\mathcal{M}$.*

3. *If $\diamond$ is an $\mathcal{M}$-implication then is also an implication for $\mathbf{L}_\mathcal{M}$.*

Corollary 3.33 Let $\mathcal{M} = \langle \mathcal{V}, \mathcal{D}, \mathcal{O} \rangle$ be a matrix for a language $\mathcal{L}$, and let $\mathcal{M}'$ be an extension of $\mathcal{M}$.

1. An $\mathcal{M}$-conjunction ($\mathcal{M}$-disjunction, $\mathcal{M}$-implication) is also a conjunction (disjunction, implication) of $\mathbf{L}_{\mathcal{M}'}$.

2. If $\mathcal{M}$ has either an $\mathcal{M}$-conjunction, or an $\mathcal{M}$-disjunction, or an $\mathcal{M}$-implication, then $\mathbf{L}_{\mathcal{M}'}$ is semi-normal. If $\mathcal{M}$ has all of them then $\mathbf{L}_{\mathcal{M}'}$ is normal.

Corollary 3.34 Let $\mathcal{M} = \langle \mathcal{V}, \mathcal{D}, \mathcal{O} \rangle$ be a matrix for $\mathcal{L}$ such that $\mathbf{L}_{\mathcal{M}}$ is $\mathbf{F}$-contained in classical logic, and let $\diamond$ be a connective of $\mathcal{L}$.

1. If $\diamond$ is an $\mathcal{M}$-conjunction then $\mathbf{F}(\diamond)$ is the classical conjunction.

2. If $\diamond$ is an $\mathcal{M}$-disjunction then $\mathbf{F}(\diamond)$ is the classical disjunction.

3. If $\diamond$ is an $\mathcal{M}$-implication then $\mathbf{F}(\diamond)$ is the classical implication.

Proof. This follows from Propositions 1.54 and 3.32 $\square$

Proposition 3.32 has a converse in a special type of matrices which is particularly important for the study of paraconsistent logics.

Definition 3.35 (f-matrices) An f-matrix is a matrix in which $\overline{\mathcal{D}}$ is a singleton. (The single non-designated element of an f-matrix will usually be denoted by f.)

Proposition 3.36 Let $\mathcal{M} = \langle \mathcal{V}, \mathcal{D}, \mathcal{O} \rangle$ be an f-matrix.

1. A connective $\wedge$ is a conjunction for $\mathbf{L}_{\mathcal{M}}$ iff it is an $\mathcal{M}$-conjunction.

2. A connective $\supset$ is an implication for $\mathbf{L}_{\mathcal{M}}$ iff it is an $\mathcal{M}$-implication.[5]

3. A connective $\vee$ is a disjunction for $\mathbf{L}_{\mathcal{M}}$ iff it is an $\mathcal{M}$-disjunction.

Proof.

1. This is identical to the first item of Proposition 3.32.

2. Assume that $\supset$ is an implication for $\mathbf{L}_{\mathcal{M}}$. This easily implies that for any two variables p and q we have:

   (a) $q \vdash_{\mathcal{M}} p \supset q$ (b) $p, p \supset q \vdash_{\mathcal{M}} q$ (c) $\vdash_{\mathcal{M}} p \supset p$

   In general, (a) implies that if $b \in \mathcal{D}$ then $a \widetilde{\supset} b \in \mathcal{D}$, while (b) implies that if $a \in \mathcal{D}$ and $b \in \overline{\mathcal{D}}$ then $a \widetilde{\supset} b \in \overline{\mathcal{D}}$. Finally, in case $\overline{\mathcal{D}}$ is a singleton (c) implies that if $a, b \in \overline{\mathcal{D}}$ then $a \widetilde{\supset} b \in \mathcal{D}$.

---

[5] Using a more complicated proof, this item can be generalized as follows: if $\mathcal{M}$ is an f-matrix, then a connective $\supset$ is a semi-implication for $\mathbf{L}_{\mathcal{M}}$ iff it is an $\mathcal{M}$-implication. See [53] for details.

3.1 Preliminaries

3. Assume that $\vee$ is a disjunction for $\mathbf{L}_\mathcal{M}$. This easily implies that for any two variables p and q we have:

   (a) $p \vdash_\mathcal{M} p \vee q$ (b) $q \vdash_\mathcal{M} p \vee q$ (c) $p \vee p \vdash_\mathcal{M} p$

   In general, the first two facts imply that if either $a \in \mathcal{D}$ or $b \in \mathcal{D}$ then $a \widetilde{\vee} b \in \mathcal{D}$. Finally, in case $\overline{\mathcal{D}}$ is a singleton (c) implies that if $a, b \in \overline{\mathcal{D}}$ then $a \widetilde{\vee} b \in \overline{\mathcal{D}}$. □

Proposition 3.37 Let $\mathcal{M} = \langle \mathcal{V}, \mathcal{D}, \mathcal{O} \rangle$ be a matrix for $\mathcal{L}$.

1. If a connective $\supset$ of $\mathcal{L}$ is an $\mathcal{M}$-implication then the $\{\supset\}$-fragment of $\mathbf{L}_\mathcal{M}$ is identical to the $\{\supset\}$-fragment of $\mathbf{CL}$.

2. Suppose $\mathcal{L}$ has connectives $\supset$, $\vee$, and $\wedge$ that are (respectively) an $\mathcal{M}$-implication, an $\mathcal{M}$-disjunction, and an $\mathcal{M}$-conjunction. Then the $\{\supset, \vee, \wedge\}$-fragment of $\mathbf{L}_\mathcal{M}$ is identical to $\mathbf{CL}^+$.

Proof. We prove the second part (the proof of the first is similar). Define a function $g : \mathcal{V} \to \{t, f\}$ by:

$$g(a) = \begin{cases} t & \text{if } a \in \mathcal{D}, \\ f & \text{if } a \notin \mathcal{D}. \end{cases}$$

Let ν be a valuation in $\mathcal{M}$. Using induction on the structure of formulas and Definition 3.29, it is easy to see that for the language of $\{\supset, \vee, \wedge\}$, $g \circ \nu$ (the composition of g and ν) is a valuation in $\mathcal{M}_{CL}$ (Definition 3.12). Obviously, $\nu \models_\mathcal{M} \varphi$ iff $g \circ \nu \models_{\mathcal{M}_{CL}} \varphi$. Now, let $\mathcal{T}$ and φ be in the language of $\{\supset, \vee, \wedge\}$, and suppose that $\mathcal{T} \vdash_{\mathcal{M}_{CL}} \varphi$ and $\nu \models_\mathcal{M} \mathcal{T}$. Then $g \circ \nu \models_{\mathcal{M}_{CL}} \mathcal{T}$, and so $g \circ \nu \models_{\mathcal{M}_{CL}} \varphi$, implying that $\nu \models_\mathcal{M} \varphi$. It follows that if $\mathcal{T} \vdash_{\mathbf{CL}^+} \varphi$ then $\mathcal{T} \vdash_{\mathbf{L}_\mathcal{M}} \varphi$. Therefore, by Proposition 1.65, $\vdash_{\mathbf{L}_\mathcal{M}}$ and $\vdash_{\mathbf{CL}^+}$ are identical for the language of $\{\supset, \vee, \wedge\}$. □

We conclude this section with a useful general construction of implications in matrices:

Definition 3.38 ($\supset_t$) Let $\mathcal{V}$ be a nonempty set, $\mathcal{D}$ a nonempty proper subset of $\mathcal{V}$, and let $t \in \mathcal{D}$. The operation $\supset_t$ on $\mathcal{V}$ is defined as follows:

$$a \supset_t b = \begin{cases} b & \text{if } a \in \mathcal{D}, \\ t & \text{otherwise}. \end{cases}$$

Proposition 3.39 Let $\mathcal{M} = \langle \mathcal{V}, \mathcal{D}, \mathcal{O} \rangle$ be a matrix with a connective $\supset_t$ defined as in 3.38 (with $t \in \mathcal{D}$). Then $\supset_t$ is an $\mathcal{M}$-implication.

Proof. Immediate from the definition of $\supset_t$. □

3.2 Paraconsistent Matrices and Their Logics

3.2.1 Matrices with Negation

In what follows we shall attribute the properties of $\mathbf{L}_\mathcal{M}$ to $\mathcal{M}$. In particular,

Definition 3.40 (¬-containment/¬-coherence of matrices) Let $\mathcal{L}$ be a propositional language with a unary connective ¬. A matrix $\mathcal{M}$ for $\mathcal{L}$ is ¬-*contained in classical logic*, or ¬-*coherent with classical logic*, if so is $\mathbf{L}_\mathcal{M}$.

Definition 3.41 (negation in a matrix) Let $\mathcal{L}$ be a propositional language with a unary connective ¬, and let $\mathcal{M}$ be a matrix for $\mathcal{L}$. We say that ¬ is a *negation in* $\mathcal{M}$, if $\mathcal{M}$ is ¬-coherent with classical logic.

Definition 3.42 ($\mathcal{M}_\mathbf{F}$) Let $\mathbf{F}$ be a bivalent ¬-interpretation of a language $\mathcal{L}$. We denote by $\mathcal{M}_\mathbf{F}$ the two-valued matrix for $\mathcal{L}$ that is induced by $\mathbf{F}$ (i.e., $\mathcal{M}_\mathbf{F} = \langle \{t,f\}, \{t\}, \{\mathbf{F}(\diamond) \mid \diamond \text{ is a connective of } \mathcal{L}\}\rangle$). Note that $\Lambda_{\mathcal{M}_\mathbf{F}} = \Lambda_{2_\mathbf{F}}$.

Proposition 3.43 *Let* $\mathcal{M} = \langle \mathcal{V}, \mathcal{D}, \mathcal{O}\rangle$ *be a matrix with negation* ¬. *There are elements* t *and* f *in* $\mathcal{V}$ *such that* $t \in \mathcal{D}$, $f \notin \mathcal{D}$, $f = \tilde{\neg}t$, *and* $\tilde{\neg}f \in \mathcal{D}$.

Proof. Obviously, it suffices to prove this in the case that $\mathcal{M}$ is ¬-contained in classical logic. So let $\mathbf{F}$ be a bivalent ¬-interpretation such that $\mathbf{L}_\mathcal{M}$ is $\mathbf{F}$-contained in classical logic. Since $p, \neg\neg p \not\vdash_{\mathcal{M}_\mathbf{F}} \neg p$, also $p, \neg\neg p \not\vdash_\mathcal{M} \neg p$, and so there is some $t \in \mathcal{D}$ such that $\tilde{\neg}t \notin \mathcal{D}$, while $\tilde{\neg}\tilde{\neg}t \in \mathcal{D}$. Let $f = \tilde{\neg}t$. Then t and f have the required properties. □

Corollary 3.44 *A negation in an f-matrix $\mathcal{M}$ is complete for $\mathbf{L}_\mathcal{M}$.*

The useful matrices are usually those in which the two classical truth-values are available in a sense which is stronger than that given in Proposition 3.43. This sense is described in the following definitions and notations.

Definition 3.45 (subclassical matrices) Let $\mathcal{M} = \langle \mathcal{V}, \mathcal{D}, \mathcal{O}\rangle$ be a matrix for a language $\mathcal{L}$, and let ¬ be a unary connective of $\mathcal{L}$. $\mathcal{M}$ is ¬-*subclassical* if there exist $t \in \mathcal{D}$ and $f \in \overline{\mathcal{D}}$ such that:

- $\tilde{\neg}t = f$; $\tilde{\neg}f = t$.

- For every $a \in \mathcal{V}$, if $a \notin \{t,f\}$ then $a \in \mathcal{D}$ iff $\neg a \in \mathcal{D}$.

- The interpretations in $\mathcal{M}$ of all the connectives are $\{t,f\}$-closed.

(We shall usually write just 'subclassical' instead of '¬-subclassical'.)

3.2 Paraconsistent Matrices and Their Logics

Notation 3.46 In a given subclassical matrix $\mathcal{M} = \langle \mathcal{V}, \mathcal{D}, \mathcal{O} \rangle$, there is obviously a unique element a of $\mathcal{V}$ such that $a \in \mathcal{D}$ while $\tilde{\neg}a \notin \mathcal{D}$. We shall henceforth denote by t this unique element, and by f the element $\tilde{\neg}t$. (Note that f is the unique element a of $\mathcal{V}$ such that $a \notin \mathcal{D}$ while $\tilde{\neg}a \in \mathcal{D}$.)

Terminology. Let $\mathcal{M} = \langle \mathcal{V}, \mathcal{D}, \mathcal{O} \rangle$ be a matrix with negation $\neg$. Suppose that there exists a *unique* $t \in \mathcal{D}$ such that $\tilde{\neg}t \in \overline{\mathcal{D}}$. Let $f = \tilde{\neg}t$.

- A connective of $\mathcal{L}$ is *classically closed* (in $\mathcal{M}$) if it is $\{t, f\}$-closed.

- $\mathcal{M}$ is *classically closed* if all its operations are classically closed.

Note that in this terminology, every subclassical matrix is classically closed.

Proposition 3.47 *Let $\mathcal{M} = \langle \mathcal{V}, \mathcal{D}, \mathcal{O} \rangle$ be a $\neg$-subclassical matrix for $\mathcal{L}$.*

1. *$\mathcal{M}$ is $\neg$-contained in classical logic.*

2. *If M has an $\mathcal{M}$-disjunction, or an $\mathcal{M}$-conjunction, or an $\mathcal{M}$-implication then $\neg$ is a negation in it.*

Proof.

1. To show that $\mathcal{M}$ is $\neg$-contained in classical logic, define a function $\mathbf{F}_{\mathcal{M}}$ on $\mathcal{L}$ by letting $\mathbf{F}_{\mathcal{M}}(\diamond) = \tilde{\diamond}_{\mathcal{M}}/\{t,f\}^n$, where n is the arity of $\diamond$, and $\tilde{\diamond}_{\mathcal{M}}/\{t,f\}^n$ is the reduction of $\tilde{\diamond}_{\mathcal{M}}$ to $\{t,f\}^n$. Since $\tilde{\neg}t = f$, $\tilde{\neg}f = t$, and $\mathcal{M}$ is classically closed, $\mathbf{F}_{\mathcal{M}}$ is a bivalent $\neg$-interpretation, and since $\{t, f\} \subseteq \mathcal{V}$, $\mathcal{M}$ is obviously $\mathbf{F}_{\mathcal{M}}$-contained in classical logic.

2. This item follows from the first, using Proposition 3.32. □

Definition 3.48 ($\mathbf{F}_{\mathcal{M}}$) Let $\mathcal{M} = \langle \mathcal{V}, \mathcal{D}, \mathcal{O} \rangle$ be a subclassical matrix for $\mathcal{L}$. Then $\mathbf{F}_{\mathcal{M}}$ from the proof of Proposition 3.47 is called the *bivalent $\neg$-interpretation induced by $\mathcal{M}$*.

Proposition 3.49 *If $\mathcal{M}$ is subclassical then $\mathbf{F}_{\mathcal{M}}$ is the only bivalent $\neg$-interpretation $\mathbf{F}$ for which $\mathbf{L}_{\mathcal{M}}$ is $\mathbf{F}$-contained in classical logic.*

Proof. Suppose for contradiction that there is some other bivalent $\neg$-interpretation $\mathbf{F} \neq \mathbf{F}_{\mathcal{M}}$ such that $\mathbf{L}_{\mathcal{M}}$ is also $\mathbf{F}$-contained in classical logic. Then there is some n-ary connective $\diamond$ of $\mathcal{L}$ such that $\mathbf{F}_{\mathcal{M}}(\diamond) \neq \mathbf{F}(\diamond)$. Since $\mathbf{F}_{\mathcal{M}}(\diamond) = \tilde{\diamond}/\{t,f\}^n$, it follows that there are elements $a_1, \ldots, a_n \in \{t, f\}$ such that $\tilde{\diamond}(a_1, \ldots, a_n) \neq \mathbf{F}(\diamond)(a_1, \ldots, a_n)$. Because $\mathbf{F}$ and $\mathbf{F}_{\mathcal{M}}$ are both bivalent $\neg$-interpretations, we may assume without loss of generality that

$\mathbf{F}(\diamond)(a_1,\ldots,a_n) = t$ and $\tilde{\diamond}(a_1,\ldots,a_n) = f$. (Otherwise we consider $\neg\diamond$ instead of $\diamond$.) Next, for $i = 1,\ldots,n$ define $\varphi_i = P_1$ if $a_i = t$ and $\varphi_i = \neg P_1$ otherwise. Since $\mathcal{M}$ is subclassical, for every $a \in \mathcal{D}$ different from t it holds that $\neg a \in \mathcal{D}$. Hence, $P_1, \diamond(\varphi_1,\ldots,\varphi_n) \vdash_\mathcal{M} \neg P_1$, while $P_1, \diamond(\varphi_1,\ldots,\varphi_n) \not\vdash_{\mathcal{M}_\mathbf{F}} \neg P_1$ (because $\nu(P_1) = t$ provides a counterexample). This contradicts the $\mathbf{F}$-containment of $\mathbf{L}_\mathcal{M}$ in classical logic. □

Corollary 3.50 *Let $\mathcal{M}$ be a $\neg$-subclassical matrix. Then:*

1. *If $\wedge$ is an $\mathcal{M}$-conjunction then $\tilde{\wedge}/\{t,f\}^2$ is the classical conjunction.*

2. *If $\vee$ is an $\mathcal{M}$-disjunction then $\tilde{\vee}/\{t,f\}^2$ is the classical disjunction.*

3. *If $\supset$ is an $\mathcal{M}$-implication then $\tilde{\supset}/\{t,f\}^2$ is the classical implication.*

4. *$\tilde{\neg}/\{t,f\}^2$ is the classical negation.*

Proof. By Proposition 3.49, and Corollary 3.34. □

The next theorem shows the important role that subclassical matrices have among matrices with negation.

Theorem 3.51 *Let $\mathcal{M} = \langle \mathcal{V}, \mathcal{D}, \mathcal{O} \rangle$ be a matrix for a language $\mathcal{L}$, and let $\neg$ be a unary connective of $\mathcal{L}$. Suppose that $\mathcal{M}$ has the following properties:*

a) *There exist $t \in \mathcal{D}$ and $f \in \overline{\mathcal{D}}$ such that:*

   i) *$\tilde{\neg} t = f$; $\tilde{\neg} f \in \{t,f\}$.*

   ii) *If $a \in \mathcal{V} - \{t,f\}$ then $a \in \mathcal{D}$ iff $\neg a \in \mathcal{D}$.*

b) *$\mathcal{M}$ has an $\mathcal{M}$-disjunction.*

Then $\mathcal{M}$ is $\neg$-subclassical iff it is $\neg$-contained in classical logic.

Proof. One direction follows from Proposition 3.47.

For the converse, Suppose that $\mathcal{M}$ is $\neg$-contained in classical logic. Then Property (b) of $\mathcal{M}$ implies (using Proposition 3.32) that $\neg$ is a negation in $\mathcal{M}$. In addition, it follows from Property (a) that t is the unique element $a \in \mathcal{V}$ such that $a \in \mathcal{D}$ while $\tilde{\neg} a \notin \mathcal{D}$. Therefore, Proposition 3.43 and Item (i) of (a) entail that $\tilde{\neg} f \in \mathcal{D} \cap \{t,f\} = \{t\}$. Hence, $\tilde{\neg} f = t$. To prove that $\mathcal{M}$ is subclassical, it remains therefore to show that it is classically closed. Assume for contradiction that it is not. Then there are elements $a_1,\ldots,a_n \in \{t,f\}$ and an n-ary connective $\diamond$ such that $\tilde{\diamond}(a_1,\ldots,a_n) \notin \{t,f\}$. For $i = 1,\ldots,n$ let $r_i = P_i$ if $a_i = t$, $r_i = \neg P_i$ if $a_i = f$. Then for every valuation $\nu \in \Lambda_\mathcal{M}$, if $\nu(P_i) = t$ for every $1 \leq i \leq n$ then $\nu(r_i) = a_i$ for every $1 \leq i \leq n$ (because $\tilde{\neg} t = f$), and so $\nu(\diamond(r_1,\ldots,r_n)) = \tilde{\diamond}(a_1,\ldots,a_n)$. Since $\tilde{\diamond}(a_1,\ldots,a_n) \notin \{t,f\}$, Property (a) implies that there are two possibilities:

3.2 Paraconsistent Matrices and Their Logics 75

- Assume that $\tilde{\diamond}(a_1,\ldots,a_n)$ and $\tilde{\neg}\tilde{\diamond}(a_1,\ldots,a_n)$ are both in $\mathcal{D}$. Then:

$$P_1,\ldots,P_n \vdash_\mathcal{M} \neg P_1 \vee \ldots \vee \neg P_n \vee \diamond(r_1,\ldots,r_n),$$

$$P_1,\ldots,P_n \vdash_\mathcal{M} \neg P_1 \vee \ldots \vee \neg P_n \vee \neg\diamond(r_1,\ldots,r_n).$$

Indeed, let ν be a model of $\{P_1,\ldots,P_n\}$. If $\nu(P_i) \neq t$ for some i then because of the uniqueness property of t, $\nu(\neg P_i) \in \mathcal{D}$, and so ν is a model of the disjunctions on the right hand sides. If $\nu(P_i) = t$ for all i then ν is a model of both $\tilde{\diamond}(r_1,\ldots,r_n)$ and $\tilde{\neg}\tilde{\diamond}(r_1,\ldots,r_n)$, and so again ν is a model of both right hand sides.

Now, since $\mathcal{M}$ is $\neg$-contained in classical logic, Corollary 3.34 entails that the above two facts remain true if we replace $\vdash_\mathcal{M}$ by $\vdash_{2\mathbf{F}}$ for some bivalent interpretation $\mathbf{F}$ of $\mathcal{L}$ that interprets $\vee$ and $\neg$ as classical disjunction and negation (respectively). This is impossible.

- Assume that $\tilde{\diamond}(a_1,\ldots,a_n)$ and $\tilde{\neg}\tilde{\diamond}(a_1,\ldots,a_n)$ are both not in $\mathcal{D}$. Then

$$\diamond(r_1,\ldots,r_n), P_1,\ldots,P_n \vdash_\mathcal{M} \neg P_1 \vee \ldots \vee \neg P_n,$$

$$\neg\diamond(r_1,\ldots,r_n), P_1,\ldots,P_n \vdash_\mathcal{M} \neg P_1 \vee \ldots \vee \neg P_n.$$

(The reason this time is that the only models of either of the left hand sides are here valuations which assign an element of $\mathcal{D} \setminus \{t\}$ to some P_i. Because of the uniqueness property of t, $\nu(\neg P_i) \in \mathcal{D}$, and so ν is a model of the disjunction on the right hand sides.) This is impossible, by an argument similar to that given in the previous case. □

3.2.2 Paraconsistent Matrices

Definition 3.52 ($\neg$-paraconsistent matrix) Let $\mathcal{M}$ be a matrix with negation $\neg$. $\mathcal{M}$ is *(boldly, strongly, pre-) $\neg$-paraconsistent* if so is $\mathbf{L}_\mathcal{M}$.

Convention: Again, henceforth we shall usually omit the $\neg$ sign, and simply refer to *(boldly, strongly, pre-) paraconsistent matrices*.

Proposition 3.53 *Every extension of a (strongly) paraconsistent matrix is (strongly) paraconsistent too.*

Proof. Immediate from Propositions 3.14 and 2.11. □

Proposition 3.54 *Supposed that $\mathcal{M}$ is pre-$\neg$-paraconsistent; $\neg$-contained in classical logic; and has an $\mathcal{M}$-conjunction, an $\mathcal{M}$-implication, and an $\mathcal{M}$-disjunction. If $\mathbf{L}_\mathcal{M}$ is fintary then $\mathbf{L}_\mathcal{M}$ is classical $\neg$-paraconsistent logic.*

Proof. This follows from Propositions 2.23, 3.32, and 3.37. □

Proposition 3.55 *Let $\mathcal{M}$ be a matrix with negation $\neg$.*

1. *$\mathcal{M}$ is pre-paraconsistent iff there is $\top \in \mathcal{D}$, such that $\tilde{\neg}\top \in \mathcal{D}$.*

2. *If $\mathcal{M}$ is paraconsistent then there are three different elements t, f, and $\top$ in $\mathcal{V}$ such that $f = \tilde{\neg}t$, $f \notin \mathcal{D}$, and $\{t, \tilde{\neg}f, \top, \tilde{\neg}\top\} \subseteq \mathcal{D}$.*

Proof.

1. $\mathcal{M}$ is pre-paraconsistent iff $p, \neg p \not\vdash_{\mathcal{M}} q$. Obviously, this happens iff $\{p, \neg p\}$ has an $\mathcal{M}$-model. The latter, in turn, is possible iff there is some $\top \in \mathcal{D}$ such that $\tilde{\neg}\top \in \mathcal{D}$.

2. Immediate from Proposition 3.43 and the first item. □

Corollary 3.56 *Any paraconsistent matrix is boldly paraconsistent. In particular: any paraconsistent matrix is strongly paraconsistent.*

Proof. The second part follows from the first by Proposition 2.20. For the first part, suppose that $\mathcal{M} = \langle \mathcal{V}, \mathcal{D}, \mathcal{O} \rangle$ is a paraconsistent matrix, σ is a formula in its language such that $\not\vdash_{\mathcal{M}} \sigma$, and q is a variable such that $q \notin \mathsf{Var}(\sigma)$. Then there is a valuation ν in $\mathcal{M}$ such that $\nu(\sigma) \notin \mathcal{D}$. Let $\top$ be an element of $\mathcal{V}$ like in the first item of Proposition 3.55. Define a valuation ν' by letting $\nu'(q) = \top$, and $\nu'(p) = \nu(p)$ for every variable $p \neq q$. Then $\nu'(\sigma) = \nu(\sigma) \notin \mathcal{D}$. Hence ν' is an $\mathcal{M}$-model of $\{\neg q, q\}$ which is not an $\mathcal{M}$-model of σ, and so $\{\neg q, q\} \not\vdash_{\mathcal{M}} \sigma$. □

Corollary 3.57

1. *Every paraconsistent matrix has at least two designated elements.[6]*

2. *No two-valued matrix can be paraconsistent.*

Proof. Immediate from the second item of Proposition 3.55. □

In the next two propositions we describe properties of a particularly important class of paraconsistent matrices: those which have the minimal possible number of designated values.

---

[6] In the literature one can find paraconsistent logics which are connected with matrices with a single designated values. (Examples can be found in [234] and [257], as well as in Section 9.4 below.) These matrices are not paraconsistent according to our definitions, and the paraconsistent logics connected with them are finite-valued logic only according to an extended meaning of this notion. (See Section 3.3 and Footnote 3.)

3.2 Paraconsistent Matrices and Their Logics

Proposition 3.58 *Let $\mathcal{M} = \langle \mathcal{V}, \mathcal{D}, \mathcal{O} \rangle$ be a paraconsistent matrix, and suppose that $|\mathcal{D}| = 2$. Then there are three different elements t, f, and $\top$ in $\mathcal{V}$ such that $\mathcal{D} = \{t, \top\}$, $\tilde{\neg}t = f$, $\tilde{\neg}f = t$, and $\tilde{\neg}\top \in \mathcal{D}$.*

Proof. From the assumption that $|\mathcal{D}| = 2$ and the second part of Proposition 3.55 it easily follows that there are elements $t, f, \top \in \mathcal{V}$ such that $\mathcal{D} = \{t, \top\}$, $\tilde{\neg}\top \in \mathcal{D}$, $f = \tilde{\neg}t$, $f \notin \mathcal{D}$, and $\tilde{\neg}f \in \mathcal{D}$. It remains to show that $\tilde{\neg}f = t$. Assume for contradiction that $\tilde{\neg}f = \top$. This implies that $\tilde{\neg}\tilde{\neg}\tilde{\neg}\top = \top$, no matter whether $\tilde{\neg}\top = \top$ or $\tilde{\neg}\top = t$. This and the facts that $\mathcal{D} = \{t, \top\}$ and $\tilde{\neg}\top \in \mathcal{D}$ imply that $p \vdash_{\mathcal{M}} \neg\neg\neg p$, which contradicts the $\neg$-coherence of $\mathcal{M}$ with classical logic. □

Proposition 3.59 *Let $\mathcal{M} = \langle \mathcal{V}, \mathcal{D}, \mathcal{O} \rangle$ be a paraconsistent matrix such that $|\mathcal{D}| = 2$. Then $\mathbf{L}_{\mathcal{M}}$ is non-exploding iff every connective of $\mathcal{M}$ is $\{\top\}$-closed (where $\top$ is as in Proposition 3.58).*

Proof. Suppose that every connective of $\mathcal{M}$ is $\{\top\}$-closed. Let $\mathcal{T}$ be a theory and q a variable such that $q \notin \mathsf{Var}(\mathcal{T})$. Let ν be a valuation in $\mathcal{M}$ such that $\nu(p) = \top$ for every $p \in \mathsf{Var}(\mathcal{T})$, while $\nu(q) \notin \mathcal{D}$. Since every connective of $\mathcal{M}$ is $\{\top\}$-closed, $\nu(\varphi) = \top$ for every $\varphi \in \mathcal{T}$. Hence ν is a model of $\mathcal{T}$ which is not a model of q. It follows that $\mathcal{T} \nvdash_{\mathcal{M}} q$.

For the converse, assume that there is an n-ary connective $\diamond$ of the language of $\mathcal{M}$ such that $\tilde{\diamond}$ is not $\{\top\}$-closed. Let

$$S = \{P_1, \neg P_1, \diamond(P_1, \ldots, P_1), \neg\diamond(P_1, \ldots, P_1)\}.$$

Then our assumption and Proposition 3.58 imply that S has no models in $\mathcal{M}$, and so $S \vdash_{\mathcal{M}} \varphi$ for every φ. Hence $\mathbf{L}_{\mathcal{M}}$ is not non-exploding. □

3.2.3 Maximality of Paraconsistent Many-Valued Logics

In the next chapter we show that every significant paraconsistent three-valued logic is strongly maximal, and if it is $\neg$-contained in classical logic then it is fully maximal. In this section we show that these properties are not confined to the three-valued case. In fact, also for every $n > 3$ there exists an extensive family of n-valued fully maximal paraconsistent logics, each of which is not equivalent to any k-valued logic with $k < n$.

Theorem 3.60 *Let $n > 3$, and let $\mathcal{M} = \langle \mathcal{V}, \mathcal{D}, \mathcal{O} \rangle$ be an n-valued matrix for a language $\mathcal{L}$ containing the unary connectives $\neg$ and $\diamond$, and the binary connective $\supset$.*

1. *Suppose that the following conditions hold in $\mathcal{M}$:*

(a) $\mathcal{V} = \{t, f, \top, \bot_1, \ldots, \bot_{n-3}\}$ and $\mathcal{D} = \{t, \top\}$;

(b) $\tilde{\neg} t = f, \tilde{\neg} f = t$, and $\tilde{\neg} x = x$ otherwise;

(c) $\tilde{\diamond} t = f, \tilde{\diamond} f = t$; $\tilde{\diamond} \top = \bot_1$, $\tilde{\diamond} \bot_i = \bot_{i+1}$ for $i < n-3$, and $\tilde{\diamond} \bot_{n-3} = \top$;

(d) $\tilde{\supset}$ is the operation $\supset_t$ from Definition 3.38 (i.e. $a \tilde{\supset} b = t$ if $a \notin \mathcal{D}$, and $a \tilde{\supset} b = b$ otherwise).

Then $\mathbf{L}_\mathcal{M}$ is a normal and boldly paraconsistent n-valued logic, which is strongly maximal, and not equivalent to any k-valued logic with $k < n$.

2. Suppose that in addition to the conditions above we have that $\tilde{\star}$ is $\{t, f\}$-closed for every other n-ary connective $\star$ of $\mathcal{M}$. Then $\mathbf{L}_\mathcal{M}$ is a fully maximal.

Proof.

1. First, we note that for any $\mathcal{M}$-valuation ν, $\nu(\diamond p \supset (p \supset p)) = t$. Hence we may assume that $\mathcal{L}$ includes propositional constants F and T such that $\nu(\mathsf{F}) = f$ and $\nu(\mathsf{T}) = t$ for any $\mathcal{M}$-valuation ν.

   Now we show that $\mathbf{L}_\mathcal{M}$ has all the stated properties.

   - $\supset$ is an $\mathcal{M}$-implication by Proposition 3.39, and so $(\varphi \supset \psi) \supset \psi$ defines an $\mathcal{M}$-disjunction by Note 3.30. It is also easy to verify that by letting $\varphi \wedge \psi = \neg(\varphi \supset \neg\psi)$ we get an $\mathcal{M}$-conjunction $\wedge$. It follows from Corollary 3.33 that $\mathbf{L}_\mathcal{M}$ is normal. Obviously, the reduction of $\mathcal{M}$ to $\mathcal{L}_{CL}$ is subclassical, and so it is paraconsistent by Propositions 3.47 and 3.55. Therefore $\mathcal{M}$ itself is boldly paraconsistent by Proposition 3.53 and Corollary 3.56.

   - Note that for any $a \in \mathcal{V} \setminus \{t, f\}$ there is $0 \leq j_a \leq n-2$ such that a valuation μ is a model in $\mathcal{M}$ of $\{\diamond^{j_a} p, \neg \diamond^{j_a} p\}$ (where $\diamond^0 p = p, \diamond^{k+1} p = \diamond(\diamond^k p)$) iff $\mu(p) = a$. ($j_\top = 0$ or $j_\top = n-2$, and $j_{\bot_i} = n-2-i$.) Let $\mathbf{L} = \langle \mathcal{L}, \vdash_{\mathbf{L}} \rangle$ be any proper simple finitary extension of $\mathbf{L}_\mathcal{M}$. Then there are some $\psi_1, \ldots, \psi_k$ and φ such that $\psi_1, \ldots, \psi_k \vdash_{\mathbf{L}} \varphi$, but $\psi_1, \ldots, \psi_k \not\vdash_\mathcal{M} \varphi$. From the latter it follows that there is a valuation μ such that $\mu(\psi_i) \in \mathcal{D}$ for every $1 \leq i \leq k$, and $\mu(\varphi) \in \overline{\mathcal{D}}$. Let $p_1, \ldots, p_m$ be the variables occurring in $\{\psi_1, \ldots, \psi_k, \varphi\}$. Since we can substitute the constant F for any p such that $\mu(p) = f$, and T for any p such that $\mu(p) = t$, we may assume that for any variable p, $\mu(p) \in \mathcal{V} \setminus \{t, f\}$. Accordingly, let $j_i = j_{\mu(p_i)}$ for $1 \leq i \leq m$. By the observations above, μ is the only model of the set $\Psi = \bigcup_{1 \leq i \leq m} \{\diamond^{j_i} p_i, \neg \diamond^{j_i} p_i\}$.

3.2 Paraconsistent Matrices and Their Logics

It follows that $\Psi \vdash_{\mathcal{M}} \psi_i$ for every $1 \leq i \leq k$, and $\Psi \cup \{\varphi\} \vdash_{\mathcal{M}} q$ (where q is a new variable). Hence $\Psi \vdash_{\mathbf{L}} q$. Now, by substituting $\diamond^{n-j_i-2}p$ for p_i (where p is different from q), we can unify Ψ to $\{\diamond^{n-2}p, \neg\diamond^{n-2}p\}$. But in $\mathbf{L}_{\mathcal{M}}$ both elements of this set follow from $\{p, \neg p\}$. Thus $p, \neg p \vdash_{\mathbf{L}} q$, and so $\mathbf{L}$ is not pre-paraconsistent. It follows that $\mathbf{L}_{\mathcal{M}}$ is strongly maximal.

- Let $\mathcal{M}' = \langle \mathcal{V}', \mathcal{D}', \mathcal{O}' \rangle$ be a matrix for $\mathcal{L}$ such that $\vdash_{\mathcal{M}'} = \vdash_{\mathcal{M}}$. We show that $\mathcal{V}'$ has at least n elements.

  Since $\vdash_{\mathcal{M}'} = \vdash_{\mathcal{M}}$, $\vdash_{\mathcal{M}'}$ has the following properties (where p and q are different variables):

  (a) $p \not\vdash_{\mathcal{M}'} \neg p$;
  (b) $\neg p \not\vdash_{\mathcal{M}'} p$;
  (c) $\neg p, p \not\vdash_{\mathcal{M}'} q$;
  (d) $p, \neg p \vdash_{\mathcal{M}'} \diamond^{n-2}p$;
  (e) $p, \neg p, \diamond^k p \vdash_{\mathcal{M}'} q$ for $1 \leq k \leq n-3$;
  (f) $p, \neg p, \neg \diamond^k p \vdash_{\mathcal{M}'} q$ for $1 \leq k \leq n-3$.

  ($\vdash_{\mathcal{M}}$ has Property (a) because of t, Property (b) because of f, Property (c) because of $\top$, and the three other properties because $\nu(p) = \top$ for every model ν in $\mathcal{M}$ of $\{\neg p, p\}$.) Now, Property (a) of $\vdash_{\mathcal{M}'}$ implies that there is an element $t' \in \mathcal{D}'$ such that $\tilde{\neg} t' \notin \mathcal{D}'$, Property (b) implies that there is an element $f' \notin \mathcal{D}'$ such that $\tilde{\neg} f' \in \mathcal{D}'$, and Property (c) implies that there is an element $\top' \in \mathcal{D}'$ such that $\tilde{\neg} \top' \in \mathcal{D}'$. These three elements of $\mathcal{V}'$ are of course different from each other.

  Let $\perp'_k = \tilde{\diamond}_{\mathcal{M}'}^k \top'$ for every $1 \leq k \leq n-3$. Then Property (e) of $\vdash_{\mathcal{M}'}$ implies that $\perp'_k \notin \mathcal{D}'$ for $1 \leq k \leq n-3$. Moreover, $\perp'_1, \ldots, \perp'_{n-3}$ are different from each other, because otherwise we would get that $\tilde{\diamond}^i \top' \notin \mathcal{D}$ for every $i > 0$, and this contradicts Property (d) of $\vdash_{\mathcal{M}'}$. Finally, Property (f) of $\vdash_{\mathcal{M}'}$ implies that for every $1 \leq k \leq n-3$ we have that also $\tilde{\neg}\perp'_k \notin \mathcal{D}$. (Otherwise, the valuation $\mu(p) = \top', \mu(q) = f'$ would contradict this property.) Hence, $\{\perp'_1, \ldots, \perp'_{n-3}\} \cap \{t', f', \top'\} = \emptyset$. It follows that $t', \top', f', \perp'_1, \ldots, \perp'_{n-3}$ are all different from each other. Hence $\mathcal{V}'$ has at least n elements.

2. Suppose $\mathcal{M}$ satisfies also the extra condition. We show that $\mathbf{L}_{\mathcal{M}}$ is $\mathbf{F}_{\mathcal{M}}$-maximal relative to classical logic (where $\mathbf{F}_{\mathcal{M}}$ is the bivalent $\neg$-interpretation given in Definition 3.48). Let φ be a formula that is not $\mathcal{M}$-valid. Suppose for contradiction that there is a classical $\mathbf{F}_{\mathcal{M}}$-tautology σ such that $\not\vdash_{\mathbf{L}} \sigma$, where $\mathbf{L}$ is obtained from $\mathbf{L}_{\mathcal{M}}$ by adding

φ as a new axiom schema. This implies that $S \not\vdash_\mathcal{M} \sigma$, where S is the set of instances of φ. Let $\mathcal{T}$ be a maximal theory extending S such that $\mathcal{T} \not\vdash_\mathcal{M} \sigma$. Then for every formula ψ, $\mathcal{T} \vdash_\mathcal{M} \psi$ iff $\psi \in \mathcal{T}$, and either $\psi \in \mathcal{T}$, or $\mathcal{T}, \psi \vdash_\mathcal{M} \sigma$ and so $\mathcal{T} \vdash_\mathcal{M} \psi \supset \sigma$.

Next, for any truth value $a \in \mathcal{V}$ and formula $\psi \in \mathcal{W}(\mathcal{L})$ define formulas $\phi_1^a(\psi)$ and $\phi_2^a(\psi)$ as follows:

$$\phi_1^t(\psi) = \psi \qquad \phi_2^t(\psi) = \neg \diamond \psi$$
$$\phi_1^f(\psi) = \neg \psi \qquad \phi_2^f(\psi) = \diamond \psi$$
$$\phi_1^\top(\psi) = \psi \qquad \phi_2^\top(\psi) = \neg \psi$$
$$\phi_1^{\perp_i}(\psi) = \diamond^{n-2-i}\psi \qquad \phi_2^{\perp_i}(\psi) = \neg \diamond^{n-2-i}\psi \quad \text{for } i = 1, \ldots, n-3$$

It is easy to check that the following holds:

($\star$) For any $a \in \mathcal{V}$, $\psi \in \mathcal{W}(\mathcal{L})$, and an $\mathcal{M}$-valuation ν, $\nu(\psi) = a$ iff ν satisfies both $\phi_1^a(\psi)$ and $\phi_2^a(\psi)$.

Now, define a valuation ν in $\mathcal{M}$ by: $\nu(\psi) = a$ if $\phi_1^a(\psi) \in \mathcal{T}$ and $\phi_2^a(\psi) \in \mathcal{T}$. We show the following facts:

(a) ν *is well-defined*. This follows from the following two facts:

   i. There must exist $a \in \mathcal{V}$ such that $\{\phi_1^a(\psi), \phi_2^a(\psi)\} \subseteq \mathcal{T}$.
   To see this, let $\phi_3^a(\psi) = \phi_1^a(\psi) \supset (\phi_2^a(\psi) \supset \sigma)$. By ($\star$), every $\mathcal{M}$-valuation satisfies $\phi_1^a(\psi)$ and $\phi_2^a(\psi)$ for some $a \in \mathcal{V}$. Hence $\{\phi_3^a(\psi) \mid a \in \mathcal{V}\} \vdash_\mathcal{M} \sigma$. Now, if $\nu(\psi) \neq a$ for some $a \in \mathcal{V}$, then $\phi_1^a(\psi) \notin \mathcal{T}$ or $\phi_2^a(\psi) \notin \mathcal{T}$, so $\mathcal{T} \cup \{\phi_1^a(\psi), \phi_2^a(\psi)\}$ is a proper extension of $\mathcal{T}$. Hence, $\mathcal{T} \cup \{\phi_1^a(\psi), \phi_2^a(\psi)\} \vdash_\mathcal{M} \sigma$, and so $\mathcal{T} \vdash_\mathcal{M} \phi_3^a(\psi)$ by Proposition 3.39. Thus, if $\nu(\psi) \neq a$ for every $a \in \mathcal{V}$ we get that $\mathcal{T} \vdash_\mathcal{M} \phi_3^a(\psi)$ for every $a \in \mathcal{V}$, and so $\mathcal{T} \vdash_\mathcal{M} \sigma$. A contradiction.

   ii. If $\{\phi_1^a(\psi), \phi_2^a(\psi)\} \subseteq \mathcal{T}$ and $\{\phi_1^b(\psi), \phi_2^b(\psi)\} \subseteq \mathcal{T}$ then $a = b$.
   Assume that $\{\phi_1^a(\psi), \phi_2^a(\psi), \phi_1^b(\psi), \phi_2^b(\psi)\} \subseteq \mathcal{T}$ for some a, b such that $a \neq b$. Then ($\star$) implies that $\mathcal{T}$ is not $\mathcal{M}$-satisfiable, and so $\mathcal{T} \vdash_\mathcal{M} \sigma$. A contradiction.

(b) ν *is an $\mathcal{M}$-valuation*.
   Let $\psi = \diamond(\psi_1, \ldots, \psi_n)$. Suppose $\nu(\psi_i) = a_i$ for $i = 1, \ldots, n$ and $\tilde{\diamond}(a_1, \ldots, a_n) = b$. Then $\phi_1^{a_i}(\psi_i) \in \mathcal{T}$ and $\phi_2^{a_i}(\psi_i) \in \mathcal{T}$ for every $1 \leq i \leq n$. Now by ($\star$), $\bigcup_{i=1}^n \{\phi_1^{a_i}(\psi_i), \phi_2^{a_i}(\psi_i)\} \vdash_\mathcal{M} \phi_j^b(\psi)$ for $j = 1, 2$. Hence $\phi_j^b(\psi) \in \mathcal{T}$ for $j = 1, 2$, and so $\nu(\psi) = b$ by the definition of ν. Therefore, $\nu(\diamond(\psi_1, \ldots, \psi_n)) = \tilde{\diamond}(a_1, \ldots, a_n)$, as required.

3.2 Paraconsistent Matrices and Their Logics   81

(c) ν is a model of $\mathcal{T}$ which is not a model of σ.

Let $\psi \in \mathcal{T}$. If $\neg\psi \in \mathcal{T}$ then $\phi_1^\top(\psi) \in \mathcal{T}$ and $\phi_2^\top(\psi) \in \mathcal{T}$, thus $\nu(\psi) = \top \in \mathcal{D}$. Otherwise, $\neg\psi \notin \mathcal{T}$, and so $\neg\psi \supset \sigma \in \mathcal{T}$. Since $\{\psi, \neg\psi \supset \sigma, \neg\Diamond\psi \supset \sigma\} \vdash_\mathcal{M} \sigma$, this implies that $\neg\Diamond\psi \supset \sigma \notin \mathcal{T}$, and so $\neg\Diamond\psi \in \mathcal{T}$. It follows that in this case $\phi_1^t(\psi) \in \mathcal{T}$ and $\phi_2^t(\psi) \in \mathcal{T}$, thus again $\nu(\psi) = t \in \mathcal{D}$.

Clearly, ν cannot be a model of σ, since if $\nu(\sigma) \in \{t, \top\}$, then in particular $\phi_1^t(\sigma) \in \mathcal{T}$ or $\phi_1^\top(\sigma) \in \mathcal{T}$. In either case $\sigma \in \mathcal{T}$, a contradiction to $\mathcal{T} \nvdash_\mathcal{M} \sigma$.

(d) ν is a valuation in $\{t, f\}$.

Assume for contradiction that there are $a \notin \{t, f\}$ and ψ_a such that $\nu(\psi_a) = a$. This implies that for every b there is a sentence ψ_b such that $\nu(\psi_b) = b$. Indeed, for $b \notin \{t, f\}$, ψ_b may be taken as a sentence of the form $\Diamond^k\psi_a$, where k is such that $\tilde{\Diamond}^k a = b$. In addition, the facts that $\nu(\sigma) \notin \mathcal{D}$ and that ν is an $\mathcal{M}$-valuation together imply (by the definition of $\tilde{\supset}$ and $\tilde{\neg}$) that we may take $\psi_t = \sigma \supset \sigma$ and $\psi_f = \neg(\sigma \supset \sigma)$. Now, let μ be an $\mathcal{M}$-valuation such that $\mu(\varphi) \notin \mathcal{D}$ (Such μ exists because φ is not valid in $\mathcal{M}$). Assume that $\mathsf{Var}(\varphi) = \{q_1, \ldots, q_k\}$ and that $\mu(q_i) = b_i$ for $i = 1, \ldots, k$. Let $\psi = \varphi\{\psi_{b_i}/q_i\}$. Then $\nu(\psi) = \mu(\varphi) \notin \mathcal{D}$. On the other hand, $\psi \in \mathcal{S} \subseteq \mathcal{T}$. Hence $\nu(\psi) \in \mathcal{D}$, since ν is a model of $\mathcal{T}$. A contradiction.

Now, since σ is a classical $\mathbf{F}_\mathcal{M}$-tautology, and ν is a valuation in $\{t, f\}$ which obviously respects the truth-tables of $\mathbf{F}_\mathcal{M}$, necessarily $\nu(\sigma) = t$. This contradicts the fact that ν is not a model of σ. □

Note 3.61 All the logics introduced in Theorem 3.60 have the further important property that $\neg\neg\psi$ is *congruent* (Definition 1.18) in them to ψ. Another (even more) important property they all have is that if p is a variable, then for every truth value x they use there is a simple set S_x such that $\mathsf{Var}(S_x) = \{p\}$ and for every valuation ν it holds that $\nu \in mod(S_x)$ iff $\nu(p) = x$. Indeed, like in the proof of the second part of Theorem 3.60, we can take $S_t = \{p, \neg\Diamond p\}$, $S_f = \{\neg p, \Diamond p\}$, $S_\top = \{p, \neg p\}$, and for every $1 \leq i \leq n-3$, $S_{\perp_i} = \{\Diamond^{n-2-i} p, \neg\Diamond^{n-2-i} p\}$ (where $\Diamond^1 p = \Diamond p$, $\Diamond^k p = \Diamond(\Diamond^{k-1} p)$). In the terminology of [59] this means that the language of these matrices is 'sufficiently expressive', and so one can apply the algorithm provided in that paper *in order to develop for each of them a Gentzen-type system which is cut-free* (though less elegant than the various Gentzen-type systems which are presented in this book). Thus, each of these logics has all the properties mentioned in the desiderata list which is presented in Section 2.4.

3.3 Bibliographical Notes and Further Reading

Matrix-based logics is a vast field with numerous publications. The survey papers of Urquhart [290] and Hähnle [177] provide a good entry to this discipline with many relevant references. Other extensive sources are [104, 105, 176, 204, 258, 304].

Theorem 3.17 has been proved in [266, 267]. Theorem 3.24 is from [200]. (See also [290, 302, 304].) Theorem 3.21 is proved in [302, 304].

Fully maximal paraconsistent logics induced by matrices and the systematic way of constructing them as considered in Section 3.2.3 have been presented in [18, 20], where logics which are both normal and fully maximal are called 'ideal'.

As mentioned in Footnote 6, in addition to the logic $\mathbf{L}_\mathcal{M}$ (to which this chapter is devoted), one can find in the literature other methods of using a given matrix $\mathcal{M} = \langle \mathcal{V}, \mathcal{D}, \mathcal{O} \rangle$ for defining or characterizing logics. (See, e.g., Chapter 1 of [304], where several logics are associated with Łukasiewicz's three-valued matrix.) Thus, following the first item of Note 1.25, one may define a tcr $\vdash_\mathcal{M}^{MP}$ by $\mathcal{T} \vdash_\mathcal{M}^{MP} \varphi$ if $\vdash_{\mathbf{L}_\mathcal{M}} \psi_1 \supset (\psi_2 \supset (\ldots \supset (\psi_n \supset \varphi)\ldots))$ for some $\psi_1, \ldots, \psi_n \in \mathcal{T}$, where $\supset$ is a connective of $\mathcal{M}$ with appropriate properties that guarantee that $\vdash_\mathcal{M}^{MP}$ is indeed a tcr (for which $\supset$ is serving as a semi-implication). Three examples (one from Chapter 9 of this book) of the use of such methods for constructing paraconsistent logics are mentioned in Footnote 6. Another famous one has been given in [273]. (See also Note 4.43 in Chapter 4.) Others can be found later in this book, especially in Part V.

Chapter 4
Three-Valued Matrices

Three-valued matrices are the most popular framework for reasoning with contradictory data. The major reason for this is that they provide the simplest semantic way of defining paraconsistent logics (cf. Corollary 3.57). Note that by Theorem 3.17, all logics defined in this way are *finitary*.

4.1 An Introduction

In every matrix there is at least one designated element, and one that is not. Therefore three-valued matrices are naturally divided into two classes: those in which there is a single designated value, and those in which there is a single non-designated value. On the basis of this simple observation, it is possible to provide a precise description of the structure of three-valued matrices with negation.

Proposition 4.1 *Let $\mathcal{M}$ be a three-valued matrix with negation $\neg$. Then $\mathcal{M}$ is isomorphic to a matrix $\mathcal{M}' = \langle \mathcal{V}, \mathcal{D}, \mathcal{O} \rangle$ in which $\mathcal{V}$ contains two elements t and f such that $t \in \mathcal{D}$, $f \notin \mathcal{D}$, $\tilde{\neg} t = f$, and $\tilde{\neg} f = t$.*

Proof. If $\mathcal{M}$ has a single designated value then the claim is immediate from Proposition 3.43. So assume that $\mathcal{M}$ has a single non-designated value, which we denote by f. Since $\neg$ is a negation in $\mathcal{M}$, $\neg p \nvDash_\mathcal{M} p$. Since f is the only non-designated element of $\mathcal{M}$, this is possible only if $\tilde{\neg} f \neq f$. Denote $\tilde{\neg} f$ by t. It remains to show that $\tilde{\neg} t = f$. Assume otherwise. Then it is easy to check that no matter how $\tilde{\neg}$ is defined for the third value of $\mathcal{M}$, necessarily $p \vDash_\mathcal{M} \neg\neg\neg p$. This is impossible, since $\neg$ is a negation in $\mathcal{M}$, while $p \nvDash_{CL} \neg\neg\neg p$. □

In what follows we assume that $\{t, f\} \subseteq \mathcal{V}$ for every three-valued matrix $\langle \mathcal{V}, \mathcal{D}, \mathcal{O} \rangle$ with negation. However, we shall use different symbols for denoting the third element, to distinguish between the two main cases noted above: a non-designated third element is denoted by $\bot$, and a designated third element is denoted by $\top$. When it does not matter whether the third element is designated or not, we shall sometimes denote it by i.

Notation 4.2 (M3$_\bot$ and M3$_\top$) In what follows we denote by M3$_\bot$ the class of three-valued matrices with one designated element, and by M3$_\top$ the class of three-valued matrices with two designated elements. We usually take the elements of M3$_\bot$ to be of the form $\langle \{t, f, \bot\}, \{t\}, \mathcal{O} \rangle$ and the elements of M3$_\top$ to be of the form $\langle \{t, f, \top\}, \{t, \top\}, \mathcal{O} \rangle$ (for some interpretation $\mathcal{O}$ of the connectives in $\mathcal{L}$), where t and f are like in Proposition 4.1. The corresponding sets of logics are denoted by $\mathbf{3}_\bot$ and $\mathbf{3}_\top$, that is: $\mathbf{3}_\bot$ consists of the logics which have the form $\langle \mathcal{L}, \vdash_\mathcal{M} \rangle$ for some $\mathcal{M} \in$ M3$_\bot$, and $\mathbf{3}_\top$ consists of the logics which have the form $\langle \mathcal{L}, \vdash_\mathcal{M} \rangle$ for some $\mathcal{M} \in$ M3$_\top$.

Next, we turn to the issue of the interpretations of the basic connectives. Below we recall some of the better known three-valued interpretations of the usual classical connectives. The most standard way of defining negation, conjunction, and disjunction in this framework is by using the total order $\leq_t$ on $\{t, f, i\}$, in which t is the maximal element and f is the minimal one. This order may be intuitively understood as reflecting differences in the amount of *truth* that each element exhibits. Conjunction and disjunction are usually interpreted as the meet and the join (respectively) of $\leq_t$, while negation is interpreted as the unique function on $\{t, f, i\}$ which is order reversing with respect to $\leq_t$. Accordingly, the most standard truth tables are the following. (Note that these interpretations do not depend on the choice whether the middle element is designated or not, so we use here the neutral notation i to denote the intermediate element.)

$\tilde{\vee}$	t	f	i
t	t	t	t
f	t	f	i
i	t	i	i

$\tilde{\wedge}$	t	f	i
t	t	f	i
f	f	f	f
i	i	f	i

	$\tilde{\neg}$
t	f
f	t
i	i

It is easy to verify that whether $\mathcal{M} \in$ M3$_\bot$ or $\mathcal{M} \in$ M3$_\top$, with these interpretations $\tilde{\vee}$ is an $\mathcal{M}$-disjunction and $\tilde{\wedge}$ is an $\mathcal{M}$-conjunction (in the sense of Definition 3.29). Note also that according to these interpretations, $\vee$ and $\wedge$ are definable in terms of each other and $\neg$ by the usual De Morgan connections, since $a \tilde{\vee} b = \tilde{\neg}(\tilde{\neg} a \tilde{\wedge} \tilde{\neg} b)$ and $a \tilde{\wedge} b = \tilde{\neg}(\tilde{\neg} a \tilde{\vee} \tilde{\neg} b)$.

The logic that is induced by the three-valued matrix with non-designated middle element and the above interpretations for conjunction, disjunction, and negation is usually called *(strong) Kleene logic* [194].

4.1 An Introduction

Note 4.3 A different common way to interpret conjunction and disjunction is by the *weak* Kleene connectives [194] (also known as *Bochvar's connectives* [103], when i is non-designated, or *Halldén connectives* [179], when i is designated), defined as follows:

$\tilde{\vee}$	t	f	i
t	t	t	i
f	t	f	i
i	i	i	i

$\tilde{\wedge}$	t	f	i
t	t	f	i
f	f	f	i
i	i	i	i

Here, an expression evaluates to t or f iff all its components evaluate to t or f. This corresponds, for instance, to call-by-value implementations of some programming languages, in which the values of $\psi \vee \varphi$ and $\psi \wedge \varphi$ cannot be evaluated unless both of the procedures for evaluating ψ and φ terminate. In such a case the third truth value intuitively represents non-terminating computations. Note that $\tilde{\vee}$ is then an $\mathcal{M}$-disjunction only for $\mathcal{M} \in \mathsf{M3}_\top$ (i.e, when the middle element i is designated) and $\tilde{\wedge}$ is an $\mathcal{M}$-conjunction only for $\mathcal{M} \in \mathsf{M3}_\bot$ (when the middle element is non-designated).

Another popular interpretation of the connectives above is by sequential evaluations. Again, the motivation for this comes from the context of programming languages: a sequential evaluation of $\psi \wedge \varphi$, say from left to right, evaluates ψ first. If ψ evaluates to f, then the processing terminates and the value of $\psi \wedge \varphi$ is f. If ψ evaluates to t, then φ is evaluated, and its value is the value assigned to $\psi \wedge \varphi$. This kind of intermediate between strong Kleene connectives and weak Kleene connectives is known as *sequential* Kleene connectives, or *McCarthy's connectives* [211]. The corresponding truth tables are the following:

$\tilde{\vee}$	t	f	i
t	t	t	t
f	t	f	i
i	i	i	i

$\tilde{\wedge}$	t	f	i
t	t	f	i
f	f	f	f
i	i	i	i

In this case too $\tilde{\vee}$ is an $\mathcal{M}$-disjunction only for $\mathcal{M} \in \mathsf{M3}_\top$ and $\tilde{\wedge}$ is an $\mathcal{M}$-conjunction only for $\mathcal{M} \in \mathsf{M3}_\bot$.

Next we consider some historically important interpretations of the implication connectives that have been considered in the three-valued context.

- First, we follow the definition of the material implication in the two values case, and denote

$$\varphi \mapsto \psi = \neg \varphi \vee \psi$$

$\mapsto$ is what is usually taken as the official implication of Kleene's three-valued logic. However, it is not an implication connective in the sense of Definition 1.23 (in either $\mathbf{3}_\perp$ or $\mathbf{3}_\top$) when $\tilde{\neg}$ and $\tilde{\vee}$ are respectively the Kleene negation and (strong) disjunction. Indeed, in a logic belonging to $\mathbf{3}_\perp$ even $p \mapsto p$ is not valid (since if $a = \perp$ then $a \mapsto a \notin \mathcal{D}$), while in a logic $\mathbf{L}$ belonging to $\mathbf{3}_\top$ we have that $p, p \mapsto q \not\vdash_{\mathbf{L}} q$ (a counter-model assigns $\top$ to p and f to q).[1]

- The first many-valued logic ever invented is Łukasiewicz three-valued logic [201, 202]. This logic uses as its official implication the connective $\to_L$, whose interpretation is defined as follows:

$$a \tilde{\to}_L b = \begin{cases} \tilde{\neg} a \tilde{\vee} b & \text{if } a >_t b, \\ t & \text{otherwise.} \end{cases}$$

Here again, $\tilde{\neg}$ and $\tilde{\vee}$ are the Kleene negation and (strong) disjunction, respectively. However, $\to_L$ too is not an implication connective in any logic $\mathbf{L}$ which is based on these interpretations of $\neg$, $\vee$, and $\to_L$: If $\mathbf{L}$ belongs to $\mathbf{3}_\top$ then $p, p \to_L q \not\vdash_{\mathbf{L}} q$ (even though $p \to_L q \vdash_{\mathbf{L}} p \to_L q$), while if $\mathbf{L}$ belongs to $\mathbf{3}_\perp$ then $p \to_L (p \to_L q) \not\vdash_{\mathbf{L}} p \to_L q$ (even though $p \to_L (p \to_L q), p \vdash_{\mathbf{L}} q$ in this case).

- In contrast to the above operations, the operation $\supset_t$ introduced in Definition 3.38 (where the concrete element t of the three-valued matrices considered here is assigned to the variable t used in Definition 3.38) is an $\mathcal{M}$-implication for every three-valued matrix $\mathcal{M}$ (by Proposition 3.39). Note that *two different* three-valued operations may be denoted by $\supset_t$ in the three-valued context: one in the case where $\mathcal{D} = \{t\}$, and the other in case $\mathcal{D} = \{t, \top\}$. Both operations have been used in the literature. The former was introduced by Słupecki's (see[271]) and has been independently reintroduced in [32, 110, 217, 303]. The other was considered by D'Ottaviano and da-Costa [127, 143] and also investigated, e.g., in [10, 32, 260]. Because of its nice properties (to be presented later in this chapter), the latter is the main interpretation of an implication connective which has been used in three-valued paraconsistent logics, as well as the main one investigated in this chapter. Therefore unless otherwise stated, the symbol $\supset$ will be interpreted in this chapter as follows:

$$a \tilde{\supset} b = \begin{cases} b & \text{if } a \neq f, \\ t & \text{if } a = f. \end{cases}$$

---

[1]Note that these observations remain valid if we replace Kleene's $\tilde{\vee}$ by any other interpretation which makes $\vee$ a disjunction connective in the sense of Definition 3.29.

Apart for introducing appropriate counterparts of the classical connectives, it is also quite common to employ in the language of a finite-valued logic also propositional constants which correspond to (some of or all) the truth-values which are used. In this chapter we shall denote by F the one for which $\forall \nu \in \Lambda\, \nu(\mathsf{F}) = f$ and by B the constant for which $\forall \nu \in \Lambda\, \nu(\mathsf{B}) = \top$.[2] (There is usually no need to consider also a constant for t, because such a constant and F are definable in terms of each other and $\neg$.) Later we shall prove (see Theorem 4.65) that we get a functionally complete set of connectives by adding these propositional constants to the central three-valued interpretations which were introduced above for the classical connectives.

4.2 Paraconsistent Three-Valued Matrices

Now we consider three-valued paraconsistent matrices. We start by characterizing the basic structure of paraconsistent matrices.

Proposition 4.4 *Let $\mathcal{M}$ be a three-valued $\neg$-paraconsistent matrix. Then $\mathcal{M}$ is an f-matrix, and it is isomorphic to a matrix $\mathcal{M}' = \langle \mathcal{V}, \mathcal{D}, \mathcal{O} \rangle$ in which $\mathcal{V} = \{t, f, \top\}$, $\mathcal{D} = \{t, \top\}$, $\tilde{\neg} t = f$, $\tilde{\neg} f = t$ and $\tilde{\neg} \top \in \mathcal{D}$.*

Proof. This is immediate from Propositions 3.55 and 3.58. □

In the rest of this chapter we assume that any three-valued paraconsistent matrix is in $\mathsf{M3}_\top$*, and has the form described in Proposition 4.4.* As our main concern in this book is paraconsistent reasoning, from now on we concentrate in this chapter on matrices of this form. In this and the next section we give some general characterizations of logics which are induced by matrices in this family, with particular emphasis on those which are actually $\neg$-contained in classical logic (and not just $\neg$-coherent with it).

Proposition 4.5 *There are exactly two possible interpretations of $\neg$ in a three-valued $\neg$-paraconsistent matrix:*

- *Kleene's negation [194]: $\tilde{\neg} t = f$, $\tilde{\neg} f = t$, $\tilde{\neg} \top = \top$.*
- *Sette's negation [265]: $\tilde{\neg} t = f$, $\tilde{\neg} f = t$, $\tilde{\neg} \top = t$.*

Proof. Immediate from Proposition 4.4. □

Corollary 4.6 *Let $\mathcal{M}$ be a three-valued $\neg$-paraconsistent matrix. Then $\neg$ is complete for $\mathbf{L}_\mathcal{M}$ and left-involutive in it.*

---
[2]This notation reflects the intuitive meaning of the constant as representing 'both' truth and falsity.

Proof. That $\neg$ is complete for $\mathbf{L}_{\mathcal{M}}$ follows from Corollary 3.44. That $\neg\neg\varphi \vdash_{\mathbf{L}_{\mathcal{M}}} \varphi$ easily follows from Proposition 4.5. □

Proposition 4.7 *Let $\mathcal{M} = \langle \mathcal{V}, \mathcal{D}, \mathcal{O} \rangle$ be a paraconsistent three-valued matrix. Then $\mathbf{L}_{\mathcal{M}}$ is non-exploding iff every connective of $\mathcal{M}$ is $\{\top\}$-closed.*

Proof. This is a particular case of Proposition 3.59. □

Proposition 4.8 *Let $\mathcal{M} = \langle \mathcal{V}, \mathcal{D}, \mathcal{O} \rangle$ be a paraconsistent 3-valued matrix.*

1. *A connective $\supset$ is a semi-implication for $\mathbf{L}_{\mathcal{M}}$ iff it is an $\mathcal{M}$-implication.*

2. *A connective $\supset$ is an implication for $\mathbf{L}_{\mathcal{M}}$ iff it is an $\mathcal{M}$-implication.*

3. *A connective $\wedge$ is a conjunction for $\mathbf{L}_{\mathcal{M}}$ iff it is an $\mathcal{M}$-conjunction.*

4. *A connective $\vee$ is a disjunction for $\mathbf{L}_{\mathcal{M}}$ iff it is an $\mathcal{M}$-disjunction.*

Proof. By Propositions 4.4, 3.36, and Footnote 5 in Chapter 3. □

Corollary 4.9 *Let $\mathcal{M} = \langle \mathcal{V}, \mathcal{D}, \mathcal{O} \rangle$ be a paraconsistent three-valued matrix. A connective $\supset$ is a semi-implication for $\mathbf{L}_{\mathcal{M}}$ iff it is an implication for $\mathbf{L}_{\mathcal{M}}$.*

Corollary 4.10 *If $\mathcal{M}$ is a paraconsistent 3-valued matrix then $\mathbf{L}_{\mathcal{M}'}$ is paraconsistent for every extension $\mathcal{M}'$ of $\mathcal{M}$, and if $\supset$ (respectively, $\vee$, $\wedge$) is a (semi-)implication (respectively, a disjunction, conjunction) for $\mathbf{L}_{\mathcal{M}}$, then it is also an implication (respectively, a disjunction, conjunction) for $\mathbf{L}_{\mathcal{M}'}$.*

Proof. Immediate from Proposition 4.8 and Corollaries 3.33 and 4.9. □

Corollary 4.11 *If $\mathcal{M}$ is a paraconsistent three-valued matrix then $\mathbf{L}_{\mathcal{M}}$ is semi-normal.*

Proof. By definition of paraconsistency, if $\mathcal{M}$ is a paraconsistent then it has a paraconsistent semi-normal fragment. Hence the claim follows from Corollary 4.10. □

Corollary 4.12 *Let $\mathcal{M}$ be a paraconsistent three-valued matrix.*

1. *If $\mathbf{L}_{\mathcal{M}}$ is $\mathbf{F}$-contained in classical logic, and $\supset$ (respectively, $\vee, \wedge$) is a (semi-)implication (respectively, a disjunction, conjunction) for $\mathbf{L}_{\mathcal{M}}$, then $\mathbf{F}(\supset)$ (respectively, $\mathbf{F}(\vee)$, $\mathbf{F}(\wedge)$) is the classical implication (respectively, disjunction, conjunction).*

2. *If the connective $\supset$ is a (semi-)implication for $\mathbf{L}_{\mathcal{M}}$, then the $\{\supset\}$-fragment of $\mathbf{L}_{\mathcal{M}}$ is identical to the $\{\supset\}$-fragment of $\mathbf{CL}$.*

4.2 Paraconsistent Three-Valued Matrices

3. If $\mathcal{M}$ is normal then the $\{\supset, \vee, \wedge\}$-fragment of $\mathbf{L}_{\mathcal{M}}$ (where $\supset, \vee,$ and $\wedge$ are respectively implication, disjunction, and conjunction for $\mathbf{L}_{\mathcal{M}}$) is identical to $\mathbf{CL}^+$.

Proof. Immediate from Proposition 4.8 and Corollaries 3.34, 3.37, 4.9. □

Corollary 4.13 *Let $\mathcal{M}$ be a three-valued pre-$\neg$-paraconsistent matrix for a language $\mathcal{L}$, and suppose that $\mathcal{L}_{CL} \subseteq \mathcal{L}$.*

1. *If $\mathbf{L}_{\mathcal{M}}$ is $\neg$-contained in classical logic and normal, with the connectives $\wedge, \vee$ and $\supset$ of $\mathcal{L}_{CL}$ being conjunction, disjunction and implication (respectively) for it, then it is a classical $\neg$-paraconsistent logic.*

2. *$\mathbf{L}_{\mathcal{M}}$ is classical $\neg$-paraconsistent logic (Definition 2.22) iff it is $\neg$-contained in classical logic, and $\tilde{\wedge}, \tilde{\vee},$ and $\tilde{\supset}$ are $\mathcal{M}$-conjunction, $\mathcal{M}$-disjunction, and $\mathcal{M}$-implication (respectively).*

Proof. Theorem 3.17, Propositions 2.23, 3.32, 4.8, and Corollary 4.12. □

Our main next goal is to find out can be the possible interpretations of the basic connectives in the framework of three-valued paraconsistent matrices. For this we need the next two important results.

Theorem 4.14 *Let $\mathcal{M} = \langle \mathcal{V}, \mathcal{D}, \mathcal{O} \rangle$ be a three-valued $\neg$-paraconsistent matrix, where $\neg$ is Kleene's negation. If $\mathcal{M}$ is $\neg$-contained in classical logic then $\mathcal{M}$ has an $\mathcal{M}$-disjunction.*

Proof. By Corollary 4.11, there are three possible cases to consider.

- Suppose that $\mathbf{L}_{\mathcal{M}}$ has a disjunction connective $\vee$. Then $\vee$ is also an $\mathcal{M}$-disjunction by Proposition 4.8.

- Suppose that $\mathbf{L}_{\mathcal{M}}$ has a (semi-)implication. By Proposition 4.8 this implies that $\mathcal{M}$ has an $\mathcal{M}$-implication. Hence $\mathcal{M}$ has an $\mathcal{M}$-disjunction too (Note 3.30).

- Suppose that $\mathbf{L}_{\mathcal{M}}$ has a conjunction $\wedge$. Then by Proposition 4.8 $\wedge$ is an $\mathcal{M}$-conjunction, and so we have:

$$(*) \quad a \tilde{\wedge} b = f \text{ iff either } a = f \text{ or } b = f.$$

First, we prove that $t \tilde{\wedge} t = t$. Assume otherwise. Then $t \tilde{\wedge} t = \top$ by $(*)$ above. Hence, if $\nu \in \Lambda_{\mathcal{M}}$ then $\nu(\neg(p \wedge p)) \in \mathcal{D}$ in case $\nu(p) = t$. By $(*)$ again, this implies that $\nu(\neg(p \wedge p)) \in \mathcal{D}$ in case $\nu(p) \in \{t, f\}$. On the other hand, if $\nu(p) = \top$ then $\nu(p) = \nu(\neg p)$, and so $\nu(\neg(p \wedge$

$p)) = \nu(\neg(p \wedge \neg p))$. It follows that $\neg(p \wedge \neg p) \vdash_{\mathcal{M}} \neg(p \wedge p)$. Since $\mathcal{M}$ is $\neg$-contained in classical logic, this and Corollary 3.34 imply that $\neg(p \wedge \neg p) \vdash_{CL} \neg(p \wedge p)$, which is false.

Next we show that using $\neg$ and $\wedge$, it is possible to define in $\mathcal{L}$ an $\mathcal{M}$-disjunction $\vee$. We have two cases to consider:

- $\mathsf{T} \tilde{\wedge} \mathsf{T} = t$:
  In this case we take $\varphi \vee \psi =_{Df} \neg(\neg(\varphi \wedge \varphi) \wedge \neg(\psi \wedge \psi))$. The fact that $t \tilde{\wedge} t = \mathsf{T} \tilde{\wedge} \mathsf{T} = t$ and (*) easily imply that this formula has the required property.

- $\mathsf{T} \tilde{\wedge} \mathsf{T} = \mathsf{T}$:
  In this case we first let $\mathbf{t}_{\varphi,\psi}$ abbreviate $\neg(\varphi \wedge \neg \varphi \wedge \psi \wedge \neg \psi)$ (where association of conjunction is taken to the right). Then $\nu(\mathbf{t}_{\varphi,\psi}) = \mathsf{T}$ in case that $\nu(\varphi) = \nu(\psi) = \mathsf{T}$, and $\nu(\mathbf{t}_{\varphi,\psi}) = t$ otherwise. Now, we take:

  $$\varphi \vee \psi =_{Df} \neg(\neg(\mathbf{t}_{\varphi,\psi} \wedge \varphi \wedge \mathbf{t}_{\varphi,\psi}) \wedge \neg(\mathbf{t}_{\varphi,\psi} \wedge \psi \wedge \mathbf{t}_{\varphi,\psi})).$$

  * Suppose first that $\nu(\varphi) = \nu(\psi) = f$. Since for every x, $x \tilde{\wedge} f = f \tilde{\wedge} x = f$, we have that $\nu(\mathbf{t}_{\varphi,\psi} \wedge \varphi \wedge \mathbf{t}_{\varphi,\psi}) = \nu(\mathbf{t}_{\varphi,\psi} \wedge \psi \wedge \mathbf{t}_{\varphi,\psi}) = f$. Since $\tilde{\neg} f = t$, $t \tilde{\wedge} t = t$, and $\tilde{\neg} t = f$, it follows that in this case $\nu(\varphi \vee \psi) = f$.
  * Suppose that $\nu(\varphi) = t$. Then $\nu(\mathbf{t}_{\varphi,\psi}) = t$. Since $t \tilde{\wedge} t = t$, $\nu(\mathbf{t}_{\varphi,\psi} \wedge \varphi \wedge \mathbf{t}_{\varphi,\psi}) = t$. Again, since $\tilde{\neg} t = f$, $f \tilde{\wedge} x = f$, and $\tilde{\neg} f = t$, we conclude that in this case $\nu(\varphi \vee \psi) = t$.
  * Suppose that $\nu(\psi) = t$. Then again $\nu(\mathbf{t}_{\varphi,\psi}) = t$. Like in the previous case, this implies that $\nu(\varphi \vee \psi) = t$.
  * Suppose that $\nu(\varphi) = \nu(\psi) = \mathsf{T}$. Then $\nu(\mathbf{t}_{\varphi,\psi}) = \mathsf{T}$. Since $\tilde{\neg} \mathsf{T} = \mathsf{T}$ and $\mathsf{T} \tilde{\wedge} \mathsf{T} = \mathsf{T}$, $\nu(\sigma) = \mathsf{T}$ for every sub-formula σ of $\varphi \vee \psi$. Hence $\nu(\varphi \vee \psi) = \mathsf{T}$ as well.
  * Suppose that $\nu(\varphi) = f$, $\nu(\psi) = \mathsf{T}$. Then $\nu(\mathbf{t}_{\varphi,\psi}) = t$, and so $\nu(\varphi \vee \psi) = \tilde{\neg}(t \tilde{\wedge} \tilde{\neg}((t \tilde{\wedge} \mathsf{T}) \tilde{\wedge} t))$. If $(t \tilde{\wedge} \mathsf{T}) \tilde{\wedge} t) = t$ (which is the case if either $t \tilde{\wedge} \mathsf{T} = t$ or $\mathsf{T} \tilde{\wedge} t = t$) then $\nu((\varphi \vee \psi) = t$, and if $(t \tilde{\wedge} \mathsf{T}) \tilde{\wedge} t) = \mathsf{T}$ ((which is the case if $t \tilde{\wedge} \mathsf{T} = \mathsf{T} \tilde{\wedge} t = \mathsf{T}$) then $\nu((\varphi \vee \psi) = \mathsf{T}$. In both cases we are done.
  * The case where $\nu(\varphi) = \mathsf{T}$ and $\nu(\psi) = f$ is similar to the previous case. □

Note 4.15 Theorem 4.14 is not always true in case $\neg$ is Sette's negation. Thus it can be shown that no disjunction is available in the $\{\neg, \wedge\}$-fragment

4.2 Paraconsistent Three-Valued Matrices

of the logic **PE3**, which is studied in Section 4.4.6 below. (That logic uses Sette's negation and Kleene's conjunction.[3])

Theorem 4.16 *Let $\mathcal{M} = \langle \mathcal{V}, \mathcal{D}, \mathcal{O} \rangle$ be a three-valued $\neg$-paraconsistent matrix. $\mathcal{M}$ is $\neg$-contained in classical logic iff it is $\neg$-subclassical.*

Proof. By Propositions 3.47 and 4.4, we only need to show that if $\mathcal{M}$ is $\neg$-contained in classical logic then $\mathcal{M}$ is classically closed. We show this for each of the two possible cases given by Proposition 4.5.

$\tilde{\neg}_\mathcal{M}$ **is Kleene's negation:** The claim immediately follows in this case from Theorems 4.14 and 3.51.

$\tilde{\neg}_\mathcal{M}$ **is Sette's negation:** Assume for contradiction that $\mathcal{M}$ is not classically closed. Then $\tilde{\neg}\top = t$, and there is an n-ary connective $\diamond$ and elements $a_1, \ldots, a_n \in \{t, f\}$ such that $\tilde{\diamond}(a_1, \ldots, a_n) = \top$. For $i = 1, \ldots, n$ let $r_i = P_i$ if $a_i = t$, $r_i = \neg P_i$ if $a_i = f$. Then for every valuation $\nu \in \Lambda_\mathcal{M}$, if $\nu(P_i) = t$ for every $1 \le i \le n$ then $\nu(\diamond(r_1, \ldots, r_n)) = \top$. Let now $S = \{\neg\neg P_1, \neg\neg P_2, \ldots \neg\neg P_n\}$. Then $\nu \models_\mathcal{M} S$ iff $\nu(P_1) = \ldots = \nu(P_n) = t$. Hence $S \vdash_\mathcal{M} \diamond(r_1, \ldots, r_n)$ and $S \vdash_\mathcal{M} \neg \diamond(r_1, \ldots, r_n)$. Since $\mathcal{M}$ is $\neg$-contained in classical logic, $S \vdash_{2_\mathbf{F}} \diamond(r_1, \ldots, r_n)$ and $S \vdash_{2_\mathbf{F}} \neg \diamond(r_1, \ldots, r_n)$ for some bivalent $\neg$-interpretation $\mathbf{F}$ for the language of $\mathcal{M}$. This means that S is classically unsatisfiable, but this is false. □

Theorem 4.17 *There are exactly 2^{13} distinct three-valued paraconsistent matrices $\mathcal{M}$ for the language $\mathcal{L}_{CL}$ such that $\mathbf{L}_\mathcal{M}$ is $\neg$-contained in classical logic and normal, with the connectives $\wedge$, $\vee$ and $\supset$ of $\mathcal{L}_{CL}$ being conjunction, disjunction and implication (respectively) for it. The corresponding matrices are those that belong to the following family* **8Kb** *of matrices from [113, 116][4] (where the notation '$x \wr y$' means that x and y are two optional values):*

$\tilde{\wedge}$	t	f	$\top$
t	t	f	$t \wr \top$
f	f	f	f
$\top$	$t \wr \top$	f	$t \wr \top$

$\tilde{\vee}$	t	f	$\top$
t	t	t	$t \wr \top$
f	t	f	$t \wr \top$
$\top$	$t \wr \top$	$t \wr \top$	$t \wr \top$

$\tilde{\supset}$	t	f	$\top$
t	t	f	$t \wr \top$
f	t	t	$t \wr \top$
$\top$	$t \wr \top$	f	$t \wr \top$

	$\tilde{\neg}$
t	f
f	t
$\top$	$t \wr \top$

---
[3] This is actually the only exception in the language of $\{\neg, \wedge\}$.
[4] In [113, 116] the language is extended with a consistency operator $\circ$, defined by $\tilde{\circ}t = t$, $\tilde{\circ}f = t$, and $\tilde{\circ}\top = f$. We shall return to this extended language later (see Note 4.63).

Proof. Propositions 4.5, 4.8, and Theorem 4.16 imply that the matrices in 8Kb exhaust all the possible cases. On the other hand, every matrix $\mathcal{M}$ in 8Kb is obviously subclassical, and the connectives $\wedge$, $\vee$ and $\supset$ are $\mathcal{M}$-conjunction, $\mathcal{M}$-disjunction and $\mathcal{M}$-implication (respectively). That such $\mathcal{M}$ has indeed the required properties follow therefore from Propositions 3.55 (first part), 3.32 and 3.47. □

Corollary 4.18 *Let $\mathcal{M}$ be a three-valued pre-$\neg$-paraconsistent matrix for a language $\mathcal{L}$, and suppose that $\mathcal{L}_{CL} \subseteq \mathcal{L}$. $\mathbf{L}_\mathcal{M}$ is classical $\neg$-paraconsistent logic iff $\mathcal{M}$ is a classically closed extension of some matrix in* 8Kb.

Proof. Immediate from Corollary 4.13 and Theorem 4.17. □

Proposition 4.19 *The logics induced by the matrices in* 8Kb *are all distinct. In other words: different matrices in* 8Kb *induce different logics.*

Proof. Suppose, for instance, that $\vdash_1$ and $\vdash_2$ are two consequence relations which are induced by two matrices in 8Kb which have different interpretations for $\vee$. Below we check the possible cases for such different interpretations and show that in each case the logics that are obtained are indeed different. First, note that the two matrices coincide on $a \tilde{\vee} b$ whenever $a \in \{t, f\}$ and $b \in \{t, f\}$. Now,

1. Suppose that $\top \tilde{\vee}_1 \top = \top$ while $\top \tilde{\vee}_2 \top = t$.
   Then $p, \neg p, q, \neg q \vdash_1 \neg (p \vee q)$, but this is not true for $\vdash_2$ (since in both cases all the models of the right-hand side assign $\top$ to p and q).

2. Suppose $\top \tilde{\vee}_1 \top = \top \tilde{\vee}_2 \top \in \{t, \top\}$, and $f \tilde{\vee}_1 \top = \top$ while $f \tilde{\vee}_2 \top = t$.
   Then $q, \neg q, \neg (p \vee q) \vdash_2 p$ (a model of the right-hand side must assign $\top$ to q, and since $f \tilde{\vee}_2 \top = t$ it cannot assign f to p), while this is not true for $\vdash_1$ (a counter-model in this case assigns f to p and $\top$ to q).

3. The remaining cases are dual to the ones in the previous cases.

We leave to the reader the cases in which two matrices in 8Kb differ in their interpretations for one of the other connectives. □

Corollary 4.20 *There are exactly 2^{13} distinct classical paraconsistent logics in the language $\mathcal{L}_{CL}$. The corresponding matrices are those that belong to the family* 8Kb.

Proof. By Theorem 4.17, Proposition 4.19, and Corollary 4.18. □

Theorem 4.21 *Let $\mathcal{M}$ be a 3-valued matrix for a language with $\neg$.*

4.2 Paraconsistent Three-Valued Matrices

1. $\mathcal{M}$ induces a $\neg$-paraconsistent logic which is $\neg$-contained in classical logic iff it is isomorphic to a matrix $\langle\{t, f, \top\}, \{t, \top\}, \mathcal{O}\rangle$ in M3$_\top$ which satisfies the following conditions:

    (a) It has as its interpretation of $\neg$ one of the two tables for $\neg$ given in Theorem 4.17;

    (b) It has a (possibly definable) connective whose interpretation is either one of the 2^3 possible interpretations for conjunction ($\wedge$) given in Theorem 4.17, or one of the 2^5 interpretations for disjunction ($\vee$) given there, or one of the 2^4 interpretations for (semi-)implication ($\supset$) given there;

    (c) All its connectives are classically closed (i.e., $\{t, f\}$-closed).

2. $\mathcal{M}$ induces a $\neg$-paraconsistent logic iff it is isomorphic to a matrix $\langle\{t, f, \top\}, \{t, \top\}, \mathcal{O}\rangle$ in M3$_\top$ which satisfies (a) and (b) above.

3. $\mathcal{M}$ induces a normal $\neg$-paraconsistent logic which is $\neg$-contained in classical logic iff it is isomorphic to a matrix $\langle\{t, f, \top\}, \{t, \top\}, \mathcal{O}\rangle$ in M3$_\top$ which satisfies the following conditions:

    (a) It is an extension of one of the matrices in **8Kb**. More precisely:

    - It has as its interpretation of $\neg$ one of the two tables for $\neg$ given in Theorem 4.17;
    - It has a (possibly definable) connective whose interpretation is one of the 8 interpretations for $\wedge$ given in Theorem 4.17, a connective whose interpretation is one of the 32 interpretations for $\vee$ given there, and a connective whose interpretation is one of the 16 interpretations for $\supset$ given there.

    (b) All its connectives are classically closed (i.e., $\{t, f\}$-closed).

Proof. That the conditions in all Parts are necessary again follows from Propositions 4.5, 4.8, and Theorem 4.16. That they are sufficient follows from Propositions 3.47, 3.55 (first part), 3.32, and Corollary 4.10. □

Note 4.22 Although all the logics which are induced by the matrices in the family **8Kb** are different from each other, some of them have the same expressive power. For instance, in any paraconsistent matrix for the language of $\{\neg, \supset\}$ in which $\tilde\neg\top = \top$ and $\tilde\supset$ is given by D'ottaviano implication described in Section 4.1 (following Definition 3.38), the formulas $\neg(\varphi \supset \neg\psi)$ and $\neg(\psi \supset \neg\varphi)$ define two different conjunctions. Hence, the corresponding matrices in the family **8Kb** are equivalent in their expressive power.

4.3 Properties of Paraconsistent 3-Valued Logics

In this section we show that every three-valued paraconsistent logic is boldly (and so strongly) paraconsistent and strongly maximal, that every three-valued paraconsistent logic which is ¬-contained in classical logic is necessarily fully maximal (see Sections 2.2 and 3.2.3), and that a lot of these logics have all the properties listed in the desiderata list given in Chapter 2.

Theorem 4.23 *If $\mathcal{M}$ is a three-valued paraconsistent matrix then $\mathbf{L}_\mathcal{M}$ is boldly (and so strongly) paraconsistent and strongly maximal.*

Proof. Let $\mathcal{M}$ be a three-valued paraconsistent matrix for a language $\mathcal{L}$. That $\mathbf{L}_\mathcal{M}$ is boldly (and so strongly) paraconsistent follows from Corollary 3.56. In addition, Theorem 4.21 implies that $\mathcal{M}$ has a classically closed binary connective $\diamond$ (from those listed in Theorem 4.17) which is either an $\mathcal{M}$-disjunction, or an $\mathcal{M}$-conjunction, or an $\mathcal{M}$-implication. Let $\Psi(p)$ be $\neg p \diamond p$ in the first case, $\neg(\neg p \diamond p)$ in the second one, and $p \diamond p$ in the third. Then $\nu(\Psi(p)) = t$ for every $\nu \in \Lambda_\mathcal{M}$ such that $\nu(p) \neq \top$.

Now, let $\langle \mathcal{L}, \vdash \rangle$ be a proper extension of $\mathbf{L}_\mathcal{M}$ by some set of rules. We show that $\langle \mathcal{L}, \vdash \rangle$ is not pre-paraconsistent. Let Γ be a finite theory and ψ a formula in $\mathcal{L}$ such that $\Gamma \vdash \psi$ but $\Gamma \not\vdash_\mathcal{M} \psi$. In particular, there is a valuation $\nu \in mod_\mathcal{M}(\Gamma)$ such that $\nu(\psi) = f$. Consider the substitution θ, defined for every $p \in \mathsf{Var}(\Gamma \cup \{\psi\})$ by

$$\theta(p) = \begin{cases} P_2 & \text{if } \nu(p) = t, \\ \neg P_2 & \text{if } \nu(p) = f, \\ P_1 & \text{if } \nu(p) = \top. \end{cases}$$

Note that $\theta(\Gamma)$ and $\theta(\psi)$ contain (at most) the variables P_1, P_2; and that $\mu(\theta(\phi)) = \nu(\phi)$ for every valuation $\mu \in \Lambda_\mathcal{M}$ such that $\mu(P_1) = \top$ and $\mu(P_2) = t$, and every formula ϕ such that $\mathsf{Var}(\{\phi\}) \subseteq \mathsf{Var}(\Gamma \cup \{\psi\})$. Thus,

($\star$) Any $\mu \in \Lambda_\mathcal{M}$ such that $\mu(P_1) = \top$ and $\mu(P_2) = t$ is an $\mathcal{M}$-model of $\theta(\Gamma)$ but not of $\theta(\psi)$.

Now, consider the following two cases:

Case I. There is a formula $\phi(p,q)$ (i.e. $\mathsf{Var}(\phi) = \{p, q\}$, where $p \neq q$) such that for every $\mu \in \Lambda_\mathcal{M}$, $\mu(\phi) \neq \top$ if $\mu(p) = \mu(q) = \top$.
In this case, let $\mathsf{tt} = \Psi(\phi(P_1, P_1))$. Note that $\mu(\mathsf{tt}) = t$ for every $\mu \in \Lambda_\mathcal{M}$ such that $\mu(P_1) = \top$. Now, as $\vdash$ is structural, $\Gamma \vdash \psi$ implies that

$$\theta(\Gamma)\,[\mathsf{tt}/P_2] \vdash \theta(\psi)\,[\mathsf{tt}/P_2]. \qquad (4.1)$$

Also, by the above property of tt and ($\star$), any $\mu \in \Lambda_\mathcal{M}$ for which $\mu(P_1) = \top$ is a model of $\theta(\Gamma)\,[\mathsf{tt}/P_2]$ but does not $\mathcal{M}$-satisfy $\theta(\psi)\,[\mathsf{tt}/P_2]$. Thus,

4.3 Properties of Paraconsistent 3-Valued Logics 95

- $P_1, \neg P_1 \vdash_{\mathcal{M}} \theta(\gamma) [\text{tt}/P_2]$ for every $\gamma \in \Gamma$. As $\langle \mathcal{L}, \vdash \rangle$ is stronger than $\langle \mathcal{L}, \vdash_{\mathcal{M}} \rangle$, this implies that

$$P_1, \neg P_1 \vdash \theta(\gamma) [\text{tt}/P_2] \text{ for every } \gamma \in \Gamma. \tag{4.2}$$

- Since the set $\{P_1, \neg P_1, \theta(\psi)[\text{tt}/P_2]\}$ is not $\mathcal{M}$-satisfiable, we have that $P_1, \neg P_1, \theta(\psi) [\text{tt}/P_2] \vdash_{\mathcal{M}} P_2$. Again this implies that

$$P_1, \neg P_1, \theta(\psi) [\text{tt}/P_2] \vdash P_2. \tag{4.3}$$

By (4.1)–(4.3), $P_1, \neg P_1 \vdash P_2$, thus $\langle \mathcal{L}, \vdash \rangle$ is not pre-paraconsistent.

Case II. For every formula $\phi(p, q)$ and for every $\mu \in \Lambda_{\mathcal{M}}$, if $\mu(p) = \mu(q) = \top$ then $\mu(\phi) = \top$.

Again, as $\vdash$ is structural, and since $\Gamma \vdash \psi$,

$$\theta(\Gamma) [\Psi(P_2)/P_2] \vdash \theta(\psi) [\Psi(P_2)/P_2]. \tag{4.4}$$

In addition, ($\star$) above entails that a valuation $\mu \in \Lambda_{\mathcal{M}}$ such that $\mu(P_1) = \top$ and $\mu(P_2) \in \{t, f\}$ is a model of $\theta(\Gamma) [\Psi(P_2)/P_2]$ which is not a model of $\theta(\psi) [\Psi(P_2)/P_2]$. Thus, the only $\mathcal{M}$-model of $\{P_1, \neg P_1, \theta(\psi) [\Psi(P_2)/P_2]\}$ is the one in which both of P_1 and P_2 are assigned the value $\top$. It follows that $P_1, \neg P_1, \theta(\psi) [\Psi(P_2)/P_2] \vdash_{\mathcal{M}} P_2$. Thus,

$$P_1, \neg P_1, \theta(\psi) [\Psi(P_2)/P_2] \vdash P_2. \tag{4.5}$$

By using ($\star$) again (for $\mu(P_2) \in \{t, f\}$) and the condition of case II (for $\mu(P_2) = \top$), we have:

$$P_1, \neg P_1 \vdash \theta(\gamma) [\Psi(P_2)/P_2] \text{ for every } \gamma \in \Gamma. \tag{4.6}$$

Again, by (4.4)–(4.6) above we have that $P_1, \neg P_1 \vdash P_2$, and so $\langle \mathcal{L}, \vdash \rangle$ is not pre-paraconsistent in this case either. $\square$

Theorem 4.24 *Let $\mathcal{M}$ be a three-valued $\neg$-paraconsistent matrix which is $\neg$-contained in classical logic. Then $\mathbf{L}_{\mathcal{M}}$ is fully maximal.*

Proof. Suppose that $\mathcal{M}$ satisfies the conditions of the theorem. Then $\mathbf{L}_{\mathcal{M}}$ is strongly maximal by Theorem 4.23. It remains to show that it is maximally paraconsistent relative to classical logic. By Theorem 4.16 and Proposition 3.49, $\mathbf{L}_{\mathcal{M}}$ is $\mathbf{F}_{\mathcal{M}}$-contained in classical logic. We end by showing that $\mathbf{L}_{\mathcal{M}}$ is $\mathbf{F}_{\mathcal{M}}$-maximal relative to classical logic. So let ψ' be a classical $\mathbf{F}_{\mathcal{M}}$-tautology not provable in $\mathbf{L}_{\mathcal{M}}$, and let $\mathcal{S}'^*$ be the set of all of its substitution instances. Let $\mathbf{L}'^*$ be the logic obtained by adding ψ' as a new axiom to $\mathbf{L}_\mathcal{M}$. Then for every theory $\mathcal{T}$ we have that $\mathcal{T} \vdash_{\mathbf{L}'^*} \phi$ iff $\mathcal{T}, \mathcal{S}'^* \vdash_\mathcal{M} \phi$. Since $\mathbf{L}_\mathcal{M}$ is strongly maximal, it follows in particular that

$$\mathcal{S}^*, \varphi, \neg\varphi \vdash_\mathcal{M} \phi \text{ for every } \varphi, \phi. \tag{4.7}$$

Suppose for contradiction that there is some classical $\mathbf{F}_\mathcal{M}$-tautology σ not provable in $\mathbf{L}'^*$. Since $\nvdash_{\mathbf{L}'^*} \sigma$, also $\mathcal{S}'^* \nvdash_\mathcal{M} \sigma$. Hence there is a valuation $\nu \in \Lambda_\mathcal{M}$ which is a model of $\mathcal{S}'^*$, but $\nu(\sigma) = f$. Suppose that there is some φ such that $\nu(\varphi) = \top$. Since ν is a model of $\mathcal{S}'^*$, it is also a model of $\mathcal{S}'^* \cup \{\varphi, \neg\varphi\}$, and so by (4.7) it is a model of σ, in contradiction to the fact that $\nu(\sigma) = f$. Hence $\nu(\varphi) \in \{t, f\}$ for all φ, and so ν is an $\mathcal{M}_{\mathbf{F}_\mathcal{M}}$-valuation which assigns f to σ. This contradicts the fact that $\vdash_{\mathcal{M}_{\mathbf{F}_\mathcal{M}}} \sigma$. It follows that all classical $\mathbf{F}_\mathcal{M}$-tautologies are provable in $\mathbf{L}'^*$, and so $\mathbf{L}_\mathcal{M}$ is $\mathbf{F}_\mathcal{M}$-maximal relative to classical logic. □

Corollary 4.25 *Let $\mathcal{M}$ be a 3-valued matrix for a language with $\neg$. $\mathbf{L}_\mathcal{M}$ is a fully maximal $\neg$-paraconsistent logic iff $\mathcal{M}$ is isomorphic to a matrix $\langle \{t, f, \top\}, \{t, \top\}, \mathcal{O}\rangle$ in $\mathsf{M3}_\top$ which satisfies the three conditions given in the first item of Theorem 4.21.*

Proof. Immediate from Theorems 4.21 and 4.24. □

Note 4.26 Let $\mathcal{M}$ be a three-valued paraconsistent matrix which is $\neg$-contained in classical logic. By Theorem 4.21, any three-valued extension of it which is obtained by enriching the language of $\mathcal{M}$ with extra classically closed connectives necessarily has the same properties. It follows from Theorem 4.24 that not only is $\mathbf{L}_\mathcal{M}$ fully maximal, but so must be all the logics which are induced by extensions of $\mathcal{M}$ with classically closed connectives.

Theorem 4.27 *Every logic which is induced by a three-valued paraconsistent matrix has a corresponding cut-free Gentzen-type system.*

Proof. By [59][5], this is true for any logic which is induced by a finite-valued matrix, provided that its language is *sufficiently expressive*, i.e. it suffices for characterizing the truth values of the matrix. This is the case here, since for every valuation ν in such a matrix we have that $\nu(\varphi) = t$ iff ν satisfies φ and does not satisfy $\neg\varphi$, $\nu(\varphi) = f$ iff ν satisfies $\neg\varphi$ and does not satisfy φ, and $\nu(\varphi) = \top$ iff ν satisfies both φ and $\neg\varphi$. (Note that this is true for both Kleene's negation and Sette's negation.) □

---

[5]In Section 4.5.1 we provide Gentzen-type systems for all the specific three-valued logics investigated in the next section, and directly prove (without relying on [59]) the cut-elimination theorem for them, as well as their soundness and completeness.

4.3 Properties of Paraconsistent 3-Valued Logics

Note 4.28 An algorithm for deriving the cut-free Gentzen-type system mentioned in the proof of Theorem 4.29 is provided in [59]. Actually, the algorithm applies to the bigger family of finite non-deterministic matrices which is described in Part III of the book. In the case of three-valued paraconsistent (non-deterministic) matrices it provides quasi-canonical systems.

The next theorem is perhaps the most important result of this chapter:

Theorem 4.29 *Every logic which is induced by one of the matrices in the family* **8Kb** *which employs Kleene's negation, or by a classically closed extension of such a matrix, has all the properties listed in the desiderata list given in Chapter 2.*

Proof. The negation connective $\neg$ of every such logic **L** is obviously involutive, and it is complete for **L** by Corollary 4.6. That each such logic is strongly paraconsistent, normal, fully maximal, and has a corresponding cut-free Gentzen-type system follows from Corollary 3.56, Theorem 4.21, Theorem 4.24, and Theorem 4.27 (respectively). □

Theorem 4.29 raises the question whether a similar theorem can be proved for the matrices in the family **8Kb** which employs Sette's negation. The problem here is that the negation $\neg$ of these matrices is complete and left-involutive (Corollary 4.6), but it is not right-involutive. However, the desiderata list given in Chapter 2 allows $\neg$ to be contrapositive rather than right-involutive. Our next goal is to show that this option is not available in the context of normal three-valued paraconsistent logics. Moreover: we prove something which is usually stronger (see Proposition 9.3): that such logics cannot be self-extensional. For this we need the next definition and proposition (which will be useful also in the next section).

Definition 4.30 Let $\mathbf{L} = \langle \mathcal{L}, \vdash_\mathbf{L} \rangle$ be a logic, where $\mathcal{L}$ includes the unary connective $\neg$. **L** is $\neg$-*extensional* if $\neg\varphi \vdash_\mathbf{L} \neg\psi$ whenever $\varphi \dashv\vdash_\mathbf{L} \psi$.

Proposition 4.31 *Let $\mathcal{M}$ be a three-valued paraconsistent matrix. The following conditions on $\mathcal{M}$ are equivalent:*

- *For every φ and ψ, if $\varphi \dashv\vdash_\mathcal{M} \psi$, then $\nu(\varphi) = \nu(\psi)$ for every $\mathcal{M}$-valuation ν.*

- $\mathbf{L}_\mathcal{M}$ *is self-extensional.*

- $\mathbf{L}_\mathcal{M}$ *is $\neg$-extensional.*

Proof. Obviously, the first condition implies the second, and the second implies the third. It remains to show that the third condition implies the first. So assume that $\mathbf{L}_\mathcal{M}$ is $\neg$-extensional, and suppose that $\nu(\varphi) \neq \nu(\psi)$ for some $\mathcal{M}$-valuation ν. If $\nu(\varphi) = f$ and $\nu(\psi) \neq f$ then $\psi \not\vdash_\mathcal{M} \varphi$. If $\nu(\varphi) = t$ and $\nu(\psi) \neq t$ then $\neg\psi \not\vdash_\mathcal{M} \neg\varphi$. Similarly, if $\nu(\psi) = f$ and $\nu(\varphi) \neq f$ then $\varphi \not\vdash_\mathcal{M} \psi$, while if $\nu(\psi) = t$ and $\nu(\varphi) \neq t$ then $\neg\varphi \not\vdash_\mathcal{M} \neg\psi$. It follows that in all cases the $\neg$-extensionality of $\mathbf{L}_\mathcal{M}$ implies that $\varphi \not\dashv\vdash_\mathbf{L} \psi$. □

Proposition 4.32 *Let $\mathcal{M}$ be a three-valued $\neg$-paraconsistent matrix, and suppose $\mathbf{L}_\mathcal{M}$ has an implication $\supset$. Then $\mathbf{L}_\mathcal{M}$ is not $\neg$-extensional.*

Proof. Assume for contradiction that $\mathbf{L}_\mathcal{M}$ is $\neg$-extensional. Since $\supset$ is an implication for $\mathbf{L}_\mathcal{M}$, $\vdash_\mathcal{M} \varphi \supset \varphi$ for every φ. Hence $P_1 \supset P_1 \dashv\vdash_\mathcal{M} P_2 \supset P_2$. Therefore it follows from Proposition 4.31 that $\nu(P_1 \supset P_1) = \nu(P_2 \supset P_2)$ for every $\mathcal{M}$-valuation ν. Since $t \widetilde{\supset} t = t$ by the second part of Theorem 4.21, this immediately implies that $a \widetilde{\supset} a = t$ for every $a \in \{t, f, \top\}$. From this it easily follows (again using Theorem 4.21) that if p is a variable then $p \dashv\vdash_\mathcal{M} \neg(p \supset \neg(p \supset p))$, while $\neg p \not\vdash_\mathcal{M} \neg\neg(p \supset \neg(p \supset p))$. (To see the latter, let $\nu(p) = \top$.) This contradicts the $\neg$-extensionality of $\mathbf{L}_\mathcal{M}$. □

Theorem 4.33 *There is no three-valued $\neg$-paraconsistent matrix $\mathcal{M}$ such that $\mathbf{L}_\mathcal{M}$ is normal and self-extensional, nor is there such $\mathcal{M}$ for which $\mathbf{L}_\mathcal{M}$ is normal while $\neg$ is contrapositive.*

Proof. Immediate from Proposition 4.32, since if $\mathbf{L}_\mathcal{M}$ is self-extensional, or if $\neg$ is contrapositive in it, then $\mathbf{L}_\mathcal{M}$ is obviously $\neg$-extensional. □

Note 4.34 A stronger form of Theorem 4.33, with a very similar proof, is given in [61]. It is shown there that no three-valued $\neg$-paraconsistent matrix $\mathcal{M}$ can have a connective $\supset$ such that $\vdash_\mathcal{M} \varphi \supset \varphi$ and $\varphi, \varphi \supset \psi \vdash_\mathcal{M} \psi$ for every φ and ψ.

4.4 Central Paraconsistent Three-Valued Logics

As shown in Theorem 4.21, there are exactly eight ways of defining conjunctions in three-valued paraconsistent matrices. Of these eight operations, only four are symmetric (and so make the formulas $\varphi \wedge \psi$ and $\psi \wedge \varphi$ congruent), and only three (all of them symmetric) seem to have some intuitive justification. These are: Sette's conjunction (Section 4.4.1), Sobociński's conjunction (Section 4.4.2), and Kleene's conjunction (described in the introduction to this chapter). To the best of our knowledge, they are indeed the only three three-valued conjunctions that have seriously been investigated in the literature on paraconsistent logics. In this section we examine

4.4 Central Paraconsistent Three-Valued Logics

the properties of the most important (and famous) paraconsistent logics in $\mathbf{3}_\top$ that are based on these three conjunctions and the two negations which are allowed by Proposition 4.5. In addition to historical importance (i.e. being investigated in the literature), our main criteria here for "importance" of three-valued paraconsistent matrices is having desirable properties (like normality or self-extensionality), or having a *natural* set of connectives that can be characterized by a combination of potentially desirable properties. Two examples of the latter are $\{t, f\}$-closure (which is equivalent to $\neg$-containment in classical logic by Theorem 4.16), and $\{\top\}$-closure (which is equivalent to being non-exploding by Proposition 4.7).

Note that by the results of the previous sections, *all* the logics studied here have the following properties (that will not be repeated in the sequel):

- They are boldly (and so strongly) paraconsistent (Theorem 4.23).
- They are strongly maximal paraconsistent logics (Theorem 4.23).
- They are semi-normal (Corollary 4.11).
- Their negation is complete and left-involutive (Corollary 4.6).

In addition, it follows from Theorem 4.27 that all these logics have corresponding cut-free Gentzen-type systems. In the next section this is shown *directly*, and we provide for each of them a quasi-canonical and cut-free Gentzen-type system, which is very close to the classical one.

4.4.1 The Logics $\mathbf{P}_1$, $\mathbf{P}_1^K$, and $\mathbf{P}_1^S$

Sette's logic $\mathbf{P}_1 = \langle \mathcal{L}_{\mathsf{P}_1}, \vdash_{\mathsf{P}_1} \rangle$ [265] is the logic that is induced by the matrix $\mathsf{P}_1 = \langle \{t, f, \top\}, \{t, \top\}, \{\tilde{\wedge}, \tilde{\neg}\} \rangle$,[6] where the operations are defined as follows:

$\tilde{\wedge}$	t	f	$\top$
t	t	f	t
f	f	f	f
$\top$	t	f	t

$\tilde{\neg}$	
t	f
f	t
$\top$	t

Proposition 4.35 $\mathbf{P}_1$ *is normal, $\neg$-contained in classical logic, and fully maximal.*

Proof. Define $\psi \vee \varphi = \neg(\neg(\psi \wedge \psi) \wedge \neg(\varphi \wedge \varphi))$ and $\psi \supset \varphi = \neg((\psi \wedge \psi) \wedge \neg(\varphi \wedge \varphi))$. The corresponding interpretations are the following:

$\tilde{\vee}$	t	f	$\top$
t	t	t	t
f	t	f	t
$\top$	t	t	t

$\tilde{\supset}$	t	f	$\top$
t	t	f	t
f	t	t	t
$\top$	t	f	t

---
[6]We note that in our notations $\mathbf{P}_1$ is also denoted $\mathbf{L}_{\mathsf{P}_1}$.

Therefore, Item 3 of Theorem 4.21 implies that $\mathbf{P_1}$ is a normal paraconsistent logic, which is $\neg$-contained in classical logic. That it is fully maximal follows from Theorem 4.24. □

Note 4.36 As a matrix for $\mathcal{L}_{CL}$, with the interpretations of $\vee$ and $\supset$ given in the proof of Proposition 4.35, $\mathbf{P_1}$ belongs to 8Kb. Hence, as a logic in $\mathcal{L}_{CL}$, $\mathbf{P_1}$ is classical paraconsistent logic (Corollary 4.18).

Note 4.37 To the best of our knowledge, $\mathbf{P_1}$ was the first paraconsistent logic for which a maximality property has been stated and proved (in [265]). Thus, it is frequently referred to as "Sette's maximal paraconsistent logic". However, the theorems of Section 4.3 show that there is nothing special about $\mathbf{P_1}$ in this respect.

Next, we characterize the expressive power of $\mathsf{P_1}$.

Proposition 4.38 *A function* $g : \{t, f, \top\}^n \to \{t, f, \top\}$ *is represented in* $\mathsf{P_1}$ *iff it is either a projection function (i.e.* $g(x_1, \ldots, x_n) = x_i$ *for some* $1 \leq i \leq n$*), or a function whose range is* $\{t, f\}$.

Proof. If ψ is a variable then F_ψ^n is a projection function. If ψ is a complex formula then the truth tables of $\mathsf{P_1}$ dictate that the range of F_ψ^n is $\{t, f\}$. Hence the condition is necessary.

Obviously, every projection function is represented in $\mathsf{P_1}$ by some variable. To prove the sufficiency of the condition, it remains to show that every function $g : \{t, f, \top\}^n \to \{t, f\}$ is represented in $\mathsf{P_1}$. For this define:

$$\psi_a(p) = \begin{cases} \neg\neg p & \text{if } a = t, \\ \neg(p \wedge p) & \text{if } a = f, \\ p \wedge \neg p & \text{if } a = \top. \end{cases}$$

It is easy to check that if ν is a valuation in $\mathbf{P_1}$, then $\nu(\psi_a(p)) = t$ if $\nu(p) = a$, and $\nu(\psi_a(p)) = f$ if $\nu(p) \neq a$. Now given a function $g : \{t, f, \top\}^n \to \{t, f\}$, it is not difficult to see that in case $n > 0$, g is represented in $\mathsf{P_1}$ by the disjunction (as defined in the proof of Proposition 4.35) of all the formulas of the form $\psi_{a_1}(P_1) \wedge \cdots \wedge \psi_{a_n}(P_n)$ such that $g(a_1, a_2, \ldots, a_n) = t$ (and by $\neg\neg P_1 \wedge \neg P_1$ if no such $a_1, a_2, \ldots, a_n$ exist). In addition, the truth-values f and t are represented in $\mathsf{P_1}$ by $\neg\neg P_1 \wedge \neg P_1$ and $\neg P_1 \vee P_1$ (respectively). □

Corollary 4.39 *Kleene's negation is not definable in* $\mathbf{P_1}$. *Hence Sette's negation is the only negation which is definable in it.*

Proof. This is immediate from Propositions 4.38 and 4.5. □

4.4 Central Paraconsistent Three-Valued Logics 101

Corollary 4.40 *The operation $\tilde{\wedge}$ of P_1, and the operations $\tilde{\vee}$ and $\tilde{\supset}$ which were defined in the proof of Proposition 4.35, are the only conjunction, disjunction and implication (respectively) which are definable in $\mathbf{P}_1$.*

Proof. This easily follows from Propositions 4.38 and 4.8. □

Note 4.41 By Proposition 4.35, $\mathbf{P}_1$ has most of the properties listed in the desiderata list given in Chapter 2. However, $\mathbf{P}_1$ also has the following severe drawbacks, which might make it less attractive than other three-valued paraconsistent logics:

1. Since the only values that complex formulas can get in P_1 are t and f, $\mathbf{P}_1$ is paraconsistent only with respect to propositional variables (that is, for a complex formula ψ we have that $\psi, \neg \psi \vdash_{\mathsf{P}_1} \varphi$ for every φ).

2. The combination of conjunction and negation does not always behave in $\mathbf{P}_1$ as one might expect or wish. For example, $\neg p \not\vdash_{\mathsf{P}_1} \neg(p \wedge q)$.

3. Its negation is neither involutive, nor contrapositive. Hence it fails to have one of the properties of the above-mentioned list.

4. Although $\varphi \wedge \varphi \dashv\vdash_{\mathbf{P}_1} \varphi$, $\varphi \wedge \varphi \not\equiv_{\mathbf{P}_1} \varphi$ (and similarly for $\varphi \vee \varphi$). Hence $\mathbf{P}_1$ makes some particularly strange distinctions.

We next examine two extensions of $\mathbf{P}_1$ which are motivated by this list.

$\mathbf{P}_1^{\mathsf{K}}$. The third drawback in the list given in Note 4.41 can be overcome if we combine Sette's conjunction with Kleene's negation instead of Sette's negation. So let $\mathsf{P}_1^{\mathsf{K}}$ be the matrix for $\{\neg, \wedge\}$ which is defined like P_1, except that the interpretation of $\neg$ is taken to be Kleene's negation. Let $\mathbf{P}_1^{\mathsf{K}}$ be the logic which is induced by $\mathsf{P}_1^{\mathsf{K}}$. Since Sette's negation is represented in $\mathsf{P}_1^{\mathsf{K}}$ by the formula $\neg P_1 \wedge \neg P_1$, while Kleene's negation is not representable in P_1 (by Corollary 4.39), $\mathsf{P}_1^{\mathsf{K}}$ is an extension of P_1, which has greater expressive power. However, the difference is minor, since the proof of Proposition 4.38 can easily be adapted to show that the only functions that are representable in $\mathsf{P}_1^{\mathsf{K}}$ but not in P_1 are the functions of the form $\lambda x_1, \ldots, x_n.\tilde{\neg}x_i$, where $1 \leq i \leq n$ and $\tilde{\neg}$ is Kleene's negation. It follows that Corollary 4.40 is true also for $\mathsf{P}_1^{\mathsf{K}}$. It is easy to show that Proposition 4.35 and Note 4.36 are true for $\mathbf{P}_1^{\mathsf{K}}$ as well; that $\mathbf{P}_1^{\mathsf{K}}$ still has the second and fourth drawbacks of $\mathbf{P}_1$ from the list in Note 4.41; and that in addition to variables, it is paraconsistent only with respect to formulas of the form $\neg \cdots \neg p$.

$\mathbf{P}_1^{\mathsf{S}}$. The forth drawback in the list given in Note 4.41 can be overcome if instead of Sette's conjunction, we combine Sette's negation with either of the other two conjunctions we use. The combination of Sette's negation with Kleene's conjunction is studied in Section 4.4.6. Here we briefly discuss the matrix and logic which are obtained by using Sette's negation and Sobociński's conjunction. (See the truth table of the latter in Section 4.4.2. It amounts to letting $a \tilde{\wedge} b$ to be f iff $a = f$ or $b = f$, and $a \tilde{\wedge} b = \top$ iff $a = b = \top$.) So let P_1^S be the matrix for $\{\neg, \wedge\}$ which is defined like P_1, except that the interpretation of $\wedge$ is taken to be Sobociński's conjunction. Let $\mathbf{P}_1^\mathsf{S}$ be the logic which is induced by P_1^S. Since Sette's conjunction is represented in P_1^K by the formula $(P_1 \wedge P_2) \wedge \neg(\neg P_1 \wedge \neg\neg P_1)$, while Sobociński's conjunction is not representable in P_1 (by Corollary 4.40), P_1^S too is an extension of P_1 which has greater expressive power. However, the difference is again minor, since the proof of Proposition 4.38 can easily be adapted to show that the only functions that are representable in P_1^S but not in P_1 are the the generalized Sobociński's conjunctions, where a function $g : \{t, f, \top\}^n \to \{t, f, \top\}$ is a generalized Sobociński's conjunction iff there is a nonempty subset A of $\{1, \ldots, n\}$ such that $g(a_1, \ldots, a_n) = f$ iff there exists $i \in A$ such that $a_i = f$, and $g(a_1, \ldots, a_n) = \top$ iff $a_i = \top$ for every $i \in A$. It follows that Sette's negation, disjunction, and implication are the only negation, disjunction, and implication (respectively) that can be defined in P_1^S, and the conjunctions of Sette's and Sobociński's are the only conjunctions that are definable there. Again, it is easy to show that Proposition 4.35 and Note 4.36 are true for P_1^S; that $\mathbf{P}_1^\mathsf{S}$ still has the second and third drawbacks of $\mathbf{P}_1$ from the list in Note 4.41; and that in addition to variables, it is paraconsistent only with respect to conjunctions of variables.

4.4.2 The Logic SRM$_{\to}$

What we call 'Sobociński's conjunction' in the previous section was first (implicitly) used by Sobociński in ([273]). He introduced there the three-valued matrix $\mathcal{A}_1 = \langle \{t, f, \top\}, \{t, \top\}, \{\tilde{\wedge}, \tilde{\neg}\}\rangle$ [7], where $\tilde{\neg}$ and $\tilde{\wedge}$ are presented below. ($\tilde{\neg}$ is Kleene's negation, and $\tilde{\wedge}$ is Sobociński's conjunction.)

$\tilde{\wedge}$	t	f	$\top$
t	t	f	t
f	f	f	f
$\top$	t	f	$\top$

$\tilde{\neg}$	
t	f
f	t
$\top$	$\top$

For reasons that are explained in Section 13.1.4 (in Chapter 13), we denote $\mathbf{L}_{\mathcal{A}_1}$ (the logic that is induced by $\mathcal{A}_1$) by $\mathbf{SRM}_{\to}$ (or $\mathbf{SRMI}^1_{\to}$).

---

[7] In Part V of this book the language of $\mathcal{A}_1$ is taken to be the language of the purely intensional (or "multiplicative") fragments of the relevant logics, with $\otimes$ used instead of $\wedge$. Actually, the language is officially taken there to be $\{\neg, \otimes, \to\}$. However, this is an insignificant difference, since $\to$ is definable in all the logics considered there in terms of $\neg$ and $\otimes$, exactly like it is defined in terms of $\neg$ and $\wedge$ in Note 4.42 of this section.

4.4 Central Paraconsistent Three-Valued Logics

Note 4.42 The official language which was used in [273] (as well as in most of the literature on relevant logic) is $\{\neg, \rightarrow\}$, and the interpretation of $\rightarrow$ there was the following *Sobociński's entailment*:[8]

$$a \tilde{\rightarrow} b = \begin{cases} \top & \text{if } a = b = \top, \\ f & \text{if } a >_t b, \\ t & \text{otherwise.} \end{cases}$$

It is easy to see that $a \tilde{\rightarrow} b = \tilde{\neg}(a \tilde{\wedge} \tilde{\neg} b)$, while $a \tilde{\wedge} b = \tilde{\neg}(a \tilde{\rightarrow} \tilde{\neg} b)$. Hence, $\mathcal{A}_1$ is equivalent to Sobociński's original matrix.

Note 4.43 It should be emphasized that $\mathbf{SRM}_{\tilde{\rightarrow}}$ is *not* identical to the logic introduced by Sobociński in [273]. That logic has only been *motivated* by the matrix $\mathcal{A}_1$. What Sobociński actually did in [273] is to axiomatize *the set of valid sentences of* $\mathcal{A}_1$ using a Hilbert-type system with Modus Ponens for $\rightarrow$ as the single rule of inference. That system is called $\mathbf{RM}_{\tilde{\rightarrow}}$ in Chapter 13 (Section 13.1.4),[9] and it has the following properties:

1. As shown in Section 13.1.4, it is *weakly* complete for $\mathcal{A}_1$. (This is the reason why it is frequently called "Sobociński's three-valued logic".) Hence $\vdash_{\mathbf{RM}_{\tilde{\rightarrow}}} \varphi$ iff $\vdash_{\mathbf{SRM}_{\tilde{\rightarrow}}} \varphi$.

2. It is *not* (strongly) complete for $\mathcal{A}_1$, and so its consequence relation is different from that of $\mathbf{SRM}_{\tilde{\rightarrow}}$. An example which is shown in Section 13.2.2 (See Proposition 13.68): $\varphi \wedge \psi \vdash_{\mathbf{SRM}_{\tilde{\rightarrow}}} \varphi$, but $\varphi \wedge \psi \nvdash_{\mathbf{RM}_{\tilde{\rightarrow}}} \varphi$.[10]

3. It induces the purely intensional fragment of the semi-relevant logic $\mathbf{RM}$, which is investigated in Chapter 15. (See Theorem 15.16.)

4. It is maximally paraconsistent (in the *weak* sense), as well as maximally paraconsistent relative to classical logic (Section 13.1.4 again). On the other hand, $\mathbf{RM}_{\tilde{\rightarrow}}$ is *not* strongly maximal, as $\mathbf{SRM}_{\tilde{\rightarrow}}$ is a proper extension of it which is still paraconsistent. (See the second item above for a concrete example.)

The connective $\rightarrow$ introduced in Note 4.42 is *not* an implication for $\mathbf{SRM}_{\tilde{\rightarrow}}$, since the schema $\varphi \rightarrow (\psi \rightarrow \varphi)$ is not valid in $\mathcal{A}_1$. Still we have:

---

[8] In this definition, $a >_t b$ means that $a \nleq_t b$, where $\leq_t$ is the order relation introduced in Section 4.1, that is: $f \leq_t \top \leq_t t$.

[9] Our axiomatization of $\mathbf{RM}_{\tilde{\rightarrow}}$ is a slight variation of Sobociński's original system, but the two systems are equivalent, and the differences in their formulations are insignificant.

[10] Note that $\wedge$ refers here to the connective denoted by $\otimes$ in Part V, while $\wedge$ refers there to another connective.

Proposition 4.44 $\mathbf{SRM}_{\to}$ *is non-exploding, normal, $\neg$-contained in classical logic, and fully maximal.*

Proof. Obviously, $\wedge$ is an $\mathcal{A}_1$-conjunction. Define $\varphi \supset \psi = \varphi \to (\varphi \wedge \psi)$, where $\varphi \to \psi = \neg(\varphi \wedge \neg\psi)$. Then $\supset$ has in $\mathcal{A}_1$ the following interpretation:

$\tilde{\supset}$	t	f	$\top$
t	t	f	t
f	t	t	t
$\top$	t	f	$\top$

It follows that $\supset$ is an $\mathcal{A}_1$-implication. Therefore (by Note 3.30) $\mathcal{A}_1$ has an $\mathcal{A}_1$-disjunction as well. Hence, Item 3 of Theorem 4.21 implies that $\mathbf{SRM}_{\to}$ is a normal paraconsistent logic, which is $\neg$-contained in classical logic. The other two properties follow from Theorem 4.24 and Proposition 4.7. $\square$

Note 4.45 With $\tilde{\wedge}$ as the interpretation of $\wedge$, and the interpretations of $\vee$ and $\supset$ as given in the proof of Proposition 4.44, $\mathcal{A}_1$ belongs to 8Kb. Hence, $\mathbf{SRM}_{\to}$ can be viewed as a classical paraconsistent logic.

Next, we characterize the expressive power of $\mathcal{A}_1$.

Proposition 4.46 *A function is represented in $\mathcal{A}_1$ iff it is both $\{\top\}$-closed and $\{\top\}$-limited.*

Proof. It is easy to verify that all the functions that are represented in $\mathcal{A}_1$ have these properties. The converse is proved in Section 13.1.5 of this book (see Corollary 13.44). $\square$

Corollary 4.47 *Sette's negation is not definable in $\mathcal{A}_1$. Hence, Kleene's negation is the only negation which is definable in it.*

Proof. This is immediate from Propositions 4.46 and 4.5, since Sette's negation is not $\{\top\}$-closed. $\square$

Corollary 4.48 *The operation $\tilde{\wedge}$ of $\mathcal{A}_1$, and the operations $\tilde{\vee}$ and $\tilde{\supset}$ which were defined in the proof of Proposition 4.44, are the* only *conjunction, disjunction and implication (respectively) which are definable in $\mathbf{SRM}_{\to}$.*

Proof. This easily follows from Propositions 4.46 and 4.8. $\square$

More information about $\mathcal{A}_1$, $\mathbf{SRM}_{\to}$, and $\mathbf{RM}_{\to}$ is given in Section 4.5, and in Part V of this book.

4.4 Central Paraconsistent Three-Valued Logics 105

4.4.3 The Logic LP and Its Main Monotonic Extensions

The most popular conjunction used in three-valued paraconsistent logics (and three-valued logics in general) is Kleene's (strong) conjunction, discussed in Section 4.1. Most of the paraconsistent logics which use it are based on Asenjo-Priest's three-valued logic **LP** [23, 244, 245, 246]. This logic is induced by the three-valued matrix $\mathsf{LP} = \langle \{t, f, \top\}, \{t, \top\}, \{\tilde{\wedge}, \tilde{\neg}\} \rangle$, where the truth tables for $\neg$ and $\wedge$ are the following:

$\tilde{\wedge}$	t	f	$\top$
t	t	f	$\top$
f	f	f	f
$\top$	$\top$	f	$\top$

$\tilde{\neg}$	
t	f
f	t
$\top$	$\top$

There are several disjunctions that can be defined in $\mathcal{M}_{LP}$. The one that was preferred by Asenjo and Priest is the strong Kleene's disjunction (see again Section 4.1), which is one of the possible interpretations of disjunction given in Theorem 4.21. It is definable in $\mathcal{M}_{LP}$ by: $\psi \vee \varphi = \neg(\neg \psi \wedge \neg \varphi)$. The truth table in $\mathcal{M}_{LP}$ of this $\vee$ is indeed precisely that of Kleene:

$\tilde{\vee}$	t	f	$\top$
t	t	t	t
f	t	f	$\top$
$\top$	t	$\top$	$\top$

Next, we introduce the simplest extensions of **LP**: those obtained by adding to its language the propositional constants from Section 4.1.

Definition 4.49 ($\mathbf{LP^F}, \mathbf{LP^B}, \mathbf{LP^{F,B}}$)

$\mathbf{LP^F}$ is the logic induced by $\mathsf{LP^F}$, the extension of the matrix LP to the language $\{\neg, \wedge, \vee, \mathsf{F}\}$ (or just $\{\neg, \wedge, \mathsf{F}\}$).

$\mathbf{LP^B}$ is the logic induced by $\mathsf{LP^B}$, the extension of the matrix LP to the language $\{\neg, \wedge, \vee, \mathsf{B}\}$.

$\mathbf{LP^{F,B}}$ is the logic induced by $\mathsf{LP^{F,B}}$, the extension of the matrix LP to the language $\{\neg, \wedge, \vee, \mathsf{F}, \mathsf{B}\}$.

Proposition 4.50

1. **LP** is $\neg$-contained in classical logic and fully maximal. The same is true for $\mathbf{LP^F}$ (but not for $\mathbf{LP^B}$ or $\mathbf{LP^{B,F}}$).

2. **LP** and $\mathbf{LP^B}$ are non-exploding, while $\mathbf{LP^F}$ and $\mathbf{LP^{B,F}}$ are not.

Proof. By Theorems 4.16, 4.24, and Proposition 4.7. □

Perhaps the most remarkable property of **LP** (and **LP^F**) is given in the next proposition.

Proposition 4.51 [244] *The tautologies of **LP** and **LP^F** are the same as those of classical logic in their languages: if ψ is in the language of $\{\neg, \wedge, \vee\}$ ($\{\neg, \wedge, \vee, F\}$) then $\vdash_{\mathcal{M}_{LP}} \psi$ ($\vdash_{\mathcal{M}^F_{LP}} \psi$) iff ψ is a classical tautology.*

Proof. One direction is trivial. For the converse, suppose that $\nu \in \Lambda_{\mathcal{M}_{LP}}$ (the proof in the case of **LP^F** is similar). Let μ be the two-valued valuation such that for every $p \in \mathsf{Var}$, $\mu(p) = t$ iff $\nu(p) \in \{t, \top\}$. It is easy to prove by induction on the complexity of ψ that if $\mu(\psi) = t$ then $\nu(\psi) \in \{t, \top\}$, and if $\mu(\psi) = f$ then $\nu(\psi) \in \{f, \top\}$. It follows that if $\mu(\psi) = t$ for every $\mu \in \Lambda_{\mathcal{M}_{CL}}$ then $\nu(\psi) \in \{t, \top\}$ for every $\nu \in \Lambda_{\mathcal{M}_{LP}}$. □

Note 4.52 Despite having the same set of valid formulas, **LP** is paraconsistent, while classical logic (in the language of $\{\neg, \wedge, \vee\}$) is not. The difference between the two is due to their *consequence relations*.

$\mathcal{THREE}_\top$ which is For characterizing the expressive power of LP and its above extensions, it is convenient to order the truth-values in another partial order, denoted here by $\leq_k$, that intuitively reflects differences in the amount of *knowledge* (or *information*) that the truth values convey. (See Section 5.2 below for motivation.) According to this relation $\top$ is the maximal element, while neither of the remaining truth-values is greater than the other. Therefore $\langle \mathcal{V}, \leq_k \rangle$ is an upper semilattice. A Double-Hasse diagram representing the structure induced by $\leq_k$ and $\leq_t$ is given in Figure 4.1. In this diagram b is an immediate $\leq_t$-successor of a iff b is on the right-hand side of a, and there is an edge between them; Similarly, b is an immediate $\leq_k$-successor of a iff b is above a, and there is an edge between them.

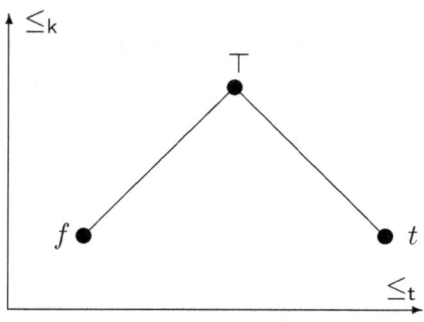

Figure 4.1: $\mathcal{THREE}_\top$

4.4 Central Paraconsistent Three-Valued Logics

Definition 4.53 A function $g : \{t, f, \top\}^n \to \{t, f, \top\}$ is $\leq_k$-*monotonic* if $g(a_1, \ldots, a_n) \leq_k g(b_1, \ldots, b_n)$ in case $a_i \leq_k b_i$ for every $1 \leq i \leq n$.

Now we are ready to characterize the expressive power of LP and its above extensions.

Theorem 4.54 [40] *Let* $g : \{t, f, \top\}^n \to \{t, f, \top\}$.

1. *g is representable in* $\mathsf{LP}^{\mathsf{F,B}}$ *iff it is $\leq_k$-monotonic.*

2. *g is representable in* LP^{B} *iff it is $\leq_k$-monotonic and $\{\top\}$-closed.*

3. *g is representable in* LP^{F} *iff it is $\leq_k$-monotonic and classically (i.e. $\{t, f\}$-) closed.*

4. *g is representable in* LP *iff it is $\leq_k$-monotonic, $\{\top\}$-closed, and classically closed.*

Proof. In all four cases it is straightforward to verify the "only if" direction (e.g. that any function which is representable in $\mathsf{LP}^{\mathsf{F,B}}$ is $\leq_k$-monotonic). Therefore we concentrate on the converse.

We start with some definitions and a lemma. A set $\mathcal{S} \subseteq \mathit{THREE}^n$ is called a *cone* in $\langle \mathit{THREE}^n, \leq_k \rangle$ if $\vec{y} \in \mathcal{S}$ whenever $\vec{y} \geq_k \vec{x}$ and $\vec{x} \in \mathcal{S}$. Obviously, a cone $\mathcal{S}$ in THREE^n is nonempty iff $\vec{\top} = (\top, \ldots, \top) \in \mathcal{S}$. An element $\vec{x} \in \mathcal{S}$ is a *stable* element of $\mathcal{S}$ if $\{\vec{y} \in \mathit{THREE}^n \mid \vec{y} \leq_k \vec{x}\} \subseteq \mathcal{S}$.

Lemma. For every cone $\mathcal{C}$ in THREE^n there is a formula $\psi_\mathcal{C}$ in $\{\neg, \wedge, \mathsf{F}\}$, so that for every $\vec{x} \in \mathit{THREE}^n$, $F^n_{\psi_\mathcal{C}}(\vec{x}) \neq f$ iff $\vec{x} \in \mathcal{C}$, and if $\vec{x}$ is a stable element of $\mathcal{C}$ then $F^n_{\psi_\mathcal{C}}(\vec{x}) = t$.

Proof of the Lemma: For every $\vec{a} \in \mathcal{C}$ and every $1 \leq i \leq n$ define a formula $\psi^i_{\vec{a}}$ as follows: If $\vec{a}$ is not a stable element of $\mathcal{C}$ then

$$\psi^i_{\vec{a}} = \begin{cases} P_i \wedge \neg P_i & a_i = \top, \\ P_i & a_i = t, \\ \neg P_i & a_i = f. \end{cases}$$

If $\vec{a}$ *is* a stable element of $\mathcal{C}$ then

$$\psi^i_{\vec{a}} = \begin{cases} \neg \mathsf{F} & a_i = \top, \\ P_i & a_i = t, \\ \neg P_i & a_i = f. \end{cases}$$

Let $\psi_{\vec{a}}$ be $\psi^1_{\vec{a}} \wedge \psi^2_{\vec{a}} \wedge \cdots \wedge \psi^n_{\vec{a}}$. It is easy to see that $F^n_{\psi_{\vec{a}}}(\vec{a}) \neq f$ for every $\vec{a} \in \mathcal{C}$, and that $F^n_{\psi_{\vec{a}}}(\vec{a}) = t$ in case $\vec{a}$ is a stable element of $\mathcal{C}$. Now we show that if $F^n_{\psi_{\vec{a}}}(\vec{x}) \neq f$ then $\vec{x} \in \mathcal{C}$. This is easy in case $\vec{a}$ is an element of $\mathcal{C}$

which is not stable, because in this case $F^n_{\psi_{\vec{a}}}(\vec{x}) \neq f$ iff $\vec{x} \geq_k \vec{a}$, and $\mathcal{C}$ is a cone. Assume that $\vec{a}$ is a stable element of $\mathcal{C}$, and that $F^n_{\psi_{\vec{a}}}(\vec{x}) \neq f$. Then $x_i \geq_k a_i$ for every i such that $a_i \neq \top$. Define:

$$c_i = \begin{cases} a_i & a_i \neq \top, \\ x_i & a_i = \top. \end{cases}$$

Then $\vec{c} \leq_k \vec{a}$, and so $\vec{c} \in \mathcal{C}$ (since $\vec{a}$ is stable in $\mathcal{C}$). But $\vec{c} \leq_k \vec{x}$ as well, and so $\vec{x} \in \mathcal{C}$ (because $\mathcal{C}$ is a cone).

Now, define $\psi_\mathcal{C}$ to be $\bigvee_{\vec{a} \in \mathcal{C}} \psi_{\vec{a}}$ in case $\mathcal{C}$ is not empty, F otherwise. Given what we have shown, it is obvious that $\psi_\mathcal{C}$ has the required properties.

Now we turn to the proof of the "if" directions of the theorem.

1. Let g be a $\leq_k$-monotonic function of arity n. Define:

$$\mathcal{C}^g_t = \{\vec{x} \in \mathit{THREE}^n \mid g(\vec{x}) \geq_k t\},$$

$$\mathcal{C}^g_f = \{\vec{x} \in \mathit{THREE}^n \mid g(\vec{x}) \geq_k f\}.$$

Since g is $\leq_k$-monotonic, both $\mathcal{C}^g_t$ and $\mathcal{C}^g_f$ are cones. Moreover: if $g(\vec{x}) = t$ then $\vec{x}$ is a stable element of $\mathcal{C}^g_t$, while if $g(\vec{x}) = f$ then $\vec{x}$ is a stable element of $\mathcal{C}^g_f$. Let ψ_t and ψ_f be, respectively, the formulas which are given by the lemma for the cones $\mathcal{C}^g_t$ and $\mathcal{C}^g_f$. Define:

$$\psi^g = (\psi_t \wedge \mathsf{B}) \vee (\neg \psi_f \wedge \mathsf{B}) \vee (\psi_t \wedge \neg \psi_f).$$

Now, if $g(\vec{x}) = t$ then $F^n_{\psi_t}(\vec{x}) = t$ (since $\vec{x}$ is a stable element of $\mathcal{C}^g_t$), while $F^n_{\psi_f}(\vec{x}) = f$ (since $\vec{x} \notin \mathcal{C}^g_f$). It follows that $F^n_{\psi^g}(\vec{x}) = t$ in this case. Similarly, if $g(\vec{x}) = f$ then $F^n_{\psi^g}(\vec{x}) = f$. Finally, if $g(\vec{x}) = \top$ then $\vec{x} \in \mathcal{C}^g_t$ and $\vec{x} \in \mathcal{C}^g_f$, and so $F^n_{\psi_t}(\vec{x}) \in \{t, \top\}$ and also $F^n_{\psi_f}(\vec{x}) \in \{t, \top\}$. This implies that $F^n_{\psi^g}(\vec{x}) = \top$ in this case. Hence $F^n_{\psi^g}(\vec{x}) = g(\vec{x})$ in all cases, and so ψ^g represents g.

2. Suppose that g is $\leq_k$-monotonic and $\{\top\}$-closed. Let ψ^g be the formula which represents g in $\{\neg, \vee, \wedge, \mathsf{F}, \mathsf{B}\}$ according to the construction given in the proof of Item 1. Obtain ψ^g_* from ψ^g by replacing every occurrence of F with the following formula:

$$\mathsf{F}_n = P_1 \wedge \neg P_1 \wedge P_2 \wedge \neg P_2 \wedge \ldots \wedge P_n \wedge \neg P_n.$$

Obviously, F_n has the following property:

$$F^n_{\mathsf{F}_n}(\vec{a}) = \begin{cases} \top & \vec{a} = \vec{\top}, \\ f & \text{otherwise.} \end{cases}$$

4.4 Central Paraconsistent Three-Valued Logics

It follows that $F^n_{\psi^g_*}(\vec{x}) = F^n_{\psi^g}(\vec{x}) = g(\vec{x})$ for every $\vec{x} \neq \vec{\top}$. Since g is $\{\top\}$-closed and ψ^g_* is in $\{\neg, \vee, \wedge, \mathsf{B}\}$, also $F^n_{\psi^g_*}(\vec{\top}) = \top = g(\top)$. Hence $F^n_{\psi^g_*}(\vec{x}) = g(\vec{x})$ for every $\vec{x} \in \mathit{THREE}^n$.

3. Suppose that g is $\leq_k$-monotonic and $\{t, f\}$-closed. Let ψ^g be the formula which represents g in $\{\neg, \vee, \wedge, \mathsf{F}, \mathsf{B}\}$ according to the construction given in the proof of Item 1. Obtain ψ^g_* from ψ^g by replacing every occurrence of B with the following formula:

$$\mathsf{B}_n = (P_1 \vee \neg P_1) \wedge (P_2 \vee \neg P_2) \wedge \ldots \wedge (P_n \vee \neg P_n).$$

It is easy to verify that B_n has the following property:

$$F^n_{\mathsf{B}_n}(\vec{a}) = \begin{cases} \top & \exists 1 \leq i \leq n \ a_i = \top, \\ t & \text{otherwise}. \end{cases}$$

It follows that $F^n_{\psi^g_*}(\vec{x}) = F^n_{\psi^g}(\vec{x}) = g(\vec{x})$ for every $\vec{x} \notin \{t, f\}^n$. Since g is $\{t, f\}$-closed and ψ^g_* is in the language $\{\neg, \vee, \wedge, \mathsf{F}\}$, the above property of $F^n_{\mathsf{B}_n}$ and the definition of ψ^g_* easily imply that $F^n_{\psi^g_*}(\vec{x}) = g(\vec{x})$ also in case $\vec{x} \in \{t, f\}^n$. Hence $F^n_{\psi^g_*} = g$.

4. Suppose that g is $\leq_k$-monotonic, $\{\top\}$-closed, and $\{t, f\}$-closed. Let ψ^g be the formula which represents g in $\{\neg, \vee, \wedge, \mathsf{F}, \mathsf{B}\}$ according to the construction given in the proof of Item 1. Obtain ψ^g_* from ψ^g by replacing every occurrence of B with B_n and every occurrence of F with F_n. Then $F^n_{\psi^g_*} = g$ (details are left to the reader). □

Corollary 4.55 *Sette's negation is not represented in* $\mathbf{LP}^{\mathsf{F},\mathsf{B}}$. *Hence Kleene's negation is the only negation which is definable in in* $\mathbf{LP}^{\mathsf{F},\mathsf{B}}$.

Proof. This is immediate from Theorem 4.54 and Proposition 4.5, since Sette's negation is not $\leq_k$-monotonic. □

Corollary 4.56 *The operation* $\tilde{\wedge}$ *of* LP *is the only conjunction which is definable in* $\mathbf{LP}$ *and* $\mathbf{LP}^{\mathsf{F}}$.

Proof. This easily follows from Theorem 4.54 and Proposition 4.8. □

Note 4.57 From Theorem 4.54 it follows that in contrast to Corollary 4.56, no less than four different disjunctions are definable in **LP**. Note that Kleene's disjunction is the $\leq_k$-minimal among them, and Bochvar-Halldén's disjunction is the $\leq_k$-maximal one. Note also that Corollary 4.56 cannot be extended to $\mathbf{LP}^{\mathsf{B}}$ and $\mathbf{LP}^{\mathsf{F},\mathsf{B}}$, since in both there are two definable conjunctions: Kleene's conjunction, and the connective $\wedge_\top$, whose interpretation is defined by $a \tilde{\wedge}_\top b = f$ if $a = f$ or $b = f$, and $a \tilde{\wedge}_\top b = \top$ otherwise.

Now comes the main drawback of the logics studied in this section.

Proposition 4.58 *Let $\mathcal{M}$ be a three-valued paraconsistent matrix which has only $\leq_k$-monotonic connectives. Then $\mathbf{L}_\mathcal{M}$ does not have an implication or even a semi-implication.*

Proof. Suppose for contradiction that $\supset$ is a definable semi-implication for $\mathbf{L}_\mathcal{M}$. It is easy to see (see the proof of Theorem 11.45) that this implies that (i) $\vdash_\mathcal{M} p \supset p$ and (ii) $p, p \supset q \vdash_\mathcal{M} q$. Now, (i) entails that $\tilde{\supset}(f, f) \in \{t, \top\}$. Therefore it follows from the $\leq_k$-monotonicity of $\supset$ that $\tilde{\supset}(\top, f) \in \{t, \top\}$. This contradicts (ii) (since it is refuted by any $\mathcal{M}$-valuation ν such that $\nu(p) = \top$ and $\nu(q) = f$). □

Corollary 4.59 *The logic $\mathbf{LP}$ is not normal. The same is true for $\mathbf{LP^B}$, $\mathbf{LP^F}$, and $\mathbf{LP^{F,B}}$.*

Proof. Immediate from Theorem 4.54 and Proposition 4.58. □

4.4.4 The Logics PAC ($\mathbf{RM_3}$) and Its Main Extensions

The most straightforward way to turn $\mathbf{LP}$ into a normal logic is to extend $\mathcal{M}_{LP}$ by the implication connective $\supset$ of D'Ottaviano and da-Costa which is mentioned in Section 4.1 (following Definition 3.38). Doing this leads to the logic **PAC** (also known as $\mathbf{RM_3}$ — see Chapter 15) [25, 32, 78, 141, 153, 260]. Thus, **PAC** is the logic in $\mathcal{L}_{CL}$ which is induced by the three-valued matrix $\mathsf{PAC} = \langle\{t, f, \top\}, \{t, \top\}, \{\tilde{\wedge}, \tilde{\vee}, \tilde{\supset}, \tilde{\neg}\}\rangle$, where $\tilde{\wedge}, \tilde{\vee}$, and $\tilde{\neg}$ are like in LP, while $\tilde{\supset}$ is given by the following truth-table:

$\tilde{\supset}$	t	f	$\top$
t	t	f	$\top$
f	t	t	t
$\top$	t	f	$\top$

Note 4.60 Let $\tilde{\rightarrow}$ be Sobociński's entailment (Note 4.42). Since $a\tilde{\rightarrow}b = (a\tilde{\supset}b)\tilde{\wedge}(\tilde{\neg}b\tilde{\supset}\tilde{\neg}a)$, while $a\tilde{\supset}b = b\tilde{\vee}(a\tilde{\rightarrow}b)$, another way that leads to **PAC** is to extend $\mathcal{A}_1$, and with it $\mathbf{SRM}_{\tilde{\rightarrow}}$, with Kleene's conjunction (which by Corollary 4.48 is not definable in $\mathcal{A}_1$).

Example 4.61 (Tweety Dilemma) We demonstrate the use of **PAC** for reasoning with inconsistent information in the case of the famous 'Tweety dilemma'. The relevant assumptions may be represented in the language of PAC as follows (where $p \mapsto q$ is an abbreviation for $\neg p \vee q$):

4.4 Central Paraconsistent Three-Valued Logics

$$\mathcal{T} = \left\{ \begin{array}{l} bird(Tweety) \mapsto fly(Tweety) \\ penguin(Tweety) \supset bird(Tweety) \\ penguin(Tweety) \supset \neg fly(Tweety) \\ bird(Tweety) \\ penguin(Tweety) \end{array} \right\}.$$

We are using different connectives here according to the strength we attach to each entailment: Penguins *never* fly. This is a characteristic property of penguins, and there are no exceptions to that. Also, every penguin is a bird and again, there are no exceptions to that fact. Thus, the second and the third facts are formulated with a real implication connective. In contrast, the first fact states only a default property of birds. Therefore we formulate it using a weaker sort of "implication". Indeed, the first assertion does not allow an automatic deduction of flying ability for any bird, since in **PAC** from ψ and $\psi \mapsto \varphi$ one cannot infer φ. (Take $\nu(\psi) = \top$, $\nu(\varphi) = f$.) However, it does give a strong connection between these two properties.[11]

Here are the six **PAC**-models of $\mathcal{T}$:

	bird	fly	penguin
ν_1	$\top$	$\top$	$\top$
ν_2	$\top$	$\top$	t
ν_3	$\top$	f	$\top$

	bird	fly	penguin
ν_4	$\top$	f	t
ν_5	t	$\top$	$\top$
ν_6	t	$\top$	t

It follows that although $\mathcal{T}$ is classically inconsistent, nontrivial conclusions can be derived from it by **PAC**: Tweety is a penguin, a bird, and it cannot fly. As expected, the complementary conclusions *cannot* be derived.

It is interesting to note that in **PAC** inconsistencies are 'localized', in the sense that non-contradictory information is not affected by them. To see this, consider the following extended theory:

$$\mathcal{T}' = \mathcal{T} \cup \left\{ \begin{array}{l} elephant(Fred) \supset \neg fly(Fred) \\ elephant(Fred) \end{array} \right\}.$$

One may conclude here, e.g., that Fred is an elephant and it cannot fly.

Next we introduce the simplest extensions of **PAC**. Again, these are the extensions that are obtained by adding to its language the propositional constants which were introduced in Section 4.1.

---

[11] In some 3-valued *nonmonotonic* logics (like Priest's three-valued logic **LPm** for reasoning with minimal inconsistency [244, 246]) $\mapsto$ *is* deductive and so in these contexts it may be considered as an implication.

Definition 4.62 (J_3, PAC^B, J_3^B)

J_3 ([141, 153]) is the logic induced by J_3, the extension of the matrix PAC to the language $\mathcal{L}_{CL}^{F\neg} = \{\neg, \wedge, \vee, \supset, F\}$.

PAC^B is the logic induced by PAC^B, the extension of the matrix PAC to the language $\mathcal{L}_{CL}^{B} = \{\neg, \wedge, \vee, \supset, B\}$.

J_3^B is the logic induced by J_3^B, the extension of the matrix PAC to the language $\mathcal{L}_{CL}^{F,B} = \{\neg, \wedge, \vee, \supset F, B\}$.

Note 4.63 Instead of the propositional constant F it is common in the literature on J_3 to use as the extra connective the *consistency* operator ○, whose interpretation $\tilde{\circ}$ is given by: $\tilde{\circ}(t) = \tilde{\circ}(f) = t$, while $\tilde{\circ}(\top) = f$. This does not make much difference, since $\tilde{\circ}(a) = (a \tilde{\wedge} \tilde{\neg} a) \tilde{\supset} f$, while $f = \tilde{\circ}(a) \tilde{\wedge} \tilde{\neg} \tilde{\circ}(a)$. As a logic in the language of $\{\neg, \wedge, \vee, \supset, \circ\}$, J_3 is equivalent to **LFI1** ([116, 113]), *the strongest logic* among those which are studied in Chapter 8.[12]

The matrix PAC is obviously an element of $8Kb$. Therefore, we have:

Proposition 4.64

1. **PAC**, J_3, PAC^B, and J_3^B *are all normal.*

2. **PAC** *and* J_3 *are fully maximal classical paraconsistent logics. (This is false for* PAC^B *and* J_3^B, *which are not $\neg$-contained in classical logic.)*

3. **PAC** *and* PAC^B *are non-exploding. This is false for* J_3 *and* J_3^B.

Proof. By Theorems 4.21, 4.24, Corollary 4.18, and Proposition 4.7. □

Next, we turn to the expressive power of PAC and its extensions.

Theorem 4.65 [40] *Let* $g : \{t, f, \top\}^n \to \{t, f, \top\}$.

1. *g is representable in* J_3^B.

2. *g is representable in* J_3 *iff it is* $\{t, f\}$-*closed (i.e. classically closed).*

3. *g is representable in* PAC^B *iff it is* $\{\top\}$-*closed.*

4. *g is representable in* PAC *iff it is both* $\{t, f\}$-*closed and* $\{\top\}$-*closed.*

---

[12] J_3 and its weaker versions have been considered also in the context of epistemic logics, where in [121, 122] it is shown that these logics can be encoded in the fragment of the modal logic **KD** which has only modal formulas without nesting.

4.4 Central Paraconsistent Three-Valued Logics

Proof. The proofs of the 'only if' directions are easy, and left for the reader. We prove the 'if' directions.

1. For $a \in \{t, f, \top\}$ we define:

$$\psi_a(p) = \begin{cases} \neg p \supset \mathsf{F} & \text{if } a = t, \\ p \supset \mathsf{F} & \text{if } a = f, \\ p \wedge \neg p & \text{if } a = \top. \end{cases}$$

It is easy to check that for every valuation ν it holds that $\nu(\psi_a(p)) \neq f$ iff $\nu(p) = a$. Next, for $\vec{a} = (a_1, \ldots, a_n) \in \{t, f, \top\}^n$ we let $\psi_{\vec{a}}$ be the formula $\psi_{a_1}(P_1) \wedge \psi_{a_2}(P_2) \wedge \cdots \wedge \psi_{a_n}(P_n)$. Then for every valuation ν, $\nu(\psi_{\vec{a}}) \neq f$ iff $\nu(P_i) = a_i$ for every $1 \leq i \leq n$.

Now, if $g(\vec{a}) = t$ for every $\vec{a} \in \{t, f, \top\}^n$ then the formula $\neg \mathsf{F}$ represents g. Otherwise, it is not difficult to check that g is represented by the $\wedge$-conjunction of all the formulas which either has the form $\psi_{\vec{a}} \supset \mathsf{F}$ where $g(\vec{a}) = f$, or the form $\psi_{\vec{a}} \supset \mathsf{B}$ where $g(\vec{a}) = \top$.

2. The proof is very similar to the proof of of the first part. The only difference is that instead of the propositional constant B (which is not available in $\mathcal{L}_{CL}^{\mathsf{F}\neg}$) we use again the formula B_n from the proof of Theorem 4.54. Recall that this formula has the following property:

$$F_{\mathsf{B}_n}^n(\vec{a}) = \begin{cases} \top & \exists 1 \leq i \leq n \ a_i = \top, \\ t & \text{otherwise.} \end{cases}$$

This implies that if g is $\{t, f\}$-closed and $g(\vec{a}) = \top$ then $\nu(\mathsf{B}_n) = \top$ in case $\nu(P_i) = a_i$ for every $1 \leq i \leq n$. Hence B_n can replace B here.

3. The proof is again similar to the proof of of the first part. However, this time we replace the propositional constant F by the formula F_n from the proof of Theorem 4.54. We leave details for the reader.

4. The proof is again similar to the proof of of the first part, only this time we replace both the propositional constant B by the formula B_n and the propositional constant F by the formula F_n from the proof of Theorem 4.54. Details are again left for the reader. □

Corollary 4.66 *Sette's negation is not represented in* PAC^B *(and* PAC*), but it is represented in* J_3.

Corollary 4.67

1. *Every three-valued paraconsistent logic can be embedded in* J_3^B.

2. **J_3** *is the strongest 3-valued paraconsistent logic which is $\neg$-contained in classical logic (i.e. every other logic with these properties, can be embedded in it). In particular: any logic which is induced by one of the matrices in* **8Kb** *can be embedded in* **J_3**.

3. **PAC^B** *is the strongest 3-valued non-exploding paraconsistent logic.*

4. **PAC** *is the strongest three-valued paraconsistent logic which is both $\neg$-contained in classical logic and non-exploding.*

Proof. By Theorems 4.65, 4.16 (or 4.21), and Propositions 4.64, 4.7. □

Note 4.68 In [40] it is shown that by adding to **PAC** *any* classically closed connective not available in it, we get a matrix in which all the classically closed connectives are available. Similarly, by adding to **PAC** *any* $\{\top\}$-closed connective not available in it, we get a matrix in which all the $\{\top\}$-closed connectives are available. Hence there is no intermediate logic between **PAC** and **J_3**, or between **PAC** and **PAC^B**. From results in [40] it also follows that there is no intermediate logic between **J_3** or **PAC^B** and **J_3^B**.

4.4.5 The Logic $PAC_\supset$

In this section we briefly study another interesting paraconsistent logic: **$PAC_\supset$**, the $\{\neg, \supset\}$-fragment of **PAC**. It is the logic induced by the matrix $PAC_\supset = \langle \{t, f, \top\}, \{t, \top\}, \{\tilde{\supset}, \tilde{\neg}\}\rangle$, where $\tilde{\supset}$ and $\tilde{\neg}$ are like in **PAC**.

Proposition 4.69 *The matrix* $PAC_\supset$ *is a proper extension of* $\mathcal{A}_1$.

Proof. It is easy to check that Sobociński's conjunction is represented in $PAC_\supset$ by the formula $\neg(\neg(P_1 \supset \neg P_2) \supset (\neg(\neg P_2 \supset P_1)))$. Hence $PAC_\supset$ is an extension of $\mathcal{A}_1$. That the extension is proper follows by Proposition 4.46 from the fact that $\tilde{\supset}$ is not $\{\top\}$-limited. □

Corollary 4.70 $PAC_\supset$ *is non-exploding, normal, $\neg$-contained in classical logic, and fully maximal.*

Proof. $PAC_\supset$ is normal by Propositions 4.69, 4.44, and Corollary 4.10. The rest follow from Theorem 4.21, 4.23, and Proposition 4.7. □

Proposition 4.71 *A function $g : \{t, f, \top\}^n \to \{t, f, \top\}$ is representable in* $PAC_\supset$ *iff it is $\{\top\}$-closed, and there is $1 \leq i \leq n$ such that $g(a_1, \ldots, a_n) = \top$ only if $a_i = \top$.*

4.4 Central Paraconsistent Three-Valued Logics

Proof. For a formula φ in $\{\neg, \supset\}$, we define $\varphi_\top$ recursively as follows: $p_\top = p$ if p is a variable, $(\neg\psi)_\top = \psi_\top$, and $(\varphi \supset \psi)_\top = \psi_\top$. It is easy to verify that for every φ, $\varphi_\top$ is a variable such that $\nu(\varphi_\top) = \top$ whenever ν is a valuation in $\mathbf{PAC}_\supset$ such that $\nu(\varphi) = \top$. This easily implies that if g representable in $\mathbf{PAC}_\supset$ then it satisfies the condition given above. Obviously, such g is also $\{\top\}$-closed. This prove the "only if" part of the proposition,

The proof of the converse is very similar to the proof of the forth part of Proposition 4.65. The only differences are that $\wedge$ should be understood as Sobociński's conjunction (which is definable in $\mathbf{PAC}_\supset$ by Proposition 4.69), and instead of the formula B_n we use for a given g the formula $P_i \supset P_i$, where i has the property that $g(a_1, \ldots, a_n) = \top$ only if $a_i = \top$. The easy details are left for the reader. □

Corollary 4.72 $\mathbf{PAC}$ *is a proper extension of* $\mathbf{PAC}_\supset$.

Proof. This follows from the previous proposition and the fact that Kleene's conjunction does not satisfy the second condition given there. □

Figure 4.2 below shows the relative expressive power of six of the three-valued logics with Kleene's negation which are considered in this section.

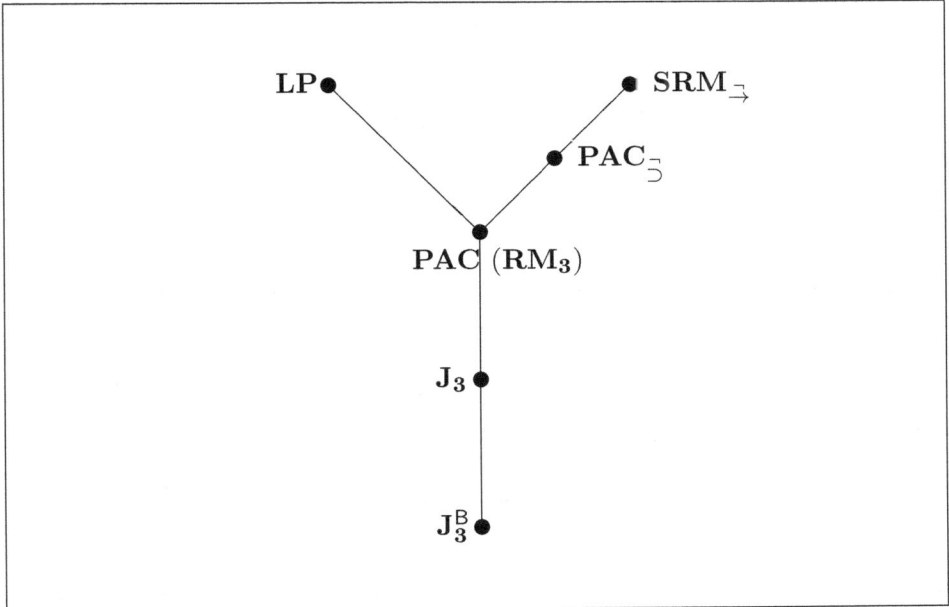

Figure 4.2: Relative strength of some logics with Kleene's negation

4.4.6 The Logic PE3

None of the logics discussed so far in this section is $\neg$-extensional (and so none of them is self-extensional or $\neg$ is contrapositive in it). For **LP** and its extensions it follows from the fact that $\neg P_1 \vee P_1 \dashv\vdash_{\mathbf{LP}} \neg P_2 \vee P_2$ while $\neg(\neg P_1 \vee P_1) \not\vdash_{\mathbf{LP}} \neg(\neg P_2 \vee P_2)$. For the others it follows from Proposition 4.32 (and other results of this chapter). However, that proposition still leaves open the possibility of a three-valued paraconsistent matrix $\mathcal{M}$ such that $\mathbf{L}_\mathcal{M}$ is semi-normal (without implication) and $\neg$-extensional (and so also self-extensional, by Proposition 4.31). In this section we show that there is exactly one such matrix for the language $\{\neg, \wedge, \vee\}$, and that $\neg$ is contrapositive in the induced logic.

Definition 4.73 $\mathbf{PE3} = \langle \{\neg, \wedge, \vee\}, \vdash_{\mathbf{PE3}} \rangle$ is the logic which is induced by the matrix $\mathsf{PE3} = \langle \{t, f, \top\}, \{t, \top\}, \mathcal{O} \rangle$ for $\{\neg, \wedge, \vee\}$, where the operations of $\mathcal{O}$ are defined as follows:

$\tilde{\neg}$	
t	f
f	t
$\top$	t

$\tilde{\wedge}$	t	f	$\top$
t	t	f	$\top$
f	f	f	f
$\top$	$\top$	f	$\top$

$\tilde{\vee}$	t	f	$\top$
t	t	t	t
f	t	f	$\top$
$\top$	t	$\top$	$\top$

Note 4.74 The matrix $\mathsf{PE3}$ is a combination of $\mathbf{P}_1$ and $\mathbf{LP}$: Like in $\mathbf{LP}$, its truth-tables for $\wedge$ and $\vee$ are those of Kleene; but its truth-table for $\neg$ is Sette's truth-table used in $\mathbf{P}_1$.

Proposition 4.75 $\mathbf{PE3}$ is $\neg$-contained in classical logic, and so fully maximal.

Proof. Left for the reader. $\square$

Now we turn to the special properties of **PE3**. For this we first need the following definition and lemma:

Definition 4.76 For $x, y \in \{t, f, \top\}$, let $x \sim_t y$ if either $x = y = t$, or both $x \neq t$ and $y \neq t$.

Lemma 4.77 Let ν_1 and ν_2 be $\mathsf{PE3}$-valuations such that $\nu_1(p) \sim_t \nu_2(p)$ for every variable p. Then $\nu_1(\varphi) \sim_t \nu_2(\varphi)$ for every formula φ.

Proof. An inspection of the truth-tables of $\mathsf{PE3}$ show that every $\mathsf{PE3}$-valuation ν satisfies the following conditions for every φ and ψ:

1. $\nu(\neg\varphi) = t$ iff $\nu(\varphi) \neq t$.

4.4 Central Paraconsistent Three-Valued Logics

2. $\nu(\varphi \wedge \psi) = t$ iff $\nu(\varphi) = t$ and $\nu(\psi) = t$.

3. $\nu(\varphi \vee \psi) = t$ iff $\nu(\varphi) = t$ or $\nu(\psi) = t$.

Now assume that ν_1 and ν_2 satisfy the condition in the lemma. Using 1.–3. above, we show by induction on the structure of φ that $\nu_1(\varphi) \sim_t \nu_2(\varphi)$.

- The case where φ is a variable holds by our assumption.

- If $\varphi = \neg \psi$, then $\nu_1(\varphi) = t$ iff $\nu_1(\psi) \neq t$, iff (by induction hypothesis) $\nu_2(\psi) \neq t$, iff $\nu_2(\varphi) = t$.

- If $\varphi = \sigma \wedge \psi$, then $\nu_1(\varphi) = t$ iff $\nu_1(\sigma) = t$ and $\nu_1(\psi) = t$, iff (by induction hypothesis) $\nu_2(\sigma) = t$ and $\nu_2(\psi) = t$, iff $\nu_2(\varphi) = t$.

- If $\varphi = \sigma \vee \psi$, then $\nu_1(\varphi) = t$ iff $\nu_1(\sigma) = t$ or $\nu_1(\psi) = t$, iff (by induction hypothesis) $\nu_2(\sigma) = t$ or $\nu_2(\psi) = t$, iff $\nu_2(\varphi) = t$. □

Theorem 4.78 $\neg$ *is contrapositive in* **PE3**.

Proof. Assume that $\neg \psi \nvdash_{\mathbf{PE3}} \neg \varphi$. Then there is an PE3-valuation ν such that $\nu(\neg \psi) \neq f$, while $\nu(\neg \varphi) = f$. By the truth-table of $\neg$ in PE3, this means that $\nu(\psi) \neq t$ and $\nu(\varphi) = t$. Let ν' be the unique PE3-valuation such that for every variable p, $\nu'(p) \in \{t, f\}$ and $\nu'(p) \sim_t \nu(p)$. Then from Lemma 4.77 it follows that $\nu'(\varphi) = t$, while $\nu'(\psi) \neq t$. Since all the operations of PE3 are $\{t, f\}$-closed, and ν' assigns to variables values in $\{t, f\}$, the fact that $\nu'(\psi) \neq t$ implies that $\nu'(\psi) = f$. This and the fact that $\nu'(\varphi) = t$ entail that $\varphi \nvdash_{\mathbf{PE3}} \psi$. □

Theorem 4.79 **PE3** *is self-extensional.*

Proof. Immediate from Theorem 4.78 and Proposition 4.31. □

Corollary 4.80 **PE3** *has no implication. Hence it is not normal.*

Proof. This follows from Theorems 4.79 and Proposition 4.32. □

Next, we show the uniqueness properties of $PE3$.

Theorem 4.81 *Let* $\mathcal{M}$ *be a three-valued paraconsistent matrix for a language* $\mathcal{L}$, *and assume that* $\mathbf{L}_{\mathcal{M}}$ *has a conjunction* $\wedge$ *and a disjunction* $\vee$. *If* $\mathbf{L}_{\mathcal{M}}$ *is self-extensional then it is an extension of* **PE3**. *Moreover:* $\mathcal{M}$'s *truth-tables for* $\neg, \wedge,$ *and* $\vee$ *are identical to those of* **PE3**.

Proof. Let $\tilde{\neg}$, $\tilde{\wedge}$, and $\tilde{\vee}$ be the interpretations in $\mathcal{M}$ of $\neg$, $\wedge$, and $\vee$ (respectively). By Proposition 4.8, $\tilde{\wedge}$ is an $\mathcal{M}$-conjunction, and $\tilde{\vee}$ is an $\mathcal{M}$-disjunction. By Proposition 4.5, $\tilde{\neg}$ is either Kleene's negation or Sette's negation. (Hence $\tilde{\neg} t = f$ and $\tilde{\neg} f = t$.) Suppose that $\mathbf{L}_\mathcal{M}$ is self-extensional. To prove our claim, we need to show the following:

1. $a \tilde{\wedge} a = a \tilde{\vee} a = a$ for every $a \in \{t, f, \top\}$.

2. $a \tilde{\wedge} f = f \tilde{\wedge} a = f$ for every $a \in \{t, f, \top\}$.

3. $t \tilde{\vee} f = f \tilde{\vee} t = t$.

4. $\tilde{\neg} \top = t$.

5. $t \tilde{\vee} \top = \top \tilde{\vee} t = t$.

6. $f \tilde{\vee} \top = \top \tilde{\vee} f = \top$.

7. $t \tilde{\wedge} \top = \top \tilde{\wedge} t = \top$.

To prove 1, note that since $\wedge$ is a conjunction for $\mathbf{L}_\mathcal{M}$ then $\varphi \wedge \varphi \dashv\vdash_L \varphi$, while since $\vee$ is a disjunction for $\mathbf{L}_\mathcal{M}$ then $\varphi \vee \varphi \dashv\vdash_L \varphi$. Therefore the equations in 1 immediately follow from Proposition 4.31.

Item 2 follows from the fact that $\tilde{\wedge}$ is an $\mathcal{M}$-conjunction.

The assumptions that $\wedge$ is a conjunction for $\mathbf{L}_\mathcal{M}$ and $\vee$ is a disjunction for it easily imply that $\varphi \vee (\varphi \wedge \neg\varphi) \dashv\vdash_\mathcal{M} \varphi$. By Proposition 4.31 it follows that $a \tilde{\vee}(a \tilde{\wedge} \tilde{\neg} a) = a$ for every $a \in \{t, f, \top\}$. In particular: $t \tilde{\vee}(t \tilde{\wedge} \tilde{\neg} t) = t$. Hence $t \tilde{\vee} f = t$ by 2 (and the fact that $\tilde{\neg} t = f$). The proof that $f \tilde{\vee} t = t$ is similar. This proves 3.

Now, note that by Corollary 4.6, $\neg P_1 \vee P_1 \dashv\vdash_\mathcal{M} \neg P_2 \vee P_2$. Therefore, by taking $\nu(P_1) = \top$, $\nu(P_2) = t$ in Proposition 4.31 we get that $\tilde{\neg}\top \tilde{\vee} \top = \tilde{\neg} t \tilde{\vee} t$. It follows by 3 (and $\tilde{\neg} t = f$) that $\tilde{\neg}\top \tilde{\vee} \top = t$. Similarly, $\top \tilde{\vee} \tilde{\neg}\top = t$. Since $\tilde{\neg}\top \neq f$, this is possible by 1 only if $\tilde{\neg}\top = t$. Hence 4 and 5.

From 3–5 it follows also that $\tilde{\neg}(\tilde{\neg} a \tilde{\vee} a) = f$ for every $a \in \{t, \top, f\}$. Hence $\neg(\neg p \vee p) \vdash_\mathcal{M} p$. Since $\vee$ is a disjunction for $\mathbf{L}_\mathcal{M}$, it follows that $p \vee \neg(\neg p \vee p) \dashv\vdash_\mathcal{M} p$. Therefore Proposition 4.31 implies that $\nu(p \vee \neg(\neg p \vee p)) = \nu(p)$ for every ν. By taking $\nu(p) = \top$ we get from this that $\top \tilde{\vee} f = \top$. The proof that $f \tilde{\vee} \top = \top$ is similar. Hence 6.

For 7, note that from 2 and 4 it follows that $\nu(\neg(\neg p \wedge \neg\neg p)) = t$ for every valuation ν in $\mathcal{M}$. Hence $\vdash_\mathcal{M} \neg(\neg p \wedge \neg\neg p)$. Since $\wedge$ is a conjunction for $\mathbf{L}_\mathcal{M}$, this implies that $p \wedge \neg(\neg p \wedge \neg\neg p) \dashv\vdash_\mathcal{M} p$. Therefore Proposition 4.31 implies that $\nu(p \wedge \neg(\neg p \wedge \neg\neg p)) = \nu(p)$ for every ν. Taking $\nu(p) = \top$ we get $\top \tilde{\wedge} t = \top$. That $t \tilde{\wedge} \top = \top$ is shown similarly. □

4.4 Central Paraconsistent Three-Valued Logics

Our final goal in this section is to determine the expressive power of PE3, that is: determine what functions are representable in it. For this we first need some definitions and a lemma.

Definition 4.82 For $x, y \in \{t, f, \top\}$, let $x \preceq y$ if either $x = y$, or $x = f$ and $y = \top$.

Definition 4.83 A set $\mathcal{S} \subseteq \{t, f, \top\}^n$ is a $\preceq$-cone if $\langle y_1, \ldots, y_n \rangle \in \mathcal{S}$ whenever $y_1, \ldots, y_n \in \{t, f, \top\}^n$, and there exist $x_1, \ldots, x_n \in \{t, f, \top\}^n$ such that $\langle x_1, \ldots, x_n \rangle \in \mathcal{S}$ and $x_i \preceq y_i$ for every $1 \leq i \leq n$.

Lemma 4.84 For every $\preceq$-cone $\mathcal{C}$ in $\{t, f, \top\}^n$ there is a formula $\psi_\mathcal{C}$ in $\{\neg, \wedge, \vee\}$ with the following properties:

1. $\mathsf{Var}(\psi_\mathcal{C}) \subseteq \{P_1, \ldots, P_n\}$.

2. For every PE3-valuation ν: $\nu \models_{\mathbf{PE3}} \psi_\mathcal{C}$ iff $\langle \nu(P_1), \ldots, \nu(P_n) \rangle \in \mathcal{C}$.

Proof. For $a \in \{t, f, \top\}$ and a variable p define:

$$\psi_a(p) = \begin{cases} \neg\neg p & \text{if } a = t \\ \neg p & \text{if } a = f \\ p \wedge \neg p & \text{if } a = \top \end{cases}$$

It is easy to check that if ν is in PE3-valuation, then $\nu \models_{\mathbf{PE3}} \psi_a(p)$ iff $a \preceq \nu(p)$.

For $\vec{a} = \langle a_1, \ldots, a_n \rangle$ let $\psi_{\vec{a}} = \psi_{a_1}(P_1) \wedge \psi_{a_2}(P_2) \wedge \cdots \wedge \psi_{a_n}(P_n)$. Then $\nu \models_{\mathbf{PE3}} \psi_{\vec{a}}$ iff $a_i \preceq \nu(P_i)$ for every $1 \leq i \leq n$. Let $\psi_\mathcal{C}$ be the disjunction of all the formulas $\psi_{\vec{a}}$ such that $\vec{a} \in \mathcal{C}$ (and $\neg\neg P_1 \wedge \neg P_1$ if no such $\vec{a}$ exists). The fact that $\mathcal{C}$ is a $\preceq$-cone implies that $\psi_\mathcal{C}$ has the desired properties. □

Definition 4.85 A function $g : \{t, f, \top\}^n \to \{t, f, \top\}$ is $\preceq$-monotonic if $g(x_1, \ldots, x_n) \preceq g(y_1, \ldots, y_n)$ whenever $x_i \preceq y_i$ for every $1 \leq i \leq n$.

Theorem 4.86 A function $g : \{t, f, \top\}^n \to \{t, f, \top\}$ is representable in $\mathcal{M}_{PE3}$ iff it is $\{t, f\}$-closed and $\preceq$-monotonic.

Proof. To show the "only if" part, we prove by induction on the structure of ψ, that if ψ represents $g : \{t, f, \top\}^n \to \{t, f, \top\}$ in $\mathcal{M}_{PE3}$, then g is $\{t, f\}$-closed and $\preceq$-monotonic. This is easy, since $\tilde{\neg}, \tilde{\wedge}$, and $\tilde{\vee}$ obviously have these properties.

We prove the "if" part of the theorem by induction on n.

If $n = 1$ then there are exactly six functions from $\{t, f, \top\}$ to $\{t, f, \top\}$ that are both $\{t, f\}$-closed and $\preceq$-monotonic. It is an easy matter to find them and to verify that each of them is represented in $\mathcal{M}_{PE3}$ by one of the following formulas: $P_1, \neg P_1, \neg\neg P_1, P_1 \wedge \neg P_1, \neg(P_1 \wedge \neg P_1), \neg\neg(P_1 \wedge \neg P_1)$.

Assume the claim is true for n, and suppose that $g : \{t, f, \top\}^{n+1} \to \{t, f, \top\}$ is $\{t, f\}$-closed and $\preceq$-monotonic.. Define $g_t : \{t, f, \top\}^n \to \{t, f, \top\}$, $g_f : \{t, f, \top\}^n \to \{t, f, \top\}$, and $\mathcal{C}_g \subseteq \{t, f, \top\}^n$ as follows:

- $g_t(x_1, \ldots, x_n) = g(x_1, \ldots, x_n, t)$ and $g_f(x_1, \ldots, x_n) = g(x_1, \ldots, x_n, f)$.
- $\mathcal{C}_g = \{\langle x_1, \ldots, x_n \rangle \in \{t, f, \top\}^n \mid g(x_1, \ldots, x_n, \top) = \top\}$.

Obviously, the fact that g is $\{t, f\}$-closed and $\preceq$-monotonic entails that so are also g_t and g_f. Hence there are by the induction hypothesis formulas ψ_{g_t} and ψ_{g_f} that respectively represent them in $\mathcal{M}_{PE3}$. The fact that g is $\preceq$-monotonic entails also that $\mathcal{C}_g$ is a $\preceq$-cone. It follows by Lemma 4.84 that there exists a formulas $\psi_{\mathcal{C}_g}$ with the two properties described there. Define:

$$\psi_g = (\neg\neg P_{n+1} \wedge \psi_{g_t}) \vee (\neg P_{n+1} \wedge \psi_{g_f}) \vee (P_{n+1} \wedge \neg P_{n+1} \wedge \psi_{\mathcal{C}_g}).$$

Now we show that ψ_g represents g in $\mathcal{M}_{PE3}$. So Let ν be an PE3-valuation. We show that $\nu(\psi_g) = g(\nu(P_1), \ldots, \nu(P_n), \nu(P_{n+1}))$.

- Suppose that $\nu(P_{n+1}) = t$. Then $\nu(\neg P_{n+1}) = f$ and $\nu(\neg\neg P_{n+1}) = t$. Hence $\nu(\psi_g) = \nu(\psi_{g_t})$. But the induction hypothesis and the fact that $\nu(P_{n+1}) = t$ together imply that

$$\nu(\psi_{g_t}) = g_t(\nu(P_1), \ldots, \nu(P_n)) = g(\nu(P_1), \ldots, \nu(P_n), \nu(P_{n+1})).$$

It follows that $\nu(\psi_g) = g(\nu(P_1), \ldots, \nu(P_n), \nu(P_{n+1}))$.

- Suppose that $\nu(P_{n+1}) = f$. Then $\nu(\neg P_{n+1}) = t$ and $\nu(\neg\neg P_{n+1}) = f$. Hence $\nu(\psi_g) = \nu(\psi_{g_f})$. But the induction hypothesis and the fact that $\nu(P_{n+1}) = f$ together imply that

$$\nu(\psi_{g_f}) = g_f(\nu(P_1), \ldots, \nu(P_n)) = g(\nu(P_1), \ldots, \nu(P_n), \nu(P_{n+1})).$$

It follows that $\nu(\psi_g) = g(\nu(P_1), \ldots, \nu(P_n), \nu(P_{n+1}))$.

- Suppose that $\nu(P_{n+1}) = \top$. Then $\nu(\neg P_{n+1}) = t$ and $\nu(\neg\neg P_{n+1}) = f$. Hence, in this case $\nu(\psi_g) = \nu(\psi_{g_f}) \widetilde{\vee} (\top \widetilde{\wedge} \nu(\psi_{\mathcal{C}_g}))$. Now there are two subcases to consider:

    - Suppose that $g(\nu(P_1), \ldots, \nu(P_n), \top) \neq \top$.
      In this case, $\langle \nu(P_1), \ldots, \nu(P_n) \rangle \notin \mathcal{C}_g$, and so $\nu(\psi_{\mathcal{C}_g}) = f$. Hence, $\nu(\psi_g) = \nu(\psi_{g_f})$. Now, by the induction hypothesis, $\nu(\psi_{g_f}) = g_f(\nu(P_1), \ldots, \nu(P_n)) = g(\nu(P_1), \ldots, \nu(P_n), f)$. This entails that $\nu(\psi_g) = g(\nu(P_1), \ldots, \nu(P_n), f)$. But the $\preceq$-monotonicity of g implies here that $g(\nu(P_1), \ldots, \nu(P_n), f) = g(\nu(P_1), \ldots, \nu(P_n), \top)$ (because $g(\nu(P_1), \ldots, \nu(P_n), \top) \neq \top$). Therefore, it follows that $\nu(\psi_g) = g(\nu(P_1), \ldots, \nu(P_n), \nu(P_{n+1}))$ in this case.

- Suppose that $g(\nu(P_1),\ldots,\nu(P_n),\mathsf{T}) = \mathsf{T}$.
  In this case, $\langle \nu(P_1),\ldots,\nu(P_n)\rangle \in \mathcal{C}_g$, and so $\nu(\psi_{\mathcal{C}_g}) \in \{t,\mathsf{T}\}$, implying that $\mathsf{T}\widetilde{\wedge}\nu(\psi_{\mathcal{C}_g}) = \mathsf{T}$. In turn, the $\preceq$-monotonicity of g implies that $g(\nu(P_1),\ldots,\nu(P_n),f) \preceq g(\nu(P_1),\ldots,\nu(P_n),\mathsf{T}) = \mathsf{T}$. It follows by the induction hypothesis for g_f that $\nu(\psi_{g_f}) \in \{f,\mathsf{T}\}$, implying that $\nu(\psi_{g_f})\widetilde{\vee}\mathsf{T} = \mathsf{T}$. Hence $\nu(\psi_{g_f})\widetilde{\vee}(\mathsf{T}\widetilde{\wedge}\nu(\psi_{\mathcal{C}_g})) = \mathsf{T}$, and so $\nu(\psi_g) = \mathsf{T}$. Hence, $\nu(\psi_g) = g(\nu(P_1),\ldots,\nu(P_n),\nu(P_{n+1}))$ in this case as well.

It follows that indeed in all cases $\nu(\psi_g) = g(\nu(P_1),\ldots,\nu(P_n),\nu(P_{n+1}))$. □

4.5 Proof Systems

This section provides proof systems for the logics presented in Section 4.4.

4.5.1 Gentzen-type Systems

From Theorem 4.27 and Note 4.28 it follows that all the logics presented in this chapter have corresponding cut-free quasi-canonical Gentzen-type systems. In this section we explicitly present such systems for a large family of logics that includes all the logics that were investigated in Section 4.4, and directly prove their strong soundness and completeness, as well as the fact that they admit strong cut-elimination.

Definition 4.87 ($3^\mathbf{G}_\mathsf{T}$) The family $3^\mathbf{G}_\mathsf{T}$ includes every logic of the form $\mathbf{L}_\mathcal{M}$, where $\mathcal{M} = \langle \{t,f,\mathsf{T}\}, \{t,\mathsf{T}\}, \mathcal{O}\rangle$ is a matrix for a language $\mathcal{L}$ such that:

- $\mathcal{L} \subseteq \mathcal{L}^{\mathsf{F},\mathsf{B}}_{CL} = \{\neg, \wedge, \vee, \supset, \mathsf{F}, \mathsf{B}\}$; $\neg \in \mathcal{L}$; and $\mathcal{L} \cap \{\wedge, \vee, \supset\} \neq \emptyset$.

- The interpretation $\widetilde{\neg}$ of $\neg$ in $\mathcal{M}$ is either Kleene's negation or Sette's negation.

- If $\wedge \in \mathcal{L}$, then its interpretation $\widetilde{\wedge}$ in $\mathcal{M}$ is either Kleene's conjunction, or Sette's conjunction, or Sobociński's conjunction. (See the beginning of Section 4.4.)

- If $\vee \in \mathcal{L}$, then its interpretation $\widetilde{\vee}$ in $\mathcal{M}$ is either Kleene's disjunction, or Sette's disjunction.

- If $\supset \in \mathcal{L}$, then its interpretation $\widetilde{\supset}$ is either D'Ottaviano and da-Costa's implication (used in J_3), or Sette's implication (used in P_1).

- If $\mathsf{F} \in \mathcal{L}$, then its interpretation $\widetilde{\mathsf{F}}$ is f.

- If $\mathsf{B} \in \mathcal{L}$, then its interpretation $\widetilde{\mathsf{B}}$ is T.

Our construction of the Gentzen-type systems for all the logics in $\mathbf{3}_{\top}^{\mathbf{G}}$ is fully modular. They all have the same axioms ($\varphi \Rightarrow \varphi$ and $\Rightarrow \neg \varphi, \varphi$), and the same classical structural rules. Now the logical rules for each connective are determined only by its own interpretation, and are completely independent of the interpretations of the other connectives (including negation). Accordingly, the logical rules of each of the systems are those which are associated with the connectives of its language according to the semantic interpretations of those connectives in the corresponding matrices. Note that in the case of one of the positive classical connectives these rules always include the rules for that connective in LK^+. However, the identity of its rules for its combination with $\neg$ depend on its chosen interpretation.

The rules associated with each interpretation we use of a connective in $\mathcal{L}_{CL}^{F,B}$ can be found in the figures below as follows:

1. The rules for Sette's negation, conjunction, disjunction, and implication can be found in Figure 4.3. That figure fully describes a Gentzen-type system $G_{\mathbf{P}_1}$ for Sette's logic $\mathbf{P}_1$.

2. The rules for Kleene's negation, conjunction, and disjunction, as well as the rules for D'Ottaviano and da-Costa's implication and the rules for the propositional constants F and B, can be found in Figure 4.4, which describes a Gentzen-type system $G_{\mathbf{J}_3^B}$ for $\mathbf{J}_3^{\mathbf{B}}$.

3. The rules for Sobociński's conjunction can be found in Figure 4.5, which describes a Gentzen-type system $G_{\mathbf{SRM}_{\rightarrow}}$ for $\mathbf{SRM}_{\rightarrow}$ in its primitive language.[13]

As an example how the modularity works, Figure 4.6 presents in full (including the rules that come from LK^+) a proof system $G_{\mathbf{PE3}}$ for $\mathbf{PE3}$. The set of rules of $G_{\mathbf{PE3}}$ is a mixture of the pure negation rules of $G_{\mathbf{P}_1}$ (Figure 4.3) and the rules for disjunction and conjunction of $G_{\mathbf{J}_3^B}$ (Figure 4.4). We leave to the reader the task of similarly constructing Gentzen-type proof systems for the other logics that were mentioned in Section 4.4. (Note that for $\mathbf{L} \in \{\mathbf{LP}, \mathbf{LP}^{\mathsf{F}}, \mathbf{LP}^{\mathsf{B}}, \mathbf{LP}^{\mathsf{F,B}}, \mathbf{PAC}, \mathbf{J_3}, \mathbf{PAC}^{\mathsf{B}}, \mathbf{PAC}_{\supset}\}$, a corresponding Gentzen system $G_{\mathbf{L}}$ is obtained from $G_{\mathbf{J}_3^B}$ by simply deleting from it the irrelevant rules, like the rules for $\supset$ and F in the case of $\mathbf{LP}^{\mathsf{B}}$.)

Here are some important notes about the Gentzen-type systems presented in this section:

---

[13] Because of its historical importance, and its crucial role in Part V, Figure 4.5 includes also rules for Sobociński's entailment $\rightarrow$. Note that it is is not difficult to add there also rules for the (definable) implication and disjunction of $\mathbf{SRM}_{\rightarrow}$ (Corollary 4.48).

4.5 Proof Systems

Axioms: $\varphi \Rightarrow \varphi \qquad \Rightarrow \neg\varphi, \varphi$

Rules: All the rules of LK^+, and the following rules for $\neg$:

$$[\neg\neg\Rightarrow] \frac{\Gamma \Rightarrow \Delta, \neg\varphi}{\Gamma, \neg\neg\varphi \Rightarrow \Delta} \qquad [\Rightarrow\neg\neg] \frac{\Gamma, \neg\varphi \Rightarrow \Delta}{\Gamma \Rightarrow \Delta, \neg\neg\varphi}$$

$$[\neg\wedge\Rightarrow] \frac{\Gamma \Rightarrow \Delta, \varphi \quad \Gamma \Rightarrow \Delta, \psi}{\Gamma, \neg(\varphi \wedge \psi) \Rightarrow \Delta} \qquad [\Rightarrow\neg\wedge] \frac{\Gamma, \varphi, \psi \Rightarrow \Delta}{\Gamma \Rightarrow \Delta, \neg(\varphi \wedge \psi)}$$

$$[\neg\vee\Rightarrow] \frac{\Gamma \Rightarrow \Delta, \varphi, \psi}{\Gamma, \neg(\varphi \vee \psi) \Rightarrow \Delta} \qquad [\Rightarrow\neg\vee] \frac{\Gamma, \varphi \Rightarrow \Delta \quad \Gamma, \psi \Rightarrow \Delta}{\Gamma \Rightarrow \Delta, \neg(\varphi \vee \psi)}$$

$$[\neg\supset\Rightarrow] \frac{\Gamma, \varphi \Rightarrow \Delta, \psi}{\Gamma, \neg(\varphi \supset \psi) \Rightarrow \Delta} \qquad [\Rightarrow\neg\supset] \frac{\Gamma \Rightarrow \Delta, \varphi \quad \Gamma, \psi \Rightarrow \Delta}{\Gamma \Rightarrow \neg(\varphi \supset \psi), \Delta}$$

Figure 4.3: The proof system $G_{\mathbf{P}_1}$

Note 4.88

- From Corollary 4.6 it immediately follows that the rule $[\Rightarrow\neg]$ of LK is valid for every logic which is induced by a three-valued paraconsistent matrix. The extra axioms $\Rightarrow \neg\varphi, \varphi$ which are used in all the of the systems introduced in this section are derivable by it from the standard identity axioms. Therefore we could have included this rule in the definition of those systems instead of the new axioms. (Note that this rule is derivable from these axioms using a cut.) We prefer our official formulation above because all of its logical rules are *invertible* (see Lemma 4.93 below). This is a very useful property in proof search and for other goals (as the proofs given below show).

- It is possible to take as axioms in all the systems above only $p \Rightarrow p$, $\neg p \Rightarrow \neg p$, and $\Rightarrow p, \neg p$, where p is a variable (and also the rules for F and B, which are really axioms, in case one or both of them are in the language). All other instances of the axioms are then derivable using the logical rules of each of these systems.

- A much more compact formulation of a Gentzen-type system for $\mathbf{P}_1$, which is easily seen to be equivalent to $G_{\mathbf{P}_1}$, is obtained from LK by restricting the applications of the rule $[\neg\Rightarrow]$ of LK (to infer $\neg\varphi, \Gamma \Rightarrow \Delta$ from $\Gamma \Rightarrow \Delta, \varphi$) to the case in which the active formula (φ) is not atomic. In contrast to $G_{\mathbf{P}_1}$, the soundness of this formulation is lost once $\mathbf{P}_1$ is extended with a new, undefinable connective.

Axioms: $\varphi \Rightarrow \varphi \qquad \Rightarrow \neg\varphi, \varphi$

Rules: All the rules of LK^+, and the following rules for $\neg$, F, and T:

$$[\neg\neg\Rightarrow] \quad \frac{\Gamma, \varphi \Rightarrow \Delta}{\Gamma, \neg\neg\varphi \Rightarrow \Delta} \qquad [\Rightarrow\neg\neg] \quad \frac{\Gamma \Rightarrow \Delta, \varphi}{\Gamma \Rightarrow \Delta, \neg\neg\varphi}$$

$$[\neg\wedge\Rightarrow] \quad \frac{\Gamma, \neg\varphi \Rightarrow \Delta \quad \Gamma, \neg\psi \Rightarrow \Delta}{\Gamma, \neg(\varphi \wedge \psi) \Rightarrow \Delta} \qquad [\Rightarrow\neg\wedge] \quad \frac{\Gamma \Rightarrow \Delta, \neg\varphi, \neg\psi}{\Gamma \Rightarrow \Delta, \neg(\varphi \wedge \psi)}$$

$$[\neg\vee\Rightarrow] \quad \frac{\Gamma, \neg\varphi, \neg\psi \Rightarrow \Delta}{\Gamma, \neg(\varphi \vee \psi) \Rightarrow \Delta} \qquad [\Rightarrow\neg\vee] \quad \frac{\Gamma \Rightarrow \Delta, \neg\varphi \quad \Gamma \Rightarrow \Delta, \neg\psi}{\Gamma \Rightarrow \Delta, \neg(\varphi \vee \psi)}$$

$$[\neg\supset\Rightarrow] \quad \frac{\Gamma, \varphi, \neg\psi \Rightarrow \Delta}{\Gamma, \neg(\varphi \supset \psi) \Rightarrow \Delta} \qquad [\Rightarrow\neg\supset] \quad \frac{\Gamma \Rightarrow \varphi, \Delta \quad \Gamma \Rightarrow \neg\psi, \Delta}{\Gamma \Rightarrow \neg(\varphi \supset \psi), \Delta}$$

$$[F\Rightarrow] \quad \Gamma, F \Rightarrow \Delta$$

$$[\Rightarrow\neg F] \quad \Gamma \Rightarrow \Delta, \neg F$$

$$[\Rightarrow B] \quad \Gamma \Rightarrow \Delta, B$$

$$[\Rightarrow\neg B] \quad \Gamma \Rightarrow \Delta, \neg B$$

Figure 4.4: The proof system $G_{\mathbf{J}_3^B}$

Our next goal is to show the strong soundness and completeness of all these Gentzen-type Systems. Our first step towards this goal is simply to recall what is the semantics of sequents in the context of matrices.

Proposition 4.89 *Let $\mathcal{M}$ be a three-valued paraconsistent matrix, and let ν be an $\mathcal{M}$-valuation. Then $\nu \models_{\mathcal{M}} \Gamma \Rightarrow \Delta$ (where $\Gamma \Rightarrow \Delta$ is a sequent in the language or $\mathcal{M}$) iff either $\nu(\varphi) = f$ for some $\varphi \in \Gamma$, or $\nu(\psi) \neq f$ (i.e., $\nu(\psi) \in \{t, \top\}$) for some $\psi \in \Delta$.*

Proof. Immediate from Definition 1.123 and Proposition 4.4. □

Next, we prove some lemmas.

Lemma 4.90 *Let $\mathcal{M}$ be a 3-valued paraconsistent matrix for a language $\mathcal{L}$, and let $\Gamma \Rightarrow \Delta$ be a sequent in $\mathcal{L}$ which consists of literals (i.e. atomic formulas or negations of atomic formulas).*

1. *$\vdash_{\mathcal{M}} \Gamma \Rightarrow \Delta$ iff either $\Gamma \cap \Delta \neq \emptyset$, or there is a variable p such that $\{p, \neg p\} \subseteq \Delta$, or (in case $F \in \mathcal{L}$) $F \in \Gamma$ or $\neg F \in \Delta$, or (in case $B \in \mathcal{L}$) $B \in \Delta$ or $\neg B \in \Delta$.*

4.5 Proof Systems

Axioms: $\Gamma, \varphi \Rightarrow \Delta, \varphi \qquad \Gamma \Rightarrow \Delta, \varphi, \neg\varphi$

Rules: Weakening, Cut, and the following logical rules:

$$\dfrac{\Gamma, \varphi \Rightarrow \Delta}{\Gamma, \neg\neg\varphi \Rightarrow \Delta} \qquad \dfrac{\Gamma \Rightarrow \Delta, \varphi}{\Gamma \Rightarrow \Delta, \neg\neg\varphi}$$

$$\dfrac{\Gamma, \varphi, \psi \Rightarrow \Delta}{\Gamma, \varphi \wedge \psi \Rightarrow \Delta} \qquad \dfrac{\Gamma \Rightarrow \Delta, \varphi \quad \Gamma \Rightarrow \Delta, \psi}{\Gamma \Rightarrow \Delta, \varphi \wedge \psi}$$

$$\dfrac{\Gamma \Rightarrow \Delta, \varphi \quad \Gamma \Rightarrow \Delta, \psi \quad \Gamma, \neg\varphi, \neg\psi \Rightarrow \Delta}{\Gamma, \neg(\varphi \wedge \psi) \Rightarrow \Delta} \qquad \dfrac{\Gamma, \varphi \Rightarrow \Delta, \neg\psi \quad \Gamma, \psi \Rightarrow \Delta, \neg\varphi}{\Gamma \Rightarrow \Delta, \neg(\varphi \wedge \psi)}$$

$$\dfrac{\Gamma \Rightarrow \Delta, \varphi \quad \Gamma \Rightarrow \Delta, \neg\psi \quad \Gamma, \neg\varphi, \psi \Rightarrow \Delta}{\Gamma, \varphi \rightarrow \psi \Rightarrow \Delta} \qquad \dfrac{\Gamma, \varphi \Rightarrow \Delta, \psi \quad \Gamma, \neg\psi \Rightarrow \Delta, \neg\varphi}{\Gamma \Rightarrow \Delta, \varphi \rightarrow \psi}$$

$$\dfrac{\Gamma, \varphi, \neg\psi \Rightarrow \Delta}{\Gamma, \neg(\varphi \rightarrow \psi) \Rightarrow \Delta} \qquad \dfrac{\Gamma \Rightarrow \Delta, \varphi \quad \Gamma \Rightarrow \Delta, \neg\psi}{\Gamma \Rightarrow \Delta, \neg(\varphi \rightarrow \psi)}$$

Figure 4.5: The proof system $G_{\mathbf{SRM}_\rightarrow}$

2. *If $\mathbf{L}$ is one of the logics dealt with in this section, and $\vdash_{\mathbf{L}} \Gamma \Rightarrow \Delta$, then $\vdash_{G_{\mathbf{L}}} \Gamma \Rightarrow \Delta$.*

Proof. Suppose Γ and Δ consist only of literals.

1. From Proposition 4.4 it follows that if Γ and Δ satisfies one of the six conditions, then $\vdash_{\mathcal{M}} \Gamma \Rightarrow \Delta$. Suppose now that $\Gamma \Rightarrow \Delta$ does not satisfy any of them. Define:

$$\nu(p) = \begin{cases} f & \text{if } p \in \Delta, \\ t & \text{if } \neg p \in \Delta, \\ \top & \text{otherwise.} \end{cases}$$

Then ν is well-defined, and $\nu \not\models_{\mathcal{M}} \Gamma \Rightarrow \Delta$. Hence $\not\vdash_{\mathcal{M}} \Gamma \Rightarrow \Delta$.

2. This follows from the first part and the fact that every sequent which satisfies the condition given in that part is an axiom of $G_{\mathbf{L}}$. □

Lemma 4.91 *Let $\mathbf{L} \in \mathbf{3}_\top^{\mathbf{G}}$, and let $\mathcal{M}$ be the three-valued paraconsistent matrix which induces $\mathbf{L}$. Then every logical rule of $G_{\mathbf{L}}$ is strongly sound for $\vdash_{\mathcal{M}}$: if S is the set of premises of (an application of) such a rule, and s is its conclusion, then $S \vdash_{\mathcal{M}} s$.*

Axioms: $\varphi \Rightarrow \varphi$ $\Rightarrow \neg\varphi, \varphi$

Rules: Cut, Weakening, and the following logical rules:

$$[\neg\neg\Rightarrow] \quad \frac{\Gamma \Rightarrow \Delta, \neg\varphi}{\Gamma, \neg\neg\varphi \Rightarrow \Delta} \qquad [\Rightarrow\neg\neg] \quad \frac{\Gamma, \neg\varphi \Rightarrow \Delta}{\Gamma \Rightarrow \Delta, \neg\neg\varphi}$$

$$[\wedge\Rightarrow] \quad \frac{\Gamma, \psi, \varphi \Rightarrow \Delta}{\Gamma, \psi \wedge \varphi \Rightarrow \Delta} \qquad [\Rightarrow\wedge] \quad \frac{\Gamma \Rightarrow \Delta, \psi \quad \Gamma \Rightarrow \Delta, \varphi}{\Gamma \Rightarrow \Delta, \psi \wedge \varphi}$$

$$[\neg\wedge\Rightarrow] \quad \frac{\Gamma, \neg\varphi \Rightarrow \Delta \quad \Gamma, \neg\psi \Rightarrow \Delta}{\Gamma, \neg(\varphi \wedge \psi) \Rightarrow \Delta} \qquad [\Rightarrow\neg\wedge] \quad \frac{\Gamma \Rightarrow \Delta, \neg\varphi, \neg\psi}{\Gamma \Rightarrow \Delta, \neg(\varphi \wedge \psi)}$$

$$[\vee\Rightarrow] \quad \frac{\Gamma, \psi \Rightarrow \Delta \quad \Gamma, \varphi \Rightarrow \Delta}{\Gamma, \psi \vee \varphi \Rightarrow \Delta} \qquad [\Rightarrow\vee] \quad \frac{\Gamma \Rightarrow \Delta, \psi, \varphi}{\Gamma \Rightarrow \Delta, \psi \vee \varphi}$$

$$[\neg\vee\Rightarrow] \quad \frac{\Gamma, \neg\varphi, \neg\psi \Rightarrow \Delta}{\Gamma, \neg(\varphi \vee \psi) \Rightarrow \Delta} \qquad [\Rightarrow\neg\vee] \quad \frac{\Gamma \Rightarrow \Delta, \neg\varphi \quad \Gamma \Rightarrow \Delta, \neg\psi}{\Gamma \Rightarrow \Delta, \neg(\varphi \vee \psi)}$$

Figure 4.6: The Gentzen-type system $G_{\mathbf{PE3}}$

Proof. Tedious, but easy. We do as an example the case of the rule for introducing $\neg(\varphi \wedge \psi)$ on the right in case the interpretation $\tilde{\wedge}$ of $\wedge$ in $\mathcal{M}$ is Sobociński's conjunction. (See Figure 4.5.) So assume that $\nu \models_{\mathcal{M}} \Gamma, \varphi \Rightarrow \Delta, \neg\psi$ and $\nu \models_{\mathcal{M}} \Gamma, \psi \Rightarrow \Delta, \neg\varphi$. We show that $\nu \models_{\mathcal{M}} \Gamma \Rightarrow \Delta, \neg(\varphi \wedge \psi)$. If $\nu \models_{\mathcal{M}} \Gamma \Rightarrow \Delta$ we are done. Otherwise, we have that either $\nu(\varphi) = f$ or $\nu(\psi) \neq t$, and either $\nu(\psi) = f$ or $\nu(\varphi) \neq t$. This leaves four possibilities. It is easy to check that in all of them $\nu(\varphi \wedge \psi) \neq t$, i.e., $\nu(\neg(\varphi \wedge \psi)) \neq f$. □

Lemma 4.92 *Let* $\mathbf{L} \in 3_{\top}^{\mathbf{G}}$. *Then* $G_{\mathbf{L}}$ *is strongly sound for* $\mathbf{L}$.

Proof. By Lemma 4.91 and Note 1.127, we need only to check that the axioms of $G_{\mathbf{L}}$ are valid in $\mathbf{L}$. This is obvious. □

Lemma 4.93 *Let* $\mathbf{L}$ *and* $\mathcal{M}$ *be like in Lemma 4.91. Every logical rule* r *of* $G_{\mathbf{L}}$ *is strongly invertible in* $\vdash_{\mathcal{M}}$: *If* s_c *is the conclusion of* r *and* s_p *is any of its premises, then* $s_c \vdash_{\mathcal{M}} s_p$.

Proof. Again we do as an example the rule of $G_{\mathbf{SRM}_\neg}$ for introducing $\neg(\varphi \wedge \psi)$ on the right in case the interpretation $\tilde{\wedge}$ of $\wedge$ in $\mathcal{M}$ is Sobociński's conjunction. So assume that $\nu \models_{\mathcal{M}} \Gamma \Rightarrow \Delta, \neg(\varphi \wedge \psi)$. We show, e.g., that $\nu \models_{\mathcal{M}} \Gamma, \psi \Rightarrow \Delta, \neg\varphi$. If $\nu \models_{\mathcal{M}} \Gamma \Rightarrow \Delta$ we are done. Otherwise $\nu(\neg(\varphi \wedge \psi)) \neq f$, and so $\nu(\varphi \wedge \psi) \neq t$. This implies that either $\nu(\varphi) = f$, or $\nu(\psi) = f$, or $\nu(\varphi) = \nu(\psi) = \top$, and so either $\nu(\psi) = f$ or $\nu(\neg\varphi) \neq f$. In both cases $\nu \models_{\mathcal{M}} \Gamma, \psi \Rightarrow \Delta, \neg\varphi$. □

4.5 Proof Systems

Lemma 4.94 *Let $\mathbf{L}$ and $\mathcal{M}$ be like in Lemma 4.91, and let s be a sequent in the language of $\mathbf{L}$. If $\vdash_{\mathcal{M}} s$ then s has a cut-free proof in $G_{\mathbf{L}}$.*

Proof. It is easy to check that by applying the logical rules of $G_{\mathbf{L}}$ backward, and using Lemma 4.93, we can construct for every sequent s a finite set $S(s)$ with the following properties:

1. Each element of $S(s)$ is a sequent which consists only of literals.

2. $s \vdash_{\mathcal{M}} s'$ for every element s' of $S(s)$.

3. There is a cut-free proof of s from $S(s)$.

Suppose now that $\vdash_{\mathcal{M}} s$. By Lemma 4.93 and the property 2 of $S(s)$ this implies that $\vdash_{\mathcal{M}} s'$ for every element s' of $S(s)$. By Lemma 4.90 and properties 1 and 3 of $S(s)$, it follows that s has a cut-free proof in $G_{\mathbf{L}}$. □

Now we are ready to prove the two main results of this section.

Theorem 4.95 (soundness and completeness of $G_{\mathbf{L}}$) *Let $\mathbf{L} \in 3^{\mathbf{G}}_{\top}$.*

1. *$G_{\mathbf{L}}$ is sound and complete for $\mathbf{L}$.*

2. *$\mathcal{T} \vdash_{G_{\mathbf{L}}} \psi$ iff $\mathcal{T} \vdash_{\mathbf{L}} \psi$.*

Proof.

1. Immediate from Lemmas 4.92 and 4.94.

2. This follows from the Part 1, Proposition 1.126, and Theorem 3.17. □

Theorem 4.96 (cut-elimination for $G_{\mathbf{L}}$) *Let $\mathbf{L} \in 3^{\mathbf{G}}_{\top}$. Then $G_{\mathbf{L}}$ admits cut-elimination (see Definition 1.120).*

Proof. Suppose that $\vdash_{G_{\mathbf{L}}} s$. By Lemma 4.92, $\vdash_{\mathbf{L}} s$. By Lemma 4.94 this implies that s has a cut-free proof in $G_{\mathbf{L}}$. □

Note 4.97

- From the last theorem and Theorem 1.122 it follows that all the Gentzen-type systems discussed in this section actually admit *strong* cut-elimination.

- By Lemma 4.92, each of the systems dealt with in this section is actually *strongly* sound with respect to its corresponding 3-valued matrix. Using the stronger proof methods used in the next part, it can be shown that each of them is also *strongly* complete (in the sense of Definition 1.123) for that matrix.

4.5.2 Hilbert-type Systems

Let **L** be a logic with an implication $\supset$, and suppose that $G_\mathbf{L}$ is a sound and complete Gentzen-type system for **L** that includes $LK_\supset$. Using the equivalences between $LK_\supset$ and $HCL_\supset$ (which are both sound and complete with respect to $\mathbf{CL}_\supset$), it is usually a standard exercise to translate $G_\mathbf{L}$ into an equivalent Hilbert-type system $H_\mathbf{L}$, which has [MP] for $\supset$ as its sole rule of inference. (The task is even easier if G contains LK^+, in which case one can use the convenient equivalences between LK^+ and HCL^+). In particular, it is a routine matter to construct a Hilbert-type system of the above sort for every three-valued paraconsistent logic **L** from the family $3_\top^G$ (Definition 4.87) that has an implication. As examples, in this section we present the results of this process in the cases of the logics with an implication that were studied in Sections 4.4.4 and 4.4.5.

Definition 4.98 ($H_\mathbf{PAC}$, $H_{\mathbf{PAC}^B}$, $H_{\mathbf{PAC}_\supset}$, $H_{\mathbf{J}_3}$, $H_{\mathbf{J}_3^B}$) Figure 4.7 contains Hilbert-type proof systems for **PAC** and $\mathbf{J}_3$. Hilbert-type proof systems $H_{\mathbf{PAC}^B}$ and $H_{\mathbf{J}_3^B}$ for $\mathbf{PAC}^B$ and $\mathbf{J}_3^B$ (respectively) are obtained by adding to $H_\mathbf{PAC}$ and $H_{\mathbf{J}_3}$ (respectively) the axioms B and ¬B. A Hilbert-type proof system $H_{\mathbf{PAC}_\supset}$ for $\mathbf{PAC}_\supset$ is obtained from $H_\mathbf{PAC}$ by replacing [t] with either of the two schemas mentioned in the first item of Note 1.91, changing $[\Rightarrow \neg \supset]$ to $\varphi \supset (\neg\psi \supset \neg(\varphi \supset \psi))$, and deleting all axioms related to $\wedge$ or $\vee$.

Note 4.99 The negation axioms in Figure 4.7 have been considered under the following different names in the literature. In what follows we shall frequently use these shorter names instead of those used in Figure 4.7.

- $[\neg\neg \Rightarrow]$ and $[\Rightarrow \neg\neg]$ are called [c] and [e] (respectively) in [113, 116].
- $[\neg \supset\Rightarrow 1]$ and $[\neg \supset\Rightarrow 2]$ are called $[n_\supset^l]_1$ and $[n_\supset^l]_2$ (respectively) in [65], and their combination is called there $[n_\supset^l]$.
- $[\Rightarrow \neg \supset]$ is called $[n_\supset^r]$ in [65].
- $[\neg\vee \Rightarrow 1]$ and $[\neg\vee \Rightarrow 2]$ are called $[n_\vee^l]_1$ and $[n_\vee^l]_2$ (respectively) in [65], and their combination is called there $[n_\vee^l]$.
- $[\Rightarrow \neg\vee]$ is called $[n_\vee^r]$ in [65].
- $[\neg\wedge \Rightarrow]$ is called $[n_\wedge^l]$ in [65].
- $[\Rightarrow \neg \wedge 1]$ and $[\Rightarrow \neg \wedge 2]$ are called $[n_\wedge^r]_1$ and $[n_\wedge^r]_2$ (respectively) in [65], and their combination is called there $[n_\wedge^r]$.

4.5 Proof Systems

Inference Rule: [MP] $\dfrac{\psi \quad \psi \supset \varphi}{\varphi}$

Axioms of $H_{\mathbf{PAC}}$: The axioms of HCL^+ and:

$[\mathsf{t}]$ $\quad\quad\quad \neg \psi \vee \psi$

$[\neg\neg\Rightarrow]$ $\quad\quad \neg\neg\varphi \supset \varphi$

$[\Rightarrow\neg\neg]$ $\quad\quad \varphi \supset \neg\neg\varphi$

$[\neg\supset\Rightarrow 1]$ $\quad \neg(\varphi \supset \psi) \supset \varphi$

$[\neg\supset\Rightarrow 2]$ $\quad \neg(\varphi \supset \psi) \supset \neg\psi$

$[\Rightarrow\neg\supset]$ $\quad\quad (\varphi \wedge \neg\psi) \supset \neg(\varphi \supset \psi)$

$[\neg\vee\Rightarrow 1]$ $\quad \neg(\varphi \vee \psi) \supset \neg\varphi$

$[\neg\vee\Rightarrow 2]$ $\quad \neg(\varphi \vee \psi) \supset \neg\psi$

$[\Rightarrow\neg\vee]$ $\quad\quad (\neg\varphi \wedge \neg\psi) \supset \neg(\varphi \vee \psi)$

$[\neg\wedge\Rightarrow]$ $\quad\quad \neg(\varphi \wedge \psi) \supset (\neg\varphi \vee \neg\psi)$

$[\Rightarrow\neg\wedge 1]$ $\quad \neg\varphi \supset \neg(\varphi \wedge \psi)$

$[\Rightarrow\neg\wedge 2]$ $\quad \neg\psi \supset \neg(\varphi \wedge \psi)$

Axioms of $H_{\mathbf{J_3}}$: The axioms of $H_{\mathbf{PAC}}$ and:

$[\mathsf{F}\supset]$ $\quad\quad\quad \mathsf{F} \supset \psi$

Figure 4.7: The proof systems $H_{\mathbf{PAC}}$ and $H_{\mathbf{J_3}}$

Theorem 4.100 $\vdash_{H_{\mathbf{L}}} = \vdash_{G_{\mathbf{L}}}$ for $\mathbf{L} \in \{\mathbf{PAC}_{\supset}, \mathbf{PAC}, \mathbf{PAC}^{\mathbf{B}}, \mathbf{J_3}, \mathbf{J_3^B}\}$.

Proof. Using cuts and the fact that $\vdash_{LK^+} \psi, \psi \supset \varphi \Rightarrow \varphi$, it is easy to show by induction on length of proofs in $H_{\mathbf{L}}$ that if $\Gamma \vdash_{H_{\mathbf{L}}} \varphi$ (where Γ is finite) then $\Gamma \vdash_{G_{\mathbf{L}}} \varphi$. All one needs to do is to show that $\vdash_{G_{\mathbf{L}}} \varphi$ for every axiom φ of $H_{\mathbf{L}}$, and this is straightforward. It immediately follows that $\vdash_{H_{\mathbf{L}}} \subseteq \vdash_{G_{\mathbf{L}}}$.

For the converse it would be more convenient to use the versions of the Gentzen-type systems which employ lists of formulas rather than finite sets (and so have all the basic structural rules — see Chapter 1), and to treat each of the six logics separately.

$\mathbf{L} = \mathbf{PAC}$. In this case it is easy to prove (either syntactically, using the cut-elimination theorem for $G_{\mathbf{PAC}}$, or semantically, using the soundness theorem for it) that a sequent $s = \varphi_1,\ldots,\varphi_n \Rightarrow \psi_1,\ldots,\psi_m$ is provable in $G_{\mathbf{PAC}}$ only if $m > 0$. For each such sequent s define a translation $Tr_{\mathbf{L}}(s)$ by $Tr_{\mathbf{L}}(s) = \varphi_1 \wedge \ldots \wedge \varphi_n \supset \psi_1 \vee \ldots \vee \psi_m$ (note that in particular: $Tr_{\mathbf{L}}(\Rightarrow \psi_1,\ldots,\psi_m) = \psi_1 \vee \ldots \vee \psi_m$). Obviously, to show that $\vdash_{G_{\mathbf{L}}} \subseteq \vdash_{H_{\mathbf{L}}}$ it suffices to prove that if $\vdash_{G_{\mathbf{L}}} s$ then $\vdash_{H_{\mathbf{L}}} Tr_{\mathbf{L}}(s)$. We prove this claim by induction on length of proofs in $G_{\mathbf{L}}$. This is a routine (though tedious) induction, and here we shall do as examples three of the various possible cases that should be considered.

- Suppose that s is an axiom of the form $\Rightarrow \neg\varphi, \varphi$. Then $Tr_{\mathbf{L}}(s)$ is an instance of the axiom [t] of $\mathbf{L}(= \mathbf{PAC})$.
- Suppose that s is inferred from s_1 and s_2 using $[\supset\Rightarrow]$. Then there are sentences φ,ψ,τ_2, and τ_1 such that $Tr_{\mathbf{L}}(s) = \tau_1 \wedge (\varphi \supset \psi) \supset \tau_2$, $Tr_{\mathbf{L}}(s_1) = \tau_1 \supset \tau_2 \vee \varphi$, and $Tr_{\mathbf{L}}(s_2) = \tau_1 \wedge \psi \supset \tau_2$ (the case where $Tr_{\mathbf{L}}(s) = (\varphi \supset \psi) \supset \tau_2$, $Tr_{\mathbf{L}}(s_1) = \tau_2 \vee \varphi$, and $Tr_{\mathbf{L}}(s_2) = \psi \supset \tau_2$ is similar, but easier). By induction hypothesis, $\vdash_{H_{\mathbf{L}}} Tr_{\mathbf{L}}(s_1)$ and $\vdash_{H_{\mathbf{L}}} Tr_{\mathbf{L}}(s_2)$. Now,

$$P_1 \supset P_2 \vee P_3,\ P_1 \wedge P_4 \supset P_2 \vdash_{CL^+} P_1 \wedge (P_3 \supset P_4) \supset P_2.$$

Since HCL^+ is complete for $\mathbf{CL^+}$, and $H_{\mathbf{L}}$ extends HCL^+, it follows (by substituting τ_1 for P_1, τ_2 for P_2, φ for P_3, and ψ for P_4) that $Tr_{\mathbf{L}}(s_1), Tr_{\mathbf{L}}(s_2) \vdash_{H_{\mathbf{L}}} Tr_{\mathbf{L}}(s)$. Hence, $\vdash_{H_{\mathbf{L}}} Tr_{\mathbf{L}}(s)$.

- Suppose that s is inferred from s_1 using $[\neg \supset \Rightarrow]$. Then there are sentences φ,ψ,τ_2, and τ_1 such that $Tr_{\mathbf{L}}(s) = \tau_1 \wedge \neg(\varphi \supset \psi) \supset \tau_2$, while $Tr_{\mathbf{L}}(s_1) = \tau_1 \wedge \varphi \wedge \neg\psi \supset \tau_2$ (again, the case where there is no τ_1 is easier). By induction hypothesis, $\vdash_{H_{\mathbf{L}}} Tr_{\mathbf{L}}(s_1)$. Now,

$$P_5 \supset P_3,\ P_5 \supset P_4,\ P_1 \wedge P_3 \wedge P_4 \supset P_2 \vdash_{CL^+} P_1 \wedge P_5 \supset P_2.$$

Since HCL^+ is complete for $\mathbf{CL^+}$ and $H_{\mathbf{L}}$ is an extension of HCL^+, it follows (by substituting τ_1 for P_1, τ_2 for P_2, φ for P_3, $\neg\psi$ for P_4, and $\neg(\varphi \supset \psi)$ for P_5) that

$$\neg(\varphi \supset \psi) \supset \varphi,\ \neg(\varphi \supset \psi) \supset \neg\psi,\ Tr_{\mathbf{L}}(s_1) \vdash_{H_{\mathbf{L}}} Tr_{\mathbf{L}}(s).$$

Using the axioms $[\neg \supset \Rightarrow 1]$ and $[\neg \supset \Rightarrow 2]$ of $H_{\mathbf{L}}$, it follows from the induction hypothesis for s_1 that $\vdash_{H_{\mathbf{L}}} Tr_{\mathbf{L}}(s)$.

The proofs in the other cases are similar. One should only note that in some of the cases (e.g. when s is inferred from s_1 using weakening on the right) there are four subcases to consider (rather than just two as in the cases handled above): that we have both τ_1 and τ_2; that we have τ_1 but not τ_2; that we have τ_2 but not τ_1; and that we have neither τ_1 nor τ_2.

4.5 Proof Systems

L = PAC^B. The proof in this case is very similar to that in the previous one, and is left to the reader.

L = J$_3$. The proof in this case is again similar to that in case of **PAC**. The main difference is that now also sequents of the form $\Gamma \Rightarrow$ may be proved in G_L, and so the translation of sequents into formulas should be extended to these type of sequents. This is done by letting $Tr_L(\varphi_1, \ldots, \varphi_n \Rightarrow)$ be $\varphi_1 \wedge \ldots \wedge \varphi_n \supset F$. Details are left to the reader.

L = J$_3^B$. The proof in this case is very similar to that in the case of **J$_3$**, and is left to the reader.

L = PAC$_\supset$. This time there is another problem: $\wedge$ and $\vee$ are not included in the language of **PAC$_\supset$**, and so we cannot employ the translation function that was used in the case of **PAC**. However, we can use the facts that $\varphi \vee \psi$ is equivalent in **CL$^+$** to $(\varphi \supset \psi) \supset \psi$ and $\varphi \wedge \psi \supset \tau$ is equivalent in **CL$^+$** to $\varphi \supset \psi \supset \tau$. With the help of this fact we can transform the definition of $Tr_{\mathbf{PAC}}$ into an equivalent (in **CL$^+$**) definition in which $\wedge$ and $\vee$ are not used:

$$Tr_{\mathbf{PAC}_\supset}(\varphi_1, \ldots, \varphi_n \Rightarrow \psi_1, \ldots, \psi_m) =$$
$$\varphi_1 \supset \ldots \varphi_n \supset (\ldots((\psi_1 \supset \psi_2) \supset \psi_2) \supset \ldots \supset \psi_m) \supset \psi_m.$$

With this definition, and using $HCL_\supset$ (which by Theorem 1.90 is sound and complete with respect to $\{\supset\}$-fragment of classical logic) instead of HCL^+, we can proceed in a way which is very similar to that used in the case **L = PAC**. □

Theorem 4.101 (soundness and completeness) H_L *is sound and complete for* **L** *for every* $\mathbf{L} \in \{\mathbf{PAC}_\supset, \mathbf{PAC}, \mathbf{PAC}^B, \mathbf{J}_3, \mathbf{J}_3^B\}$. *In other words:* $\mathcal{T} \vdash_{H_L} \psi$ *iff* $\mathcal{T} \vdash_L \psi$ *for each such* **L**.

Proof. This is a direct corollary of Theorems 4.100 and 4.95.

It is also possible to provide an alternative *direct* proof of the completeness part of the theorem by using Theorem 1.98. (The easy direct proof of the soundness of **L** for the corresponding 3-valued matrix is left to the reader.) As an example, we present such a proof for the case **L = PAC**. So assume that $\mathcal{T} \nvdash_{H_L} \psi$. We construct a model of $\mathcal{T}$ which is not a model of ψ. By Theorem 1.98 there is a theory $\mathcal{T}^*$ in $\mathcal{L}$ with the following properties:

1. $\mathcal{T} \subseteq \mathcal{T}^*$.

2. $\mathcal{T}^* \nvdash_{H_L} \psi$.

3. For every φ and τ, $\mathcal{T}^* \vdash_{H_L} \varphi \supset \tau$ iff either $\mathcal{T}^* \nvdash_{H_L} \varphi$ or $\mathcal{T}^* \vdash_{H_L} \tau$.

4. For every φ and τ, $\mathcal{T}^* \vdash_{H_L} \varphi \wedge \tau$ iff $\mathcal{T}^* \vdash_{H_L} \varphi$ and $\mathcal{T}^* \vdash_{H_L} \tau$.

5. For every φ and τ, $\mathcal{T}^* \vdash_{H_L} \varphi \vee \tau$ iff either $\mathcal{T}^* \vdash_{H_L} \varphi$ or $\mathcal{T}^* \vdash_{H_L} \tau$.

6. For every φ, either $\mathcal{T}^* \vdash_{H_L} \varphi$, or $\mathcal{T}^* \vdash_{H_L} \neg\varphi$.

Now define a valuation ν as follows:

$$\nu(\varphi) = \begin{cases} \top & \text{if } \mathcal{T}^* \vdash_{H_L} \varphi \text{ and } \mathcal{T}^* \vdash_{H_L} \neg\varphi, \\ t & \text{if } \mathcal{T}^* \vdash_{H_L} \varphi \text{ and } \mathcal{T}^* \nvdash_{H_L} \neg\varphi, \\ f & \text{if } \mathcal{T}^* \nvdash_{H_L} \varphi. \end{cases}$$

Obviously, ν is well-defined. By Properties 1 and 2 of $\mathcal{T}^*$, $\nu(\psi) = f$, while $\nu(\varphi)$ is designated in case $\varphi \in \mathcal{T}$. Thus, it remains to show that ν is actually a valuation. Below we show that $\nu(\neg\varphi) = \widetilde{\neg}\nu(\varphi)$ and that $\nu(\varphi \supset \tau) = \nu(\varphi)\widetilde{\supset}\nu(\tau)$, leaving the other two cases for the reader.

The case of $\neg$:

- The fact that if $\nu(\varphi) = t$ then $\nu(\neg\varphi) = f$ directly follows from the definition of ν

- Suppose that $\nu(\varphi) = f$. Then $\mathcal{T}^* \nvdash_{H_L} \varphi$. Therefore, Property 6 of $\mathcal{T}^*$ and the axiom $[\neg\neg \Rightarrow]$ respectively imply that $\mathcal{T}^* \vdash_{H_L} \neg\varphi$ and $\mathcal{T}^* \nvdash_{H_L} \neg\neg\varphi$. Hence $\nu(\neg\varphi) = t$.

- Suppose that $\nu(\varphi) = \top$. Then $\mathcal{T}^* \vdash_{H_L} \varphi$ and $\mathcal{T}^* \vdash_{H_L} \neg\varphi$. By the axiom $[\Rightarrow \neg\neg]$, also $\mathcal{T}^* \vdash_{H_L} \neg\neg\varphi$. It follows that $\nu(\neg\varphi) = \top$.

The case of $\supset$:

- Suppose that $\nu(\varphi) = f$. Then $\mathcal{T}^* \nvdash_{H_L} \varphi$. By Property 3 of $\mathcal{T}^*$ and by axiom $[\neg \supset\Rightarrow 1]$ (respectively), this entails that $\mathcal{T}^* \vdash_{H_L} \varphi \supset \tau$ but $\mathcal{T}^* \nvdash_{H_L} \neg(\varphi \supset \tau)$. Hence, $\nu(\varphi \supset \tau) = t = \nu(\varphi)\widetilde{\supset}\nu(\tau)$.

- Suppose that $\nu(\varphi) \in \{t, \top\}$. Then $\mathcal{T}^* \vdash_{H_L} \varphi$. Hence Property 3 of $\mathcal{T}^*$ implies that $\mathcal{T}^* \vdash_{H_L} \varphi \supset \tau$ iff $\mathcal{T}^* \vdash_{H_L} \tau$, while $[\neg \supset\Rightarrow 2]$ and $[\Rightarrow \neg \supset]$ entail that $\mathcal{T}^* \vdash_{H_L} \neg(\varphi \supset \tau)$ iff $\mathcal{T}^* \vdash_{H_L} \neg\tau$. It follows that $\nu(\varphi \supset \tau) = \nu(\tau)$. But $\nu(\varphi)\widetilde{\supset}\nu(\tau) = \nu(\tau)$ too, because $\nu(\varphi) \in \{t, \top\}$. $\square$

Note 4.102 It is as straightforward to transform the system $G_{\mathbf{P}_1}$ (Figure 4.3) into an equivalent Hilbert-type system $H_{\mathbf{P}_1}$ as it has been in the case of the systems connected with Figure 4.4. An alternative Hilbert-type system $H^*_{\mathbf{P}_1}$ for $\mathbf{P}_1$ with fewer axioms is presented in Figure 4.8. It was obtained by the same method from the compact Gentzen-type system for that logic that was described in Note 4.88.

Inference Rule: $\quad$ [MP] $\dfrac{\psi \quad \psi \supset \varphi}{\varphi}$

Axioms: The axioms of HCL^+ and:

$\quad$ [t] $\qquad\qquad \neg \psi \vee \psi$

$\quad$ [$\neg \supset \Rightarrow$] $\qquad (\varphi \supset \psi) \supset \neg(\varphi \supset \psi) \supset \tau$

$\quad$ [$\neg \vee \Rightarrow$] $\qquad (\varphi \vee \psi) \supset \neg(\varphi \vee \psi) \supset \tau$

$\quad$ [$\neg \wedge \Rightarrow$] $\qquad (\varphi \wedge \psi) \supset \neg(\varphi \wedge \psi) \supset \tau$

$\quad$ [$\neg \Rightarrow$] $\qquad \neg\varphi \supset \neg\neg\varphi \supset \tau$

Figure 4.8: The proof system $H^*_{\mathbf{P}_1}$

The proofs of the soundness and completeness of $H^*_{\mathbf{P}_1}$ for $\mathbf{P}_1$ are very similar to the proofs given above of the soundness and completeness of $H_{\mathbf{PAC}}$ for PAC. However, in the direct proof the effect of the axioms in Figures 4.8 other than [t] is here to exclude the possibility that both $\mathcal{T}^* \vdash_{H_L} \varphi$ and $\mathcal{T}^* \vdash_{H_L} \neg\varphi$ in case φ is not atomic. The fact that the function ν constructed in the proof of Theorem 4.101 is a valuation here too easily follows from this observation and Properties 1–6 of $\mathcal{T}^*$. Details are left for the reader.

Note 4.103 By using the definable implication of $\mathbf{SRM}_{\neg\rightarrow}$ which was described in the proof of Theorem 4.44, it is not difficult to construct for $\mathbf{SRM}_{\neg\rightarrow}$ too a sound and complete Hilbert-type system (in the language of $\{\neg, \wedge, \supset\}$) which employs MP for $\supset$ as the only inference rule. An alternative one, which reflects the connections of $\mathbf{SRM}_{\neg\rightarrow}$ with the family of relevance logics, will be given in Chapter 13. (See Corollary 13.71 and Note 13.72.)

4.6 Bibliographical Notes and Further Reading

Three-valued matrices is an old subject which may be traced back to the work of Łukasiewicz [201] (see also [202]). They have been found useful in many areas, like database theory [123, 171] (the database query language SQL implements a 3-valued logic as a means of handling comparisons with null field content; See also [222]), logic programming [134, 137, 156, 292], verification of programs [77], fuzzy and approximative reasoning [133], design issues in digital circuits, different aspects of AI and non-monotonic reasoning [135, 136], semantics for argumentation frameworks [305], and so forth.

As this chapter demonstrates, three-valued matrices have been extensively investigated and used also in the context of paraconsistent reasoning. To the many references given in the body of this chapter we should add here the following:

- [113, 205, 208], a series of papers by Carnielli, Coniglio and Marcos in which the 2^{13} LFIs of the family **8Kb** are investigated and shown to be maximally paraconsistent relative to classical logic;

- [18, 19], where some results of this chapter, including Theorem 4.29, have first been presented;

- [56, 61], from which the material on **PE3** and self-extensionality in paraconsistent three-valued logics is taken. (Note that both of these last two papers use a notion of "paraconsistent logic" which is much more general than the one used in this book.)

- Papers like [234] and [257], in which three-valued matrices with a single designated values are used for paraconsistent reasoning. (See Section 3.3 and footnote 6.)

Beyond the matrices investigated in this chapter, two other three-valued matrices are studied and used for paraconsistent reasoning in Chapters 8 (Corollary 8.109 and Note 8.110). Chapter 15 contains further information about the logic **RM$_3$** (see Notes 15.18 and 15.46).

This Chapter is an extensively extended and improved version of [15].

Chapter 5
Paradefinite Matrices

Uncertainty of information involves two main phenomena: inconsistency and incompleteness. Three-valued matrices (the subject of the previous chapter) can be used to handle each of these two types of uncertainty alone, but not to handle both of them simultaneously. The subject of this chapter are logics that can do that, and are adequate for reasoning with information that might be *both* contradictory and partial. Such logics are called paradefinite. Here is the precise definition:

Definition 5.1 (paracomplete/paradefinite logics) Let $\mathbf{L} = \langle \mathcal{L}, \vdash_\mathbf{L} \rangle$ be a propositional logic for a language $\mathcal{L}$ with a unary connective $\neg$.

- $\mathbf{L}$ is $\neg$-*paracomplete* if $\neg$ is a negation for $\mathbf{L}$ which is not complete for $\mathbf{L}$ (see Definition 2.5).

- $\mathbf{L}$ is $\neg$-*paradefinite* if it is both $\neg$-paraconsistent and $\neg$-paracomplete.[1]

We say that a matrix $\mathcal{M}$ is $\neg$-*paracomplete* ($\neg$-*paradefinite*) if so is $\mathbf{L}_\mathcal{M}$. As usual, we shall sometimes omit the prefix $\neg$ from the above notions.

As we show below, the availability of at least four different truth values is needed for developing paradefinite logics in the framework of matrices. This fact provides the primary motivation for using four-valued matrices. In this chapter we consider the most useful paradefinite four-valued matrices, as well as their natural generalizations to algebraic structures called *bilattices*.

5.1 The Four-Valued Paradefinite Framework

First we show that any paradefinite matrix should be at least four-valued.

---
[1] Paradefinite logics are called 'non-alethic' by da Costa, and 'paranormal' by Béziau (see [93]). We think that our terminology is a bit more definite than these alternatives.

Proposition 5.2 Let $\mathcal{M} = \langle \mathcal{V}, \mathcal{D}, \mathcal{O} \rangle$ be a matrix for a language with $\neg$.

1. If $\mathcal{M}$ is $\neg$-paracomplete then there is an element $\bot \in \mathcal{V}$ such that $\bot \notin \mathcal{D}$ and $\tilde{\neg}\bot \notin \mathcal{D}$.

2. If $\mathcal{M}$ has an $\mathcal{M}$-disjunction such that the $\{\neg, \vee\}$-fragment of $\mathcal{M}$ is $\neg$-contained in classical logic, and there is an element $\bot \in \mathcal{V}$ such that $\bot \notin \mathcal{D}$ and $\tilde{\neg}\bot \notin \mathcal{D}$, then $\mathcal{M}$ is $\neg$-paracomplete.

Proof.

1. Suppose that $\mathcal{M}$ is paracomplete. Then there is a set of formulas Γ and formulas ψ, φ such that (i) $\Gamma, \psi \vdash_\mathcal{M} \varphi$, (ii) $\Gamma, \neg\psi \vdash_\mathcal{M} \varphi$, and (iii) $\Gamma \nvdash_\mathcal{M} \varphi$. From (iii) it follows that there is a four-valued valuation $\nu \in mod_\mathcal{M}(\Gamma) \setminus mod_\mathcal{M}(\varphi)$. Thus, in order to satisfy conditions (i) and (ii), necessarily $\nu(\psi) \notin \mathcal{D}$ and $\tilde{\neg}\nu(\psi) = \nu(\neg\psi) \notin \mathcal{D}$. Hence, $\nu(\psi)$ is an element $\bot$ as required.

2. Suppose that the two conditions are satisfied. Then Proposition 3.32 easily implies that $\neg$ is a negation for $\mathcal{M}$, and that $p \vdash_\mathcal{M} \neg p \vee p$ and $\neg p \vdash_\mathcal{M} \neg p \vee p$. But if $\nu(p) = \bot$ then $\nu(\neg p \vee p) \notin \mathcal{D}$ by the definitions of $\bot$ and of an $\mathcal{M}$-disjunction. Hence $\mathcal{M}$ is paracomplete. □

Proposition 5.3 Let $\mathcal{M} = \langle \mathcal{V}, \mathcal{D}, \mathcal{O} \rangle$ be a matrix for a language with $\neg$. If $\mathcal{M}$ is $\neg$-paradefinite then there are four elements $t, f, \top, \bot$ in $\mathcal{D}$ such that:

1. $t \in \mathcal{D}$ and $\tilde{\neg}t \notin \mathcal{D}$.

2. $f \notin \mathcal{D}$ and $\tilde{\neg}f \in \mathcal{D}$.

3. $\top \in \mathcal{D}$ and $\tilde{\neg}\top \in \mathcal{D}$.

4. $\bot \notin \mathcal{D}$ and $\tilde{\neg}\bot \notin \mathcal{D}$.

5. $\tilde{\neg}t = f$.

Proof. This follows from Propositions 3.55 and 5.2. □

Proposition 5.4 Any $\neg$-paradefinite four-valued matrix $\mathcal{M}$ is isomorphic to a matrix of the form $\mathcal{M}' = \langle \{t, f, \top, \bot\}, \{t, \top\}, \mathcal{O} \rangle$, in which $\tilde{\neg}t = f$, $\tilde{\neg}f = t$, $\tilde{\neg}\top \in \{t, \top\}$, and $\tilde{\neg}\bot \in \{f, \bot\}$.

Proof. This easily follows from Propositions 5.3 and 3.58. □

Because of Proposition 5.4, in the rest of this chapter we shall assume that the four-valued matrices we study have the form described in that proposition. In other words: we continue to denote by $t, f, \top$ the three elements that any paraconsistent matrix should have according to Proposition 3.55, and we denote the fourth element by $\bot$.

5.1 The Four-Valued Paradefinite Framework

Note 5.5 By Corollary 3.57, any four-valued paraconsistent matrix can have either two or three designated elements. The former option is much more popular in the literature (see, e.g., [4, 11, 12, 32, 86, 87, 151, 159]). Proposition 5.4 provides an explanation to this fact, and is the reason why we too concentrate in this chapter on such matrices. However, in Chapter 13 a use is made of an important four-valued paraconsistent matrix which has *three* designated elements.

Proposition 5.4 leaves exactly four possible interpretations for $\neg$ in four-valued paradefinite matrices. However, the next proposition and its corollary show that one of them is by far more natural than the others.

Proposition 5.6 *Let $\mathcal{M}$ be a $\neg$-paradefinite four-valued matrix. Then:*

1. *If $\neg$ is left involutive for $\mathbf{L}_{\mathcal{M}}$ then $\tilde{\neg}\bot = \bot$.*

2. *If $\neg$ is right involutive for $\mathbf{L}_{\mathcal{M}}$ then $\tilde{\neg}\top = \top$.*

Proof. If $\neg$ is left involutive for $\mathbf{L}_{\mathcal{M}}$ then $\neg\neg p \vdash_{\mathcal{M}} p$. Hence $\tilde{\neg}\bot \neq f$ (otherwise $\nu(p) = \bot$ would have been a counter-model by Proposition 5.4), and so $\tilde{\neg}\bot = \bot$. If $\neg$ is right involutive then $p \vdash_{\mathcal{M}} \neg\neg p$. Hence $\tilde{\neg}\top \neq t$ (otherwise $\nu(p) = \top$ would have been a counter-model by Proposition 5.4), and so $\tilde{\neg}\top = \top$. □

Corollary 5.7 *The only involutive negation of paradefinite four-valued matrices is Dunn-Belnap negation: $\tilde{\neg}t = f$, $\tilde{\neg}f = t$, $\tilde{\neg}\top = \top$, and $\tilde{\neg}\bot = \bot$.*

The converse of Proposition 5.4 is of course not always true. However, the addition of just one very natural demand suffices:

Definition 5.8 Let $\mathcal{L}$ be a propositional language which includes the unary connective $\neg$. We denote by $\mathsf{M4}^{\mathcal{L}}$ the set of four-valued matrices $\mathcal{M}$ for $\mathcal{L}$ of the form $\langle \{t, f, \top, \bot\}, \{t, \top\}, \mathcal{O} \rangle$ which satisfy the conditions on $\tilde{\neg}$ given in Proposition 5.4. By $\mathsf{M4}$ we denote the class of matrices $\mathcal{M}$ such that $\mathcal{M}$ belongs to $\mathsf{M4}^{\mathcal{L}}$ for some language $\mathcal{L}$, and $\mathcal{M}$ has a classically closed (i. e. $\{t, f\}$-closed) $\mathcal{M}$-disjunction $\vee$.

Proposition 5.9 *Let $\mathcal{M}$ be an element of $\mathsf{M4}$.*

1. $\mathbf{L}_{\mathcal{M}}$ *is finitary.*

2. $\mathcal{M}$ *is $\neg$-contained in classical logic iff it is classically closed.*

3. $\mathcal{M}$ *is semi-normal, paradefinite, and boldly paraconsistent (thus it is also strongly paraconsistent).*

4. *If $\mathcal{M}$ is classically closed, and has an $\mathcal{M}$-conjunction, and an $\mathcal{M}$-implication, Then $\mathbf{L}_\mathcal{M}$ is classical $\neg$-paraconsistent logic.*

5. *$\mathcal{M}$ is non-exploding iff every connective of $\mathcal{M}$ is $\{\top\}$-closed.*

Proof.

1. This is an instance of Theorem 3.17.

2. Obviously, $\mathcal{M} \in \mathbf{M4}$ is $\neg$-subclassical iff it is $\{t, f\}$-closed. Hence this item follows from Theorem 3.51.

3. Let $\vee$ be a classically closed $\mathcal{M}$-disjunction. The $\{\neg, \vee\}$-fragment of $\mathcal{M}$ is obviously $\neg$-subclassical. Therefore, it is paraconsistent by the second item of Proposition 3.47, and the first item of Proposition 3.55. The rest of the claim follows from Proposition 3.32, Corollary 3.56, and the second part of Proposition 5.2.

4. This follows from the previous items and Proposition 3.54.

5. This follows from Proposition 3.59. □

In what follows we shall concentrate on matrices in $\mathbf{M4}$. Unlike the previous chapter, we shall make no attempt to characterize all elements of this family. Instead, we study the most useful and well-motivated ones. Our two formal criteria for usefulness of a four-valued paradefinite matrix is the existence of an illuminating and easy-to-use cut-free corresponding proof system which is as close as possible to LK and $G_{\mathbf{J_3}}$, and the expressive power of its language: we use only languages whose set of definable connectives can be characterized by a simple property that has clear significance for paraconsistent reasoning. For each such property we present a small set of connectives which have this property and together suffice for generating all other connectives which have it. This set is selected from a small stock of connectives, which is presented and motivated in the next section. It should be noted that the language of one of the matrices we study is functionally complete for four-valued matrices, and so the logic it induces is a conservative extension of any other paradefinite four-valued logic. Still another logic we study is similarly a conservative extension of any paradefinite four-valued logic which is $\neg$-contained in classical logic.

5.2 Dunn-Belnap's Matrix $\mathcal{FOUR}$

5.2.1 Motivation

By the definition of $\mathbf{M4}$, the first step in constructing useful elements of it is to choose the interpretations of $\neg$ and $\vee$. For doing so, the best motivation

5.2 Dunn-Belnap's Matrix $\mathcal{FOUR}$

we know was given by Belnap in [86, 87]. There Belnap suggested a four-valued framework for collecting and processing information.[2] His framework was later generalized in [60]. In this generalized framework there is a set S of information sources and a processor P. The sources provide information about formulas over $\{\neg, \vee\}$, and we assume that for each such formula ψ, a source $s \in S$ can say one of the following: that ψ is true (i.e., it assigns it the truth-value 1), that ψ is false (i.e., it assigns it the truth-value 0), or that it has no knowledge about ψ. In turn, the processor collects information from the sources, combines it according to some strategy and defines the resulting combined valuation of ψ in terms of a subset $d(\psi) \subseteq \{0, 1\}$. Now in Belnap's original model the sources are allowed to provide information only about *propositional variables*.[3] The processor then assigns to a variable p a subset $d(p)$ of $\{0, 1\}$ according to the following two simple principles: $1 \in d(p)$ iff some source claims that p is true, and $0 \in d(p)$ iff some source claims that p is false. The processor's evaluation of complex formulas is then derived by using the following basic principles (induced by the classical truth-tables of $\neg$ and $\vee$) for assigning a subset $d(\varphi)$ of $\{0, 1\}$ to a complex formula φ over $\{\neg, \vee\}$:

db1 $0 \in d(\neg\varphi)$ iff $1 \in d(\varphi)$,

db2 $1 \in d(\neg\varphi)$ iff $0 \in d(\varphi)$,

db3 $1 \in d(\varphi \vee \psi)$ iff $1 \in d(\varphi)$ or $1 \in d(\psi)$,

db4 $0 \in d(\varphi \vee \psi)$ iff $0 \in d(\varphi)$ and $0 \in d(\psi)$.

Now, in this model $d(\varphi) = \{0, 1\}$ means that φ is known to be true and also known to be false (i.e., the information about φ is inconsistent). $d(\varphi) = \{1\}$ means that φ is only known to be true, while $d(\varphi) = \{0\}$ means that φ is only known to be false. Finally, $d(\varphi) = \emptyset$ means that there is no knowledge about φ. This observation leads to the following identification of the four truth-values used in **M4** with the subsets of $\{0, 1\}$: $t = \{1\}$, $f = \{0\}$, $\top = \{0, 1\}$, $\bot = \emptyset$. Accordingly, the truth tables for $\neg$ and $\vee$ that the above principles lead to are the following (where the connective $\wedge$ is defined by: $\varphi \wedge \psi =_{Df} \neg(\neg\varphi \vee \neg\psi)$):

$\tilde{\vee}$	t	f	$\top$	$\bot$
t	t	t	t	t
f	t	f	$\top$	$\bot$
$\top$	t	$\top$	$\top$	t
$\bot$	t	$\bot$	t	$\bot$

$\tilde{\wedge}$	t	f	$\top$	$\bot$
t	t	f	$\top$	$\bot$
f	f	f	f	f
$\top$	$\top$	f	$\top$	f
$\bot$	$\bot$	f	f	$\bot$

$\tilde{\neg}$	
t	f
f	t
$\top$	$\top$
$\bot$	$\bot$

---

[2] The corresponding lattice $\mathcal{FOUR}$ which is described below was first introduced by Dunn in [146] (see also [148]), following an observation of Smiley (see [4]).

[3] This assumption is dropped in Chapter 7.

Definition 5.10 (*FOUR* and $\mathcal{FOUR}$) *Dunn-Belnap basic matrix* $\mathcal{FOUR}$ is the matrix $\langle FOUR, \mathcal{D}, \mathcal{O}\rangle$ for the language $\mathcal{L}_{\mathcal{FOUR}} = \{\neg, \vee, \wedge\}$ (or just $\{\neg, \vee\}$) in which $FOUR = \{t, f, \top, \bot\}$, $\mathcal{D} = \{t, \top\}$, and the interpretations of the connectives are given by the truth tables above.

Proposition 5.11 *Let $\mathcal{M}$ be an extension of $\mathcal{FOUR}$. Then $\vee$ is an $\mathcal{M}$-disjunction and $\wedge$ is an $\mathcal{M}$-conjunction. In particular: $\mathcal{M} \in \mathsf{M4}$, and $\mathbf{L}_{\mathcal{M}}$ is semi-normal, paradefinite, and boldly paraconsistent.*

Proof. This easily follows from the definitions, Corollary 3.33, and Proposition 5.9. □

Note 5.12 Let $\leq_t$ be the partial order on $FOUR$ in which t is the maximal element, f is the minimal one, and $\top, \bot$ are (incomparable) intermediate elements. This order is a natural extension to the four-valued case of the total order $\leq_t$ of the three-valued case, considered in Section 4.1. Again, the conjunction and disjunction are interpreted here, respectively, as the meet and the join of $\leq_t$, and negation is interpreted by a $\leq_t$-reversing function. Note that this interpretation of $\neg$ is exactly the unique involutive negation of paradefinite four-valued logics given in Corollary 5.7.

Note 5.13 Another useful dual representation of $\mathcal{FOUR}$ (to which we shall return in Section 5.5 and in later Chapters) uses pairs from $\{1, 0\} \times \{1, 0\}$. Given such a pair $\langle a, b\rangle$, the first component intuitively represents the information about the truth of a formula, and the second one represents the information about its falsity, or the truth of its negation. According to this representation, we have that $t = \langle 1, 0\rangle$, $f = \langle 0, 1\rangle$, $\top = \langle 1, 1\rangle$, $\bot = \langle 0, 0\rangle$. The operations of $\mathcal{FOUR}$ are then defined by: $\langle a_1, b_1\rangle \vee \langle a_2, b_2\rangle = \langle max(a_1, b_1), min(a_2, b_2)\rangle$, $\langle a_1, b_1\rangle \wedge \langle a_2, b_2\rangle = \langle min(a_1, b_1), max(a_2, b_2)\rangle$, and $\neg\langle a, b\rangle = \langle b, a\rangle$. This representation is used e.g. in [7, 21, 62, 124, 160].

5.2.2 Other Useful Connectives

As Theorem 5.22 and Note 5.23 below show, the language of $\mathcal{FOUR}$ is rather limited, even if we add to it propositional constants for the two classical truth-values. Therefore, now we introduce several other useful and natural connectives on $FOUR$ that (by Theorem 5.22) cannot be defined in the language of $\mathcal{FOUR}$:

- First of all, the logic $\mathbf{L}_{\mathcal{FOUR}}$ of $\mathcal{FOUR}$ is not normal: no implication connective is available in it (see Proposition 5.40 below). To overcome this problem, the most natural and useful choice (both for the proof systems we introduce later in this chapter, and for keeping close

5.2 Dunn-Belnap's Matrix $\mathcal{FOUR}$

connections with the implication connective of **PAC** and **J₃**) is the connective $\supset_t$ introduced in Definition 3.38. By Proposition 3.39 this is an $\mathcal{M}$-implication for every matrix $\mathcal{M}$. In what follows we shall denote this connective simply by $\supset$. We note that its interpretation is in the present context the following:

$$a \tilde{\supset} b = \begin{cases} b & \text{if } a \in \{t, \top\}, \\ t & \text{if } a \in \{f, \bot\}. \end{cases}$$

- As in the three-valued case, for characterizing the expressive power of $\mathcal{FOUR}$ one needs to order its four truth-values by another partial order, denoted again by $\leq_k$. Again, $\leq_k$ reflects differences in the amount of information that the the truth-values exhibit (as explained below, this makes it very important for Belnap's model). Since $\top$ and $\bot$ intuitively represent inconsistent and incomplete information (respectively), they are taken as the maximal and the minimal elements of $\leq_k$, while this time t and f are the intermediate elements. This means that $\leq_k$ is here simply the subset relation $\subseteq$ according to Belnap's identification of $\mathcal{FOUR}$'s elements with the subsets of $\{0, 1\}$.

Together, the lattices $\langle \mathcal{V}, \leq_t \rangle$ and $\langle \mathcal{V}, \leq_k \rangle$ form a single four-valued structure (denoted again by $\mathcal{FOUR}$) known as Belnap's bilattice [86, 87]. It is represented in the double-Hasse diagram of Figure 5.1 (which is strongly related to Figure 4.1). In Section 5.5 we shall consider algebraic structures which generalize it.

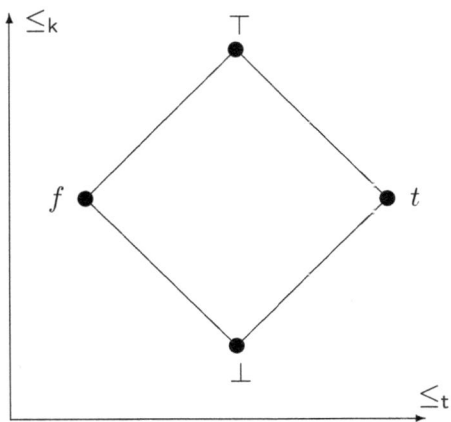

Figure 5.1: $\mathcal{FOUR}$

Following Fitting's notations, we shall denote the join and the meet of $\leq_k$ by $\oplus$ and $\otimes$ (respectively). The $\leq_k$-reversing function on $FOUR$

which is dual to $\tilde{\neg}$ is called *conflation* [158], and the corresponding connective is usually denoted by $-$. The truth tables of these $\leq_k$-connectives are given below.

$\tilde{\oplus}$	t	f	$\top$	$\bot$
t	t	$\top$	$\top$	t
f	$\top$	f	$\top$	f
$\top$	$\top$	$\top$	$\top$	$\top$
$\bot$	t	f	$\top$	$\bot$

$\tilde{\otimes}$	t	f	$\top$	$\bot$
t	t	$\bot$	t	$\bot$
f	$\bot$	f	f	$\bot$
$\top$	t	f	$\top$	$\bot$
$\bot$	$\bot$	$\bot$	$\bot$	$\bot$

$\tilde{-}$	
t	t
f	f
$\top$	$\bot$
$\bot$	$\top$

- Another group of connectives that it is natural to introduce in the present context are propositional constants for the four truth-values of $\mathcal{FOUR}$. Below we use T, F, B (read as "both" true and false) and N (for "neither" true nor false) as such constants, to be interpreted (respectively) by the truth-values $t, f, \top,$ and $\bot$ (that is, $\nu(\mathsf{B}) = \top$ for every valuation, and similarly for the other three constants).

Definition 5.14 (4All) Let $\mathcal{L}_{All} = \{\neg, \vee, \wedge, \supset, \mathsf{F}, \mathsf{T}, \mathsf{B}, \mathsf{N}, -, \oplus, \otimes\}$. $\mathcal{M}_{All}$ is the extension of $\mathcal{FOUR}$ to $\mathcal{L}_{All}$, and **4All** is the logic induced by $\mathcal{M}_{All}$.

Note 5.15 If $\mathcal{M} \in \mathsf{M4}$, and either of the $\leq_k$-meet $\otimes$, the $\leq_k$-join $\oplus$, the propositional constant B, or the propositional constant N is definable in the languages of $\mathcal{M}$, then the language of $\mathcal{M}$ is not classically closed. (Indeed, $a \tilde{\oplus} b \notin \{t, f\}$ and $a \tilde{\otimes} b \notin \{t, f\}$ for any $a \neq b \in \{t, f\}$). It follows by Proposition 5.9 that $\mathcal{M}$ induces in such a case a logic which is only $\neg$-*coherent* with classical logic but not $\neg$-*contained* in it. (This is again one of the exceptional cases mentioned in Note 2.21.)

5.2.3 Properties of $\mathsf{L}_{\mathcal{FOUR}}$

As usual, a major step in investigating the logic induced by $\mathcal{FOUR}$ is to determine the expressive power of its language.

Definition 5.16 (characterization by a formula) Let $\mathcal{M} \in \mathsf{M4}$ and let ψ be a formula in $\mathcal{L}_\mathcal{M}$ such that $\mathsf{Var}(\psi) \subseteq \{P_1, \ldots, P_n\}$. The subset S_ψ^n of $FOUR^n$ that is *characterized* in $\mathcal{M}$ by ψ is (see Definition 3.10 for the notation):

$$S_\psi^n = \{(a_1, a_2, \ldots, a_n) \in FOUR^n \mid F_\psi^n(a_1, a_2, \ldots, a_n) \in \mathcal{D}\}.$$

A subset $C \subseteq FOUR^n$ is *characterizable* in $\mathcal{M}$ iff there exists a formula ψ of $\mathcal{L}_\mathcal{M}$ such that $C = S_\psi^n$.

5.2 Dunn-Belnap's Matrix $\mathcal{FOUR}$

Convention. Let $\mathcal{L} \subseteq \mathcal{L}_{all}$. We say that a function is *representable* (by a formula ψ) in $\mathcal{L}$ if it is representable (by ψ) in some (or every) extension of $\mathcal{FOUR}$ whose language contains $\mathcal{L}$, and its interpretations of the connectives in $\mathcal{L}$ are those given in the previous section. Similarly, a set is *characterizable* (by a formula ψ) in $\mathcal{L}$ if it is characterizable (by ψ) in some (or every) extension of $\mathcal{FOUR}$ with the same properties.

Lemma 5.17 *Given* $\mathcal{M} \in \mathcal{M}4$ *we have:*

1. $\bigcup_{i=1}^{k} S_{\psi_i}^n = S_{\psi_1 \vee \psi_2 \vee \cdots \vee \psi_k}^n$.

2. $\bigcap_{i=1}^{k} S_{\psi_i}^n = S_{\psi_1 \wedge \psi_2 \wedge \cdots \wedge \psi_k}$.

Proof. This easily follows from Proposition 5.11. □

Definition 5.18 (cone) Let $\mathcal{V} = FOUR$. A subset $\mathcal{S}$ of V^n is called a *cone* if $\vec{y} \in \mathcal{S}$ whenever $\vec{y} \geq_k \vec{x}$ (i.e. $y_i \geq_k x_i$ for every $1 \leq i \leq n$, $\vec{y} \in V^n$), and $\vec{x} \in \mathcal{S}$. If $\mathcal{S} = V^n$ then the cone is called *trivial*.

Proposition 5.19 *Every non-trivial and non-empty cone* $\mathcal{S}$ *in* $FOUR^n$ *can be characterized by a formula* $\psi_\mathcal{S}$ *in* $\{\neg, \vee, \wedge\}$.

Proof. It is easy to see that given $\vec{a} \in FOUR^n$ such that $\vec{a} \neq \vec{\bot}$, the set $\{\vec{x} \in FOUR^n \mid \vec{x} \geq_k \vec{a}\}$ is characterized by $\psi_{i_1} \wedge \psi_{i_2} \wedge \cdots \wedge \psi_{i_l}$, where $\{i_1, \ldots, i_l\} = \{i \mid 1 \leq i \leq n, a_i \neq \bot\}$ and for $i \in \{i_1, \ldots, i_l\}$:

$$\psi_i = \begin{cases} P_i \wedge \neg P_i & \text{if } a_i = \top, \\ P_i & \text{if } a_i = t, \\ \neg P_i & \text{if } a_i = f. \end{cases}$$

Now, assume that $\mathcal{S}$ is a non-trivial and non-empty cone in $FOUR^n$. Then $\vec{\top} \in \mathcal{S}$, and $\vec{\bot} \notin \mathcal{S}$. Hence $\mathcal{S}$ is the union of the non-empty set of all the subsets of $FOUR^n$ of the form $\{\vec{x} \in FOUR^n \mid \vec{x} \geq_k \vec{a}\}$, where $\vec{a} \in \mathcal{S}$, and $\vec{a} \neq \vec{\bot}$. Hence the claim follows from what we have just shown, and from Lemma 5.17. □

Definition 5.20 (monotonic functions) Let $f : FOUR^n \to FOUR$. f is *monotonic* if $f(a_1, \ldots, a_n) \leq_k f(b_1, \ldots, b_n)$ whenever $a_i \leq_k b_i$ for $1 \leq i \leq n$ (i.e. if it is $\leq_k$-monotonic).

Notation 5.21 Let $g : FOUR^n \to FOUR$. Denote:

$$g_t = \{\vec{x} \in FOUR^n \mid g(\vec{x}) \geq_k t\},$$

$$g_f = \{\vec{x} \in FOUR^n \mid g(\vec{x}) \geq_k f\}.$$

Theorem 5.22 *Let* $g : FOUR^n \to FOUR$. g *is representable in* $\{\neg, \vee, \wedge\}$ *(or just* $\{\neg, \vee\}$*) iff it satisfies the following conditions:*

1. *It is* monotonic.

2. *It commutes with conflation, i.e.* $g(\tilde{-}a_1, \ldots, \tilde{-}a_n) = \tilde{-}g(a_1, \ldots, a_n)$ *for every* $a_1, \ldots, a_n \in FOUR$.

3. *It is* $\{\top\}$-*closed (i.e.,* $g(\top, \ldots, \top) = \top$*).*

Proof. It is easy to verify that the three conditions are preserved under compositions of functions, and that the projection functions, $\tilde{\neg}$, $\tilde{\vee}$, and $\tilde{\wedge}$ satisfy all of the three. It follows that every function which is representable in $\{\neg, \vee, \wedge\}$ satisfies the three conditions.

For the converse, assume that $g : FOUR^n \to FOUR$ satisfies the three conditions. Then $g(\top, \ldots, \top) = \top$, and so $g(\bot, \ldots, \bot) = \bot$, because g commutes with conflation. These two equations and the monotonicity of g imply that g_t is a nontrivial and nonempty cone. It follows by Proposition 5.19 that g_t is characterized by some formula ψ_t^g in the language $\{\neg, \vee, \wedge\}$. This means that for every $\vec{a} \in FOUR^n$ we have:

$$(*) \quad g(\vec{a}) \in \{t, \top\} \quad \text{iff} \quad F^n_{\psi_t^g}(\vec{a}) \in \{t, \top\}.$$

Now we show that actually $F^n_{\psi_t^g} = g$.

- Suppose that $g(a_1, \ldots, a_n) = \top$. Then $g(\tilde{-}a_1, \ldots, \tilde{-}a_n) = \bot$. By $(*)$ it follows that $F^n_{\psi_t^g}(a_1, \ldots, a_n) \in \{t, \top\}$ and $F^n_{\psi_t^g}(\tilde{-}a_1, \ldots, \tilde{-}a_n) \in \{f, \bot\}$. Since $F^n_{\psi_t^g}$ commutes with conflation by the direction we have already proved, this is possible only if $F^n_{\psi_t^g}(a_1, \ldots, a_n) = \top$ too.

- Suppose that $g(a_1, \ldots, a_n) = t$. Then $g(\tilde{-}a_1, \ldots, \tilde{-}a_n) = t$. By $(*)$ it follows that $F^n_{\psi_t^g}(a_1, \ldots, a_n) \in \{t, \top\}$ and $F^n_{\psi_t^g}(\tilde{-}a_1, \ldots, \tilde{-}a_n) \in \{t, \top\}$. Since $F^n_{\psi_t^g}$ commutes with conflation, necessarily $F^n_{\psi_t^g}(a_1, \ldots, a_n) = t$.

- Suppose that $g(a_1, \ldots, a_n) = f$. Then $g(\tilde{-}a_1, \ldots, \tilde{-}a_n) = f$. By $(*)$, $F^n_{\psi_t^g}(a_1, \ldots, a_n) \in \{f, \bot\}$ and $F^n_{\psi_t^g}(\tilde{-}a_1, \ldots, \tilde{-}a_n) \in \{f, \bot\}$. Since $F^n_{\psi_t^g}$ commutes with conflation, necessarily $F^n_{\psi_t^g}(a_1, \ldots, a_n) = f$.

- Suppose that $g(a_1, \ldots, a_n) = \bot$. Then $g(\tilde{-}a_1, \ldots, \tilde{-}a_n) = \top$. By $(*)$ it follows that $F^n_{\psi_t^g}(a_1, \ldots, a_n) \in \{f, \bot\}$ and $F^n_{\psi_t^g}(\tilde{-}a_1, \ldots, \tilde{-}a_n) \in \{t, \top\}$. Since $F^n_{\psi_t^g}$ commutes with conflation, necessarily $F^n_{\psi_t^g}(a_1, \ldots, a_n) = \bot$.

It follows that indeed $F^n_{\psi_t^g} = g$, and so ψ_t^g represents g. □

5.2 Dunn-Belnap's Matrix $\mathcal{FOUR}$

Note 5.23 It is not difficult to check that the constant functions $\lambda \vec{a}.t$ and $\lambda \vec{a}.f$ are the only functions from $FOUR^n$ to $FOUR$ which satisfy the first two conditions given in Theorem 5.22 but not the third. It follows that every function from $FOUR^n$ to $FOUR$ which is monotonic and commute with conflation is representable in $\{\neg, \vee, \wedge, \mathsf{F}\}$, where F is a propositional constant whose interpretation is always the truth-value f.

Proposition 5.24 *In addition to the properties listed in Theorem 5.22, every function which is representable in $\{\neg, \vee, \wedge\}$ is also:*

1. $\{t, f, \top\}$-closed.

2. $\{t, f, \bot\}$-closed.

3. $\{t, f\}$-closed (i.e., classically closed).

4. $\{\bot\}$-closed.

Proof. The third item follows from the second property listed in Theorem 5.22. The first two items follow from the third and the monotonicity of g (Property 1 in Theorem 5.22). The last one follows from Properties 2 and 3 in Theorem 5.22. (All the items can easily be shown directly too.) □

Note 5.25 The properties listed in Proposition 5.24 do not suffice for replacing the second condition given in Theorem 5.22 (i.e. commuting with conflation). Indeed, the following function g is monotonic, $\{t, f, \top\}$-closed, $\{t, f, \bot\}$-closed, $\{t, f\}$-closed, $\{\top\}$-closed, and $\{\bot\}$-closed. Yet, it does not always commute with conflation (thus $g(\top, \bot) = \top$, but $g(\bot, \top) \neq \bot$).

$$g(x, y) = \begin{cases} \top & \text{if } x = \top \text{ or } y = \top, \\ \bot & \text{if } x = y = \bot. \\ t & \text{otherwise.} \end{cases}$$

Now we are able to list the main properties of $\mathbf{L}_{\mathcal{FOUR}}$:

Proposition 5.26

1. $\mathbf{L}_{\mathcal{FOUR}}$ is a semi-normal, paradefinite, and boldly paraconsistent logic, which is $\neg$-contained in classical logic and non-exploding.

2. $\mathbf{L}_{\mathcal{FOUR}}$ is not normal.

3. $\mathbf{L}_{\mathcal{FOUR}}$ is neither maximally paraconsistent, nor maximally paraconsistent relative to classical logic.

4. $\neg$ is contrapositive in $\mathbf{L}_{\mathcal{FOUR}}$, and $\mathbf{L}_{\mathcal{FOUR}}$ is self-extensional.

Proof.

1. This easily follows from Proposition 5.9.

2. This is a special case of Proposition 5.40 below.

3. The first item in Proposition 5.24 easily implies that the paraconsistent logic **LP** is an extension of $\mathbf{L}_{\mathcal{FOUR}}$. Since $\neg\varphi \vee \varphi$ is valid in **LP**, it follows that one can add this schema to $\mathbf{L}_{\mathcal{FOUR}}$ and get by this a proper extension of the latter which is still paraconsistent. Hence $\mathbf{L}_{\mathcal{FOUR}}$ is not maximally paraconsistent. Since **LP** is $\neg$-contained in classical logic, it follows by Proposition 1.54 that $\mathbf{L}_{\mathcal{FOUR}}$ is also not maximally paraconsistent relative to classical logic.[4]

4. This is shown in Proposition 5.66 below. □

5.3 Important Extensions of $\mathcal{FOUR}$

Not all the properties of the functions which are representable in $\mathcal{FOUR}$ according to Theorem 5.22 and Proposition 5.24 are significant for paraconsistent reasoning. Thus it is difficult to see why the connection with conflation is desirable. Also the importance of being $\{\bot\}$-closed is debatable. Thus it seems rather possible to have knowledge about the truth or falsity of a disjunction without knowing anything about the truth or falsity of its disjuncts. However, each of the other properties, or at least a combination of them, *is* significant — for reasons that will be explained below. In this section we determine what subsets of the connectives which were introduced in the previous section correspond to each of the significant properties, determining by this what we take to be the most important extensions of $\mathcal{FOUR}$.

Note 5.27 The various connectives we investigate are not independent of each other. Thus we obviously do not need to consider both F and T, because each of them is definable in terms of the other and $\neg$ (which we always assume of course to be in the language)[5]. However, there are many other, more complicated dependencies among our connectives, and some of them are mentioned and used below.

---

[4]Using e.g. the Gentzen-type systems for **LP** (as given in the previous chapter) and $\mathbf{L}_{\mathcal{FOUR}}$ (to be given later in this chapter) it is easy to see that by adding $\neg\varphi \vee \varphi$ as an axiom to $\mathbf{L}_{\mathcal{FOUR}}$ one gets **LP**.

[5]The same applies of course to $\vee$ and $\wedge$, so in principle we need to include only one of them in the various sets we consider below. Still, for convenience we take here both of them as primitive.

5.3 Important Extensions of $\mathcal{FOUR}$

5.3.1 A Maximal Extension

We begin by showing that the set of connectives we use (and actually a proper subset of it) suffices for defining any operation on $\mathcal{FOUR}$. Our first step in this direction is given by the following useful lemma:

Lemma 5.28 *Let* $\{\neg, \vee, \wedge, \mathsf{B}, \mathsf{N}\} \subseteq \mathcal{L} \subseteq \mathcal{L}_{all}$, *and let* $g: FOUR^n \to FOUR$. *Assume that g_t and g_f (Notation 5.21) are characterized by formulas in $\mathcal{L}$. Then g itself is representable in $\mathcal{L}$.*

Proof. Suppose that g_t and g_f are characterized in $\mathcal{L}$ by ψ_t^g and ψ_f^g (respectively). Let $\psi = (\psi_t^g \wedge \mathsf{B}) \vee (\neg \psi_f^g \wedge \mathsf{N})$. Then for every $\vec{a} \in FOUR^n$:

$$(*) \quad F_\psi^n(\vec{a}) = (F_{\psi_t^g}^n(\vec{a}) \,\tilde{\wedge}\, \top) \,\tilde{\vee}\, (\tilde{\neg} F_{\psi_f^g}^n(\vec{a}) \,\tilde{\wedge}\, \bot).$$

Now we show that $F_\psi^n(\vec{a}) = g(\vec{a})$ for every $\vec{a} \in FOUR^n$.

- Let $g(\vec{a}) = \top$. Then $\vec{a} \in g_t$ and $\vec{a} \in g_f$. It follows that $F_{\psi_t^g}^n(\vec{a}) \in \{\top, t\}$ and $F_{\psi_f^g}^n(\vec{a}) \in \{\top, t\}$. This and (*) entail that $F_\psi^n(\vec{a}) = \top$.

- Let $g(\vec{a}) = t$. Then $\vec{a} \in g_t$ and $\vec{a} \notin g_f$. It follows that $F_{\psi_t^g}^n(\vec{a}) \in \{\top, t\}$ and $F_{\psi_f^g}^n(\vec{a}) \in \{\bot, f\}$. This and (*) entail that $F_\psi^n(\vec{a}) = t$.

- Let $g(\vec{a}) = f$. Then $\vec{a} \notin g_t$ and $\vec{a} \in g_f$. It follows that $F_{\psi_t^g}^n(\vec{a}) \in \{\bot, f\}$ and $F_{\psi_f^g}^n(\vec{a}) \in \{\top, t\}$. This and (*) entail that $F_\psi^n(\vec{a}) = f$.

- Let $g(\vec{a}) = \bot$. Then $\vec{a} \notin g_t$ and $\vec{a} \notin g_f$. It follows that $F_{\psi_t^g}^n(\vec{a}) \in \{\bot, f\}$ and $F_{\psi_f^g}^n(\vec{a}) \in \{\bot, f\}$. This and (*) entail that $F_\psi^n(\vec{a}) = \bot$. □

Proposition 5.29

1. A subset S of $FOUR^n$ is characterizable by some formula in the language $\mathcal{L}_{CL}$ iff $\langle \top, \top, \ldots, \top \rangle \in S$.

2. Every subset of $FOUR^n$ is characterizable in $\mathcal{L}_{CL}^{\mathsf{F}\neg}$.

Proof.

1. The necessity of the condition is easy. For the converse define:

$$\mathsf{F}_n = \neg(P_1 \supset P_1) \wedge \ldots \wedge \neg(P_n \supset P_n).$$

Then F_n has the following property:

$$F_{\mathsf{F}_n}^n(\vec{a}) = \begin{cases} \top & \vec{a} = \vec{\top}, \\ f & \text{otherwise.} \end{cases}$$

Let $\vec{a} = \langle a_1, \ldots, a_n \rangle \in FOUR^n$. Define, for every $1 \leq i \leq n$,

$$\psi_i^{\vec{a}} = \begin{cases} P_i \wedge \neg P_i & \text{if } a_i = \top, \\ P_i \wedge (\neg P_i \supset \mathsf{F}_n) & \text{if } a_i = t, \\ \neg P_i \wedge (P_i \supset \mathsf{F}_n) & \text{if } a_i = f, \\ (\neg P_i \supset \mathsf{F}_n) \wedge (P_i \supset \mathsf{F}_n) & \text{if } a_i = \bot. \end{cases}$$

Using the second part of Lemma 5.17, it is easy to see that $\psi^{\vec{a}} = \psi_1^{\vec{a}} \wedge \psi_2^{\vec{a}} \wedge \ldots \psi_n^{\vec{a}}$ characterizes $\{\vec{\top}, \vec{a}\}$, where $\vec{\top} = \langle \top, \top, \ldots, \top \rangle$. Hence the proposition follows from the first part of Lemma 5.17.

2. All we need to change in the last proof is to use F instead of F_n in the definition of $\psi_i^{\vec{a}}$. After this change the $\wedge$-conjunction of the new $\psi_i^{\vec{a}}$'s characterizes $\{\vec{a}\}$ and not $\{\vec{\top}, \vec{a}\}$. This suffices (using $\vee$) for the characterization of every nonempty set. The empty set itself is characterized by F. □

Theorem 5.30 *The language of $\{\neg, \vee, \wedge, \supset, \mathsf{B}, \mathsf{N}\}$ is functionally complete for FOUR.*

Proof. Since F is defined in $\{\neg, \vee, \wedge, \supset, \mathsf{B}, \mathsf{N}\}$ by the formula $\mathsf{B} \wedge \mathsf{N}$, this theorem is an immediate corollary of Lemma 5.28 and Proposition 5.29. □

Corollary 5.31 *Every logic which is induced by some element of* **M4** *is equivalent to some fragment of* **4All** *(and of its $\{\neg, \vee, \wedge, \supset, \mathsf{B}, \mathsf{N}\}$-fragment).*

Note 5.32 Since $\bot = f \tilde{\otimes} \neg f$ while $\top = f \tilde{\oplus} \neg f$ we have that the language of $\{\neg, \vee, \wedge, \supset, \otimes, \oplus, \mathsf{F}\}$ is also functionally complete for $FOUR$. The use of this language has a certain advantage of modularity over the use of $\{\neg, \vee, \wedge, \supset, \mathsf{B}, \mathsf{N}\}$, since it has been proved in [40] that if Ξ is a subset of $\{\otimes, \oplus, \mathsf{F}\}$, then a function $g : FOUR^n \to FOUR$ is representable in $\{\neg, \wedge, \supset\} \cup \Xi$ iff it is S-closed for every $S \in \{\{\top\}, \{t, f, \top\}, \{t, f, \bot\}\}$ for which all the (functions that directly correspond to the) connectives in Ξ are S-closed. In other words, g is:

1. representable in $\mathcal{L}_{CL} = \{\neg, \vee, \wedge, \supset\}$ iff it is $\{\top\}$-closed, $\{t, f, \bot\}$-closed, and $\{t, f, \top\}$-closed.

2. representable in $\mathcal{L}_{CL}^{\mathsf{F}\neg} = \{\neg, \vee, \wedge, \supset, \mathsf{F}\}$ iff it is $\{t, f, \bot\}$-closed and $\{t, f, \top\}$-closed.

3. representable in $\{\neg, \vee, \wedge, \supset, \oplus\}$ iff it is $\{\top\}$-closed and $\{t, f, \top\}$-closed.

4. representable in $\{\neg, \vee, \wedge, \supset, \otimes\}$ iff it is $\{\top\}$-closed and $\{t, f, \bot\}$-closed.

5.3 Important Extensions of $\mathcal{FOUR}$ 149

5. representable in $\{\neg, \vee, \wedge, \supset, \otimes, \mathsf{F}\}$ iff it is $\{t, f, \bot\}$-closed.

6. representable in $\{\neg, \vee, \wedge, \supset, \oplus, \otimes\}$ iff it is $\{\top\}$-closed.

7. representable in $\{\neg, \vee, \wedge, \supset, \oplus, \mathsf{F}\}$ iff it is $\{t, f, \top\}$-closed.

8. representable in $\{\neg, \vee, \wedge, \supset, \oplus, \otimes, \mathsf{F}\}$.

As noted above, the last item in this list is equivalent to Theorem 5.30. In addition we prove below theorems which are easily seen to be equivalent to items 1, 2, and 6. The other four items can be proved according to similar lines, but they are not important for us here. It is also worth noting that it is easy to find examples that show that the eight fragments above are different from each other (see [12] and [40]).

Proposition 5.33

1. **4All** *is a paradefinite, normal and boldly paraconsistent logic.*

2. **4All** *is neither $\neg$-contained in classical logic nor non-exploding.*

3. **4All** *has no proper simple extension, and so it is strongly maximal.*

Proof. The first two items follow easily from Propositions 5.9 and 3.39. The third is immediate from Proposition 3.15. □

5.3.2 A Maximal Monotonic Extension

In [87] Belnap suggested to use in the sources-processor model only languages with monotonic interpretations of the connectives. The reason is that the assignments of truth-values to formulas of such languages (and *only* such languages) are stable in the sense that the arrival of new data from new sources does not change previous knowledge about truth and falsity: Suppose that an assignment ν which is based on the information collected from some set S of sources recognizes a formula φ as "true" (i.e., $1 \in \nu(\varphi)$). Then any assignment which is based on a superset of S would still recognize φ as "true" (though it might recognize it also as "false"). The same holds if ν recognizes φ as "false" (i.e., $0 \in \nu(\varphi)$). In other words: new data might only increase our knowledge about a formula, but not reject existing one. From Belnap's point of view an optimal language for information processing is therefore a language in which it is possible to represent *all* monotonic functions, and only monotonic functions. Next we show that not much should be added to the basic language of $\{\neg, \vee, \wedge\}$ (or just $\{\neg, \vee\}$) in order to obtain such a language.

Definition 5.34 (4Mon) Let $\mathcal{L}_{Mon} = \{\neg, \vee, \wedge, \mathsf{B}, \mathsf{N}\}$. $\mathcal{M}_{Mon}$ is the extension of $\mathcal{FOUR}$ to $\mathcal{L}_{Mon}$, and **4Mon** is the logic that is induced by $\mathcal{M}_{Mon}$.

Theorem 5.35 *A function* $g : FOUR^n \to FOUR$ *is representable in* $\mathcal{L}_{Mon}$ *iff it is monotonic.*

Proof. It is easy to see that every function which is representable in $\mathcal{L}_{Mon}$ is monotonic. For the converse, assume that $g : FOUR^n \to FOUR$ is monotonic. This implies that g_t and g_f are cones. Since the empty set is characterized by N and $FOUR^n$ is characterized by B, it follows from Proposition 5.19 that g_t and g_f are characterizable in $\mathcal{L}_{Mon}$. Hence, g is representable in $\mathcal{L}_{Mon}$ by Lemma 5.28. □

Corollary 5.36 *The logic* **4Mon** *contains every logic which is induced by an element of* **M4** *that employs only monotonic functions.*

Proposition 5.37

1. **4Mon** *is a semi-normal, paradefinite, and boldly paraconsistent logic.*

2. **4Mon** *is neither* $\neg$*-contained in classical logic nor non-exploding.*

3. **4Mon** *has no proper simple extension, and so it is strongly maximal.*

Proof. Similar to the proof of Propositions 5.33. □

Example 5.38 (Fitting's guard connective) The meet and the join in $\mathcal{FOUR}$ with respect to $\leq_t$ correspond to the conjunction and disjunction of strong Kleene's logic, in the sense that their restrictions to $\{t, f, \top\}$ and to $\{t, f, \bot\}$ coincide, respectively, with the corresponding connectives on $\mathcal{THREE}_\top$ and on $\mathcal{THREE}_\bot$ (recall Section 4.1). In order to represent the connectives of the other Kleene's three-valued logics (weak-Kleene and sequential-Kleene considered in Note 4.3) Fitting [161] introduced the *guard* connective. This connective is denoted there by a colon, and is evaluated as follows: if p is assigned a designated value (t or $\top$), the value of $p : q$ has the value of q, otherwise $p : q$ has the value $\bot$. The guard connective is monotonic, and so it is definable in $\mathcal{L}_{mon}$ as follows:[6]

$$p : q = ((p \wedge q) \wedge \mathsf{B}) \vee ((\neg p \vee q) \wedge \mathsf{N}).$$

---

[6] Fitting himself provides the following definition for the guard connective: $p : q = ((p \otimes \mathsf{T}) \oplus \neg(p \otimes \mathsf{T})) \otimes q$. Still another representation is $p : q = (p \supset q) \otimes \neg(p \supset \neg q)$. The latter is perhaps the easier to grasp, but it uses the non-monotonic implication $\supset$.

5.3 Important Extensions of $\mathcal{FOUR}$

Example 5.39 The operations $\tilde{\oplus}$ and $\tilde{\otimes}$ on $\mathcal{FOUR}$ are monotonic. Hence they are representable in $\mathcal{L}_{mon}$. Here are their simplest representations:

$$\varphi \oplus \psi = (\varphi \wedge \mathsf{B}) \vee (\psi \wedge \mathsf{B}) \vee (\varphi \wedge \psi),$$

$$\varphi \otimes \psi = (\varphi \wedge \mathsf{N}) \vee (\psi \wedge \mathsf{N}) \vee (\varphi \wedge \psi).$$

On the other hand, the connections given in Note 5.32 imply that the language of $\{\neg, \vee, \wedge, \oplus, \otimes, \mathsf{F}\}$ is also complete for the monotonic functions.

Important as **4Mon** is, it has at least two serious drawbacks. One drawback is that it is not ¬-contained in classical logic (Proposition 5.37). Another important drawback is that it is not normal — and neither is any of its fragments (like $\mathbf{L}_{\mathcal{FOUR}}$). This is shown in the next proposition.

Proposition 5.40 *Let $\mathcal{M} \in \mathbf{M4}$. Suppose that the interpretations of all the connectives of $\mathcal{M}$ are monotonic. Then $\mathbf{L}_{\mathcal{M}}$ has no semi-implication, and so it is not normal. In particular:* **4Mon** *and $\mathbf{L}_{\mathcal{FOUR}}$ are not normal.*

Proof. The proof is similar to that of Proposition 4.58: Suppose for contradiction that $\supset$ is a definable semi-implication for $\mathbf{L}_{\mathcal{M}}$. Then $\tilde{\supset}$ is monotonic. Now since $\varphi \supset \varphi$ is valid for any semi-implication, $\tilde{\supset}(f,f) \in \mathcal{D}$. This and the monotonicity of $\tilde{\supset}$ imply that $\tilde{\supset}(\top,f) \in \mathcal{D}$. It follows that $p, p \supset q \nvDash_{\mathcal{M}} q$ (because $\nu(p) = \top, \nu(q) = f$ provides a counterexample). This contradicts the assumption that $\supset$ is a semi-implication for $\mathbf{L}_{\mathcal{M}}$. □

5.3.3 A Maximal Classical Paraconsistent Extension

Definition 5.41 (4CC) Let $\mathcal{L}_{CC} = \{\neg, \vee, \wedge, \supset, -\}$. $\mathcal{M}_{CC}$ is the extension of $\mathcal{FOUR}$ to $\mathcal{L}_{CC}$, and **4CC** is the logic that is induced by $\mathcal{M}_{CC}$.

Theorem 5.42 *A function $g : FOUR^n \to FOUR$ is representable in $\mathcal{L}_{CC}$ iff it is classically closed.*

Proof. It is easy to see that every function which is representable in $\mathcal{L}_{CC}$ is classically closed. For the converse, note first that F is defined in $\mathcal{L}_{CC}$ by $P_1 \wedge --\neg P_1$. Therefore it follows from Proposition 5.29 that every subset of $FOUR^n$ is characterizable in $\mathcal{L}_{CC}$. Next we define for $n > 0$:

$$\mathsf{B}_n^* = (P_1 \supset P_1) \wedge \ldots \wedge (P_n \supset P_n) \wedge (-P_1 \supset -P_1) \wedge \ldots \wedge (-P_n \supset -P_n),$$

$$\mathsf{N}_n^* = -\mathsf{B}_n^*.$$

It is easy to see that for every $\vec{a} \in FOUR^n$ we have:

$$F_{\mathsf{B}_n^*}^n(\vec{a}) = \begin{cases} \top & \exists 1 \leq i \leq n \; a_i \notin \{t, f\}, \\ t & \text{otherwise.} \end{cases}$$

$$F_{\mathsf{N}_n^*}^n(\vec{a}) = \begin{cases} \bot & \exists 1 \leq i \leq n \; a_i \notin \{t, f\}, \\ t & \text{otherwise.} \end{cases}$$

Now, assume that $g : FOUR^n \to FOUR$ is classically closed. The proof that g is representable in $\mathcal{L}_{CC}$ is similar to the proof of Lemma 5.28, except that instead of using B and N we use B_n^* and N_n^* (respectively): Suppose that g_t and g_f are characterized in $\mathcal{L}_{CC}$ by some formulas ψ_t^g and ψ_f^g (respectively). Let $\psi = (\psi_t^g \wedge \mathsf{B}_n^*) \vee (\neg \psi_f^g \wedge \mathsf{N}_n^*)$. By the equations above the following holds for every $\vec{a} \in FOUR^n$:

$$(*) \quad F_\psi^n(\vec{a}) = \begin{cases} (F_{\psi_t^g}^n(\vec{a}) \tilde{\wedge} \top) \tilde{\vee} (\tilde{\neg} F_{\psi_f^g}^n(\vec{a}) \tilde{\wedge} \bot) & \exists 1 \leq i \leq n \; a_i \notin \{t, f\}, \\ F_{\psi_t^g}^n(\vec{a}) \tilde{\vee} \tilde{\neg} F_{\psi_f^g}^n(\vec{a}) & \text{otherwise.} \end{cases}$$

Now, if there is $i \leq n$ such that $a_i \notin \{t, f\}$, then (*) implies that $F_\psi^n(\vec{a}) = g(\vec{a})$, exactly like in the proof of Lemma 5.28. So, assume that $a_i \in \{t, f\}$ for every $i \leq n$. Since g, $F_{\psi_t^g}^n$, and $F_{\psi_f^g}^n$ are all classically closed, this implies that $g(\vec{a})$, $F_{\psi_t^g}^n(\vec{a})$, and $F_{\psi_f^g}^n(\vec{a})$ are all in $\{t, f\}$. This and the definitions of ψ_f^g imply, e.g., that $g(\vec{a}) = t \Leftrightarrow g(\vec{a}) \notin \{f, \top\} \Leftrightarrow F_{\psi_f^g}^n(\vec{a}) \notin \{t, \top\} \Leftrightarrow F_{\psi_f^g}^n(\vec{a}) = f$. Similarly, $g(\vec{a}) = t \Leftrightarrow F_{\psi_t^g}^n(\vec{a}) = t$, $g(\vec{a}) = f \Leftrightarrow F_{\psi_t^g}^n(\vec{a}) = f$, and $g(\vec{a}) = f \Leftrightarrow F_{\psi_f^g}^n(\vec{a}) = t$. By (*), these facts entail that again $F_\psi^n(\vec{a}) = g(\vec{a})$. Hence, $F_\psi^n(\vec{a}) = g(\vec{a})$ for every $\vec{a} \in FOUR^n$, and so ψ represents g. □

Corollary 5.43 *The logic* **4CC** *contains every logic which is induced by an element of* **M4** *and is $\neg$-contained in classical logic. In particular: it contains every classical paraconsistent induced by an element of* **M4**.

Proof. Immediate from Proposition 5.9 and Theorem 5.42. □

Proposition 5.44

1. **4CC** *is a paradefinite, classical (and so normal) paraconsistent logic.*

2. **4CC** *is boldly paraconsistent, but not non-exploding.*

3. **4CC** *is fully maximal and is not equivalent to any three-valued logic.*

Proof. The first two items follow from Propositions 5.9, 5.11, 3.39, and Theorem 5.42. The third follows from the second part of Theorem 3.60, since with $\diamond = \neg -$, all the conditions given there are satisfied by $\mathcal{M}_{CC}$. □

Note 5.45 $\mathcal{M}_{CC}$ is subclassical, and Theorem 5.42 implies that it is an extension of every other subclassical element of **M4**.

5.3.4 A Maximal Non-Exploding Extension

Definition 5.46 (4Nex) Let $\mathcal{L}_{Nex} = \{\neg, \vee, \wedge, \supset, \oplus, \otimes\}$. $\mathcal{M}_{Nex}$ is the extension of $\mathcal{FOUR}$ to $\mathcal{L}_{Nex}$, and **4Nex** is the logic that is induced by $\mathcal{M}_{Nex}$.

Theorem 5.47 *A function* $g : FOUR^n \to FOUR$ *is representable in* $\mathcal{L}_{Nex}$ *iff it is* $\{\top\}$*-closed.*

Proof. It is easy to see that every function which is representable in $\mathcal{L}_{Nex}$ is $\{\top\}$-closed. For the converse, note first that B is defined in $\mathcal{L}_{Nex}$ by $(P_1 \supset P_1) \oplus \neg(P_1 \supset P_1)$. Next we define for $n > 0$:

$$\mathsf{N}_n = P_1 \otimes \neg P_1 \otimes P_2 \otimes \neg P_2 \otimes \cdots P_n \otimes \neg P_n.$$

It is easy to see that for any valuation ν in $\mathcal{M}_{Nex}$ we have:

$$F_{\mathsf{N}_n}^n(\vec{a}) = \begin{cases} \top & \vec{a} = \vec{\top}, \\ \bot & \text{otherwise.} \end{cases}$$

Now, assume that $g : FOUR^n \to FOUR$ is $\{\top\}$-closed. Then $\vec{\top} \in g_t$ and $\vec{\top} \in g_f$. Therefore it follows from Proposition 5.29 that g_t and g_f are characterizable in $\mathcal{L}_{Nex}$ by some formulas ψ_t^g and ψ_f^g (respectively). The proof that g is representable in $\mathcal{L}_{Nex}$ is now similar to that of Lemma 5.28, except that N_n is used instead of N: Let $\psi = (\psi_t^g \wedge \mathsf{B}) \vee (\neg \psi_f^g \wedge \mathsf{N}_n)$. By the equation above, the following holds for every $\vec{a} \in FOUR^n$:

$$(*) \quad F_\psi^n(\vec{a}) = \begin{cases} (F_{\psi_t^g}^n(\vec{a}) \,\tilde{\wedge}\, \top) \,\tilde{\vee}\, (\tilde{\neg} F_{\psi_f^g}^n(\vec{a}) \,\tilde{\wedge}\, \bot) & \exists 1 \leq i \leq n\ a_i \neq \top, \\ \top & \forall 1 \leq i \leq n\ a_i = \top. \end{cases}$$

Now, if there exists $i \leq n$ such that $a_i \neq \top$ then (*) implies that $F_\psi^n(\vec{a}) = g(\vec{a})$ exactly like in the proof of Lemma 5.28. On the other hand, (*) implies that $F_\psi^n(\vec{a}) = g(\vec{a})$ in the case $\vec{a} = \vec{\top}$ as well, because g is $\{\top\}$-closed. It follows that $F_\psi^n(\vec{a}) = g(\vec{a})$ for every $\vec{a} \in FOUR^n$, and so ψ represents g. □

Corollary 5.48 *The logic* **4Nex** *contains every logic which is induced by an element of* **M4** *and is non-exploding.*

Proof. Immediate from Proposition 5.9 and Theorem 5.47. □

Proposition 5.49

1. **4Nex** *is a paradefinite and boldly paraconsistent logic.*

2. **4Nex** *is non-exploding, but not* $\neg$*-contained in classical logic.*

3. **4Nex** *has no proper simple extension, and so it is strongly maximal.*

Proof. We leave the proofs of the the first two items to the reader. For the third item, define T_n and F_n respectively as $\mathsf{B} \vee \mathsf{N}_n$ and $\mathsf{B} \wedge \mathsf{N}_n$ (where the definitions of B and N_n are like in the proof of Theorem 5.47). Then for every $\vec{a} \in FOUR^n$ we have:

$$F_{\mathsf{T}_n}^n(\vec{a}) = \begin{cases} \top & \vec{a} = \vec{\top}, \\ t & \text{otherwise.} \end{cases} \qquad F_{\mathsf{F}_n}^n(\vec{a}) = \begin{cases} \top & \vec{a} = \vec{\top}, \\ f & \text{otherwise.} \end{cases}$$

Let $\mathbf{L} = \langle \mathcal{L}, \vdash_{\mathbf{L}} \rangle$ be a proper simple finitary extension of **4Nex**, and suppose that $\psi_1, \ldots, \psi_k \vdash_{\mathbf{L}} \varphi$, but $\psi_1, \ldots, \psi_k \nvdash_{\mathbf{4Nex}} \varphi$. From the latter it follows that there is a valuation μ in $\mathcal{M}_{Nex}$, such that $\mu(\psi_i) \in \{t, \top\}$ for every $1 \leq i \leq k$, while $\mu(\varphi) \in \{f, \bot\}$. Without loss of generality, we may assume that for some n the variables occurring in $\{\psi_1, \ldots, \psi_k, \varphi\}$ are $P_1, \ldots, P_n$. Given a formula θ, denote by $\theta^\#$ the formula that resulted by substituting in θ T_n for every p such that $\mu(p) = t$, F_n for every p such that $\mu(p) = f$, B for every p such that $\mu(p) = \top$, and N_n for every p such that $\mu(p) = \bot$. Then the properties noted above of $\mathsf{T}_n, \mathsf{F}_n$, and N_n imply that $\nu(\theta^\#) = \mu(\theta)$ for every valuation ν and every θ such that $\mathsf{Var}(\theta) \subseteq \{P_1, \ldots, P_n\}$ (because all connectives of $\mathcal{M}_{Nex}$ are $\{\top\}$-closed). It follows that $\psi_1^\#, \ldots, \psi_k^\#$ and $\varphi^\# \supset P_1$ are all valid in $\mathcal{M}_{Nex}$ and so are theorems of $\mathbf{L}$. These facts and the fact that $\psi_1^\#, \ldots, \psi_k^\# \vdash_{\mathbf{L}} \varphi^\#$ (because $\psi_1, \ldots, \psi_k \vdash_{\mathbf{L}} \varphi$ and $\mathbf{L}$ is structural) imply that $\vdash_{\mathbf{L}} P_1$. This contradicts the assumption that $\mathbf{L}$ is a logic (and so not trivial). □

5.3.5 A Maximal Flexible Extension

Among the properties listed in Proposition 5.24, the combination of the first two ($\{t, f, \top\}$-closure and $\{t, f, \bot\}$-closure) seems very desirable. The reason is that it allows flexibility in the use of our four basic truth-values. Obviously, there is no point in using $\top$ in case no contradiction in our knowledge base is expected, while in the dual case there is no point in using $\bot$. The use of connectives which have both of the above properties ensures that one can easily switch from the use of the four-valued framework to the use of the appropriate 3-valued framework. Note also that this combination is a natural strengthening of the condition of classical closure. These considerations motivate the four-valued logic introduced next.

Definition 5.50 (flexible functions) $g : FOUR^n \to FOUR$ *is called* flexible *iff it is both $\{t, f, \top\}$-closed and $\{t, f, \bot\}$-closed.*

Note 5.51 Obviously, every flexible function is classically closed, but the converse is not true.

5.3 Important Extensions of $\mathcal{FOUR}$

Definition 5.52 (4Flex) $\mathcal{M}_{Flex}$ is the extension of $\mathcal{FOUR}$ to $\mathcal{L}_{CL}^{\mathsf{F}\neg}$, and **4Flex** is the logic that is induced by $\mathcal{M}_{Flex}$.

Theorem 5.53 *A function $g : FOUR^n \to FOUR$ is representable in $\mathcal{L}_{CL}^{\mathsf{F}\neg}$ iff it is flexible.*

Proof. It is easy to see that every function which is representable in $\mathcal{L}_{CL}^{\mathsf{F}\neg}$ is flexible. For the converse, note first that from Proposition 5.29 it follows that every subset of $FOUR^n$ is characterizable in $\mathcal{L}_{CL}^{\mathsf{F}\neg}$. Next we define for $n > 0$:

$$\mathsf{B}_n^{**} = (P_1 \supset P_1) \land (P_2 \supset P_2) \land \ldots \land (P_n \supset P_n),$$

$$\mathsf{N}_n^{**} = \neg((P_1 \land (P_1 \supset \mathsf{F})) \lor ((P_2 \land (P_2 \supset \mathsf{F})) \lor \ldots \lor ((P_n \land (P_n \supset \mathsf{F})))).$$

It is easy to see that for any $\vec{a} \in FOUR^n$ we have:

$$F_{\mathsf{B}_n^{**}}^n(\vec{a}) = \begin{cases} \top & \exists 1 \le i \le n \; a_i = \top, \\ t & \text{otherwise.} \end{cases}$$

$$F_{\mathsf{N}_n^{**}}^n(\vec{a}) = \begin{cases} \bot & \exists 1 \le i \le n \; a_i = \bot, \\ t & \text{otherwise.} \end{cases}$$

Now, assume that $g : FOUR^n \to FOUR$ is flexible. The proof that g is representable in $\mathcal{L}_{CL}^{\mathsf{F}\neg}$ is basically similar to the proof of Lemma 5.28, except that we need to normalize ψ_t^g and ψ_f^g, and instead of B and N we use B_n^{**} and N_n^{**} (respectively): Suppose that ψ_t^g and ψ_f^g (respectively) characterize g_t and g_f in $\mathcal{L}_{CL}^{\mathsf{F}\neg}$. Let $\psi = (\neg(\psi_t^g \supset \mathsf{F}) \land \mathsf{B}_n^{**}) \lor ((\psi_f^g \supset \mathsf{F}) \land \mathsf{N}_n^{**})$. Now, we show that $F_\psi^n(\vec{a}) = g(\vec{a})$ for every $\vec{a} \in FOUR^n$.

- Assume that $g(\vec{a}) = \top$. Then $\vec{a} \in g_t$ and $\vec{a} \in g_f$. It follows that $F_{\psi_t^g}^n(\vec{a}) \in \{\top, t\}$ and $F_{\psi_f^g}^n(\vec{a}) \in \{\top, t\}$. Moreover: since g is $\{t, f, \bot\}$-closed and $g(\vec{a}) = \top$, necessarily $a_i = \top$ for some $1 \le i \le n$. Hence, the first equation above implies that $F_{\mathsf{B}_n^{**}}^n(\vec{a}) = \top$. These facts entail that $F_\psi^n(\vec{a}) = \top$ in this case.

- Assume that $g(\vec{a}) = t$. Then $\vec{a} \in g_t$ and $\vec{a} \notin g_f$. It follows that $F_{\psi_t^g}^n(\vec{a}) \in \{\top, t\}$ and $F_{\psi_f^g}^n(\vec{a}) \in \{\bot, f\}$. This, the fact that $t \tilde{\lor} x = x \tilde{\lor} t = \top \tilde{\lor} \bot = t$, and the two equations above concerning B_n^{**} and N_n^{**}, entail that $F_\psi^n(\vec{a}) = t$ in this case.

- Assume that $g(\vec{a}) = f$. Then $\vec{a} \notin g_t$ and $\vec{a} \in g_f$. It follows that $F_{\psi_t^g}^n(\vec{a}) \in \{\bot, f\}$ and $F_{\psi_f^g}^n(\vec{a}) \in \{\top, t\}$. This entails that $F_\psi^n(\vec{a}) = f$ in this case.

- Assume that $g(\vec{a}) = \perp$. Then $\vec{a} \notin g_t$ and $\vec{a} \notin g_f$. It follows that $F^n_{\psi^g_t}(\vec{a}) \in \{\perp, f\}$ and $F^n_{\psi^g_f}(\vec{a}) \in \{\perp, f\}$. Moreover: since g is $\{t, f, \top\}$-closed and $g(\vec{a}) = \perp$, necessarily $a_i = \perp$ for some $1 \leq i \leq n$. Thus, the second equation above implies that $F^n_{\mathsf{N}^{**}_n}(\vec{a}) = \perp$. These facts entail that $F^n_\psi(\vec{a}) = \perp$ in this case.

It follows that $F^n_\psi = g$, and so ψ represents g in $\mathcal{L}^{\mathsf{F}\neg}_{CL}$. □

Corollary 5.54 *The logic* **4Flex** *contains every logic which is induced by an element of* **M4** *and employs only flexible connectives.*

Proposition 5.55

1. **4Flex** *is a paradefinite, classical (and so normal) paraconsistent logic.*

2. **4Flex** *is boldly paraconsistent, but not non-exploding.*

3. **4Flex** *is is neither maximally paraconsistent, nor maximally paraconsistent relative to classical logic.*

Proof. The proof of the first two items is like in the proof of Proposition 5.44. For the third item note that the fact that every connective of $\mathcal{L}^{\mathsf{F}\neg}_{CL}$ is $\{t, f, \top\}$-closed entails that the three-valued paraconsistent logic $\mathbf{J_3}$ is an extension of **4Flex**. Since again $\neg\varphi \vee \varphi$ is valid in $\mathbf{J_3}$, it follows that one can add this schema to **4Flex** and get by this a proper extension of the latter which is still paraconsistent. Since the logic obtained by adding $\neg\varphi \vee \varphi$ to **4Flex** is $\neg$-contained in classical logic, it again follows by Proposition 1.54 that $\mathbf{L}_{\mathcal{FOUR}}$ is also not maximally paraconsistent relative to classical logic.[7] □

5.3.6 The Classical Extension

The final extension of $\mathbf{L}_{\mathcal{FOUR}}$ we present is the one we believe to be the most useful and natural, since it is the maximal one which is *both* non-exploding and flexible, its language is exactly the classical language of $\mathcal{L}_{CL}$, and also its Gentzen-type proof system is very natural, and similar to some well-known quasi-canonical versions of **LK**.

Definition 5.56 (4CL) $\mathcal{M}_{4CL}$ is the extension of $\mathcal{FOUR}$ to $\mathcal{L}_{CL}$ (i.e. $\{\neg, \vee, \wedge, \supset\}$), and **4CL** is the logic that is induced by $\mathcal{M}_{4CL}$.

---

[7]Like in the case of $\mathbf{L}_{\mathcal{FOUR}}$, one can use, e.g., the Gentzen-type systems for $\mathbf{J_3}$ (as given in the previous chapter) and **4Nex** (to be given later in this chapter) to show that by adding $\neg\varphi \vee \varphi$ as an axiom to **4Nex** one gets $\mathbf{J_3}$.

5.3 Important Extensions of $\mathcal{FOUR}$

Theorem 5.57 *A function $g : FOUR^n \to FOUR$ is representable in $\mathcal{L}_{CL}$ iff it is flexible and $\{\top\}$-closed.*

Proof. It is easy to see that every function which is representable in $\mathcal{L}_{CL}$ is flexible and $\{\top\}$-closed. For the converse, assume that $g : FOUR^n \to FOUR$ is flexible and $\{\top\}$-closed. Then by Theorem 5.53 there is a formula ψ' of $\mathcal{L}_{CL}^{\mathsf{F},\neg}$ that represents g. Let ψ be the formula in $\mathcal{L}_{CL}$ which is obtained from ψ' by replacing every occurrence of F in ψ' with the formula F_n from the proof of Proposition 5.29. Now from the property of F_n described in that proof it follows that if $\vec{a} \neq \vec{\top}$, then $F_\psi^n(\vec{a}) = F_{\psi'}^n(\vec{a}) = g(\vec{a})$. On the other hand $F_\psi^n(\vec{\top}) = \top = g(\vec{\top})$, because g is $\{\top\}$-closed and ψ is $\mathcal{L}_{CL}$. It follows that $F_\psi^n(\vec{a}) = g(\vec{a})$ for every $\vec{a}$, and so ψ represents g. □

Corollary 5.58 *The logic* **4CL** *contains every non-exploding logic which is induced by an element of* **M4** *and employs only flexible connectives.*

Proof. Immediate from Proposition 5.9 and Theorem 5.57. □

Proposition 5.59

1. **4CL** *is a paradefinite, classical (and so normal) paraconsistent logic.*

2. **4CL** *is boldly paraconsistent and non-exploding.*

3. **4CL** *is neither maximally paraconsistent, nor maximally paraconsistent relative to classical logic.*

Proof. We leave the proof to the reader. □

Note 5.60 It seems natural to look for the maximal extension of $\mathbf{L}_{\mathcal{FOUR}}$ which is both $\neg$-contained in classical logic and non-exploding. In other words: the maximal extension of $\mathbf{L}_{\mathcal{FOUR}}$ that employs only connectives that are both classically closed and $\{\top\}$-closed. Unfortunately, among our basic set of natural four-valued connectives, only those in $\mathcal{L}_{CL}$ have both properties, and from Theorem 5.57 it easily follows that not all connectives that have both properties are representable in it. It follows that one has to introduce new connectives in order to capture the class of connectives that are both classically closed and $\{\top\}$-closed. Now, we introduce two additional connectives that suffice for the job:

$$\Delta(a) = \begin{cases} a & a \neq \bot, \\ \top & a = \bot. \end{cases} \qquad \#(a,b) = \begin{cases} \top & a = b = \top, \\ t & a \in \{t, f\} \text{ and } b \in \{t, f\}, \\ \bot & \text{otherwise.} \end{cases}$$

To prove that a function $g : FOUR^n \to FOUR$ is representable in $\mathcal{L}_{CL} \cup \{\Delta, \#\}$ iff it is both classically closed and $\{\top\}$-closed, we imitate the proof of Theorem 5.42. In doing so the only problem is to find appropriate substitutes for the formulas B_n^* and N_n^* used there. So, we define B_n' like B_n^*, only using Δ instead of the conflation $-$. We also define N_n' by induction as follows: $\mathsf{N}_1' = P_1$, and $\mathsf{N}_n' = \mathsf{N}_{n-1}' \# P_n$ for $n \geq 1$. It is easy to check that for every $\vec{a} \in FOUR^n$ we have:

$$F_{\mathsf{B}_n'}^n(\vec{a}) = \begin{cases} \top & \exists 1 \leq i \leq n\ a_i \notin \{t, f\}, \\ t & \text{otherwise.} \end{cases}$$

$$F_{\mathsf{N}_n'}^n(\vec{a}) = \begin{cases} \top & \vec{a} = \vec{\top}, \\ t & \forall 1 \leq i \leq n\ a_i \in \{t, f\}, \\ \bot & \text{otherwise.} \end{cases}$$

From this point, the proof is similar to that of Theorem 5.42, with B_n' and N_n' replacing B_n^* and N_n^* (respectively).

5.4 Proof Systems

Next, we consider proof systems for the logics presented in this chapter.

5.4.1 Gentzen-type Systems

A proof which is very similar to that given in the case of Theorem 4.29 (using Note 4.28) shows that again all the logics presented in this chapter have corresponding cut-free ¬-quasi-canonical Gentzen-type systems (Section 2.3). In this section we explicitly present such systems. Note that since all of them are fragments of the logic **4All**, it actually suffices to provide a modular one for that logic.

Definition 5.61 (Gentzen systems for 4All and its fragments)

- $G_{\mathbf{4All}}$ is the Gentzen-type system which is obtained from $G_{\mathbf{J}_3^\mathsf{B}}$ (Figure 4.4) by:

  1. Deleting the axiom $\Rightarrow \varphi, \neg\varphi$.
  2. Adding the axioms and rules given in Figure 5.2.[8]

---

[8]Note that the positive rules for $\wedge$ and $\otimes$ are identical. Both of them behave as classical conjunction. The difference is with respect to the negations of $p \wedge q$ and $p \otimes q$. Unlike the conjunction of classical logic, the negation of $p \otimes q$ is equivalent to $\neg p \otimes \neg q$. This follows from the fact that $p \leq_k q$ iff $\neg p \leq_k \neg q$. The difference between $\vee$ and $\oplus$ is similar.

5.4 Proof Systems

$$[-\Rightarrow] \quad \frac{\Gamma \Rightarrow \Delta, \neg\psi}{\Gamma, -\psi \Rightarrow \Delta} \qquad [\Rightarrow -] \quad \frac{\Gamma, \neg\psi \Rightarrow \Delta}{\Gamma \Rightarrow \Delta, -\psi}$$

$$[\neg - \Rightarrow] \quad \frac{\Gamma \Rightarrow \Delta, \psi}{\Gamma, \neg-\psi \Rightarrow \Delta} \qquad [\Rightarrow \neg-] \quad \frac{\Gamma, \psi \Rightarrow \Delta}{\Gamma \Rightarrow \Delta, \neg-\psi}$$

$$[\otimes \Rightarrow] \quad \frac{\Gamma, \psi, \varphi \Rightarrow \Delta}{\Gamma, \psi \otimes \varphi \Rightarrow \Delta} \qquad [\Rightarrow \otimes] \quad \frac{\Gamma \Rightarrow \Delta, \psi \quad \Gamma \Rightarrow \Delta, \varphi}{\Gamma \Rightarrow \Delta, \psi \otimes \varphi}$$

$$[\neg\otimes \Rightarrow] \quad \frac{\Gamma, \neg\psi, \neg\varphi \Rightarrow \Delta}{\Gamma, \neg(\psi \otimes \varphi) \Rightarrow \Delta} \qquad [\Rightarrow \neg\otimes] \quad \frac{\Gamma \Rightarrow \Delta, \neg\psi \quad \Gamma \Rightarrow \Delta, \neg\varphi}{\Gamma \Rightarrow \Delta, \neg(\psi \otimes \varphi)}$$

$$[\oplus \Rightarrow] \quad \frac{\Gamma, \psi \Rightarrow \Delta \quad \Gamma, \varphi \Rightarrow \Delta}{\Gamma, \psi \oplus \varphi \Rightarrow \Delta} \qquad [\Rightarrow \oplus] \quad \frac{\Gamma \Rightarrow \Delta, \psi, \varphi}{\Gamma \Rightarrow \Delta, \psi \oplus \varphi}$$

$$[\neg\oplus \Rightarrow] \quad \frac{\Gamma, \neg\psi \Rightarrow \Delta \quad \Gamma, \neg\varphi \Rightarrow \Delta}{\Gamma, \neg(\psi \oplus \varphi) \Rightarrow \Delta} \qquad [\Rightarrow \neg\oplus] \quad \frac{\Gamma \Rightarrow \Delta, \neg\psi, \neg\varphi}{\Gamma \Rightarrow \Delta, \neg(\psi \oplus \varphi)}$$

$$[\mathsf{N}\Rightarrow] \quad \Gamma, \mathsf{N} \Rightarrow \Delta$$

$$[\neg\mathsf{N}\Rightarrow] \quad \Gamma, \neg\mathsf{N} \Rightarrow \Delta$$

Figure 5.2: Inference rules for $-$, N, $\otimes$ and $\oplus$

- For $\mathbf{L} \in \{\mathbf{4All}, \mathbf{L}_{\mathcal{FOUR}}, \mathbf{4Mon}, \mathbf{4CC}, \mathbf{4Nex}, \mathbf{4Flex}, \mathbf{4CL}\}$ we denote by $\mathbf{G_L}$ the restriction of $G_{\mathbf{4All}}$ to the language of $\mathbf{L}$ (i.e. the Gentzen-type system in the language of $\mathbf{L}$ whose axioms and rules are the axioms and rules of $G_{\mathbf{4All}}$ which are relevant to that language).

Theorem 5.62 (completeness of $G_\mathbf{L}$, Cut-elimination) *Let $\mathbf{L}$ be one of the logics mentioned in Definition 5.61. Then:*

1. *$G_\mathbf{L}$ is sound and complete for $\mathbf{L}$.*

2. *$\mathcal{T} \vdash_{G_\mathbf{L}} \psi$ iff $\mathcal{T} \vdash_\mathbf{L} \psi$.*

3. *$G_\mathbf{L}$ admits cut-elimination.*

Proof. The main adjustments which are needed in the four-valued case to the proofs of the corresponding three-valued results are in Proposition 4.89 and Lemma 4.90. The former should be formalized according the following semantic of sequents in the 4-valued case: $\nu \models_\mathcal{M} \Gamma \Rightarrow \Delta$ for $\mathcal{M} \in \mathbf{M4}$ and $\nu \in \Lambda_\mathcal{M}$ iff either $\nu(\varphi) \in \{f, \bot\}$ for some $\varphi \in \Gamma$, or $\nu(\psi) \in \{t, \top\}$ for some $\psi \in \Delta$. As for Lemma 4.90, in the 4-valued case it is reformulated as Lemma 5.63 below. Using that lemma, the rest of the proof of Theorem 5.62

is similar to the proofs of Theorems 4.95 and 4.96 (with appropriate modifications of Lemmas 4.92, 4.94, and the other results that they are based on). We omit the details. □

Lemma 5.63 *Let $\mathcal{M} \in \mathbf{M4}$, and let $\Gamma \Rightarrow \Delta$ be a sequent which consists of literals (i.e. atomic formulas or negations of atomic formulas). Then*

1. $\vdash_\mathcal{M} \Gamma \Rightarrow \Delta$ *iff either $\Gamma \cap \Delta \neq \emptyset$, or at least one of the following holds:*
   $\mathsf{F} \in \Gamma$, $\mathsf{N} \in \Gamma$, $\neg \mathsf{N} \in \Gamma$, $\neg \mathsf{F} \in \Delta$, $\mathsf{B} \in \Delta$, $\neg \mathsf{B} \in \Delta$.

2. *If $\mathbf{L}$ is any fragment of $\mathbf{4All}$ which includes negation (in particular: if $\mathbf{L}$ is mentioned in Definition 5.61) and $\vdash_\mathbf{L} \Gamma \Rightarrow \Delta$, then $\vdash_{G_\mathbf{L}} \Gamma \Rightarrow \Delta$.*

Proof. For the first part, suppose that $\Gamma \Rightarrow \Delta$ consists only of literals. It is easy to check that if Γ and Δ have a literal in common or one of the other conditions in the lemma holds, then $\vdash_\mathcal{M} \Gamma \Rightarrow \Delta$. If not, then consider the following valuation $\nu \in \Lambda_{\mathsf{M4}}$:

$$\nu(p) = \begin{cases} \top & \text{if both } p \text{ and } \neg p \text{ are in } \Gamma, \\ \bot & \text{if both } p \text{ and } \neg p \text{ are in } \Delta, \\ t & \text{if } (p \in \Gamma \text{ and } \neg p \notin \Gamma) \text{ or } (p \notin \Delta \text{ and } \neg p \in \Delta), \\ f & \text{otherwise.} \end{cases}$$

Obviously, ν is a well defined valuation which assigns values in $\{\top, t\}$ to the literals in Γ, and values in $\{\bot, f\}$ to the literals in Δ. Hence $\nu \not\models_\mathcal{M} \Gamma \Rightarrow \Delta$, and so $\not\vdash_\mathcal{M} \Gamma \Rightarrow \Delta$ in this case.

The second part of the lemma follows from the first, and the fact that every sequent which satisfies the condition there is provable in $G_\mathbf{L}$. □

Note 5.64 As noted at the beginning of this section, the Gentzen-type systems considered in it can be obtained by the algorithm for deriving cut-free Gentzen-type system mentioned in Note 4.28. As is shown in [59], this construction too guarantees the validity of Theorem 5.62.

Note 5.65

- From Theorem 5.62 and Theorem 1.122 it follows that all the Gentzen-type systems discussed in this section admit *strong* cut-elimination. The proof of Theorem 5.62 also implies that each of them is *strongly* sound with respect to its corresponding 4-valued matrix. Moreover, with the help of the stronger proof methods used in Part III it can be shown that each of them is also *strongly* complete (in the sense of Definition 1.123) for its matrix.

5.4 Proof Systems

- The proof of Theorem 5.62 (as well as the previous item) shows that the following hold for every logic **L** considered in that theorem:

  - All the rules of $G_\mathbf{L}$ are invertible.[9]
  - Given any sequent $\Gamma \Rightarrow \Delta$, one can construct a finite set S of clauses (i.e., sequents that contain only literals) such that $\Gamma \Rightarrow \Delta$ is provable in $G_\mathbf{L}$ iff every $s \in S$ is provable in $G_\mathbf{L}$.

This means that one can develop tableaux proof systems for the logics discussed in this section which are almost identical to that of classical logic.[10] The main difference is that unlike classical logic, here a clause $\Gamma \Rightarrow \Delta$ without propositional constants is valid iff $\Gamma \cap \Delta \neq \emptyset$. One should note also that it is impossible here to translate a clause $\Gamma \Rightarrow \Delta$ in which $\Gamma \neq \emptyset$ into a sentence of the language without using the implication connective $\supset$.

We end this section with an interesting application of Theorem 5.62.

Proposition 5.66 $\mathbf{L}_{\mathcal{FOUR}}$ *is self-extensional, and $\neg$ is contrapositive in it.*

Proof. Using an induction on the structure of cut-free proofs in $\mathbf{G}_{\mathbf{L}_{\mathcal{FOUR}}}$, it is not difficult to see that if $\Gamma \Rightarrow \Delta$ has a proof in this system, then so does $\neg\Delta \Rightarrow \neg\Gamma$ (where $\neg\{\varphi_1, \ldots, \varphi_n\} = \{\neg\varphi_1, \ldots, \neg\varphi_n\}$). It follows in particular that if $\varphi \Rightarrow \psi$ has a cut-free proof in $\mathbf{G}_{\mathbf{L}_{\mathcal{FOUR}}}$, then so does $\neg\psi \Rightarrow \neg\varphi$. Hence Theorem 5.62 implies that $\neg$ is contrapositive in $\mathbf{L}_{\mathcal{FOUR}}$. Using this fact it is easy to show (using again $\mathbf{G}_{\mathbf{L}_{\mathcal{FOUR}}}$ and Theorem 5.62) that $\mathbf{L}_{\mathcal{FOUR}}$ is self-extensional. □

Note 5.67 In contrast to Proposition 5.66, if **L** is one of the logic considered in this chapter, and $\supset$ is in in the language of **L**, then **L** is not self-extensional, and $\neg$ is not contrapositive in it. Thus, $P_1 \supset P_1 \dashv\vdash_\mathbf{L} P_2 \supset P_2$, but $\neg(P_2 \supset P_2) \not\vdash_\mathbf{L} \neg(P_1 \supset P_1)$.

5.4.2 Hilbert-type Systems

Definition 5.68 (Hilbert-type systems for fragments of 4All)

- $H_\mathbf{4All}$ is the Hilbert-type system obtained from $H_{\mathbf{J}_3}$ (Figure 4.7) by:

---

[9] For example, the rule $[\Rightarrow\neg\supset]$ is invertible in any system which contains $[\neg\supset\Rightarrow]$, since both $\neg(\psi \supset \varphi) \Rightarrow \psi$ and $\neg(\psi \supset \varphi) \Rightarrow \neg\varphi$ are easily derivable using that rule.

[10] Such a system was introduced in [157], but in that paper only *validity* of *signed* formulas is considered there and not the consequence relation. Moreover, only $\leq_k$-monotonic operators are dealt with in that paper.

1. Deleting the axiom [t] (excluded middle),
2. Adding the axiom $\psi \supset \neg \mathsf{F}$,
3. Adding the axioms and rules given in Figure 5.3.

$[\Rightarrow \otimes]\ \psi \supset \varphi \supset \psi \otimes \varphi$ $\qquad$ $[\otimes \Rightarrow]\ \psi \otimes \varphi \supset \psi,\ \psi \otimes \varphi \supset \varphi$

$[\Rightarrow \oplus]\ \psi \supset \psi \oplus \varphi,\ \varphi \supset \psi \oplus \varphi$ $\qquad$ $[\oplus \Rightarrow]\ (\psi \supset \tau) \supset (\varphi \supset \tau) \supset (\psi \oplus \varphi \supset \tau)$

$[\Rightarrow \neg \oplus]\ \neg \psi \oplus \neg \varphi \supset \neg (\psi \oplus \varphi)$ $\qquad$ $[\neg \oplus \Rightarrow]\ \neg (\psi \oplus \varphi) \supset \neg \psi \oplus \neg \varphi$

$[\Rightarrow \neg \otimes]\ \neg \psi \otimes \neg \varphi \supset \neg (\psi \otimes \varphi)$ $\qquad$ $[\neg \otimes \Rightarrow]\ \neg (\psi \otimes \varphi) \supset \neg \psi \otimes \neg \varphi$

$[\Rightarrow -]\ \neg \psi \vee -\psi$ $\qquad$ $[-\Rightarrow]\ (\neg \psi \wedge -\psi) \supset \varphi$

$[\Rightarrow -\neg]\ \psi \vee \neg -\psi$ $\qquad$ $[-\neg \Rightarrow]\ (\psi \wedge \neg -\psi) \supset \varphi$

$[\Rightarrow \mathsf{B}]\ \psi \supset \mathsf{B}$ $\qquad$ $[\mathsf{N} \Rightarrow]\ \mathsf{N} \supset \psi$

$[\Rightarrow \neg \mathsf{B}]\ \psi \supset \neg \mathsf{B}$ $\qquad$ $[\neg \mathsf{N} \Rightarrow]\ \neg \mathsf{N} \supset \psi$

Figure 5.3: Axioms for $\oplus$, $\otimes$, $-$, N, B

- For $\mathbf{L} \in \{\mathbf{4All}, \mathbf{4CC}, \mathbf{4Nex}, \mathbf{4Flex}, \mathbf{4CL}\}$ we denote by $\mathbf{H_L}$ the restriction of $H_{\mathbf{4All}}$ to the language of $\mathbf{L}$ (i.e., the Hilbert-type system in the language of $\mathbf{L}$ whose axioms and rules are the axioms and rules of $H_{\mathbf{4All}}$ which are relevant to that language).

Theorem 5.69 $\vdash_{H_\mathbf{L}} = \vdash_{G_\mathbf{L}}$ *for each logic* $\mathbf{L}$ *mentioned in Definition 5.68.*

Proof. Similar to that of Theorem 4.100. $\qquad \square$

Theorem 5.70 (soundness and completeness of $H_\mathbf{L}$) *For each logic* $\mathbf{L}$ *in Definition 5.68, $H_\mathbf{L}$ is strongly sound and complete for* $\mathbf{L}$.

Proof. The theorem can easily be derived from Theorems 5.62 and 5.69. Below we give an alternative direct proof.

Since $\wedge$, $\vee$, and $\supset$ are $\mathcal{M}_{All}$-conjunction, $\mathcal{M}_{All}$-disjunction, and $\mathcal{M}_{All}$-implication (respectively), HCL^+ is strongly sound for $\mathbf{L}$ by Proposition 3.37 and Theorem 1.90. In addition, it is easy to check that with the exception of [t], each axiom in Figure 4.7 or Figure 5.3 is valid in $\mathcal{M}_{All}$. These facts imply that $H_\mathbf{L}$ is strongly sound for $\mathbf{L}$.

The proof of the strong completeness of $H_\mathbf{L}$ is again similar to the direct proof given in Chapter 4 to Theorem 4.101. Assume that $\mathcal{T} \not\vdash_{H_\mathbf{L}} \psi$. We construct a model $\mathcal{T}$ which is not a model of ψ. By Theorem 1.98 there exists a theory $\mathcal{T}^*$ in $\mathcal{L}$ with the following properties:

5.4 Proof Systems

1. $\mathcal{T} \subseteq \mathcal{T}^*$.

2. $\mathcal{T}^* \not\vdash_{H_\mathbf{L}} \psi$.

3. For every φ and τ, $\mathcal{T}^* \vdash_{H_\mathbf{L}} \varphi \supset \tau$ iff either $\mathcal{T}^* \not\vdash_{H_\mathbf{L}} \varphi$ or $\mathcal{T}^* \vdash_{H_\mathbf{L}} \tau$.

4. For every φ and τ, $\mathcal{T}^* \vdash_{H_\mathbf{L}} \varphi \wedge \tau$ iff $\mathcal{T}^* \vdash_{H_\mathbf{L}} \varphi$ and $\mathcal{T}^* \vdash_{H_\mathbf{L}} \tau$.

5. For every φ and τ, $\mathcal{T}^* \vdash_{H_\mathbf{L}} \varphi \vee \tau$ iff either $\mathcal{T}^* \vdash_{H_\mathbf{L}} \varphi$ or $\mathcal{T}^* \vdash_{H_\mathbf{L}} \tau$.

Now, define a valuation ν as follows:

$$\nu(\varphi) = \begin{cases} \top & \text{if } \mathcal{T}^* \vdash_{H_\mathbf{L}} \varphi \text{ and } \mathcal{T}^* \vdash_{H_\mathbf{L}} \neg\varphi, \\ \bot & \text{if } \mathcal{T}^* \not\vdash_{H_\mathbf{L}} \varphi \text{ and } \mathcal{T}^* \not\vdash_{H_\mathbf{L}} \neg\varphi, \\ t & \text{if } \mathcal{T}^* \vdash_{H_\mathbf{L}} \varphi \text{ and } \mathcal{T}^* \not\vdash_{H_\mathbf{L}} \neg\varphi, \\ f & \text{if } \mathcal{T}^* \not\vdash_{H_\mathbf{L}} \varphi \text{ and } \mathcal{T}^* \vdash_{H_\mathbf{L}} \neg\varphi. \end{cases}$$

Obviously, $\nu(\varphi)$ is designated whenever $\mathcal{T}^* \vdash_{H_\mathbf{L}} \varphi$, while $\nu(\psi)$ is not. It remains to show that ν is actually a valuation. Here we shall show that $\nu(\varphi \supset \tau) = \nu(\varphi)\tilde{\supset}\nu(\tau)$, and that $\nu(\varphi \vee \tau) = \nu(\varphi)\tilde{\vee}\nu(\tau)$, leaving the other cases for the reader.

- To show that $\nu(\varphi \vee \tau) = \nu(\varphi)\tilde{\vee}\nu(\tau)$, we note that Property 4 of $\mathcal{T}^*$ and Axioms [n$^l_\vee$] and [n$^r_\vee$] (see Note 4.99) entail that $\mathcal{T}^* \vdash_{H_\mathbf{L}} \neg(\varphi \vee \tau)$ iff both $\mathcal{T}^* \vdash_{H_\mathbf{L}} \neg\varphi$ and $\mathcal{T}^* \vdash_{H_\mathbf{L}} \neg\tau$. The desired equation easily follows from this fact and from Property 5 of $\mathcal{T}^*$.

- For showing that $\nu(\varphi \supset \tau) = \nu(\varphi)\tilde{\supset}\nu(\tau)$, we distinguish between two cases:
  1. $\nu(\varphi) \in \{f, \bot\}$. On one hand this means that $\nu(\varphi)\tilde{\supset}\nu(\tau) = t$. On the other it is equivalent to $\mathcal{T}^* \not\vdash_{H_\mathbf{L}} \varphi$. By Property 3 above the latter entails that $\mathcal{T}^* \vdash_{H_\mathbf{L}} \varphi \supset \tau$, while by axiom $[\neg \supset \Rightarrow 1]$ it implies that $\mathcal{T}^* \not\vdash_{H_\mathbf{L}} \neg(\varphi \supset \tau)$. Hence $\nu(\varphi \supset \tau) = t$ too.
  2. $\nu(\varphi) \in \{t, \top\}$. Then $\nu(\varphi)\tilde{\supset}\nu(\tau) = \nu(\tau)$. In addition, it means that $\mathcal{T}^* \vdash_{H_\mathbf{L}} \varphi$. Hence Property 3 implies that $\mathcal{T}^* \vdash_{H_\mathbf{L}} \varphi \supset \tau$ iff $\mathcal{T}^* \vdash_{H_\mathbf{L}} \tau$, while axioms $[\neg \supset \Rightarrow 2]$ and $[\Rightarrow \neg \supset]$ imply that $\mathcal{T}^* \vdash_{H_\mathbf{L}} \neg(\varphi \supset \tau)$ iff $\mathcal{T}^* \vdash_{H_\mathbf{L}} \neg\tau$. It follows that $\nu(\varphi \supset \tau) = \nu(\tau)$, and so $\nu(\varphi \supset \tau) = \nu(\varphi)\tilde{\supset}\nu(\tau)$. □

Note 5.71 Other proof systems for paradefinite logics have been considered in the literature. One of them is Bou and Rivieccio's Hilbert-style proof system [106]. This system has 23 rules for the language of $\{\neg, \vee, \wedge, \otimes, \oplus\}$, and no axioms. In [106] it is shown that it is equivalent to the corresponding fragment of $G_{4\mathrm{All}}$, and it is not difficult to see that it is obtained by a straightforward translation of that system.

5.5 Bilattices and Their Logics

Fitting, The four-elements structure $\mathcal{FOUR}$, shown in Figure 5.1, may be viewed as a particular case of general algebraic structures called *bilattices* [172, 173]. In bilattices the elements are simultaneously organized in two partial orders $\leq_t$ and $\leq_k$ that may have the same intuitive meanings as in the four-valued case. The resulting structures have been used by Ginsberg, and others for providing a unified platform for a diversity of applications in AI [166, 173, 261], semantics of logic programing [131, 157, 158, 160, 163], non-classical reasoning [11, 14, 21, 168, 189], different types of fuzzy sets [124, 138, 155, 168], and so forth. In this section we briefly review the matrices and the logics that are induced by these structures, and show the primary role that four-valued logics have among bilattice-based logics.

5.5.1 Bilattice-Valued Matrices

Definition 5.72 (bilattices) A *bilattice* is a structure $\mathcal{B} = \langle B, \leq_t, \leq_k, \neg \rangle$ which satisfies the following conditions:

a) B is a nonempty set containing at least two elements,

b) $\langle B, \leq_t \rangle$ and $\langle B, \leq_k \rangle$ are complete lattices,

c) $\neg$ is a unary operation on B that has the following properties:

　i) if $a \leq_t b$ then $\neg a \geq_t \neg b$,

　ii) if $a \leq_k b$ then $\neg a \leq_k \neg b$,

　iii) $\neg\neg a = a$.

Note 5.73 There are several definitions in the literature of bilattices. Here we follow Ginsberg's original definition [173], in which the two lattices are connected trough $\neg$. In other definitions, like those of Fitting (e.g., [158, 164]), they are connected by interlacing conditions or by distributive rules.

For convenience, in this section we use the same symbols for the unary and binary connectives of $\mathcal{L}_{all}$ and their interpretations in bilattices (as we did already with $\neg$ in Definition 5.72). Thus $\vee$ and $\wedge$ represent in it, respectively, also the $\leq_t$-lub and the $\leq_t$-glb, and $\oplus$ and $\otimes$ represent, respectively, also the $\leq_k$-lub and the $\leq_k$-glb.[11] In addition, $t_\mathcal{B}, f_\mathcal{B}, \top_\mathcal{B}$ and

---

[11]While $\wedge$ and $\vee$ can be associated with their usual intuitive meanings of 'and' and 'or', one may understand $\otimes$ and $\oplus$ as the 'consensus' and the 'gullibility' ('accept all') operators, respectively: $p \otimes q$ is the most that p and q can agree on, while $p \oplus q$ accepts the combined knowledge of p and of q. (See [159, 161] for a further discussion on the meaning of these connectives.)

5.5 Bilattices and Their Logics

$\perp_\mathcal{B}$ will denote the $\leq_t$-maximal, $\leq_t$-minimal, $\leq_k$-maximal, and $\leq_k$-minimal elements (respectively) of a bilattice $\mathcal{B}$. It is easy to verify that these four elements are different from each other, and that $t_\mathcal{B} = \neg f_\mathcal{B}$, $f_\mathcal{B} = \neg t_\mathcal{B}$, $\top_\mathcal{B} = \neg \perp_\mathcal{B}$, $\perp_\mathcal{B} = \neg \top_\mathcal{B}$ (see Lemma 5.77 below for their characterization in particular constructions of bilattices). When the underlying bilattice is known, we shall sometimes omit the subscripts from the notations.

Example 5.74 Belnap's $\mathcal{FOUR}$ [86, 87], presented in Figure 5.1, is the smallest bilattice. Two other known bilattices appear in Figure 5.4. Ginsberg's $\mathcal{DEFAULT}$ was introduced in [173] as a tool for non-monotonic reasoning. Its truth values with a prefix 'd' in their names are designed to represent values of default assumptions (dt = true by default, etc.). The negations in $\mathcal{DEFAULT}$ of $\top, t, f, \perp$ are the same as those in $\mathcal{FOUR}$, while $\neg df = dt$, $\neg dt = df$, and $\neg d\top = d\top$. The bilattice $\mathcal{NINE}$ has two more truth values, ot and of, where $\neg of = ot$ and $\neg ot = of$. We note that in $\mathcal{FOUR}$ and $\mathcal{NINE}$ all 12 possible distributive laws concerning $\wedge, \vee, \otimes$, and $\oplus$ hold, while in $\mathcal{DEFAULT}$ this is not the case. (For instance, $df \vee (d\top \otimes f) = d\top$ but $(df \vee d\top) \otimes (df \vee f) = df$.)

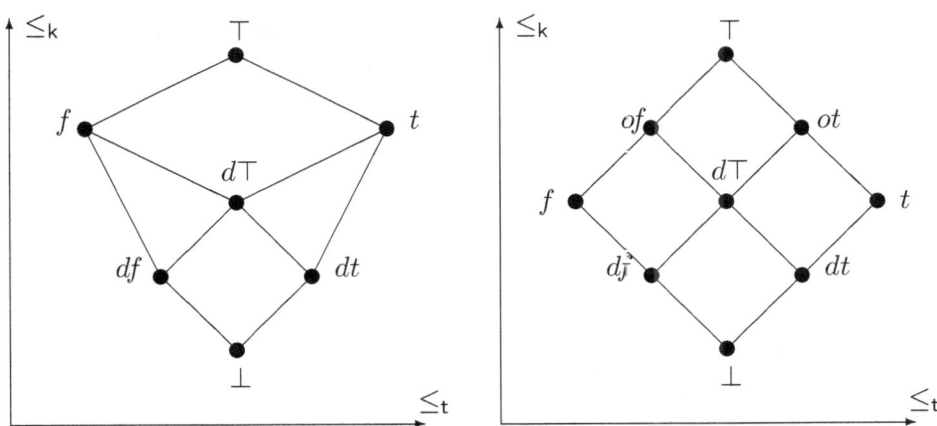

Figure 5.4: $\mathcal{DEFAULT}$ (left) and $\mathcal{NINE}$ (right)

In [173], Ginsberg proposed the following simple way of constructing bilattices out of lattices.

Definition 5.75 Let $\langle L, \leq_L \rangle$ be a complete lattice. The structure $L \odot L = \langle L \times L, \leq_t, \leq_k, \neg \rangle$ is defined as follows:
$\langle b_1, b_2 \rangle \geq_t \langle a_1, a_2 \rangle$ iff $b_1 \geq_L a_1$ and $b_2 \leq_L a_2$,
$\langle b_1, b_2 \rangle \geq_k \langle a_1, a_2 \rangle$ iff $b_1 \geq_L a_1$ and $b_2 \geq_L a_2$,
$\neg \langle a_1, a_2 \rangle = \langle a_2, a_1 \rangle$.

It is easy to verify that $L\odot L$ is indeed a bilattice. A pair $\langle x,y\rangle \in L\odot L$ may intuitively be understood so that x represents the amount of belief *for* some assertion, and y is the amount of belief *against* it. Following [173], $L\odot L$ was examined, e.g., in [11, 13, 37, 124, 138, 157, 158, 159].

Example 5.76 Let $\mathcal{TWO} = \langle \{0,1\}, <_2\rangle$ and $\mathcal{THREE} = \langle \{0,\frac{1}{2},1\}, <_3\rangle$, where $0 <_2 1$ and $0 <_3 \frac{1}{2} <_3 1$. Then $\mathcal{FOUR}$ is isomorphic to $\mathcal{TWO}\odot\mathcal{TWO}$, and $\mathcal{NINE}$ is isomorphic to $\mathcal{THREE}\odot\mathcal{THREE}$.

Lemma 5.77 [173] *Let $\langle L,\leq_L\rangle$ be a complete lattice with a join $\sqcap_L$ and a meet $\sqcup_L$. Then $L\odot L$ is a bilattice with in which:*

$\langle a,b\rangle \vee \langle c,d\rangle = \langle a\sqcup_L c, b\sqcap_L d\rangle,\quad \langle a,b\rangle \wedge \langle c,d\rangle = \langle a\sqcap_L c, b\sqcup_L d\rangle,$
$\langle a,b\rangle \oplus \langle c,d\rangle = \langle a\sqcup_L c, b\sqcup_L d\rangle,\quad \langle a,b\rangle \otimes \langle c,d\rangle = \langle a\sqcap_L c, b\sqcap_L d\rangle,$
$\neg\langle a,b\rangle = \langle b,a\rangle.$

$\bot_{L\odot L} = \langle \inf(L), \inf(L)\rangle, \qquad \top_{L\odot L} = \langle \sup(L), \sup(L)\rangle,$
$t_{L\odot L} = \langle \sup(L), \inf(L)\rangle, \qquad f_{L\odot L} = \langle \inf(L), \sup(L)\rangle.$

In [173] it is shown that if L is a distributive lattice then $L\odot L$ is a distributive bilattice, and that every distributive bilattice is isomorphic to $L\odot L$ for some complete distributive lattice L. These results are respectively extended in [157] and [37] to the larger family of *interlaced* bilattices ([157]).

Now we turn to *bilattice-valued matrices*:

Definition 5.78 (bifilters)

- Let $\langle L, \leq\rangle$ be a lattice. A nonempty proper subset $\mathbb{F}$ of L is called a *filter*, if for all $a,b \in L$, $a\wedge b \in \mathbb{F}$ iff $a\in\mathbb{F}$ and $b\in\mathbb{F}$. A filter $\mathbb{F}$ is called *prime*, if for all $a,b\in L$, $a\vee b\in\mathbb{F}$ iff $a\in\mathbb{F}$ or $b\in\mathbb{F}$.

- Let $\mathcal{B} = \langle B, \leq_t, \leq_k, \neg\rangle$ be a bilattice. A nonempty proper subset $\mathbb{F}$ of B is a [prime] *bifilter* of $\mathcal{B}$ if it is a [prime] filter with respect to both of $\langle B, \leq_t\rangle$ and $\langle B, \leq_k\rangle$.

Example 5.79 The bilattices $\mathcal{FOUR}$ and $\mathcal{DEFAULT}$ contain exactly one bifilter, $\mathbb{F} = \{\top, t\}$, which is prime in both. This set is also the only bifilter of $\mathcal{FIVE}$ (Figure 5.5), but it is *not* prime there: $d\top \vee \bot = t \in \mathbb{F}$, while $d\top \notin \mathbb{F}$, and $\bot \notin \mathbb{F}$. The bilattice $\mathcal{NINE}$ contains two bifilters: $\{\top, ot, t\}$ and $\{\top, ot, t, of, d\top, dt\}$, both of which are prime.

Note 5.80 Clearly, a [bi]filter $\mathbb{F}$ in a [bi]lattice is upward-closed with respect to [both of] the [bi]lattice order[s]. Also, $t\in\mathbb{F}$ [and $\top\in\mathbb{F}$], while $f\notin\mathbb{F}$ [and $\bot\notin\mathbb{F}$]. A detailed discussion on bifilters and their construction can be found in [11, 13]. Here we recall one such result.

5.5 Bilattices and Their Logics

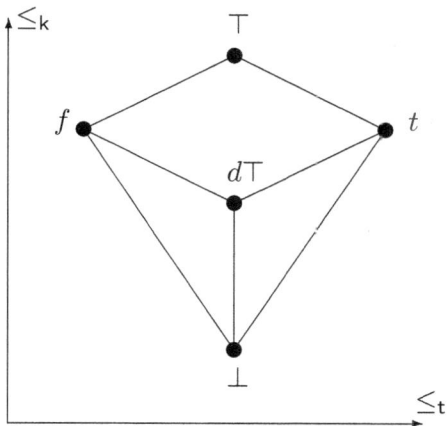

Figure 5.5: $\mathcal{FIVE}$

Lemma 5.81 [13] $\mathcal{D}_L$ *is a (prime) filter in a lattice L iff $\mathcal{D}_L \times L$ is a (prime) filter in the bilattice $L \odot L$.*

Given a bifilter $\mathbb{F}$ in a bilattice $\mathcal{B}$, a corresponding implication connective may be defined just as in Definition 3.38, namely:[12]

$$a \supset b = \begin{cases} b & \text{if } a \in \mathbb{F}, \\ t_\mathcal{B} & \text{otherwise.} \end{cases}$$

Definition 5.82 (bilattice-valued matrices)

- Let $\mathcal{L}_\mathcal{B} = \mathcal{L}_{All} - \{-\}$. In other words: $\mathcal{L}_\mathcal{B}$ is the language obtained from $\mathcal{L}_{All}$ (Definition 5.14) by deleting the conflation connective. [13]

- Let $\mathcal{L}_{\mathcal{FOUR}} \subseteq \mathcal{L} \subseteq \mathcal{L}_\mathcal{B}$ (Definition 5.10). A *bilattice-valued matrix* for $\mathcal{L}$ is a matrix $\langle \mathcal{V}, \mathcal{D}, \mathcal{O} \rangle$ where the elements in $\mathcal{V}$ form a bilattice, the interpretations in $\mathcal{O}$ of the connectives in $\mathcal{L}$ are the corresponding bilattice operations on that bilattice, and $\mathcal{D}$ is a *bifilter* on it.

5.5.2 Characterization in $\mathcal{FOUR}$

In this section we show that from a strictly logical point of view, the use of paradefinite bilattice-valued consequence relations does not provide anything beyond what can be obtained by using the matrices in Section 5.3,

---

[12] To simplify the reading, we have omitted $\mathbb{F}$ from the notation for $\supset$, although the identity of the latter depends on it. This will not cause any confusion in what follows.

[13] By Theorem 5.30, for $FOUR$ $\mathcal{L}_\mathcal{B}$ is equivalent to the functionally complete $\mathcal{L}_{all}$. However, unlike the conflation $-$ in $\mathcal{FOUR}$, a $\leq_k$-reversing function which is dual to $\neg$ is not definable for every bilattice. Therefore, the connective $-$ is excluded from $\mathcal{L}_\mathcal{B}$.

which extend the Dunn-Belnap matrix $\mathcal{FOUR}$. This does not mean, of course, that bilattices have no value. (Thus Boolean algebras are useful even though their logic can be characterized in $\{t, f\}$.) However, it does demonstrate the fundamental role of $\mathcal{FOUR}$ and its four-valued extensions.

The next proposition is crucial for showing the correspondence between bilattice-based consequence relations and four-valued consequence relations:

Proposition 5.83 *Let $\mathcal{L}_{\mathcal{FOUR}} \subseteq \mathcal{L} \subseteq \mathcal{L}_{\mathcal{B}}$, and let $\mathcal{M} = \langle \mathcal{V}, \mathcal{D}, \mathcal{O} \rangle$ be a bilattice-valued matrix for $\mathcal{L}$. If $\mathcal{M}^4 = \langle FOUR, \{t, \top\}, \mathcal{O} \rangle$ is the extension of $\mathcal{FOUR}$ to $\mathcal{L}$ then the following function $h : \mathcal{V} \to FOUR$ is a homomorphism from $\mathcal{M}$ onto $\mathcal{M}^4$:*

$$h(a) = \begin{cases} \top & \text{if } a \in \mathcal{D}, \neg a \in \mathcal{D}, \\ t & \text{if } a \in \mathcal{D}, \neg a \notin \mathcal{D}, \\ f & \text{if } a \notin \mathcal{D}, \neg a \in \mathcal{D}, \\ \bot & \text{if } a \notin \mathcal{D}, \neg a \notin \mathcal{D}, \end{cases}$$

Moreover, $\nu \in \mathrm{mod}_{\mathcal{M}}(\mathcal{T})$ iff $h \circ \nu \in \mathrm{mod}_{\mathcal{M}^4}(\mathcal{T})$ for every $\mathcal{L}$-theory $\mathcal{T}$ and a valuation $\nu \in \Lambda_{\mathcal{M}}$.

Proof. Note first that h is obviously an homomorphism with respect to the propositional constants $\mathsf{T}, \mathsf{F}, \mathsf{B}, \mathsf{N}$, and with respect to $\neg$. It thus remains to show that h is also a homomorphism with respect to $\wedge, \vee, \otimes, \oplus,$ and $\supset$.

a) The case of $\wedge$:

- If $a \wedge b \in \mathcal{D}$ and $\neg(a \wedge b) \in \mathcal{D}$, then by the former $a \in \mathcal{D}$ and $b \in \mathcal{D}$, and by the latter $\neg a \vee \neg b \in \mathcal{D}$. Since $\mathcal{D}$ is prime, $\neg a \in \mathcal{D}$ or $\neg b \in \mathcal{D}$. It follows that $\{a, \neg a\} \subseteq \mathcal{D}$ or $\{b, \neg b\} \subseteq \mathcal{D}$, hence either $h(a) = \top$ or $h(b) = \top$. Since both $h(a)$ and $h(b)$ are in $\{\top, t\}$, and $\top \wedge \top = \top \wedge t = \top$, it follows that $h(a) \wedge h(b) = \top = h(a \wedge b)$.
- If $a \wedge b \in \mathcal{D}$ but $\neg(a \wedge b) \notin \mathcal{D}$, then $a \in \mathcal{D}$ and $b \in \mathcal{D}$, but $\neg a \vee \neg b \notin \mathcal{D}$, and so neither $\neg a$ nor $\neg b$ is in $\mathcal{D}$. It follows that $h(a) = h(b) = t$, so this time $h(a) \wedge h(b) = t = h(a \wedge b)$.
- If $a \wedge b \notin \mathcal{D}$ and $\neg(a \wedge b) \in \mathcal{D}$, then either $\neg a \in \mathcal{D}$ or $\neg b \in \mathcal{D}$. Assume, e.g., that $\neg a \in \mathcal{D}$. If $a \notin \mathcal{D}$ then $h(a) = f$ and so $h(a) \wedge h(b) = f = h(a \wedge b)$. If $a \in \mathcal{D}$, then $h(a) = \top$. In addition, $b \notin \mathcal{D}$ (otherwise, $a \wedge b \in \mathcal{D}$), and so $h(b) \in \{f, \bot\}$. Since in $\mathcal{FOUR}$ $\top \wedge f = \top \wedge \bot = f$, we have that $h(a) \wedge h(b) = f = h(a \wedge b)$.
- If $a \wedge b \notin \mathcal{D}$ and $\neg(a \wedge b) \notin \mathcal{D}$, then $\neg a \notin \mathcal{D}$, $\neg b \notin \mathcal{D}$, and either $a \notin \mathcal{D}$ or $b \notin \mathcal{D}$. It follows that either $h(a) = \bot$ or $h(b) = \bot$. Assume, without loss of generality, the former. Since $\neg b \notin \mathcal{D}$, then $h(b) \in \{t, \bot\}$. But $\bot \wedge t = \bot \wedge \bot = \bot$. Hence we have that $h(a) \wedge h(b) = \bot = h(a \wedge b)$.

5.5 Bilattices and Their Logics   169

b) The case of $\vee$ follows from the identity $a \vee b = \neg(\neg a \wedge \neg b)$.

c) The case of $\otimes$:

- If $a \otimes b \in \mathcal{D}$ and $\neg(a \otimes b) \in \mathcal{D}$, then since $\neg(a \otimes b) = \neg a \otimes \neg b$, we have that $a, b, \neg a, \neg b \in \mathcal{D}$. Hence $h(a) = h(b) = \top$, and so $h(a) \otimes h(b) = \top = h(a \otimes b)$.

- If $a \otimes b \in \mathcal{D}$ and $\neg(a \otimes b) \notin \mathcal{D}$, then $a \in \mathcal{D}$, $b \in \mathcal{D}$, and either $\neg a \notin \mathcal{D}$ or $\neg b \notin \mathcal{D}$. It follows that $\{t\} \subseteq \{h(a), h(b)\} \subseteq \{\top, t\}$. Hence, $h(a) \otimes h(b) = t = h(a \otimes b)$.

- If $a \otimes b \notin \mathcal{D}$ and $\neg(a \otimes b) \in \mathcal{D}$, the argument is similar.

- If $a \otimes b \notin \mathcal{D}$ and $\neg(a \otimes b) \notin \mathcal{D}$ then either $a \notin \mathcal{D}$ or $b \notin \mathcal{D}$, and also either $\neg a \notin \mathcal{D}$ or $\neg b \notin \mathcal{D}$. Assume, e.g., that $a \notin \mathcal{D}$. If also $\neg a \notin \mathcal{D}$, then $h(a) = \bot$, and so $h(a) \otimes h(b) = \bot = h(a \otimes b)$. If $\neg a \in \mathcal{D}$, then $\neg b \notin \mathcal{D}$, and so we have that $h(a) = f$, and $h(b) \in \{t, \bot\}$. Since in $\mathcal{FOUR}$ $f \otimes t = f \otimes \bot = \bot$, we have again that $h(a) \otimes h(b) = \bot = h(a \otimes b)$.

d) The case of $\oplus$:

- Assume that $a \oplus b \in \mathcal{D}$ and $\neg(a \oplus b) \in \mathcal{D}$. Then $a \in \mathcal{D}$ or $b \in \mathcal{D}$, and $\neg a \in \mathcal{D}$ or $\neg b \in \mathcal{D}$. Assume, e.g., that $a \in \mathcal{D}$. Then $h(a) \in \{\top, t\}$. If, in addition, $\neg a \in \mathcal{D}$, then $h(a) = \top$, and so $h(a) \oplus h(b) = \top = h(a \oplus b)$. Otherwise, $\neg b \in \mathcal{D}$, and so $h(b) \in \{\top, f\}$. Since in $\mathcal{FOUR}$, $\top \oplus \top = \top \oplus t = \top \oplus f = t \oplus f = \top$, we have that $h(a) \oplus h(b) = \top = h(a \oplus b)$.

- If $a \oplus b \in \mathcal{D}$ and $\neg(a \oplus b) \notin \mathcal{D}$, then $a \in \mathcal{D}$ or $b \in \mathcal{D}$, and neither $\neg a$ nor $\neg b$ in in $\mathcal{D}$. It follows that $h(a)$ and $h(b)$ are both in $\{t, \bot\}$, and at least one of then is t. Hence, $h(a) \oplus h(b) = t = h(a \oplus b)$.

- If $a \oplus b \notin \mathcal{D}$ and $\neg(a \oplus b) \in \mathcal{D}$, the argument is similar.

- If $a \oplus b \notin \mathcal{D}$ and $\neg(a \oplus b) \notin \mathcal{D}$, then $a, \neg a, b, \neg b$ are all not in $\mathcal{D}$, and so $h(a) = h(b) = \bot$. It follows that $h(a) \oplus h(b) = \bot = h(a \oplus b)$.

e) The case of $\supset$:

- If $a \in \mathcal{D}$ then $a \supset b = b$, so $h(a \supset b) = h(b) = h(a) \supset h(b)$, since $h(a) \in \{\top, t\}$ when $a \in \mathcal{D}$.

- if $a \notin \mathcal{D}$, then $a \supset b = t$ and so $h(a \supset b) = h(t) = t$. But since in this case $h(a) \in \{\bot, f\}$, necessarily $h(a) \supset h(b) = t$ too. $\square$

Note 5.84 It is easy to see that the function used in the proof of Proposition 5.83 is the unique homomorphism $h : \mathcal{V} \to \mathcal{FOUR}$ such that $h(b) \in$

$\{\top, t\}$ iff $b \in \mathcal{D}$. For Boolean algebras we have, in fact, only weaker theorem: Given x from a Boolean algebra B, and a filter $\mathbb{F} \subseteq B$ s.t. $x \notin \mathbb{F}$, we have a homomorphism $h_x : B \to \mathcal{TWO}$ with respect to $\neg, \wedge, \vee$ such that $h_x(x) \notin \mathbb{F}(\mathcal{TWO})$, and $h_x(y) \in \mathbb{F}(\mathcal{TWO})$ for every $y \in \mathbb{F}$. In our case, the same h is good for all x. On the other hand, in Boolean algebras we have the property that if $x, y \in B$ and $x \neq y$, then there is a homomorphism $h : B \to \mathcal{TWO}$ which separates them. This further implies that equalities which hold in $\mathcal{TWO}$ are valid in any Boolean algebra. In contrast, bilattices and $\mathcal{FOUR}$ do not enjoy this property. Thus, the distributive law $a \wedge (b \vee c) = (a \wedge b) \vee (a \wedge c)$ is valid in $\mathcal{FOUR}$, but *not* in every bilattice-based matrix in general (take, e.g., a matrix with the values in $\mathcal{DEFAULT}$).

To examine some basic properties of bilattice-valued matrices and their logics, we need the following definitions.

Definition 5.85 (types of truth values) Given a matrix $\mathcal{M} = \langle \mathcal{V}, \mathcal{D}, \mathcal{O} \rangle$ for a language with a unary connective $\neg$, the elements in $\mathcal{V}$ are divided to four types as follows:

$$\mathfrak{T}_t^{\mathcal{M}} = \{x \in \mathcal{V} \mid x \in \mathcal{D}, \neg x \notin \mathcal{D}\}, \quad \mathfrak{T}_f^{\mathcal{M}} = \{x \in \mathcal{V} \mid x \notin \mathcal{D}, \neg x \in \mathcal{D}\},$$
$$\mathfrak{T}_\top^{\mathcal{M}} = \{x \in \mathcal{V} \mid x \in \mathcal{D}, \neg x \in \mathcal{D}\}, \quad \mathfrak{T}_\bot^{\mathcal{M}} = \{x \in \mathcal{V} \mid x \notin \mathcal{D}, \neg x \notin \mathcal{D}\}.$$

We shall sometimes omit the superscripts, and just write $\mathfrak{T}_t, \mathfrak{T}_f, \mathfrak{T}_\top, \mathfrak{T}_\bot$.[14]

Definition 5.86 (similar valuations) Let $\mathcal{M}_1 = \langle \mathcal{V}_1, \mathcal{D}_1, \mathcal{O} \rangle$ and $\mathcal{M}_2 = \langle \mathcal{V}_2, \mathcal{D}_2, \mathcal{O} \rangle$ be two matrices for a language $\mathcal{L}$ with negation.

- Two elements $a_1 \in \mathcal{V}_1$ and $a_2 \in \mathcal{V}_2$ are called *similar* (or *of the same type*) if, in the notations of Definition 5.85, there is an $x \in FOUR$ such that $a_1 \in \mathfrak{T}_x^{\mathcal{M}_1}$ and $a_2 \in \mathfrak{T}_x^{\mathcal{M}_2}$.

- Two valuations $\nu_1 \in \Lambda_{\mathcal{M}_1}$ and $\nu_2 \in \Lambda_{\mathcal{M}_2}$ are called *similar* (or *of the same time*), if for every variable p, $\nu_1(p)$ and $\nu_2(p)$ are similar.

Example 5.87 The proof of Proposition 5.83 shows that (in the notations of that proposition and under its assumption) the $\mathcal{M}$-valuation ν and the $\mathcal{M}^4$-valuation $h \circ \nu$ are similar.

Note 5.88 The types of valuations depend on the identity of the bifilter, so two valuations might not be of the same type even in case they are identical and the underlying bilattice is the same. (Consider, e.g., a valuation ν on $\mathcal{NINE}$ such that $\nu(p) = ot$ for some variable p.)

---
[14]Note that, if $h_\mathcal{M}$ is the homomorphism defined in Proposition 5.83 for a bilattice-based matrix $\mathcal{M}$, then for every $x \in FOUR$ it holds that $\mathfrak{T}_x^{\mathcal{M}} = h_\mathcal{M}^{-1}(x)$.

5.5 Bilattices and Their Logics

Proposition 5.89 *Let $\mathcal{M}$ be a bilattice-valued matrix. Then $t \in \mathfrak{T}_t^{\mathcal{M}}$, $f \in \mathfrak{T}_f^{\mathcal{M}}$, $\top \in \mathfrak{T}_\top^{\mathcal{M}}$ and $\bot \in \mathfrak{T}_\bot^{\mathcal{M}}$.*

Proof. This follows from Note 5.80. □

Corollary 5.90 *Every bilattice-valued matrix is paradefinite.*

Proof. Let $\mathcal{M}$ be a bilattice-valued matrix. Since $\top \in \mathfrak{T}_\top^{\mathcal{M}}$, by Proposition 3.55-1 $\mathcal{M}$ is pre-paraconsistent; Since $\bot \in \mathfrak{T}_\bot^{\mathcal{M}}$ and $\vee$ is a $\{t, f\}$-closed $\mathcal{M}$-disjunction, $\mathcal{M}$ is $\neg$-paracomplete by Proposition 5.2-2. □

Proposition 5.91 *Let $\mathcal{L}_{\mathcal{FOUR}} \subseteq \mathcal{L} \subseteq \mathcal{L}_{\mathcal{B}}$, and let $\mathcal{M}_1 = \langle \mathcal{V}_1, \mathcal{D}_1, \mathcal{O} \rangle$ and $\mathcal{M}_2 = \langle \mathcal{V}_2, \mathcal{D}_2, \mathcal{O} \rangle$ be two bilattice-based matrices for $\mathcal{L}$. Suppose that ν_1 and ν_2 are two valuations on $\Lambda_{\mathcal{M}_1}$ and $\Lambda_{\mathcal{M}_2}$ (respectively), which are of the same type. Then for every formula ψ, $\nu_1(\psi)$ and $\nu_2(\psi)$ are of the same type.*

Proof. By an induction on the structure of ψ (The fact that $\mathcal{D}_1$ and $\mathcal{D}_2$ are *prime* is crucial here!). □

We can now show the main result of this section.

Theorem 5.92 (characterization in $\mathcal{FOUR}$) *Let $\mathcal{L}$, $\mathcal{M}$, and $\mathcal{M}^4$ be like in Proposition 5.83. Then $\mathcal{T} \vdash_{\mathcal{M}} \psi$ iff $\mathcal{T} \vdash_{\mathcal{M}^4} \psi$.*

Proof. Suppose that $\mathcal{T} \vdash_{\mathcal{M}} \psi$. We show that $\mathcal{T} \vdash_{\mathcal{M}^4} \psi$. So let ν^4 be an $\mathcal{M}^4$-model of $\mathcal{T}$. Define a valuation $\nu^{\mathcal{M}} \in \Lambda_{\mathcal{M}}$ by: $\nu^4(p) = x$ iff $\nu^{\mathcal{M}}(p) = x_{\mathcal{B}}$ (where $\mathcal{B}$ is the bilattice on which $\mathcal{M}$ is based) for every variable p and $x \in \mathcal{FOUR}$. By Proposition 5.89, $\nu^{(4)}$ and $\nu^{\mathcal{M}}$ are similar, and so by Proposition 5.91, $\nu^4(\varphi)$ and $\nu^{\mathcal{M}}(\varphi)$ are similar for every φ.[15] Hence $\nu^{\mathcal{M}}$ is an $\mathcal{M}$-model of $\mathcal{T}$, and so $\nu^{\mathcal{M}} \models \psi$, implying (by the similarity) that $\nu^4 \models \psi$ as well.

For the converse, suppose that $\mathcal{T} \not\vdash_{\mathcal{M}} \psi$. Then there is a valuation $\nu^{\mathcal{M}}$ that is an $\mathcal{M}$-model of $\mathcal{T}$ but $\nu^{\mathcal{M}}(\psi) \notin \mathcal{D}$. Let $\nu^4 = h \circ \nu^{\mathcal{M}}$. From Propositions 5.91 and 5.83 it follows that ν^4 is an $\mathcal{M}^4$-model of $\mathcal{T}$ but $\nu^4(\psi) \notin \{t, \top\}$. Thus, $\mathcal{T} \not\vdash_{\mathcal{M}^4} \psi$. □

We conclude this section with some easy consequences of Theorem 5.92.

Proposition 5.93 *Let $\mathcal{L}_{\mathcal{FOUR}} \subseteq \mathcal{L} \subseteq \mathcal{L}_{Nex}$ and let $\mathcal{M} = \langle \mathcal{V}, \mathcal{D}, \mathcal{O} \rangle$ be a bilattice-valued matrix for $\mathcal{L}$. Then $\mathbf{L}_{\mathcal{M}}$ is non-exploding.*

---

[15]In the specific case where the underlying bilattice $\mathcal{B}$ is of the form of Definition 5.75 (or is interlaced, [157]), it is shown in Proposition 3.1 of [158] that $\mathcal{FOUR}$ is actually a sub-bilattice of $\mathcal{B}$. Therefore, in this case $\nu^4(\psi)$ and $\nu^{\mathcal{M}}(\psi)$ are not only similar, but are actually identical.

Proof. By Theorem 5.92 it is enough to show that $\mathbf{L}_{\mathcal{M}^4}$ is non-exploding. This follows from Proposition 5.49 (see also Proposition 3.59). □

Proposition 5.94 *Let $\mathcal{M}$ be a bilattice-valued matrix for a language $\mathcal{L} \subseteq \{\vee, \wedge, \oplus, \otimes, \neg\}$. Then $\mathbf{L}_{\mathcal{M}}$ does not have tautologies.*

Proof. Again, by Theorem 5.92 it is enough to show the proposition for $\mathbf{L}_{\mathcal{M}^4}$. This follows from the fact that $\mathcal{L}$ is $\{\bot\}$-closed in $\mathcal{M}^4$, and so for every formula ψ in $\mathcal{L}$ and a valuation ν for which $\nu(p) = \bot$ for every $p \in \mathsf{Var}(\psi)$ we have that $\nu(\psi) = \bot \notin \mathcal{D}$. □

Proposition 5.95 *Let $\mathbf{L} \in \{\mathbf{4CC}, \mathbf{4Nex}, \mathbf{4Flex}, \mathbf{4CL}\}$ and let $\mathcal{M}$ be a bilattice-valued matrix for the corresponding language. Then $\vdash_{\mathcal{M}}$, $\vdash_{\mathbf{G_L}}$, and $\vdash_{H_\mathbf{L}}$ coincide.*

Proof. By Theorems 5.62, 5.69, and 5.92. □

5.6 Bibliographical Notes and Further Reading

Like three-valued logics, four-valued reasoning may be traced back to the work of Łukasiewicz in the early 1920's [201], which is followed by a series of works in the 1950's under different names (e.g., quasi-Boolean algebras [97, 98] and distributive i-lattices [188]). The seminal papers of Belnap on four-valued computerized reasoning [86, 87], that continued the work of Dunn on this subject [146, 148], have stimulated a revived interest in this topic, which has been considered in many different contexts. A special volume of Studia Logica dedicated to the 40th anniversary of Belnap's papers has recently appeared in [232].

A non-exhaustive list of areas where logics of this type have been applied includes relations to two-valued logics [21, 165], knowledge-base integration [60], fuzzy logic and preferential modeling [140, 285], relevant logics [101, 253], semantics for argumentation frameworks [8, 9], self reference [293], and so forth.

Good surveys on bilattices and their applications can be found in Fitting's [163, 164] and Rivieccio's [255]. In addition to the references given in Section 5.5, some further applications and bilattice-based formalisms are presented in [138, 155, 189, 216]. Extensions of bilattices to structures with higher dimensions are described, e.g., in [190, 229, 268, 269, 296, 306].

This chapter is mainly based on the papers in [11, 12, 13, 40]. An early version of the first four sections of this chapter, which includes most of the new results in these sections, has appeared in [16]. Recent related works on four-valued logics appear, e.g., in [132, 231].

Part III
Non-Determinism

Chapter 6

Non-Deterministic Matrices

6.1 Motivation

All the semantic methods for handling paraconsistency introduced so far are based on the principle of *truth-functionality*, according to which the truth-value of a complex formula is uniquely determined by the truth-values of its subformulas. Unfortunately, this principle is not adequate for providing semantic characterization for one of the most famous families of paraconsistent logics: that of da Costa's Brazilian school. Nevertheless, it is still possible to provide in a modular way simple and useful multiple-valued semantics for these logics by relaxing the principle of truth-functionality. (See Chapter 8.) One way this can be done is by using *non-deterministic matrices* (Nmatrices). Nmatrices are a natural generalization of ordinary multiple-valued matrices, in which the truth-value of a complex formula can be chosen *non-deterministically* out of some non-empty set of options. The semantics of Nmatrices shares with the semantics of matrices such attractive properties as compactness and decidability in the finite case — and even more importantly: analyticity. Recall that the latter property trivially holds for matrices (Proposition 3.9), and is crucial (among other things) for the decidability of logics which are induced by finite matrices (see the proof of Theorem 3.27). The fact that the semantics of Nmatrices is analytic is the main feature that distinguishes it from the two other types of non-truth-functional semantics which have been used for paraconsistent logics: bivaluation semantics and possible translations semantics (see [111, 113]). In these types analyticity is not a priori guaranteed, and it needs to be proved from scratch for every new instance of the semantics. Another significant advantage of the semantic framework of Nmatrices is its rich general theory, which includes special useful operations, not available for matrices or for the other types of non-deterministic semantics. This theory is presented in this chapter, and is applied in Chapters 7, 8, and 10.

6.2 Basic Theory of Nmatrices

In this section and in the next one we review basic notions and facts about Nmatrices in general. We omit the proofs of propositions that appear elsewhere in the literature. (See Section 6.6 for references.)

Definition 6.1 (Nmatrices) A (multi-valued) *non-deterministic matrix* (*Nmatrix*) for $\mathcal{L}$ is a triple $\mathcal{M} = \langle \mathcal{V}, \mathcal{D}, \mathcal{O} \rangle$, where

- $\mathcal{V}$ is a non-empty set of truth values;

- $\mathcal{D}$ is a non-empty proper subset of $\mathcal{V}$ (whose elements are called the *designated* elements of $\mathcal{V}$);

- $\mathcal{O}$ is a function that associates an n-ary function $\tilde{\diamond} : \mathcal{V}^n \to 2^{\mathcal{V}} \setminus \{\emptyset\}$ with every n-ary connective $\diamond$ of $\mathcal{L}$.

Notation 6.2 $\overline{\mathcal{D}} = \mathcal{V} \setminus \mathcal{D}$.

Definition 6.3 (non-deterministic connectives, proper Nmatrices) Let $\mathcal{M} = \langle \mathcal{V}, \mathcal{D}, \mathcal{O} \rangle$ be an Nmatrix for $\mathcal{L}$.

- An n-ary connective $\diamond$ of $\mathcal{L}$ is *non-deterministic in* $\mathcal{M}$, if there are some $x_1, \ldots, x_n \in \mathcal{V}$, such that $\tilde{\diamond}(x_1, \ldots, x_n)$ is not a singleton.

- $\mathcal{M}$ is a *proper* Nmatrix if at least one of the connectives of $\mathcal{L}$ is non-deterministic in $\mathcal{M}$. $\mathcal{M}$ is *strictly proper* if that connective has arity $n > 0$. $\mathcal{M}$ is *deterministic* if it is not proper.

Note 6.4 In deterministic Nmatrices each $\tilde{\diamond}$ is a function taking singleton values only. In practice such a function can be seen as a function $\tilde{\diamond} : \mathcal{V}^n \to \mathcal{V}$. Hence ordinary matrices may be identified with the corresponding deterministic Nmatrices.

The rest of this section is very similar to the basic theory of ordinary matrices given in Section 3.1. Most of the definitions and propositions in it are practically identical to those presented there. It is very important, though, to note the few (but crucial) differences that do exist.

Definition 6.5 (extension/reduction) Let $\mathcal{M} = \langle \mathcal{V}, \mathcal{D}, \mathcal{O} \rangle$ be an Nmatrix for $\mathcal{L}$, and let $\mathcal{M}_e = \langle \mathcal{V}_e, \mathcal{D}_e, \mathcal{O}_e \rangle$ be an Nmatrix for $\mathcal{L}_e$. $\mathcal{M}_e$ is an *extension* of $\mathcal{M}$ to $\mathcal{L}_e$, and $\mathcal{M}$ is a *reduction* of $\mathcal{M}_e$ to $\mathcal{L}$, if $\mathcal{L} \subseteq \mathcal{L}_e$, $\mathcal{V} = \mathcal{V}_e$, $\mathcal{D} = \mathcal{D}_e$, and $\mathcal{O}_e(\diamond) = \mathcal{O}(\diamond)$ for every connective $\diamond$ of $\mathcal{L}$.

Definition 6.6 (finite and effective Nmatrices) Let $\mathcal{M} = \langle \mathcal{V}, \mathcal{D}, \mathcal{O} \rangle$ be an Nmatrix for $\mathcal{L}$.

6.2 Basic Theory of Nmatrices

- We say that $\mathcal{M}$ is *(in)finite* if so is $\mathcal{V}$.

- We call $\mathcal{M}$ *effective*, if it satisfies the following conditions:

  a) $\mathcal{V}$ can effectively be enumerated (and so is at most countable).
  b) The relation $\{\langle a_1, \ldots, a_n, b\rangle \in \mathcal{V}^{n+1} \mid b \in \tilde{\diamond}(a_1, \ldots, a_n)\}$ is decidable for each connective $\diamond$ of $\mathcal{L}$ (where n is the arity of $\diamond$).
  c) The set $\mathcal{D}$ of the designated truth values is decidable.

Definition 6.7 (valuations, satisfiability) Let $\mathcal{M} = \langle \mathcal{V}, \mathcal{D}, \mathcal{O}\rangle$ be an Nmatrix for $\mathcal{L}$.

- A *partial $\mathcal{M}$-valuation* for $\mathcal{L}$ is a function $\nu: \mathcal{W} \to \mathcal{V}$ for some subset $\mathcal{W}$ of $\mathcal{W}(\mathcal{L})$ that satisfies the following conditions:

  - The set $\mathcal{W}$ is closed under subformulas.
  - For every n-ary connective $\diamond$ of $\mathcal{L}$ and every $\psi_1, \ldots, \psi_n$ such that $\diamond(\psi_1, \ldots, \psi_n) \in \mathcal{W}$,

  $$(*) \quad \nu(\diamond(\psi_1, \ldots, \psi_n)) \in \tilde{\diamond}(\nu(\psi_1), \ldots, \nu(\psi_n)).$$

- A partial $\mathcal{M}$-valuation is a (full) $\mathcal{M}$-*valuation* if its domain is $\mathcal{W}(\mathcal{L})$. We denote the set of all the $\mathcal{M}$-valuations by $\Lambda_\mathcal{M}$.

- A valuation $\nu \in \Lambda_\mathcal{M}$ is an $\mathcal{M}$-*model* of a formula ψ, or ν $\mathcal{M}$-*satisfies* ψ (notation: $\nu \models_\mathcal{M} \psi$), if $\nu(\psi) \in \mathcal{D}$. We say that ν is an $\mathcal{M}$-model of a theory $\mathcal{T}$ (notation: $\nu \models_\mathcal{M} \mathcal{T}$), if it is an $\mathcal{M}$-model of every element of $\mathcal{T}$. We denote the set of $\mathcal{M}$-models of ψ by $mod_\mathcal{M}(\psi)$, and the set of $\mathcal{M}$-models of $\mathcal{T}$ by $mod_\mathcal{M}(\mathcal{T})$. (So $mod_\mathcal{M}(\psi) = \{\nu \in \Lambda_\mathcal{M} \mid \nu(\psi) \in \mathcal{D}\}$, and $mod_\mathcal{M}(\mathcal{T}) = \cap_{\psi \in \mathcal{T}} mod_\mathcal{M}(\psi)$.)

- A formula ψ is $\mathcal{M}$-*satisfiable* if $mod_\mathcal{M}(\psi) \neq \emptyset$; it is an $\mathcal{M}$-*tautology* if $mod_\mathcal{M}(\psi) = \Lambda_\mathcal{M}$. A theory $\mathcal{T}$ is $\mathcal{M}$-satisfiable if $mod_\mathcal{M}(\mathcal{T}) \neq \emptyset$.

Note 6.8 The only difference between Definitions 6.7 and 3.7 is in Condition $(*)$. Recall that in the case of matrices we have '=' rather than '$\in$' in the corresponding condition. (This amounts to the same thing when a matrix $\mathcal{M}$ is viewed as an Nmatrix in which $\tilde{\diamond}$ always returns a singleton.)

Note 6.9 Like in the case of matrices, for every Nmatrix $\mathcal{M}$ the pair $\langle \Lambda_\mathcal{M}, \models_\mathcal{M}\rangle$ is a denotational semantics in the sense of Definition 1.37.

Notation 6.10 To simplify reading, in what follows we shall sometimes omit the prefix '$\mathcal{M}$' from the notions above.

Proposition 6.11 (basic properties of the semantics) *The semantics induced by an Nmatrix $\mathcal{M}$ for $\mathcal{L}$ has the following properties:*

ANALYTICITY. *Any partial $\mathcal{M}$-valuation can be extended to a full valuation.*

SUBSTITUTION-CLOSURE. *If $\nu \in \Lambda_\mathcal{M}$ and θ is an $\mathcal{L}$-substitution, then the composition $\nu \circ \theta$ of ν and θ is in $\Lambda_\mathcal{M}$ as well.*

The following is again a special case of schema (1.1) in Definition 1.38:

Definition 6.12 ($\vdash_\mathcal{M}$, $\mathbf{L}_\mathcal{M}$) Let $\mathcal{M}$ be an Nmatrix for $\mathcal{L}$. $\vdash_\mathcal{M}$, the consequence relation that is induced by $\mathcal{M}$, is defined by: $\mathcal{T} \vdash_\mathcal{M} \psi$ if $mod_\mathcal{M}(\mathcal{T}) \subseteq mod_\mathcal{M}(\psi)$.[1] We denote by $\mathbf{L}_\mathcal{M}$ the logic $\langle \mathcal{L}, \vdash_\mathcal{M} \rangle$ which is induced by $\mathcal{M}$.

Proposition 6.13 *For every propositional language $\mathcal{L}$ and every Nmatrix $\mathcal{M}$ for $\mathcal{L}$, $\mathbf{L}_\mathcal{M}$ is a propositional logic.*

Note 6.14 Following Note 3.16, we can now considerably extend the standard notion of an n-valued logic as follows: a logic $\mathbf{L}$ is called n-valued if $\mathbf{L} = \mathbf{L}_\mathcal{M}$ for some Nmatrix $\mathcal{M}$ that has n elements. (Recall that a matrix with n elements can be considered a special type of an Nmatrix in which the interpretations of the logical connectives are deterministic.)

Definition 3.18 and Proposition 3.19 can also be generalized for Nmatrices in a straightforward way:

Definition 6.15 ($\vdash_\mathcal{F}$) Let $\mathcal{F}$ be a set of Nmatrices for $\mathcal{L}$. The relation $\vdash_\mathcal{F}$ that is induced by $\mathcal{F}$ is defined by: $\mathcal{T} \vdash_\mathcal{F} \psi$ if $\mathcal{T} \vdash_\mathcal{M} \psi$ for every $\mathcal{M} \in \mathcal{F}$.

Proposition 6.16 $\langle \mathcal{L}, \vdash_\mathcal{F} \rangle$ *is a propositional logic for every propositional language $\mathcal{L}$ and a set $\mathcal{F}$ of Nmatrices for $\mathcal{L}$.*

Next come generalizations of some important facts about matrices.

Proposition 6.17 *Suppose that $\mathcal{M} = \langle \mathcal{V}, \mathcal{D}, \mathcal{O} \rangle$ is an Nmatrix for $\mathcal{L}$, and $\mathcal{M}' = \langle \mathcal{V}', \mathcal{D}', \mathcal{O}' \rangle$ is an extension of $\mathcal{M}$ to $\mathcal{L}'$. Then $\vdash_{\mathcal{M}'}$ is a conservative extension of $\vdash_\mathcal{M}$: if $\mathcal{T} \cup \{\varphi\} \subseteq \mathcal{W}(\mathcal{L})$ then $\mathcal{T} \vdash_{\mathcal{M}'} \varphi$ iff $\mathcal{T} \vdash_\mathcal{M} \varphi$.*

Theorem 6.18 (compactness) *For every propositional language $\mathcal{L}$ and a finite Nmatrix $\mathcal{M}$ for $\mathcal{L}$, $\mathbf{L}_\mathcal{M}$ is a finitary propositional logic.*

The notions of soundness and completeness are again defined exactly as in the case of matrices (Definition 3.22):

---

[1] A Scott (multiple-conclusion) consequence relation is defined similarly by $\mathcal{T} \vdash_\mathcal{M} \mathcal{S}$ if for every $\nu \in mod_\mathcal{M}(\mathcal{T})$ there is some $\psi \in \mathcal{S}$ such that $\nu \in mod_\mathcal{M}(\psi)$.

6.2 Basic Theory of Nmatrices

Definition 6.19 Let $\mathbf{L} = \langle \mathcal{L}, \vdash_\mathbf{L} \rangle$ be a propositional logic, and let $\mathcal{M}$ be an Nmatrix for $\mathcal{L}$.

- If $\mathbf{L}_\mathcal{M}$ is an extension of $\mathbf{L}$ we say that $\mathbf{L}$ is *sound* for $\mathcal{M}$.
- If $\mathbf{L}$ is an extension of $\mathbf{L}_\mathcal{M}$ we say that $\mathbf{L}$ is *complete* for $\mathcal{M}$.
- $\mathcal{M}$ is a *characteristic* Nmatrix for $\mathbf{L}$ if $\mathbf{L} = \mathbf{L}_\mathcal{M}$ (i.e, if $\mathbf{L}$ is both sound and complete for $\mathbf{L}_\mathcal{M}$).
- $\mathbf{L}_\mathcal{M}$ is *weakly sound* for $\mathbf{L}$ if for every $\psi \in \mathcal{W}(\mathcal{L})$, $\vdash_\mathcal{M} \psi$ implies that $\vdash_\mathbf{L} \psi$, and $\mathbf{L}_\mathcal{M}$ is *weakly complete* for $\mathbf{L}$ if $\vdash_\mathbf{L} \psi$ implies that $\vdash_\mathcal{M} \psi$.
- $\mathcal{M}$ is a *weakly characteristic* Nmatrix for $\mathbf{L}$ if $\mathbf{L}$ is both weakly sound and weakly complete for $\mathbf{L}_\mathcal{M}$ (i.e., $\vdash_{\mathbf{L}_\mathcal{M}} \psi$ iff $\vdash_\mathbf{L} \psi$)).

Proposition 6.20 *The logic $\mathbf{L}_\mathcal{M}$ induced by an Nmatrix $\mathcal{M}$ is uniform.*

Corollary 6.21 *If $\mathcal{M}$ is an Nmatrix then $\mathbf{L}_\mathcal{M}$ has a characteristic* matrix.[2]

Proof. Immediate from Proposition 6.20 and Theorem 3.24. □

The proof of the next theorem is identical to that of Theorem 3.27.

Theorem 6.22 (decidability) *Let Γ be a finite theory, ψ a formula, and $\mathcal{M}$ an effective Nmatrix for $\mathcal{L}$.*

1. *The question whether $\Gamma \vdash_\mathcal{M} \psi$ is co-semi-decidable.*
2. *If $\mathcal{M}$ is finite, then the question whether $\Gamma \vdash_\mathcal{M} \psi$ is decidable.*
3. *If $\vdash_\mathcal{M}$ has a sound and complete effective proof system, then $\vdash_\mathcal{M}$ is decidable even if $\mathcal{M}$ is infinite.*

It is natural to ask whether finite Nmatrices can be used for characterizing logics that cannot be characterized by finite ordinary matrices. The next theorem provides a positive answer to this question for the two-valued case. Many more examples will be given in later chapters.

Theorem 6.23 *Let $\mathcal{M}$ be a two-valued, strictly proper Nmatrix. Then there is no finite family $\mathcal{F}$ of finite ordinary matrices such that $\vdash_\mathcal{M} = \vdash_\mathcal{F}$. If in addition $\mathcal{M}$ includes the classical implication, then there is no finite family $\mathcal{F}$ of ordinary matrices such that $\vdash_\mathcal{M} \psi$ iff $\vdash_\mathcal{F} \psi$.*

---
[2] Note that this does *not* mean that if $\mathcal{M}$ is finite or effective, then $\mathbf{L}_\mathcal{M}$ has a characteristic matrix with the same property.

Next is a characterization of the basic connectives (those considered in Definition 1.23) in Nmatrix-based logics.

Definition 6.24 (basic $\mathcal{M}$-connectives) Let $\mathcal{M} = \langle \mathcal{V}, \mathcal{D}, \mathcal{O} \rangle$ be an Nmatrix for a language $\mathcal{L}$.

- A connective $\wedge$ in $\mathcal{L}$ is called an $\mathcal{M}$-*conjunction* if for every $a, b \in \mathcal{V}$, if $a \in \mathcal{D}$ and $b \in \mathcal{D}$ then $a \tilde{\wedge} b \subseteq \mathcal{D}$, and if $a \in \overline{\mathcal{D}}$ or $b \in \overline{\mathcal{D}}$ then $a \tilde{\wedge} b \subseteq \overline{\mathcal{D}}$.

- A connective $\vee$ in $\mathcal{L}$ is called an $\mathcal{M}$-*disjunction* if for every $a, b \in \mathcal{V}$, if $a \in \mathcal{D}$ or $b \in \mathcal{D}$ then $a \tilde{\vee} b \subseteq \mathcal{D}$, and if $a \in \overline{\mathcal{D}}$ and $b \in \overline{\mathcal{D}}$ then $a \tilde{\vee} b \subseteq \overline{\mathcal{D}}$.

- A connective $\supset$ in $\mathcal{L}$ is called an $\mathcal{M}$-*implication* if for every $a, b \in \mathcal{V}$, if $a \in \overline{\mathcal{D}}$ or $b \in \mathcal{D}$ then $a \tilde{\supset} b \subseteq \mathcal{D}$, and if $a \in \mathcal{D}$ and $b \in \overline{\mathcal{D}}$ then $a \tilde{\supset} b \subseteq \overline{\mathcal{D}}$.

Note 6.25 It is again easy to verify that if $\supset$ is an $\mathcal{M}$-implication then the connective $\vee$ defined by $\varphi \vee \psi = (\varphi \supset \psi) \supset \psi$ is an $\mathcal{M}$-disjunction.

The next three results are as easy to prove as their counterparts in Chapter 3 (Proposition 3.32 and Corollaries 3.33, 3.34).

Proposition 6.26 *Let $\mathcal{M} = \langle \mathcal{V}, \mathcal{D}, \mathcal{O} \rangle$ be an Nmatrix for a language $\mathcal{L}$, and let $\diamond$ be a connective of $\mathcal{L}$.*

1. *$\diamond$ is an $\mathcal{M}$-conjunction iff it is a conjunction for $\mathbf{L}_{\mathcal{M}}$.*

2. *If $\diamond$ is an $\mathcal{M}$-disjunction then it is also a disjunction for $\mathbf{L}_{\mathcal{M}}$.*

3. *If $\diamond$ is an $\mathcal{M}$-implication then is also an implication for $\mathbf{L}_{\mathcal{M}}$.*

Corollary 6.27 *Let $\mathcal{M} = \langle \mathcal{V}, \mathcal{D}, \mathcal{O} \rangle$ be an Nmatrix for a language $\mathcal{L}$, and let $\mathcal{M}'$ be an extension of $\mathcal{M}$.*

1. *An $\mathcal{M}$-conjunction ($\mathcal{M}$-disjunction, $\mathcal{M}$-implication) is also a conjunction (disjunction, implication) of $\mathbf{L}_{\mathcal{M}'}$.*

2. *If $\mathcal{M}$ has one of the three connectives defined in Definition 6.24, then $\mathbf{L}_{\mathcal{M}'}$ is semi-normal. If $\mathcal{M}$ has them all then $\mathbf{L}_{\mathcal{M}'}$ is normal.*

Corollary 6.28 *Let $\mathcal{M} = \langle \mathcal{V}, \mathcal{D}, \mathcal{O} \rangle$ be an Nmatrix for $\mathcal{L}$ such that $\mathbf{L}_{\mathcal{M}}$ is $\mathbf{F}$-contained in classical logic, and let $\diamond$ be a connective of $\mathcal{L}$.*

1. *If $\diamond$ is an $\mathcal{M}$-conjunction then $\mathbf{F}(\diamond)$ is the classical conjunction.*

2. *If $\diamond$ is an $\mathcal{M}$-disjunction then $\mathbf{F}(\diamond)$ is the classical disjunction.*

3. *If $\diamond$ is an $\mathcal{M}$-implication then $\mathbf{F}(\diamond)$ is the classical implication.*

Proposition 6.26 again has a converse in a special type of Nmatrices which is particularly important for the study of paraconsistent logics.

Definition 6.29 (f-Nmatrices) An *f-Nmatrix* is an Nmatrix in which $\overline{\mathcal{D}}$ is a singleton. (Again, we shall usually take $\overline{\mathcal{D}} = \{f\}$ in such a case.)

Proposition 6.30 *Let $\mathcal{M} = \langle \mathcal{V}, \mathcal{D}, \mathcal{O} \rangle$ be an f-Nmatrix.*

1. *A connective $\wedge$ is a conjunction for $\mathbf{L}_\mathcal{M}$ iff it is an $\mathcal{M}$-conjunction.*

2. *A connective $\supset$ is an implication for $\mathbf{L}_\mathcal{M}$ iff it is an $\mathcal{M}$-implication.*

3. *A connective $\vee$ is a disjunction for $\mathbf{L}_\mathcal{M}$ iff it is an $\mathcal{M}$-disjunction.*

Proof. Similar to the proof of Proposition 3.36. □

Proposition 6.31 *Let $\mathcal{M} = \langle \mathcal{V}, \mathcal{D}, \mathcal{O} \rangle$ be an Nmatrix for $\mathcal{L}$.*

1. *If a connective $\supset$ of $\mathcal{L}$ is an $\mathcal{M}$-implication then the $\{\supset\}$-fragment of $\mathbf{L}_\mathcal{M}$ is identical to the $\{\supset\}$-fragment of $\mathbf{CL}$.*

2. *Suppose $\mathcal{L}$ has connectives $\supset$, $\vee$, and $\wedge$ that are (respectively) an $\mathcal{M}$-implication, an $\mathcal{M}$-disjunction, and an $\mathcal{M}$-conjunction. Then the $\{\supset, \vee, \wedge\}$-fragment of $\mathbf{L}_\mathcal{M}$ is identical to $\mathbf{CL}^+$.*

Proof. The proof is identical to that of Proposition 3.37. □

6.3 Operations on Nmatrices

In this section we introduce operations on Nmatrices which will be used extensively in the next chapters.

The first operation is completely peculiar to Nmatrices, and non-trivial applications of it to deterministic matrices produce strictly proper Nmatrices. Intuitively, applying it to a given Nmatrix $\mathcal{M}$ means "duplicating" its elements, i.e., constructing a new Nmatrix, which is completely equivalent to $\mathcal{M}$, by replacing each element of $\mathcal{M}$ by some nonempty set of "copies", and defining the operations in the new Nmatrix to be "the same" as in $\mathcal{M}$, without distinguishing between two copies of the same element.

Definition 6.32 (expansion) Let $\mathcal{M} = \langle \mathcal{V}, \mathcal{D}, \mathcal{O} \rangle$ be an Nmatrix for $\mathcal{L}$.

- An *expansion function for $\mathcal{M}$* is a function E on $\mathcal{V}$ such that $E(x) \neq \emptyset$ for every $x \in \mathcal{V}$, and $E(x_1) \cap E(x_2) = \emptyset$ in case $x_1 \neq x_2$.

- Let E be an expansion function for $\mathcal{M}$. The *E-expansion* of $\mathcal{M}$ is the following Nmatrix $\mathcal{M}_E = \langle \mathcal{V}_E, \mathcal{D}_E, \mathcal{O}_E \rangle$:

  - $\mathcal{V}_E = \bigcup_{x \in \mathcal{V}} E(x)$,
  - $\mathcal{D}_E = \bigcup_{x \in \mathcal{D}} E(x)$,
  - For every n-ary connective $\diamond$ of $\mathcal{L}$ and every $y_1, \ldots, y_n \in \mathcal{V}_E$,

    $$\tilde{\diamond}_{\mathcal{M}_E}(y_1, \ldots, y_n) = \bigcup_{z \in \tilde{\diamond}_{\mathcal{M}}(x_1, \ldots, x_n)} E(z),$$

    where for $1 \leq i \leq, n$, x_i is the element of $\mathcal{V}$ such that $y_i \in E(x_i)$.

- We say that $\mathcal{M}_1$ is an expansion of $\mathcal{M}_2$ if $\mathcal{M}_1$ is the E-expansion of $\mathcal{M}_2$ for some function E.

Proposition 6.33 *If $\mathcal{M}_1$ is an expansion of $\mathcal{M}_2$, then $\vdash_{\mathcal{M}_1} = \vdash_{\mathcal{M}_2}$.*

Example 6.34 Let $E(t) = \{t, \top\}$ and $E(f) = \{f\}$. The E-expansion of the matrix $\mathcal{M}_{CL}^+$ is the Nmatrix $\mathcal{M}_\top = \langle \{t, f, \top\}, \{t, \top\}, \{\tilde{\wedge}, \tilde{\vee}, \tilde{\supset}\} \rangle$, in which:

$$a \tilde{\vee} b = \begin{cases} \mathcal{D} & \text{if either } a \in \mathcal{D} \text{ or } b \in \mathcal{D}, \\ \{f\} & \text{if } a = b = f. \end{cases}$$

$$a \tilde{\wedge} b = \begin{cases} \mathcal{D} & \text{if } a, b \in \mathcal{D}, \\ \{f\} & \text{if either } a = f \text{ or } b = f. \end{cases}$$

$$a \tilde{\supset} b = \begin{cases} \mathcal{D} & \text{if either } a = f \text{ or } b \in \mathcal{D}, \\ \{f\} & \text{if } a \in \mathcal{D} \text{ and } b = f. \end{cases}$$

By Proposition 6.33, $\mathbf{L}_{\mathcal{M}_\top} = \mathbf{CL}^+$.

The second operation is used for reducing the degree of nondeterminism of a given Nmatrix. It generalizes the notion of submatrices (Definition 3.4), but it includes a component which is again peculiar to Nmatrices.

Definition 6.35 (refinement) Let $\mathcal{M}_1$ and $\mathcal{M}_2$ be Nmatrices for $\mathcal{L}$, where $\mathcal{M}_1 = \langle \mathcal{V}_1, \mathcal{D}_1, \mathcal{O}_1 \rangle$ and $\mathcal{M}_2 = \langle \mathcal{V}_2, \mathcal{D}_2, \mathcal{O}_2 \rangle$.

- $\mathcal{M}_2$ is a *refinement* of $\mathcal{M}_1$ if $\mathcal{V}_2 \subseteq \mathcal{V}_1$, $\mathcal{D}_2 = \mathcal{V}_2 \cap \mathcal{D}_1$, and for every $a_1, \ldots, a_n \in \mathcal{V}_2$, $\tilde{\diamond}_{\mathcal{M}_2}(a_1, \ldots, a_n) \subseteq \tilde{\diamond}_{\mathcal{M}_1}(a_1, \ldots, a_n)$.

- $\mathcal{M}_2$ is a *simple* refinement of $\mathcal{M}_1$ if it is a refinement in which $\mathcal{V}_1 = \mathcal{V}_2$ (and so also $\mathcal{D}_1 = \mathcal{D}_2$).

- $\mathcal{M}_2$ is a *determinization* of $\mathcal{M}_1$ if it is a deterministic simple refinement of $\mathcal{M}_1$.

6.3 Operations on Nmatrices

Example 6.36 Let $\mathbf{PAC}^+$ be the reduction to $\{\wedge, \vee, \supset\}$ of the matrix $\mathbf{PAC}$ (and its extensions which are discussed in Section 4.4.4). Then $\mathbf{PAC}^+$ is a determinization of the Nmatrix $\mathcal{M}_\top$ from Example 6.34.

Proposition 6.37 *If $\mathcal{M}_2$ is a refinement of $\mathcal{M}_1$, then $\vdash_{\mathcal{M}_1} \subseteq \vdash_{\mathcal{M}_2}$.*

The two operations are frequently done successively. Therefore, it is useful to combine them into one operation, which we call *rexpansion*.

Definition 6.38 Let $\mathcal{M}_1 = \langle \mathcal{V}_1, \mathcal{D}_1, \mathcal{O}_1 \rangle$ and $\mathcal{M}_2 = \langle \mathcal{V}_2, \mathcal{D}_2, \mathcal{O}_2 \rangle$ be Nmatrices for $\mathcal{L}$, and let E be an expansion function for $\mathcal{M}_1$.

- $\mathcal{M}_2$ is a (simple) *E-rexpansion of* $\mathcal{M}_1$ if it is a (simple) refinement of an E-expansion of $\mathcal{M}_1$.

- $\mathcal{M}_2$ is a *preserving E-rexpansion of* $\mathcal{M}_1$ if it is an E-rexpansion of $\mathcal{M}_1$, and $E(x) \cap \mathcal{V}_2 \neq \emptyset$ for every $x \in \mathcal{V}_1$.

- $\mathcal{M}_2$ is a *strongly preserving E-rexpansion of* $\mathcal{M}_1$ if it is a preserving E-rexpansion of $\mathcal{M}_1$, and for every n-ary $\diamond \in \mathcal{L}$, $x_1, \ldots, x_n \in \mathcal{V}_2$, and $y, z_1, \ldots, z_n \in \mathcal{V}_1$: if $x_i \in E(z_i)$ for every $1 \leq i \leq n$, and $y \in \mathcal{O}_1(\diamond)(z_1, \ldots, z_n)$, then $E(y) \cap \mathcal{O}_2(\diamond)(x_1, \ldots, x_n) \neq \emptyset$.

- $\mathcal{M}_2$ is a *(strongly) preserving rexpansion* of $\mathcal{M}_1$ if it is a (strongly) preserving E-rexpansion of it for some expansion function E for $\mathcal{M}_1$.

Example 6.39 By Examples 6.34 and 6.36, the matrix $\mathbf{PAC}^+$ is a rexpansion of $\mathcal{M}_{CL}^+$. It is easy to see that it is strongly preserving one.

Proposition 6.40 *Let $\mathcal{M} = \langle \mathcal{V}, \mathcal{D}, \mathcal{O} \rangle$ be an Nmatrix for $\mathcal{L}$. If $\wedge$ is an $\mathcal{M}$-conjunction ($\mathcal{M}$-disjunction, $\mathcal{M}$-implication) then it is an $\mathcal{M}'$-conjunction ($\mathcal{M}'$-disjunction, $\mathcal{M}'$-implication) in any rexpansion $\mathcal{M}'$ of $\mathcal{M}$.*

Lemma 6.41 *A preserving rexpansion of a matrix is strongly preserving.*

Theorem 6.42 *If the Nmatrix $\mathcal{M}_2$ is a strongly preserving rexpansion of the Nmatrix $\mathcal{M}_1$ then $\vdash_{\mathcal{M}_1} = \vdash_{\mathcal{M}_2}$.*

Corollary 6.43 *If the Nmatrix $\mathcal{M}_2$ is a preserving rexpansion of the (deterministic) matrix $\mathcal{M}_1$ then $\vdash_{\mathcal{M}_1} = \vdash_{\mathcal{M}_2}$.*

Example 6.44 By Example 6.39 and Corollary 6.43, the positive fragment of the logics $\mathbf{PAC}$ (and $\mathbf{J_3}$) is identical to $\mathbf{CL}^+$.

6.4 Paraconsistent Nmatrices

Like in the deterministic case, we shall attribute to an Nmatrix $\mathcal{M}$ the properties of $\mathbf{L}_\mathcal{M}$. In particular,

Definition 6.45 ($\neg$-coherence with classical logic) Let $\neg$ be a unary connective of the language $\mathcal{L}$. An Nmatrix $\mathcal{M}$ for $\mathcal{L}$ is $\neg$-*contained in classical logic*, or $\neg$-*coherent with classical logic* (Definition 2.2) if so is $\mathbf{L}_\mathcal{M}$.

Proposition 6.46 *Let $\mathcal{M} = \langle \mathcal{V}, \mathcal{D}, \mathcal{O} \rangle$ be an Nmatrix for a language with $\neg$. If $\mathcal{M}$ is $\neg$-coherent with classical logic then there are elements t and f in $\mathcal{V}$ such that $t \in \mathcal{D}$, $f \in \tilde{\neg} t \cap \overline{\mathcal{D}}$, and $\tilde{\neg} f \cap \mathcal{D} \neq \emptyset$.*

Proof. Obviously, it suffices to prove this in the case that $\mathcal{M}$ is $\neg$-contained in classical logic. So Let $\mathbf{F}$ be a bivalent $\neg$-interpretation such that $\mathbf{L}_\mathcal{M}$ is $\mathbf{F}$-contained in classical logic. Since $p, \neg\neg p \not\vdash_{\mathcal{M}_\mathbf{F}} \neg p$, also $p, \neg\neg p \not\vdash_\mathcal{M} \neg p$. Hence there is a valuation ν in $\mathcal{M}$ such that $\nu(p) \in \mathcal{D}$, $\nu(\neg p) \notin \mathcal{D}$, and $\nu(\neg\neg p) \in \mathcal{D}$. Then $t = \nu(p)$ and $f = \nu(\neg p)$ are as required. $\square$

Definition 6.47 ($\neg$-paraconsistent Nmatrix) Let $\mathcal{L}$ be a propositional language with a unary connective $\neg$. An Nmatrix $\mathcal{M}$ for $\mathcal{L}$ is *(boldly, strongly, pre-) $\neg$-paraconsistent* if so is $\mathbf{L}_\mathcal{M}$.

Convention: Again, henceforth we shall usually omit the $\neg$ sign, and simply refer to *(boldly, strongly, pre-) paraconsistent Nmatrices*.

The following proposition and two corollaries should be compared with Proposition 3.55 and Corollary 3.56:

Proposition 6.48 *Let $\mathcal{M} = \langle \mathcal{V}, \mathcal{D}, \mathcal{O} \rangle$ be an Nmatrix for a language with $\neg$. $\mathcal{M}$ is pre-paraconsistent iff there is $\top \in \mathcal{D}$ such that $\tilde{\neg}\top \cap \mathcal{D} \neq \emptyset$.*

Proof. $\mathcal{M}$ is pre-paraconsistent iff $p, \neg p \not\vdash_\mathcal{M} q$. Obviously, this happens iff $\{p, \neg p\}$ has an $\mathcal{M}$-model. The latter, in turn, is possible iff there is some $\top \in \mathcal{D}$ such that $\tilde{\neg}\top \cap \mathcal{D} \neq \emptyset$.

Corollary 6.49 *Let $\mathcal{L}$ be a language with $\neg$, and let $\mathcal{M} = \langle \mathcal{V}, \mathcal{D}, \mathcal{O} \rangle$ be a paraconsistent Nmatrix for $\mathcal{L}$. Then $\mathcal{M}$ has the following properties:*

t-C *There is some $t \in \mathcal{D}$ such that $\tilde{\neg} t \cap \overline{\mathcal{D}} \neq \emptyset$.*

f-C *There is some $f \in \tilde{\neg} t \cap \overline{\mathcal{D}}$ such that $\tilde{\neg} f \cap \mathcal{D} \neq \emptyset$.*

$\top$-C *There is some $\top \in \mathcal{D}$ such that $\tilde{\neg}\top \cap \mathcal{D} \neq \emptyset$.*[3]

---
[3] Note that unlike in the case of ordinary matrices, here it is possible that $t = \top$.

Proof. Immediate from Propositions 6.46 and 6.48. □

Corollary 6.50 *Any paraconsistent Nmatrix is boldly paraconsistent. In particular: any paraconsistent Nmatrix is strongly paraconsistent.*

Proof. The second part follows from the first by Proposition 2.20. For the first part, suppose that $\mathcal{M} = \langle \mathcal{V}, \mathcal{D}, \mathcal{O} \rangle$ is a paraconsistent Nmatrix, σ is a formula in its language such that $\not\vdash_\mathcal{M} \sigma$, and q is a propositional variable such that $q \notin \mathsf{Var}(\sigma)$. Then there is a valuation ν in $\mathcal{M}$ such that $\nu(\sigma) \notin \mathcal{D}$. Let $\top$ be an element of $\mathcal{V}$ like in Proposition 6.48. Define a valuation ν' in $\mathcal{M}$ by letting $\nu'(q) = \top$, $\nu'(\neg q) \in \tilde{\neg}\top \cap \mathcal{D}$, $\nu'(\varphi) = \nu(\varphi)$ for every subformula φ of σ, and then extending the resulting partial valuation to a full valuation by analyticity. Then $\nu'(\sigma) = \nu(\sigma) \notin \mathcal{D}$. Hence, ν' is an $\mathcal{M}$-model of $\{\neg q, q\}$ which is not an $\mathcal{M}$-model of σ, and so $\{\neg q, q\} \not\vdash_\mathcal{M} \sigma$. □

Note 6.51 Note that despite Corollary 6.49, and in contrast to the deterministic case, Corollary 3.57 does *not* hold for Nmatrices. As we show below, there *are* Nmatrices, and even two-valued ones, in which $\mathcal{D}$ is a singleton.

We end this section with a characterization of $\neg$-completeness in Nmatrices. We leave its easy proof to the reader.

Proposition 6.52 *Let $\mathcal{L}$ be a language with $\neg$, and let $\mathcal{M} = \langle \mathcal{V}, \mathcal{D}, \mathcal{O} \rangle$ be an Nmatrix for $\mathcal{L}$. If $\tilde{\neg}a \subseteq \mathcal{D}$ for every $a \in \overline{\mathcal{D}}$ then $\neg$ is complete for $\mathbf{L}_\mathcal{M}$. If $\mathcal{M}$ has an $\mathcal{M}$-disjunction then the condition is also necessary.*

6.5 Maximality and Pre-Maximality

In this section we look at the maximality properties of paraconsistent Nmatrices. Its main conclusion is that although the expressive power of Nmatrices is in general greater than that of ordinary matrices (recall Theorem 6.23), this is not the case in the context of strong maximal paraconsistency.

Theorem 6.53 *Let $\mathcal{M}$ be a $\neg$-paraconsistent Nmatrix for a language $\mathcal{L}$. If $\mathbf{L}_\mathcal{M}$ is strongly maximal, then there is a determinization $\mathcal{M}^*$ of $\mathcal{M}$ such that $\Gamma \vdash_{\mathcal{M}^*} \varphi$ iff $\Gamma \vdash_\mathcal{M} \varphi$ whenever $\Gamma \cup \{\varphi\}$ is a finite set of sentences of $\mathcal{L}$.*

Proof. By Proposition 6.48, there is $\top \in \mathcal{D}$ such that $\tilde{\neg}\top \cap \mathcal{D} \neq \emptyset$. Let $\mathcal{M}^*$ be any determinization of $\mathcal{M}$ for which $\tilde{\neg}_{\mathcal{M}^*}\top \in \tilde{\neg}\top \cap \mathcal{D}$. Define a finitary logic $\mathbf{L}^* = \langle \mathcal{L}, \vdash_{\mathbf{L}^*} \rangle$ by letting $\mathcal{T} \vdash_{\mathbf{L}^*} \varphi$ iff there is a finite $\Gamma \subseteq \mathcal{T}$ such that $\Gamma \vdash_{\mathcal{M}^*} \varphi$. Then $\mathbf{L}^*$ is still pre-paraconsistent by Proposition 3.55, while $\vdash_\mathcal{M} \subseteq \vdash_{\mathbf{L}^*}$ by Proposition 6.37, and the fact that $\mathbf{L}_\mathcal{M}$ is finitary (by definition of strong maximality). Since $\mathbf{L}_\mathcal{M}$ is strongly maximal, this implies that $\mathbf{L}_\mathcal{M} = \mathbf{L}^*$. Hence $\mathcal{M}^*$ has the required properties. □

Corollary 6.54 *Every finite paraconsistent Nmatrix $\mathcal{M}$ such that $\mathbf{L}_{\mathcal{M}}$ is strongly maximal has a determination which induces the same logic.*

Proof. Immediate from Theorem 6.53 and the compactness theorem for Nmatrices (Theorem 6.18). □

Proposition 6.55 *Let $\mathcal{M} = \langle \mathcal{D}, \mathcal{V}, \mathcal{O} \rangle$ be a $\neg$-paraconsistent Nmatrix for a language $\mathcal{L}$ with $\neg$. If $\mathcal{D}$ is a singleton, then $\mathbf{L}_{\mathcal{M}}$ is not strongly maximal.*

Proof. Suppose that $\mathcal{M}$ is $\neg$-paraconsistent, and $\mathcal{D} = \{t\}$ for some $t \in \mathcal{V}$. By Proposition 6.48, $t \in \tilde{\neg} t$. Let $\mathcal{M}^*$ be some determinization of $\mathcal{M}$. If $\tilde{\neg}_{\mathcal{M}^*}(t) = \{t\}$, then $p \vdash_{\mathcal{M}^*} \neg p$. On the other hand, if $\tilde{\neg}_{\mathcal{M}^*}(t) \neq \{t\}$ (and so $\tilde{\neg}_{\mathcal{M}^*}(t) \subseteq \overline{\mathcal{D}}$) then $\neg p, p \vdash_{\mathcal{M}^*} q$. In contrast, since $\mathcal{M}$ is $\neg$-paraconsistent, both $p \not\vdash_{\mathcal{M}} \neg p$ and $\neg p, p \not\vdash_{\mathcal{M}} q$. Therefore, it follows from Theorem 6.53 that $\mathbf{L}_{\mathcal{M}}$ cannot be strongly maximal. □

Theorem 6.53 implies that it is not useful to look for Nmatrices which induce fully maximal paraconsistent logics. Nevertheless, we shall see in the next chapter that Nmatrices can be used for representing in a clear and concise way big families of (deterministic) matrices which induce fully maximal paraconsistent logics. This is done by weakening the notion of (fully) maximal paraconsistency to the following notion of *(fully) pre-maximal paraconsistency*, which captures the "core" of (fully) maximal paraconsistency of all possible determinizations of certain non-deterministic matrices, thus representing what is really essential for the (fully) maximal paraconsistency of those determinizations.

Definition 6.56 *A paraconsistent Nmatrix is strongly/fully pre-maximal if its determinizations induce strongly/fully maximal paraconsistent logics.*[4]

6.6 Bibliographical Notes and Further Reading

Nmatrices were introduced by Avron and Lev in [67, 68]. Independently, non-deterministic truth-tables were used in [125] and [184]. Special cases of the idea were practically anticipated already by Schütte [262], Tait [277], Quine [249], Girard [175], and Batens [79]. For a comprehensive survey on Nmatrices and their applications, we refer the reader to [72]. Proof theory for finite Nmatrices was developed in [59, 62]. For a detailed discussion on analyticity, see [47]. Compactness of Nmatrices (Theorem 6.18) was proved in [67]. The material on the rexpansion operation is due to [74, 75]. Properties of paraconsistent Nmatrices were studied in [19], and Section 6.5 is based on material from there, but with improved results and proofs.

---
[4]Notice that the notion of pre-maximality which was used in [19] was a little bit different: There, only *paraconsistent* determinizations were taken into account.

Chapter 7
Using the Basic Four Truth-Values

Non-deterministic semantics provide an excellent tool for investigating properties of negation. This is due to the remarkable property of *modularity* that such a semantics usually has: given a set of axioms, the semantic effect of each axiom can frequently be analyzed separately, and expressed as a condition imposed on the underlying non-deterministic semantics. In this chapter this approach is demonstrated by using the four basic truth-values from Chapter 5 for investigating ¬-paraconsistent logics, especially *classical* ones (Definition 2.22).

We start the chapter with some general results about the power of *proper* paraconsistent Nmatrices that are based on the four basic truth-values. After that we turn to study the case in which just the two classical truth values t and f (associated with truth and falsity) are used. Then we add to them the truth value $\top$, that represents inconsistency, and in many cases also $\bot$, the truth value which represents incompleteness.

7.1 ¬-Fundamental Nmatrices

Recall that the idea behind Dunn-Belnap's basic matrix $\mathcal{FOUR}$ is that its truth values correspond to what Belnap has described as the four basic states of information: they represent the support we have concerning the truth and falsity of a formula ψ. (See in particular Note 5.13.) Assuming that ¬ represents the idea of falsehood within the language, this means:

- $\nu(\psi) = t$ if the truth of ψ is supported and that of $\neg\psi$ is not.
- $\nu(\psi) = f$ if the truth of $\neg\psi$ is supported and that of ψ is not.
- $\nu(\psi) = \top$ if both the truth of ψ and the truth of $\neg\psi$ are supported.
- $\nu(\psi) = \bot$ if neither the truth of ψ nor the truth of $\neg\psi$ is supported.

Given the truth value of ψ, what do these principles tell us about the truth-value of its negation? They imply that in general $\nu(\psi) \in \{t, \top\}$ iff the truth of ψ is supported, and $\nu(\psi) \in \{f, \bot\}$ iff it is not. Now the only information about $\neg\psi$ that follows from the fact that $\nu(\psi) = t$ (according to the first principle above) is that its truth is not supported, hence $\nu(\neg\psi) \in \{f, \bot\}$. Similar considerations in the other cases lead to the following derived principles (and nothing stronger, as long as we do not introduce additional assumptions concerning supporting the truth of $\neg\psi$):

- If $\nu(\psi) = t$, then $\nu(\neg\psi) \in \{f, \bot\}$.
- If $\nu(\psi) = f$, then $\nu(\neg\psi) \in \{t, \top\}$.
- If $\nu(\psi) = \top$, then $\nu(\neg\psi) \in \{t, \top\}$.
- If $\nu(\psi) = \bot$, then $\nu(\neg\psi) \in \{f, \bot\}$.

It follows that the truth-value of ψ does not fully determine the truth-value of $\neg\psi$, and so the use of non-deterministic semantics is appropriate. The general structure of all the Nmatrices considered in this chapter is determined by the above four principles, together with the following one: for any $a \in \{t, f, \top, \bot\}$, the value in $\mathcal{FOUR}$ of $\tilde{\neg}a$ should always be one of the options. These principles lead to the following very important class of paraconsistent Nmatrices.

Definition 7.1 ($\neg$-fundamental Nmatrix) Let $\mathcal{M} = \langle \mathcal{V}, \mathcal{D}, \mathcal{O} \rangle$ be an Nmatrix for a language which includes the unary connective $\neg$. $\mathcal{M}$ is called $\neg$-*fundamental* if it has the following properties:

(i) $\mathcal{V} \subseteq \{t, f, \top, \bot\}$, and $\mathcal{D} = \mathcal{V} \cap \{t, \top\}$.

(ii) The following hold for the operation $\tilde{\neg}$ of $\mathcal{O}$:

- If $t \in \mathcal{V}$ then $\{f\} \subseteq \tilde{\neg}t \subseteq \{f, \bot\}$.
- If $f \in \mathcal{V}$ then $\{t\} \subseteq \tilde{\neg}f \subseteq \{t, \top\}$.
- If $\top \in \mathcal{V}$ then $\{\top\} \subseteq \tilde{\neg}\top \subseteq \{t, \top\}$.
- If $\bot \in \mathcal{V}$ then $\{\bot\} \subseteq \tilde{\neg}\bot \subseteq \{f, \bot\}$.

Note 7.2 It follows from Definition 7.1 that if $\langle \mathcal{V}, \mathcal{D}, \mathcal{O} \rangle$ is $\neg$-fundamental, and $\neg$ is a negation for it, then $\{t, f\} \subseteq \mathcal{V}$, $t \in \mathcal{D}$, and $f \notin \mathcal{D}$.

Note 7.3 The representation of the four basic truth-values given in Note 5.13 can be very illuminating here. Recall that according to this representation, the set $FOUR$ is identified with $\{0, 1\}^2$ (where $t = \langle 1, 0 \rangle$, $f = \langle 0, 1 \rangle$, $\top = \langle 1, 1 \rangle$, $\bot = \langle 0, 0 \rangle$), and if ν is a valuation in $FOUR$ then the intended intuitive meaning of $\nu(\varphi) = \langle x, y \rangle$ is the following:

7.1 ¬-Fundamental Nmatrices 189

- $x = 1$ iff φ is "true" (i.e. $\nu(\varphi) \in \mathcal{D}$).

- $y = 1$ iff $\neg \varphi$ is "true" (i.e. $\nu(\neg\varphi) \in \mathcal{D}$).

From this point of view, an Nmatrix $\langle \mathcal{V}, \mathcal{D}, \mathcal{O} \rangle$ is ¬-fundamental iff the following three conditions are satisfied:

(i) $\mathcal{V} \subseteq \{1,0\} \times \{1,0\}$.

(ii) If $\langle a, b \rangle \in \mathcal{V}$ then $\langle a, b \rangle \in \mathcal{D}$ iff $a = 1$.

(iii) $\{\langle b, a \rangle\} \subseteq \tilde{\neg}\langle a, b \rangle \subseteq \{\langle b, 1 \rangle, \langle b, 0 \rangle\}$ whenever $\langle a, b \rangle \in \mathcal{V}$.

Another way to understand the notion of a ¬-fundamental Nmatrix is the following. Given a valuation ν in $FOUR$, let $\nu^+(\varphi) = \Pi_1(\nu(\varphi))$ and $\nu^-(\varphi) = \Pi_2(\nu(\varphi))$ (where $\Pi_1(\langle x_1, x_2 \rangle) = x_1$, and $\Pi_2(\langle x_1, x_2 \rangle) = x_2$). Then the four principles above concerning $\nu(\psi)$ can be summarized by the following constraint on ν, that this interpretation of the truth-values dictates:

$$\text{(NEG)} \quad \nu^+(\neg\psi) = \nu^-(\psi).$$

Moreover, ν is a model of a formula ψ (i.e., $\nu(\psi) \in \mathcal{D}$) iff $\nu^+(\psi) = 1$. Together with condition (NEG) this observation allows us to present in many cases the semantics provided by ¬-fundamental Nmatrices in terms of pairs $\langle \nu^+, \nu^- \rangle$ of *bivaluations* (that is, functions from the set of formulas to $\{0, 1\}$, or to $\{t, f\}$) which satisfy (NEG).

Our first theorem shows that the use of proper ¬-fundamental Nmatrices cannot be replaced by the use of finite ordinary matrices.

Theorem 7.4 *Let* $\mathcal{M} = \langle \mathcal{V}, \mathcal{D}, \mathcal{O} \rangle$ *be a proper ¬-fundamental Nmatrix.*

1. *There is no finite ordinary matrix that is characteristic for* $\mathbf{L}_\mathcal{M}$.

2. *If* $\mathcal{M}$ *is strictly proper, then there is no finite family* $\mathcal{F}$ *of finite ordinary matrices such that* $\vdash_\mathcal{M} = \vdash_\mathcal{F}$.

Proof. Assume first that there is a propositional constant C of the language such that $\mathcal{O}(\mathsf{C})$ is not a singleton. Suppose for contradiction that $\mathcal{M}' = \langle \mathcal{V}', \mathcal{D}', \mathcal{O}' \rangle$ is a finite matrix which is characteristic for $\mathbf{L}_\mathcal{M}$. Let n be number of elements of $\mathcal{V}'$. Since $\mathcal{M}'$ is an ordinary matrix, then either $\vdash_{\mathcal{M}'} \mathsf{C}$ or $\mathsf{C} \vdash_{\mathcal{M}'} p$ for every propositional variable p. Similarly, either $\vdash_{\mathcal{M}'} \neg\mathsf{C}$ or $\neg\mathsf{C} \vdash_{\mathcal{M}'} p$ for every propositional variable p. The first fact and (i) of Definition 7.1 easily exclude the possibility that both $\mathcal{O}(\mathsf{C}) \cap \{f, \bot\} \neq \emptyset$ and $\mathcal{O}(\mathsf{C}) \cap \{t, \top\} \neq \emptyset$. The second fact, together with (ii) and (i) of

Definition 7.1, exclude the two remaining possibilities (i.e., that $\mathcal{O}(\mathsf{C}) = \{t, \top\}$ or $\mathcal{O}(\mathsf{C}) = \{f, \bot\}$). Hence we get a contradiction.

Now, assume that there is a connective $\diamond$ with arity $k > 0$ in the language, and $a_1, \ldots, a_k \in \mathcal{V}$ such that $\tilde{\diamond}(a_1, \ldots, a_k)$ is not a singleton. Suppose for contradiction that $\mathcal{F}$ is a finite family of finite ordinary matrices such that $\vdash_{\mathcal{M}} = \vdash_{\mathcal{F}}$. Then there is some natural number n which is greater than the number of truth-values in every matrix in $\mathcal{F}$. Let $p_1, \ldots, p_k$ be k distinct propositional variables. Define inductively: $\psi_0 = p_1$, $\psi_{i+1} = \neg\neg\psi_i$. Then for every assignment ν' in some matrix $\mathcal{M}' \in \mathcal{F}$ the following holds:

$$(*) \qquad \nu'(\psi_n) \in \{\nu'(\psi_0), \ldots, \nu'(\psi_{n-1})\}.$$

This is trivially true if $\{\nu'(\psi_0), \ldots, \nu'(\psi_{n-1})\} = \mathcal{V}'$. Otherwise, $\nu'(\psi_i) = \nu'(\psi_j)$ for some $0 \leq i < j \leq n-1$, implying (since $\mathcal{M}'$ is deterministic) that $\nu'(\psi_n) = \nu'(\psi_{n-j+i}) \in \{\nu'(\psi_0), \ldots, \nu'(\psi_{n-1})\}$.

For $0 \leq i \leq n$ let $\varphi_i = \diamond(\psi_i, p_2, \ldots, p_k)$. Since the elements of $\mathcal{F}$ are deterministic, $(*)$ implies that $\nu'(\varphi_n) \in \{\nu'(\varphi_0), \ldots, \nu'(\varphi_{n-1})\}$ and $\nu'(\neg\varphi_n) \in \{\nu'(\neg\varphi_0), \ldots, \nu'(\neg\varphi_{n-1})\}$ for every assignment ν' in some matrix $\mathcal{M}' \in \mathcal{F}$. Hence $\varphi_0, \ldots, \varphi_{n-1} \vdash_{\mathcal{M}} \varphi_n$, and $\neg\varphi_0, \ldots, \neg\varphi_{n-1} \vdash_{\mathcal{M}} \neg\varphi_n$.

Now, there are again three cases to consider.

- For some $c, d \in \mathcal{V}$: $c \in \tilde{\diamond}(a_1, \ldots, a_k) \cap \{t, \top\}$, $d \in \tilde{\diamond}(a_1, \ldots, a_k) \cap \{f, \bot\}$. In this case, $\varphi_0, \ldots, \varphi_{n-1} \not\vdash_{\mathcal{M}} \varphi_n$. To see this, take any assignment ν in $\mathcal{M}$ with the following properties:
  - $\nu(p_i) = a_i$ for every $1 \leq i \leq k$.
  - $\nu(\psi_i) = a_1$ for every $0 \leq i \leq n$. (This is possible because of condition (ii) in Definition 7.1 concerning $\tilde{\neg}$.)
  - $\nu(\varphi_i) = c$ for every $0 \leq i \leq n - 1$.
  - $\nu(\varphi_n) = d$.

- $\tilde{\diamond}(a_1, \ldots, a_k) = \{t, \top\}$.
  In this case, $\neg\varphi_0, \ldots, \neg\varphi_{n-1} \not\vdash_{\mathcal{M}} \neg\varphi_n$. To see this, take any assignment ν in $\mathcal{M}$ with the following properties:
  - $\nu(p_i) = a_i$ for every $1 \leq i \leq k$.
  - $\nu(\psi_i) = a_1$ for every $0 \leq i \leq n$.
  - $\nu(\varphi_i) = \nu(\neg\varphi_i) = \top$ for every $0 \leq i \leq n - 1$.
  - $\nu(\varphi_n) = t$ and $\nu(\neg\varphi_n) = f$.

- $\tilde{\diamond}(a_1, \ldots, a_k) = \{f, \bot\}$.
  In this case too, $\neg\varphi_0, \ldots, \neg\varphi_{n-1} \not\vdash_{\mathcal{M}} \neg\varphi_n$. To see this, take any assignment ν in $\mathcal{M}$ with the following properties:

7.1 ¬-Fundamental Nmatrices

- $\nu(p_i) = a_i$ for every $1 \leq i \leq k$.
- $\nu(\psi_i) = a_1$ for every $0 \leq i \leq n$.
- $\nu(\varphi_i) = f$ and $\nu(\neg \varphi_i) = t$ for every $0 \leq i \leq n-1$.
- $\nu(\varphi_n) = \bot$ and $\nu(\neg \varphi_n) = \bot$.

In all three cases we got a contradiction. Hence no such $\mathcal{F}$ exists. □

Next, we turn to the topic of congruence (Definition 1.18) in the framework of proper ¬-fundamental Nmatrices.

Proposition 7.5 *Let $\mathcal{M} = \langle \mathcal{V}, \mathcal{D}, \mathcal{O} \rangle$ be a ¬-fundamental Nmatrix for $\mathcal{L}$. If $\varphi \equiv_{\mathbf{L}_\mathcal{M}} \psi$ then $\nu(\varphi) = \nu(\psi)$ for every valuation ν in $\mathcal{M}$.*

Proof. Suppose that $\nu(\varphi) \neq \nu(\psi)$ for some valuation ν in $\mathcal{M}$. If $\nu(\varphi) \in \mathcal{D}$ and $\nu(\psi) \notin \mathcal{D}$ then $\varphi \not\vdash_\mathcal{M} \psi$. If $\nu(\psi) \in \mathcal{D}$ and $\nu(\varphi) \notin \mathcal{D}$ then $\psi \not\vdash_\mathcal{M} \varphi$. If $\nu(\varphi) = t$ and $\nu(\psi) = \top$ then $\nu(\neg\psi) \in \mathcal{D}$ while $\nu(\neg\varphi) \notin \mathcal{D}$, and so $\neg\psi \not\vdash_\mathcal{M} \neg\varphi$. If $\nu(\varphi) = \bot$ and $\nu(\psi) = f$ then again $\nu(\neg\psi) \in \mathcal{D}$ while $\nu(\neg\varphi) \notin \mathcal{D}$, and so $\neg\psi \not\vdash_\mathcal{M} \neg\varphi$. Similarly, if $\nu(\psi) = t$ and $\nu(\varphi) = \top$, or $\nu(\psi) = \bot$ and $\nu(\varphi) = f$, then $\neg\varphi \not\vdash_\mathcal{M} \neg\psi$. Thus in all cases $\varphi \not\equiv_{\mathbf{L}_\mathcal{M}} \psi$. □

For our next theorem concerning strictly proper ¬-fundamental Nmatrices we need the following definition.

Definition 7.6 (isolated formula, sparse logic) *Let $\mathbf{L}$ be a logic.*

- *A formula φ of the language of $\mathbf{L}$ is isolated in $\mathbf{L}$, if the only formula which is congruent to φ in $\mathbf{L}$ (Definition 1.13) is φ itself.*

- *$\mathbf{L}$ is called sparse, if every formula of its language is isolated in $\mathbf{L}$.*

Theorem 7.7 *Let $\mathcal{M} = \langle \mathcal{V}, \mathcal{D}, \mathcal{O} \rangle$ be a strictly proper, ¬-fundamental Nmatrix for $\mathcal{L}$. Suppose that there is a connective $\supset$ of $\mathcal{L}$ such that:*

- *$(a \tilde{\supset} a) \cap \mathcal{D} \neq \emptyset$ for every $a \in \mathcal{V}$.*

- *If $a \in \mathcal{D}$ and $b \in \mathcal{V}$ then $b \in (a \tilde{\supset} b)$.*

Then $\mathbf{L}_\mathcal{M}$ is sparse.

Proof. Since $\mathcal{M}$ is strictly proper, there is $\diamond \in \mathcal{C}(\mathcal{L})$ of arity $n > 0$, and elements $a_1, \ldots, a_n, b, c \in \mathcal{V}$ such that and $\{b, c\} \subseteq \tilde{\diamond}(a_1, \ldots, a_n)$ and $b \neq c$.

Let $\varphi, \psi \in \mathcal{W}(\mathcal{L})$ be distinct, and let $q_1, \ldots, q_n$ be propositional variables that do not occur in φ or ψ. Define: $\sigma^* = \diamond((\sigma \supset \sigma) \supset q_1, q_2, \ldots, q_n)$ for every $\sigma \in \mathcal{W}(\mathcal{L})$. Let $\mathcal{W}$ be the set of subformulas of φ and ψ, $\mathcal{W}^*$ the set of subformulas of φ^* and ψ^*, and $\nu : \mathcal{W} \to \mathcal{V}$ a partial $\mathcal{M}$-valuation. Suppose first that neither $\nu(\varphi \supset \varphi)$ nor $\nu(\psi \supset \psi)$ is defined. Using the assumed properties of $\tilde{\supset}$, we extend ν to a partial $\mathcal{M}$-valuation ν^* on $\mathcal{W}^*$ by letting:

- $\nu^*(q_i) = a_i$ $(i = 1, \ldots, n)$.
- $\nu^*(\varphi \supset \varphi) \in (\nu(\varphi) \tilde{\supset} \nu(\varphi)) \cap \mathcal{D}$.
- $\nu^*(\psi \supset \psi) \in (\nu(\psi) \tilde{\supset} \nu(\psi)) \cap \mathcal{D}$.
- $\nu^*((\varphi \supset \varphi) \supset q_1) = a_1$.
- $\nu^*((\psi \supset \psi) \supset q_1) = a_1$.
- $\nu^*(\varphi^*) = b$.
- $\nu^*(\psi^*) = c$.

Since $b \neq c$, it followed by the analyticity of the semantics (Proposition 6.11) and Proposition 7.5 that $\varphi^* \not\equiv_{\mathbf{L}_\mathcal{M}} \psi^*$. Hence $\varphi \not\equiv_{\mathbf{L}_\mathcal{M}} \psi$ as well.

Now, assume that $\nu(\varphi \supset \varphi)$ (say) is defined. Then $\varphi \supset \varphi$ is a subformula of ψ. Hence $\nu(\psi \supset \psi)$ is not defined, and we can choose $\nu^*(\psi \supset \psi)$ as above, and $\nu^*(\varphi \supset \varphi) = \nu(\varphi \supset \varphi)$. If $\nu(\varphi \supset \varphi) \in (\nu(\varphi) \tilde{\supset} \nu(\varphi)) \cap \mathcal{D}$ we continue as above. If not, then $\psi \supset \psi \not\vdash_{\mathbf{L}_\mathcal{M}} \varphi \supset \varphi$, and so again $\varphi \not\equiv_{\mathbf{L}_\mathcal{M}} \psi$. □

7.2 Bivalence

The simplest kind of paraconsistent Nmatrices is *two-valued*. This is in contrast to the deterministic case, where the simplest kind of paraconsistent logics must be based on at least three truth-values (Corollary 3.57).

We begin by showing that there is only one possible (and natural) candidate for a two-valued classical ¬-paraconsistent logic. First, recall that the normality of a logic (Definition 1.32) depends on the availability in it of implication, conjunction and disjunction. This implies the following interpretations (assuming, as usual, that $\mathcal{V} = \{t, f\}$ and $\mathcal{D} = \{t\}$):

Proposition 7.8 *Let $\mathcal{M} = \langle \{t, f\}, \{t\}, \mathcal{O} \rangle$ be a two-valued Nmatrix for a language $\mathcal{L}$. The connectives $\wedge, \vee,$ and $\supset$ of $\mathcal{L}$ are conjunction, disjunction and implication for $\mathbf{L}_\mathcal{M}$ (respectively) iff they have in $\mathcal{M}$ the following interpretations:*

$$a \tilde{\supset} b = \begin{cases} \{f\} & \text{if } a = t \text{ and } b = f, \\ \{t\} & \text{otherwise.} \end{cases}$$

$$a \tilde{\wedge} b = \begin{cases} \{t\} & \text{if } a = t \text{ and } b = t, \\ \{f\} & \text{otherwise.} \end{cases}$$

$$a \tilde{\vee} b = \begin{cases} \{t\} & \text{if } a = t \text{ or } b = t, \\ \{f\} & \text{otherwise.} \end{cases}$$

7.2 Bivalence

Proof. It is easy to verify that if $\supset$ ($\wedge$, $\vee$) is an $\mathcal{M}$-implication ($\mathcal{M}$-conjunction, $\mathcal{M}$-disjunction) iff it has in $\mathcal{M}$ the above interpretation. Since $\mathcal{M}$ is obviously an f-Nmatrix, the claim follows from Proposition 6.30. □

Next, we consider the interpretation of negation. For $\neg$-paraconsistency, Corollary 6.49 implies that $t \in \tilde{\neg}t$, $t \in \tilde{\neg}f$ and $f \in \tilde{\neg}t$. These constraints allow two possible interpretations of $\neg$. One of them is $\lambda x \in \{t,f\}.\{t,f\}$. But with it $\neg\psi$ behaves just like an propositional variable for any $\psi \in \mathcal{W}(\mathcal{L})$. The effect of creating new propositional variables surely cannot justify the name 'negation'. Therefore, we are left with the following Nmatrix as the only reasonable choice for a two-valued, $\neg$-paraconsistent Nmatrix for $\mathcal{L}_{CL}$:

Definition 7.9 ($\mathcal{M}_2$, CLuN)

- The Nmatrix $\mathcal{M}_2 = (\{t,f\},\{t\},\mathcal{O})$ for $\mathcal{L}_{CL}$ is defined by:

		$\tilde{\vee}$	$\tilde{\wedge}$	$\tilde{\supset}$
t	t	$\{t\}$	$\{t\}$	$\{t\}$
t	f	$\{t\}$	$\{f\}$	$\{f\}$
f	t	$\{t\}$	$\{f\}$	$\{t\}$
f	f	$\{f\}$	$\{f\}$	$\{t\}$

	$\tilde{\neg}$
t	$\{t,f\}$
f	$\{t\}$

- The logic $\mathbf{L}_{\mathcal{M}_2}$ is known as Batens' **CLuN** ([80]).

Proposition 7.10 (properties of CLuN)

1. **CLuN** *is finitary and normal.*
2. *The* $\{\supset, \vee, \wedge\}$-*fragment of* **CLuN** *is identical to* $\mathbf{CL}^+$.
3. **CLuN** *is contained (and so $\neg$-contained) in classical logic.*
4. **CLuN** *is boldly (and so strongly) $\neg$-paraconsistent.*
5. **CLuN** *is sparse.*

Proof. The first two items follow from Theorem 6.18, Proposition 7.8 (and its proof) and Proposition 6.31. The third item easily follows from Proposition 6.37, and the fact that the classical two-valued matrix is obviously a determinization of $\mathcal{M}_2$. Item 4 follows from Item 3, Proposition 6.48, and Corollary 6.50. Finally, the last item follows from Theorem 7.7. □

Corollary 7.11 **CLuN** *is a classical $\neg$-paraconsistent logic.*

Proof. Immediate from Propositions 7.10 and 2.23. □

The next proposition shows that the use of non-deterministic semantics is essential for **CLuN**.

Proposition 7.12 CLuN *has no finite weakly characteristic (ordinary) matrix. What is more, there is no finite family $\mathcal{F}$ of finite ordinary matrices such that for every ψ, $\vdash_{\mathcal{M}_2} \psi$ iff $\vdash_{\mathcal{F}} \psi$.*

Proof. This is a special case of Theorem 6.23. □

Turning to proof systems for **CLuN**, we note that this logic was originally formulated in terms of a Hilbert-style calculus:

Definition 7.13 (*HCLuN*) *HCLuN* is the system obtained from HCL^+ (Figure 1.1) by adding the axiom [t] $\neg\varphi \vee \varphi$.

Theorem 7.14 *Suppose that $\mathcal{M} = \langle \mathcal{V}, \mathcal{D}, \mathcal{O} \rangle$ is an Nmatrix for $\mathcal{L}_{CL}$ which satisfies the following conditions:*

- *If $a \in \mathcal{D}$ then $\tilde{\neg}a \cap \mathcal{D} \neq \emptyset$ and $\tilde{\neg}a \cap \overline{\mathcal{D}} \neq \emptyset$.*

- *If $a \in \overline{\mathcal{D}}$ then $\tilde{\neg}a \subseteq \mathcal{D}$.*

- *$\wedge$, $\vee$, and $\supset$ are $\mathcal{M}$-disjunction, $\mathcal{M}$-implication, and $\mathcal{M}$-conjunction, respectively.*

Then $\vdash_{HCLuN} = \vdash_{\mathcal{M}}$.

Proof. Since $\wedge$, $\vee$, and $\supset$ are $\mathcal{M}$-conjunction, $\mathcal{M}$-disjunction, and $\mathcal{M}$-implication (respectively), HCL^+ is sound for $\mathcal{M}$ by Proposition 6.31 and Theorem 1.90. In addition, the second condition on $\mathcal{M}$ implies that [t] is valid in $\mathcal{M}_2$. Hence, *HCLuN* is sound for $\mathcal{M}$.

The proof of the completeness of *HCLuN* for $\mathcal{M}$ is similar to the direct proof given in Chapter 4 to Theorem 4.101: Assume that $\mathcal{T} \nvdash_{HCLuN} \varphi_0$. We construct a model of $\mathcal{T}$ in $\mathcal{M}$ which is not a model of ψ. By Theorem 1.98 there exists a theory $\mathcal{T}^*$ in $\mathcal{L}$ with the following properties:

1. $\mathcal{T} \subseteq \mathcal{T}^*$, and for every φ: $\varphi \in \mathcal{T}^*$ iff $\mathcal{T}^* \vdash_{HCLuN} \varphi$.

2. $\varphi_0 \notin \mathcal{T}^*$.

3. $\varphi \vee \psi \in \mathcal{T}^*$ iff either $\varphi \in \mathcal{T}^*$ or $\psi \in \mathcal{T}^*$.

4. $\varphi \wedge \psi \in \mathcal{T}^*$ iff both $\varphi \in \mathcal{T}^*$ and $\psi \in \mathcal{T}^*$.

5. $\varphi \supset \psi \in \mathcal{T}^*$ iff either $\varphi \notin \mathcal{T}^*$ or $\psi \in \mathcal{T}^*$.

6. If $\varphi \notin \mathcal{T}^*$ then $\neg\varphi \in \mathcal{T}^*$.

Now, we recursively define a valuation ν as follows:

7.2 Bivalence

- If $p \in \mathsf{Var}$ we let $\nu(p) \in \mathcal{D}$ if $p \in \mathcal{T}^*$, and $\nu(p) \in \overline{\mathcal{D}}$ if $p \notin \mathcal{T}^*$.
- If $\varphi = \varphi_1 \diamond \varphi_2$ where $\diamond \in \{\wedge, \vee, \supset\}$, we let $\nu(\varphi) \in \nu(\varphi_1)\tilde{\diamond}\nu(\varphi_2)$.
- If $\varphi = \neg\psi$ and $\nu(\psi) \in \overline{\mathcal{D}}$ we let $\nu(\varphi) \in \tilde{\neg}\nu(\psi)$.
- If $\varphi = \neg\psi$, $\nu(\psi) \in \mathcal{D}$, and $\neg\psi \in \mathcal{T}^*$, we let $\nu(\varphi) \in \tilde{\neg}\nu(\psi) \cap \mathcal{D}$.
- If $\varphi = \neg\psi$, $\nu(\psi) \in \mathcal{D}$, and $\neg\psi \notin \mathcal{T}^*$, we let $\nu(\varphi) \in \tilde{\neg}\nu(\psi) \cap \overline{\mathcal{D}}$.

The first property of $\mathcal{M}$ implies that it is indeed possible to define such a function ν, and from its definition it immediately follows that ν is an $\mathcal{M}$-valuation. An easy induction on the structure of φ, which uses Properties 3–6 of $\mathcal{T}^*$ and the second and third properties of $\mathcal{M}$, shows that $\nu(\varphi) \in \mathcal{D}$ iff $\mathcal{T}^* \vdash_{HCLuN} \varphi$. Therefore Properties 1 and 2 of $\mathcal{T}^*$ imply that $\nu(\varphi_0) \notin \mathcal{D}$, while $\nu(\varphi) \in \mathcal{D}$ in case $\varphi \in \mathcal{T}$. Hence, ν is indeed a model in $\mathcal{M}$ of $\mathcal{T}$ which is not a model of φ_0. □

Corollary 7.15 (completeness of HCLuN) $\mathbf{L}_{HCLuN} = \mathbf{CLuN}$.

Proof. That $\vdash_{HCLuN} = \vdash_{\mathcal{M}_2}$ is a special case of Theorem 7.14. □

Corollary 7.16 (minimality of CLuN) CLuN *is the minimal extension of* $\mathbf{CL}^+$ *for which* $\neg$ *is complete.*

Proof. Immediate from Corollary 7.15, the second item of Proposition 7.10, Proposition 1.16, and Note 2.6. □

The next proposition shows that by using Nmatrices with a single designated value we cannot go beyond **CLuN**.

Proposition 7.17 *Let* $\mathcal{M} = \langle \mathcal{V}, \mathcal{D}, \mathcal{O} \rangle$ *be a paraconsistent Nmatrix for* $\mathcal{L}_{CL}$ *such that* $\neg$ *is complete for* $\mathbf{L}_\mathcal{M}$, *and* $\wedge$, $\vee$, *and* $\supset$ *are* $\mathcal{M}$-*disjunction,* $\mathcal{M}$-*implication, and* $\mathcal{M}$-*conjunction, respectively. If* $\mathcal{D}$ *is a singleton then* $\mathbf{L}_\mathcal{M} = \mathbf{CLuN}$.

Proof. From Corollary 6.49 and Proposition 6.52 it easily follows that $\mathcal{M}$ necessarily satisfies all the conditions of Theorem 7.14. □

Definition 7.18 (*GCLuN*) The Gentzen-style system *GCLuN* is obtained from LK^+ (Figure 1.2) by adding the rule $[\Rightarrow \neg]$.

Theorem 7.19 (completeness and cut-elimination for *GCLuN*)

1. $\vdash_{GCLuN} = \vdash_{HCLuN}$, *and so* $\mathbf{L}_{GCLuN} = \mathbf{CLuN}$.

2. G_{CLuN} enjoys cut-admissibility.

Proof. The proof of the first half of the first item is very similar to (but shorter than) the proof of Theorem 4.100 in the case of **PAC**. The second half then follows from Proposition 7.15. We leave the details to the reader. The second item can be proved exactly like in Gentzen's original proof of cut-elimination for **LK** (Theorem 1.133).[1]

An alternative direct semantic proofs for both items can also be given. It is similar to (but easier than) that given to Theorem 7.60 below in the case $S = \{[t]\}$. We leave this too as an exercise for the reader. □

Despite its nice properties, **CLuN** has also several serious drawbacks:

Proposition 7.20

1. The negation of **CLuN** is neither left involutive nor right involutive.

2. **CLuN** is neither strongly maximal, nor maximally ¬-paraconsistent relative to classical logic.

Proof.

1. For left involutivity, we show that $p \not\vdash_{\mathbf{CLuN}} \neg\neg p$ by taking a partial $\mathcal{M}_2$-valuation ν such that $\nu(p) = t$, $\nu(\neg p) = t$ and $\nu(\neg\neg p) = f$. By analyticity of Nmatrices (Proposition 6.11), ν can be extended to a full refuting $\mathcal{M}_2$-valuation. The proof for right involutivity is similar.

2. By Corollary 7.15, Theorem 4.101, and the first item of this proposition, the paraconsistent logic **PAC** discussed in Section 4.4.4 is a proper extension of **CLuN**, and it is obtained from **CLuN** by extending it with classical tautologies. Using Proposition 1.54, this immediately implies the claim. □

7.3 Maximality and Pre-maximality

Next we start to study logics which are induced by three-valued Nmatrices. For now we focus on the subject of maximality properties, leaving other issues (like proof systems for such logics) to the next section and chapter. We start by determining the three-valued proper Nmatrices that (despite Corollary 6.54) induce strongly maximal paraconsistent logics.

---

[1] Both Theorem 4.100 and the second part of Theorem 7.19 are particular cases of the general cut-elimination theorem for canonical sequent systems which is proved in [69].

7.3 Maximality and Pre-maximality

Theorem 7.21 *Let $\mathcal{M}$ be a three-valued properly non-deterministic Nmatrix such that $\mathbf{L}_\mathcal{M}$ is strongly maximal $\neg$-paraconsistent. Then $\mathcal{M}$ is isomorphic to an Nmatrix $\langle \mathcal{V}, \mathcal{D}, \mathcal{O}\rangle$, in which $\mathcal{V} = \{t, \top, f\}$, $\mathcal{D} = \{t, \top\}$, and:*

1. *$\tilde{\neg} t = \{f\}$, $\tilde{\neg}\top = \{t, f\}$ and $\tilde{\neg} f = \{t\}$.*

2. *The interpretation of any other connective $\diamond$ of $\mathcal{M}$ is deterministic, gets values only in $\{t, f\}$, and does not distinguish between t and $\top$ (i.e., $\tilde{\diamond}(x_1, \ldots, x_{j-1}, t, x_{j+1}, \ldots, x_n) = \tilde{\diamond}(x_1, \ldots, x_{j-1}, \top, x_{j+1}, \ldots, x_n)$ in case $\diamond$ is n-ary, $1 \leq j \leq n$, and $x_1, \ldots, x_{j-1}, x_{j+1}, \ldots, x_n \in \mathcal{V}$).*

3. *$\mathcal{M}$ has an $\mathcal{M}$-conjunction or an $\mathcal{M}$-disjunction.*

Proof. Let $\mathcal{M} = \langle \mathcal{V}, \mathcal{D}, \mathcal{O}\rangle$. We start by showing that there is no $x \in \mathcal{D}$ such that both $x \in \tilde{\neg} x$ and $\tilde{\neg} x \cap \overline{\mathcal{D}} \neq \emptyset$. Suppose for contradiction that there is such x. By Theorem 6.55, there is some $y \in \mathcal{D} \setminus \{x\}$. Since $\mathcal{M}$ is three-valued, necessarily $\mathcal{D} = \{x, y\}$. We define $\mathcal{M}^*$ as follows:

- If $\tilde{\neg} y \cap \mathcal{D} \neq \emptyset$, we let $\mathcal{M}^*$ be some determinization of $\mathcal{M}$ for which $\tilde{\neg}_{\mathcal{M}^*}(x) \in \overline{\mathcal{D}}$ and $\tilde{\neg}_{\mathcal{M}^*}(y) \in \mathcal{D}$.

- If $\tilde{\neg} y \subseteq \overline{\mathcal{D}}$, we let $\mathcal{M}^*$ be some determinization of $\mathcal{M}$ for which $\tilde{\neg}_{\mathcal{M}^*}(x) = x$ (and necessarily $\tilde{\neg}_{\mathcal{M}^*}(y) \in \overline{\mathcal{D}}$).

By Proposition 6.37, $\mathbf{L}_{\mathcal{M}^*}$ extends $\mathbf{L}_\mathcal{M}$. It is easy to check that in both cases, $\mathbf{L}_{\mathcal{M}^*}$ is pre-$\neg$-paraconsistent, and $p, \neg p, \neg\neg p \vdash_{\mathcal{M}^*} \neg\neg\neg p$. On the other hand, $p, \neg p, \neg\neg p \not\vdash_\mathcal{M} \neg\neg\neg p$, since we may take $\nu(p) = \nu(\neg p) = \nu(\neg\neg p) = x$, and $\nu(\neg\neg\neg p) \in \mathcal{D}$. Thus, the pre-$\neg$-paraconsistent logic $\mathbf{L}_{\mathcal{M}^*}$ is a proper simple finitary (Theorem 6.18) extension of $\mathbf{L}_\mathcal{M}$, in contradiction to the strong maximality of $\mathbf{L}_\mathcal{M}$.

Theorem 6.55, Corollary 6.49, and what we just have proved together imply that $\mathcal{V}$ consists of three elements t, f, and $\top$, such that $\mathcal{D} = \{t, \top\}$, $\tilde{\neg} t$ is either $\{f\}$ or $\{f, \top\}$, $\tilde{\neg} f \cap \mathcal{D} \neq \emptyset$, $\{f, \top\} \not\subseteq \tilde{\neg}\top$, and if $\tilde{\neg} t = \{f\}$, then either $\tilde{\neg}\top \subseteq \mathcal{D}$, or $\tilde{\neg}\top = \{f, t\}$.[2] Moreover, we may further assume that in all cases $\tilde{\neg}\top \subseteq \mathcal{D}$, or $\tilde{\neg}\top = \{f, t\}$, because in the one possible exceptional case, in which $\tilde{\neg}\top = \{f\}$ and $\tilde{\neg} t = \{f, \top\}$, we can exchange the roles of t and $\top$ (renaming t as $\top$ and $\top$ as t).

We next prove some facts about $\mathcal{M}$.

---
[2] Note that in Corollary 6.49 it is possible that $t = \top$, while here $t \neq \top$. So, although for convenience we have already used here the names t and $\top$ used in Corollary 6.49, it is not clear yet that what we denote here by $\top$ has the properties of what is denoted by $\top$ in Corollary 6.49. In particular: if $\tilde{\neg} t = \{f, \top\}$ then it is possible that $\tilde{\neg}\top = \{f\}$.

1. $\tilde{\neg} t = \{f\}$.

   Assume otherwise. Then $\tilde{\neg} t = \{f, \top\}$. Let $\mathcal{M}^*$ be the simple refinement of $\mathcal{M}$ which is obtained by letting $\tilde{\neg}_{\mathcal{M}^*} t = \{\top\}$ and $\tilde{\neg}_{\mathcal{M}^*}(\top) = \tilde{\neg}\top \cap \mathcal{D}$. Obviously, $\mathcal{M}^*$ is still pre-$\neg$-paraconsistent by Proposition 6.48, and $p \vdash_{\mathcal{M}^*} \neg p$. Since $p \not\vdash_{\mathcal{M}} \neg p$ (because $\neg$ is a negation of $\mathcal{M}$), we get that $\vdash_{\mathcal{M}^*}$ properly extends $\vdash_{\mathcal{M}}$. This contradicts the strong maximal paraconsistency of $\mathbf{L}_{\mathcal{M}}$.

2. $\tilde{\neg} f$ is a singleton.

   Assume that $\tilde{\neg} f$ is not a singleton. In this case, $\tilde{\neg} f \cap \{t, f\} \neq \emptyset$. Hence, we can get a simple refinement $\mathcal{M}^*$ of $\mathcal{M}$ s.t. $\neg\neg p \vdash_{\mathcal{M}^*} p$ by letting $\tilde{\neg}_{\mathcal{M}^*} t = \{f\}$, and either defining $\tilde{\neg}_{\mathcal{M}^*} f = \{f\}$ or $\tilde{\neg}_{\mathcal{M}^*} f = \{t\}$. Now, since $\tilde{\neg} f$ is not a singleton, either $f \in \tilde{\neg} f$ or $\top \in \tilde{\neg} f$. In the first case we let $\nu(p) = \nu(\neg p) = f$ and $\nu(\neg\neg p)$ be some element of $\tilde{\neg} f \cap \mathcal{D}$. In the second case we take $\nu(p) = f$, $\nu(\neg p) = \top$, and $\nu(\neg\neg p) \in \tilde{\neg}\top \cap \mathcal{D}$. (Recall that we are assuming that $\tilde{\neg}\top \subseteq \mathcal{D}$, or $\tilde{\neg}\top = \{f, t\}$.) In both cases we get an $\mathcal{M}$-model of $\neg\neg p$ which is not a model of p. It follows that $\neg\neg p \not\vdash_{\mathcal{M}} p$, and so $\vdash_{\mathcal{M}^*}$ properly extends $\vdash_{\mathcal{M}}$. On the other hand, $\mathcal{M}^*$ is still pre-$\neg$-paraconsistent by Proposition 6.48, since $\tilde{\neg}_{\mathcal{M}^*}(\top) = \tilde{\neg}_{\mathcal{M}}(\top)$, and so $\tilde{\neg}_{\mathcal{M}^*}(\top) \cap \mathcal{D} \neq \emptyset$. This contradicts the strong maximal paraconsistency of $\mathbf{L}_{\mathcal{M}}$.

3. If $\diamond \neq \neg$ then $\tilde{\diamond}$ is deterministic.

   Suppose for contradiction that there are a connective $\diamond \neq \neg$ and $a_1, \ldots, a_n \in \mathcal{V}$ such that $S = \tilde{\diamond}(a_1, \ldots, a_n)$ is not a singleton. Let $\psi = \diamond(q_1, \ldots, q_n)$, where $q_i = p_1$ if $a_i = \top$, $q_i = p_2$ if $a_i = t$, and $q_i = \neg p_2$ if $a_i = f$. Then $\nu(\psi) \in S$ for every ν such that $\nu(p_1) = \top$ and $\nu(p_2) = t$ (and so $\nu(\neg p_2) = f$ by Fact 1). Moreover, any element of S can be chosen to be $\nu(\psi)$ in this case.

   (a) Suppose that $\top \in S$. In this case, $p_1, \neg p_1, p_2, \psi, \neg\psi \not\vdash_{\mathcal{M}} \neg p_2$, since by taking $\nu(p_1) = \top$, $\nu(\neg p_1) \in \mathcal{D}$, $\nu(p_2) = t$, $\nu(\neg p_2) = f$, $\nu(\psi) = \top$, and $\nu(\neg\psi) \in \mathcal{D}$, we get a counter-model. Let $\mathcal{M}^*$ be the simple refinement of $\mathcal{M}$ in which $\tilde{\diamond}_{\mathcal{M}^*}(a_1, \ldots, a_n) = S \setminus \{\top\}$, and $\tilde{\neg}_{\mathcal{M}^*}(\top) = \tilde{\neg}\top \cap \mathcal{D}$. ($S \setminus \{\top\} \neq \emptyset$, because S is not a singleton.) Then $p_1, \neg p_1, p_2, \psi, \neg\psi \vdash_{\mathcal{M}^*} \neg p_2$. Indeed, if ν is a counter-model then necessarily $\nu(p_1) = \top$, and $\nu(p_2) = t$. But in this case $\nu(\psi) \in S \setminus \{\top\}$, and so ν cannot be a model of $\{\psi, \neg\psi\}$. It follows that $\mathbf{L}_{\mathcal{M}^*}$, which is obviously pre-$\neg$-paraconsistent, properly extends $\mathbf{L}_{\mathcal{M}}$. Hence $\mathbf{L}_{\mathcal{M}}$ is not strongly maximal. A contradiction.

   (b) Suppose that $S = \{t, f\}$. In this case, $p_1, \neg p_1, p_2, \psi \not\vdash_{\mathcal{M}} \neg p_2$, since by taking $\nu(p_1) = \top$, $\nu(\neg p_1) \in \mathcal{D}$, $\nu(p_2) = t$, $\nu(\neg p_2) = f$,

7.3 Maximality and Pre-maximality

and $\nu(\psi) = t$ we get a counter-model. Let $\mathcal{M}^*$ be the simple refinement of $\mathcal{M}$ in which $\tilde{\diamond}_{\mathcal{M}^*}(a_1,\ldots,a_n) = \{f\}$, and $\tilde{\neg}_{\mathcal{M}^*}(\top) = \tilde{\neg}\top \cap \mathcal{D}$. Then $p_1, \neg p_1, p_2, \psi \vdash_{\mathcal{M}^*} \neg p_2$. Indeed, if ν is a countermodel then again necessarily $\nu(p_1) = \top$, and $\nu(p_2) = t$. But in this case $\nu(\psi) = f$, and so ν cannot be a model of ψ. Since $\mathbf{L}_{\mathcal{M}^*}$ is obviously pre-$\neg$-paraconsistent, this again contradicts the strong maximality of $\mathbf{L}_{\mathcal{M}}$.

4. $\tilde{\neg}\top = \{t, f\}$.

   Since $\mathcal{M}$ is properly non-deterministic, it follows from Facts 1–3 that $\tilde{\neg}\top$ is not a singleton. Since we assume that either $\tilde{\neg}\top \subseteq \mathcal{D}$, or $\tilde{\neg}\top = \{f, t\}$, it suffices to show that $\tilde{\neg}\top \neq \{t, \top\}$. Assume otherwise. Then $p \not\vdash_{\mathcal{M}} \neg\neg p$, since by taking $\nu(p) = \top$, $\nu(\neg p) = t$, and $\nu(\neg\neg p) = f$, we get a counter-model. Let $\mathcal{M}^*$ be the simple refinement of $\mathcal{M}$ in which $\tilde{\neg}_{\mathcal{M}^*}(\top) = \top$. Then $\mathbf{L}_{\mathcal{M}^*}$ is obviously pre-$\neg$-paraconsistent. It also properly extends $\mathbf{L}_{\mathcal{M}}$, since it is easy to check that $p \vdash_{\mathcal{M}^*} \neg\neg p$. This contradicts the strong maximality of $\mathbf{L}_{\mathcal{M}}$.

5. $\tilde{\neg}f = \{t\}$. Since $\tilde{\neg}f \cap \mathcal{D} \neq \emptyset$, it follows by Fact 2 that either $\tilde{\neg}f = \{t\}$, or $\tilde{\neg}f = \{\top\}$. Assume the latter. Using Fact 4, let $\mathcal{M}^*$ be the simple refinement of $\mathcal{M}$ in which $\tilde{\neg}\top = \{t\}$. Obviously, $\mathcal{M}^*$ is still pre-$\neg$-paraconsistent. Now in $\mathcal{M}^*$ we have that $p, \neg p \vdash_{\mathcal{M}^*} \neg\neg\neg p$ (ν is a model of $\{p, \neg p\}$ only if $\nu(p) = \top$, and in $\mathcal{M}^*$ $\nu(\neg\neg\neg p) = t$ for such ν). However, $p, \neg p \not\vdash_{\mathcal{M}} \neg\neg\neg p$, since we get a counter-model by taking $\nu(p) = \top$, $\nu(\neg p) = t$, $\nu(\neg\neg p) = f$, $\nu(\neg\neg\neg p) = \top$, and $\nu(\neg\neg\neg\neg p) = f$. Again, this contradicts the strong maximal paraconsistency of $\mathbf{L}_{\mathcal{M}}$. Hence, $\tilde{\neg}f \neq \{\top\}$, and so $\tilde{\neg}f = \{t\}$.

Given Facts 1, 3, 4, and 5, to prove that $\mathcal{M}$ has Properties 1 and 2 (from the formulation of the theorem) it only remains to show that if $\diamond \neq \neg$, then $\tilde{\diamond}$ gets values only in $\{t, f\}$, and does not distinguish between t and $\top$. For this we use $\mathcal{M}'$, the determinization of $\mathcal{M}$ in which $\tilde{\neg}\top = \{t\}$. Obviously, $\mathbf{L}_{\mathcal{M}'}$ is still pre-$\neg$-paraconsistent (and by Proposition 6.37 it extends $\mathbf{L}_{\mathcal{M}}$).

Assume first that $\diamond$ does not get values only in $\{t, f\}$. Then by Fact 3, $\tilde{\diamond}(x_1,\ldots,x_n) = \{\top\}$ for some $x_1,\ldots,x_n$. Like in the proof of Fact 3, this implies the existence of a formula ψ such that $\mathsf{Var}(\psi) \subseteq \{p_1, p_2\}$, and $\nu(\psi) = \top$ for every ν such that $\nu(p_1) = \top$ and $\nu(p_2) = t$. Therefore $p_1, \neg p_1, p_2, \neg\neg p_2 \not\vdash_{\mathcal{M}} \neg\psi$, because a counterexample is provided by taking $\nu(p_1) = \top$, $\nu(\neg p_1) = t$, $\nu(p_2) = t$ (and so $\nu(\neg\neg p_2) = t$), $\nu(\psi) = \top$, and $\nu(\neg\psi) = f$. On the other hand $p_1, \neg p_1, p_2, \neg\neg p_2 \vdash_{\mathcal{M}'} \neg\psi$, and so $\vdash_{\mathcal{M}'}$ properly extends $\vdash_{\mathcal{M}}$. Hence $\mathcal{M}$ is not strongly maximal. A contradiction.

Now, assume that $\diamond$ distinguishes between t and $\top$. So, there are e.g. $x_1,\ldots,x_{n-1}$ such that $\tilde{\diamond}(x_1,\ldots,x_{n-1},\top) \neq \tilde{\diamond}(x_1,\ldots,x_{n-1},t)$. Since $\tilde{\diamond}$ is deterministic and gets values only in $\{t,f\}$, we may assume (using, if necessary, $\neg \diamond(q_1,\ldots,q_{n-1},q)$ instead of $\diamond(q_1,\ldots,q_{n-1},q)$, and the fact that $\tilde{\neg} t = \{f\}, \tilde{\neg} f = \{t\}$) that $\tilde{\diamond}(x_1,\ldots,x_{n-1},\top) = \{t\}$, while $\tilde{\diamond}(x_1,\ldots,x_{n-1},t) = \{f\}$. Let $\psi = \diamond(q_1,\ldots,q_{n-1},q)$, where for $1 \leq i \leq n-1$, $q_i = p_1$ if $x_i = \top$, $q_i = p_2$ if $x_i = t$, and $q_i = \neg p_2$ if $x_i = f$. Then $\nu(\psi) = t$ for every assignment ν such that $\nu(p_1) = \top, \nu(p_2) = t$ and $\nu(q) = \top$, while $\nu(\psi) = f$ for every assignment ν such that $\nu(p_1) = \top, \nu(p_2) = t$ and $\nu(q) = t$. It follows that $p_1, \neg p_1, p_2, \neg\neg p_2, q, \psi \not\vdash_{\mathcal{M}} \neg q$, (take $\nu(p_1) = \nu(q) = \top, \nu(p_2) = t, \nu(\neg p_1) = t, \nu(\neg p_2) = f, \nu(\neg\neg p_2) = t, \nu(\psi) = t$, and $\nu(\neg q) = f$). On the other hand, it is easy to see that $p_1, \neg p_1, p_2, \neg\neg p_2, q, \psi \vdash_{\mathcal{M}'} \neg q$. Hence again $\vdash_{\mathcal{M}'}$ properly extends $\vdash_{\mathcal{M}}$, and so $\mathcal{M}$ is not strongly maximal.

Finally, since $\mathcal{M}$ is paraconsistent, it is $\neg$-coherent with classical logic. By Proposition 6.30 and Note 6.25, this and the fact that $\mathcal{M}$ is an f-Nmatrix (which is immediate from what we have already proved) imply that $\mathcal{M}$ also has the third required property, that is: it has an $\mathcal{M}$-conjunction or an $\mathcal{M}$-disjunction. □

The following theorem provides a converse for Theorem 7.21:

Theorem 7.22 *Let* $\mathcal{M} = \langle \{t, \top, f\}, \{t, \top\}, \mathcal{O} \rangle$ *be a three-valued Nmatrix which satisfies the three conditions specified in Theorem 7.21. Then* $\mathbf{L}_{\mathcal{M}}$ *is a strongly maximal paraconsistent logic.*

Proof. The conditions obviously imply that $\mathbf{L}_{\mathcal{M}}$ is $\neg$-paraconsistent. To show that it is strongly maximal, let $\mathcal{M}'$ be the determinization of $\mathcal{M}$ in which $\tilde{\neg}\top = \{t\}$. Obviously $\mathcal{M}'$ is paraconsistent, and so $\mathbf{L}_{\mathcal{M}'}$ is strongly maximal by Theorem 4.23. Therefore it suffices to show that $\vdash_{\mathcal{M}} = \vdash_{\mathcal{M}'}$. By Proposition 6.37, $\vdash_{\mathcal{M}} \subseteq \vdash_{\mathcal{M}'}$. For the converse, assume $\Gamma \not\vdash_{\mathcal{M}} \psi$. Let $\nu \in \Lambda_{\mathcal{M}}$ be a model of Γ in $\mathcal{M}$ such that $\nu(\psi) = f$. Define $\nu' \in \Lambda_{\mathcal{M}'}$ as follows: $\nu'(p) = t$ in case p is propositional variable such that $\nu(p) = \top$ and $\nu(\neg p) = f$, $\nu'(\varphi) = \nu(\varphi)$ for any other φ. (Note that $\nu(\varphi) = \top$ only if φ is a propositional variable.) It is easy to see that ν' is indeed in $\Lambda_{\mathcal{M}'}$, and that for every formula φ, $\nu'(\varphi)$ is designated iff $\nu(\varphi)$ is designated. In particular, ν' is a model of Γ in $\mathcal{M}'$ which is not a model of ψ. It follows that $\Gamma \not\vdash_{\mathcal{M}'} \psi$. Hence $\vdash_{\mathcal{M}'} \subseteq \vdash_{\mathcal{M}}$. □

Example 7.23 Theorem 7.21 allows exactly one three-valued properly nondeterministic matrix for $\mathcal{L}_{CL}$ which induces a normal and strongly maximal paraconsistent logic. It is the one that defers from the matrix P_1 (viewed as a deterministic Nmatrix) by having $\tilde{\neg} t = \{t, f\}$ rather than $\tilde{\neg} t = \{f\}$. By Theorem 7.22 and its proof, the logic induced by this Nmatrix is Sette's logic P_1. Therefore, among the logics which are studied in Section 4.4, P_1 is the

7.3 Maximality and Pre-maximality

only one which in addition to its characteristic three-valued ordinary matrix has also a characteristic three-valued properly non-deterministic matrix.

The above results show that three-valued properly non-deterministic matrices usually do not induce strongly maximal paraconsistent logics. However, now we show that there are many such Nmatrices which are (fully) pre-maximal. The importance of this is due to the fact that by the results of Chapter 4 (in particular: Theorems 4.24 and 4.17), for $\mathcal{L}_{CL}$ alone there is already a huge number of normal, fully maximal paraconsistent three-valued matrices. The vast variety of (fully) maximal paraconsistent logics can be systematized by using pre-maximal non-deterministic bases. The point is that each (fully) pre-maximal Nmatrix represents the family of its (fully) maximal determinizations, up to the point in which choices based on other considerations should be made.

The most important fully pre-maximal three-valued Nmatrix is implicit in Theorem 4.17.

Definition 7.24 ($\mathcal{M}_{8Kb}$) $\mathcal{M}_{8Kb}$ is the following three-valued Nmatrix:

$\tilde{\wedge}$	t	f	$\top$
t	$\{t\}$	$\{f\}$	$\{t,\top\}$
f	$\{f\}$	$\{f\}$	$\{f\}$
$\top$	$\{t,\top\}$	$\{f\}$	$\{t,\top\}$

$\tilde{\vee}$	t	f	$\top$
t	$\{t\}$	$\{t\}$	$\{t,\top\}$
f	$\{t\}$	$\{f\}$	$\{t,\top\}$
$\top$	$\{t,\top\}$	$\{t,\top\}$	$\{t,\top\}$

$\tilde{\supset}$	t	f	$\top$
t	$\{t\}$	$\{f\}$	$\{t,\top\}$
f	$\{t\}$	$\{t\}$	$\{t,\top\}$
$\top$	$\{t,\top\}$	$\{f\}$	$\{t,\top\}$

	$\tilde{\neg}$
t	$\{f\}$
f	$\{t\}$
$\top$	$\{t,\top\}$

Theorem 7.25 *Let $\mathcal{M}$ be a three-valued $\neg$-paraconsistent Nmatrix for $\mathcal{L}$.*

1. *$\mathcal{M}$ is strongly pre-maximal iff there is a (possibly definable) binary connective $\diamond \in \{\wedge, \vee, \supset\}$ of $\mathcal{L}$ such that $\mathcal{M}$ is isomorphic to an Nmatrix of the form $\langle \{t, f, \top\}, \{t, \top\}, \mathcal{O} \rangle$ whose reduction (Definition 6.5) to the language of $\{\neg, \diamond\}$ is a simple refinement of the reduction of $\mathcal{M}_{8Kb}$ to that language.*

2. *$\mathcal{M}$ is fully pre-maximal iff it is isomorphic to some Nmatrix which is of the type described in Item 1, and satisfies also the following condition:*

   - *$\tilde{*}(x_1, \ldots, x_n) \subseteq \{t, f\}$ whenever $\{x_1, \ldots, x_n\} \subseteq \{t, f\}$, and $*$ is an n-ary connective of $\mathcal{L}$.*

3. $\mathcal{M}$ is normal and strongly pre-maximal iff it is isomorphic to an Nmatrix $\langle \{t, f, \top\}, \{t, \top\}, \mathcal{O}\rangle$ that has (possibly definable) binary connectives $\wedge$, $\vee$, and $\supset$ such that the reduction of $\mathcal{M}$ to $\mathcal{L}_{CL}$ is a simple refinement of $\mathcal{M}_{8Kb}$. It is also fully pre-maximal iff in addition $\mathcal{M}$ satisfies the condition given in the second item.

Proof. Use Theorem 4.23, Theorem 4.21, and Corollary 4.25. □

Corollary 7.26 *Every simple refinement of $\mathcal{M}_{8Kb}$ (including $\mathcal{M}_{8Kb}$ itself) is fully pre-maximal.*

Note 7.27 The set of determinations of $\mathcal{M}_{8Kb}$ is exactly the family 8Kb from Theorem 4.17.

Note 7.28 From Theorem 7.7 it follows that the logic induced by $\mathcal{M}_{8Kb}$ is sparse. That theorem also suffices for showing that the same is true for most of the proper simple refinements of $\mathcal{M}_{8Kb}$. It is not difficult to determine which of the simple refinements of $\mathcal{M}_{8Kb}$ to which Theorem 7.7 does not apply is nevertheless sparse too, but we shall not do it here.

7.4 Minimality and Modularity

The set of axioms for negation given in Figure 4.7 (which are the negation axioms of the logic **PAC** mentioned in the proof of Proposition 7.20) reflect the most important properties of classical negation and its combinations with other connectives that might be considered as desirable or undesirable in the context of paraconsistent logics. In this section we characterize for each subset of these properties the minimal paraconsistent extension of $\mathbf{CL}^+$ which has all the properties in that subset. (Thus, the focus in this section is on *minimality* with respect to sets of negation properties.) This is done in a modular way, using three-valued and four-valued $\neg$-fundamental Nmatrices in which the principles that govern the support (or not) of truth and falsity of implications, conjunctions, and disjunctions are exactly those that stand behind positive classical logic. These principle lead to the following four-valued $\neg$-fundamental Nmatrix for $\mathcal{L}_{CL}$, of which Dunn-Belnap's (deterministic) matrix $\mathcal{FOUR}$ is just one of many possible refinements.

Definition 7.29 ($\mathcal{M}_4^b$, $\mathcal{M}_{4F}^b$) Let $\mathcal{M}_4^b = \langle \mathcal{V}, \mathcal{D}, \mathcal{O}\rangle$ be the Nmatrix for $\mathcal{L}_{CL}$ in which $\mathcal{V} = FOUR = \{t, \top, \bot, f\}$, $\mathcal{D} = \{t, \top\}$, and $\mathcal{O}$ is defined as:

$\tilde{\vee}$	t	$\top$	f	$\bot$
t	$\{t,\top\}$	$\{t,\top\}$	$\{t,\top\}$	$\{t,\top\}$
$\top$	$\{t,\top\}$	$\{t,\top\}$	$\{t,\top\}$	$\{t,\top\}$
f	$\{t,\top\}$	$\{t,\top\}$	$\{f,\bot\}$	$\{f,\bot\}$
$\bot$	$\{t,\top\}$	$\{t,\top\}$	$\{f,\bot\}$	$\{f,\bot\}$

$\tilde{\supset}$	t	$\top$	f	$\bot$
t	$\{t,\top\}$	$\{t,\top\}$	$\{f,\bot\}$	$\{f,\bot\}$
$\top$	$\{t,\top\}$	$\{t,\top\}$	$\{f,\bot\}$	$\{f,\bot\}$
f	$\{t,\top\}$	$\{t,\top\}$	$\{t,\top\}$	$\{t,\top\}$
$\bot$	$\{t,\top\}$	$\{t,\top\}$	$\{t,\top\}$	$\{t,\top\}$

7.4 Minimality and Modularity

$\tilde{\wedge}$	t	$\top$	f	$\bot$
t	$\{t,\top\}$	$\{t,\top\}$	$\{f,\bot\}$	$\{f,\bot\}$
$\top$	$\{t,\top\}$	$\{t,\top\}$	$\{f,\bot\}$	$\{f,\bot\}$
f	$\{f,\bot\}$	$\{f,\bot\}$	$\{f,\bot\}$	$\{f,\bot\}$
$\bot$	$\{f,\bot\}$	$\{f,\bot\}$	$\{f,\bot\}$	$\{f,\bot\}$

$\tilde{\neg}$	
t	$\{f,\bot\}$
$\top$	$\{t,\top\}$
f	$\{t,\top\}$
$\bot$	$\{f,\bot\}$

In other words:

$$a \,\tilde{\wedge}\, b = \begin{cases} \mathcal{D} & \text{if } a \in \mathcal{D} \text{ and } b \in \mathcal{D}, \\ \overline{\mathcal{D}} & \text{otherwise.} \end{cases}$$

$$a \,\tilde{\vee}\, b = \begin{cases} \mathcal{D} & \text{if } a \in \mathcal{D} \text{ or } b \in \mathcal{D}, \\ \overline{\mathcal{D}} & \text{otherwise.} \end{cases}$$

$$a \,\tilde{\supset}\, b = \begin{cases} \mathcal{D} & \text{if } a \notin \mathcal{D} \text{ or } b \in \mathcal{D}, \\ \overline{\mathcal{D}} & \text{otherwise.} \end{cases}$$

$$\tilde{\neg} a = \begin{cases} \mathcal{D} & \text{if } a \in \{f,\top\}, \\ \overline{\mathcal{D}} & \text{otherwise.} \end{cases}$$

$\mathcal{M}^b_{4F}$ is the Nmatrix for the language $\mathcal{L}^{F\neg}_{CL}$ of $\{\wedge, \vee, \supset, \neg, F\}$, in which $\tilde{\wedge}, \tilde{\vee}, \tilde{\supset}, \tilde{\neg}$ are defined as in $\mathcal{M}^b_4$, and in addition $\tilde{F} = \mathcal{V} - \mathcal{D}$.

Note 7.30 Like in Note 7.3, the representation of the four basic truth-values given in Note 5.13 can be very illuminating here too. This time the behavior of the connectives other than $\neg$ of $\mathcal{L}^{F\neg}_{CL}$ can be characterized by the following principles (where $\tilde{\diamond}_2$ is the standard two-valued interpretation of $\diamond$ in $\{0,1\}$):

(iv) $\langle a_1, b_1 \rangle \tilde{\diamond} \langle a_2, b_2 \rangle \subseteq \{v \in \{0,1\}^2 \mid \Pi_1(v) = a_1 \tilde{\diamond}_2 a_2\}$.

(v) $\tilde{F} \subseteq \{v \in \{0,1\}^2 \mid \Pi_1(v) = 0\}$.

These conditions, together with Conditions (i)–(iii) given in Note 7.3, apply to all the refinements of $\mathcal{M}^b_4$ and $\mathcal{M}^b_{4F}$ we consider below. In $\mathcal{M}^b_4$ and $\mathcal{M}^b_{4F}$ themselves the interpretations of the connectives are the maximal sets which are allowed by Conditions (i)–(v).

Finally, we note that in terms of pairs $\langle \nu^+, \nu^- \rangle$ (Note 7.3), Conditions (iv) and (iv) can respectively be formulated as follows:

(iv) $\nu^+(\varphi \diamond \psi) = \nu^+(\varphi) \,\tilde{\diamond}_2\, \nu^+(\psi)$.

(v) $\nu^+(F) = 0$.

Note that in refinements of $\mathcal{M}_4^b$ and $\mathcal{M}_{4\mathsf{F}}^b$ the valuation ν^+ (but not ν^-!) is deterministic in the sense that the value assigned by ν^+ to a formula is completely determined by the values ν^+ and ν^- assign to its subformulas. Moreover: ν^+ behaves classically with respect to the positive connectives and F (with 1 and 0 replacing t and f).

Note 7.31 The interpretations of the positive connectives $\wedge, \vee, \supset$ in $\mathcal{M} = \mathcal{M}_4^b$ (and in $\mathcal{M}_{4\mathsf{F}}^b$) are the most general (non-deterministic) interpretations for which these connectives are $\mathcal{M}$-conjunction, $\mathcal{M}$-disjunction, and $\mathcal{M}$-implication (respectively). This is made precise in the next proposition.

Proposition 7.32 *Let $\{\wedge, \vee, \supset\} \subseteq \mathcal{L}$, and let $\mathcal{M} = \langle \mathcal{V}, \mathcal{D}, \mathcal{O} \rangle$ be an Nmatrix for $\mathcal{L}$ such that $\mathcal{V} \subseteq \{t, \top, \bot, f\}$ and $\mathcal{D} \subseteq \{t, \top\}$. Then $\wedge, \vee, \supset$ are $\mathcal{M}$-conjunction, $\mathcal{M}$-disjunction, and $\mathcal{M}$-implication (respectively) iff the reduction of $\mathcal{M}$ to $\{\wedge, \vee, \supset\}$ is a refinement of the reduction of $\mathcal{M}_4^b$ ($\mathcal{M}_{4\mathsf{F}}^b$) to $\{\wedge, \vee, \supset\}$.*

Proof. Immediate from the definitions. □

Corollary 7.33 $\wedge$, $\vee$, *and* $\supset$ *are $\mathcal{M}_4^b$ ($\mathcal{M}_{4\mathsf{F}}^b$)-conjunction, $\mathcal{M}_4^b$ ($\mathcal{M}_{4\mathsf{F}}^b$)-disjunction, and $\mathcal{M}_4^b$ ($\mathcal{M}_{4\mathsf{F}}^b$)-implication (respectively).*

Proposition 7.34 *Let $\mathcal{M}$ be an extension (Definition 6.5) of $\mathcal{M}_4^b$ ($\mathcal{M}_{4\mathsf{F}}^b$). Then HCL^+ (HCL^F) is sound for any refinement of $\mathcal{M}$.*

Proof. By Corollary 7.33, the second part of Proposition 6.31, Proposition 6.40, Theorem 1.90 and the last item of Note 1.91. □

Our next goal is to use the four-valued non-deterministic framework for investigating the semantic effects of the most standard axioms for negation.

Definition 7.35 (A_{PAC}, $HCL^+(S)$, $HCL^\mathsf{F}(S)$) Let A_{PAC} be the set of the negation axioms of **PAC** from Figure 4.7. (See Note 4.99 for their other names.) For $S \subseteq \mathsf{A}_{PAC}$, the system $HCL^+(S)$ ($HCL^\mathsf{F}(S)$) is the axiomatic extension in $\mathcal{L}_{CL}$ (in $\mathcal{L}_{CL}^{\mathsf{F}\neg}$) of HCL^+ (HCL^F) with the axioms in S.

Notation: For $HCL^+(\{a_1, \ldots, a_n\})$ ($HCL^\mathsf{F}(\{a_1, \ldots, a_n\})$) we omit the brackets and write $HCL^+(a_1, \ldots, a_n)$ ($HCL^\mathsf{F}(a_1, \ldots, a_n)$) instead. We shall also frequently employ as names of axioms those that are described in Note 4.99.

Proposition 7.36 *For every $S \subseteq \mathsf{A}_{PAC}$, $\mathbf{L}_{HCL^+(S)}$ ($\mathbf{L}_{HCL^\mathsf{F}(S)}$) is classical (and so normal) paraconsistent logic, which is boldly paraconsistent.*

7.4 Minimality and Modularity

Proof. Let $S \subseteq \mathbf{A}_{PAC}$. Since $\mathbf{A}_{PAC}$ consists of classical tautologies, $\mathbf{L}_{HCL^+(S)}$ ($\mathbf{L}_{HCL^F(S)}$) is an axiomatic extension of $\mathbf{CL}^+$, which is contained in classical logic. That it is boldly paraconsistency follows from Theorem 4.23, since $\mathbf{L}_{HCL^+(S)}$ and $\mathbf{L}_{HCL^F(S)}$ are contained in $\mathbf{PAC}$ and $\mathbf{J}_3$ respectively. Hence it is classical paraconsistent logic by Definition 2.22. □

The family $\{\mathbf{L}_{HCL^+(S)} \mid S \subseteq \mathbf{A}_{PAC}\}$ includes some particularly important logics. Three of them were studied above. As was hinted at the proof of Proposition 7.36, $\mathbf{L}_{HCL^+(\mathbf{A}_{PAC})}$ is just the logic $\mathbf{PAC}$ studied in Chapter 4. Similarly, $\mathbf{L}_{HCL^+(\mathbf{A}_{PAC}-\{[t]\})}$ is the logic $\mathbf{4CL}$ studied in Chapter 5. Finally, $\mathbf{L}_{HCL^+([t])}$ is the logic $\mathbf{CLuN}$ studied in Section 7.2.

Another very important member of this family is the following:

Definition 7.37 ($\mathbf{C}_{min}$) $\mathbf{C}_{min}$ is the logic $\mathbf{L}_{HCL^+(\{[t],[c]\})}$.

Proposition 7.38 $\mathbf{C}_{min}$ *is the minimal classical paraconsistent logic which has a negation which is both complete and left-involutive.*[3]

Proof. Immediate from Note 2.6 (and Definition 2.22). □

Note 7.39 $\mathbf{C}_{min}$ is one of the most basic systems in the family which has been developed by da Costa's school. It is further investigated in Chapter 8, and plays an important role in Chapter 9 as well.

Note 7.40 Proposition 7.38 is just one example of a general principle: For each $S \subseteq \mathbf{A}_{PAC}$, $HCL^+(S)$ induces the minimal classical paraconsistent logic in which all the axioms in S are valid. This means that it is the minimal classical paraconsistent logic that has all the properties that the axioms in S correspond to. Here are some other examples of this principle:

1. $\mathbf{L}_{HCL^+([c])}$ is the minimal classical ¬-paraconsistent logic in which ¬ is left-involutive.

2. $\mathbf{L}_{HCL^+([e])}$ is the minimal classical ¬-paraconsistent logic in which ¬ is right-involutive.

3. $\mathbf{L}_{HCL^+(\{[t],[c],[e]\})}$ is the minimal classical ¬-paraconsistent logic in which ¬ is both complete and involutive.

---
[3]Recall that by Proposition 2.14, in the context of paraconsistent logics one should give up having a negation which is complete, right-involutive, and contrapositive. In this chapter we do not consider contrapositivity, but we will return to it in Chapter 9.

Now we turn to provide non-deterministic semantics for all the logics in the two families defined in Definition 7.35. The idea is that each of the axioms in A_{PAC} corresponds to a condition which leads to a certain refinement of the basic Nmatrices $\mathcal{M}_4^b$ and $\mathcal{M}_{4F}^B$ given in Definition 7.29. To see how these conditions are obtained, consider, e.g., the axiom [c] $\neg\neg\varphi \supset \varphi$. This axiom can be refuted in $\mathcal{M}_4^b$ in one of the following ways: (i) assigning f to φ, $\top$ to $\neg\varphi$, and either $\top$ or t to $\neg\neg\varphi$, or (ii) assigning $\bot$ to φ, f to $\neg\varphi$ and again either $\top$ or t to $\neg\neg\varphi$. Such refutations can only be excluded by requiring that (i) $\tilde{\neg}f \subseteq \{t\}$ and (ii) $\tilde{\neg}\bot \subseteq \{\bot\}$. A similar analysis can be made for the each of the other axioms, resulting in the following list of conditions:

Definition 7.41 (refining conditions) The refining conditions $C(x)$ for every $x \in A_{PAC}$ are defined as follows:

$C([\mathsf{t}])$: the truth-value $\bot$ is deleted,

$C([\mathsf{c}])$: $\tilde{\neg}f = \{t\}$ and $\tilde{\neg}\bot = \{\bot\}$,

$C([\mathsf{e}])$: $\tilde{\neg}t = \{f\}$ and $\tilde{\neg}\top = \{\top\}$,

$C([\mathsf{n}_\supset^l]_1)$: $x\tilde{\supset}y \subseteq \{t, \bot\}$ for $x \in \{f, \bot\}$,

$C([\mathsf{n}_\supset^l]_2)$: $x\tilde{\supset}y \subseteq \{t, \bot\}$ for $y \in \{t, \bot\}$,

$C([\mathsf{n}_\supset^r])$: $x\tilde{\supset}y = \{y\}$ for $x \in \{t, \top\}$ and $y \in \{f, \top\}$,

$C([\mathsf{n}_\vee^l]_1)$: $x\tilde{\vee}y = \{sup_t(x, y)\}$ for $x \in \{t, \bot\}$,

$C([\mathsf{n}_\vee^l]2)$: $x\tilde{\vee}y = \{sup_t(x, y)\}$ for $y \in \{t, \bot\}$,

$C([\mathsf{n}_\vee^r])$: $x\tilde{\vee}y = \{sup_t(x, y)\}$ for $x, y \in \{f, \top\}$,

$C([\mathsf{n}_\wedge^l])$: $x\tilde{\wedge}y = \{inf_t(x, y)\}$ for $x, y \in \{t, \bot\}$,

$C([\mathsf{n}_\wedge^r]_1)$: $x\tilde{\wedge}y = \{inf_t(x, y)\}$ for $x \in \{f, \top\}$,

$C([\mathsf{n}_\wedge^r]_2)$: $x\tilde{\wedge}y = \{inf_t(x, y)\}$ for $y \in \{f, \top\}$.

As one more example of the derivation of the refining conditions given in Definition 7.41, let us show how $C([\mathsf{n}_\supset^r])$ was derived. Recall that $[\mathsf{n}_\supset^r]$ is $(\varphi \wedge \neg\psi) \supset \neg(\varphi \supset \psi)$. This can be refuted in $\mathcal{M}_4^b$ only by assigning to φ a value in $\{t, \top\}$, to ψ a value in $\{f, \top\}$, and to $\varphi \supset \psi$ a value in $\{t, \bot\}$. To prevent this, we should demand that if $x \in \{t, \top\}$ and $y \in \{f, \top\}$, then

7.4 Minimality and Modularity

$x \tilde{\supset} y \subseteq \{f, \top\}$.[4] However, since we are looking for refinements of $\mathcal{M}_4^b$, this condition implies that if $x \in \{t, \top\}$ and $y = f$ then $x \tilde{\supset} y = \{f\}$, while if $x \in \{t, \top\}$ and $y = \top$ then $x \tilde{\supset} y = \{\top\}$. This is equivalent to demanding that if $x \in \{t, \top\}$ and $y \in \{f, \top\}$, then $x \tilde{\supset} y = \{y\}$.

Note 7.42 An analysis similar to that we did for $C([\mathsf{n}_\supset^r])$ shows that the conditions for $[\mathsf{n}_\supset^l]_1$ and $[\mathsf{n}_\supset^l]_2$ can also be strengthened to the following ones:

- $C([\mathsf{n}_\supset^l]_1)$: $x \tilde{\supset} y = \{t\}$ for $x \in \{f, \bot\}$,

- $C([\mathsf{n}_\supset^l]_2)$: $x \tilde{\supset} y = \{t\}$ if either $y = t$, or $x \in \{f, \bot\}$ and $y = \bot$; $x \tilde{\supset} y = \{y\}$ if either $y = t$, or $x \in \{t, \top\}$ and $y = \bot$.

In this chapter it is more convenient to directly use these stricter forms rather than those given in Definition 7.41. However, the latter are valid also for the semantics of the constructive systems which are studied in Chapter 10, while this is not the case for the stronger forms.

Definition 7.43 (weakest refinement) Let $\mathcal{M}$ be an Nmatrix, and let $\mathcal{R}$ be a non-empty set of refinements of $\mathcal{M}$. An element $\mathcal{M}^*$ of $\mathcal{R}$ is the *weakest refinement of $\mathcal{M}$ in $\mathcal{R}$* if every element of $\mathcal{R}$ is a refinement of $\mathcal{M}^*$.

Definition 7.44 $(C(S), \mathcal{M}_4^b(S), \mathcal{M}_{4F}^b(S))$ For $S \subseteq \mathbf{A}_{PAC}$, $\mathcal{M}_4^b(S)$ and $\mathcal{M}_{4F}^b(S)$ are the weakest refinements of $\mathcal{M}_4^b$ and $\mathcal{M}_{4F}^b$ (respectively) in which the conditions in $C(S) = \{C(x) \mid x \in \mathbf{A}_{PAC}\}$ are satisfied.

Proposition 7.45 $\mathcal{M}_4^b(S)$ and $\mathcal{M}_{4F}^b(S)$ $(S \subseteq \mathbf{A}_{PAC})$ are well-defined.

Proof. The matrices PAC and J_3 (Section 4.4.4) are refinements of $\mathcal{M}_4^b$ and $\mathcal{M}_{4F}^b$ (respectively), and they satisfy all the conditions which are induced by the axioms in $\mathbf{A}_{PAC}$. This implies the non-emptiness of the sets of refinements of $\mathcal{M}_4^b$ and $\mathcal{M}_{4F}^b$ in which the conditions in $C(S)$ are satisfied. It is easy now to see that each of these two sets has a weakest element. Thus, $\mathcal{M}_4^b(S) = \langle \mathcal{V}_S, \mathcal{D}_S, \mathcal{O}_S \rangle$ where:

- $\mathcal{V}_S = \begin{cases} \{t, f, \top\} & [t] \in S, \\ \{t, f, \bot, \top\} & \text{otherwise.} \end{cases}$

- $\mathcal{D}_S = \{t, \top\}$.

---

[4] A shorter way to see this is to use the representation of the semantics in terms of the pairs $\langle \nu^+, \nu^- \rangle$ that was given in Notes 5.13, 7.3, and 7.30. $[\mathsf{n}_\supset^r]$ is valid iff for each such pair, if $\nu^+(\varphi) = 1$ and $\nu^-(\psi) = 1$ then $\nu^-(\varphi \supset \psi) = 1$. This is equivalent to what we have just got.

- For $* \in \{\vee, \wedge, \supset\}$ denote by $*_S$ the operation in $\mathcal{O}_S$ which corresponds to $*$, and denote by $*_4$ the interpretation of $*$ in $\mathcal{M}_4^b$. We let $*_S(x,y) = \{z\}$ iff $*(x,y) = \{z\}$ is dictated by one of the conditions that correspond to the axioms in S (including those described in Note 7.42), and $*_S(x,y) = *_4(x,y)$ otherwise. The operation in $\mathcal{O}_S$ corresponding to $\neg$ is defined similarly.

$\mathcal{M}_{4F}^b(S)$ is defined similarly, with the addition of $\tilde{\mathsf{F}} = \mathcal{V}_S - \mathcal{D}_S$. □

Example 7.46 It is easy to check that $\mathcal{M}_4^b(\mathsf{A}_{PAC})$ is the matrix PAC from Section 4.4.4, $\mathcal{M}_{4F}^b(\mathsf{A}_{PAC})$ is J_3 (Definition 4.62), and $\mathcal{M}_4^b(\mathsf{A}_{PAC} - \{[\mathsf{t}]\})$ is the matrix $\mathcal{M}_{4CL}$ from Chapter 5. Hence, the semantics of the logics **PAC**, **J$_3$**, and **4CL** given by Definition 7.44 is the same as that given to these logics in Chapter 4 and Chapter 5. Note that these three matrices are the only matrices (i.e., deterministic Nmatrices) in the family of Nmatrices which is introduced in Definition 7.44. On the other hand, $\mathcal{M}_{4F}^b(\mathsf{A}_{PAC} - \{[\mathsf{t}]\})$ is *not* identical with $\mathcal{M}_{Flex}$, since in the former $\tilde{\mathsf{F}} = \{f, \bot\}$, while in the matrix $\mathcal{M}_{Flex}$ $\tilde{\mathsf{F}} = \{f\}$. To be able to force the latter, we could have included in A_{PAC} the axiom $\neg \mathsf{F}$. We would have got then that $\tilde{\mathsf{F}} = \{f\}$ in $\mathcal{M}_{4F}^b(S)$ iff either $[\mathsf{t}] \in S$ or $\neg \mathsf{F} \in S$.

Example 7.47 The semantics for $\mathbf{C}_{min}$ that is given by Definition 7.44 is the Nmatrix $\mathcal{M}_4^b(\{[\mathsf{t}], [\mathsf{c}]\}) = \langle \mathcal{V}, \mathcal{D}, \mathcal{O} \rangle$, where $\mathcal{V} = \{t, \top, f\}$ ($\bot$ is deleted by the condition $C([\mathsf{t}])$), $\mathcal{D} = \{t, \top\}$, and the connectives are interpreted as below. (Note that the one difference from the basic Nmatrix $\mathcal{M}_4^b$ which is not due to $[\mathsf{t}]$ is in the interpretation of $\neg f$, which due to $[\mathsf{c}]$ is $\{t\}$ rather than $\{t, \top\}$.)

$\tilde{\vee}$	t	$\top$	f
t	$\{t, \top\}$	$\{t, \top\}$	$\{t, \top\}$
$\top$	$\{t, \top\}$	$\{t, \top\}$	$\{t, \top\}$
f	$\{t, \top\}$	$\{t, \top\}$	$\{f\}$

$\tilde{\supset}$	t	$\top$	f
t	$\{t, \top\}$	$\{t, \top\}$	$\{f\}$
$\top$	$\{t, \top\}$	$\{t, \top\}$	$\{f\}$
f	$\{t, \top\}$	$\{t, \top\}$	$\{t, \top\}$

$\tilde{\wedge}$	t	$\top$	f
t	$\{t, \top\}$	$\{t, \top\}$	$\{f\}$
$\top$	$\{t, \top\}$	$\{t, \top\}$	$\{f\}$
f	$\{f\}$	$\{f\}$	$\{f\}$

$\tilde{\neg}$	
t	$\{f\}$
$\top$	$\{t, \top\}$
f	$\{t\}$

Note 7.48 The Nmatrix $\mathcal{M}_4^b(\{[\mathsf{t}]\})$ for *HCLuN* that is given by Definition 7.44 is three-valued, and so is *different* from the two-valued Nmatrix $\mathcal{M}_2$ for which *HCLuN* is shown to be sound and complete in Section 7.2. Still, by the next theorem $\mathcal{M}_4^b(\{[\mathsf{t}]\})$ too is characteristic for *HCLuN*. This is not an unusual phenomenon when the semantic framework of Nmatrices is used. Note that $\mathcal{M}_4^b(\{[\mathsf{t}]\})$ differs from the Nmatrix for $\mathbf{C}_{min}$ given in Example 7.47 only in having $\neg f = \{t, \top\}$ rather than $\{t\}$.

7.4 Minimality and Modularity

Proposition 7.49 *For $S \subseteq \mathbf{A}_{PAC}$, $\wedge$ is an $\mathcal{M}_4^b(S)$-conjunction (and an $\mathcal{M}_{4F}^b(S)$-conjunction); $\vee$ is an $\mathcal{M}_4^b(S)$-disjunction ($\mathcal{M}_{4F}^b(S)$-disjunction); and $\supset$ is an $\mathcal{M}_4^b(S)$-implication ($\mathcal{M}_{4F}^b(S)$-implication).*

Proof. Immediate by Proposition 7.32. □

Theorem 7.50 (soundness and completeness) *Let $S \subseteq \mathbf{A}_{PAC}$. Then $HCL^+(S)$ ($HCL^\mathsf{F}(S)$) is sound and complete for $\mathbf{L}_{\mathcal{M}_4^b(S)}$ ($\mathbf{L}_{\mathcal{M}_{4F}^b(S)}$).*

Proof. We prove for the case of $\mathbf{L}_{\mathcal{M}_4^b(S)}$, leaving the similar proof for $\mathbf{L}_{\mathcal{M}_{4F}^b(S)}$ to the reader. For soundness, it suffices by Proposition 7.34 to show that for any axiom $ax \in \mathbf{A}_{PAC}$, ax is valid in any refinement of $\mathcal{M}_4^b$ in which the condition $C(ax)$ is satisfied. We show this, by way of example, for $ax = [\mathsf{n}_\vee^!]_1$. Let $\mathcal{M} = \langle \mathcal{V}, \mathcal{D}, \mathcal{O} \rangle$ be some refinement of $\mathcal{M}_4^b$ in which the condition $C([\mathsf{n}_\vee^!]_1)$ is satisfied. So assume that ν is a valuation in $\mathcal{M}$ such that $\nu(\neg\varphi) \notin \mathcal{D}$. Then $\nu(\varphi) \in \{t, \bot\}$. Condition $C([\mathsf{n}_\vee^!]_1)$ entails that in this case also $\nu(\varphi \vee \psi) \in \{t, \bot\}$, and so $\nu(\neg(\varphi \vee \psi)) \notin \mathcal{D}$. Hence $\neg(\varphi \vee \psi) \vdash_\mathcal{M} \neg\psi$. The proofs for the rest of the axioms are similar.

For completeness, suppose that $\mathcal{T} \nvdash_{HCL^+(S)} \varphi_0$. We show that there exists a model of $\mathcal{T}$ in $\mathcal{M}_4^b(S)$ which is not a model of φ_0. Like in the proof of Theorem 7.14, we use for this Theorem 1.98 to get an extension $\mathcal{T}^*$ of $\mathcal{T}$ that has the following properties:

1. $\psi \in \mathcal{T}^*$ iff $\mathcal{T}^* \vdash_{HCL^+(S)} \psi$.

2. $\varphi_0 \notin \mathcal{T}^*$.

3. $\varphi \vee \psi \in \mathcal{T}^*$ iff either $\varphi \in \mathcal{T}^*$ or $\psi \in \mathcal{T}^*$.

4. $\varphi \wedge \psi \in \mathcal{T}^*$ iff both $\varphi \in \mathcal{T}^*$ and $\psi \in \mathcal{T}^*$.

5. $\varphi \supset \psi \in \mathcal{T}^*$ iff either $\varphi \notin \mathcal{T}^*$ or $\psi \in \mathcal{T}^*$.

Define a valuation ν in $\mathcal{M}_4^b(S)$ as follows:

$$\nu(\psi) = \begin{cases} \bot & \psi \notin \mathcal{T}^*, \neg\psi \notin \mathcal{T}^*, \\ f & \psi \notin \mathcal{T}^*, \neg\psi \in \mathcal{T}^*, \\ t & \psi \in \mathcal{T}^*, \neg\psi \notin \mathcal{T}^*, \\ \top & \psi \in \mathcal{T}^*, \neg\psi \in \mathcal{T}^*. \end{cases}$$

Now, we show that ν is an $\mathcal{M}_4^b(S)$-valuation, i.e., it respects the interpretations of the connectives in $\mathcal{M}_4^b(S)$. Properties 3–5 of $\mathcal{T}^*$ easily imply that ν respects the basic constraints concerning the positive connectives. As for the basic constraints concerning negation, we have:

- Assume that $\nu(\psi) \in \{t, \bot\}$. By definition, this implies $\neg\psi \notin \mathcal{T}^*$, and so $\nu(\neg\psi) \in \{f, \bot\}$.

- Assume that $\nu(\psi) \in \{f, \top\}$. By definition, this implies $\neg\psi \in \mathcal{T}^*$, and so $\nu(\neg\psi) \in \{t, \top\}$.

We now show that ν respects the conditions induced by the axioms in S.

- $C([\mathsf{t}])$: Assume that $[\mathsf{t}] \in \mathsf{S}$. Then $\varphi \vee \neg\varphi \in \mathcal{T}^*$, and so Property 3 above entails that for every φ, either $\varphi \in \mathcal{T}^*$, or $\neg\varphi \in \mathcal{T}^*$. Hence there is no φ such that $\nu(\varphi) = \bot$.

- $C([\mathsf{c}])$: Assume that $[\mathsf{c}] \in \mathsf{S}$. Then $\neg\neg\varphi \notin \mathcal{T}^*$ whenever $\varphi \notin \mathcal{T}^*$. This easily entails (by the definition of ν) that $\nu(\varphi) = f$ implies $\nu(\neg\varphi) = t$, while $\nu(\varphi) = \bot$ implies $\nu(\neg\varphi) = \bot$.

- $C([\mathsf{e}])$: Assume that $[\mathsf{e}] \in \mathsf{S}$. Then $\neg\neg\varphi \in \mathcal{T}^*$ whenever $\varphi \in \mathcal{T}^*$. This easily implies (by the definition of ν) that if $\nu(\varphi) = t$ then $\nu(\neg\varphi) = f$, while if $\nu(\varphi) = \top$ then $\nu(\neg\varphi) = \top$.

- $C([\mathsf{n}^!_\vee]_1)$: Assume that $[\mathsf{n}^!_\vee]_1 \in \mathsf{S}$. Then $\neg(\varphi \vee \psi) \notin \mathcal{T}^*$ if $\neg\varphi \notin \mathcal{T}^*$. Hence, if $\nu(\varphi) \in \{t, \bot\}$ then $\neg(\varphi \vee \psi) \notin \mathcal{T}^*$. Now, if $\nu(\varphi) = t$ then $\varphi \in \mathcal{T}^*$, and so $\varphi \vee \psi \in \mathcal{T}^*$. Hence, in this case $\nu(\varphi \vee \psi) = t$. If $\nu(\varphi) = \bot$ then $\varphi \notin \mathcal{T}^*$, and so $\varphi \vee \psi \in \mathcal{T}^*$ iff $\psi \in \mathcal{T}^*$ (iff $\nu(\psi) \in \mathcal{D}$). It follows that in this case $\nu(\varphi \vee \psi) = t$ if $\nu(\psi) \in \mathcal{D}$, and $\nu(\varphi \vee \psi) = \bot$ otherwise. In all cases we find that $\nu(\varphi \vee \psi) = sup_t(\nu(\varphi), \nu(\psi))$ if $\nu(\varphi) \in \{t, \bot\}$.

We leave the proofs of the conditions corresponding to the remaining axioms to the reader. Now, $\nu(\psi) \in \mathcal{D}$ iff $\psi \in \mathcal{T}^*$. Hence, $\nu(\psi) \in \mathcal{D}$ for every $\psi \in \mathcal{T}$, while $\nu(\varphi_0) \notin \mathcal{D}$ by property 2 of $\mathcal{T}^*$. It follows that ν is indeed a model of $\mathcal{T}$ which is not a model of $\nu(\varphi_0)$. □

Note 7.51 By Example 7.46, the soundness and completeness of $H_{\mathbf{PAC}}$, $H_{\mathbf{J_3}}$, and $H_{\mathbf{4CL}}$ (which were proved in Chapter 4 and Chapter 5) are special cases of Theorem 7.50.

Corollary 7.52 *For $S \subseteq \mathsf{A}_{PAC}$, $\mathbf{L}_{HCL^+(S)}$ and $\mathbf{L}_{HCL^F(S)}$ are decidable.*

Proof. By Theorems 7.50 and 6.22. □

Now we show that using non-deterministic semantics is essential here.

Theorem 7.53 *Let $S \subseteq \mathsf{A}_{PAC}$. If $S \neq \mathsf{A}_{PAC}$ and $S \neq \mathsf{A}_{PAC} - \{[\mathsf{t}]\}$ then there is no finite family $\mathcal{F}$ of finite ordinary matrices such that $\vdash_{HCL^+(S)} = \vdash_{\mathcal{F}}$ or $\vdash_{HCL^F(S)} = \vdash_{\mathcal{F}}$.*

7.4 Minimality and Modularity

Proof. Immediate from Theorems 7.50 and 7.4. □

Example 7.54 As noted in Example 7.46, three of the four systems left out in Theorem 7.53 have finite characteristic matrices. By Theorem 7.4 the fourth, $HCL^F(A_{PAC})$ does not (see Example 7.46 again). However, it can easily be seen that it is characterized by the family $\mathcal{F}$ which consists of the two possible determinizations of $\mathcal{M}^b_{4F}(A_{PAC})$: the one in which $\tilde{\mathsf{F}} = \bot$, and $\mathcal{M}_{Flex}$ (Definition 5.52), in which $\tilde{\mathsf{F}} = \{f\}$.

Corollary 7.55 *Let $S \subseteq A_{PAC}$. If $S \neq A_{PAC}$ and $S \neq A_{PAC} - \{[t]\}$ then $\mathbf{L}_{HCL^+(S)}$ and $\mathbf{L}_{HCL^F(S)}$ are sparse.*

Proof. By Theorems 7.50 and 7.7. □

Next, we show that the logics considered in this section induce different paraconsistent logics.

Theorem 7.56 *Let S_1 and S_2 be two different subsets of A_{PAC}. Then $\mathbf{L}_{HCL^+(S_1)} \neq \mathbf{L}_{HCL^+(S_2)}$; $\mathbf{L}_{HCL^+(S_1)} \neq \mathbf{L}_{HCL^F(S_2)}$; $\mathbf{L}_{HCL^F(S_1)} \neq \mathbf{L}_{HCL^F(S_2)}$.*

Proof. In order to prove that $\mathbf{L}_{HCL^+(S_1)} \neq \mathbf{L}_{HCL^+(S_2)}$ for two different subsets S_1 and S_2 of A_{PAC}, we show that if $ax \in S$ then ax is not provable in $HCL^+(S - \{ax\})$. By Theorem 7.50, this is equivalent to showing that ax is not valid in $\mathcal{M}^b_4(S')$ where $S' = A_{PAC} - \{ax\}$. For this, it suffices (by Proposition 6.11) to provide an appropriate partial $\mathcal{M}^b_4(S')$-valuation ν which refutes ax. Here are the refuting partial valuations in each case:

- $ax = [\mathsf{t}]$:
$$\nu(p) = \bot \quad \nu(\neg p) = \bot$$

- $ax = [\mathsf{c}]$:
$$\nu(p) = f \quad \nu(\neg p) = \top \quad \nu(\neg\neg p) = \top$$

- $ax = [\mathsf{e}]$:
$$\nu(p) = \top \quad \nu(\neg p) = t \quad \nu(\neg\neg p) = f$$

- $ax = [\mathsf{n}^l_\vee]_1$:
$$\nu(p) = t \quad \nu(q) = \top \quad \nu(p \vee q) = \top \quad \nu(\neg(p \vee q)) = \top \quad \nu(\neg p) = f$$

- $ax = [\mathsf{n}^l_\vee]_2$:
$$\nu(p) = \top \quad \nu(q) = t \quad \nu(p \vee q) = \top \quad \nu(\neg(p \vee q)) = \top \quad \nu(\neg q) = f$$

- $ax=[n_\vee^r]$:

$$\nu(p) = \nu(q) = \top \quad \nu(p \vee q) = t \quad \nu(\neg p) = \nu(\neg q) = \top$$
$$\nu(\neg p \wedge \neg q) = \top \quad \nu(\neg(p \vee q)) = f$$

- $ax=[n_\wedge^l]$:

$$\nu(p) = \nu(q) = t \quad \nu(\neg p) = \nu(\neg q) = f \quad \nu(p \wedge q) = \top$$
$$\nu(\neg(p \wedge q)) = \top \quad \nu(\neg p \vee \neg q) = f$$

- $ax=[n_\wedge^r]_1$:

$$\nu(p) = \top \quad \nu(q) = t \quad \nu(p \wedge q) = t \quad \nu(\neg p) = \top \quad \nu(\neg(p \wedge q)) = f$$

- $ax=[n_\wedge^r]_2$:

$$\nu(p) = t \quad \nu(q) = \top \quad \nu(p \wedge q) = t \quad \nu(\neg q) = \top \quad \nu(\neg(p \wedge q)) = f$$

- $ax=[n_\supset^l]_1$:

$$\nu(p) = f \quad \nu(q) = \top \quad \nu(p \supset q) = \top \quad \nu(\neg(p \supset q)) = \top$$

- $ax=[n_\supset^l]_1$:

$$\nu(p) = t \quad \nu(q) = t \quad \nu(p \supset q) = \top \quad \nu(\neg(p \supset q)) = \top \quad \nu(\neg q) = f$$

- $ax=[n_\supset^r]$:

$$\nu(p) = \top \quad \nu(q) = \top \quad \nu(\neg q) = \top \quad \nu(p \wedge \neg q) = \top$$
$$\nu(p \supset q) = t \quad \nu(\neg(p \supset q)) = f$$

The fact that if $ax \in S$, ax is not provable in $HCL^F(S - \{ax\})$, is shown similarly; We leave the rest of the proof to the reader. □

Now we turn to providing quasi-canonical Gentzen-type systems for all the logics of the form $\mathbf{L}_{HCL^+(S)}$. This is easy, since if $S \subseteq A_{PAC}$, then $HCL^+(S)$ is a subsystem of $HCL^+(A_{PAC}) = H_{\mathbf{PAC}}$. Now, $H_{\mathbf{PAC}}$ itself is equivalent to the Gentzen-type system $G_{\mathbf{PAC}}$ (Section 4.5.1). $G_{\mathbf{PAC}}$, in turn, is easily seen to be equivlent to the system $G^*_{\mathbf{PAC}}$ that is obtained from it by splitting each of $[\Rightarrow \neg\wedge]$, $[\neg\vee \Rightarrow]$, and $[\neg \supset\Rightarrow]$ into two rules in the obvious way, and replacing the axioms $\Rightarrow \neg\varphi, \varphi$ by the rule $[\Rightarrow \neg]$ of LK. (See first item of Note 4.88.) $G^*_{\mathbf{PAC}}$ has 12 negation-related rules, and each of them corresponds to exactly one of the negation-related axioms

7.4 Minimality and Modularity

	ax	$R(ax)$
$[t] = [\Rightarrow \neg]$	$\neg\varphi \vee \varphi$	$\dfrac{\Gamma, \varphi \Rightarrow \Delta}{\Gamma \Rightarrow \neg\varphi, \Delta}$
$[c] = [\neg\neg \Rightarrow]$	$\neg\neg\varphi \supset \varphi$	$\dfrac{\Gamma, \varphi \Rightarrow \Delta}{\Gamma, \neg\neg\varphi \Rightarrow \Delta}$
$[e] = [\Rightarrow \neg\neg]$	$\varphi \supset \neg\neg\varphi$	$\dfrac{\Gamma \Rightarrow \Delta, \varphi}{\Gamma \Rightarrow \Delta, \neg\neg\varphi}$

Figure 7.1: Gentzen rules induced by the negation axioms of A_{PAC}

of $H_{\mathbf{PAC}}$. (Those axioms have actually been obtained in this way. See the beginning of Section 4.5.2.) The Gentzen-type systems for the various $\mathbf{L}_{HCL^+(S)}$ ($S \subseteq \mathsf{A}_{PAC}$) are then modularly obtained by collecting the rules of $G^*_{\mathbf{PAC}}$ which correspond to the axioms in S. For treating the logics of the form $\mathbf{L}_{HCL^F(S)}$, we repeat the process, using $G_{\mathbf{J_3}}$ instead of $G_{\mathbf{PAC}}$.

For the reader's convenience, in Figures 7.1 and 7.2 we present for every $ax \in \mathsf{A}_{PAC}$ the Gentzen-type rule $R(ax)$ which corresponds to it.

Note 7.57 An alternative syntactic method for deriving the rules in Figures 7.1 and 7.2 (which reverses the order used in Sections 4.5.1 and 4.5.2) is to use the approach of [119]. There, an algorithm is provided for transforming Hilbert axioms (of a certain general form, which all of the axioms in A_{PAC} have) into equivalent Gentzen-type rules. That approach is again completely modular: for each $ax \in \mathsf{A}_{PAC}$, an equivalent Gentzen-type rule $R(ax)$ is obtained, so that for every s: $\vdash_{LK^+ \cup \{\Rightarrow ax\}} s$ iff $\vdash_{LK^+ \cup \{R(ax)\}} s$ and $\vdash_{LK^F \cup \{\Rightarrow ax\}} s$ iff $\vdash_{LK^F \cup \{R(ax)\}} s$.

Still another approach for the same task, this time semantic one, is presented and used in the next chapter. There the Gentzen-type rules are directly derived from corresponding semantic conditions.

Definition 7.58 ($LK^+(S)$, $LK^F(S)$) Let $S \subseteq \mathsf{A}_{PAC}$. $LK^+(S)$ ($LK^F(S)$) is the Gentzen-type system obtained from LK^+ (LK^F) by adding the rules in the set $\{R(ax) \mid ax \in S\}$.

$[n^r_\wedge]_1 = [\Rightarrow \neg \wedge 1]$	$\neg\varphi \supset \neg(\varphi \wedge \psi)$	$\dfrac{\Gamma \Rightarrow \Delta, \neg\varphi}{\Gamma \Rightarrow \Delta, \neg(\varphi \wedge \psi)}$
$[n^r_\wedge]_2 = [\Rightarrow \neg \wedge 2]$	$\neg\psi \supset \neg(\varphi \wedge \psi)$	$\dfrac{\Gamma \Rightarrow \Delta, \neg\psi}{\Gamma \Rightarrow \Delta, \neg(\varphi \wedge \psi)}$
$[n^l_\wedge] = [\neg\wedge \Rightarrow]$	$\neg(\varphi \wedge \psi) \supset (\neg\varphi \vee \neg\psi)$	$\dfrac{\Gamma, \neg\varphi \Rightarrow \Delta \quad \Gamma, \neg\psi \Rightarrow \Delta}{\Gamma, \neg(\varphi \wedge \psi) \Rightarrow \Delta}$
$[n^r_\vee] = [\Rightarrow \neg\vee]$	$(\neg\varphi \wedge \neg\psi) \supset \neg(\varphi \vee \psi)$	$\dfrac{\Gamma \Rightarrow \Delta, \neg\varphi \quad \Gamma \Rightarrow \Delta, \neg\psi}{\Gamma \Rightarrow \Delta, \neg(\varphi \vee \psi)}$
$[n^l_\vee]_1 = [\neg\vee \Rightarrow 1]$	$\neg(\varphi \vee \psi) \supset \neg\varphi$	$\dfrac{\Gamma, \neg\varphi \Rightarrow \Delta}{\Gamma, \neg(\varphi \vee \psi) \Rightarrow \Delta}$
$[n^l_\vee]_2 = [\neg\vee \Rightarrow 2]$	$\neg(\varphi \vee \psi) \supset \neg\psi$	$\dfrac{\Gamma, \neg\psi \Rightarrow \Delta}{\Gamma, \neg(\varphi \vee \psi) \Rightarrow \Delta}$
$[n^r_\supset] = [\Rightarrow \neg \supset]$	$(\varphi \wedge \neg\psi) \supset \neg(\varphi \supset \psi)$	$\dfrac{\Gamma \Rightarrow \Delta, \varphi \quad \Gamma \Rightarrow \Delta, \neg\psi}{\Gamma \Rightarrow \Delta, \neg(\varphi \supset \psi)}$
$[n^l_\supset]_1 = [\neg\supset\Rightarrow 1]$	$\neg(\varphi \supset \psi) \supset \varphi$	$\dfrac{\Gamma, \varphi \Rightarrow \Delta}{\Gamma, \neg(\varphi \supset \psi) \Rightarrow \Delta}$
$[n^l_\supset]_2 = [\neg\supset\Rightarrow 2]$	$\neg(\varphi \supset \psi) \supset \neg\psi$	$\dfrac{\Gamma, \neg\psi \Rightarrow \Delta}{\Gamma, \neg(\varphi \supset \psi) \Rightarrow \Delta}$

Figure 7.2: Gentzen rules induced by other axioms of $\mathbf{A}_{PAC}$

7.4 Minimality and Modularity

Example 7.59 The system $LK^+(\{[\text{t}],[\text{c}]\})$ for $\mathbf{C}_{min}$ is given in Figure 7.3.

Axioms: $\psi \Rightarrow \psi$

Rules: Cut, Weakening, and the following logical rules:

$$[\wedge \Rightarrow] \quad \frac{\Gamma, \psi, \varphi \Rightarrow \Delta}{\Gamma, \psi \wedge \varphi \Rightarrow \Delta} \qquad [\Rightarrow \wedge] \quad \frac{\Gamma \Rightarrow \Delta, \psi \quad \Gamma \Rightarrow \Delta, \varphi}{\Gamma \Rightarrow \Delta, \psi \wedge \varphi}$$

$$[\vee \Rightarrow] \quad \frac{\Gamma, \psi \Rightarrow \Delta \quad \Gamma, \varphi \Rightarrow \Delta}{\Gamma, \psi \vee \varphi \Rightarrow \Delta} \qquad [\Rightarrow \vee] \quad \frac{\Gamma \Rightarrow \Delta, \psi, \varphi}{\Gamma \Rightarrow \Delta, \psi \vee \varphi}$$

$$[\supset \Rightarrow] \quad \frac{\Gamma \Rightarrow \psi, \Delta \quad \Gamma, \varphi \Rightarrow \Delta}{\Gamma, \psi \supset \varphi \Rightarrow \Delta} \qquad [\Rightarrow \supset] \quad \frac{\Gamma, \psi \Rightarrow \varphi, \Delta}{\Gamma \Rightarrow \psi \supset \varphi, \Delta}$$

$$[\neg\neg \Rightarrow] \quad \frac{\Gamma, \psi \Rightarrow \Delta}{\Gamma, \neg\neg\psi \Rightarrow \Delta} \qquad [\Rightarrow \neg] \quad \frac{\Gamma, \psi \Rightarrow \Delta}{\Gamma \Rightarrow \neg\psi, \Delta}$$

Figure 7.3: The system $LK^+(\{[\text{t}],[\text{c}]\})$ for $\mathbf{C}_{min}$

Theorem 7.60 (completeness and cut-elimination) *Let $S \subseteq \mathbf{A}_{PAC}$.*

1. $\mathcal{T} \vdash_{LK^+(S)} \psi$ *iff* $\mathcal{T} \vdash_{\mathcal{M}_4^b(S)} \psi$, *and* $\mathcal{T} \vdash_{LK^\mathsf{F}(S)} \psi$ *iff* $\mathcal{T} \vdash_{\mathcal{M}_{4F}^B(S)}$.

2. $LK^+(S)$ *and* $LK^\mathsf{F}(S)$ *admit cut-elimination.*

Proof. We do first the case of $LK^+(S)$.

It is straightforward to show the soundness of $LK^+(S)$ for $\mathcal{M}_4^b(S)$. Using Theorem 6.18, we prove its completeness for $\mathcal{M}_4^b(S)$ and the cut-elimination theorem for it together, by showing that any sequent in $\mathcal{L}_{LC}$ either has a cut-free proof in $LK^+(S)$, or it is not valid in $\mathcal{M}_4^b(S)$.

So, suppose that $\Gamma_0 \Rightarrow \Delta_0$ has no cut-free proof in $LK^+(S)$. Then $\Gamma_0 \Rightarrow \Delta_0$ can obviously be extended to a *saturated* sequent $\Gamma \Rightarrow \Delta$ that still has no cut-free proof in $LK^+(S)$, where we call $\Gamma \Rightarrow \Delta$ saturated (with respect to S) if it is closed under the rules of $LK^+(S)$ applied backwards. This means that $\Gamma \Rightarrow \Delta$ satisfies the following conditions:

- If $[\text{t}] \in S$ and $\neg\varphi \in \Delta$, then $\varphi \in \Gamma$.

- If $[\text{c}] \in S$ and $\neg\neg\varphi \in \Gamma$, then $\varphi \in \Gamma$.

- If $[e] \in S$ and $\neg\neg\varphi \in \Delta$, then $\varphi \in \Delta$.
- If $\varphi \wedge \psi \in \Gamma$, then $\varphi \in \Gamma$ and $\psi \in \Gamma$.
- If $\varphi \wedge \psi \in \Delta$, then $\varphi \in \Delta$ or $\psi \in \Delta$.
- If $[n_\wedge^l] \in S$ and $\neg(\varphi \wedge \psi) \in \Gamma$, then either $\neg\varphi \in \Gamma$ or $\neg\psi \in \Gamma$.
- If $[n_\wedge^r]_1 \in S$ and $\neg(\varphi \wedge \psi) \in \Delta$, then $\neg\varphi \in \Delta$.
- If $[n_\wedge^r]_2 \in S$ and $\neg(\varphi \wedge \psi) \in \Delta$, then $\neg\psi \in \Delta$.
- Similar conditions concerning the rules for $\vee$ and $\supset$.

Now we recursively define a valuation ν in $\mathcal{M}_4^b(S)$ which refutes $\Gamma_0 \Rightarrow \Delta_0$.

- If φ is a propositional variable:

$$\nu(\varphi) = \begin{cases} \top & \varphi \in \Gamma \text{ and } \neg\varphi \in \Gamma, \\ \bot & \varphi \in \Delta \text{ and } \neg\varphi \in \Delta, \\ t & (\varphi \in \Gamma \text{ and } \neg\varphi \notin \Gamma) \text{ or } (\varphi \notin \Delta \text{ and } \neg\varphi \in \Delta), \\ f & \text{otherwise.} \end{cases}$$

- If $\varphi = \diamond(\varphi_1, \ldots, \varphi_n)$ where $\diamond \in \{\neg, \wedge, \vee, \supset\}$ and n is the arity of $\diamond$:

$$\nu(\varphi) = \begin{cases} a & \tilde{\diamond}(\nu(\varphi_1), \ldots, \nu(\varphi_n)) = \{a\}, \\ \top & \tilde{\diamond}(\nu(\varphi_1), \ldots, \nu(\varphi_n)) = \{t, \top\} \text{ and } \neg\varphi \in \Gamma, \\ t & \tilde{\diamond}(\nu(\varphi_1), \ldots, \nu(\varphi_n)) = \{t, \top\} \text{ and } \neg\varphi \notin \Gamma, \\ \bot & \tilde{\diamond}(\nu(\varphi_1), \ldots, \nu(\varphi_n)) = \{f, \bot\} \text{ and } \neg\varphi \in \Delta, \\ f & \tilde{\diamond}(\nu(\varphi_1), \ldots, \nu(\varphi_n)) = \{f, \bot\} \text{ and } \neg\varphi \notin \Delta. \end{cases}$$

Obviously, ν is well-defined valuation in $\mathcal{M}_4^b(S)$. (In the case of variables this follows from the fact that $\Gamma \cap \Delta = \emptyset$, since $\Gamma \Rightarrow \Delta$ has no cut-free proof.) Now we prove that ν has the following four properties:

1. If $\varphi \in \Gamma$ then $\nu(\varphi) \in \{t, \top\}$.
2. If $\neg\varphi \in \Gamma$ then $\nu(\varphi) \in \{f, \top\}$.
3. If $\varphi \in \Delta$ then $\nu(\varphi) \in \{f, \bot\}$.
4. If $\neg\varphi \in \Delta$ then $\nu(\varphi) \in \{t, \bot\}$.

The proofs of 1-4 is by simultaneous induction on the structure of φ.

7.4 Minimality and Modularity 217

1. Suppose that $\varphi \in \Gamma$. Then $\varphi \notin \Delta$. We show that $\nu(\varphi) \in \{t, \top\}$.

   - When φ is a variable, this is immediate from the definition of ν.
   - If $\varphi = \neg\psi$ then $\nu(\psi) \in \{f, \top\}$ by the induction hypothesis for ψ (Property 2). Hence, $\nu(\varphi) \in \{t, \top\}$.
   - Suppose $\varphi = \varphi_1 \wedge \varphi_2$. Since $\Gamma \Rightarrow \Delta$ is saturated, also $\varphi_1 \in \Gamma$ and $\varphi_2 \in \Gamma$. By the induction hypothesis for φ_1 and φ_2, this implies that $\nu(\varphi_1) \in \{t, \top\}$ and $\nu(\varphi_2) \in \{t, \top\}$. Hence, $\nu(\varphi) \in \{t, \top\}$.
   - Suppose $\varphi = \varphi_1 \supset \varphi_2$. Since $\Gamma \Rightarrow \Delta$ is saturated, this implies that either $\varphi_2 \in \Gamma$ or $\varphi_1 \in \Delta$. It follows by the induction hypothesis for φ_1 and φ_2 that either $\nu(\varphi_2) \in \{t, \top\}$, or $\nu(\varphi_1) \in \{f, \bot\}$. In either case, $\nu(\varphi) \in \{t, \top\}$.
   - The cases where $\varphi = \varphi_1 \vee \varphi_2$ is similar to the previous one, and is left for the reader.

2. Suppose that $\neg\varphi \in \Gamma$. Then $\neg\varphi \notin \Delta$. We show that $\nu(\varphi) \in \{f, \top\}$.

   - When φ is a variable, this is immediate from the definition of ν.
   - Suppose $\varphi = \diamond(\varphi_1, \ldots, \varphi_n)$. We show that $\nu(\varphi) \notin \{t, \bot\}$.
     - $\nu(\varphi) \neq t$.
       By the definition of ν, the fact that $\neg\varphi \in \Gamma$ implies that $\nu(\varphi)$ can be t only if $\tilde{\diamond}(\nu(\varphi_1), \ldots, \nu(\varphi_n)) = \{t\}$. We check all the cases in which this might happens, and show that none of them is applicable here.
       * Suppose $\varphi = \neg\psi$, [c] $\in$ S, and $\nu(\psi) = f$. Since $\Gamma \Rightarrow \Delta$ is saturated, the facts that $\neg\varphi \in \Gamma$ and [c] $\in$ S imply that $\psi \in \Gamma$. Hence $\nu(\psi) \neq f$ by the induction hypothesis for ψ. A contradiction.
       * Suppose $\varphi = \varphi_1 \wedge \varphi_2$, $[\mathsf{n}^!_\wedge] \in$ S, and $\nu(\varphi_1) = \nu(\varphi_2) = t$. Since $\Gamma \Rightarrow \Delta$ is saturated, the facts that $\neg\varphi \in \Gamma$ and $[\mathsf{n}^!_\wedge] \in$ S imply that either $\neg\varphi_1 \in \Gamma$ or $\neg\varphi_2 \in \Gamma$. By the induction hypothesis for φ_1 and φ_2, it follows that either $\nu(\varphi_1) \neq t$ or $\nu(\varphi_2) \neq t$. A contradiction.
       * Suppose $\varphi = \varphi_1 \vee \varphi_2$, $[\mathsf{n}^!_\vee]_1) \in$ S, and either $\nu(\varphi_1) = t$, or $\nu(\varphi_1) = \bot$ and $\nu(\varphi_2) \in \{t, \top\}$. Since $\Gamma \Rightarrow \Delta$ is saturated, the facts that $\neg\varphi \in \Gamma$ and $[\mathsf{n}^!_\vee]_1) \in S$ imply that $\neg\varphi_1 \in \Gamma$. By the induction hypothesis for φ_1, it follows that $\nu(\varphi_1) \notin \{t, \bot\}$. A contradiction.
       * Similar analysis applies in the remaining three cases that we need to check here (connected with the rules $[\mathsf{n}^!_\vee]_2$, $[\mathsf{n}^!_\supset]_1$, and $[\mathsf{n}^!_\supset]_2$).

- $\nu(\varphi) \neq \bot$.
  Since $\neg\varphi \in \Gamma$, $\neg\varphi \notin \Delta$. By the definition of ν, this implies that $\nu(\varphi)$ can be $\bot$ only if $\tilde{\diamond}(\nu(\varphi_1),\ldots,\nu(\varphi_n)) = \{\bot\}$. This is of course impossible in case $[t] \in S$. Otherwise the analysis is very similar to that made in showing that $\nu(\varphi) \neq t$. Details are left to the reader.

3. The proof of Property 3 is similar (and dual) to that of Property 1. Details are left to the reader.

4. Suppose that $\neg\varphi \in \Delta$. We show that $\nu(\varphi) \in \{t, \bot\}$.
   - When φ is a variable, this is immediate from the definition of ν.
   - Suppose $\varphi = \diamond(\varphi_1,\ldots,\varphi_n)$. We show that $\nu(\varphi) \notin \{f, \top\}$.
     - $\nu(\varphi) \neq \top$.
       The proof of this is similar to the proof that $\nu(\varphi) \neq \bot$ in case $\neg\varphi \in \Gamma$. (See the proof of property 2.) We leave the analysis to the reader.
     - $\nu(\varphi) \neq f$.
       If $[t] \notin S$ then the proof of this is similar to the proof that $\nu(\varphi) \neq t$ in case $\neg\varphi \in \Gamma$. If $[t] \in S$ then the fact that $\neg\varphi \in \Delta$ implies (since $\Gamma \Rightarrow \Delta$ is saturated) that $\varphi \in \Gamma$. Hence, $\nu(\varphi) \neq f$ by property 1 of φ (that we have already proved using the induction hypothesis).

Since $\Gamma_0 \subseteq \Gamma$ and $\Delta_0 \subseteq \Delta$, it follows from the above four properties of ν that it is indeed a refutation of $\Gamma_0 \Rightarrow \Delta_0$.

The proof in the case of $LK^F(F)$ is very similar to that for $LK^+(S)$. We only have to add the following clause to the definition of ν:

$$\nu(F) = \begin{cases} \bot & \neg\varphi \in \Delta, \\ f & \text{otherwise.} \end{cases}$$

Then we have to show two things: (i) $\nu(F) \in \{f, \bot\}$, and $\nu(F) = f$ in case $[t] \in S$; (ii) Properties 1.–4. above obtain now also in the case where $\varphi = F$. For this, note first that since $\Gamma \Rightarrow \Delta$ has no cut free proof, necessarily $F \notin \Gamma$, and if $[t] \in S$ also $\neg F \notin \Delta$. (i) and (ii) easily follow from these two facts and the definition of $\nu(F)$. □

Corollary 7.61 *If $S \subseteq A_{PAC}$ then $\mathcal{T} \vdash_{LK^+(S)} \psi$ iff $\mathcal{T} \vdash_{HCL^+(S)} \psi$, and $\mathcal{T} \vdash_{LK^F(S)} \psi$ iff $\mathcal{T} \vdash_{HCL^F(S)} \psi$.*

Proof. Immediate from Theorem 7.60 and 7.50.[5] □

---

[5]Alternatively, a purely syntactic proof can be given which is similar to the proof of Theorem 4.100 (in which the case $S = A_{PAC}$ is shown).

7.5 Processing Information from a Set of Sources

In the previous section we studied the effect of different negation axioms. Accordingly, our starting point was the most general intuitive interpretation of negation. In this section we turn to one important concrete application of the basic four truth-values: their use for solving the problem of collecting and processing information from a set of sources. This problem has already been tackled in Section 5.2, where Dunn-Belnap's *matrix* $\mathcal{FOUR}$ has been used for this task. However, as explained in Section 5.2.1, in Belnap's model, on which this use is based, the sources are allowed to provide information only about *atomic formulas*. In other words: it is adequate for the case in which the sources are simple relational databases. However, it does not capture all the situations encountered in practice. In particular, knowledge bases and disjunctive databases can provide information also about *complex* formulas. We show that the framework of Nmatrices allows for a more general framework, in which the more complex situations can also be handled.

In the framework described below, a source may provide the processor with information (in the form of a truth value from $\{0, 1, i\}$, where i means "don't know") about *arbitrary* formulas over $\{\neg, \vee, \wedge\}$. The assignment of subsets of $\{0, 1\}$ to formulas is carried out in two stages. In the first stage, the processor follows Belnap's model by *initially* including a value $x \in \{0, 1\}$ in the value of a formula ψ iff some source assigns x to ψ. In the second stage, the processor expands the subset of $\{0, 1\}$ which was initially assigned to a formula in order to close it under certain *coherence* constraints. Here the crucial assumption we make is that the final assignment ν developed by the processor should include everything that can be derived from the classical truth tables without assuming consistency or full knowledge. Practically this means that we use Conditions **db1**–**db4** from Section 5.2.1 (together with the analogs for $\wedge$ of **db3** and **db4**), except that Condition **db3** (and its analog for $\wedge$) should be weakened, so that only the "if" part is retained (but not the "only if").

Now, the point of introducing this general framework is to allow situations in which a source may know (for example) that a certain disjunction holds, without knowing which of the disjuncts is the true one. This does not mean, of course, that a reasonable source may arbitrarily assign truth-values to formulas. This time we should expect it to be able to derive everything that can be derived from the classical truth tables without assuming full knowledge (but a reasonable source should be consistent!). This means, e.g., that if either φ or ψ is assigned 1, then so should be $\varphi \vee \psi$, and also that $\varphi \vee \psi$ may be assigned 1 only if either φ or ψ is assigned 1, or if both of them are assigned i.

7.5.1 The logic EIP

We turn to corresponding formal definitions. *At the rest of this section* $\mathcal{L} = \{\neg, \vee, \wedge\}$, *and* $\mathcal{W} = \mathcal{W}(\mathcal{L})$.

Definition 7.62 (source valuation) A *source valuation* is a function $s : \mathcal{W} \Rightarrow \{0, 1, i\}$ for which the following hold:

(s1) $s(\neg\varphi) = 0$ iff $s(\varphi) = 1$;

(s2) $s(\neg\varphi) = 1$ iff $s(\varphi) = 0$;

(s3) If $s(\varphi) = 1$ or $s(\psi) = 1$ then $s(\varphi \vee \psi) = 1$;

(s4) If $s(\varphi \vee \psi) = 1$ and $s(\varphi) = 0$ then $s(\psi) = 1$;

(s5) If $s(\varphi \vee \psi) = 1$ and $s(\psi) = 0$ then $s(\varphi) = 1$;

(s6) $s(\varphi \vee \psi) = 0$ iff $s(\varphi) = 0$ and $s(\psi) = 0$;

(s7) $s(\varphi \wedge \psi) = 1$ iff $s(\varphi) = 1$ and $s(\psi) = 1$;

(s8) If $s(\varphi) = 0$ or $s(\psi) = 0$ then $s(\varphi \wedge \psi) = 0$;

(s9) If $s(\varphi \wedge \psi) = 0$ and $s(\psi) = 1$ then $s(\varphi) = 0$;

(s10) If $s(\varphi \wedge \psi) = 0$ and $s(\varphi) = 1$ then $s(\psi) = 0$.

Note 7.63 It is not difficult to see that a function $s : \mathcal{W} \Rightarrow \{0, 1, i\}$ is a source valuation iff it is an $\mathcal{M}_r^3$-valuation, where the Nmatrix $\mathcal{M}_r^3 = \langle \mathcal{V}, \mathcal{D}, \mathcal{O} \rangle$ is defined as follows:

- $\mathcal{V} = \{0, 1, i\}$ and $\mathcal{D} = \{1\}$. (So $\mathcal{M}_r^3 \in M3_\perp$.)

- The operations in $\mathcal{O}$ are given by the following tables:

$\tilde{\neg}$	
0	$\{1\}$
1	$\{0\}$
i	$\{i\}$

$\tilde{\vee}$	0	1	i
0	$\{0\}$	$\{1\}$	$\{i\}$
1	$\{1\}$	$\{1\}$	$\{1\}$
i	$\{i\}$	$\{1\}$	$\{i, 1\}$

$\tilde{\wedge}$	0	1	i
0	$\{0\}$	$\{0\}$	$\{0\}$
1	$\{0\}$	$\{1\}$	$\{i\}$
i	$\{0\}$	$\{i\}$	$\{0, i\}$

Definition 7.64 (g_S, d_S, EIP-valuation) Let S be a non-empty set of source valuations.

- g_S, the *existential collecting function of S*, is the function from $\mathcal{W}$ to $\mathcal{P}(\{0, 1\})$ defined by: $g(\varphi) = \{x \in \{0, 1\} \mid \exists s \in S.\ s(\varphi) = x\}$.

- d_S, *The existential information processing valuation* (EIP-valuation, in short) induced by S, is the function from $\mathcal{W}$ to $\mathcal{P}(\{0, 1\})$ which is inductively defined as follows:

7.5 Processing Information from a Set of Sources

(i) $x \in g_S(p) \Rightarrow x \in d_S(p)$ for any $x \in \{0,1\}$ and any $p \in \mathsf{Var}$.
(ii) $x \in d_S(\varphi) \Rightarrow 1 - x \in d_S(\neg\varphi)$ for any $x \in \{0,1\}$.
(iii) $1 \in d_S(\varphi)$ or $1 \in d_S(\psi)$ or $1 \in g_S(\varphi \vee \psi) \Rightarrow 1 \in d_S(\varphi \vee \psi)$.
(iv) $0 \in d_S(\varphi)$ and $0 \in d_S(\psi) \Rightarrow 0 \in d_S(\varphi \vee \psi)$.
(v) $1 \in d_S(\varphi)$ and $1 \in d_S(\psi) \Rightarrow 1 \in d_S(\varphi \wedge \psi)$.
(vi) $0 \in d_S(\varphi)$ or $0 \in d_S(\psi)$ or $0 \in g_S(\varphi \wedge \psi) \Rightarrow 0 \in d_S(\varphi \wedge \psi)$.

The following proposition characterizes the EIP-valuation induced by S in terms of its main properties.

Proposition 7.65 *Let S be a non-empty set of source valuations. Then d_S is the minimal function d from $\mathcal{W}$ to $\mathcal{P}(\{0,1\})$ (where $h_1 \leq h_2$ if $h_1(\varphi) \subseteq h_2(\varphi)$ for every φ) that satisfies the following conditions for every φ and ψ:*

(d0) $g_S(\varphi) \subseteq d(\varphi)$;

(d1) $0 \in d(\neg\varphi)$ iff $1 \in d(\varphi)$;

(d2) $1 \in d(\neg\varphi)$ iff $0 \in d(\varphi)$;

(d3) $1 \in d(\varphi \vee \psi)$ if $1 \in d(\varphi)$ or $1 \in d(\psi)$;

(d4) $0 \in d(\varphi \vee \psi)$ iff $0 \in d(\varphi)$ and $0 \in d(\psi)$;

(d5) $1 \in d(\varphi \wedge \psi)$ iff $1 \in d(\varphi)$ and $1 \in d(\psi)$;

(d6) $0 \in d(\varphi \wedge \psi)$ if $0 \in d(\varphi)$ or $0 \in d(\psi)$.

Proof. Obviously, every function $d : \mathcal{W} \to \mathcal{P}(\{0,1\})$ which satisfies (d0)–(d6) also satisfies (i)–(vi) from Definition 7.64, and so $d_S \leq d$. It remains to show that d_S itself satisfies (d0)–(d6). That it satisfies (d1)–(d6) (including the 'only if' direction in (d2), (d3), (d4) and (d5)) easily follows from the inductive definition of D_S. We show that (d0) too is satisfied by d_S by induction on the structure of φ.

- The case where $\varphi \in \mathsf{Var}$ directly follows from the definition of d_S.

- Suppose that $\varphi = \neg\psi$, and $x \in g_S(\varphi)$. Then there exists $s \in S$ such that $s(\varphi) = x$. Since s is a source valuation, this means that $s(\psi) = 1 - x$. Hence $1 - x \in g_S(\psi)$, and so $1 - x \in d_S(\psi)$ by the induction hypothesis for ψ. It follows by the definition of d_S that $x \in d_S(\varphi)$.

- Suppose that $\varphi = \psi_1 \vee \psi_2$, and $x \in g_S(\varphi)$.

- If $x = 1$ then $x \in d_S(\varphi)$ by definition.
- If $x = 0$ then there exists $s \in S$ such that $s(\varphi) = 0$. Since s is a source valuation, this implies that $s(\psi_1) = s(\psi_2) = 0$. Hence $0 \in g_S(\psi_1)$ and $0 \in g_S(\psi_2)$. It follows by induction hypothesis for ψ_1 and ψ_2 that $0 \in d_S(\psi_1)$ and $0 \in d_S(\psi_2)$, and so $0 \in d_S(\varphi)$ by the definition of d_S.

- The proof in the case where $\varphi = \psi_1 \wedge \psi_2$ is similar to the proof in the case where $\varphi = \psi_1 \vee \psi_2$. □

Now we turn to the logic that is induced by the set of EIP-valuations.

Definition 7.66 ($\models_{\mathcal{EIP}}$, $\vdash_{\mathcal{EIP}}$, The logic EIP)

- Let S be a non-empty set of source valuations, and let φ be a formula of $\mathcal{L}$. $S \models_{\mathcal{EIP}} \varphi$ if $1 \in d_S(\varphi)$.

- **EIP** is the logic $\langle \mathcal{L}, \vdash_{\mathcal{EIP}} \rangle$, where $\mathcal{T} \vdash_{\mathcal{EIP}} \varphi$ if $S \models_{\mathcal{EIP}} \varphi$ for every non-empty set S of source valuations such that $S \models_{\mathcal{EIP}} \mathcal{T}$.

Note 7.67 It is easy to see that if EIP denotes the set of all EIP-valuations which are induced by some non-empty set S of source valuations, then $\langle \text{EIP}, \models_{\mathcal{EIP}} \rangle$ is a denotational semantics according to Definition 1.37, and $\vdash_{\mathcal{EIP}}$ is the corresponding tcr, as given in Definition 1.38.

Our next goal is to show that the logic **EIP** has a four-valued characteristic Nmatrix. The identity of that Nmatrix is rather obvious, once we recall the identification made in Section 5.2.1 of the four truth-values of $\mathcal{FOUR}$ with the subsets of $\{0, 1\}$: $t = \{1\}$, $f = \{0\}$, $\top = \{0, 1\}$, $\bot = \emptyset$.

Definition 7.68 ($\mathcal{M}_E^4$) $\mathcal{M}_E^4 = \langle \mathcal{V}, \mathcal{D}, \mathcal{O} \rangle$ is the $\neg$-fundamental Nmatrix for the language of $\{\neg, \vee, \wedge\}$, which is defined as follows:

- Like in Dunn-Belnap matrix $\mathcal{FOUR}$, $\mathcal{V} = \{f, \bot, \top, t\}$ and $\mathcal{D} = \{\top, t\}$.

- $\tilde{\neg}$ is again defined like in $\mathcal{FOUR}$ (where the latter is viewed an Nmatrix), that is: $\tilde{\neg} t = \{f\}$, $\tilde{\neg} f = \{t\}$, $\tilde{\neg} \bot = \{\bot\}$, and $\tilde{\neg} \top = \{\top\}$.

- $\tilde{\vee}$ and $\tilde{\wedge}$ are given by the following tables:

$\tilde{\vee}$	f	$\bot$	$\top$	t
f	$\{f, \top\}$	$\{t, \bot\}$	$\{\top\}$	$\{t\}$
$\bot$	$\{t, \bot\}$	$\{t, \bot\}$	$\{t\}$	$\{t\}$
$\top$	$\{\top\}$	$\{t\}$	$\{\top\}$	$\{t\}$
t	$\{t\}$	$\{t\}$	$\{t\}$	$\{t\}$

$\tilde{\wedge}$	f	$\bot$	$\top$	t
f	$\{f\}$	$\{f\}$	$\{f\}$	$\{f\}$
$\bot$	$\{f\}$	$\{f, \bot\}$	$\{f\}$	$\{f, \bot\}$
$\top$	$\{f\}$	$\{f\}$	$\{\top\}$	$\{\top\}$
t	$\{f\}$	$\{f, \bot\}$	$\{\top\}$	$\{t, \top\}$

7.5 Processing Information from a Set of Sources

Proposition 7.69 *A function $d : \mathcal{W} \to \mathcal{P}(\{0,1\})$ satisfies Conditions (d1)–(d6) from Proposition 7.65 iff it is an $\mathcal{M}_E^4$-valuation.*

Proof. A simple check suffices. □

Theorem 7.70 $\mathbf{EIP} = \mathbf{L}_{\mathcal{M}_E^4}$ *(that is: $\mathcal{T} \vdash_{\mathcal{EIP}} \varphi$ iff $\mathcal{T} \vdash_{\mathcal{M}_E^4} \varphi$ for all $\mathcal{T}, \varphi$).*

Proof. Proposition 7.69, Proposition 7.65, Definition 7.66, and the fact that $1 \in d_S(\varphi)$ iff $d_S(\varphi) \in \{t, \top\}$ together imply that if $\mathcal{T} \vdash_{\mathcal{M}_E^4} \varphi$ then $\mathcal{T} \vdash_{\mathcal{EIP}} \varphi$.

For the converse, it obviously suffices to construct for every $\mathcal{M}_E^4$-valuation ν a non-empty set S_ν of source valuations such that $\nu = d_{S_\nu}$. To do this, we first split ν into a family S_ν^* of partial function from $\mathcal{W}$ to $\{0,1\}$. S_ν^* includes the following partial functions:

1. For any variable $p \in \mathsf{Var}$, the following functions s_p^0, s_p^1 on $\{p\}$:

$$s_p^0(p) = \begin{cases} 0 & \text{if } 0 \in \nu(p), \\ i & \text{otherwise.} \end{cases} \qquad s_p^1(p) = \begin{cases} 1 & \text{if } 1 \in \nu(p), \\ i & \text{otherwise.} \end{cases}$$

2. For any formula of the form $\varphi \vee \psi$ such that $1 \in \nu(\varphi \vee \psi)$, the function $s_{\varphi \vee \psi}$ which is defined on the set of subformulas of $\varphi \vee \psi$ by:

$$s_{\varphi \vee \psi}(\sigma) = \begin{cases} 1 & \sigma = \varphi \vee \psi, \\ i & \text{otherwise.} \end{cases}$$

3. For any formula of the form $\varphi \wedge \psi$ such that $0 \in \nu(\varphi \wedge \psi)$, the function $s_{\varphi \wedge \psi}$ which is defined on the set of subformulas of $\varphi \wedge \psi$ by:

$$s_{\varphi \wedge \psi}(\sigma) = \begin{cases} 0 & \sigma = \varphi \wedge \psi, \\ i & \text{otherwise.} \end{cases}$$

The set S_ν is then the set of all functions from $\mathcal{W}$ to $\{0,1\}$ which are obtained from the partial functions in S_ν^* by first assigning the value i to all the variables for which those functions are not defined, and then extending the resulting new partial valuations to all other formulas in $\mathcal{W}$ (for which they are not yet defined) according to the truth tables of the three-valued matrix LP (Section 4.4). It is not difficult to see that all the elements of S_ν are $\mathcal{M}_r^3$-valuations (Note 7.63), and so S_ν is a non-empty set of source valuations. It remains to prove that $d_S(\varphi) = \nu(\varphi)$ for every $\varphi \in \mathcal{W}$. We show this by induction on the structure of φ.

Suppose first that φ is a propositional variable p. Then $d_S(p) = g_S(p)$. Now for any $x \in \{0,1\}$, $x \in g_S(p)$ iff $x = s(p)$ for some $s \in S_\nu$. However, by the definition of the valuations in S_v, $x = s(p)$ for $s \in S_v$ iff $s = s_p^x$ and $x \in \nu(p)$. Thus $g_S(p) = \nu(p)$. It follows that $d_S(p) = \nu(p)$.

Assume now that the equation holds for φ_1 and φ_2. We prove that it holds also for $\neg\varphi_1$, $\varphi_1 \vee \varphi_2$, and $\varphi_1 \wedge \varphi_2$.

$\neg\varphi_1$: Substituting $1-y$ for x in (ii) of Definition 7.64, we conclude that, for any $y \in \{0,1\}$, $y \in d_S(\neg\varphi_1)$ iff $1-y \in d_S(\varphi_1)$. Since by the inductive hypothesis $d_S(\varphi_1) = \nu(\varphi_1)$, this yields $y \in d_S(\neg\varphi_1)$ iff $1-y \in \nu(\varphi_1)$. As ν is an $\mathcal{M}_E^4$-valuation, the latter holds iff $y = 1-(1-y) \in \nu(\neg\varphi_1)$, whence the equation holds for $\neg\varphi_1$.

$\varphi_1 \vee \varphi_2$: By (iv) of Definition 7.64, we have $0 \in d_S(\varphi_1 \vee \varphi_2)$ iff $0 \in d_S(\varphi_1)$ and $0 \in d_S(\varphi_2)$, which, by the inductive hypothesis, holds iff $0 \in \nu(\varphi_1)$ and $0 \in \nu(\varphi_2)$. As ν is an $\mathcal{M}_E^4$-valuation, the latter in turn holds iff $0 \in \nu(\varphi_1 \vee \varphi_2)$.

Assume now $1 \in d_S(\varphi_1 \vee \varphi_2)$. Then, by (iii) of Definition 7.64, either $1 \in d_S(\varphi_1)$ or $1 \in d_S(\varphi_2)$ or $1 \in g_S(\varphi_1 \vee \varphi_2)$. By the inductive hypothesis, the first two cases are equivalent to $1 \in \nu(\varphi_1)$ and $1 \in \nu(\varphi_2)$, respectively, and both of the latter options yield $1 \in \nu(\varphi_1 \vee \varphi_2)$ by the truth table for $\vee$ of $\mathcal{M}_E^4$. Finally, if $1 \in g_S(\varphi_1 \vee \varphi_2)$, then $s(\varphi_1 \vee \varphi_2) = 1$ for some $s \in S_v$. Then one of the following must hold:

1. $s(\varphi_i) = 1$ for some i,
2. $s(\varphi_1) = s(\varphi_2) = \mathrm{i}$ and $s(\varphi_1 \vee \varphi_2) = 1$.

In Case 1, we have $1 \in g_S(\varphi_i)$, whence $1 \in d_S(\varphi_i)$ by (i) of Definition 7.64, and $1 \in \nu(\varphi_i)$ by the inductive hypothesis. Consequently, $1 \in \nu(\varphi_1 \vee \varphi_2)$ by the definition of $\mathcal{M}_E^4$. In Case 2, we must clearly have $s = s_{\varphi_1 \vee \varphi_2}$, and so $1 \in \nu(\varphi_1 \vee \varphi_2)$ by the definition of $s_{\varphi_1 \vee \varphi_2}$.

For the converse, suppose $1 \in \nu(\varphi_1 \vee \varphi_2)$. Then, by the definition of S_v, S_v contains a source $s = s_{\varphi_1 \vee \varphi_2}$ such that $s(\varphi_1 \vee \varphi_2) = 1$. Hence, $1 \in g_S(\varphi_1 \vee \varphi_2)$, whence $1 \in d_S(\varphi_1 \vee \varphi_2)$ by (i) of Definition 7.64.

$\varphi_1 \wedge \varphi_2$: The proof is again analogous to that for disjunction, with $s_{\varphi \vee \psi}$ replaced by $s_{\varphi \wedge \psi}$, (iv) of Definition 7.64 replaced by (v), and (iii) of that definition replaced by (vi). □

Corollary 7.71 *EIP is finitary and decidable.*

Proof. Immediate from Theorems 6.18, 6.22, and 7.70. □

Proposition 7.72 *EIP is semi-normal, $\neg$-contained in classical logic, and boldly paraconsistent.*

7.5 Processing Information from a Set of Sources

Proof. A check of the corresponding truth-table reveals that $\wedge$ is an $\mathcal{M}_E^4$-conjunction. It follows by Theorem 7.70 and Proposition 6.26 that $\wedge$ is a conjunction for **EIP**. Hence **EIP** is semi-normal.

It is also easy to check that $\mathcal{FOUR}$ is a simple refinement of $\mathcal{M}_E^4$. Therefore, it follows from Theorem 7.70, Proposition 6.37, and Proposition 5.26 that **EIP** is $\neg$-contained in classical logic, and boldly paraconsistent. $\square$

Note 7.73 While $\wedge$ is a conjunction for **EIP**, $\vee$ is *not* a disjunction for **EIP**. Indeed, it is easy to see that $\sigma, \varphi \vdash_{\mathcal{EIP}} (\sigma \wedge \varphi) \vee (\sigma \wedge \psi)$ and $\sigma, \psi \vdash_{\mathcal{EIP}} (\sigma \wedge \varphi) \vee (\sigma \wedge \psi)$. However, $\sigma, \varphi \vee \psi \nvdash_{\mathcal{EIP}} (\sigma \wedge \varphi) \vee (\sigma \wedge \psi)$. (Take $\nu(\sigma) = t$, $\nu(\varphi) = \nu(\psi) = \nu(\sigma \wedge \varphi) = \nu(\sigma \wedge \psi) = \nu((\sigma \wedge \varphi) \vee (\sigma \wedge \psi)) = f$, $\nu(\varphi \vee \psi) = \top$.)

Proposition 7.74 **EIP** *is sparse.*

Proof. Suppose φ and ψ are distinct formulas of $\mathcal{L}$. Let q be some propositional variable that does not occur in φ or ψ. Using Proposition 6.11, define an $\mathcal{M}_E^4$-valuation ν such that $\nu(\sigma) \in \{t, f\}$ for every subformula of either φ or ψ, $\nu(\neg \varphi \vee \varphi) = \nu(\neg \psi \vee \psi) = \nu(q) = t$, $\nu((\neg \varphi \vee \varphi) \wedge q) = t$, and $\nu((\neg \psi \vee \psi) \wedge q) = \top$. Since $\mathcal{M}_E^4$ is $\neg$-fundamental, it follows from Proposition 7.5 that $(\neg \varphi \vee \varphi) \wedge q \not\equiv_{\mathbf{L}_{\mathcal{M}_E^4}} (\neg \psi \vee \psi) \wedge q$. Hence $\varphi \not\equiv_{\mathbf{L}_{\mathcal{M}_E^4}} \psi$. Therefore it follows from Theorem 7.70 that **EIP** is sparse. $\square$

Next, we prove that the use of a finite non-deterministic matrix for capturing **EIP** cannot be replaced by using finite ordinary matrices.

Theorem 7.75 *There is no finite family $\mathcal{F}$ of finite ordinary matrices such that $\vdash_{\mathcal{EIP}} = \vdash_{\mathcal{F}}$. In particular:* **EIP** *has no finite characteristic matrix.*

Proof. Since $\mathcal{M}_E^4$ is $\neg$-fundamental and strictly proper, this follows from Theorem 7.4 and Theorem 7.70. $\square$

Note 7.76 There is another significant difference between Belnap's model of information processing (as described in Section 5.2) and the one presented here. It is not difficult to show (see [63]) that the same matrix ($\mathcal{FOUR}$) and logic is obtained if we allow a processor only to process information that comes from at most two different Belnap's sources. The situation is completely different with respect to the type of sources considered here: Let $\vdash_{\mathcal{EIP}_{\leq n}}$ be defined similarly to $\vdash_{\mathcal{EIP}}$ (Definition 7.66), except that only sets of source valuations with at most n elements are allowed. Then it is proved in [63] that $\vdash_{\mathcal{EIP}} \neq \vdash_{\mathcal{EIP}_{\leq n}}$ for every n, and $\vdash_{\mathcal{EIP}_{\leq n}} \neq \vdash_{\mathcal{EIP}_{\leq k}}$ for $n \neq k$.

The next proposition lists some problems with the expressive power of the language of **EIP**.

Proposition 7.77

1. **EIP** *has no tautologies.*

2. **EIP** *is not normal. In particular: it has no implication.*

3. *There is no definable unary connective $\diamond$ of $\mathcal{L}$ such that for every assignment ν in $\mathcal{M}_E^4$ and for every formula φ: $\nu \models \diamond\varphi$ iff $\nu \not\models \varphi$.*

Proof.

1. This is immediate from the fact that by letting $\nu_\bot(\varphi) = \bot$ for every formula φ we get an $\mathcal{M}_E^4$-assignment.

2. This follows from the first item, since every logic with an implication $\supset$ has tautologies (like $\varphi \supset \varphi$).

3. This follows from the fact that by letting $\nu_\top(\varphi) = \top$ for every formula φ we get an $\mathcal{M}_E^4$-assignment. □

7.5.2 A Gentzen-type System for EIP

In view of Proposition 7.77 and Note 7.73, finding an appropriate Gentzen-type system for **EIP** is a particularly important task:

- Since **EIP** has no tautologies, it has no decent Hilbert-type system.

- The use of sequents partially compensates for the lack of implication in $\mathcal{L}$ for **EIP**, since sequents provide non-nestable ("first-degree", in the terminology of [4]) version of entailment.

- The expressive power of *formulas* of $\mathcal{L}$ is too weak. Thus the third item of Proposition 7.77 means that there is no way to express that a certain formula φ is *not true* (meaning that $1 \notin d_S(\varphi)$). Similarly, there is no way to express *disjunctive knowledge* of the form "one of the sentences φ and ψ is known to be true" (meaning that either $1 \in d_S(\varphi)$ or $1 \in d_S(\psi)$), because it is possible that $1 \in d_S(\varphi \vee \psi)$ but neither $1 \in d_S(\varphi)$ nor $1 \in d_S(\psi)$. With the help of our next definition (which actually contains two instances of Definition 1.123), these problems can partially be overcome by using sequents.

Definition 7.78

- Let $s = \Gamma \Rightarrow \Delta$ be a sequent in $\mathcal{L}$.
    - If S is a non-empty set of source valuations then $S \models_{EIP} s$ if either $1 \notin d_S(\varphi)$ for some $\varphi \in \Gamma$, or $1 \in d_S(\psi)$ for some $\psi \in \Delta$.

7.5 Processing Information from a Set of Sources

- If ν is an assignment in $\mathcal{M}_E^4$ then $\nu \models_{\mathcal{M}_E^4} s$ if either $\nu(\varphi) \in \{f, \bot\}$ for some $\varphi \in \Gamma$, or $\nu(\psi) \in \{t, \top\}$ for some $\psi \in \Delta$.

- Let $\mathcal{S} \cup \{s\}$ be a set of sequents.
  - $\mathcal{S} \vdash_{\mathcal{EIP}} s$ if $S \models_{EIP} s$ for every non-empty set of source valuations S such that $S \models_{EIP} \mathcal{S}$.
  - $\mathcal{S} \vdash_{\mathcal{M}_E^4} s$ if $\nu \models_{\mathcal{M}_E^4} s$ for every assignment ν in $\mathcal{M}_E^4$ such that $\nu \models_{\mathcal{M}_E^4} \mathcal{S}$.

Proposition 7.79 *Let $\mathcal{S} \cup \{s\}$ be a set of sequents. $\mathcal{S} \vdash_{\mathcal{EIP}} s$ if $\mathcal{S} \vdash_{\mathcal{M}_E^4} s$.*

Proof. Immediate from Theorem 7.70 and its proof. □

Definition 7.80 ($\mathcal{G}_{\mathbf{EIP}}$) $\mathcal{G}_{\mathbf{EIP}}$ is the Gentzen-type system which is obtained from $\mathcal{G}_{\mathbf{L}_{\mathcal{FOUR}}}$ (Definition 5.61) by deleting $[\vee \Rightarrow]$ and $[\neg \wedge \Rightarrow]$. (For the reader's convenience, $\mathcal{G}_{\mathbf{EIP}}$ is presented in Figure 7.4 below.)

Axioms: $\psi \Rightarrow \psi$

Rules: Cut, Weakening, and the following logical rules:

$$[\wedge \Rightarrow] \; \frac{\Gamma, \psi, \varphi \Rightarrow \Delta}{\Gamma, \psi \wedge \varphi \Rightarrow \Delta} \qquad [\Rightarrow \wedge] \; \frac{\Gamma \Rightarrow \Delta, \psi \quad \Gamma \Rightarrow \Delta, \varphi}{\Gamma \Rightarrow \Delta, \psi \wedge \varphi}$$

$$[\Rightarrow \vee] \; \frac{\Gamma \Rightarrow \Delta, \psi, \varphi}{\Gamma \Rightarrow \Delta, \psi \vee \varphi}$$

$$[\neg\neg \Rightarrow] \; \frac{\Gamma, \varphi \Rightarrow \Delta}{\Gamma, \neg\neg\varphi \Rightarrow \Delta} \qquad [\Rightarrow \neg\neg] \; \frac{\Gamma \Rightarrow \Delta, \varphi}{\Gamma \Rightarrow \Delta, \neg\neg\varphi}$$

$$[\Rightarrow \neg\wedge] \; \frac{\Gamma \Rightarrow \Delta, \neg\varphi, \neg\psi}{\Gamma \Rightarrow \Delta, \neg(\varphi \wedge \psi)}$$

$$[\neg\vee \Rightarrow] \; \frac{\Gamma, \neg\varphi, \neg\psi \Rightarrow \Delta}{\Gamma, \neg(\varphi \vee \psi) \Rightarrow \Delta} \qquad [\Rightarrow \neg\vee] \; \frac{\Gamma \Rightarrow \Delta, \neg\varphi \quad \Gamma \Rightarrow \Delta, \neg\psi}{\Gamma \Rightarrow \Delta, \neg(\varphi \vee \psi)}$$

Figure 7.4: The Gentzen-type system $\mathcal{G}_{\mathbf{EIP}}$

Next, we show the adequacy of $\mathcal{G}_{\mathbf{EIP}}$ for **EIP**. Note that because of the use noted above of sequents to enhance the expressive power of $\mathcal{L}$, we cannot be satisfied with just ordinary soundness and completeness of $\mathcal{G}_{\mathbf{EIP}}$, but we need *strong* soundness and completeness (Definition 1.125).

Notation. $\mathcal{S} \vdash^{cf}_{G_{\mathbf{EIP}}} s$ means that there is a proof in $G_{\mathbf{EIP}}$ of the sequent s from the set of sequents $\mathcal{S}$ in which each cut is on a formula φ such that $\varphi \in \Gamma \cup \Delta$ for some sequent $\Gamma \Rightarrow \Delta$ in $\mathcal{S}$.

Proposition 7.81 (strong soundness) *Let $\mathcal{S} \cup \{s\}$ be a set of sequents of $\mathcal{L}$. If $\mathcal{S} \vdash_{G_{\mathbf{EIP}}} s$ then $\mathcal{S} \vdash_{\mathcal{M}^4_E} s$.*

Proof. It is straightforward to check that if ν is an assignment in $\mathcal{M}^4_E$ that satisfies all the premises of some rule of $G_{\mathbf{EIP}}$, then ν satisfies also the conclusion of that rule. This easily implies the proposition. □

Proposition 7.82 (strong completeness) *Let $\mathcal{S} \cup \{s\}$ be a set of sequents of $\mathcal{L}$. If $\mathcal{S} \vdash_{\mathcal{M}^4_E} s$ then $\mathcal{S} \vdash^{cf}_{G_{\mathbf{EIP}}} s$.*

Proof. Suppose that $\mathcal{S} \not\vdash^{cf}_{G_{\mathbf{EIP}}} s$. We show that $\mathcal{S} \not\vdash_{\mathcal{M}^4_E} s$ by constructing a model ν of $\mathcal{S}$ in $\mathcal{M}^4_E$ which is not a model of s.

Suppose that $s = \Gamma_0 \Rightarrow \Delta_0$. Let $\varphi_1, \varphi_2, \ldots$ be an enumeration of the formulas of $\mathcal{L}$. Starting with $\Gamma_0 \Rightarrow \Delta_0$, we inductively define a sequence of sequents $\Gamma_n \Rightarrow \Delta_n$ ($n = 1, 2, \ldots$) as follows:

- If $\mathcal{S} \not\vdash^{cf}_{G_{\mathbf{EIP}}} \varphi_n, \Gamma_{n-1} \Rightarrow \Delta_{n-1}$ then $\Gamma_n = \Gamma_{n-1} \cup \{\varphi_n\}$, $\Delta_n = \Delta_{n-1}$.
- If $\mathcal{S} \vdash^{cf}_{G_{\mathbf{EIP}}} \varphi_n, \Gamma_{n-1} \Rightarrow \Delta_{n-1}$ while $\mathcal{S} \not\vdash^{cf}_{G_{\mathbf{EIP}}} \Gamma_{n-1} \Rightarrow \Delta_{n-1}, \varphi_n$ then $\Gamma_n = \Gamma_{n-1}$, $\Delta_n = \Delta_{n-1} \cup \{\varphi_n\}$.
- Otherwise $\Gamma_n = \Gamma_{n-1}$ and $\Delta_n = \Delta_{n-1}$.

The assumption that $\mathcal{S} \not\vdash^{cf}_{G_{\mathbf{EIP}}} s$ and the way the above sequence is defined imply that the following hold for every $n \geq 0$:

(i) $\Gamma_n \subseteq \Gamma_{n+1}$ and $\Delta_n \subseteq \Delta_{n+1}$

(ii) $\mathcal{S} \not\vdash^{cf}_{G_{\mathbf{EIP}}} \Gamma_n \Rightarrow \Delta_n$.

Let $\mathcal{T}_l = \bigcup_{n=0}^{\infty} \Gamma_n$, $\mathcal{T}_r = \bigcup_{n=0}^{\infty} \Delta_n$. We show that $\mathcal{T}_l$ and $\mathcal{T}_r$ have the following properties

1. $\Gamma_0 \subseteq \mathcal{T}_l$, and $\Delta_0 \subseteq \mathcal{T}_r$.

2. If $\Gamma \subseteq \mathcal{T}_l$, $\Delta \subseteq \mathcal{T}_r$, and Γ, Δ are finite, then $\mathcal{S} \not\vdash^{cf}_{G_{\mathbf{EIP}}} \Gamma \Rightarrow \Delta$.

3. If $\psi \notin \Gamma$ then there exist $\Gamma \subseteq \mathcal{T}_l$, $\Delta \subseteq \mathcal{T}_r$ such that $\mathcal{S} \vdash^{cf}_{G_{\mathbf{EIP}}} \psi, \Gamma \Rightarrow \Delta$.

4. If $\psi \notin \Delta$ then there exist $\Gamma \subseteq \mathcal{T}_l$, $\Delta \subseteq \mathcal{T}_r$ such that $\mathcal{S} \vdash^{cf}_{G_{\mathbf{EIP}}} \Gamma \Rightarrow \Delta, \psi$.

7.5 Processing Information from a Set of Sources

5. If $\Gamma \Rightarrow \Delta \in \mathcal{S}$, then $\Gamma \cup \Delta \subseteq \mathcal{T}_l \cup \mathcal{T}_r$.

6. $\mathcal{T}_l$ and $\mathcal{T}_r$ are closed under the logical rules of $G_{\mathbf{EIP}}$ applied backward:

   (a) If $\neg\neg\psi \in \mathcal{T}_l$, then $\psi \in \mathcal{T}_l$.
   (b) If $\neg\neg\psi \in \mathcal{T}_r$, then $\psi \in \mathcal{T}_r$.
   (c) If $\varphi \wedge \psi \in \mathcal{T}_l$, then $\varphi \in \mathcal{T}_l$ and $\psi \in \mathcal{T}_l$.
   (d) If $\varphi \wedge \psi \in \mathcal{T}_r$, then $\varphi \in \mathcal{T}_r$ or $\psi \in \mathcal{T}_r$.
   (e) If $\neg(\varphi \wedge \psi) \in \mathcal{T}_r$, then $\neg\varphi \in \mathcal{T}_r$ and $\neg\psi \in \mathcal{T}_r$.
   (f) If $\neg(\varphi \vee \psi) \in \mathcal{T}_l$, then $\neg\varphi \in \mathcal{T}_l$ and $\neg\psi \in \mathcal{T}_l$.
   (g) If $\varphi \vee \psi \in \mathcal{T}_r$, then $\varphi \in \mathcal{T}_r$ and $\psi \in \mathcal{T}_r$.
   (h) If $\neg(\varphi \vee \psi) \in \mathcal{T}_r$, then $\neg\varphi \in \mathcal{T}_r$ or $\neg\psi \in \mathcal{T}_r$.

Property 1 is trivial. Property 2 follows from (ii) above, because if Γ and Δ are finite subsets of $\mathcal{T}_l$ and $\mathcal{T}_r$ (respectively) then it follows from (i) that there exists n such that $\Gamma \subseteq \Gamma_n$ and $\Delta \subseteq \Delta_n$.

To prove Property 4, assume that $\psi \notin \Delta$. Suppose that $\psi = \varphi_n$ (where $n \geq 1$). Then $\psi \notin \Delta_n$. By the construction of $\Gamma_n \Rightarrow \Delta_n$, this implies that either $\psi \in \Gamma_n$ (and so $\psi \in \mathcal{T}_l$), or $\mathcal{S} \vdash^{cf}_{G_{\mathbf{EIP}}} \Gamma_{n-1} \Rightarrow \Delta_{n-1}, \psi$. In the first case we take $\Gamma = \{\psi\}$, $\Delta = \emptyset$. In the second case we take $\Gamma = \Gamma_{n-1}$, $\Delta = \Delta_{n-1}$.

The proof of Property 3 is similar to the proof of Property 4 (but is simpler).

To prove Property 5, assume that $\psi \in \Gamma \cup \Delta$ for some $\Gamma \Rightarrow \Delta \in \mathcal{S}$, but $\psi \notin \mathcal{T}_l \cup \mathcal{T}_r$. By properties 3 and 4 this implies that there are finite $\Gamma_1 \subseteq \mathcal{T}_l$, $\Gamma_2 \subseteq \mathcal{T}_l$, $\Delta_1 \subseteq \mathcal{T}_r$, and $\Delta_2 \subseteq \mathcal{T}_r$ such that $\mathcal{S} \vdash^{cf}_{G_{\mathbf{EIP}}} \Gamma_1 \Rightarrow \Delta_1, \psi$ and $\mathcal{S} \vdash^{cf}_{G_{\mathbf{EIP}}} \psi, \Gamma_2 \Rightarrow \Delta_2$. It follows that $\mathcal{S} \vdash^{cf}_{G_{\mathbf{EIP}}} \Gamma_1, \Gamma_2 \Rightarrow \Delta_1, \Delta_2$, using weakenings and an allowed cut on ψ. This contradicts Property 2.

The proofs of the various parts of Property 6 are similar. We do the last one, (h), as an example. So suppose that $\neg(\varphi \vee \psi) \in \mathcal{T}_r$, but neither $\neg\varphi \in \mathcal{T}_r$ nor $\neg\psi \in \mathcal{T}_r$. By Property 4 this implies that there are finite $\Gamma_1 \subseteq \mathcal{T}_l$, $\Gamma_2 \subseteq \mathcal{T}_l$, $\Delta_1 \subseteq \mathcal{T}_r$, and $\Delta_2 \subseteq \mathcal{T}_r$ such that $\mathcal{S} \vdash^{cf}_{G_{\mathbf{EIP}}} \Gamma_1 \Rightarrow \Delta_1, \neg\varphi$ and $\mathcal{S} \vdash^{cf}_{G_{\mathbf{EIP}}} \Gamma_2 \Rightarrow \Delta_2, \neg\psi$. But then $\mathcal{S} \vdash^{cf}_{G_{\mathbf{EIP}}} \Gamma_1, \Gamma_2 \Rightarrow \Delta_1, \Delta_2, \neg(\varphi \vee \psi)$ (using weakenings and an application of $[\Rightarrow \neg\vee]$). This contradicts Property 2.

Using $\mathcal{T}_l$, $\mathcal{T}_r$, and the identification of the truth-values of $\mathcal{M}^4_E$ with the subsets of $\{0,1\}$, we now inductively define the desired ν as follows:

(v0) For a variable p, $1 \in \nu(p)$ iff $p \in \mathcal{T}_l$, $0 \in \nu(p)$ iff $\neg p \in \mathcal{T}_l$;

(v1) $1 \in \nu(\neg\varphi)$ iff $0 \in \nu(\varphi)$;

(v2) $0 \in \nu(\neg\varphi)$ iff $1 \in \nu(\varphi)$;

(v3) $1 \in \nu(\varphi \vee \psi)$ iff $1 \in \nu(\varphi)$ or $1 \in \nu(\psi)$ or $(\varphi \vee \psi) \in \mathcal{T}_l$;

(v4) $0 \in \nu(\varphi \vee \psi)$ iff $0 \in \nu(\varphi)$ and $0 \in \nu(\psi)$;

(v5) $1 \in \nu(\varphi \wedge \psi)$ iff $1 \in \nu(\varphi)$ and $1 \in \nu(\psi)$;

(v6) $0 \in \nu(\varphi \wedge \psi)$ iff $0 \in \nu(\varphi)$ or $0 \in \nu(\psi)$ or $\neg(\varphi \wedge \psi) \in \mathcal{T}_l$;

From Proposition 7.69 and (v1)–(v6) it follows that ν is indeed an $\mathcal{M}_E^4$-assignment. Next we show that it has the following two properties:

(A) $1 \in \nu(\psi)$ for every $\psi \in \mathcal{T}_l$ (B) $1 \notin \nu(\psi)$ for every $\psi \in \mathcal{T}_r$

We argue by induction on the complexity of ψ.

- To prove (A), assume that $\psi \in \mathcal{T}_l$.

    - Assume that ψ is a variable. Then $1 \in \nu(\psi)$ by (v0).
    - Assume that $\psi = \neg\psi'$. There are four subcases to consider:

        $\psi' = p$ (where p is a variable): Then $\neg p = \psi \in \mathcal{T}_l$, and by (v0) in the definition of ν we have $0 \in \nu(p)$, whence by (v1) of that definition we get $1 \in \nu(\psi)$;

        $\psi' = \neg\varphi$: Then $\neg\neg\varphi = \psi \in \mathcal{T}_l$. By (a) of Property 6 above we have $\varphi \in \mathcal{T}_l$, whence by the induction hypothesis $1 \in \nu(\varphi)$, and so $1 \in \nu(\neg\neg\varphi) = \nu(\psi)$ by applications of (v2) and (v1);

        $\psi' = \varphi \wedge \tau$: Then $\neg(\varphi \wedge \tau) = \psi \in \mathcal{T}_l$. Hence $0 \in \nu(\varphi \wedge \tau)$ by (v6), and so $1 \in \nu(\neg(\varphi \wedge \tau)) = \nu(\psi)$.

        $\psi' = \varphi \vee \tau$: Then $\neg(\varphi \vee \tau) = \psi \in \mathcal{T}_l$. By (f) of Property 6 we have $\neg\varphi, \neg\psi \in \mathcal{T}_l$, whence by the induction hypothesis $1 \in \nu(\neg\varphi)$ and $1 \in \nu(\neg\tau)$. Thus $0 \in \nu(\varphi), 0 \in \nu(\tau)$ by (v1), and so $0 \in \nu(\varphi \vee \tau)$ by (v4), and $1 \in \nu(\neg(\varphi \vee \tau)) = \nu(\psi)$ by (v1).

    - Assume that $\psi = \psi_1 \vee \psi_2$. Then $\psi_1 \vee \psi_2 = \psi \in \mathcal{T}_l$, so by (v3) we have $1 \in \nu(\psi_1 \vee \psi_2) = \nu(\psi)$.
    - Assume that $\psi = \psi_1 \wedge \psi_2$. Since $\psi \in \mathcal{T}_l$, it follows by (c) of Property 6 that $\psi_1, \psi_2 \in \mathcal{T}_l$. Therefore the induction hypothesis implies that $1 \in \nu(\psi_1), 1 \in \nu(\psi_2)$. Hence, $1 \in \nu(\psi_1 \wedge \psi_2) = \nu(\psi)$ by (v5).

7.5 Processing Information from a Set of Sources

- To prove (B), assume that $\psi \in \mathcal{T}_r$.

  Assume first that $\psi = p$ where p is a variable. As $\psi \in \mathcal{T}_r$, Property 2 above of $\mathcal{T}_l$ and $\mathcal{T}_r$ implies that $p \notin \mathcal{T}_l$. Hence $1 \notin \nu(\psi)$ by (v0). Similarly, if $\psi = \neg p$, then $\neg p \notin \mathcal{T}_l$. Thus by (v0) $0 \notin \nu(p)$, whence by (v1) we have $1 \notin \nu(\neg p) = \nu(\psi)$.

  Now, assume that ψ is not a literal. The proof that (B) holds for ψ in this case is by an induction which is similar to that which was carried out in the proof of (A), using the relevant items of Property 6 above. Details are left for the reader.

That $\nu \not\models_{\mathcal{M}_E^4} \mathcal{S}$ immediately follows from (A) and (B) because of Property 1 above of $\mathcal{T}_l$ and $\mathcal{T}_r$. Therefore it only remains to show that $\nu \models_{\mathcal{M}_E^4} \mathcal{S}$, that is: that if $\Gamma \Rightarrow \Delta \in \mathcal{S}$ then $\nu \models_{\mathcal{M}_E^4} \Gamma \Rightarrow \Delta$. So let $\Gamma \Rightarrow \Delta \in \mathcal{S}$. Then $\Gamma \cup \Delta \subseteq \mathcal{T}_l \cup \mathcal{T}_r$ by Property 5 of $\mathcal{T}_l$ and $\mathcal{T}_r$. On the other hand, Property 2 of $\mathcal{T}_l$ and $\mathcal{T}_r$ implies that either $\Gamma \not\subseteq \mathcal{T}_l$ or $\Delta \not\subseteq \mathcal{T}_r$ (because if $\Gamma \Rightarrow \Delta \in \mathcal{S}$ then trivially $\mathcal{S} \vdash_{G_{\mathbf{EIP}}}^{cf} \Gamma \Rightarrow \Delta$). It follows that there is either $\varphi \in \Gamma$ such that $\varphi \in \mathcal{T}_r$, or $\varphi \in \Delta$ such that $\varphi \in \mathcal{T}_l$. By (A) and (B) this implies that there is either $\varphi \in \Gamma$ such that $1 \notin \nu(\varphi)$, or $\varphi \in \Delta$ such that $1 \in \nu(\varphi)$. In either case $\nu \models_{\mathcal{M}_E^4} \Gamma \Rightarrow \Delta$ (by Definition 7.78). □

Theorem 7.83 (strong soundness and completeness) *If $\mathcal{S} \cup \{s\}$ is a set of sequents of $\mathcal{L}$, then $\mathcal{S} \vdash_{\mathcal{EIP}} s$ iff $\mathcal{S} \vdash_{G_{\mathbf{EIP}}} s$. In other words: $G_{\mathbf{EIP}}$ is strongly sound and complete for* **EIP**.

Proof. Immediate from Propositions 7.81, 7.82, and 7.79. □

Theorem 7.84 (compactness) *If $\mathcal{S} \cup \{s\}$ is a set of sequents of $\mathcal{L}$, then $\mathcal{S} \vdash_{\mathcal{EIP}} s$ iff there is a finite subset $\mathcal{S}'$ of $\mathcal{S}$ such that $\mathcal{S}' \vdash_{\mathcal{EIP}} s$.*

Proof. Immediate from Theorem 7.83. □

Theorem 7.85 (cut-elimination) $G_{\mathbf{EIP}}$ *admits strong cut-elimination.*

Proof. Immediate from Propositions 7.81 and 7.82. □

Corollary 7.86 *Let $\mathcal{T} \cup \varphi$ be a set of formulas of $\mathcal{L}$. $\mathcal{T} \vdash_{\mathcal{EIP}} \varphi$ iff there is a finite subset Γ of $\mathcal{T}$ such that $\Gamma \Rightarrow \varphi$ has a cut-free proof in $G_{\mathcal{EIP}}$.*

Proof. This follows from Corollary 7.71, Theorems 7.83 and 7.85, and the fact that by Definition 7.78, if Γ is finite then $\Gamma \vdash_{\mathcal{EIP}} \varphi$ iff $\vdash_{\mathcal{EIP}} \Gamma \Rightarrow \varphi$. □

Corollary 7.87 *It is decidable whether $\mathcal{S} \vdash_{\mathcal{EIP}} s$ in case $\mathcal{S} \cup \{s\}$ is a finite set of sequents of $\mathcal{L}$.*

Proof. Derivations of s from $\mathcal{S}$ in which all cuts are on formulas which occur in $\mathcal{S}$ may contain only subformulas of formulas in $\mathcal{S} \cup \{s\}$. If the latter is finite, then so is the set of such subformulas. Hence finding such a derivation involves only a finite search, and so the corollary follows from Theorems 7.83 and 7.85. □

Note 7.88 Theorem 7.84 and Corollary 7.87 strengthen Corollary 7.71, because $\mathcal{T} \vdash_{\mathcal{EIP}} \varphi$ iff $\{\Rightarrow \psi \mid \psi \in \mathcal{T}\} \vdash_{\mathcal{EIP}} \Rightarrow \varphi$ (where $\mathcal{T} \cup \{\varphi\}$ is a set of formulas). The converse is not true, because (as noted above) sequents in $\mathcal{L}$ cannot be translated into formulas of $\mathcal{L}$.

7.6 Bibliographical Notes and Further Reading

Many of the logics discussed above have already been studied in the literature. The logic **CLuN** was first introduced by Batens in [78] under the name of **PI**, and was further investigated in [79, 80, 83] as the basic adaptive logic. The logic $\mathbf{C}_{min}$ is used by Carnielli, Coniglio and Marcos in [113, 116] as the basis for the taxonomy of C-systems. The logic $\mathbf{L}_{HCL^+(\{[t],[\Rightarrow\neg\neg],[\Rightarrow\neg\vee]\})}$ was also introduced in [78], under the name **PI***.

The material in Section 7.1 and all the results in the later sections whose proofs rely on it (like Theorem 7.53) are based on [57].

The material in Section 7.3 is based on [18, 19]. The material in Section 7.4 is based on [44, 45].

The material in Section 7.5 is based on (and improves) [60] and [63]. However, Proposition 7.74, Theorem 7.84, and and Corollary 7.87 are new here, while Proposition 7.81 and Theorem 7.83 were previously proved only for the case where $\mathcal{S}$ is finite.

Chapter 8

Logics of Formal (In)consistency

8.1 Introduction

The Brazilian logician Newton da Costa is the father of what is known as the "Brazilian school" of paraconsistent logics, which is one of the oldest and currently most well-known approaches to study of paraconsistency. Starting with the hierarchy of paraconsistent calculi $\{\mathbf{C}_n \mid 0 < n < \omega\}$, which was developed by him in the sixties ([127, 129, 142]), his ideas led to the development of a large family of paraconsistent logics called Logics of Formal (In)consistency (LFIs, [111, 113, 116]). This chapter is devoted to this class, and especially to its most important subclass: the family of C-systems.

The main idea of the Brazilian school is that propositions may be divided into two sorts: "normal" (or "consistent") and "abnormal" (or "inconsistent"); Contradictions are acceptable for the latter, but not the former. LFIs are logics that have resources to express this meta-notion of consistency of formulas within the object language. Thus, an LFI is able to separate between propositions for which explosion holds (the normal ones), and those for which it does not hold (the abnormal ones). A C-system is a special kind of LFI which uses a unary (primitive or defined) connective ∘ for this purpose. Thus in da Costa's most basic system $\mathbf{C}_1$ the "normality" of a sentence ψ is expressed as $\neg(\psi \wedge \neg\psi)$.

We start our study of LFIs with a precise definition of this concept.[1]

Definition 8.1 (LFI) Let **L** be a logic for a language $\mathcal{L}$ with negation $\neg$. **L** is a Logic of Formal Inconsistency (LFI) with respect to $\neg$ if there exists a non-empty set $\bigcirc(p)$ of formulas in $\mathcal{L}$ (depending on a single propositional variable p, that is: $\mathsf{Var}(\varphi) = \{p\}$ for every $\varphi \in \bigcirc(p)$) with the following properties (where $\bigcirc(\psi) = \{\varphi[\psi/p] \mid \varphi \in \bigcirc(p)\}$):

---
[1]There are other definitions that have been proposed. See [111, 116] for variations.

- **L** is ¬-paraconsistent.
- There are formulas ψ_0, φ_0, such that:
  - $\circ(\psi_0), \psi_0 \nvdash_\mathbf{L} \varphi_0$,
  - $\circ(\psi_0), \neg\psi_0 \nvdash_\mathbf{L} \varphi_0$.
- $\circ(\psi), \psi, \neg\psi \vdash_\mathbf{L} \varphi$ for every ψ and φ.

Like the notion of an LFI, the notion of a C-system has several definitions. Usually it is defined with respect to some base logic, which such a C-system extends (like $\mathbf{CL}^+$ or $\mathbf{IL}^+$). In this chapter our focus is on C-systems extending $\mathbf{CL}^+$, and among them we restrict our attention only to classical ¬-paraconsistent logics (Definition 2.22 and Proposition 2.23).

Definition 8.2 (consistency operator) Let $\mathbf{L} = \langle \mathcal{L}, \vdash_\mathbf{L} \rangle$ be a logic such that $\mathcal{L}_{CL} \subseteq \mathcal{L}$ and $\mathbf{CL}^+ \subseteq \mathbf{L}$, and let $\circ$ be a unary connective of $\mathcal{L}$. We say that $\circ$ is a *consistency operator* for ¬ (in $\mathbf{L}$) if:

- $[\mathsf{n}_0]$ $\vdash_\mathbf{L} (\circ\psi \wedge \neg\psi \wedge \psi) \supset \varphi$ for every $\psi, \varphi \in \mathcal{W}(\mathcal{L})$.
- $[\mathsf{n}_1]$ $\nvdash_\mathbf{L} (\circ p \wedge \neg p) \supset q$ whenever p and q are distinct variables.
- $[\mathsf{n}_2]$ $\nvdash_\mathbf{L} (\circ p \wedge p) \supset q$ whenever p and q are distinct variables.

Definition 8.3 (C-system) Let $\mathbf{L}$ be a like in Definition 8.2, and let $\circ$ be a unary connective of $\mathcal{L}$.

- $\mathbf{L}$ is a $\circ$-*C-system* if it is classical ¬-paraconsistent logic in which $\circ$ is a consistency operator for ¬. If $\circ$ is a primitive connective of $\mathcal{L}$ then $\mathbf{L}$ is called *strict* $\circ$-C-system for ¬.
- $\mathbf{L}$ is a (strict) *C-system* if it is a (strict) $\circ$-C-system for some $\circ$.

Proposition 8.4 *Every C-system is normal.*

Proof. This is a special case of Proposition 2.23. □

Convention *As usual, unless there is a risk of confusion about the identity of $\circ$ or ¬, or if the exact identity of $\circ$ does not matter, we shall write just* "C-system" *instead of* "$\circ$-C-system for ¬".

Note 8.5 Because of condition $[\mathsf{n}_0]$, the conditions $[\mathsf{n}_1]$, $[\mathsf{n}_2]$, and the condition of pre-paraconsistency can respectively be reformulated as follows:

$[\mathsf{n}_1\text{'}]$ $\nvdash_\mathbf{L} (\circ p \wedge \neg p) \supset p$.

8.1 Introduction

[n$_2$'] $\not\vdash_{\mathbf{L}} (\circ p \wedge p) \supset \neg p$.

[n$_3$'] $\not\vdash_{\mathbf{L}} (\neg p \wedge p) \supset \circ p$.

Taken together, conditions [n$_1$'], [n$_2$'], [n$_3$'], and [n$_0$] mean that every formula follows from the set $\{p, \neg p, \circ p\}$, but no element of this set follows from the other two. Thus, the set which consists of these four conditions is completely symmetric with respect to $\neg$ and $\circ$. What nevertheless distinguishes between $\neg$ and $\circ$ in C-systems is the demand that $\mathbf{L}$ is a *classical* $\neg$-paraconsistent logic. In particular: the $\mathcal{L}_{CL}$-fragment of $\mathbf{L}$ is contained in **CL**. The next definition extends this condition to $\{\wedge, \vee, \supset, \neg, \circ\}$ in a way which further breaks the symmetry between $\neg$ and $\circ$, and reflects the intuition that classical logic is based on the assumption that all formulas are consistent (or "normal").

Definition 8.6 (regular C-system) Let $\mathbf{L}$ and $\circ$ be like in Definition 8.2.

- We say that $\circ$ is a *regular consistency operator* (for $\neg$) if it satisfies condition [n$_0$] from Definition 8.2, as well as the following one:

    - [c$_0$] $\mathbf{L}$ is **F**-contained in **CL** (Definition 1.45) for some bivalent $\neg$-interpretation **F** such that $F(\circ) = \lambda x.t$ (i.e., $(\mathbf{F}(\circ))(t) = (\mathbf{F}(\circ))(f) = t$).[2]

- $\mathbf{L}$ is a *regular $\circ$-C-system* if it is a pre-$\neg$-paraconsistent axiomatic extension of $\mathbf{CL}^+$ in which $\circ$ is a regular consistency operator, and $\neg$ is complete.[3]

- $\mathbf{L}$ is a *regular C-system* if it is regular $\circ$-C-system for some $\circ$.

The next theorem provides the most important property of regular consistency operators. It demonstrates the basic intuition (noted above) that classical logic is based on the assumption that all formulas are consistent. What is more: it shows that in the presence of a regular consistency operator, any classical proof can be reproduced in any regular C-system by adding assumptions concerning consistency of formulas.

---

[2] Recall that by Proposition 1.54, the interpretations that such **F** assigns to the connectives of $\mathcal{L}_{CL}$ are necessarily the classical ones.

[3] The last condition (the completeness of $\neg$) has not been included before in any definition of a C-system. Nevertheless, all the C-systems investigated in [113, 116], including what are taken there to be the most basic C-systems, do satisfy it. Moreover, this property is needed for the proof of the crucial Theorem 8.7 below (even though a weaker demand, that $\circ\varphi \vdash \neg\varphi \vee \varphi$, would do). Thus, we have found it convenient to include it in the definition of our new notion of a regular C-system.

Theorem 8.7 *Let* **L** *and* ∘ *be like in Definition 8.2. Suppose that* ∘ *is a regular consistency operator for* ¬*. Let* $\mathcal{T} \cup \{\psi\}$ *be a set of formulas in* $\mathcal{L}_{CL}$. *Then* $\mathcal{T} \vdash_{\mathbf{CL}} \psi$ *iff there exists a finite set* Δ *of subformulas of* $\mathcal{T} \cup \{\psi\}$ *such that* $\circ \Delta \cup \mathcal{T} \vdash_{\mathbf{L}} \psi$ *(where* $\circ \Delta = \{\circ \psi \mid \psi \in \Delta\}$*).*

Proof. Suppose that $\circ \Delta \cup \mathcal{T} \vdash_{\mathbf{L}} \psi$ for some set Δ of formulas in $\mathcal{L}_{CL}$. Let **F** be a bivalent ¬-interpretation that satisfies condition [c₀] (from Definition 8.6). Since **L** is **F**-contained in **CL**, $\circ \Delta \cup \mathcal{T} \vdash_{\mathbf{2_F}} \psi$. Given the value of **F**(∘), this means (see footnote 2) that $\mathcal{T} \vdash_{\mathbf{CL}} \psi$.

For the converse, suppose that $\mathcal{T} \vdash_{\mathbf{CL}} \psi$. Then there is a derivation $\pi = \varphi_1, \ldots, \varphi_n$ in *HCL* of ψ from $\mathcal{T}$. This may not be a derivation in **L**, because instances of the axiom [¬ ⊃] may not be theorems of **L**. Let $\varphi_{j_1}, \ldots, \varphi_{j_m}$ be all the instances of [¬ ⊃] in π, so that $\varphi_{j_i} = \neg \theta_{j_i} \supset (\theta_{j_i} \supset \theta'_{j_i})$ for some θ_{j_i}, θ'_{j_i}. Then $\{\varphi_{j_1}, \ldots, \varphi_{j_m}\} \cup \mathcal{T} \vdash_{\mathbf{L}} \psi$. But for every $1 \leq i \leq m$, $\circ \theta_{j_i} \vdash_{\mathbf{L}} \varphi_{j_i}$ by condition [n₀]. Hence $\circ \Delta \cup \mathcal{T} \vdash_{\mathbf{L}} \psi$, where $\Delta = \{\theta_{j_1}, \ldots, \theta_{j_m}\}$.

To show that Δ can be confined to consist of subformulas of $\mathcal{T} \cup \{\psi\}$, we can use cut-free derivations in *LK* rather than derivations in *HCL*. Indeed, it is not too difficult to show that if $\Gamma \Rightarrow \psi_1, \ldots, \psi_n$ has a cut-free proof in *LK* then there is a set Δ, which consists of subformulas of $\Gamma \cup \{\psi_1, \ldots, \psi_n\}$, such that $\circ \Delta, \Gamma \vdash_{\mathbf{L}} \psi_1 \vee \cdots \vee \psi_n$. We omit the details, since this direction of the theorem is an immediate consequence of Propositions 8.29, 8.48, and Theorem 8.46 below. □

Corollary 8.8 *Let* **L** *and* ∘ *be like in Definition 8.2.*

1. *If* ∘ *is a regular consistency operator then it is a consistency operator.*

2. *If* **L** *is a regular (∘-)C-system then it is a (∘-)C-system.*

Proof.

1. We have to show that if ∘ is a regular consistency operator then it satisfies conditions [n₁] and [n₂]. Suppose for example that it does not satisfy [n₁]. (The case of [n₂] is similar.) Then $\vdash_{\mathbf{L}} (\circ p \wedge \neg p) \supset q$. From Theorem 8.7 it easily follows that this implies that $\vdash_{\mathbf{CL}} \neg p \supset q$, which is false. A contradiction.

2. Condition [c₀] implies ¬-containment in classical logic. Hence, every regular (∘-)C-system is a classical ¬-paraconsistent logic. That it is also a (∘-)C-system now follows from the first item. □

Note 8.9 To the best of our knowledge, all the C-systems that have ever been studied in the literature are regular. (This includes of course all the C-systems studied in this chapter.)

8.1 Introduction

As argued in [64], there is yet another (and stronger) condition which is natural to impose on $\circ$ and $\neg$.

Definition 8.10 (strong consistency operator) Let **L** and $\circ$ be like in Definition 8.2. $\circ$ is a *strong consistency operator* (for $\neg$) if every instance of the following axioms is a theorem of **L**:

[b] $(\circ\varphi \wedge \neg\varphi \wedge \varphi) \supset \psi$,

[k] $\circ\varphi \vee (\neg\varphi \wedge \varphi)$.

Note 8.11 Under the intuitive meaning of $\circ\psi$ as 'ψ is consistent', Axiom [b] implies that no formula is both consistent and contradictory. Axiom [k] complements [b] by saying that every formula is *either* consistent or contradictory. Together, these axioms capture the essence of the intended meaning of $\circ$, and so requiring their combination is quite natural.

Proposition 8.12 *Let* **L** *and* $\circ$ *be like in Definition 8.2. Suppose that* **L** *is* $\neg$*-contained in* **CL**. *If* $\circ$ *is a strong consistency operator for* $\neg$ *in* **L**, *then it is also a regular consistency operator for* $\neg$ *in* **L**.

Proof. The validity of Axiom [b] in **L** ensures that $\circ$ satisfies condition [n$_0$]. We show that it satisfies [c$_0$] as well. By our assumptions, **L** is **F**-contained in **CL** for some bivalent $\neg$-interpretation **F**. Because of Axiom [k], this implies that $\circ\varphi \vee (\neg\varphi \wedge \varphi)$ is valid in 2$_\mathbf{F}$. But $\mathbf{F}(\wedge) = \mathbf{F}_{CL}(\wedge)$ and $\mathbf{F}(\vee) = \mathbf{F}_{CL}(\vee)$. (This is shown by the same argument as in the proof of Proposition 1.54.) It follows that $\circ\varphi$ is valid in 2$_\mathbf{F}$, and so $\mathbf{F}(\circ) = \lambda x.t$. □

Definition 8.13 (strong C-system) Let **L** and $\circ$ be like in Definition 8.2.

- **L** is a *strong* $\circ$-*C-system* (for $\neg$) if it is a classical $\neg$-paraconsistent logic in which $\circ$ is a strong consistency operator for $\neg$, and $\neg$ is complete.

- **L** is a *strong C-system* if it is strong $\circ$-C-system for some $\circ$.

Proposition 8.14 *A logic* **L** *is a strong* $(\supset$-$)$*C-system iff it is a regular* $(\circ$-$)$*C-system in which every instance of* [k] *is a theorem.*

Proof. The 'if' direction follows from the definitions and the second item of Corollary 8.8. The converse easily follows from Proposition 8.12. □

The next proposition shows that we do not lose much if we restrict our attention to strong C-systems.

Proposition 8.15 *Every C-system* **L** *in which* $\neg$ *is complete is also a strong C-system.*

Proof. Let **L** be a ○-C-system. Then it is classical ¬-paraconsistent. Define:

$$\otimes\varphi =_{Df} (\neg\varphi \wedge \varphi) \supset \circ\varphi.$$

Then the fact that ○ satisfies condition [n₀] easily implies that so does ⊗. This means that every instance of [b], in which ○ is replaced by ⊗, is a theorem of **L**. The same is true for [k], because $(\sigma \supset \psi) \vee \sigma$ is a positive classical tautology. Hence, ⊗ is a strong consistency operator for ¬. □

Note 8.16 The proof of Proposition 8.15 actually shows something stronger: that **L** is a strong C-system iff it is a classical ¬-paraconsistent logic in which ¬ is complete, and there is an operation ○ for which [b] is valid.

Proposition 8.17 *Let **L** be a classical ¬-paraconsistent logic, and suppose that ○ is a consistency operator for ¬ in **L**, while ⊗ is a strong consistency operator for ¬ in **L**. Then $\vdash_\mathbf{L} \circ\varphi \supset \otimes\varphi$.*

Proof. $\vdash_\mathbf{L} \otimes\varphi \supset (\circ\varphi \supset \otimes\varphi)$ (since $\mathbf{CL}^+ \subseteq \mathbf{L}$), and $\vdash_\mathbf{L} (\neg\varphi \wedge \varphi) \supset (\circ\varphi \supset \otimes\varphi)$ (because ○ is a consistency operator for ¬). Using reasoning in $\mathbf{CL}^+$, it follows that $\vdash_\mathbf{L} (\otimes\varphi \vee (\neg\varphi \wedge \varphi)) \supset (\circ\varphi \supset \otimes\varphi)$. But $\vdash_\mathbf{L} \otimes\varphi \vee (\neg\varphi \wedge \varphi)$ (because ⊗ is a strong consistency operator for ¬). Hence $\vdash_\mathbf{L} \circ\varphi \supset \otimes\varphi$. □

Corollary 8.18 *Let **L** be a classical ¬-paraconsistent logic. A strong consistency operator is unique in **L** up to equivalence[4], i.e., if $\circ_1$ and $\circ_2$ are both strong consistency operators in **L** then $\vdash_\mathbf{L} \circ_1 p \supset \circ_2 p$ and $\vdash_\mathbf{L} \circ_2 p \supset \circ_1 p$.*

Proof. Immediate from Propositions 8.12, 8.17, and Corollary 8.8. □

Next, we provide another characterization of (strong) C-systems.

Definition 8.19 (bottom element) A formula F is a *bottom element* for a logic **L** in $\mathcal{L}$ if $\vdash_\mathbf{L} \mathsf{F} \supset \varphi$ for every $\varphi \in \mathcal{W}(\mathcal{L})$.[5]

Proposition 8.20 *A classical ¬-paraconsistent logic **L** has a (strong) consistency operator iff it has a bottom element.*

Proof. Suppose that ○ is a consistency operator in **L**. Then for every formula φ, $\circ\varphi \wedge \neg\varphi \wedge \varphi$ is a bottom element for **L**.

For the converse, assume that **L** has a bottom element F. Define: $\circ\varphi =_{Df} \neg\varphi \wedge \varphi \supset \mathsf{F}$. The facts that $\mathbf{CL}^+ \subseteq \mathbf{L}$ and that F is a bottom element easily imply (again because $(\sigma \supset \psi) \vee \sigma$ is a positive classical tautology) that ○ is a strong consistency operator in **L**. □

---

[4] Equivalence does not imply here congruence. In other words: that $\circ_1$ and $\circ_2$ are equivalent in **L** does not imply that one can be substituted for the other in all contexts.

[5] In previous chapters the symbol 'F' was used as a special propositional constant that serves as a bottom element in many of the logics we were studying. Note that here it is used in a somewhat generalized, but closely related, sense.

Corollary 8.21 *Let* **L** *be a classical $\neg$-paraconsistent logic in which $\neg$ is complete.* **L** *is a (strong) C-system iff it has a bottom element.*

Corollary 8.22

1. $\mathbf{L}_{HCL^F(S)}$ *(Section 7.4) is a C-system for every $S \subseteq A_{PAC}$.*

2. $\mathbf{L}_{HCL^F(S)}$ *is a strong C-system for every $S \subseteq A_{PAC}$ such that $[t] \in S$.*

Proposition 8.23 *A logic* **L** *is a strong C-system for $\neg$ iff it is axiomatic extension of* $\mathbf{CL}^F$, *$\neg$-contained in classical logic, pre-$\neg$-paraconsistent, and $\neg$ is complete in it.*

Proof. From the fourth item of Note 1.91 and Proposition 1.65 it follows that if a classical $\neg$-paraconsistent logic has a bottom element F, then its $\{\vee, \wedge, \supset, F\}$-fragment (where F is used as a propositional constant) is $\mathbf{CL}^F$. The proposition is an immediate corollary of this fact, the definitions of a classical $\neg$-paraconsistent logic and of a C-system, and Proposition 8.20. □

Proposition 8.23 and Corollary 8.22 show that according to its formal definition the family of C-systems is rather big, and includes many of the logics that have already been investigated in previous chapters. (This includes, e.g, $\mathbf{J_3}$. See Note 4.63.) However, when a logic is viewed (and is intended to be used) as a C-system, it is the consistency operator that is of interest. Accordingly, we concentrate in this chapter on the behavior of this operator in the various logics studied in it. In fact, answering the question: "What combinations of properties of ○ should be taken as desirable?" is the main motivation behind the introduction of these logics in the first place.

8.2 The Basic C-systems

In this section we start our exploration of the family of regular C-systems. We begin with the two most basic systems in this family.

Definition 8.24 ($\mathcal{L}_C$) $\mathcal{L}_C = \{\wedge, \vee, \supset, \neg, \circ\}$.

Definition 8.25 (*HB*, *HBk*, B, Bk) The Hilbert-type system *HB* is the extension in $\mathcal{L}_C$ of *HCLuN* (Definition 7.13) with the axiom [b]. The system *HBk* is the extension of *HB* with [k]. We denote by **B** and **Bk** the logics in $\mathcal{L}_C$ which are induced by *HB* and *HBk*, respectively. (In other words: $\mathbf{B} = \mathbf{L}_{HB}$, $\mathbf{Bk} = \mathbf{L}_{HBk}$.)

Note 8.26 **B** is called **mbC** in [111, 113, 116], while **Bk** is called **mbCciw** in [111], and **BK** in [64, 65].

Proposition 8.27 *Let* **L** *be a logic in* $\mathcal{L}$, *where* $\mathcal{L}_C \subseteq \mathcal{L}$. *The following conditions on* **L** *are equivalent:*

- **L** *is a strong* ∘-*C*-*system for* ¬.

- **L** *is an axiomatic extension of* **Bk** *which is pre-*¬-*paraconsistent, and* ¬-*contained in classical logic.*

Proof. That strong ∘-C-systems satisfy the second condition is immediate from the definitions of such systems and of a classical ¬-paraconsistent logic, and Corollaries 7.15, 7.16. For the converse, assume that **L** satisfies the second condition. Using Theorem 1.90, this implies that **L** is a classical ¬-paraconsistent logic. From this it follows (using axioms [t], [b], and [k]) that **L** is a strong ∘-C-system. □

Proposition 8.28

1. **B** *is a regular* ∘-*C*-*system (for* ¬*).*

2. **Bk** *is a strong* ∘-*C*-*system (for* ¬*).*

Proof.

1. It is easy to see that **B** satisfies the conditions [n₀] and [c₀], and that its negation is complete. To show that it is a regular ∘-C-system for ¬ it remains therefore to show that it is ¬-paraconsistent. This follows from the fact that the logic **J₃** is ¬-paraconsistent (Proposition 4.64), since by defining $\circ\varphi =_{Df} \neg\varphi \wedge \varphi \supset F$ we make **J₃** an extension of **B**.

2. The proof that **Bk** is a regular ∘-C-system (for ¬) is identical to the proof of the first item. That it is a a strong ∘-C-system (for ¬) follows therefore from Proposition 8.14. □

Proposition 8.29 *The logic* **B** *is the minimal* ∘-*C*-*system (for* ¬*) in which* ¬ *is complete. (That is:* **B** *has these properties, and any other logic which has them is an axiomatic extension of* **B**.*)*

Proof. This follows from Proposition 8.28, together with Corollaries 7.15 and 7.16. □

Corollary 8.30 *The logic* **B** *is the minimal regular* ∘-*C*-*system, and* **Bk** *is the minimal strong* ∘-*C*-*system.*

Proof. Immediate from Propositions 8.29, 8.28 and 8.27. □

8.2 The Basic C-systems

Proposition 8.31 *Let* **L** *be a logic in* $\mathcal{L}$, *where* $\mathcal{L}_C \subseteq \mathcal{L}$. *The following conditions on* **L** *are equivalent:*

- **L** *is a regular* $\circ$-*C-system for* $\neg$.

- **L** *is an axiomatic extension of* **B** *which is pre-*$\neg$-*paraconsistent and* $\neg$-*contained in classical logic, and satisfies condition* [c$_0$].

Proof. Immediate from Corollary 8.30 and the relevant definitions. □

Note 8.32 In contrast to the minimality of **Bk** which is described in Proposition 8.27, there are $\neg$-paraconsistent axiomatic extensions of **B** which are $\neg$-contained in classical logic, but in which $\circ$ is not a consistency operator. An example is provided by the $\mathcal{L}_C$-fragment of **J$_3$**, if we define $\circ\varphi =_{Df} \mathsf{F}\wedge\varphi$. We leave the verification of this claim to the reader.

Proposition 8.33 **B** *and* **Bk** *are conservative extensions of* **CLuN**.

Proof. Extend the Nmatrix $\mathcal{M}_2$ (Definition 7.9) to the Nmatrix $\mathcal{M}_2^\mathsf{F}$ for $\mathcal{L}_{CL}^{\mathsf{F}\neg}$ by letting $\widetilde{\mathsf{F}} = f$. Using Theorem 6.18 and Proposition 2.23, it is easy to see that like $\mathbf{L}_{\mathcal{M}_2}$ (i.e., **CLuN**), $\mathbf{L}_{\mathcal{M}_2^\mathsf{F}}$ is a classical $\neg$-paraconsistent logic, and it obviously has a bottom element. Hence it is a strong C-system by Proposition 8.20. Since $\neg$ is complete in $\mathbf{L}_{\mathcal{M}_2^\mathsf{F}}$, this logic is an extension of **Bk** by Proposition 8.27. It follows that the $\mathcal{L}_{LC}$-fragments of **B** and **Bk** are between **CLuN** and the corresponding fragment of $\mathbf{L}_{\mathcal{M}_2^\mathsf{F}}$. The latter is identical to $\mathbf{L}_{\mathcal{M}_2}$, which is **CLuN**. Therefore both fragments are identical to **CLuN**. □

Corollary 8.34 *There is no finite family* $\mathcal{F}$ *of finite ordinary matrices such that* $\vdash_\mathcal{F} = \vdash_\mathbf{B}$ *or* $\vdash_\mathcal{F} = \vdash_\mathbf{Bk}$. *In particular,* **B** *and* **Bk** *have no finite weakly characteristic matrix.*

Proof. Immediate from Propositions 8.33 and 7.12. □

Next we show that unlike **CLuN**, even the use of a two-valued Nmatrix is not sufficient for characterizing **B** and **Bk**. Actually, we prove something much stronger.

Proposition 8.35 *A strict C-system cannot have a weakly characteristic two-valued Nmatrix.*

Proof. Let **L** be a strict C-system. Suppose that $\mathcal{M} = \langle\{t,f\},\{t\},\mathcal{O}\rangle$ is a two-valued Nmatrix which is weakly sound for **L**. A proof which is similar to that of Proposition 1.54 shows that the interpretations $\mathcal{O}$ assigns to $\wedge, \vee, \supset$ are the classical ones. Moreover, the validity of [t] obviously enforces the condition $\tilde{\neg} f = \{t\}$. Hence there are two cases to consider:

- If $\tilde{\neg} t = \{f\}$, then $\neg p \supset (p \supset q))$ is valid in $\mathcal{M}$. Since **L** is paraconsistent, $\mathcal{M}$ is not weakly characteristic for **L**.

- If $t \in \tilde{\neg} t$, then it cannot be the case that also $t \in \tilde{\circ} t$, since otherwise [b] would not be valid in $\mathcal{M}$. It follows that $\tilde{\circ} t = \{f\}$, and so $p \wedge \circ p \supset q$ is valid in $\mathcal{M}$. Since $\circ$ is a consistency operator for $\neg$ in **L**, by definition this formula is not a theorem of **L**. Hence $\mathcal{M}$ is not weakly characteristic for **L** in this case too. □

Note 8.36 The condition of strictness in the last theorem is necessary. Thus the logic $\mathbf{L}_{\mathcal{M}_2^{\mathrm{F}}}$ (see the proof of Proposition 8.33) is a (strong) C-system by Corollary 8.21, and it has a characteristic two-valued Nmatrix.

Corollary 8.37 **B** and **Bk** *do not have a weakly characteristic two-valued Nmatrix.*

Proof. Immediate from Propositions 8.35 and 8.28. □

Given Corollaries 8.34 and 8.37, the simplest semantic characterization we can hope for in the case of **B** and **Bk** is in terms of three-valued Nmatrices. Next we provide such semantic characterizations.

Definition 8.38 ($\mathcal{M}_B$, $\mathcal{M}_{Bk}$) The Nmatrix $\mathcal{M}_B$ is the extension to $\mathcal{L}_C$ of the Nmatrix $\mathcal{M}_4^b(\{[\mathsf{t}]\})$ (Definition 7.44) in which $\tilde{\circ}\top = \{f\}$, while $\tilde{\circ} a = \{t, f, \top\}$ if $a \in \{t, f\}$. In other words: $\mathcal{M}_B$ is the Nmatrix $\langle \mathcal{V}, \mathcal{D}, \mathcal{O} \rangle$ for $\mathcal{L}_C$ in which $\mathcal{V} = \{t, f, \top\}$, $\mathcal{D} = \{t, \top\}$, and $\mathcal{O}$ is defined as follows:

	$\tilde{\neg}$	$\tilde{\circ}$
t	$\{f\}$	$\{t, \top, f\}$
$\top$	$\{t, \top\}$	$\{f\}$
f	$\{t, \top\}$	$\{t, \top, f\}$

$\tilde{\wedge}$	t	$\top$	f
t	$\{t, \top\}$	$\{t, \top\}$	$\{f\}$
$\top$	$\{t, \top\}$	$\{t, \top\}$	$\{f\}$
f	$\{f\}$	$\{f\}$	$\{f\}$

$\tilde{\vee}$	t	$\top$	f
t	$\{t, \top\}$	$\{t, \top\}$	$\{t, \top\}$
$\top$	$\{t, \top\}$	$\{t, \top\}$	$\{t, \top\}$
f	$\{t, \top\}$	$\{t, \top\}$	$\{f\}$

$\tilde{\supset}$	t	$\top$	f
t	$\{t, \top\}$	$\{t, \top\}$	$\{f\}$
$\top$	$\{t, \top\}$	$\{t, \top\}$	$\{f\}$
f	$\{t, \top\}$	$\{t, \top\}$	$\{t, \top\}$

$\mathcal{M}_{Bk}$ is the simple refinement of $\mathcal{M}_B$ in which $\tilde{\circ} t = \tilde{\circ} f = \{t, \top\}$.

Proposition 8.39 $\wedge, \vee$ *and* $\supset$ *are* $\mathcal{M}_{Bk}(\mathcal{M}_B)$*-conjunction,* $\mathcal{M}_{Bk}(\mathcal{M}_B)$*-disjunction and* $\mathcal{M}_{Bk}(\mathcal{M}_B)$*-implication respectively.*

Proof. By Corollary 7.49. □

8.2 The Basic C-systems

Theorem 8.40 $\mathcal{M}_B$ *is characteristic for* **B**, *and* $\mathcal{M}_{Bk}$ — *for* **Bk**.

Proof. It is easy to verify that [b] is valid in $\mathcal{M}_B$ (and so also in $\mathcal{M}_{Bk}$), and that [k] is valid in $\mathcal{M}_{Bk}$. These facts and the soundness part of Theorem 7.50 imply the soundness part here. The proof of completeness closely follows that of Theorem 7.50 in the case of $\mathcal{M}_4^b(\{[t]\})$. The only needed addition is to show that the assignment ν defined there respects $\tilde{\circ}$.

- Assume $\nu(\psi) = \top$. By definition, this means that $\{\psi, \neg\psi\} \subseteq \mathcal{T}^*$. By the axiom [b], this implies that $\circ\psi \notin \mathcal{T}^*$, and so $\nu(\circ\psi) = f$.

- Assume $\nu(\psi) \in \{t, f\}$. Then either $\psi \notin \mathcal{T}^*$, or $\neg\psi \notin \mathcal{T}^*$. By property 4 of $\mathcal{T}^*$ (see the proof of Theorem 7.50), this implies that $\neg\psi \wedge \psi \notin \mathcal{T}^*$. Hence it follows from property 3 of $\mathcal{T}^*$ that in the presence of axiom [k] necessarily $\circ\psi \in \mathcal{T}^*$, and so $\nu(\circ\psi) \in \{t, \top\}$ by definition of ν. □

Proposition 8.41 **B** *and* **Bk** *are normal, boldly $\neg$-paraconsistent, and decidable.*

Proof. By Theorem 8.40, with the help of Corollary 6.50 in the case of bold paraconsistency, and Propositions 8.4 and 8.28 (or Proposition 8.39) in the case of normality. □

We turn to the construction of quasi-canonical Gentzen-type systems for **B** and **Bk**. The main tool used in this chapter for such a task[6] is the following proposition, whose easy verification we leave to the reader.

Proposition 8.42 *Let $\mathcal{M}$ be an extension of some refinement of $\mathcal{M}_4^b(\{[t]\})$, and let ν be an $\mathcal{M}$-valuation. With the notion of satisfaction of sequents given in Definition 1.123 we have:*

1. $\nu(\psi) = t$ iff ν satisfies $\neg\psi \Rightarrow$.

2. $\nu(\psi) = f$ iff ν satisfies $\psi \Rightarrow$.

3. $\nu(\psi) = \top$ iff ν satisfies both $\Rightarrow \psi$ and $\Rightarrow \neg\psi$.

4. $\nu(\psi) \in \{f, \top\}$ iff ν satisfies $\Rightarrow \neg\psi$.

5. $\nu(\psi) \in \{t, \top\}$ iff ν satisfies $\Rightarrow \psi$.

6. $\nu(\psi) \in \{t, f\}$ iff ν satisfies $\psi, \neg\psi \Rightarrow$.

---

[6] Actually, an obvious generalization to the four-valued case could have been used to construct also the various quasi-canonical Gentzen-type systems which were used in the previous chapters. See also Note 7.57.

Definition 8.43 (*GB*, *GBk*) *GB* is the Gentzen-style system obtained from LK^+ by adding the rule $[\Rightarrow \neg]$ (Figure 1.3) and the rule $[\circ \Rightarrow]$ below. *GBk* is the system obtained from *GB* by adding to it the rule $[\Rightarrow \circ]$ below.

$$[\circ \Rightarrow] \quad \frac{\Gamma \Rightarrow \psi, \Delta \quad \Gamma \Rightarrow \neg\psi, \Delta}{\Gamma, \circ\psi \Rightarrow \Delta} \qquad [\Rightarrow \circ] \quad \frac{\Gamma, \psi, \neg\psi \Rightarrow \Delta}{\Gamma \Rightarrow \circ\psi, \Delta}$$

Note 8.44 *GB* and *GBk* are obtained by translating the truth tables of $\mathcal{M}_B$ and $\mathcal{M}_{Bk}$ into sequent rules, using Proposition 8.42. Take, for example, the tables for $\circ$ in $\mathcal{M}_{Bk}$. According to them: (1) if $v(\varphi) = \top$ then $\nu(\circ\varphi) = f$, and (2) if $\nu(\varphi) \in \{t, f\}$ then $\nu(\circ\varphi) \in \{t, \top\}$. Using Facts 3 and 2 in Proposition 8.42, (1) translates (after adding contexts) to the rule $[\circ \Rightarrow]$ above. Similarly, (2) translates to the rule $[\Rightarrow \circ]$ using Facts 5 and 6.[7]

Proposition 8.45 *The rules $[\Rightarrow \circ]$ and $[\circ \Rightarrow]$ are invertible in GBk.*

Proof. For the invertibility of $[\circ \Rightarrow]$, we show that $\Gamma, \circ\psi \Rightarrow \Delta \vdash_{GBk} \Gamma \Rightarrow \psi, \Delta$ and $\Gamma, \circ\psi \Rightarrow \Delta \vdash_{GBk} \Gamma \Rightarrow \neg\psi, \Delta$. Here are the derivations:

$$\frac{\Gamma, \circ\psi \Rightarrow \Delta \quad \dfrac{\Gamma, \psi, \neg\psi \Rightarrow \psi, \Delta}{\Gamma \Rightarrow \circ\psi, \psi, \Delta}}{\Gamma \Rightarrow \psi, \Delta} \qquad \frac{\Gamma, \circ\psi \Rightarrow \Delta \quad \dfrac{\Gamma, \psi, \neg\psi \Rightarrow \neg\psi, \Delta}{\Gamma \Rightarrow \circ\psi, \neg\psi, \Delta}}{\Gamma \Rightarrow \neg\psi, \Delta}$$

The proof in the case of $[\Rightarrow \circ]$ is similar (using $[\circ \Rightarrow]$ instead of $[\Rightarrow \circ]$). □

Theorem 8.46 (completeness and cut admissibility of *GB*, *GBk*)

1. *GB is sound and complete for* **B**, *and GBk — for* **Bk**.

2. *GB and GBk enjoy cut-admissibility.*[8]

Proof. The proofs for both systems rely on Theorem 8.40, and closely follow the proof of the corresponding theorem for $LK^+(\{[t]\})$ and $\mathcal{M}_4^b(\{[t]\})$. (See Theorem 7.60.)

It is easy to check that $[\circ \Rightarrow]$ is valid in $\mathcal{M}_B$ (and so also in $\mathcal{M}_{Bk}$), and that $[\Rightarrow \circ]$ is valid in $\mathcal{M}_{Bk}$. Given the soundness of $LK^+(\{[t]\})$ for $\mathcal{M}_4^b(\{[t]\})$, this implies the soundness of *GB* for $\mathcal{M}_B$ and of *GBk* for $\mathcal{M}_{Bk}$.

Using Theorem 6.18, we again prove together the completeness part and cut-elimination by showing that any sequent in $\mathcal{L}_C$ either has a cut-free proof in *GB* (*GBk*) or it is not valid in $\mathcal{M}_B$ ($\mathcal{M}_{Bk}$).

We start with the case of *GB*. So suppose that $\Gamma_0 \Rightarrow \Delta_0$ has no cut-free proof in *GB*. We first extend it to a saturated sequent (see the proof of Theorem 7.60) $\Gamma \Rightarrow \Delta$ that still has no cut-free proof in *GB*. Then $\Gamma \Rightarrow \Delta$ satisfies the following conditions:

---

[7]This approach constitutes an application of the general method provided in [62] for generating cut-free sequent calculi for logics induced by finite Nmatrices.

[8]Again, this part can easily be seen to be an instance of Theorem 64 from [50].

8.2 The Basic C-systems

- If $\neg\varphi \in \Delta$, then $\varphi \in \Gamma$.
- If $\circ\varphi \in \Gamma$ then either $\varphi \in \Delta$ or $\neg\varphi \in \Delta$.
- If $\varphi \wedge \psi \in \Gamma$, then $\varphi \in \Gamma$ and $\psi \in \Gamma$.
- If $\varphi \wedge \psi \in \Delta$, then $\varphi \in \Delta$ or $\psi \in \Delta$.
- Similar conditions concerning the rules for $\vee$ and $\supset$.

Next we recursively define a valuation ν in $\mathcal{M}_B$ which refutes $\Gamma_0 \Rightarrow \Delta_0$.

- If φ is a propositional variable:

$$\nu(\varphi) = \begin{cases} f & \varphi \in \Delta, \\ t & \neg\varphi \in \Delta, \\ \top & \text{otherwise.} \end{cases}$$

- If $\varphi = \diamond(\varphi_1, \ldots, \varphi_n)$ where $\diamond \in \{\neg, \wedge, \vee, \supset\}$ and n is the arity of $\diamond$:

$$\nu(\varphi) = \begin{cases} a & \tilde{\diamond}(\nu(\varphi_1), \ldots, \nu(\varphi_n)) = \{a\}, \\ \top & \tilde{\diamond}(\nu(\varphi_1), \ldots, \nu(\varphi_n)) = \{t, \top\} \text{ and } \neg\varphi \in \Gamma, \\ t & \tilde{\diamond}(\nu(\varphi_1), \ldots, \nu(\varphi_n)) = \{t, \top\} \text{ and } \neg\varphi \notin \Gamma, \end{cases}$$

- If $\varphi = \circ\psi$:

$$\nu(\varphi) = \begin{cases} f & \nu(\psi) = \top \text{ or } \varphi \in \Delta, \\ t & \nu(\psi) \neq \top \text{ and } \neg\varphi \in \Delta, \\ \top & \text{otherwise.} \end{cases}$$

Since the presence of the rule for [t] (i.e. [$\Rightarrow \neg$]) excludes the possibility that both $\varphi \in \Delta$ and $\neg\varphi \in \Delta$, ν is well-defined. From its definition it is also immediate that it is a valuation in $\mathcal{M}_B$. Now we prove that it has the following properties:

1. If $\varphi \in \Gamma$ then $\nu(\varphi) \in \{t, \top\}$.
2. If $\neg\varphi \in \Gamma$ then $\nu(\varphi) \in \{f, \top\}$.
3. If $\varphi \in \Delta$ then $\nu(\varphi) = f$.
4. If $\neg\varphi \in \Delta$ then $\nu(\varphi) = t$.

If φ is a propositional variable then this follow from the fact that $\Gamma \cap \Delta = \emptyset$. If $\varphi = \diamond(\varphi_1, \ldots, \varphi_n)$ where $\diamond \in \{\neg, \wedge, \vee, \supset\}$, then the argument is similar to that given for this case in the proof Theorem 7.60 (but is simpler, because we deal only with t, f, and $\top$). Finally, assume that $\varphi = \circ\psi$.

- Suppose that $\varphi \in \Gamma$. Since $\Gamma \Rightarrow \Delta$ is saturated, this implies that $\psi \in \Delta$ or $\neg\psi \in \Delta$. Therefore it follows by the induction hypothesis for ψ that either $\nu(\psi) = f$ or $\nu(\psi) = t$, implying that $\nu(\psi) \neq \top$. In addition, $\varphi \notin \Delta$ because $\Gamma \cap \Delta = \emptyset$. Hence $\nu(\varphi) \neq f$, and so $\nu(\varphi) \in \{t, \top\}$.

- If $\neg\varphi \in \Gamma$ then $\neg\varphi \notin \Delta$. Hence $\nu(\varphi) \neq t$, and so $\nu(\varphi) \in \{f, \top\}$.

- Suppose that $\varphi \in \Delta$. Then $\nu(\varphi) = f$ by the definition of ν.

- Suppose that $\neg\varphi \in \Delta$. Since $\Gamma \Rightarrow \Delta$ is saturated, it follows that $\varphi \in \Gamma$. As shown above, this implies that $\nu(\psi) \neq \top$. Hence $\nu(\varphi) = t$ by definition of ν.

Since $\Gamma_0 \subseteq \Gamma$ and $\Delta_0 \subseteq \Delta$, it again follows from the above four properties of ν that it is indeed a refutation of $\Gamma_0 \Rightarrow \Delta_0$.

The proof of the theorem in the case of GBk is almost identical to the proof just given for the case of GB. The only difference is that in the case where $\varphi = \circ\psi$ in the definition of ν we demand $\nu(\varphi)$ to be f only in case $\nu(\psi) = \top$. As a result, the fact that $\nu(\varphi) = f$ in case $\varphi = \circ\psi$ and $\varphi \in \Delta$ does not directly follow anymore from the definition of ν. However, in the case of GBk the fact that $\Gamma \Rightarrow \Delta$ is saturated implies that if $\circ\psi \in \Delta$ then both $\psi \in \Gamma$ and $\neg\psi \in \Gamma$. Therefore it follows from the induction hypothesis for ψ that if $\circ\psi \in \Delta$ then necessarily $\nu(\psi) = \top$, and so $\nu(\circ\psi) = f$. □

One important property of GB (shared by GBk and any other extension of GB) is that it allows a stronger reduction of **CL** to **B** (and so, by Corollary 8.30 to any regular $\circ$-C-system) than that given in Theorem 8.7.

Lemma 8.47 *The following rule is derivable in GB without using cut:*

$$[\circ\neg\Rightarrow] \quad \frac{\Gamma \Rightarrow \psi, \Delta}{\Gamma, \circ\psi, \neg\psi \Rightarrow \Delta}.$$

Proof. By applying $[\circ \Rightarrow]$ to $\Gamma, \neg\psi \Rightarrow \psi, \Delta$ and $\Gamma, \neg\psi \Rightarrow \neg\psi, \Delta$ we get $\Gamma, \circ\psi, \neg\psi \Rightarrow \Delta$. □

Proposition 8.48 *Let $\mathcal{T} \cup \{\varphi\}$ be a set of formulas in $\mathcal{L}_{CL}$. If $\mathcal{T} \vdash_{\mathbf{CL}} \varphi$ then there is a cut-free proof in GB of a sequent of the form $\circ\Delta, \Gamma \Rightarrow \varphi$ in which $\Gamma \subseteq \mathcal{T}$ and Δ consists of subformulas of $\Gamma \cup \{\varphi\}$.*

Proof. Let GB' be the system obtained from GB by replacing the rule $[\circ \Rightarrow]$ with the rule $[\circ\neg\Rightarrow]$ from Lemma 8.47.
Suppose $\mathcal{T} \vdash_{\mathbf{CL}} \varphi$. Then for some finite $\Gamma \subseteq \mathcal{T}$ there is a cut-free proof in LK of $\Gamma \Rightarrow \varphi$. Replace in that proof every application of $[\neg \Rightarrow]$ by an

8.2 The Basic C-systems

application of $[\circ \neg \Rightarrow]$. We get by this a cut-free proof in GB' of a sequent of the form $\circ\Delta, \Gamma \Rightarrow \varphi$ in which Δ consists of subformulas of $\Gamma \cup \{\varphi\}$. The proposition follows therefore from Lemma 8.47. □

From Propositions 8.27 and 8.29 it follows that the study of (strong) C-systems is the study of certain axiomatic extensions of **B** (**Bk**). The simplest such extensions are those which are obtained by adding axioms from the set A_{PAC} (Definition 7.35). The main results of this section can easily be generalized to these extensions.

Definition 8.49 (**B**(S), **Bk**(S), $HB(S)$, $HBk(S)$, $GB(S)$, $GBk(S)$)
Let $S \subseteq \mathcal{W}(\mathcal{L}_C)$.

- $HB(S)$ and $HBk(S)$ are the axiomatic extensions of HB and HBk (respectively) with the axioms in S.

- **B**$(S) = \mathbf{L}_{HB(S)}$, **Bk**$(S) = \mathbf{L}_{HBk(S)}$. (That is: **B**$(S)$ and **Bk**(S) are the logics which are induced by $HB(S)$ and $HBk(S)$, respectively.)

- Suppose R is a function that associates a Gentzen-type rule with each $ax \in S$. Then $GB_R(S)$ and $GBk_R(S)$ (or just $GB(S)$ and $GBk(S)$ respectively, in case the identity of R is clear from the context) are the Gentzen-type system obtained from GB and GBk (respectively) by adding to them the rules in the set $\{R(ax) \mid ax \in S\}$.

Definition 8.50 ($\mathcal{M}_B(S)$, $\mathcal{M}_{Bk}(S)$) Let $S \subseteq \mathcal{W}(\mathcal{L}_C)$, and suppose that C is a function that associates semantic conditions on $\mathcal{M}_B$ ($\mathcal{M}_{Bk}$) with each $ax \in S$. $\mathcal{M}_B^C(S)$ ($\mathcal{M}_{Bk}^C(S)$) is the weakest refinement (Definition 7.43), if such exists, of $\mathcal{M}_B$ (of $\mathcal{M}_{Bk}(S)$) in which all the conditions in $C(S) = \{C(x) \mid x \in S\}$ are satisfied. (Since the identity of C will always be clear from the context, in what follows we shall write just $\mathcal{M}_B(S)$ ($\mathcal{M}_{Bk}(S)$) instead of $\mathcal{M}_B^C(S)$ ($\mathcal{M}_{Bk}^C(S)$).)

Note 8.51 For every S and C, the Nmatrices $\mathcal{M}_B^C(S)$ and $\mathcal{M}_{Bk}^C(S)$ are $\neg$-fundamental (Definition 7.1) whenever they exist.

Proposition 8.52 $\mathcal{M}_B^C(S)$ and $\mathcal{M}_{Bk}^C(S)$ exist for every $S \subseteq A_{PAC}$ (where C is as given in Definition 7.41).

Proof. It is easy to see that $\mathcal{M}_B(S)$ ($\mathcal{M}_{Bk}(S)$) is just the extension of $\mathcal{M}_4^b(S \cup \{[\mathsf{t}]\})$ (Definition 7.44) to $\mathcal{L}_C$ in which the interpretation $\tilde{\circ}$ of $\circ$ is like in $\mathcal{M}_B$ ($\mathcal{M}_{Bk}$). □

Theorem 8.53 Let $S \subseteq A_{PAC}$, and let R be the function given in Figures 7.1 and 7.2.

1. $\mathbf{B}(S)$ is a regular $\circ$-C-system, and $\mathbf{Bk}(S)$ is a strong $\circ$-C-system.

2. $\wedge, \vee$ and $\supset$ are $\mathcal{M}_{Bk}(S)$ ($\mathcal{M}_B(S)$)-conjunction, $\mathcal{M}_{Bk}(S)$ ($\mathcal{M}_B(S)$)-disjunction and $\mathcal{M}_{Bk}(S)$ ($\mathcal{M}_B(S)$)-implication, respectively.

3. $\mathcal{M}_B(S)$ is characteristic for $\mathbf{B}(S)$, and $\mathcal{M}_{Bk}(S)$ — for $\mathbf{Bk}(S)$.

4. $\mathbf{B}(S)$ and $\mathbf{Bk}(S)$ are normal, boldly $\neg$-paraconsistent, and decidable.

5. $GB(S)$ is sound and complete for $\mathbf{B}(S)$, and $GBk(S)$ — for $\mathbf{Bk}(S)$. Both systems enjoy cut-admissibility.

6. $\mathbf{B}(S)$ and $\mathbf{Bk}(S)$ have no finite characteristic matrix (or even a finite family $\mathcal{F}$ of finite ordinary matrices that induces it), and no weakly characteristic two-valued Nmatrix.

7. $\mathbf{B}(S)$ and $\mathbf{Bk}(S)$ are conservative extensions of $\mathbf{L}_{HCL^+(S \cup \{[t]\})}$.

8. $\mathbf{B}(S)$ and $\mathbf{Bk}(S)$ are sparse.

Proof.

1. Similar to the proof of Proposition 8.28.

2. Similar to the proof of Proposition 8.39.

3. The proof is similar to the proof of Theorem 8.40, only this time it closely follows that of Theorem 7.50 in the case of $\mathcal{M}_4^b(S \cup \{[t]\})$ (rather than the case of $\mathcal{M}_4^b(\{[t]\})$).

4. Similar to the proof of Proposition 8.41.

5. Similar to the proof of Theorem 8.46, closely following the proof of corresponding theorem for $LK^+(S \cup \{[t]\})$ and $\mathcal{M}_4^b(S \cup \{[t]\})$. (See Theorem 7.60.)

6. The first part follows from the third item and Theorem 7.4. The second part follows from the first item and Proposition 8.35.

7. This easily follows from the third item of this theorem, and from Theorem 7.50, using the observation made in the proof of Proposition 8.52. (Alternatively, the claim follows from the fifth item of this theorem, and from Corollary 7.61.)

8. Immediate from the third item and Theorem 7.7. □

8.3 Propagation of Consistency

Because of Proposition 8.29 (as well as Corollary 8.30 and Proposition 8.31), **B** is often considered to be 'the most basic C-system', and so it is called **mbC** in [111, 113, 116]. However, for several reasons **Bk** is in our opinion more appropriate for this role. First, as explained in Note 8.11, while [b] means that no formula is both consistent and contradictory, [k] complements this by saying that every formulas is either consistent or contradictory. This principle seems to be no less essential for the intended meaning of $\circ\varphi$ than that expressed by [b]. Second, Note 8.32 shows that **Bk** has a certain completeness property that **B** lacks, which again indicates that something is missing in the latter. Third, [k] is anyway a theorem of almost every important C-system ever studied. This is due to the fact that it is derivable in **B** from three important axioms related to $\circ$, which are studied in the sequel: [i], [l], and [d]. (See Proposition 8.84.) Forth, the invertibility of the rules for $\circ$ in the Gentzen-type system GBk (Proposition 8.45) also indicates that **Bk** is the more natural choice for the basic C-system. Therefore, from now on we focus on extensions of **Bk**.

Already in the previous section we investigate extensions of **Bk** with the most standard axioms concerning negation. Next we study the effects of axioms which are related to properties of the consistency operator $\circ$. (Note that except for [b] and [k], which together may be viewed as providing a "definition" of $\circ$, **Bk** does not include any such axioms.) This section is devoted to the study of a special (and particularly important) type of such axioms: those that deal with consistency propagation. (Other important axioms connected with $\circ$ are dealt with in the next section.) The list of these axioms is provided in Figure 8.1. It is divided between two sets: the a-axioms, and the o-axioms. These sets describe different ways of consistency propagation. The safer (for retaining strong paraconsistency) a-axioms require *every* immediate proper subformula of a complex formula φ to be consistent in order to force φ to be consistent, while in the o-axioms it suffices for this that *some* immediate proper subformula of φ is consistent.

One of the most important application of the a-axioms is that they allow to considerably strengthen the crucial Theorem 8.7.

Theorem 8.54 *Let* **L** *be a regular $\circ$-C-system in which all the a-axioms (with the possible exception of $a_\circ$) are provable, and let $\mathcal{T} \cup \{\psi\}$ be a set of formulas in $\mathcal{L}_{CL}$. Then $\mathcal{T} \vdash_{\mathbf{CL}} \psi$ iff there exists a finite set Δ of propositional variables which occur in $\mathcal{T} \cup \{\psi\}$ such that $\circ\Delta \cup \mathcal{T} \vdash_{\mathbf{L}} \psi$.*

Proof. Using the a-axioms, it is easy to prove by induction on the structure of $\varphi \in \mathcal{L}_{CL}$, that $\{\circ p \mid p \in \mathsf{Var}(\varphi)\} \vdash_{\mathbf{L}} \circ\varphi$. The theorem is an immediate corollary of this observation and Theorem 8.7. □

$$
\begin{array}{ll}
[\mathsf{a}_\neg] & \circ\varphi \supset \circ\neg\varphi \\
[\mathsf{a}_\wedge] & (\circ\varphi \wedge \circ\psi) \supset \circ(\varphi \wedge \psi) \\
[\mathsf{a}_\vee] & (\circ\varphi \wedge \circ\psi) \supset \circ(\varphi \vee \psi) \\
[\mathsf{a}_\supset] & (\circ\varphi \wedge \circ\psi) \supset \circ(\varphi \supset \psi) \\
[\mathsf{a}_\circ] & \circ\varphi \supset \circ\circ\varphi \\
[\mathsf{o}^1_\wedge] & \circ\varphi \supset \circ(\varphi \wedge \psi) \\
[\mathsf{o}^2_\wedge] & \circ\psi \supset \circ(\varphi \wedge \psi) \\
[\mathsf{o}^1_\vee] & \circ\varphi \supset \circ(\varphi \vee \psi) \\
[\mathsf{o}^2_\vee] & \circ\psi \supset \circ(\varphi \vee \psi) \\
[\mathsf{o}^1_\supset] & \circ\varphi \supset \circ(\varphi \supset \psi) \\
[\mathsf{o}^2_\supset] & \circ\psi \supset \circ(\varphi \supset \psi) \\
\end{array}
$$

Figure 8.1: Propagation axioms for the connectives of $\mathcal{L}_C$

Notation 8.55 ([o],[a])

- For $\sharp \in \{\wedge, \vee, \supset\}$, the axioms $[\mathsf{o}^1_\sharp]$ and $[\mathsf{o}^2_\sharp]$ are sometimes combined into one axiom: $[\mathsf{o}_\sharp]$ $(\circ\varphi \vee \circ\psi) \supset \circ(\varphi \sharp \psi)$.

- We denote by $[\mathsf{o}]$ the set $\{[\mathsf{o}_\sharp] \mid \sharp \in \{\wedge, \vee, \supset\}\}$, and by $[\mathsf{a}]$ the set $\{[\mathsf{a}_\sharp] \mid \sharp \in \{\wedge, \vee, \supset\}\}$.

- Let $\mathbf{A} = [\mathsf{a}] \cup \{[\mathsf{a}_\neg], [\mathsf{a}_\circ]\}$, and let $\mathbf{O} = [\mathsf{o}] \cup \{[\mathsf{a}_\neg], [\mathsf{a}_\circ]\}$.

Note 8.56 Using the instance $\circ\circ\varphi \vee (\circ\varphi \wedge \neg \circ \varphi)$ of [k], it is easy to show that $\circ\varphi \supset \circ\circ\varphi$ is equivalent in **Bk** to $\circ\circ\varphi$. Hence the latter can be adopted as an axiom instead of $[\mathsf{a}_\circ]$.

In the rest of this section we investigate the family of strong C-systems which are obtained by extending **Bk** with combinations of axioms from $\mathbf{A}_{PAC}$ (Definition 7.35) and the propagation axioms above.

Definition 8.57 ($\mathbf{A}_C$, $\mathbf{A}_{CPAC}$)

- $\mathbf{A}_C$ is the set of axioms given in Figure 8.1.

- $\mathbf{A}_{CPAC} = \mathbf{A}_C \cup \mathbf{A}_{PAC}$.

8.3 Propagation of Consistency

In the next two propositions we determine the connections between the axioms in A_{CPAC}. First, we observe that two of the [a]-axioms are equivalent in **Bk** to axioms in A_{PAC}. (Hence they are not really needed in A_{CPAC}.)

Proposition 8.58 *In* **Bk** $[a_\neg]$ *is equivalent to* $[c]$, *and* $[a_\wedge]$ — *to* $[n^!_\wedge]$.

Proof. Using Theorem 8.40 and $\mathcal{M}_{Bk}$, it can easily be checked that for every particular φ, the corresponding instances of $[c]$ and $[a_\neg]$ imply each other in **Bk**. Similarly, for every φ and ψ the corresponding instances of $[a_\wedge]$ and $[n^!_\wedge]$ imply each other in **Bk**. We show for example that

$$\neg(\varphi \wedge \psi) \supset (\neg\varphi \vee \neg\psi) \vdash_{\mathbf{Bk}} (\circ\varphi \wedge \circ\psi) \supset \circ(\varphi \wedge \psi)$$

So, let ν be a valuation in $\mathcal{M}_{Bk}$ for which $\nu((\circ\varphi \wedge \circ\psi) \supset \circ(\varphi \wedge \psi)) = f$. Then $\nu(\circ\varphi) \neq f$, $\nu(\circ\psi) \neq f$, and $\nu(\circ(\varphi \wedge \psi)) = f$. Hence $\nu(\varphi) \in \{t, f\}$, $\nu(\psi) \in \{t, f\}$, and $\nu(\varphi \wedge \psi) = \top$. This is possible in $\mathcal{M}_{Bk}$ only if $\nu(\varphi) = \nu(\psi) = t$ and $\nu(\varphi \wedge \psi) = \top$. But in this case $\nu(\neg(\varphi \wedge \psi)) \neq f$, while $\nu(\neg\varphi) = \nu(\neg\psi) = f$, implying that $\nu(\neg(\varphi \wedge \psi) \supset (\neg\varphi \vee \neg\psi)) = f$. □

Proposition 8.59 *The following dependencies hold in* **Bk**:

1. $[a_\sharp]$ *follows from* $[o^i_\sharp]$ *for* $\sharp \in \{\wedge, \vee, \supset\}$ *and* $i \in \{1, 2\}$.

2. $[a_\vee]$ *follows from* $\{[n^!_\vee]_1, [n^!_\vee]_2\}$.

3. $[n^!_\vee]_1$ *follows from* $[o^1_\vee]$.

4. $[n^!_\vee]_2$ *follows from* $[o^2_\vee]$.

5. $[o^2_\supset]$ *follows from* $\{[n^!_\supset]_1, [n^!_\supset]_2\}$.

6. $[n^!_\supset]_1$ *follows from* $[o^1_\supset]$.

7. $[n^!_\supset]_2$ *follows from* $[o^2_\supset]$.

Proof. Similar to that of Proposition 8.58. □

Note 8.60 None of the converses of the dependencies noted in Proposition 8.59 hold, and there are no other dependencies among the axioms in A_{CPAC} beyond those that follow from Propositions 8.58 and 8.59. However, to show these facts is more difficult than the proofs of these two propositions. Thus to show that $[n^!_\vee]_1$ does not follow in **Bk** from $[a_\vee]$ it is *not* sufficient to show that $\neg(\varphi \vee \psi) \supset \neg\varphi$ does not follow in **Bk** from $(\circ\varphi \wedge \circ\psi) \supset \circ(\varphi \vee \psi)$. The reason is that the addition of $[a_\vee]$ to **Bk** means the addition of *all* its instances to **Bk**, and so we have to show that $\neg(\varphi \vee \psi) \supset \neg\varphi$ does not follow in **Bk** from the set of all these instances. Still, the task does become straightforward with the help of the semantic characterizations of the various axioms and logics. These characterizations are our next topic.

Our semantic characterizations of the family of logics which are induced by the axioms in Definition 8.57 again employ the framework of Nmatrices. Again each of the axioms in A_{CPAC} corresponds to a condition which this time leads to a certain refinement of the basic Nmatrix $\mathcal{M}_{Bk}$. The list of these conditions is given in the next definition. Note that the refining conditions given there for the axioms of A_{PAC} are just the simplifications of those given in Definition 7.41 for the case in which $\mathcal{V} = \{t, f, \top\}$.

Definition 8.61 (refining conditions for A_{CPAC}) For $ax \in A_{CPAC}$, the refining condition $C(ax)$ is defined as follows (where $x \in \{f, t, \top\}$):

$C([c])$: $\tilde{\neg} f = \{t\}$.

$C([e])$: $\tilde{\neg} \top = \{\top\}$

$C([n_\wedge^l])$: $t \tilde{\wedge} t = \{t\}$.

$C([n_\wedge^r]_1)$: $\top \tilde{\wedge} t = \top \tilde{\wedge} \top = \{\top\}$.

$C([n_\wedge^r]_2)$: $t \tilde{\wedge} \top = \top \tilde{\wedge} \top = \{\top\}$.

$C([n_\vee^l]_1)$: $t \tilde{\vee} x = \{t\}$.

$C([n_\vee^l]_2)$: $x \tilde{\vee} t = \{t\}$.

$C([n_\vee^r])$: $f \tilde{\vee} \top = \top \tilde{\vee} f = \top \tilde{\vee} \top = \{\top\}$.

$C([n_\supset^l]_1)$: $f \tilde{\supset} x = \{t\}$.

$C([n_\supset^l]_2)$: $x \tilde{\supset} t = \{t\}$.

$C([n_\supset^r])$: $t \tilde{\supset} \top = \top \tilde{\supset} \top = \{\top\}$.

$C([a_\neg])$: $\tilde{\neg} f = \{t\}$.

$C([a_\circ])$: $\tilde{\circ} t = \tilde{\circ} f = \{t\}$.

$C([a_\wedge])$: $t \tilde{\wedge} t = \{t\}$.

$C([a_\vee])$: $t \tilde{\vee} t = t \tilde{\vee} f = f \tilde{\vee} t = \{t\}$.

$C([a_\supset])$: $f \tilde{\supset} f = f \tilde{\supset} t = t \tilde{\supset} t = \{t\}$.

$C([o_\wedge^1])$: $t \tilde{\wedge} t = t \tilde{\wedge} \top = \{t\}$.

$C([o_\wedge^2])$: $t \tilde{\wedge} t = \top \tilde{\wedge} t = \{t\}$.

$C([o_\vee^1])$: $t \tilde{\vee} x = f \tilde{\vee} t = f \tilde{\vee} \top = \{t\}$.

8.3 Propagation of Consistency

$C([\mathsf{o}_\vee^2])$: $x\tilde{\vee}t = t\tilde{\vee}f = \mathsf{T}\tilde{\vee}f = \{t\}$.

$C([\mathsf{o}_\supset^1])$: $f\tilde{\supset}x = t\tilde{\supset}t = t\tilde{\supset}\mathsf{T} = \{t\}$.

$C([\mathsf{o}_\supset^2])$: $x\tilde{\supset}t = f\tilde{\supset}f = \{t\}$.

Note 8.62 Unsurprisingly (see Proposition 8.58), $C([\mathsf{a}_\neg]) = \mathsf{C}([\mathsf{c}])$ and $C([\mathsf{a}_\wedge]) = \mathsf{C}([\mathsf{n}_\wedge^l])$. We have included all of them in the list above for the reader convenience.[9]

Note 8.63 Examples how the above refining conditions are derived have been given before and after Definition 7.41. As another example, this time of an axiom in A_C, consider the axiom $[\mathsf{a}_\vee]$. To refute it, we should have an assignment ν such that $\nu(\circ\varphi) \in \{t, \mathsf{T}\}$, $\nu(\circ\psi) \in \{t, \mathsf{T}\}$, and $\nu(\circ(\varphi\vee\psi)) = f$. In $\mathcal{M}_{Bk}$ this is possible iff $\nu(\varphi) \in \{t, f\}$, $\nu(\psi) \in \{t, f\}$, and $\nu(\varphi \vee \psi) = \mathsf{T}$. Since in $\mathcal{M}_{Bk}$ $\mathsf{T} \notin f\tilde{\vee}f$, to guarantee that no such refutation exists we should ensure that $\mathsf{T} \notin t\tilde{\vee}t$, $\mathsf{T} \notin t\tilde{\vee}f$, and $\mathsf{T} \notin f\tilde{\vee}t$. In view of the truth table of $\vee$ in $\mathcal{M}_{Bk}$, this is equivalent to $C([\mathsf{a}_\vee])$ above.

Using the refining conditions given above, we can now extend the function R given in Figures 7.1 and 7.2 (in Chapter 7) to the whole of A_{CPAC}. This extension is presented in Figures 8.2 and 8.3, which list the Gentzen-type rules that correspond to the consistency propagation axioms.

Note 8.64 To see how these Gentzen-type rules are derived, consider, e.g., the axiom $(\mathsf{a}_\vee)$. The refining conditions induced by this axioms mean that for $b \in \{t, f\}$: (i) $t \vee b = \{t\}$, and (ii) $b \vee t = \{t\}$. Using Proposition 8.42, (i) can be reformulated as follows: if $\neg\varphi \Rightarrow$ and $\psi, \neg\psi \Rightarrow$ are true, then $\neg(\varphi \vee \psi) \Rightarrow$ is true. By adding context we obtain:

$$\frac{\Gamma, \neg\varphi \Rightarrow \Delta \quad \Gamma, \psi, \neg\psi \Rightarrow \Delta}{\Gamma, \neg(\varphi \vee \psi) \Rightarrow \Delta}$$

Similarly, (ii) translates into the other rule of $R([\mathsf{a}_\vee])$.[10]

Definition 8.65 ($HBk(S)$, **Bk**(S), $\mathcal{M}_{Bk}(S)$, $G\mathcal{E}k(S)$) For $S \subseteq \mathsf{A}_{CPAC}$, $HBk(S)$, **Bk**(S), $\mathcal{M}_{Bk}(S)$, and $GBk(S)$ are defined like in Definitions 8.49 and 8.50, using the extended functions C and R introduced above.

---

[9]Actually, observing these equalities is what has suggested Proposition 8.58, and they can indeed be used (with the help of the soundness and completeness theorem proved below) for providing an alternative proof of that proposition.

[10]Note that what we have just described is again a (simplified) application of the general method provided in [62] for generation of cut-free sequent calculi.

	ax	$R(ax)$
$[a_\circ]$	$\circ\varphi \supset \circ\circ\varphi$	$\dfrac{\Gamma, \neg\varphi, \varphi \Rightarrow \Delta}{\Gamma, \neg\circ\varphi \Rightarrow \Delta}$
$[a_\neg]$	$\circ\varphi \supset \circ\neg\varphi$	$\dfrac{\Gamma, \varphi \Rightarrow \Delta}{\Gamma, \neg\neg\varphi \Rightarrow \Delta}$
$[a_\wedge]$	$(\circ\varphi \wedge \circ\psi) \supset \circ(\varphi \wedge \psi)$	$\dfrac{\Gamma, \neg\varphi \Rightarrow \Delta \quad \Gamma, \neg\psi \Rightarrow \Delta}{\Gamma, \neg(\varphi \wedge \psi) \Rightarrow \Delta}$
$[a_\vee]$	$(\circ\varphi \wedge \circ\psi) \supset \circ(\varphi \vee \psi)$	$\dfrac{\Gamma, \neg\psi \Rightarrow \Delta \quad \Gamma, \neg\varphi, \varphi \Rightarrow \Delta}{\Gamma, \neg(\varphi \vee \psi) \Rightarrow \Delta}$ $\dfrac{\Gamma, \neg\varphi \Rightarrow \Delta \quad \Gamma, \neg\psi, \psi \Rightarrow \Delta}{\Gamma, \neg(\varphi \vee \psi) \Rightarrow \Delta}$
$[a_\supset]$	$(\circ\varphi \wedge \circ\psi) \supset \circ(\varphi \supset \psi)$	$\dfrac{\Gamma, \neg\varphi, \varphi \Rightarrow \Delta \quad \Gamma, \neg\psi \Rightarrow \Delta}{\Gamma, \neg(\varphi \supset \psi) \Rightarrow \Delta}$ $\dfrac{\Gamma, \neg\psi, \psi \Rightarrow \Delta \quad \Gamma, \varphi \Rightarrow \Delta}{\Gamma, \neg(\varphi \supset \psi) \Rightarrow \Delta}$

Figure 8.2: Gentzen rules induced by the a-axioms

Notation 8.66 In the sequel we shall frequently follow standard conventions made in the relevant literature concerning the names of systems. This means that we might omit parentheses of all sorts in such names in case there is no danger of confusion, and also employ only boldface letters. Thus we shall write **BkA** instead of **Bk(A)** (see Notation 8.55), **BkeA** instead of

8.3 Propagation of Consistency 255

	ax	R(ax)
$[\circ_\wedge^1]$	$\circ\varphi \supset \circ(\varphi \wedge \psi)$	$\dfrac{\Gamma, \neg\varphi \Rightarrow \Delta \quad \Gamma \Rightarrow \Delta, \psi}{\Gamma, \neg(\varphi \wedge \psi) \Rightarrow \Delta}$
$[\circ_\wedge^2]$	$\circ\psi \supset \circ(\varphi \wedge \psi)$	$\dfrac{\Gamma, \neg\psi \Rightarrow \Delta \quad \Gamma \Rightarrow \Delta, \varphi}{\Gamma, \neg(\varphi \wedge \psi) \Rightarrow \Delta}$
$[\circ_\vee^1]$	$\circ\varphi \supset \circ(\varphi \vee \psi)$	$\dfrac{\Gamma, \varphi \Rightarrow \Delta \quad \Gamma \Rightarrow \Delta, \psi}{\Gamma, \neg(\varphi \vee \psi) \Rightarrow \Delta}$ $\dfrac{\Gamma, \neg\varphi \Rightarrow \Delta}{\Gamma, \neg(\varphi \vee \psi) \Rightarrow \Delta}$
$[\circ_\vee^2]$	$\circ\psi \supset \circ(\varphi \vee \psi)$	$\dfrac{\Gamma, \psi \Rightarrow \Delta \quad \Gamma \Rightarrow \Delta, \varphi}{\Gamma, \neg(\varphi \vee \psi) \Rightarrow \Delta}$ $\dfrac{\Gamma, \neg\psi \Rightarrow \Delta}{\Gamma, \neg(\varphi \vee \psi) \Rightarrow \Delta}$
$[\circ_\supset^1]$	$\circ\varphi \supset \circ(\varphi \supset \psi)$	$\dfrac{\Gamma, \varphi \Rightarrow \Delta}{\Gamma, \neg(\varphi \supset \psi) \Rightarrow \Delta}$ $\dfrac{\Gamma, \neg\varphi \Rightarrow \Delta \quad \Gamma \Rightarrow \Delta, \psi}{\Gamma, \neg(\varphi \supset \psi) \Rightarrow \Delta}$
$[\circ_\supset^2]$	$\circ\psi \supset \circ(\varphi \supset \psi)$	$\dfrac{\Gamma, \neg\psi \Rightarrow \Delta}{\Gamma, \neg(\varphi \supset \psi) \Rightarrow \Delta}$ $\dfrac{\Gamma, \varphi \Rightarrow \Delta \quad \Gamma, \psi \Rightarrow \Delta}{\Gamma, \neg(\varphi \supset \psi) \Rightarrow \Delta}$

Figure 8.3: Gentzen rules induced by the o-axioms

Bk({[e]} ∪ A), **Bka** instead of **Bk**([a]), and **Bkca** instead of **Bk**({[c]} ∪ [a]).

Note 8.67 The names we employ here for the various C-systems are different from those that are used in [111, 113, 116] and in other works on the subject (including our own). First of all, the system called here **B**({[c]}) (or just **Bc** – see Notation 8.66) is (implicitly) denoted in [111, 113, 116] by **C**. In addition the set {[k], [a$_\circ$]} is (implicitly) denoted there by **i**. (See Proposition 8.85 below for the reason.) Therefore, the system called here **Bk**({[c], [a$_\circ$]}) (which by Proposition 8.58 is equivalent to **Bk**({[a$_\neg$], [a$_\circ$]})) is called in the previous literature **Ci**, while **BkA** is called **Cia**.

As opposed to the case of A$_{PAC}$ (Proposition 7.45), not all combinations of axioms from A$_{CPAC}$ are compatible with the demand of strong paraconsistency. Next we characterize all those that are.

Definition 8.68 (Bk-coherence) Let $S \subseteq \mathcal{W}(\mathcal{L}_C)$. We say that S is **Bk**-*coherent* (or just *coherent*), if the axiomatic extension of **Bk** by the formulas of S is strongly paraconsistent.

Theorem 8.69 *The following conditions on $S \subseteq$ A$_{CPAC}$ are equivalent:*

1. *S is **Bk**-coherent.*

2. *S does not contain any of the following pairs of axioms:*

    (a) [o$_\wedge^1$], [n$_\wedge^r$]$_2$;

    (b) [o$_\wedge^2$], [n$_\wedge^r$]$_1$;

    (c) [o$_\vee^1$], [n$_\vee^r$];

    (d) [o$_\vee^2$], [n$_\vee^r$];

    (e) [o$_\supset^1$], [n$_\supset^r$].

3. $\mathcal{M}_{Bk}(S)$ *exists.*

Proof. To show that the first condition implies the second, it suffices to show that if $S \subseteq$ A$_{CPAC}$ contains any of the pairs listed above then it is not strongly ¬-paraconsistent. This immediately follows from the following five facts (that can easily be verified using $\mathcal{M}_{Bk}$):

(a) $\circ\varphi \supset \circ(\varphi \wedge \psi), \neg\psi \supset \neg(\varphi \wedge \psi), \psi, \neg\psi \vdash_{\mathbf{Bk}} \neg\varphi$.

(b) $\circ\psi \supset \circ(\varphi \wedge \psi), \neg\varphi \supset \neg(\varphi \wedge \psi), \varphi, \neg\varphi \vdash_{\mathbf{Bk}} \neg\psi$.

(c) $\circ\varphi \supset \circ(\varphi \vee \psi), (\neg\varphi \wedge \neg\psi) \supset \neg(\varphi \vee \psi), \psi, \neg\psi \vdash_{\mathbf{Bk}} \varphi$.

(d) $\circ\psi \supset \circ(\varphi \vee \psi), (\neg\varphi \wedge \neg\psi) \supset \neg(\varphi \vee \psi), \varphi, \neg\varphi \vdash_{\mathbf{Bk}} \psi$.

8.3 Propagation of Consistency 257

(e) $\circ\varphi \supset \circ(\varphi \supset \psi), (\varphi \wedge \neg\psi) \supset \neg(\varphi \supset \psi), \psi, \neg\psi \vdash_{\mathbf{Bk}} \neg\varphi$.

Showing that the second condition implies the third involves just a straightforward (though tedious) mechanical check.

To complete the circle, supposed that $\mathcal{M}_{Bk}(S)$ exists. Then $\mathcal{M}_{Bk}(S)$ is a simple refinement of $\mathcal{M}_{Bk}$, and it is easy to check that $HBk(S)$ is sound for it. Since every simple refinement of $\mathcal{M}_{Bk}$ induces a strongly ¬-paraconsistent logic, it follows that so is $\mathbf{Bk}(S)$. □

Note 8.70 As can be seen from Theorem 8.69, the o-axioms are involved in all the conflicts among the semantic conditions that correspond to the axioms in A_{CPAC}. This fact and Theorem 8.69 demonstrate our claim above, that the a-axioms are safer and more natural than the o-axioms. (They were indeed the axioms used by da Costa in his original C-systems.)

Note 8.71 From the proof of Theorem 8.69 it immediately follows that $\mathbf{Bk}(S)$ is not even ¬-paraconsistent in the third and forth cases. It is not too difficult to show that the same is true in the fifth case as well. [11] This is not necessarily what happens in the first two cases. Take for example $S = \{[\mathsf{o}_\wedge^1], [\mathsf{o}_\wedge^2], [\mathsf{n}_\wedge^r]_1, [\mathsf{n}_\wedge^r]_2\}$. (Note that this S falls under *both* of the first two cases in the list given in Theorem 8.69.) It is easy to see that in this case $HBk(S)$ is sound for the following two-valued matrices (in both of which $\vee, \wedge$, and $\supset$ get their classical interpretations):

1. The matrix in which $\tilde{\mathsf{o}} = \lambda x.t$, and $\tilde{\neg}$ is the classical negation;

2. The matrix in which $\tilde{\mathsf{o}}$ is the classical negation, while $\tilde{\neg} = \lambda x.t$.

Now, the soundness with respect to the first matrix ensures that $\mathbf{Bk}(S)$ is ¬-contained in classical logic, while the soundness with respect to the second one ensures that it is pre-¬-paraconsistent. Hence $\mathbf{Bk}(S)$ is a strong C-system by Proposition 8.27. Nevertheless, as stated in Chapter 2, in this book we investigate only strongly (and usually boldly) paraconsistent logics. Therefore, in the rest of this section we focus on the extensions of $\mathbf{Bk}$ which are induced by *coherent* subsets of A_{CPAC}.

The next theorem generalizes the first six items of Theorem 8.53. The eighth (and last) item of that theorem (concerning sparseness) will be generalized in Theorem 8.74, while a certain counterpart of the seventh is given at the end of this section (Note 8.81).

---

[11] $\circ\varphi \supset \circ(\varphi \supset \psi), (\varphi \wedge \neg\psi) \supset \neg(\varphi \supset \psi), \circ\varphi, \varphi, \psi, \neg\psi \vdash_{\mathbf{Bk}} \sigma$. By substituting $\mathsf{F} \supset \mathsf{F}$ for φ (where F is any bottom element for $\mathbf{Bk}$), we get $\psi, \neg\psi \vdash_{HBk(S)} \sigma$ in case $\{[\mathsf{o}_\supset^1], [\mathsf{n}_\supset^r]\} \subseteq S$. (Note that $\vdash_{\mathbf{Bk}} \circ\mathsf{F}$, and so $\vdash_{HBk(S)} \circ(\mathsf{F} \supset \mathsf{F})$ in case $[\mathsf{o}_\supset^1] \in S$.)

Theorem 8.72 *Let S be a coherent subset of A_{CPAC}.*

1. $\mathbf{Bk}(S)$ *is a strong (and so regular) $\circ$-C-system.*

2. *$\wedge, \vee$ and $\supset$ are respectively $\mathcal{M}_{Bk}(S)$-conjunction, $\mathcal{M}_{Bk}(S)$-disjunction, and $\mathcal{M}_{Bk}(S)$-implication.*

3. *$\mathcal{M}_{Bk}(S)$ is characteristic for $\mathbf{Bk}(S)$.*

4. *$\mathbf{Bk}(S)$ is normal, boldly $\neg$-paraconsistent, and decidable.*

5. *$GBk(S)$ is sound and complete for $\mathbf{Bk}(S)$, and the cut-elimination theorem obtains for it.*

6. *There is no finite family $\mathcal{F}$ of finite ordinary matrices such that $\vdash_{\mathbf{Bk}(S)}$ is identical to $\vdash_{\mathcal{F}}$. (In particular: $\mathbf{Bk}(S)$ has no finite characteristic matrix.) $\mathbf{Bk}(S)$ also has no weakly characteristic two-valued Nmatrix.*

Proof.

1. By definition of coherence, $\mathbf{Bk}(S)$ is $\neg$-paraconsistent. Like in the case of $\mathbf{Bk}$ (see the proof of Proposition 8.28), this fact implies that $\mathbf{Bk}(S)$ is a strong $\circ$-C-system.

2. Similar to the proof of Proposition 8.39.

3. The proof again closely follows that of Theorem 7.50 in the case of $\mathcal{M}_4^b(S \cup \{[\mathsf{t}]\})$. What should be added here to that proof, beyond what has already been added to it in the proof of Theorem 8.40, are demonstrations that ν (as defined in the proof of Theorem 7.50) respects $C(ax)$ for every $ax \in S \cap A_C$. We do here as an example the case of $[\mathsf{a}_\vee]$. So let $[\mathsf{a}_\vee] \in S$, and assume that $\nu(\varphi) \neq \top$, and $\nu(\psi) \neq \top$. As in the proof of Theorem 8.40, these assumptions imply that $\circ\varphi \in \mathcal{T}^*$ and $\circ\psi \in \mathcal{T}^*$. It follows by the presence of $[\mathsf{a}_\vee]$ that $\circ(\varphi \vee \psi) \in \mathcal{T}^*$ as well. Again like in the proof of Theorem 8.40, this implies that $\nu(\varphi \vee \psi) \neq \top$. It easily follows that ν respects $C([\mathsf{a}_\vee])$.

4. Similar to the proof of Proposition 8.41.

5. The proof is similar to the proof of Theorem 8.53, closely following the proof of Theorem 7.60 in the case of $LK^+(S\cup\{[\mathsf{t}]\})$ and $\mathcal{M}_4^b(S\cup\{[\mathsf{t}]\})$. The only difference is that there are now more cases to be checked when we show that if $\neg\varphi \in \Gamma$ then $\nu(\varphi) \in \{f, \top\}$ (that is, $\nu(\varphi) \neq t$). We do as an example one of them. So, suppose that $\varphi = \varphi_1 \vee \varphi_2$, $[\mathsf{a}_\vee] \in S$, $\nu(\varphi_1) = t$ and $\nu(\varphi_2) \neq \top$. (The case where $\nu(\varphi_2) = t$ and $\nu(\varphi_1) \neq \top$ is treated similarly.) Since $\Gamma \Rightarrow \Delta$ is saturated, the facts

8.3 Propagation of Consistency

that $\neg\varphi \in \Gamma$ and $[a_\vee] \in S$ imply (by the second rule associated with $[a_\vee]$) that either $\neg\varphi_1 \in \Gamma$, or $\{\varphi_2, \neg\varphi_2\} \subseteq \Gamma$. In the first case $\nu(\varphi_1) \neq t$ by the induction hypothesis for φ_1. In the second case $\nu(\varphi_2) = \top$ by the induction hypothesis for φ_2. In both cases we got a contradiction.

6. Again, the first part follows from the third item and Theorem 7.4, and the second part — from the first item and Proposition 8.35. □

Example 8.73 The set **A** is coherent by Theorem 8.69. Hence, by Theorem 8.72, $\mathcal{M}_{Bk}(A)$ is characteristic for the logic **BkA** (also known as **Cia**, see Note 8.67). Here are the operations of this Nmatrix:

	$\tilde{\neg}$	$\tilde{\circ}$
t	$\{f\}$	$\{t\}$
$\top$	$\{t, \top\}$	$\{f\}$
f	$\{t\}$	$\{t\}$

$\tilde{\wedge}$	t	$\top$	f
t	$\{t\}$	$\{t, \top\}$	$\{f\}$
$\top$	$\{t, \top\}$	$\{t, \top\}$	$\{f\}$
f	$\{f\}$	$\{f\}$	$\{f\}$

$\tilde{\vee}$	t	$\top$	f
t	$\{t\}$	$\{t, \top\}$	$\{t\}$
$\top$	$\{t, \top\}$	$\{t, \top\}$	$\{t, \top\}$
f	$\{t\}$	$\{t, \top\}$	$\{f\}$

$\tilde{\supset}$	t	$\top$	f
t	$\{t\}$	$\{t, \top\}$	$\{f\}$
$\top$	$\{t, \top\}$	$\{t, \top\}$	$\{f\}$
f	$\{t\}$	$\{t, \top\}$	$\{t\}$

Turning to the topic of sparseness, we note that here Theorem 7.7 allows us only to provide *sufficient* criteria for the sparseness of **Bk**(S), when S is a coherent subset of $\mathbf{A}_{CPAC}$. (For example: that if $[o^1_\supset] \notin S$ and $\mathcal{M}_{Bk}(S)$ is a proper Nmatrix then **Bk**(S) is sparse.) Nevertheless, the next theorem provides an effective criterion which is both sufficient and necessary.

Proposition 8.74 *Let S be a coherent subset of $\mathbf{A}_{CPAC}$. **Bk**(S) is sparse iff $\mathcal{M}_{Bk}(S)$ is a proper Nmatrix, and there is an n-ary primitive connective $\diamond$ of $\mathcal{L}_C$ ($n \in \{1, 2\}$) and $a_1, \ldots, a_n \in \{t, f, \top\}$ such that $a_i \neq \top$ for some $1 \leq i \leq n$, and in $\mathcal{M}_{Bk}(S)$ $\tilde{\diamond}(a_1, \ldots, a_n) \in \{\{\top\}, \{t, \top\}\}$.*

Proof. For sufficiency, Suppose that the condition is satisfied. Then there are two cases to consider:

- $\tilde{\diamond}(a_1, \ldots, a_n) = \{t, \top\}$ for some n-ary connective $\diamond$ of $\mathcal{L}_C$ ($n \in \{1, 2\}$) and $a_1, \ldots, a_n \in \{t, f, \top\}$ such that a_1 (say) is in $\{t, f\}$. Let φ and ψ be two distinct formulas of $\mathcal{L}$, and let $q_2, \ldots, q_n$ be propositional variables that do not occur in φ or ψ. For $\tau \in \{\varphi, \psi\}$, let $\tau' = \tau \supset \tau$ in case $a_1 = t$, $\tau' = \neg(\tau \supset \tau)$ in case $a_1 = f$, and $\tau^* = \diamond(\tau', q_2, \ldots, q_n)$. Let $\mathcal{W}$ be the set of subformulas of φ and ψ, and let $\mathcal{W}^*$ be the set of subformulas of φ^* and ψ^*. Finally, let $\nu : \mathcal{W} \to \{t, f\}$ be a partial $\mathcal{M}_{Bk}(S)$-valuation. (It is easy to see that such ν always exists). Using the assumed property of $\tilde{\diamond}$, we extend ν to a partial $\mathcal{M}_{Bk}(S)$-valuation ν^* on $\mathcal{W}^*$ by letting:

- $\nu^*(q_i) = a_i$ $(i = 2, \ldots, n)$.
- $\nu^*(\varphi \supset \varphi) = t$ and $\nu^*(\psi \supset \psi) = t$.
- $\nu^*(\neg(\varphi \supset \varphi)) = f$ and $\nu^*(\neg(\psi \supset \psi)) = f$.
- $\nu^*(\varphi^*) = t$.
- $\nu^*(\psi^*) = \top$.

By the analyticity of the semantics and Proposition 7.5 it follows that $\varphi^* \not\equiv_{\mathbf{Bk}(S)} \psi^*$. Hence $\varphi \not\equiv_{\mathbf{Bk}(S)} \psi$ as well.

- The first case does not obtain. Since $\{t, \top\}$ is the only value which is not a singleton that a primitive operation of $\mathcal{M}_{Bk}(S)$ can get, this assumption implies that there are an n-ary connective $\diamond$ ($n \in \{1, 2\}$), $a_1, \ldots, a_n \in \{t, f, \top\}$, and a k-ary connective $\#$ ($k \in \{1, 2\}$) such that a_1 (say) is in $\{t, f\}$, $\tilde{\diamond}(a_1, \ldots, a_n) = \{\top\}$ and $\tilde{\#}(\top, \ldots, \top) = \{t, \top\}$. Now, define $\varphi^*, \psi^*, \mathcal{W}^*$, and ν^* as in the first case, except that now we take $\nu^*(\varphi^*) = \top$ (rather than $\nu^*(\varphi^*) = t$). Extend ν^* to $\mathcal{W}^* \cup \{\#(\varphi^*, \ldots, \varphi^*), \#(\psi^*, \ldots, \psi^*)\}$ by letting $\nu^*(\#(\varphi^*, \ldots, \varphi^*)) = t$ and $\nu^*(\#(\psi^*, \ldots, \psi^*)) = \top$. Again, it follows by the analyticity of the semantics and Proposition 7.5 that $\varphi \not\equiv_{\mathbf{Bk}(S)} \psi$.

For necessity of the condition, assume that $\mathcal{M}_{Bk}(S)$ does not satisfy it. Then again there are two cases to consider:

- In $\mathcal{M}_{Bk}(S)$ $\tilde{\diamond}(a_1, \ldots, a_n) = \{t\}$ or $\tilde{\diamond}(a_1, \ldots, a_n) = \{f\}$ for every n-ary primitive connective $\diamond$ of $\mathcal{L}_C$ ($n \in \{1, 2\}$) and $a_1, \ldots, a_n \in \{t, f, \top\}$ such that $a_i \neq \top$ for some $1 \leq i \leq n$. Using an easy induction on the structure of σ, this implies that the following holds for all formulas φ, ψ, σ, every $p \in \mathsf{Var}(\sigma)$, and every valuation ν in $\mathcal{M}_{Bk}(S)$:

  1. If $\nu(\varphi) \in \{t, f\}$ then $\nu(\sigma[\varphi/p]) \in \{t, f\}$.
  2. If $\nu(\varphi) \in \{t, f\}$ and $\nu(\psi) = \nu(\varphi)$ then $\nu(\sigma[\varphi/p]) = \nu(\sigma[\psi/p])$.

  Since $\nu(\circ\varphi \wedge \neg\varphi \wedge \varphi) = f$ for every formula φ and valuation ν in $\mathcal{M}_{Bk}(S)$, it follows from (1) and (2) that $\circ\varphi \wedge \neg\varphi \wedge \varphi$ is congruent in $\mathbf{L}_{\mathcal{M}_{Bk}(S)}$ to $\circ\psi \wedge \neg\psi \wedge \psi$ for any φ and ψ.

- $\mathcal{M}_{Bk}(S)$ is deterministic. Since $\nu(\circ\varphi \wedge \neg\varphi \wedge \varphi) = f$ for every formula φ and valuation ν in $\mathcal{M}_{Bk}(S)$, this again implies that $\circ\varphi \wedge \neg\varphi \wedge \varphi$ is congruent in $\mathbf{L}_{\mathcal{M}_{Bk}(S)}$ to $\circ\psi \wedge \neg\psi \wedge \psi$ for any φ and ψ.

8.3 Propagation of Consistency

It follows that in both cases $\mathbf{Bk}(S) = \mathbf{L}_{\mathcal{M}_{Bk}(S)}$ is not sparse. □

Theorem 8.75 *There are exactly five logics of the form $\mathbf{Bk}(S)$ (where S is a coherent subset of A_{CPAC}) which are not sparse.*[12] *With $\mathsf{A}'_{PAC} = \mathsf{A}_{PAC} \cup \{[\mathsf{a}_\circ]\}$, these are (see Notation 8.66): (i) $\mathbf{BkO}$; (ii) $\mathbf{BkeO}$; (iii) $\mathbf{Bk}(\mathsf{A}'_{PAC})$; (iv) $\mathbf{Bk}((\mathsf{A}'_{PAC} - \{[\mathsf{n}^r_\wedge]_2\}) \cup \{[\mathsf{o}^1_\wedge]\})$; (v) $\mathbf{Bk}((\mathsf{A}'_{PAC} - \{[\mathsf{n}^r_\wedge]_1\}) \cup \{[\mathsf{o}^2_\wedge]\})$.*

Proof. From Proposition 8.74 it follows that if S is a coherent subset of A_{CPAC}, then $\mathbf{Bk}(S)$ is not sparse iff one of the following conditions obtains:

- In $\mathcal{M}_{Bk}(S)$ $\tilde{\diamond}(a_1, \ldots, a_n) = \{t\}$ or $\tilde{\diamond}(a_1, \ldots, a_n) = \{f\}$ for every n-ary primitive connective $\diamond$ of $\mathcal{L}_C$ ($n \in \{1, 2\}$) and $a_1, \ldots, a_n \in \{t, f, \top\}$ such that $a_i \neq \top$ for some $1 \leq i \leq n$. This happens iff all the axioms in $\mathbf{O}$ are valid in $\mathcal{M}_{Bk}(S)$. Since from Propositions 8.58, 8.59 and Theorem 8.68 it follows that [e] is the only axiom in A_{CPAC} that neither follows from $\mathbf{O}$ nor is incoherent with it, it follows that there are exactly two Nmatrices of the form $\mathcal{M}_{Bk}(S)$ (where S is a coherent subset of A_{CPAC}) that satisfies this condition: $\mathcal{M}_{\mathbf{BkO}}$ and $\mathcal{M}_{\mathbf{BkeO}}$.

- Not all elements of $\mathbf{O}$ are valid in $\mathcal{M}_{Bk}(S)$, and $\mathcal{M}_{Bk}(S)$ is deterministic. A careful check of the various refining conditions for A_{CPAC} (Definition 8.61) reveals that this is the case iff all the elements of one of the other three sets listed in the theorem are valid in $\mathcal{M}_{Bk}(S)$. □

Note 8.76 Given the soundness and completeness of $\mathbf{Bk}(S)$ for $\mathcal{M}_{Bk}(S)$, checking whether $\mathbf{Bk}(S)$ is sparse reduces to checking whether $\mathcal{M}_{Bk}(S)$ is identical to one of the five Nmatrices that correspond to the logics listed in Theorem 8.75. Note that of these five Nmatrices, three are deterministic matrices, and these are the only matrices in the family of Nmatrices of the form $\mathcal{M}_{Bk}(S)$, where S is a coherent subset of A_{CPAC}.

Example 8.77 From Theorem 8.75 it easily follows that $\mathbf{BkA}$ is sparse.

Next we show that even in $\mathbf{BkO}$ and $\mathbf{BkeO}$ there is a major class of isolated formulas.

Proposition 8.78 *Every formula of $\mathcal{L}_{CL}$ is isolated in $\mathbf{BkO}$ and $\mathbf{BkeO}$.*

Proof. Obviously, it suffices to prove the claim for $\mathbf{BkeO}$.

First we note that in $\mathcal{M}_{\mathbf{BkeO}}$, $\nu(\psi) \neq \top$ for every valuation ν and every formula ψ which contains $\circ$. Since no formula φ of $\mathcal{L}_{CL}$ has this property, such φ cannot be congruent to any formula ψ which does contain $\circ$.

---
[12] Note that because of Propositions 8.58 and 8.59, the number of coherent subsets S of A_{CPAC} such that $\mathbf{Bk}(S)$ is not sparse is somewhat greater than five.

Now, suppose that φ and ψ are two distinct formulas of $\mathcal{L}_{CL}$. let $\mathcal{W}^*$ be the set of subformulas of $\varphi \vee \varphi$ and $\psi \vee \psi$. Since φ and ψ are in $\mathcal{L}_{CL}$, there is in $\mathcal{M}_{\mathbf{BkeO}}$ a partial valuation ν on $\mathcal{W}^*$ such that $\nu(\sigma) = \top$ for any subformula of $\varphi \vee \varphi$ or ψ, while $\nu(\psi \vee \psi) = t$. It follows by the analyticity of the semantics and Proposition 7.5 that $\varphi \vee \varphi \not\models_{\mathbf{Bk}(S)} \psi \vee \psi$. Hence $\varphi \not\models_{\mathbf{Bk}(S)} \psi$ as well. □

We end this section with some remarks about the $\mathcal{L}_{CL}$-fragments of the logics which were investigated in it. We start with the most important one, which implies that none of these fragments is a C-system itself. (This is in sharp contrast with the logics which are studied in the next section.)

Proposition 8.79 *The $\mathcal{L}_{CL}$-fragment of $\mathbf{Bk}(S)$, where S is a coherent subset of A_{CPAC}, has no consistency operator.*

Proof. The function ν which assigns $\top$ to every formula of $\mathcal{L}_{CL}$ is a partial $\mathcal{M}_{Bk}(S)$-valuation. Hence, the soundness and completeness of $\mathbf{Bk}(S)$ for $\mathcal{M}_{Bk}(S)$ imply that the $\mathcal{L}_{CL}$-fragment of $\mathbf{Bk}(S)$ has no bottom element. It follows by Proposition 8.20 that it has no consistency operator either. □

Next, we turn to proof systems for these fragments.

Proposition 8.80 *If S is a coherent subset of A_{CPAC} then the $\circ$-free fragment of $GBk(S)$ is sound and complete for the $\mathcal{L}_{CL}$-fragment of $\mathbf{Bk}(S)$, and it admits cut-elimination.*

Proof. This is immediate from the fifth item of Theorem 8.72, since the cut-elimination theorem ensures that any sequent in $\mathcal{L}_{CL}$ which is derivable in $GBk(S)$ has there a $\circ$-free proof. □

Note 8.81 Using Proposition 8.80, it is not difficult to obtain also Hilbert-style axiomatizations for the $\mathcal{L}_{CL}$-fragments under question. For this purpose, one can employ a standard method of using an implication for translating Gentzen-type rules into corresponding axioms. (See the introduction to Section 4.5.2.) Take for instance the first rule in Figure 8.3 that corresponds to $[\mathrm{o}_\vee^1]$. By letting $\Gamma = \{\psi\}$ and $\Delta = \{\varphi\}$ in this rule, we get a derivation of the sequent $\psi, \neg(\varphi \vee \psi) \Rightarrow \varphi$. (Note that the provability of this sequent is actually equivalent to the rule, since the rule can be derived from it using cuts.) Using the rules for implication, this sequent, in turn, is equivalent to $\Rightarrow \neg(\varphi \vee \psi) \supset (\psi \supset \varphi)$. By applying a similar process to the other rule in Figure 8.3 which corresponds to $[\mathrm{o}_\vee^1]$, we obtain that adding $[\mathrm{o}_\vee^1]$ to $\mathbf{Bk}$ is equivalent to adding to it the axioms $\neg(\varphi \vee \psi) \supset \neg \varphi$ and $\neg(\varphi \vee \psi) \supset (\psi \supset \varphi)$. The other axioms in Figures 8.3 and 8.2 can be treated similarly. In this way we can get from every cut-free Gentzen-type system for the $\mathcal{L}_{CL}$-fragment of $\mathbf{Bk}(S)$ an equivalent Hilbert-style system (and a $\circ$-free equivalent for each axiom in Figures 8.3 and 8.2).

8.4 da Costa's Consistency Operator(s)

8.4.1 The Strict Case

LFI is a short for "Logics of Formal Inconsistency". However, in all the systems studied in this chapter the focus is on an operator for *formal consistency*, not formal inconsistency.[13] The reason for this is that the inconsistency of φ means that both φ and $\neg\varphi$ are true, and this is easily expressed by the sentence $\varphi \wedge \neg\varphi$. The latter can therefore be taken as the definition of formal inconsistency.[14] The content of the two negation axioms of HBk is indeed that each of $\circ\varphi$ and $\varphi \wedge \neg\varphi$ is equivalent to the *classical negation* of the other: always one of them is true, but never both of them are true. The main idea behind da Costa's main system $\mathbf{C}_1$ (see Example 8.141 below) was that with respect to the question of consistency/inconsistency of formulas, the negation $\neg$ of the logic behaves as classical negation, and so $\neg(\varphi \wedge \neg\varphi)$ can serve as a definition inside $\mathcal{L}_{CL}$ of the consistency operator. Now the schemes $\circ\varphi \supset \neg(\varphi \wedge \neg\varphi)$ and $(\varphi \wedge \neg\varphi) \supset \neg\circ\varphi$ are easily seen to be theorems already of the most basic system $\mathbf{B}$. Therefore, what is needed for making $\circ\varphi$ equivalent to the negation of $\varphi \wedge \neg\varphi$, and vice versa, are the following two axioms:

[i] $\neg\circ\varphi \supset (\varphi \wedge \neg\varphi)$,

[l] $\neg(\varphi \wedge \neg\varphi) \supset \circ\varphi$.

Note 8.82 Instead of $\varphi \wedge \neg\varphi$ one can of course take $\neg\varphi \wedge \varphi$ as expressing the inconsistency of φ. Choosing this would have led to the use of $\neg(\neg\varphi \wedge \varphi)$ as the consistency operator. In this case [l] should be replaced by the following axiom (which by Item 8 of Theorem 8.53 is *not* equivalent in $\mathbf{Bk}$ to [l]):

[d] $\neg(\neg\varphi \wedge \varphi) \supset \circ\varphi$.

Obviously, there is no rational reason to prefer either [l] or [d] over the other. For historical reasons, we follow da Costa and mainly focus on [l]. But switching instead to [d], or to the use of both, is straightforward.

Note 8.83 It is very easy to verify that both $(\neg P_1 \supset P_2) \supset (P_1 \vee P_2)$ and $(\neg P_2 \supset P_1) \supset (P_1 \vee P_2)$ are theorems of **CLuN**. (Either use its characteristic Nmatrix $\mathcal{M}_2$, or provide a short direct derivation.) Hence, both [i] and [l] (as

---

[13] The use of a special operator for formal inconsistency (usually denoted by •) is also considered in the literature on LFIs (see [111, 113]). However, it plays there a very secondary role, and is usually defined in terms of ∘.

[14] In contrast, that φ is consistent means that either φ or $\neg\varphi$ is *false*. But the falsity of a sentence cannot be expressed in $\mathcal{L}_{CL}$, unless we assume that φ is false iff $\neg\varphi$ is true — but precisely this assumption is what is rejected in paraconsistent logics.

well as [d]) directly strengthen [k] already over **CLuN**. Thus, we have the following proposition, which allows us to use **B** as our base system (rather than **Bk**) whenever either of [i], [l], or [d] is present.

Proposition 8.84 $\mathbf{Bki} = \mathbf{Bi}$, $\mathbf{Bkl} = \mathbf{Bl}$, and $\mathbf{Bkd} = \mathbf{Bd}$.[15]

Proof. Immediate from Note 8.83. □

Our next observation is that the axiom [i] does not involve anything new, since it is equivalent over **B** to the combination of [k] and [a$_\circ$].

Proposition 8.85 $\mathbf{Bi} = \mathbf{Bka_\circ}$.

Proof. Using $\mathcal{M}_{Bk}$, it is easy to check that:

$$\vdash_{\mathbf{Bk}} (\neg \circ \varphi \supset (\varphi \wedge \neg \varphi)) \supset (\circ \varphi \supset \circ \circ \varphi),$$

$$\vdash_{\mathbf{Bk}} (\circ \varphi \supset \circ \circ \varphi) \supset (\neg \circ \varphi \supset (\varphi \wedge \neg \varphi)).$$

It follows that $\mathbf{Bki} = \mathbf{Bka_\circ}$. Hence, $\mathbf{Bi} = \mathbf{Bka_\circ}$ by Proposition 8.84. □

Note 8.86 From Proposition 8.85 it follows that every instance of [a$_\circ$] is derivable in **Bi**. Note that this is so despite the fact that

$$\nvdash_{\mathbf{B}} (\neg \circ \varphi \supset (\varphi \wedge \neg \varphi)) \supset (\circ \varphi \supset \circ \circ \varphi).$$

(This fact can e.g. be seen by taking an assignment ν in $\mathcal{M}_B$ for which $\nu(\varphi) \in \{t, f\}$, $\nu(\circ \varphi) = t$, and $\nu(\circ \circ \varphi) = f$.) To infer [a$_\circ$] in **Bi** one indeed needs *two different* instances of [i], not just one.

Note 8.87 As noted above, $\circ \varphi$ and $\neg(\varphi \wedge \neg \varphi)$ are equivalent in **Bl**. As we show below, they are nevertheless not congruent there, as well as in most of the extensions of **Bl** that we study below (even in the presence of [i]). Hence, $\circ \varphi$ and $\neg(\varphi \wedge \neg \varphi)$ cannot be taken to "have the same meaning" in these logics. Still, their equivalence will be exploited in the sequel to provide a Gentzen-type rule which corresponds to [l].

Because of Proposition 8.85 we shall not dwell on [i] any further in this section. Therefore, its main subject will be extensions of **Bk** by those subsets of the following set of axioms which include either [l] or [d].

Definition 8.88 (A$_{ld}$) $A_{ld} = A_{CPAC} \cup \{[l], [d]\}$.

---

[15]Note that we are employing here and later the conventions which where introduced in Notation 8.66 and used in Note 8.67.

8.4 da Costa's Consistency Operator(s)

Let $S \subseteq \mathbf{A}_{ld}$, $S \cap \{[l], [d]\} \neq \emptyset$, and suppose that $\mathbf{Bk}(S)$ ($= \mathbf{B}(S)$ by Proposition 8.84) is strongly paraconsistent. When we try to develop appropriate semantics for $\mathbf{B}(S)$ we face a new major obstacle. Unlike the case of axioms in $\mathbf{A}_{CPAC}$, the semantic effect of adding [l] or [d] to HBk cannot be formulated as a refining condition on $\mathcal{M}_{Bk}$. This is due to the fact that both [l] and [d] involve a conjunction of a formula with its negation. Therefore, to handle them we must be able to distinguish between the conjunction of an "inconsistent" formula ψ with its negation, and the conjunction of ψ with other formulas. As is proved in Theorem 8.108, this necessarily requires an *infinite* number of truth-values. Accordingly, from now we use as our basic Nmatrix not $\mathcal{M}_{Bk}$, but a certain infinite preserving rexpansion (Definition 6.38) of it, in which the truth values t and $\top$ are replaced by infinite sets $\mathcal{E}_t$, $\mathcal{E}_\top$ (respectively). As will be clear from the propositions that follow its definition, for the systems without [l] and [d] its use is completely equivalent to the use of $\mathcal{M}_{Bk}$.

Definition 8.89 ($\mathcal{M}^\infty_{Bk}$) $\mathcal{M}^\infty_{Bk}$ is the f-Nmatrix $\langle \mathcal{V}, \mathcal{D}, \mathcal{O} \rangle$ for $\mathcal{L}_C$, in which $\mathcal{V} = \mathcal{E}_t \cup \mathcal{E}_\top \cup \{f\}$, where $\mathcal{E}_t = \{t_i^j \mid i \geq 0, j \geq 0\}$, $\mathcal{E}_\top = \{\top_i^j \mid i \geq 0, j \geq 0\}$, $\mathcal{D} = \mathcal{E}_t \cup \mathcal{E}_\top$, and $\mathcal{O}$ is defined as follows:

$$a \tilde{\vee} b = \begin{cases} \mathcal{D} & \text{if either } a \in \mathcal{D} \text{ or } b \in \mathcal{D}, \\ \{f\} & \text{if } a = b = f. \end{cases}$$

$$a \tilde{\supset} b = \begin{cases} \mathcal{D} & \text{if either } a = f \text{ or } b \in \mathcal{D}, \\ \{f\} & \text{if } a \in \mathcal{D} \text{ and } b = f. \end{cases}$$

$$a \tilde{\wedge} b = \begin{cases} \{f\} & \text{if either } a = f \text{ or } b = f, \\ \mathcal{D} & \text{otherwise.} \end{cases}$$

$$\tilde{\neg} a = \begin{cases} \{f\} & \text{if } a \in \mathcal{E}_t, \\ \mathcal{D} & \text{if } a = f, \\ \{\top_i^{j+1}, t_i^{j+1}\} & \text{if } a = \top_i^j. \end{cases}$$

$$\tilde{\circ} a = \begin{cases} \mathcal{D} & \text{if } a \in \mathcal{E}_t \cup \{f\}, \\ \{f\} & \text{if } a \in \mathcal{E}_\top. \end{cases}$$

Proposition 8.90 $\mathcal{M}^\infty_{Bk}$ *is a strongly preserving rexpansion of* $\mathcal{M}_{Bk}$.

Proof. Define the following expansion function for $\mathcal{M}_{Bk}$:

$$E = \lambda x \in \{t, f, \top\}. \begin{cases} \mathcal{E}_t & x = t, \\ \{f\} & x = f, \\ \mathcal{E}_\top & x = \top. \end{cases}$$

It is easy to check that $\mathcal{M}_{Bk}^{\infty}$ is a strongly preserving E-rexpansion of $\mathcal{M}_{Bk}$. (Note that the $\neg$-free reduction of $\mathcal{M}_{Bk}^{\infty}$ is simply an E-expansion of the $\neg$-free reduction of $\mathcal{M}_{Bk}$.) □

Corollary 8.91 $\mathbf{Bk} = \mathbf{L}_{\mathcal{M}_{Bk}^{\infty}}$.

Proof. Immediate from Propositions 8.90 and Theorem 6.42. □

The use of $\mathcal{M}_{Bk}^{\infty}$ instead of $\mathcal{M}_{Bk}$ will not affect the *modularity* of the semantics: Again, each axiom of $\mathbf{A}_{ld}$ corresponds to a certain semantic condition, this time on refinements of $\mathcal{M}_{Bk}^{\infty}$. These conditions are calculated like in Note 8.63 (or in Note 8.95 below). As can be seen in the next definition, for the axioms in $\mathbf{A}_{CPAC}$ these semantic conditions are just the obvious generalizations of those provided in Definition 8.61. The only differences are that now $\mathcal{E}_t$ and $\mathcal{E}_\top$ replace $\{t\}$ and $\{\top\}$ (respectively), and $\subseteq$ replaces $=$.

Definition 8.92 (general refining conditions for $\mathbf{A}_{CPAC}$) Given $ax \in \mathbf{A}_{CPAC}$, the general refining condition $GC(ax)$ for it is defined as follows:

$GC([\mathsf{c}])$: $\tilde{\neg} f \subseteq \mathcal{E}_t$.

$GC([\mathsf{e}])$: $\tilde{\neg} a \subseteq \mathcal{E}_\top$ if $a \in \mathcal{E}_\top$. (In other words: $\tilde{\neg} \mathsf{T}_i^j = \{\mathsf{T}_i^{j+1}\}$.)

$GC([\mathsf{n}_\wedge^l])$: $a \tilde{\wedge} b \subseteq \mathcal{E}_t$ if $\{a,b\} \subseteq \mathcal{E}_t$.

$GC([\mathsf{n}_\wedge^r]_1)$: $a \tilde{\wedge} b \subseteq \mathcal{E}_\top$ if $a \in \mathcal{E}_\top$ and $b \neq f$.

$GC([\mathsf{n}_\wedge^r]_2)$: $a \tilde{\wedge} b \subseteq \mathcal{E}_\top$ if $b \in \mathcal{E}_\top$ and $a \neq f$.

$GC([\mathsf{n}_\vee^l]_1)$: $a \tilde{\vee} b \subseteq \mathcal{E}_t$ if $a \in \mathcal{E}_t$.

$GC([\mathsf{n}_\vee^l]_2)$: $a \tilde{\vee} b \subseteq \mathcal{E}_t$ if $b \in \mathcal{E}_t$.

$GC([\mathsf{n}_\vee^r])$: $a \tilde{\vee} b \subseteq \mathcal{E}_\top$ if $\{a,b\} \subseteq \mathcal{E}_\top \cup \{f\}$, and either $a \neq f$ or $b \neq f$.

$GC([\mathsf{n}_\supset^l]_1)$: $f \tilde{\supset} b \subseteq \mathcal{E}_t$.

$GC([\mathsf{n}_\supset^l]_2)$: $a \tilde{\supset} b \subseteq \mathcal{E}_t$ if $b \in \mathcal{E}_t$.

$GC([\mathsf{n}_\supset^r])$: $a \tilde{\supset} b \subseteq \mathcal{E}_\top$ if $a \neq f$ and $b \in \mathcal{E}_\top$.

$GC([\mathsf{a}_\neg])$: $\tilde{\neg} f \subseteq \mathcal{E}_t$.

$GC([\mathsf{a}_\circ])$: $\tilde{\circ} a \subseteq \mathcal{E}_t$ if $a \in \mathcal{E}_t \cup \{f\}$.

$GC([\mathsf{a}_\wedge])$: $a \tilde{\wedge} b \subseteq \mathcal{E}_t$ if $\{a,b\} \subseteq \mathcal{E}_t$.

$GC([\mathsf{a}_\vee])$: $a \tilde{\vee} b \subseteq \mathcal{E}_t$ if $\{a,b\} \subseteq \mathcal{E}_t \cup \{f\}$, and either $a \neq f$ or $b \neq f$.

8.4 da Costa's Consistency Operator(s)

$GC([a_\supset])$: $a\tilde{\supset}b \subseteq \mathcal{E}_t$ if $\{a,b\} \subseteq \mathcal{E}_t \cup \{f\}$, and either $a = f$ or $b \neq f$.

$GC([o_\wedge^1])$: $a\tilde{\wedge}b \subseteq \mathcal{E}_t$ if $a \in \mathcal{E}_t$ and $b \neq f$.

$GC([o_\wedge^2])$: $a\tilde{\wedge}b \subseteq \mathcal{E}_t$ if $b \in \mathcal{E}_t$ and $a \neq f$.

$GC([o_\vee^1])$: $a\tilde{\vee}b \subseteq \mathcal{E}_t$ if $a \in \mathcal{E}_t \cup \{f\}$, and either $a \neq f$ or $b \neq f$.

$GC([o_\vee^2])$: $a\tilde{\vee}b \subseteq \mathcal{E}_t$ if $b \in \mathcal{E}_t \cup \{f\}$, and either $a \neq f$ or $b \neq f$.

$GC([o_\supset^1])$: $a\tilde{\supset}b \subseteq \mathcal{E}_t$ if $a \in \mathcal{E}_t \cup \{f\}$, and either $a = f$ or $b \neq f$.

$GC([o_\supset^2])$: $a\tilde{\supset}b \subseteq \mathcal{E}_t$ if $b \in \mathcal{E}_t \cup \{f\}$, and either $a = f$ or $b \neq f$.

Note 8.93 Again (see Proposition 8.58 and Note 8.62), we have of course that $GC([a_\neg]) = \mathsf{GC}([c])$ and $GC([a_\wedge]) = \mathsf{GC}([n_\wedge^l])$.

Now we turn to the problematic axioms [l] and [d], and define the refining conditions induced by them in the context of refinements of $\mathcal{M}_{Bk}^\infty$.

Definition 8.94 (general semantic conditions for [l], [d])

$GC([l])$: $a\tilde{\wedge}b \subseteq \mathcal{E}_t$ if $a \in \mathcal{E}_\top$ and $b \in \tilde{\neg}a$.

$GC([d])$: $a\tilde{\wedge}b \subseteq \mathcal{E}_t$ if $b \in \mathcal{E}_\top$ and $a \in \tilde{\neg}b$.

Note 8.95 To demonstrate how the conditions in Definition 8.94 are derived, consider e.g [l]. To refute it in an Nmatrix of the type we deal with, we should have an assignment ν such that $\nu(\circ\varphi) = f$, while $\nu(\neg(\varphi \wedge \neg\varphi)) \neq f$. To guarantee that no such refutation exists is equivalent to making sure that $\nu(\varphi \wedge \neg\varphi) \in \mathcal{E}_t$ whenever $\nu(\varphi) \in \mathcal{E}_\top$. This, in turn, is equivalent to imposing $GC([l])$ above.

Definition 8.96 ($\mathcal{M}_{Bk}^\infty(S)$) Let $S \subseteq \mathsf{A}_{ld}$. $\mathcal{M}_{Bk}^\infty(S)$ is the weakest refinement of $\mathcal{M}_{Bk}^\infty$ (if such a refinement exists) in which the condition $GC(ax)$ is satisfied for every $ax \in S$.

The conditions $GC([l])$ and $GC([d])$ given in Definition 8.94 are *not* refining conditions, as they involve a non-simple connection between the interpretations of two connectives. In the next proposition we show how to replace them by strict refining conditions in each case in which $S \subseteq \mathsf{A}_{ld}$.

Proposition 8.97 Let $S \subseteq \mathsf{A}_{CPAC}$ be coherent, $G(S) = \{GC(x) \mid x \in S\}$.

1. $G(S) \cup \{GC([l])\}$ *is not satisfiable in case* $[n_\wedge^r]_1 \in S$ *or* $\{[e], [n_\wedge^r]_2\} \subseteq S$.
   $G(S) \cup \{GC([d])\}$ *is not satisfiable in case* $[n_\wedge^r]_2 \in S$ *or* $\{[e], [n_\wedge^r]_1\} \subseteq S$.

2. *The conditions $GC([l])$ and $GC([d])$ may be reformulated as follows in the context of $G(S)$ and $\mathcal{M}^{\infty}_{Bk}(S)$:*

$GC([l])$:

$[e] \in S$: $\quad\quad\quad\quad T_i^j \tilde{\wedge} T_i^{j+1} \subseteq \mathcal{E}_t.$

$[n_{\wedge}^r]_2 \in S$: $\quad\quad \tilde{\neg} T_i^j = \{t_i^{j+1}\}$ and $T_i^j \tilde{\wedge} t_i^{j+1} \subseteq \mathcal{E}_t.$

$S \cap \{[e], [n_{\wedge}^r]_2\} = \emptyset$: $\quad$ if $b \in \{T_i^{j+1}, t_i^{j+1}\}$ then $T_i^j \tilde{\wedge} b \subseteq \mathcal{E}_t.$

$GC([d])$:

$[e] \in S$: $\quad\quad\quad\quad T_i^{j+1} \tilde{\wedge} T_i^j \subseteq \mathcal{E}_t.$

$[n_{\wedge}^r]_1 \in S$: $\quad\quad \tilde{\neg} T_i^j = \{t_i^{j+1}\}$ and $t_i^{j+1} \tilde{\wedge} T_i^j \subseteq \mathcal{E}_t.$

$S \cap \{[e], [n_{\wedge}^r]_1\} = \emptyset$: $\quad$ if $a \in \{T_i^{j+1}, t_i^{j+1}\}$ then $a \tilde{\wedge} T_i^j \subseteq \mathcal{E}_t.$

Proof. We show the claims concerning $GC([l])$. (The proofs of those concerning $GC([d])$ are similar.)

1. We leave the easy check of this claim to the reader.

2. Since $\tilde{\neg} T_i^j \subseteq \{t_i^{j+1}, T_i^{j+1}\}$, the condition given for the third (and last) case in the reformulation of $GC([l])$ always implies $GC([l])$. The converse is true in case the conditions in $G(S) \cup \{GC([l])\}$ do not force $\tilde{\neg} T_i^j$ to be a proper subset of $\{t_i^{j+1}, T_i^{j+1}\}$. A careful (but easy) check reveals that this is the case unless $S \cap \{[e], [n_{\wedge}^r]_1\} \neq \emptyset$. (And, by the first item, neither $[n_{\wedge}^r]_1 \in S$ nor $\{[e], [n_{\wedge}^r]_2\} \subseteq S$.) It remains to check the cases in which $[e] \in S$ and in which $[n_{\wedge}^r]_2 \in S$.

   - The claim is obvious in case $[e] \in S$, since $GC([e])$ reduces here to the condition: $\tilde{\neg} T_i^j = \{T_i^{j+1}\}$.

   - Suppose that $[n_{\wedge}^r]_2 \in S$. Then from $GC([n_{\wedge}^r]_2)$ (as given in Definition 8.92) it follows that $T_i^j \tilde{\wedge} T_i^{j+1} \subseteq \mathcal{E}_T$ in $\mathcal{M}^{\infty}_{Bk}(S)$. However, $GC([l])$ implies that if $T_i^{j+1} \in \tilde{\neg} T_i^j$ then $T_i^j \tilde{\wedge} T_i^{j+1} \subseteq \mathcal{E}_t$. Hence $T_i^{j+1} \not\in \tilde{\neg} T_i^j$ in case $\{[l], [n_{\wedge}^r]_2\} \subseteq S$, and so $\tilde{\neg} T_i^j = \{t_i^{j+1}\}$ in this case. This immediately implies that the reformulation of $[l]$ given above for the present case indeed obtains in $\mathcal{M}^{\infty}_{Bk}(S)$ if $\{[l], [n_{\wedge}^r]_2\} \subseteq S$. It is obvious that the converse (i.e. that if this reformulation is satisfied then so does $GC([l])$) is true for *any* Nmatrix of the form $\mathcal{M}^{\infty}_{Bk}(S)$. $\square$

It can easily be proved that Proposition 8.90 and Corollary 8.91 can be extended to extensions of **Bk** with subsets of $\mathbf{A}_{CPAC}$: If $S \subseteq \mathbf{A}_{CPAC}$ is coherent, then $\mathcal{M}^{\infty}_{Bk}(S)$ exists, and is a strongly preserving E-rexpansion of $\mathcal{M}_{Bk}(S)$, where E is the expansion function defined in the proof of

8.4 da Costa's Consistency Operator(s)

Proposition 8.90. Hence $\mathbf{Bk}(S) = \mathbf{L}_{\mathcal{M}^\infty_{\tilde{B}k}(S)}$ (by Proposition 6.42 and Theorem 8.72). However, the situation with respect to *arbitrary* subsets of A_{ld} is more complicated. First, we determine which of them is **Bk**-coherent. The keys to the solution of this question have been given in Theorem 8.69 and the first part of Proposition 8.97.

Theorem 8.98 *The following conditions on $S \subseteq \mathsf{A}_{ld}$ are equivalent:*

1. *S is **Bk**-coherent.*

2. *S does not contain any of the following sets of axioms:*

    (a) $\{[\mathsf{o}^1_\wedge], [\mathsf{n}^r_\wedge]_2\}$,
    (b) $\{[\mathsf{o}^2_\wedge], [\mathsf{n}^r_\wedge]_1\}$,
    (c) $\{[\mathsf{o}^1_\vee], [\mathsf{n}^r_\vee]\}$,
    (d) $\{[\mathsf{o}^2_\vee], [\mathsf{n}^r_\vee]\}$,
    (e) $\{[\mathsf{o}^1_\supset], [\mathsf{n}^r_\supset]\}$,
    (f) $\{[\mathsf{l}], [\mathsf{n}^r_\wedge]_1\}$,
    (g) $\{[\mathsf{l}], [\mathsf{e}], [\mathsf{n}^r_\wedge]_2\}$,
    (h) $\{[\mathsf{d}], [\mathsf{n}^r_\wedge]_2\}$,
    (i) $\{[\mathsf{d}], [\mathsf{e}], [\mathsf{n}^r_\wedge]_1\}$.

3. $\mathcal{M}^\infty_{Bk}(S)$ *exists.*

Proof. To show that the first condition implies the second, it suffices to show that if $S \subseteq \mathsf{A}_{ld}$ contains any of the sets listed above, then $\mathbf{Bk}(S)$ is not strongly $\neg$-paraconsistent. For the sets given in (a)–(e) this was already shown in the proof of Theorem 8.69.

For (f), suppose that $\{[\mathsf{l}], [\mathsf{n}^r_\wedge]_1\} \subseteq S$. Then $\neg\varphi \supset \neg(\varphi \wedge \neg\varphi)$ and $\neg(\varphi \wedge \neg\varphi) \supset \mathsf{o}\varphi$ are theorems of $\mathbf{Bk}(S)$. (The first is an instance of $[\mathsf{n}^r_\wedge]_1$, the second is an instance of $[\mathsf{l}]$.) That $\varphi, \neg\varphi \vdash_{\mathbf{Bk}(S)} \psi$ for every φ, ψ follows therefore from the following easily shown fact:

$$\neg\varphi \supset \neg(\varphi \wedge \neg\varphi), \neg(\varphi \wedge \neg\varphi) \supset \mathsf{o}\varphi, \varphi, \neg\varphi \vdash_{\mathbf{B}} \psi.$$

For (g), suppose that $\{[\mathsf{l}], [\mathsf{e}], [\mathsf{n}^r_\wedge]_2\} \subseteq S$. Then $\neg\neg\varphi \supset \neg(\varphi \wedge \neg\varphi)$, $\neg(\varphi \wedge \neg\varphi) \supset \mathsf{o}\varphi$, and $\varphi \supset \neg\neg\varphi$ are theorem of $\mathbf{Bk}(S)$. (The first is this time an instance of $[\mathsf{n}^r_\wedge]_2$.) That $\varphi, \neg\varphi \vdash_{\mathbf{Bk}(S)} \psi$ for every φ, ψ follows therefore from the following easily shown fact:

$$\varphi \supset \neg\neg\varphi, \neg\neg\varphi \supset \neg(\varphi \wedge \neg\varphi), \neg(\varphi \wedge \neg\varphi) \supset \mathsf{o}\varphi, \varphi, \neg\varphi \vdash_{\mathbf{B}} \psi.$$

For (h) and (i), the argument is similar to that for (f) and (g) (respectively).

Showing that the second condition implies the third again involves just an exhaustive check. This check is more complicated than in the proof of Theorem 8.69, but with the help of Proposition 8.97 it is still mechanical.

To complete the circle, it is again not difficult to show that if $\mathcal{M}_{Bk}^{\infty}(S)$ exists, then $\mathbf{Bk}(S)$ is sound for it. Since $\mathcal{M}_{Bk}^{\infty}(S)$ obviously induces a strongly ¬-paraconsistent logic, $\mathbf{Bk}(S)$ is strongly ¬-paraconsistent as well. □

Example 8.99

- Let $S_1 = A \cup \{[\mathsf{l}]\}$. Then S_1 is coherent, and so $\mathcal{M}_{Bk}^{\infty}(S_1)$ exists. It is the Nmatrix $\langle \mathcal{V}, \mathcal{D}, \mathcal{O} \rangle$ for $\mathcal{L}_C$, where $\mathcal{V} = \mathcal{E}_t \cup \mathcal{E}_\top \cup \{f\}$, $\mathcal{D} = \mathcal{E}_t \cup \mathcal{E}_\top$, and $\mathcal{O}$ is defined as follows:

$$a \widetilde{\vee} b = \begin{cases} \{f\} & \text{if } a = b = f, \\ \mathcal{E}_t & a \in \mathcal{E}_t \text{ and } b \in \mathcal{E}_t \cup \{f\}, \text{ or } a = f, \text{ and } b \in \mathcal{E}_t, \\ \mathcal{D} & \text{otherwise.} \end{cases}$$

$$a \widetilde{\supset} b = \begin{cases} \{f\} & \text{if } a \in \mathcal{D} \text{ and } b = f, \\ \mathcal{E}_t & \text{if } a = f \text{ and } b \in \{f\} \cup \mathcal{E}_t, \text{ or } \{a,b\} \subseteq \mathcal{E}_t, \\ \mathcal{D} & \text{otherwise.} \end{cases}$$

$$a \widetilde{\wedge} b = \begin{cases} \{f\} & \text{if either } a = f \text{ or } b = f, \\ \mathcal{E}_t & \text{if } a = \mathsf{T}_i^j \text{ and } b \in \{\mathsf{T}_i^{j+1}, t_i^{j+1}\}, \text{ or } \{a,b\} \subseteq \mathcal{E}_t, \\ \mathcal{D} & \text{otherwise.} \end{cases}$$

$$\widetilde{\neg} a = \begin{cases} \{f\} & \text{if } a \in \mathcal{E}_t, \\ \mathcal{E}_t & \text{if } a = f, \\ \{\mathsf{T}_i^{j+1}, t_i^{j+1}\} & \text{if } a = \mathsf{T}_i^j. \end{cases}$$

$$\widetilde{\circ} a = \begin{cases} \{f\} & \text{if } a \in \mathcal{E}_\top, \\ \mathcal{E}_t & \text{if } a \in \mathcal{E}_t \cup \{f\}. \end{cases}$$

- Let S_2 be obtained from S_1 by deleting $[a_\circ]$. (So S_2 is equivalent to $[a] \cup \{[\mathsf{l}], [c]\}$.) Then the only difference between $\mathcal{M}_{Bk}^{\infty}(S_2)$ and $\mathcal{M}_{Bk}^{\infty}(S_1)$ is that in the former $\widetilde{\circ}$ is defined by:

$$\widetilde{\circ} a = \begin{cases} \{f\} & \text{if } a \in \mathcal{E}_\top, \\ \mathcal{D} & \text{if } a \in \mathcal{E}_t \cup \{f\}. \end{cases}$$

- Let $S_3 = S_1 \cup \{[\mathsf{n}_\wedge^r]_2\}$. This addition causes changes in the definitions of $\widetilde{\neg}$ and $\widetilde{\wedge}$, which are defined in $\mathcal{M}_{Bk}^{\infty}(S_3)$ as follows:

$$\widetilde{\neg} a = \begin{cases} \{f\} & \text{if } a \in \mathcal{E}_t, \\ \mathcal{E}_t & \text{if } a = f, \\ \{t_i^{j+1}\} & \text{if } a = \mathsf{T}_i^j. \end{cases}$$

8.4 da Costa's Consistency Operator(s)

$$a \tilde{\wedge} b = \begin{cases} \{f\} & \text{if either } a = f \text{ or } b = f, \\ \mathcal{E}_t & \text{if } a = \top_i^j \text{ and } b = t_i^{j+1}, \text{ or } \{a,b\} \subseteq \mathcal{E}_t, \\ \mathcal{E}_\top & \text{if } a \in \mathcal{D} \text{ and } b \in \mathcal{E}_\top, \\ \mathcal{D} & \text{otherwise.} \end{cases}$$

Note that $\mathcal{M}_{Bk}^\infty(S_3)$ is *not* a refinement of $\mathcal{M}_{Bk}^\infty(S_1)$, even though $\mathbf{Bk}(S_3)$ is an axiomatic extension of $\mathbf{Bk}(S_1)$.

Proposition 8.100 *If S is a coherent subset of A_{ld} then $\mathbf{Bk}(S)$ is a strong (and so regular) $\circ$-C-system.*

Proof. Similar to the proof of the first item of Theorem 8.72 □

Corollary 8.101 *If S is a coherent subset of A_{ld} then $\mathbf{Bk}(S)$ is normal.*

Proof. Immediate from Propositions 8.4 and 8.100.[16] □

Notation 8.102 $\mathbf{Bl}(S) = \mathbf{B}(S \cup \{[l]\})$, $\mathbf{Bd}(S) = \mathbf{B}(S \cup \{[d]\})$, $\mathbf{Bld}(S) = \mathbf{B}(S \cup \{[l],[d]\})$.

Corollary 8.103 *If $S \cup \{[l]\}$ is coherent then $\mathbf{Bl}(S)$ is a strong (and so regular and normal) $\circ$-C-system. Similar results apply to $\mathbf{Bd}(S)$ and $\mathbf{Bld}(S)$.*

Proof. Immediate from Propositions 8.84 and 8.100 □

Theorem 8.104 *If S is a coherent subset of A_{ld} then $\mathcal{M}_{Bk}^\infty(S)$ is characteristic for $\mathbf{Bk}(S)$.*

Proof. The soundness part, which has already been noted in the proof of Theorem 8.98, is easy. As noted immediately after Definition 8.96, the completeness is also easy in case $S \subseteq \mathsf{A}_{CPAC}$, because $\mathcal{M}_{Bk}^\infty(S)$ is then a strongly preserving rexpansion of $\mathcal{M}_{Bk}(S)$. Accordingly, in the rest of the proof we assume that $[l] \in S$. (The case where $[d] \in S$ is treated similarly.)

So we suppose (for the proof of completeness) that $\mathcal{T} \not\vdash_{HBk(S)} \varphi_0$. We show that there exists a model of $\mathcal{T}$ in $\mathcal{M}_{Bk}^\infty(S)$ which is not a model of φ_0. Like in the proofs of Theorems 7.14 and 8.40, we use for this Theorem 1.98 to get an extension $\mathcal{T}^*$ of $\mathcal{T}$ that has the following properties:

1. $\psi \in \mathcal{T}^*$ iff $\mathcal{T}^* \vdash_{HBk(S)} \psi$.

2. $\varphi_0 \notin \mathcal{T}^*$.

3. $\varphi \vee \psi \in \mathcal{T}^*$ iff either $\varphi \in \mathcal{T}^*$ or $\psi \in \mathcal{T}^*$.

---

[16]The corollary follows also from Theorem 8.104 below, and the fact that $\wedge, \vee$ and $\supset$ are $\mathcal{M}_{Bk}^\infty(S)$-conjunction, $\mathcal{M}_{Bk}^\infty(S)$-disjunction, and $\mathcal{M}_{Bk}^\infty(S)$-implication (respectively).

4. $\varphi \wedge \psi \in \mathcal{T}^*$ iff both $\varphi \in \mathcal{T}^*$ and $\psi \in \mathcal{T}^*$.

5. $\varphi \supset \psi \in \mathcal{T}^*$ iff either $\varphi \notin \mathcal{T}^*$ or $\psi \in \mathcal{T}^*$.

6. Either $\varphi \in \mathcal{T}^*$ or $\neg \varphi \in \mathcal{T}^*$ for every $\varphi \in \mathcal{W}(\mathcal{L}_C)$.

7. $\circ \varphi \in \mathcal{T}^*$ iff either $\varphi \notin \mathcal{T}^*$ or $\neg \varphi \notin \mathcal{T}^*$.

(Property 6 is due to Axiom [t], Property 7 is due to [b] and [k].)

Let $\lambda i.\sigma_i$ be an enumeration of all the formulas of $\mathcal{L}_C$ that do not begin with $\neg$. Then for any formula ψ of $\mathcal{L}_C$ there are unique $n(\psi), k(\psi)$ such that $\psi = \neg_{k(\psi)} \sigma_{n(\psi)}$, where $\neg_k \sigma$ is σ preceded by k negation symbols. Define a valuation ν in $\mathcal{M}^{\infty}_{Bk}(S)$ as follows:

$$\nu(\psi) = \begin{cases} f & \psi \notin \mathcal{T}^*, \\ t^{k(\psi)}_{n(\psi)} & \neg\psi \notin \mathcal{T}^*, \\ \top^{k(\psi)}_{n(\psi)} & \psi \in \mathcal{T}^*, \neg\psi \in \mathcal{T}^*. \end{cases}$$

That ν is well-defined follows from Property 6 of $\mathcal{T}^*$. Now we show that it is an $\mathcal{M}^{\infty}_{Bk}(S)$-valuation, i.e., it respects the interpretations in $\mathcal{M}^{\infty}_{Bk}(S)$ of the connectives of $\mathcal{L}_C$. We do the cases of $\neg$ and $\wedge$, leaving the proofs for the other connectives for the reader. (The demonstrations in these cases are essentially identical to the corresponding ones in the proof of Item 3 of Theorem 8.72, with $t^{k(\psi)}_{n(\psi)}$ and $\top^{k(\psi)}_{n(\psi)}$ replacing t and $\top$, respectively.)

First, we note that since we assume that $[l] \in S$, Theorem 8.98 implies that $[n^r_\wedge]_1 \notin S$, and either $[e] \notin S$, or $[n^r_\wedge]_2 \notin S$.

The case of $\neg$: We show that $\nu(\neg\psi) \in \tilde{\neg}\nu(\psi)$.

- Suppose that $\nu(\psi) = f$. Then $\psi \notin \mathcal{T}^*$, and so $\neg\psi \in \mathcal{T}^*$, implying that $\nu(\neg\psi) \in \{t^{k(\neg\psi)}_{n(\neg\psi)}, \top^{k(\neg\psi)}_{n(\neg\psi)}\} \subseteq \mathcal{D}$. If in addition $[c] \in S$ (or, equivalently, $[a_\neg] \in S$) then $\neg\neg\psi \notin \mathcal{T}^*$, and so $\nu(\neg\psi) = t^{k(\neg\psi)}_{n(\neg\psi)} \in \mathcal{E}_t$, as needed.

- Suppose that $\nu(\psi) = t^{k(\psi)}_{n(\psi)}$. By definition of ν this means that $\neg\psi \notin \mathcal{T}^*$. Hence $\nu(\neg\psi) = f$.

- Suppose that $\nu(\psi) = \top^{k(\psi)}_{n(\psi)}$. By definition of ν this means that $\psi \in \mathcal{T}^*$ and $\neg\psi \in \mathcal{T}^*$. The latter implies that $\nu(\neg\psi) \in \{t^{k(\neg\psi)}_{n(\neg\psi)}, \top^{k(\neg\psi)}_{n(\neg\psi)}\}$. But $n(\neg\psi) = n(\psi)$, $k(\neg\psi) = k(\psi) + 1$. Hence $\nu(\neg\psi) \in \{t^{k(\psi)+1}_{n(\psi)}, \top^{k(\psi)+1}_{n(\psi)}\}$, as needed. Since $[n^r_\wedge]_1 \notin S$, Proposition 8.97 implies that in addition we have in this case exactly two special subcases to consider:

8.4 da Costa's Consistency Operator(s)

- Suppose that $[e] \in S$. Since $\psi \in \mathcal{T}^*$, this means that $\neg\neg\psi \in \mathcal{T}^*$. But $\neg\psi \in \mathcal{T}^*$ as well. Hence $\nu(\neg\psi) = \top_{n(\neg\psi)}^{k(\neg\psi)} = \top_{n(\psi)}^{k(\psi)+1}$, as needed.
- Suppose that $[n_\wedge^r]_2 \in S$. Then $\vdash_{\mathbf{Bk}(S)} \neg\neg\psi \supset \neg(\psi \wedge \neg\psi)$. Using the axioms [l] and [b], this implies that $\vdash_{\mathbf{Bk}(S)} \neg\neg\psi \wedge \neg\psi \wedge \psi \supset \varphi_0$. Since ψ and $\neg\psi$ are in $\mathcal{T}^*$, it follows that $\neg\neg\psi \notin \mathcal{T}^*$. Hence $\nu(\neg\psi) = t_{n(\neg\psi)}^{k(\neg\psi)} = t_{n(\psi)}^{k(\psi)+1}$, as needed.

The case of $\wedge$: We show that $\nu(\psi_1 \wedge \psi_2) \in \nu(\psi_1)\tilde{\wedge}\nu(\psi_2)$.

- Suppose that $\nu(\psi_1) = \top_i^j$, $\nu(\psi_2) \in \{\top_i^{j+1}, t_i^{j+1}\}$. By definition of ν, this is possible only if ψ_1 and ψ_2 are both in $\mathcal{T}^*$, and $\psi_2 = \neg\psi_1$. By the axioms [l] and [b], this implies that $\neg(\psi_1 \wedge \psi_2) \notin \mathcal{T}^*$, and so $\nu(\psi_1 \wedge \psi_2) \in \mathcal{E}_t$. By Proposition 8.97 this is exactly what we need to show. (This is obvious in case $S \cap \{[e], [n_\wedge^r]_2\} = \emptyset$. It is true also if not, since if $[e] \in S$ then $\nu(\psi_2) = \nu(\neg\psi_1) \neq t_i^{j+1}$, while if $[n_\wedge^r]_2 \in S$ then $\nu(\psi_2) = \nu(\neg\psi_1) \neq \top_i^{j+1}$.)

- Suppose that $\nu(\psi_2) = \top_i^j$, $\nu(\psi_1) \in \{\top_i^{j+1}, t_i^{j+1}\}$, and $[d] \in S$. Using Proposition 8.97, the proof in this case is similar to the proof in the previous one, except that since we assume that $[l] \in S$, we do not need to take into account the possibility that $[n_\wedge^r]_1 \in S$.

- The proofs in all of the other cases is again essentially identical to what is done in the proof of Item 3 of Theorem 8.72. Details are left for the reader.

Finally, since $\mathcal{T} \subseteq \mathcal{T}^*$, $\nu(\psi) \in \mathcal{D}$ for every $\psi \in \mathcal{T}$, while $\nu(\varphi_0) = f$. Hence $\mathcal{T} \not\vdash_{\mathcal{M}_{Bk}^\infty(S)} \varphi_0$. □

Corollary 8.105 *If $S \subseteq \mathsf{A}_{ld}$ is coherent, $\mathbf{Bk}(S)$ is boldly paraconsistent.*

Proof. By Theorem 8.104 and Corollary 6.50. □

Corollary 8.106 *If $S \subseteq \mathsf{A}_{ld}$ is coherent then $\mathbf{Bk}(S)$ is decidable.*

Proof. Since $\mathbf{Bk}(S)$ is easily seen to be effective, this follows simply from Theorems 8.104 and 6.22. However, a better decision procedure can be derived from the proof of Theorem 8.104. It is clear from this proof (and the theorem itself) that in order to check whether a given formula φ is provable in $\mathbf{Bk}(S)$, it suffices to check all partial $\mathcal{M}_{Bk}^\infty(S)$-valuations which assign to subformulas of φ values in $\{f\} \cup \bigcup \{\{t_i^j, \top_i^j\} \mid 0 \leq i \leq n^*(\varphi), 0 \leq j \leq k^*(\varphi)\}$, where $n^*(\varphi)$ is the number of subformulas of φ which do not begin with $\neg$, and $k^*(\varphi)$ is the maximal number of consecutive negation symbols occurring within φ. This is a finite process. □

Note 8.107 In [46] it is shown that the simplest refinement of $\mathcal{M}_{Bk}^\infty(S)$ in which the set of truth values is just $\{f\} \cup \{t_0^j \mid 0 \leq j\} \cup \{I_0^j \mid 0 \leq j)\}$ is already *weakly* characteristic for $\mathbf{Bk}(S)$ (if S is coherent). Hence this simpler Nmatrix suffices for deciding theoremhood in $\mathbf{Bk}(S)$. Note, however, that this Nmatrix is *not* characteristic for $\mathbf{Bk}(S)$.

Next, we show that in the presence of [l] or [d], most logics of the form $\mathbf{B}(S)$ $(=\mathbf{Bk}(S))$[17] lack finite-valued semantics in a way which is much stronger than that given in Item 6 of Theorem 8.72.

Theorem 8.108 *Suppose $S \subseteq \mathbf{A}_{ld}$ is coherent, and $S \cap \{[l], [d]\} \neq \emptyset$.*

1. *If either $\{[l], [\mathsf{n}_\wedge^r]_2, [\mathsf{o}_\wedge^2]\} \subseteq S$ or $\{[d], [\mathsf{n}_\wedge^r]_1, [\mathsf{o}_\wedge^1]\} \subseteq S$, then $\mathbf{B}(S)$ has a characteristic three-valued Nmatrix.*

2. *If neither $\{[l], [\mathsf{n}_\wedge^r]_2, [\mathsf{o}_\wedge^2]\} \subseteq S$ nor $\{[d], [\mathsf{n}_\wedge^r]_1, [\mathsf{o}_\wedge^1]\} \subseteq S$, then there is no finite family $\mathcal{F}$ of finite Nmatrices such that $\vdash_{\mathbf{B}(S)} = \vdash_{\mathcal{F}}$.*

Proof.

1. Suppose, e.g., that $\{[l], [\mathsf{n}_\wedge^r]_2, [\mathsf{o}_\wedge^2]\} \subseteq S$. Let $S' = S - \{[l]\}$. Define a three-valued Nmatrix $\mathcal{M} = \langle\{t, f, \top\}, \{t, \top\}, \mathcal{O}\rangle$ for $\mathcal{L}_C$ as follows:

   - For $\diamond \in \{\vee, \supset, \circ\}$, $\tilde{\diamond}_\mathcal{M} = \tilde{\diamond}_{\mathcal{M}_{Bk}(S')}$.
   - $\tilde{\neg}_\mathcal{M}(t) = \{f\}$; $\tilde{\neg}_\mathcal{M}(\top) = \{t\}$; $\tilde{\neg}_\mathcal{M}(f) = \{t\}$ in case [c] $\in S$ or $[a_\neg] \in S$, and $\tilde{\neg}_\mathcal{M}(f) = \{t, \top\}$ otherwise.
   - $\tilde{\wedge}_\mathcal{M}(f, x) = f$, and $\tilde{\wedge}_\mathcal{M}(t, x) = \tilde{\wedge}_\mathcal{M}(\top, x) = x$.

   It is now a straightforward matter to check that $\mathcal{M}_{Bk}^\infty(S)$ is a strongly preserving E-rexpansion of $\mathcal{M}$, where E is the expansion function introduced in Proposition 8.90. Therefore, it follows from Theorems 6.42 and 8.104 that $\mathcal{M}$ is characteristic for $\mathbf{B}(S)$.

2. Suppose, e.g., that [l] $\in S$ and $\{[\mathsf{n}_\wedge^r]_2, [\mathsf{o}_\wedge^2]\} \not\subseteq S$. (Note that by Theorem 8.98, $\{[d], [\mathsf{n}_\wedge^r]_1, [\mathsf{o}_\wedge^1]\} \not\subseteq S$ in case [l] $\in S$.) Assume for contradiction that there is a finite family $\mathcal{F}$ of finite ordinary Nmatrices such that $\vdash_{\mathbf{B}(S)} = \vdash_\mathcal{F}$. Let n be the maximal number of truth-values in one of the Nmatrices of $\mathcal{F}$. Define:

$$\mathcal{T} = \{p_i \mid 1 \leq i \leq n+1\} \cup \{\neg p_i \mid 1 \leq i \leq n+1\} \cup$$
$$\cup \{\neg(p_i \wedge \neg p_j) \mid i \neq j, 1 \leq i \leq n+1, 1 \leq j \leq n+1\}.$$

   Let ν be a valuation in $\mathcal{M}_B^\infty(S)$ such that:

---

[17]Recall that if $S \cap \{[l], [d]\} \neq \emptyset$ then $\mathbf{Bk}(S) = \mathbf{B}(S)$ (See Proposition 8.84).

8.4 da Costa's Consistency Operator(s)

- $\nu(p_{n+2}) = f$, and $\nu(p_i) = \mathsf{T}_i^0$ for $1 \leq i \leq n+1$.
- $\nu(\neg p_i) = t_i^1$ if $[\mathsf{n}_\wedge^r]_2 \in S$, and $\nu(\neg p_i) = \mathsf{T}_i^1$ otherwise;
- For $i \neq j, 1 \leq i \leq n+1, 1 \leq j \leq n+1$, $\nu(p_i \wedge \neg p_j) = \mathsf{T}_0^0$ and $\nu(\neg(p_i \wedge \neg p_j)) \in \{\mathsf{T}_0^1, t_0^1\}$.

(Such a valuation exists, since here $[\mathsf{o}_\wedge^2] \notin S$ in case $[\mathsf{n}_\wedge^r]_2 \in S$.) Obviously, ν is a model of $\mathcal{T}$ which is not a model of p_{n+2}. It follows that $\mathcal{T} \not\vdash_{\mathbf{B}(S)} p_{i+2}$. Since $\vdash_{\mathbf{B}(S)} = \vdash_{\mathcal{F}}$, there is an Nmatrix $\mathcal{M} = \langle \mathcal{V}, \mathcal{D}, \mathcal{O} \rangle \in \mathcal{F}$ and an $\mathcal{M}$-valuation ν_0 such that ν_0 is a model of $\mathcal{T}$ but not of p_{n+2}. By the pigeonhole principle, there are $1 \leq i_0 < j_0 \leq n+1$ such that $\nu_0(p_{i_0}) = \nu_0(p_{j_0})$. Define a partial valuation ν_1 in $\mathcal{M}$ by letting $\nu_1(p_{n+2}) = \nu_0(p_{n+2})$, $\nu_1(p_{i_0}) = \nu_0(p_{i_0})$, $\nu_1(\neg p_{i_0}) = \nu_0(\neg p_{j_0})$, $\nu_1(p_{i_0} \wedge \neg p_{i_0}) = \nu_0(p_{i_0} \wedge \neg p_{j_0})$, and $\nu_1(\neg(p_{i_0} \wedge \neg p_{i_0})) = \nu_0(\neg(p_{i_0} \wedge \neg p_{j_0}))$. Since ν_0 is an $\mathcal{M}$-valuation and $\nu_0(p_{i_0}) = \nu_0(p_{j_0})$, ν_1 is indeed a partial $\mathcal{M}$-valuation. Now, ν_0 is a model of $\mathcal{T}$, and so it is a model of $\{p_{i_0}, \neg p_{j_0}, \neg(p_{i_0} \wedge \neg p_{j_0})\}$. Hence ν_1 is a model of $\{p_{i_0}, \neg p_{i_0}, \neg(p_{i_0} \wedge \neg p_{i_0})\}$. On the other hand, ν_1 is not a model of p_{n+2} (because ν_0 is not a model of p_{n+2}). Since $\mathbf{B}(S)$ is sound for $\mathcal{M}$, it follows that $\{p_{i_0}, \neg p_{i_0}, \neg(p_{i_0} \wedge \neg p_{i_0})\} \not\vdash_\mathbf{L} p_{n+2}$. Since $\mathbf{B}(S)$ is an extension of $\mathbf{Bl}$, this contradicts the fact that $\{p_{i_0}, \neg p_{i_0}, \neg(p_{i_0} \wedge \neg p_{i_0})\} \vdash_{\mathbf{Bl}} p_{n+2}$. □

Corollary 8.109 *Let $S^l = (\mathcal{A}_{PAC} \cup \{[\mathsf{o}_\wedge^2], [\mathsf{a}_\mathsf{c}]\}) - \{[\mathsf{e}], [\mathsf{n}_\wedge^r]_1\}$. Then $\mathbf{Bl}(S^l)$ is the only C-system of the form $\mathbf{Bl}(S)$ (where $S \subseteq \mathcal{A}_{ld}$) that has a characteristic matrix. Similarly, if $S^d = (\mathcal{A}_{PAC} \cup \{[\mathsf{o}_\wedge^1], [\mathsf{a}_\circ]\}) - \{[\mathsf{e}], [\mathsf{n}_\wedge^r]_2\}$, then $\mathbf{Bd}(S^d)$ is the only such C-system of the form $\mathbf{Bd}(S)$ ($S \subseteq \mathcal{A}_{ld}$).*

Proof. A careful check of the Nmatrices that according to Theorem 8.108 characterize some C-systems of the form $\mathbf{Bl}(S)$ ($S \subseteq \mathcal{A}_{ld}$) reveals that exactly one of them is a matrix: that which characterizes $\mathbf{Bl}(S^l)$. The case of logics of the form $\mathbf{Bd}(S)$ ($S \subseteq \mathcal{A}_{ld}$) is similar. □

Note 8.110 The matrix that characterizes $\mathbf{Bl}(S^l)$ is of course three-valued. Its operations are given by the following tables:

	$\tilde{\neg}$	$\tilde{\circ}$
t	f	t
$\top$	t	f
f	t	t

$\tilde{\wedge}$	t	$\top$	f
t	t	$\top$	f
$\top$	t	$\top$	f
f	f	f	f

$\tilde{\vee}$	t	$\top$	f
t	t	t	t
$\top$	t	$\top$	$\top$
f	t	$\top$	f

$\tilde{\supset}$	t	$\top$	f
t	t	$\top$	f
$\top$	t	$\top$	f
f	t	t	t

The operations in the matrix that characterizes $\mathbf{Bd}(S^d)$ are almost identical, except that there $\tilde{\wedge}(t, \top) = t$, while $\tilde{\wedge}(\top, t) = \top$.

Our next task is to provide in a modular way a Gentzen-type system for $\mathbf{Bk}[S]$ for every coherent $S \subseteq \mathsf{A}_{ld}$. For this we need first of all to translate the axioms [l] and [d] too into Gentzen-type rules. Using the intuitive meaning of $\circ\psi$ given in Note 8.87, this translation is obtained by substituting in $[\circ \Rightarrow]$ the formulas $\neg(\psi \wedge \neg\psi)$ and $\neg(\neg\psi \wedge \psi)$ (respectively) for $\circ\psi$. (Note that the rules that are obtained by applying the same procedure to $[\Rightarrow \circ]$ are derivable in $\mathbf{Bk}$.)

Definition 8.111 $(R([\mathsf{l}]), R([\mathsf{d}]))$ The Gentzen-type rules $R([\mathsf{l}])$ and $R([\mathsf{d}])$ that corresponds to [l] and [d] (respectively) are defined as follows:

$$[R([\mathsf{l}])] \quad \frac{\Gamma \Rightarrow \varphi, \Delta \quad \Gamma \Rightarrow \neg\varphi, \Delta}{\Gamma, \neg(\varphi \wedge \neg\varphi) \Rightarrow \Delta} \qquad [R([\mathsf{d}])] \quad \frac{\Gamma \Rightarrow \varphi, \Delta \quad \Gamma \Rightarrow \neg\varphi, \Delta}{\Gamma, \neg(\neg\varphi \wedge \varphi) \Rightarrow \Delta}$$

Note 8.112 Recall that the Gentzen-type rules corresponding to the axioms in A_{CPAC} are given in Figures 8.2, 8.3, 7.1, and 7.2. It is worth noting that they could also have been derived directly from the general refining conditions which are associated with those axioms in Proposition 8.92. This derivation uses the following easily seen facts (that are analogous to those given in Proposition 8.42) about an $\mathcal{M}$-valuation ν, in case $\mathcal{M}$ is a simple refinement of $\mathcal{M}_{Bk}^\infty$:

1. $\nu(\psi) \in \mathcal{E}_t$ iff ν satisfies $\neg\psi \Rightarrow$.

2. $\nu(\psi) = f$ iff ν satisfies $\psi \Rightarrow$.

3. $\nu(\psi) \in \mathcal{E}_\top$ iff ν satisfies both $\Rightarrow \psi$ and $\Rightarrow \neg\psi$.

4. $\nu(\psi) \in \{f\} \cup \mathcal{E}_\top$ iff ν satisfies $\Rightarrow \neg\psi$.

5. $\nu(\psi) \in \mathcal{E}_t \cup \mathcal{E}_\top$ iff ν satisfies $\Rightarrow \psi$.

6. $\nu(\psi) \in \{f\} \cup \mathcal{E}_t$ iff ν satisfies $\psi, \neg\psi \Rightarrow$.

Let us take $a_\vee$ as an example. Its corresponding semantic condition $\mathsf{GC}(a_\vee)$ from Definition 8.92 can be reformulated as follows: If $a \in \mathcal{E}_t$ and $b \in \mathcal{T} \cup \{f\}$ then: (i) $a \vee b \subseteq \mathcal{E}_t$ and (ii) $b \vee a \subseteq \mathcal{E}_t$. Now, (i) implies that if $\nu(\varphi) \in \mathcal{E}_t$ and $\nu(\psi) \in \mathcal{E}_t \cup \{f\}$, then $\nu(\varphi \vee \psi) \in \mathcal{E}_t$. Using the above facts, we rewrite this as: if ν satisfies $\neg\varphi \Rightarrow$ and $\neg\psi, \psi \Rightarrow$, then ν satisfies $\neg(\varphi \vee \psi) \Rightarrow$. By adding context we obtain the second rule in $R(a_\vee)$ as given in Figure 8.2. The first one is similarly obtained from (ii).

8.4 da Costa's Consistency Operator(s)

Proposition 8.113 *Let S be a coherent subset of A_{ld}, and let R be the function given in Figures 7.1, 7.2, 8.2 and 8.3, and in Definition 8.111. Then $GBk(S)$ (Definition 8.49) is sound and complete for $\mathbf{Bk}(S)$.*

Proof. Since $R([\mathsf{l}])$ and $R([\mathsf{d}])$ are obviously equivalent to $[\mathsf{l}]$ and $[\mathsf{d}]$ (respectively), it easily follows from Theorem 8.72 that $\vdash_{HBk(S)} \subseteq \vdash_{GBk(S)}$. It is also not difficult to show that $GBk(S)$ is sound for $\mathcal{M}^\infty_{Bk}(S)$. Hence the proposition follows from Theorem 8.104. □

Things become more complicated when we turn to the issue of cut-elimination for $GBk(S)$ ($S \subseteq \mathsf{A}_{ld}$).

Example 8.114 Let $\{[\mathsf{l}], [\mathsf{n}^r_\wedge]_2\} \subseteq S$. From Proposition 8.97 it follows that $P_1, \neg P_1, \neg\neg P_1 \Rightarrow$ is valid in $\mathcal{M}^\infty_{Bk}(S)$. Therefore it is derivable in $GBk(S)$ by Proposition 8.113. (It is indeed not difficult to provide a direct derivation of this sequent in $GBk(S)$ using cuts.) However, it is obvious that no cut-free proof of this sequent is possible in $GBk(S)$.

The next definition overcomes the problem noted in Example 8.114.

Definition 8.115 ($GBk^*(S)$) Let S be a coherent subset of A_{ld}.

- If either $\{[\mathsf{l}], [\mathsf{n}^r_\wedge]_2\} \subseteq S$ or $\{[\mathsf{d}], [\mathsf{n}^r_\wedge]_1\} \subseteq S$, then $GBk^*(S)$ is the system obtained from $GBk(S)$ by adding to it the following rule:

$$[R([\mathsf{n}^r_\wedge])^*] \quad \frac{\Gamma \Rightarrow \varphi, \Delta \quad \Gamma \Rightarrow \neg\varphi, \Delta}{\Gamma, \neg\neg\varphi \Rightarrow \Delta}$$

- Otherwise, $GBk^*(S) = GBk(S)$.

Theorem 8.116 (completeness of $GBk^*(S)$, cut elimination) *If $S \subseteq \mathsf{A}_{ld}$ is coherent, then $GBk^*(S)$ is sound and complete for $\mathbf{Bk}(S)$, and it admits cut-elimination.*

Proof. The case in which $S \subseteq \mathsf{A}_{CPAC}$ was shown in Item 5 of Theorem 8.72. So here we do the proof under the assumption that $[\mathsf{l}] \in S$.[18] Note that in this case it follows from Theorem 8.98 that $[\mathsf{n}^r_{\wedge 1}] \notin S$, $\{[\mathsf{e}], [\mathsf{n}^r_\wedge]_2\} \nsubseteq S$, $\{[\mathsf{d}], [\mathsf{n}^r_\wedge]_2\} \nsubseteq S$, $\{[\mathsf{o}^1_\wedge], [\mathsf{n}^r_\wedge]_2\} \nsubseteq S$, $\{[\mathsf{o}^1_\vee], [\mathsf{n}^r_\vee]\} \nsubseteq S$, $\{[\mathsf{o}^2_\vee], [\mathsf{n}^r_\vee]\} \nsubseteq S$, and $\{[\mathsf{o}^1_\supset], [\mathsf{n}^r_\supset]\} \nsubseteq S$. Therefore (using Proposition 8.97), in $\mathcal{M}^\infty_{Bk}(S)$ we have that if $a = \top^j_i$ then $\tilde{\neg} a$ is either $\{\top^{j+1}_i\}$ (in case $[\mathsf{e}] \in S$), or $\{t^{j+1}_i\}$ (in

---

[18]This does not rule out the possibility that $[\mathsf{d}] \in S$ as well. The case in which $[\mathsf{d}] \in S$ while $[\mathsf{l}] \notin S$ is completely similar to the case in which $[\mathsf{l}] \in S$ while $[\mathsf{d}] \notin S$. Note also that the proof given here can easily be adapted to the case where $S \subseteq \mathsf{A}_{CPAC}$.

case $[\mathsf{n}_\wedge^r]_2 \in S$), or $\{\mathsf{T}_i^{j+1}, t_i^{j+1}\}$ (if $S \cap \{[\mathsf{e}], [\mathsf{n}_\wedge^r]_2\} = \emptyset$). In any other case $\tilde{\diamond}(a_1, \ldots, a_n)$ is either $\{f\}$, or $\mathcal{E}_t$, or $\mathcal{E}_\top$, or $\mathcal{D} = \mathcal{E}_t \cup \mathcal{E}_\top$. (Here $\diamond \in \mathcal{C}(\mathcal{L}_C)$, n is the arity of $\diamond$ ($n \in \{1, 2\}$), and $a_1, \ldots, a_n \in \mathcal{E}_t \cup \mathcal{E}_\top \cup \{f\}$.)

The soundness (and in fact also the completeness) of $GBk^*(S)$ for $\mathbf{Bk}(S)$ is immediate from Proposition 8.113 and Example 8.114. As usual in this book, what we will show is that if a sequent has no cut-free proof in $GBk^*(S)$, then it is not valid in $\mathcal{M}_{Bk}^\infty(S)$. Given the soundness of $GBk^*(S)$ for $\mathcal{M}_{Bk}^\infty(S)$, this will prove the cut-elimination theorem for $GBk^*(S)$, and at the same time will provide a new, direct proof of the completeness of $GBk^*(S)$ for $\mathcal{M}_{Bk}^\infty(S)$.

So, suppose that $\Gamma_0 \Rightarrow \Delta_0$ has no cut-free proof in $GBk(S)$. We construct a valuation ν that refutes it in $\mathcal{M}_{Bk}^\infty(S)$. As in the proof of Theorem 7.60, it is a standard matter to show that $\Gamma_0 \Rightarrow \Delta_0$ can be extended to a saturated (with respect to S) sequent $\Gamma \Rightarrow \Delta$ such that $\Gamma_0 \subseteq \Gamma$, $\Delta_0 \subseteq \Delta$, and $\Gamma \Rightarrow \Delta$ has no cut-free proof in $GBk(S)$. Here, as usual, $\Gamma \Rightarrow \Delta$ being saturated (with respect to S) means that it is closed under the rules of $GBk(S)$ applied backwards. In other words, $\Gamma \Rightarrow \Delta$ has the following properties:

(S0) $\Gamma \cap \Delta = \emptyset$. (Since $\Gamma \Rightarrow \Delta$ has no cut-free proof.)

(S1) Suppose $\sigma = \varphi \# \psi$ where $\# \in \{\wedge, \vee, \supset\}$.

- If $\# = \wedge$ and $\sigma \in \Gamma$, then $\varphi, \psi \in \Gamma$.
- If $\# = \wedge$ and $\sigma \in \Delta$, then either $\varphi \in \Delta$ or $\psi \in \Delta$.
- Similar conditions for the cases $\# = \vee$; $\# = \supset$.

(S2) Suppose $\sigma = \circ\varphi$.

- If $\sigma \in \Gamma$, then either $\varphi \in \Delta$ or $\neg\varphi \in \Delta$.
- If $\sigma \in \Delta$, then $\varphi \in \Gamma$ and $\neg\varphi \in \Gamma$.

(S3) If $\neg\varphi \in \Delta$, then $\varphi \in \Gamma$.

(S4) If $\neg(\varphi \wedge \neg\varphi) \in \Gamma$, then either $\varphi \in \Delta$ or $\neg\varphi \in \Delta$.

(S5) If $[\mathsf{d}] \in S$ and $\neg(\neg\varphi \wedge \varphi) \in \Gamma$, then either $\varphi \in \Delta$ or $\neg\varphi \in \Delta$.

(S6) If $[\mathsf{c}] \in S$ or $[\mathsf{a}_\neg] \in S$, and $\neg\neg\varphi \in \Gamma$, then $\varphi \in \Gamma$.

(S7) If $[\mathsf{e}] \in S$ and $\neg\neg\varphi \in \Delta$, then $\varphi \in \Delta$.

(S8) If $[\mathsf{n}_\wedge^r]_2 \in S$ and $\neg\neg\varphi \in \Gamma$, then either $\varphi \in \Delta$ or $\neg\varphi \in \Delta$.

(S9) If $[\mathsf{a}_\circ] \in S$ and $\neg\circ\varphi \in \Gamma$, then $\neg\varphi, \varphi \in \Gamma$.

8.4 da Costa's Consistency Operator(s)

(S10) If $[n_\wedge^l] \in S$ or $[a_\wedge] \in S$, and $\neg(\varphi \wedge \psi) \in \Gamma$, then $\neg\varphi \in \Gamma$ or $\neg\psi \in \Gamma$.

(S11) If $[n_\wedge^r]_2 \in S$ and $\neg(\varphi \wedge \psi) \in \Delta$ then $\neg\psi \in \Delta$.

(S12) If $[a_\vee] \in S$ and $\neg(\varphi_1 \vee \varphi_2) \in \Gamma$ then:

- Either $\neg\varphi_2 \in \Gamma$, or $\{\varphi_1, \neg\varphi_1\} \subseteq \Gamma$.
- Either $\neg\varphi_1 \in \Gamma$, or $\{\varphi_2, \neg\varphi_2\} \subseteq \Gamma$.

(S13) Similarly for the remaining axioms of A_{ld}.

Let $\lambda i.\sigma_i$, $n(\psi)$, and $k(\psi)$ be like in the proof of Theorem 8.104. (That is: $\lambda i.\sigma_i$ is an enumeration of all the formulas in $\mathcal{L}_C$ that do not begin with $\neg$, and $\psi = \neg_{k(\psi)}\sigma_{n(\psi)}$ for every formula ψ of $\mathcal{L}_C$.) We define a valuation ν in $\mathcal{M}^\infty_{Bk}(S)$ which refutes $\Gamma \Rightarrow \Delta$ as follows:

- If p is a propositional variable, then

$$\nu(p) = \begin{cases} f & p \in \Delta, \\ t^0_{n(p)} & \neg p \in \Delta, \\ \top^0_{n(p)} & otherwise. \end{cases}$$

This is well-defined because of (S3) and (S0).

- If $\varphi = \diamond(\psi_1, \ldots, \psi_j)$ (where $j \in \{1, 2\}$), then

$$\nu(\varphi) = \begin{cases} f & \tilde{\diamond}(\nu(\psi_1), \ldots, \nu(\psi_j)) = \{f\}, \\ t^{k(\varphi)}_{n(\varphi)} & \text{either } \tilde{\diamond}(\nu(\psi_1), \ldots, \nu(\psi_j)) \subseteq \mathcal{E}_t \text{ or} \\ & \neg\varphi \in \Delta \text{ and } t^{k(\varphi)}_{n(\varphi)} \in \tilde{\diamond}(v(\psi_1), \ldots, v(\psi_j)), \\ \top^{k(\varphi)}_{n(\varphi)} & otherwise. \end{cases}$$

Note that from this definition of ν it immediately follows that $\nu(\varphi) \in \{f, t^{k(\varphi)}_{n(\varphi)}, \top^{k(\varphi)}_{n(\varphi)}\}$ for every $\varphi \in W(\mathcal{L})$. This fact will be freely used in what follows, without any further notice.

First of all, we prove that ν is an $\mathcal{M}^\infty_{Bk}(S)$-valuation, that is: for every j-ary connective $\diamond$, $\nu(\diamond(\psi_1, \ldots, \psi_j)) \in \tilde{\diamond}(\nu(\psi_1), \ldots, \nu(\psi_j))$.

$\diamond = \neg$: In this case we should show that $\nu(\neg\psi_1) \in \tilde{\neg}\nu(\psi_1)$.

- Suppose that $\nu(\psi_1) = f$. If $[c] \in S$ (or $[a_\neg] \in S$) then $\tilde{\neg}\nu(\psi_1) = \mathcal{E}_t$, and so $\nu(\neg\psi_1) = t^{k(\neg\psi_1)}_{n(\neg\psi_1)} \in \mathcal{E}_t$ by definition of ν. Otherwise $\tilde{\neg}\nu(\psi_1) = \mathcal{D}$, and so $\nu(\neg\psi_1) \neq f$ by definition of ν, implying that $\nu(\neg\psi_1) \in \tilde{\neg}\nu(\psi_1)$ in this case too.

- Suppose that $\nu(\psi_1) = t_{n(\psi_1)}^{k(\psi_1)}$. Then $\tilde{\neg}\nu(\psi_1) = \{f\}$. Hence $\nu(\neg\psi_1) = f$ by definition of ν, and so $\nu(\neg\psi_1) \in \tilde{\neg}\nu(\psi_1)$.

- Suppose that $\nu(\psi_1) = \mathsf{T}_{n(\psi_1)}^{k(\psi_1)}$. Since $n(\neg\psi_1) = n(\psi_1)$ and $k(\neg\psi_1) = k(\psi_1)+1$, the definition of $\tilde{\neg}$ in $\mathcal{M}_{Bk}^{\infty}(S)$ implies that in this case $\tilde{\neg}\nu(\psi_1)$ is either $\{\mathsf{T}_{n(\neg\psi_1)}^{k(\neg\psi_1)}\}$, or $\{t_{n(\neg\psi_1)}^{k(\neg\psi_1)}\}$, or $\{\mathsf{T}_{n(\neg\psi_1)}^{k(\neg\psi_1)}, t_{n(\neg\psi_1)}^{k(\neg\psi_1)}\}$. From the definition of ν it easily follows that in all three cases we have that $\nu(\neg\psi_1) \in \tilde{\neg}\nu(\psi_1)$.

$\diamond = \circ$: In this case that $\tilde{\circ}(\nu(\psi_1))$ is either $\{f\}$, or $\mathcal{E}_t$, or $\mathcal{D}$. An examination of the definition of ν easily reveals that in all three cases we get that $\nu(\diamond(\psi_1)) \in \tilde{\circ}(\nu(\psi_1))$.

$\diamond \in \{\land, \lor, \supset\}$: In this case the value of of $\tilde{\diamond}(\nu(\psi_1), \nu(\psi_2))$ is either $\{f\}$, or $\mathcal{E}_t$, or $\mathcal{E}_\top$, or $\mathcal{D}$. An examination of the definition of ν easily reveals that in all four cases we get that $\nu(\diamond(\psi_1, \psi_2)) \in \tilde{\diamond}(\nu(\psi_1), \nu(\psi_2))$.

It remains to show that ν is a refuting valuation for $\Gamma \Rightarrow \Delta$. For this we prove by induction on the structure of φ the following four properties of ν:

1. If $\varphi \in \Delta$ then $\nu(\varphi) = f$.

2. If $\varphi \in \Gamma$ then $\nu(\varphi) = t_{n(\varphi)}^{k(\varphi)}$ or $\nu(\varphi) = \mathsf{T}_{n(\varphi)}^{k(\varphi)}$.

3. If $\neg\varphi \in \Delta$ then $\nu(\varphi) = t_{n(\varphi)}^{k(\varphi)}$.

4. If $\neg\varphi \in \Gamma$ then $\nu(\varphi) = f$ or $\nu(\varphi) = \mathsf{T}_{n(\varphi)}^{k(\varphi)}$.

- If φ is a variable then 1-4 are immediate from the definition of ν, the fact that if $\varphi \in \Gamma$ then $\varphi \notin \Delta$, and the fact that $k(\varphi) = 0$ in this case.

- Suppose that $\varphi = \neg\psi$.

    1. Suppose that $\varphi = \neg\psi \in \Delta$. By the induction hypothesis for ψ, we have $\nu(\psi) = t_{n(\psi)}^{k(\psi)}$. Since $\tilde{\neg}t_{n(\psi)}^{k(\psi)} = \{f\}$, $\nu(\varphi) = f$ by the definition of v.

    2. Suppose that $\varphi = \neg\psi \in \Gamma$. By the induction hypothesis for ψ, either $\nu(\psi) = f$ or $\nu(\psi) = \mathsf{T}_{n(\psi)}^{k(\psi)}$. Hence $\tilde{\neg}\nu(\psi) \neq \{f\}$. By the definition of ν, $\nu(\varphi) \neq f$, and so $\nu(\varphi) = t_{n(\varphi)}^{k(\varphi)}$ or $\nu(\varphi) = \mathsf{T}_{n(\varphi)}^{k(\varphi)}$.

    3. Suppose that $\neg\varphi = \neg\neg\psi \in \Delta$. By (**S3**), this implies that $\neg\psi \in \Gamma$. Hence it follows from the induction hypothesis for ψ that either $\nu(\psi) = f$ or $\nu(\psi) = \mathsf{T}_{n(\psi)}^{k(\psi)}$. Now there are two possibilities:

8.4 da Costa's Consistency Operator(s) 281

- [e] ∈ S. Then $\psi \in \Delta$ by **(S7)**, and so $\nu(\psi) = f$ by the induction hypothesis. Hence, in this case $t_{n(\varphi)}^{k(\varphi)} \in \tilde{\neg}\nu(\psi)$ (because $\tilde{\neg} f$ can be either $\mathcal{E}_t$ or $\mathcal{D}$).
- [e] ∉ S. Then each of $\tilde{\neg}\top_{n(\psi)}^{k(\psi)}$ and $\tilde{\neg} f$ can be either $\mathcal{D}$, or $\mathcal{E}_t$, or $\{\top_{n(\psi)}^{k(\psi)+1}, t_{n(\psi)}^{k(\psi)+1}\}$, or $\{t_{n(\psi)}^{k(\psi)+1}\}$. Hence, $t_{n(\psi)}^{k(\psi)+1} \in \tilde{\neg}\nu(\psi)$. Since $k(\psi)+1 = k(\varphi)$ and $n(\psi) = n(\varphi)$, we again have $t_{n(\varphi)}^{k(\varphi)} \in \tilde{\neg}\nu(\psi)$.

Thus, in both cases, $t_{n(\varphi)}^{k(\varphi)} \in \tilde{\neg}\nu(\psi)$. Since $\neg\varphi \in \Delta$, $\nu(\varphi) = t_{n(\varphi)}^{k(\varphi)}$ by the definition of ν.

4. Suppose that $\neg\varphi \in \Gamma$. Then $\neg\varphi \notin \Delta$, and so $\nu(\varphi) = t_{n(\varphi)}^{k(\varphi)}$ only if $\tilde{\neg}\nu(\psi) \subseteq \mathcal{E}_t$. This may happen only in one of the following cases:
   - $\nu(\psi) = f$, and [c] ∈ S (or [a$_\neg$] ∈ S). But by **(S6)**, the assumption that $\neg\varphi = \neg\neg\psi \in \Gamma$ implies in this case that $\psi \in \Gamma$, and so $\nu(\psi) \neq f$ by the induction hypothesis.
   - $\nu(\psi) = \top_{n(\psi)}^{k(\psi)}$, and $[n^r_\wedge]_2 \in S$. But by **(S8)**, the assumption that $\neg\varphi = \neg\neg\psi \in \Gamma$ implies in this case that either $\psi \in \Delta$ or $\neg\psi \in \Delta$. In both cases the induction hypothesis implies that $\nu(\psi) \neq \top_{n(\psi)}^{k(\psi)}$.

It follows that $\tilde{\neg}\nu(\psi) \neq t_{n(\varphi)}^{k(\varphi)}$, and so $\nu(\varphi) = f$ or $\nu(\varphi) = \top_{n(\varphi)}^{k(\varphi)}$.

- Suppose that $\varphi = \circ\psi$.

  1. Suppose that $\varphi \in \Delta$. By **(S2)**, this implies that $\psi, \neg\psi \in \Gamma$. Therefore, $\nu(\psi) = \top_{n(\psi)}^{k(\psi)}$ by the induction hypothesis for ψ. Since $\tilde{o}(\top_{n(\psi)}^{k(\psi)}) = \{f\}$, $\nu(\varphi) = f$ by the definition of ν.
  2. Suppose that $\varphi \in \Gamma$. By **(S2)**, either $\psi \in \Delta$ or $\neg\psi \in \Delta$. By the induction hypothesis for ψ, it follows that $\nu(\psi) \notin \mathcal{E}_\top$. Therefore, $\tilde{o}(\nu(\psi)) \neq \{f\}$, and so $\nu(\varphi) = t_{n(\varphi)}^{k(\varphi)}$ or $\nu(\varphi) = \top_{n(\varphi)}^{k(\varphi)}$.
  3. Suppose that $\neg\varphi \in \Delta$. By **(S3)**, $\varphi = \circ\psi \in \Gamma$. By **(S2)**, either $\psi \in \Delta$ or $\neg\psi \in \Delta$. Hence $\nu(\psi) \notin \mathcal{E}_\top$ by the induction hypothesis for ψ. Thus $\tilde{o}(\nu(\psi))$ is either $\mathcal{E}_t$ or $\mathcal{D}$, and so $t_{n(\varphi)}^{k(\varphi)} \in \tilde{o}(\nu(\psi))$. Since $\neg\varphi \in \Delta$, this and ν's definition imply that $\nu(\varphi) = t_{n(\varphi)}^{k(\varphi)}$.
  4. Suppose that $\neg\varphi \in \Gamma$. Then $\neg\varphi \notin \Delta$. Hence, $\nu(\varphi) = t_{n(\varphi)}^{k(\varphi)}$ only if $\tilde{o}\nu(\psi) = \mathcal{E}_t$. This may happen only if $[a_\circ] \in S$, and $\nu(\psi) \neq \top_{n(\psi)}^{k(\psi)}$. However, by **(S9)**, if $[a_\circ] \in S$ and $\neg\circ\psi \in \Gamma$ then $\psi, \neg\psi \in \Gamma$, and so $\nu(\psi) = \top_{n(\psi)}^{k(\psi)}$ by the induction hypothesis for ψ.

- Suppose that $\varphi = \psi_1 \wedge \psi_2$.

  1. Suppose that $\varphi \in \Delta$. By **(S1)**, $\psi_1 \in \Delta$ or $\psi_2 \in \Delta$. By the induction hypothesis, $\nu(\psi_1) = f$ or $\nu(\psi_2) = f$. Hence, $\tilde{\wedge}(\nu(\psi_1), \nu(\psi_2)) = \{f\}$, and $\nu(\varphi) = f$ by the definition of ν.

  2. Suppose that $\varphi \in \Gamma$. By **(S1)**, $\psi_1, \psi_2 \in \Gamma$. By the induction hypothesis, $\nu(\psi_i) = t_{n(\psi_i)}^{k(\psi_i)}$ or $\nu(\psi_i) = \top_{n(\psi_i)}^{k(\psi_i)}$ for $i \in \{1, 2\}$. Hence, $\tilde{\wedge}(\nu(\psi_1), \nu(\psi_2)) \neq \{f\}$, and by the definition of ν, $\nu(\varphi) = t_{n(\varphi)}^{k(\varphi)}$ or $\nu(\varphi) = \top_{n(\varphi)}^{k(\varphi)}$.

  3. Suppose that $\neg\varphi \in \Delta$. By **(S3)**, $\varphi \in \Gamma$. Hence, $\psi_1, \psi_2 \in \Gamma$ by **(S1)**. It follows by the induction hypothesis for ψ_1 and ψ_2 that $\nu(\psi_1), \nu(\psi_2) \in \mathcal{D}$. Thus, $\tilde{\wedge}(\nu(\psi_1), \nu(\psi_2)) \neq \{f\}$. Suppose for contradiction that $\tilde{\wedge}(\nu(\psi_1), \nu(\psi_2)) = \mathcal{E}_\top$. Since $[\mathsf{n}_\wedge^r]_1 \notin S$, this is possible only if $[\mathsf{n}_\wedge^r]_2 \in S$, and $\nu(\psi_2) \in \mathcal{E}_\top$. But by **(S11)**, if $\neg\varphi \in \Delta$ and $[\mathsf{n}_\wedge^r]_2 \in S$ then $\neg\psi_2 \in \Delta$, and so $\nu(\psi_2) \in \mathcal{E}_t$ by the induction hypothesis for ψ_2. A contradiction. It follows that $\tilde{\wedge}(\nu(\psi_1), \nu(\psi_2))$ can be either $\mathcal{E}_t$ or $\mathcal{D}$, and so $t_{n(\varphi)}^{k(\varphi)} \in \tilde{\wedge}(\nu(\psi_1), \nu(\psi_2))$. Since $\neg\varphi \in \Delta$, the definition of ν implies that $\nu(\varphi) = t_{n(\varphi)}^{k(\varphi)}$.

  4. Suppose that $\neg\varphi \in \Gamma$. We show that none of the cases in which $\tilde{\wedge}(\nu(\psi_1), \nu(\psi_2)) \subseteq \mathcal{E}_t$ is applicable under this assumption:

     - Suppose that $\nu(\psi_1) = \top_j^i$ and $\nu(\psi_2) \in \{t_j^{i+1}, \top_j^{i+1}\}$. By the definition of ν, we must have $\psi_2 = \neg\psi_1$. Then by **(S4)** either $\psi_1 \in \Delta$ or $\psi_2 \in \Delta$. By the induction hypothesis, either $\nu(\psi_1) = f$ or $\nu(\psi_2) = f$. A contradiction.
     - Suppose that $[\mathsf{d}] \in S$, $\nu(\psi_2) = \top_j^i$ and $\nu(\psi_1) \in \{t_j^{i+1}, \top_j^{i+1}\}$. Using **(S5)**, we get a contradiction like in the previous case.
     - Suppose that $[\mathsf{n}_\wedge^l] \in S$ or $[\mathsf{a}_\wedge] \in S$, and $\nu(\psi_1), \nu(\psi_2) \in \mathcal{E}_t$. Then by **(S10)** either $\neg\psi_1 \in \Gamma$ or $\neg\psi_2 \in \Gamma$. By the induction hypothesis, either $\nu(\psi_1) \notin \mathcal{E}_t$ or $\nu(\psi_2) \notin \mathcal{E}_t$. A contradiction.
     - Suppose that $[\mathsf{o}_\wedge^1] \in S$, $\nu(\psi_1) \in \mathcal{E}_t$ and $\nu(\psi_2) \in \mathcal{D}$. By **(S13)**, either $\neg\psi_1 \in \Gamma$, or $\psi_2 \in \Delta$. By the induction hypothesis, either $\nu(\psi_1) \notin \mathcal{E}_t$, or $\nu(\psi_2) \notin \mathcal{D}$. A contradiction.
     - Suppose that $[\mathsf{o}_\wedge^2] \in S$, $\nu(\psi_1) \in \mathcal{D}$ and $\nu(\psi_2) \in \mathcal{E}_t$. Similarly to the previous case, we get a contradiction.

     It follows that $\tilde{\wedge}(\nu(\psi_1), \nu(\psi_2)) \not\subseteq \mathcal{E}_t$. As $\neg\varphi \in \Gamma$, $\neg\varphi \notin \Delta$. Hence, the definition of ν implies that either $\nu(\varphi) = f$ or $\nu(\varphi) = \top_{n(\varphi)}^{k(\varphi)}$.

- Suppose that $\varphi = \psi_1 \vee \psi_2$.

8.4 da Costa's Consistency Operator(s)

1. The case where $\varphi \in \Delta$ is left to the reader.
2. The case where $\varphi \in \Gamma$ is left to the reader.
3. Suppose that $\neg\varphi \in \Delta$. By **(S3)**, $\varphi \in \Gamma$. Hence, by **(S1)**, either $\psi_1 \in \Gamma$ or $\psi_2 \in \Gamma$. It follows by the induction hypothesis for ψ_1 and ψ_2 that either $\nu(\psi_1) \in \mathcal{D}$, or $\nu(\psi_2) \in \mathcal{D}$. In both cases $\tilde{V}(\nu(\psi_1), \nu(\psi_2)) \neq \{f\}$. Suppose for contradiction that $\tilde{V}(\nu(\psi_1), \nu(\psi_2)) = \mathcal{E}_\top$. This is possible only if $[n_V^r] \in S$, and $\{\nu(\psi_1), \nu(\psi_2)\} \cap \mathcal{E}_t = \emptyset$. But if $[n_V^r] \in S$ then by **(S13)**, either $\neg\psi_1 \in \Delta$, or $\neg\psi_2 \in \Delta$. By the induction hypothesis for ψ_1 and ψ_2, this implies that either $\nu(\psi_1) \in \mathcal{E}_t$, or $\nu(\psi_2) \in \mathcal{E}_t$. A contradiction. It follows that $\tilde{V}(\nu(\psi_1), \nu(\psi_2))$ can be either $\mathcal{E}_t$ or $\mathcal{D}$, and so $t_{n(\varphi)}^{k(\varphi)} \in \tilde{V}(\nu(\psi_1), \nu(\psi_2))$. Since $\neg\varphi \in \Delta$, the definition of ν implies that $\nu(\varphi) = t_{n(\varphi)}^{k(\varphi)}$.
4. Suppose that $\neg\varphi \in \Gamma$. We show that none of the cases in which $\tilde{V}(\nu(\psi_1), \nu(\psi_2)) \subseteq \mathcal{E}_t$ is applicable under this assumption:
   - Suppose that $[a_\vee] \in S$, and that either $\nu(\psi_1) \in \mathcal{E}_t$ and $\nu(\psi_2) \in \mathcal{E}_t \cup \{f\}$, or $\nu(\psi_2) \in \mathcal{E}_t$ and $\nu(\psi_1) \in \mathcal{E}_t \cup \{f\}$. Assume e.g. the former. Since $[a_\vee] \in S$, it follows by **(S12)** that either $\neg\psi_1 \in \Gamma$, or $\{\psi_2, \neg\psi_2\} \subseteq \Gamma$. By the induction hypothesis for ψ_1 and ψ_2, in the first case $\nu(\psi_1) \notin \mathcal{E}_t$, while in the second case $\nu(\psi_2) \notin \mathcal{E}_t \cup \{f\}$. In both cases we got a contradiction.
   - We leave to the reader the remaining four cases (corresponding to the cases: $[o_\vee^1] \in S$, $[o_\vee^2] \in S$, $[n_\vee^l]_1 \in S$, $[n_\vee^l]_2 \in S$).

   It follows that $\tilde{V}(\nu(\psi_1), \nu(\psi_2)) \not\subseteq \mathcal{E}_t$. As $\neg\varphi \in \Gamma$, $\neg\varphi \notin \Delta$. Hence the definition of ν implies that either $\nu(\varphi) = f$ or $\nu(\varphi) = \top_{n(\varphi)}^{k(\varphi)}$.

- We leave the cases where $\varphi = \psi_1 \supset \psi_2$ to the reader.

It follows that for every $\psi \in \Gamma$, $\nu(\psi) \in \mathcal{D}$ and for every $\varphi \in \Delta$, $\nu(\varphi) = f$. Since $\Gamma_0 \subseteq \Gamma$ and $\Delta_0 \subseteq \Delta$, $\Gamma_0 \Rightarrow \Delta_0$ is not valid in $\mathcal{M}_{Bk}^\infty(S)$. □

We end this section by determining what effect the axioms [l] and [d] have on the sparseness of a system from the family we investigate in it.

Theorem 8.117 *Let S be a coherent subset of $\mathbf{A}_{ld}$ such that $S \cap \{[l], [d]\} \neq \emptyset$. Then $\mathbf{Bk}(S)$ is not sparse iff either it has a characteristic (deterministic) matrix, or all the axioms in $\mathbf{O}$ are derivable in it.*

Proof. If one of the two conditions is satisfied then $\circ\varphi \wedge \neg\varphi \wedge \varphi$ is congruent in $\mathbf{L}_{\mathcal{M}_{Bk}(S)}$ to $\circ\psi \wedge \neg\psi \wedge \psi$ for any φ and ψ. The proof of this is practically identical to the proof of the necessity direction of Proposition 8.74 (using $\mathcal{E}_t$ instead of $\{t\}$, $\mathcal{E}_t \cup \{f\}$ instead of $\{t,f\}$, '$\not\subseteq \mathcal{E}_\top$' instead of '$\neq \top$' etc.). We leave the details to the reader.

For the converse, let φ and ψ be two distinct formulas, and assume that neither of the two conditions is satisfied. By Corollary 8.109 the latter means that there are at least one axiom from $\mathbf{O}$, one axiom from S^l, and one axiom from S^d that are not derivable in $\mathbf{Bk}(S)$. The first of these three facts implies that there are two cases to consider concerning $\mathcal{M}_{Bk}^\infty(S)$:

- $\tilde{\diamond}(a_1, \ldots, a_n) = \mathcal{D}$ for some n-ary connective $\diamond$ of $\mathcal{L}_C$ ($n \in \{1, 2\}$) and $a_1, \ldots, a_n$ such that a_1 (say) is in $\mathcal{E}_t \cup \{f\}$. In this case we define φ^*, ψ^*, and ν^* exactly like in the first item in the proof of the sufficiency direction of Proposition 8.74 (using t_0^0 instead of t, and $\top_0^0$ instead of $\top$), and use them like in that proof to show that $\varphi \not\equiv_{\mathbf{Bk}(S)} \psi$.

- $\tilde{\diamond}'(x_1, \ldots, x_n) \neq \mathcal{D}$ for every n-ary connective $\diamond'$ of $\mathcal{L}_C$ ($n \in \{1, 2\}$) and $x_1, \ldots, x_n$ such that x_i is in $\mathcal{E}_t \cup \{f\}$ for some $1 \leq i \leq n$; but $\tilde{\diamond}(a_1, \ldots, a_n) = \mathcal{E}_\top$ for some n-ary connective $\diamond$ of $\mathcal{L}_C$ and $a_1, \ldots, a_n$ such that a_1 (say) is in $\mathcal{E}_t \cup \{f\}$. A careful check shows that together with our assumptions concerning the relations between $\mathbf{Bk}(S)$ and S^l or S^d, this implies that $\{t_0^1, \top_0^1\} \subseteq \tilde{\#}(\top_0^0, \ldots, \top_0^0)$ for some connective $\#$. Now, we define φ^*, ψ^*, and ν^* as in the first case, except that now we take $\nu^*(\varphi^*) = \top_0^0$. Similarly to what is done in the second item in the proof of the sufficiency direction of Proposition 8.74, we extend ν^* to $\mathcal{W}^* \cup \{\#(\varphi^*, \ldots, \varphi^*), \#(\psi^*, \ldots, \psi^*)\}$ by letting $\nu^*(\#(\varphi^*, \ldots, \varphi^*)) = t_0^1$ and $\nu^*(\#(\psi^*, \ldots, \psi^*)) = \top_0^1$. Using this ν^* it is again easy to show that $\varphi \not\equiv_{\mathbf{Bk}(S)} \psi$ in this case too. □

Corollary 8.118 *With eight exceptions, all the logics of the form $\mathbf{Bk}(S)$ are sparse, in case S is a coherent subset of $\mathbf{A}_{ld}$ such that $S \cap \{[l], [d]\} \neq \emptyset$. The exceptions are:* $\mathbf{BlO}$, $\mathbf{BleO}$, $\mathbf{BdO}$, $\mathbf{BdeO}$, $\mathbf{BldO}$, $\mathbf{BldeO}$, *and the logics* $\mathbf{Bl}(S^l)$ *and* $\mathbf{Bd}(S^d)$ *from Corollary 8.109.*

Proof. This directly follows Theorem 8.117 and Corollary 8.109, using an analysis which is similar to that done in the proof of Theorem 8.75. □

Example 8.119 From Theorem 8.117 (or Corollary 8.118) it follows that $\mathbf{BlA}$ (=$\mathbf{BklA}$ by Proposition 8.84) is sparse.

Note 8.120 In continuation to Note 8.67, we note that again, and for the same reasons as there, the systems mentioned in Corollary 8.118 and Example 8.119 have different names in [111, 113, 116] and elsewhere. Thus, $\mathbf{BlA}$ is (equivalent to what is) called there $\mathbf{Cila}$, while $\mathbf{BlO}$ is (equivalent to what is) called there $\mathbf{Cilo}$.

8.4 da Costa's Consistency Operator(s)

Note 8.121 Proposition 8.78 cannot be extended to the case in which [l] (or [d]) is present. The reason is that from the proof of Theorem 8.117 it easily follows that if $\mathbf{Bl}(S)$ is not sparse, then any two bottom elements are congruent in it. Now in $\mathbf{Bl}$ any formula of the form $\neg(\varphi \wedge \neg\varphi) \wedge (\varphi \wedge \neg\varphi)$ is a bottom element. Hence no formula of this form is isolated in $\mathbf{Bl}(S)$ in case this system is not sparse, even if it a formula of $\mathcal{L}_{CL}$. (A similar argument holds for $\mathbf{Bd}(S)$.)

8.4.2 The Non-Strict Case

All the C-systems treated above include $\circ$ as a *primitive* connective. As explained at the beginning of this section, this has not been the case in da Costa's original C-systems, like his major system $\mathbf{C}_1$ (Example 8.141 below). The language of those logics has been $\mathcal{L}_{CL}$, and $\circ$ has been a *defined* connective of it. Our next proposition shows that the $\mathcal{L}_{CL}$-fragments of the C-systems which were studied in this section are C-systems themselves. Hence, they are all C-systems of da Costa's original type.

Definition 8.122 (L↾CL) Let $\mathbf{L}$ be a logic in a language which includes $\mathcal{L}_{CL}$. We denote by $\mathbf{L} \upharpoonright CL$ the $\mathcal{L}_{CL}$-fragment of $\mathbf{L}$.

Proposition 8.123 *Let S be a coherent subset of $\mathbf{A}_{ld}$ such that $S \cap \{[l], [d]\}$ is not empty. Then $\mathbf{B}(S) \upharpoonright CL$ ($= \mathbf{Bk}(S) \upharpoonright CL$) is a strong (and so regular) C-system. In particular:*

1. *If $[l] \in S$ then $\otimes \varphi =_{Df} \neg(\varphi \wedge \neg\varphi)$ is a strong consistency operator.*

2. *If $[d] \in S$ then $\otimes \varphi =_{Df} \neg(\neg\varphi \wedge \varphi)$ is a strong consistency operator.*

Proof. If $S \subseteq \mathbf{A}_{ld}$ is coherent, then by Proposition 8.100, $\mathbf{Bk}(S)$ is a classical $\neg$-paraconsistent logic in which $\neg$ is complete. Hence so is its $\mathcal{L}_{CL}$-fragment. It remains to show, e.g., that if $[l] \in S$ then $\otimes \varphi =_{Df} \neg(\varphi \wedge \neg\varphi)$ is a strong consistency operator. (The proof in case $[d] \in S$ is similar.) This is easy, since the derivability of [b] for $\otimes$ is immediate from the axioms [b] and [l] of $\mathbf{Bl}$, while axiom [k] for $\otimes$ is an instance of [t]. □

Note 8.124 If $\{[l], [d]\} \subseteq S$, then *both* $\neg(\varphi \wedge \neg\varphi)$ and $\neg(\neg\varphi \wedge \varphi)$ define strong consistency operators in $\mathbf{B}(S)$. Hence these formulas are equivalent in $\mathbf{B}(S)$ by Corollary 8.18. It follows that their disjunction and conjunction are equivalent to both, and so they too define strong consistency operators.

Corollary 8.125 *Let S be a coherent subset of $\mathbf{A}_{ld}$ such that $S \cap \{[l], [d]\}$ is not empty. Then $\mathbf{B}(S) \upharpoonright CL$ is normal.*

Proof. Immediate from Propositions 8.4 and 8.123. □

Next, we present some important properties that $\mathbf{B}(S)\!\upharpoonright\!CL$ inherits from $\mathbf{B}(S)$, in case S is a coherent subset of A_{ld}, and $S \cap \{[\mathsf{l}], [\mathsf{d}]\} \neq \emptyset$.

Theorem 8.126 *If S is a coherent subset of A_{ld}, and $S \cap \{[\mathsf{l}], [\mathsf{d}]\} \neq \emptyset$, then the reduction of $\mathcal{M}^\infty_{Bk}(S)$ to $\mathcal{L}_{CL}$ is characteristic for $\mathbf{B}(S)\!\upharpoonright\!CL$.*

Proof. Immediate from Theorem 8.104 (and Proposition 8.84). □

Corollary 8.127 *If S is a coherent subset of A_{ld}, $S \cap \{[\mathsf{l}], [\mathsf{d}]\} \neq \emptyset$, and $S' = S \cup \{[\mathsf{a}_\circ]\}$, then $\mathbf{B}(S')\!\upharpoonright\!CL = \mathbf{B}(S)\!\upharpoonright\!CL$.*

Proof. This easily follows Theorem 8.126, since the reductions of $\mathcal{M}^\infty_{Bk}(S)$ and $\mathcal{M}^\infty_{Bk}(S')$ to $\mathcal{L}_{CL}$ are identical in case $S' = S \cup \{[\mathsf{a}_\circ]\}$. □

Note 8.128 Corollary 8.127 implies that the inclusion of the axiom $[\mathsf{a}_\circ]$ (or $[\mathsf{i}]$ — see Proposition 8.85) in any of the systems studied in this section does not affect the $\circ$-free fragment of that system.[19]

Proposition 8.129 *If S is a coherent subset of A_{ld}, and $S \cap \{[\mathsf{l}], [\mathsf{d}]\} \neq \emptyset$, then $\mathbf{B}(S)\!\upharpoonright\!CL$ is boldly paraconsistent and decidable.*

Proof. Immediate from Corollaries 8.105 and 8.106. □

Theorem 8.130 *Suppose that $S \subseteq \mathsf{A}_{ld}$ is coherent, and $S \cap \{[\mathsf{l}], [\mathsf{d}]\} \neq \emptyset$.*

1. *If either $\{[\mathsf{l}], [\mathsf{n}^r_\wedge]_2, [\mathsf{o}^2_\wedge]\} \subseteq S$ or $\{[\mathsf{d}], [\mathsf{n}^r_\wedge]_1, [\mathsf{o}^1_\wedge]\} \subseteq S$, then $\mathbf{B}(S)\!\upharpoonright\!CL$ has a characteristic three-valued Nmatrix.*

2. *If neither $\{[\mathsf{l}], [\mathsf{n}^r_\wedge]_2, [\mathsf{o}^2_\wedge]\} \subseteq S$ nor $\{[\mathsf{d}], [\mathsf{n}^r_\wedge]_1, [\mathsf{o}^1_\wedge]\} \subseteq S$, then there is no finite family $\mathcal{F}$ of finite Nmatrices such that $\vdash_{\mathbf{B}(S)\upharpoonright CL} = \vdash_\mathcal{F}$.*

Proof. The proof is identical to that of Theorem 8.108. □

Corollary 8.131 *Let S^l and S^d be the sets defined in Corollary 8.109. Then $\mathbf{Bl}(S^l)\!\upharpoonright\!CL$ and $\mathbf{Bd}(S^d)\!\upharpoonright\!CL$ are the only C-systems of the form $\mathbf{Bl}(S)\!\upharpoonright\!CL$ or $\mathbf{Bd}(S)\!\upharpoonright\!CL$ (where $S \subseteq \mathsf{A}_{ld}$) which have characteristic matrices.*

Proof. The proof is similar to that of Corollary 8.109. □

---

[19]This is actually true, with a similar proof, for *any* logic of the form $\mathbf{Bk}(S)$, where S is a coherent subset of A_{ld} (not only those for which $S \cap \{[\mathsf{l}], [\mathsf{d}]\} \neq \emptyset$).

8.4 da Costa's Consistency Operator(s)

Theorem 8.132 *Let S be a coherent subset of A_{ld} such that $S \cap \{[\mathsf{l}], [\mathsf{d}]\} \neq \emptyset$. Then $\mathbf{Bk}(S) \restriction CL$ is not sparse iff either it has a characteristic three-valued matrix, or all the axioms in $\mathbf{O} - \{[\mathsf{a}_\circ]\}$ are derivable in $\mathbf{Bk}(S)$.*

Proof. The proof is practically identical to that of Theorem 8.117. We only have to use the *defined* strong consistency operator of $\mathbf{Bk}(S) \restriction CL$ instead of the primitive connective $\circ$. (That is: we can use $\neg(\varphi \wedge \neg\varphi)$ instead of $\circ\varphi$ in case $[\mathsf{l}] \in S$, and $\neg(\neg\varphi \wedge \varphi)$ in case $[\mathsf{d}] \in S$.) □

Corollary 8.133 *Let S be a coherent subset of A_{ld} such that $S \cap \{[\mathsf{l}], [\mathsf{d}]\} \neq \emptyset$. Then $\mathbf{Bk}(S) \restriction CL$ is not sparse iff $\mathbf{Bk}(S \cup \{[\mathsf{a}_\circ]\})$ is one of the eight non-sparse logics listed in Corollary 8.118.*

Proof. By Corollary 8.118, what we have to prove is that $\mathbf{Bk}(S) \restriction CL$ is not sparse iff $\mathbf{Bk}(S \cup \{[\mathsf{a}_\circ]\})$ is not sparse.

- Suppose $\mathbf{Bk}(S) \restriction CL$ is not sparse. By Theorem 8.132, this implies that either all the axioms in $\mathbf{O} - \{[\mathsf{a}_\circ]\}$ are derivable in $\mathbf{Bk}(S)$, or $\mathbf{Bk}(S) \restriction CL$ has a characteristic three-valued matrix. In the first case all axioms of $\mathbf{O}$ are derivable in $\mathbf{Bk}(S \cup \{[\mathsf{a}_\circ]\})$. In the second case it is easy to see that $\mathbf{Bk}(S \cup \{[\mathsf{a}_\circ]\})$ has a characteristic three-valued matrix too. Therefore, in both cases it follows from Theorem 8.117 that $\mathbf{Bk}(S \cup \{[\mathsf{a}_\circ]\})$ is not sparse.

- Suppose $\mathbf{Bk}(S \cup \{[\mathsf{a}_\circ]\})$ is not sparse. Using Theorem 8.132 and Theorem 8.117 it similarly follows from that that $\mathbf{Bk}(S \cup \{[\mathsf{a}_\circ]\}) \restriction CL$ is not sparse. Hence $\mathbf{Bk}(S) \restriction CL$ is not sparse by Corollary 8.127. □

Next, we turn to proof-systems for the logics studied in this section. In the case of Gentzen-type systems this is an easy task.

Definition 8.134 ($GBk^*(S) \restriction CL$) Let $S \subseteq \mathsf{A}_{ld}$ be coherent. $GBk^*(S) \restriction CL$ is the system obtained from $GBk^*(S)$ (Definition 8.115) by discarding the rules for $\circ$ (i.e. $[\circ \Rightarrow]$, $[\Rightarrow \circ]$, and if $[\mathsf{a}_\circ] \in S$ then also $R([\mathsf{a}_\circ])$).

Proposition 8.135 *Let $S \subseteq \mathsf{A}_{ld}$. If S is coherent, then $GBk^*(S) \restriction CL$ is sound and complete for $\mathbf{Bk}(S) \restriction CL$, and it admits cut-elimination.*

Proof. That $GBk^*(S) \restriction CL$ admits cut-elimination immediately follows from the fact that $GBk^*(S)$ does (by Theorem 8.116). Now we prove its soundness and completeness for $\mathbf{Bk}(S) \restriction CL$. Let $\mathcal{T} \cup \{\psi\}$ be a set of formulas over $\mathcal{L}_{CL}$. Clearly, if $\mathcal{T} \vdash_{GBk^*(S) \restriction CL} \psi$ then $\mathcal{T} \vdash_{GBk^*(S)} \psi$. By Theorem 8.116 this implies that $\mathcal{T} \vdash_{\mathbf{Bk}(S)} \psi$. For the converse, suppose that $\mathcal{T} \vdash_{\mathbf{Bk}(S)} \psi$ for some $\mathcal{T} \cup \{\psi\}$ in $\mathcal{L}_{CL}$. By Theorem 8.116 this implies that for some $\Gamma \subseteq \mathcal{T}$, $\Gamma \Rightarrow \psi$ has a cut-free derivation P in $GBk^*(S)$. Since $\Gamma \Rightarrow \psi$ does not contain $\circ$, P does not contain applications of any rule which introduces $\circ$ (since $\circ$ can only be eliminated using cuts). Hence P is also a cut-free derivation in $GBk^*(S){\restriction}CL$, implying that $\mathcal{T} \vdash_{GBk^*(S){\restriction}CL} \psi$. □

Note 8.136 Since $GBk^*(S \cup \{[a_\circ]\}) \restriction CL = GBk^*(S) \restriction CL$ by definition, Proposition 8.135 can be used for an alternative proof of Corollary 8.127.

Turning to Hilbert-type systems, we first note that it is easy to translate the Gentzen-type systems given in Proposition 8.135 into equivalent Hilbert-type systems by using the method explained in Note 8.81. However, some of the axioms that this method produces look rather strange and unintuitive. More intuitive Hilbert-type systems for logics of the form $\mathbf{B}k(S) \restriction CL$ are obtained by using the method described in the next definition.

Definition 8.137 ($HBk^\sigma(S)$) Let $S \subseteq \mathbf{A}_{ld}$ be a coherent set of axioms, and let σ be a formula such that $\mathsf{Var}(\sigma) = \{P_1\}$. $HBk^\sigma(S)$ is the Hilbert-style system obtained from $HBk(S)$ by deleting [k], [l], [d], and [a$_\circ$]; and replacing in the remaining axioms every formula of the form $\circ\varphi$ by $\sigma[\varphi/P_1]$.

Definition 8.138 (σ_l, σ_d) Let $\sigma_l = \neg(P_1 \wedge \neg P_1)$, $\sigma_d = \neg(\neg P_1 \wedge P_1)$.

Proposition 8.139 Let $S \subseteq \mathbf{A}_{ld}$ be a coherent set of axioms. $HBk^\sigma(S)$ is sound and complete for $\mathbf{B}k(S){\restriction}CL$ whenever:

1. $[l] \in S$, and $\sigma = \sigma_l$.

2. $[d] \in S$, and $\sigma = \sigma_d$.

Proof. By Corollary 8.127, we may assume that $[a_\circ] \notin S$. Now it is easy to check that in the two cases considered in the above theorem, both $\sigma[\varphi/P_1] \supset \circ\varphi$ and $\circ\varphi \supset \sigma[\varphi/P_1]$ are theorems of $HBk(S)$. Since $HBk(S)$ is an extension of the classical positive logic, this fact and the form of the elements of S imply that by replacing $\circ\varphi$ with $\sigma[\varphi/P_1]$ we get a theorem of $HBk(S)$ from any axiom of $HBk(S)$.[20] Consequently, if $\mathcal{T} \vdash_{HBk^\sigma(S)} \psi$ then $\mathcal{T} \vdash_{HBk(S)} \psi$. For the converse, define $tr_\sigma : \mathcal{W}(\mathcal{L}_C) \to \mathcal{W}(\mathcal{L}_{CL})$ as follows: $tr_\sigma(p) = p$ if p is a variable, $tr_\sigma(\neg\varphi) = \neg tr_\sigma(\varphi)$, $tr_\sigma(\varphi \sharp \psi) = tr_\sigma(\varphi) \sharp tr_\sigma(\psi)$ (for $\sharp \in \{\wedge, \vee, \supset\}$), and $tr_\sigma(\circ\varphi) = \sigma(tr_\sigma(\varphi))$. Using induction on the length of derivations in $HBk(S)$ and our assumption that $[a_\circ] \notin S$, we can easily show that in both cases, if $\mathcal{T} \vdash_{HBk(S)} \psi$ then $\{tr_\sigma(\varphi) \mid \varphi \in \mathcal{T}\} \vdash_{HBk^\sigma(S)} tr_\sigma(\psi)$. Since $tr_\sigma(\varphi) = \varphi$ if φ is $\circ$-free, this in particular implies that if all formulas of $\mathcal{T} \cup \{\psi\}$ are in $\mathcal{L}_{CL}$ and $T \vdash_{HBk(S)} \psi$, then $\mathcal{T} \vdash_{HBk^\sigma(S)} \psi$. □

---

[20] Unless $[c] \in S$ this is *not* true if $[a_\circ] \in S$ (or if $[i]$ is used instead of $[a_\circ]$ — see Proposition 8.85). Hence, our ability to assume that $[a_\circ] \notin S$ is crucial here.

8.4 da Costa's Consistency Operator(s)

Note 8.140 It should be emphasized that the converse of what is shown about tr_σ in the second part of the proof of Proposition 8.139 is *not true*, unless $\mathbf{Bk}(S)$ is one of the eight logics listed in Corollary 8.118. Indeed, from that corollary it follows that $\mathbf{Bk}(S)$ is sparse in any other case. Therefore $\circ P_1$ and σ_l are then not congruent in $\mathbf{Bk}(S)$, even though $tr_{\sigma_l}(\circ P_1)$ and $tr_{\sigma_l}(\sigma_l)$ are congruent in $\mathbf{Bk}(S){\restriction}\mathcal{CL}$, since $tr_{\sigma_l}(\circ P_1) = tr_{\sigma_l}(\sigma_l) = \sigma_l$.

Example 8.141 da Costa's main paraconsistent logic, $\mathbf{C}_1$, was originally introduced in [127] using the Hilbert-type system HC_1, which is just (see Notation 8.55) $HBk^{\sigma_l}(\{[l], [c]\} \cup [a]) = HBk^{\sigma_l}(\{[l], [c], [a_\circ]\} \cup [a])$.[21] In other words: the system HC_1 for $\mathcal{L}_{CL}$ is obtained by adding to HCL^+ the axioms [t], [c] and the versions of axioms [b] and [a] in which $\circ\psi =_{Df} \neg(\psi \wedge \neg\psi)$. From Proposition 8.139 (and Proposition 8.58) it follows that $\mathbf{C}_1 = \mathbf{BlA}{\restriction} CL = \mathbf{Blca}{\restriction}CL$, i.e. $\mathbf{C}_1$ is the $\circ$-free fragment of $\mathbf{BlA}$ and $\mathbf{Blca}$.[22] However, from Note 8.140 it follows that it cannot be identified with either.

From the results of this section it follows that $\mathbf{C}_1$ is a strong C-system, which is boldly paraconsistent, decidable, normal, and sparse. It also follows from them that there is no finite family $\mathcal{F}$ of finite Nmatrices such that $\vdash_{\mathbf{C}_1} = \vdash_{\mathcal{F}}$. In addition, these results also provide a semantic characterization of $\mathbf{C}_1$, and a cut-free Gentzen-type system for it. The latter is obtained by discarding $[\circ \Rightarrow]$ and $[\Rightarrow \circ]$ from $GBk(\{[c], [l], [a]\})$. The semantics is given by the Nmatrix $\mathcal{M}_{\mathbf{C}_1} = \langle \mathcal{V}, \mathcal{D}, \mathcal{O} \rangle$ for $\mathcal{L}_{CL}$, in which $\mathcal{V} = \mathcal{E}_t \cup \mathcal{E}_\top \cup \{f\}$, $\mathcal{D} = \mathcal{E}_t \cup \mathcal{E}_\top$, and $\mathcal{O}$ is defined as follows:

$$a \widetilde{\vee} b = \begin{cases} \{f\} & \text{if } a, b = f, \\ \mathcal{E}_t & a \in \mathcal{E}_t \text{ and } b \in \mathcal{E}_t \cup \{f\}, \text{ or } a \in \{f\} \text{ and } b \in \mathcal{E}_t, \\ \mathcal{D} & \text{otherwise.} \end{cases}$$

$$a \widetilde{\supset} b = \begin{cases} \{f\} & \text{if } a \in \mathcal{D} \text{ and } b = f, \\ \mathcal{E}_t & \text{if } a = f \text{ and } b \in \{f\} \cup \mathcal{E}_t, \text{ or } a \in \mathcal{E}_t \cup \{f\} \text{ and } b \in \mathcal{E}_t, \\ \mathcal{D} & \text{otherwise.} \end{cases}$$

$$a \widetilde{\wedge} b = \begin{cases} \{f\} & \text{if either } a = f \text{ or } b = f, \\ \mathcal{E}_t & \text{if } a = \top_i^j \text{ and } b \in \{\top_i^{j+1}, t_i^{j+1}\}, \text{ or } a, b \in \mathcal{E}_t, \\ \mathcal{D} & \text{otherwise.} \end{cases}$$

$$\widetilde{\neg} a = \begin{cases} \{f\} & \text{if } a \in \mathcal{E}_t, \\ \mathcal{E}_t & \text{if } a = f, \\ \{\top_i^{j+1}, t_i^{j+1}\} & \text{if } a = \top_i^j. \end{cases}$$

---

[21] The original HC_1 has actually employed a more complicated version of [b], which is easily seen to be equivalent to [b] in **CLuN**.

[22] Recall that **BlA** and **Blca** are called in [111, 113, 116] **Cila** and **Cla**, respectively.

8.5 Bibliographical Notes and Further Reading

The ideas that led to the family of LFIs in general, and C-systems in particular were first introduced in [126, 127]. Earlier studies include (among others) [88, 89, 129, 218, 250, 286, 287, 288]; see [130, 142] for surveys on the early development of C-systems.

The most comprehensive sources (so far) concerning LFIs are the surveys [113, 116], and very recently the book [111]. We refer the readers to them for a lot of further information about the subject. This includes the study and the use of alternative semantic frameworks for C-systems that are not discussed here, like bivaluation semantics and possible translation semantics. (Note, however, that the present chapter also includes many results that are not presented in these three sources.)

The use of the framework of Nmatrices for providing effective semantics to C-systems was first made in [46, 71]. (See [111] for its relations with the other types of semantics for these systems that are mentioned above.) The semantic study which is presented in this chapter is based on these two papers, but includes significant improvements, corrections, and extensions of what has been done in them.

Early efforts for providing analytic calculi for C-systems focused on the historical system C_1 of da Costa (see Example 8.141): sequent calculi in [89, 96, 250] and tableau calculi in [113, 114, 115]. Later, some analytic calculi have been introduced also for a few other C-systems (see, e.g., [169, 226]). A systematic construction of cut-free sequent calculi for C-systems (using the general method that was introduced in [62]) was first made in [64, 65]. The proof-theoretical part of this chapter is based on these two papers, but again includes significant improvements, corrections, and extensions.

Corollary 8.34, and item (6) of Theorems 8.53 and 8.72 are due to [57]. (Many particular cases of them have been proven before by Marcos and others. See [113].)

Part IV
Possible Worlds

Chapter 9

Negation as a Modal Operator

9.1 Replacement in Paraconsistent Logics

None of the *normal* strongly paraconsistent logics considered so far in this book has a negation which is both complete and contrapositive. What is more, none of them has both a complete negation and the quite desirable replacement property (Definition 1.18).[1] Can we construct reasonable normal strongly paraconsistent logics which are self-extensional (that is, do enjoy the replacement property), and satisfy all the properties in the desiderata list given in Section 2.4? In this chapter we provide a positive answer to this question. However, as Note 1.19 indicates, and the next proposition shows, our goal cannot be achieved by the usual method of adding axioms that force the stronger substitution (or strong replacement) property, which a logic **L** with an implication $\supset$ has if the following holds for every propositional variable p and formulas φ, ψ, σ: $\varphi \supset \psi, \psi \supset \varphi \vdash_\mathbf{L} \sigma[\psi/p] \supset \sigma[\varphi/p]$.

Proposition 9.1 *Let* $\mathbf{CAR}^I$ *be the logic which is obtained from* $\mathbf{IL}^+$ *by adding to it* $\neg\varphi \vee \varphi$ *and* $(\psi \supset \varphi) \wedge (\varphi \supset \psi) \supset (\neg\psi \supset \neg\varphi)$ *as axiom schemas. Then no extension of* $\mathbf{CAR}^I$ *is strongly paraconsistent.*

Proof. $\vdash_{\mathbf{IL}^+} (\tau \vee \varphi) \supset (((\psi \supset \varphi) \wedge (\varphi \supset \psi) \supset (\sigma \supset \tau)) \supset (\sigma \supset (\psi \supset \tau)))$. Our claim easily follows from this schema by substituting in it $\neg\varphi$ for τ and $\neg\psi$ for σ, and then using the two additional schemas of $\mathbf{CAR}^I$. □

Example 9.2 Let **CAR** be the logic which is obtained from **CLuN** by adding to it the second schema mentioned in Proposition 9.1. Then Proposition 9.1 implies that **CAR** is not strongly paraconsistent. Note that it

---

[1] In fact, if **L** is one of the normal strongly paraconsistent logics considered so far, and the negation $\neg$ of **L** is complete, then **L** is not even $\neg$-extensional (Definition 4.30), since $P_1 \supset P_1 \dashv\vdash_\mathbf{L} P_2 \supset P_2$, $P_1 \dashv\vdash_\mathbf{L} (P_1 \supset P_1) \supset P_1$, but either $\neg(P_1 \supset P_1) \not\vdash_\mathbf{L} \neg(P_2 \supset P_2)$, or $\neg P_1 \not\vdash_\mathbf{L} \neg((P_1 \supset P_1) \supset P_1)$.

is not difficult to show that **CAR** is equivalent to the logic that is called by the same name in [128]. In Chapter 3 of [227] the same logic (with yet another axiomatization) is called **Le**; an extensive study of this logic is provided there, including its history as well as many references.

Proposition 9.1 entails that in order to develop strongly paraconsistent extensions of $\mathbf{IL}^+$ which have both a complete negation and the replacement property (and in which $\supset$ is still an implication), the inference of $\neg\varphi \supset \neg\psi$ from $\varphi \supset \psi$ and $\psi \supset \varphi$ should be forced only in the case where the premises are theorems of the logic. This can be done by including this rule in the corresponding proof systems not as a rule of derivation, but just as a *rule of proof*, that is: a rule that is used only to define the set of axioms of the system, but not its consequence relation. However, since we would like $\neg$ to satisfy the desiderata list given in Section 2.4, it would be much better to adopt as a rule of proof the inference of $\neg\varphi \supset \neg\psi$ from $\psi \supset \varphi$ alone. This would make $\neg$ *contrapositive* (Definition 2.5 and Note 2.6). The next proposition implies that as long as we use the language of classical logic $\mathcal{L}_{CL}$, doing this would also suffice for forcing the replacement property.

Proposition 9.3 *Let* **L** *be a logic in* $\mathcal{L}_{CL}$ *which extends* $\mathbf{IL}^+$. *Suppose that* $\supset$ *is an implication for* **L**, *and that* $\neg$ *is contrapositive in* **L**. *Then* **L** *is self-extensional (i.e., it has the replacement property).*

Proof. Suppose $\psi \dashv\vdash_\mathbf{L} \varphi$ (Definition 1.18). Since $\supset$ is an implication for **L**, this means that $\vdash_\mathbf{L} \varphi \supset \psi$ and $\vdash_\mathbf{L} \psi \supset \varphi$. We prove by induction on the structure of σ that $\vdash_\mathbf{L} \sigma[\psi/p] \supset \sigma[\varphi/p]$ and $\vdash_\mathbf{L} \sigma[\varphi/p] \supset \sigma[\psi/p]$ for every formula σ and variable p. This is obvious if σ is a variable. The case where $\sigma = \neg\sigma'$ is taken care by our assumption that $\neg$ is contrapositive, while the case where $\sigma = \sigma_1 \diamond \sigma_2$ ($\diamond \in \{\wedge, \vee, \supset\}$) follows from the following fact:

$$\sigma_1 \supset \sigma_1', \sigma_1' \supset \sigma_1, \sigma_2 \supset \sigma_2', \sigma_2' \supset \sigma_2 \vdash_{\mathbf{IL}^+} (\sigma_1 \diamond \sigma_2) \supset (\sigma_1' \diamond \sigma_2').$$

Since $\supset$ is an implication for **L**, the replacement property follows. □

9.2 The Modal Logic B as a Paraconsistent Logic

In this section we introduce a $\neg$-classical paraconsistent logic, which has a negation which is both complete and contrapositive, and satisfies all the conditions in the desiderata list of Section 2.4 (the last one – by being a minimal logic among those which satisfy all the other conditions that we have just listed). Moreover, this logic is a C-system which is closely related to one of the most basic systems in da Costa's school: $\mathbf{C}_{min}$ (see Definition 7.37). Recall that by Proposition 7.38, $\mathbf{C}_{min}$ is the minimal

9.2 The Modal Logic B as a Paraconsistent Logic

extension of $\mathbf{CL}^+$ which has a negation which is both complete and left-involutive.

9.2.1 The Logic NB and Its Proof Systems

Definition 9.4 (NB)

- $Th(\mathbf{NB})$ is the minimal set S of formulas in $\mathcal{L}_{CL}$ such that:

  1. S includes all axioms of $\mathbf{C}_{min}$ (Definition 7.37).
  2. S is closed under the following two rules:

     - [MP] $\dfrac{\varphi \quad \varphi \supset \psi}{\psi}$

     - [CP] $\dfrac{\vdash \psi \supset \varphi}{\vdash \neg\varphi \supset \neg\psi}$

- HNB is the Hilbert-type system whose set of axioms is $Th(\mathbf{NB})$ and has [MP] for $\supset$ as its sole rule of inference.

- **NB** is the logic in $\mathcal{L}_{CL}$ which is induced by HNB.

Note 9.5 Obviously, $\vdash_{\mathbf{NB}} \varphi$ iff $\varphi \in Th(\mathbf{NB})$.

Note 9.6 We would like to emphasize that [CP] is *not* a rule of HNB. (Otherwise $\supset$ would not be an implication for it, and the whole rationale behind the construction of HNB would be lost). It is only used for defining its set of axioms. By its definition, **NB** is in fact an *axiomatic* extension of $\mathbf{C_{min}}$ (and so of $\mathbf{CL}^+$, etc). The best way to describe and understand HNB is that it is the system which is obtained from HC_{min} by adding [CP] to it as a *rule of proof* (see the explanation in the previous section). This is similar to the role that the necessitation rule (from ψ infer $\Box\psi$) usually has in Hilbert-type systems in modal logics. More precisely, whether necessitation is taken in modal logics as a rule of proof or a rule of derivation depends on the intended consequence relation. If the *local* one of preserving truth in worlds is used, then the rule should be taken only as a rule of proof. In contrast, if the *global* one of preserving validity in frames (that is, truth in all worlds of a frame) is used, then the rule should be taken as a usual rule of derivation (like [MP]).

Definition 9.7 $\varphi \equiv \psi =_{Df} (\varphi \supset \psi) \wedge (\psi \supset \varphi)$.

The following lemma will be useful in the sequel:

Lemma 9.8

1. *If* $\vdash_{\mathbf{NB}} \varphi$ *then for every* ψ, $\neg\varphi \vdash_{\mathbf{NB}} \psi$.

2. $\vdash_{\mathbf{NB}} \neg(\varphi \wedge \psi) \equiv (\neg\varphi \vee \neg\psi)$.

3. $\vdash_{\mathbf{NB}} \neg\neg\neg\varphi \equiv \neg\varphi$.

Proof.

1. Suppose $\vdash_{\mathbf{NB}} \varphi$. Then $\vdash_{\mathbf{NB}} \neg\psi \supset \varphi$ (since **NB** extends $\mathbf{CL}^+$). It follows by [CP] that $\vdash_{\mathbf{NB}} \neg\varphi \supset \neg\neg\psi$, and so (using the axiom [c] of HC_{min}) $\vdash_{\mathbf{NB}} \neg\varphi \supset \psi$, implying that $\neg\varphi \vdash_{\mathbf{NB}} \psi$.

2. Obviously, $\vdash_{\mathbf{NB}} \neg\varphi \supset (\neg\varphi \vee \neg\psi)$. Hence an application of [CP] yields that $\vdash_{\mathbf{NB}} \neg(\neg\varphi \vee \neg\psi) \supset \neg\neg\varphi$. Using axiom [c] of HC_{min} and $\mathbf{CL}^+$ we get: $\vdash_{\mathbf{NB}} \neg(\neg\varphi \vee \neg\psi) \supset \varphi$. Similarly, $\vdash_{\mathbf{NB}} \neg(\neg\varphi \vee \neg\psi) \supset \psi$. It follows (again using $\mathbf{CL}^+$) that $\vdash_{\mathbf{NB}} \neg(\neg\varphi \vee \neg\psi) \supset (\varphi \wedge \psi)$. By applying [CP] once more we get: $\vdash_{\mathbf{NB}} \neg(\varphi \wedge \psi) \supset \neg\neg(\neg\varphi \vee \neg\psi)$. Hence $\vdash_{\mathbf{NB}} \neg(\varphi \wedge \psi) \supset (\neg\varphi \vee \neg\psi)$, by $\mathbf{CL}^+$ and another use of [c]. For the converse, note that $\vdash_{\mathbf{NB}} \varphi \wedge \psi \supset \varphi$. It follows by [CP] that $\vdash_{\mathbf{NB}} \neg\varphi \supset \neg(\varphi \wedge \psi)$. Similarly, $\vdash_{\mathbf{NB}} \neg\psi \supset \neg(\varphi \wedge \psi)$. Hence $\vdash_{\mathbf{NB}} (\neg\varphi \vee \neg\psi) \supset \neg(\varphi \wedge \psi)$.

3. $\neg\neg\neg\varphi \supset \neg\varphi$ is an instance of the axiom $\neg\neg\varphi \supset \varphi$ of **NB**. On the other hand, an application of [CP] to $\neg\neg\varphi \supset \varphi$ yields $\neg\varphi \supset \neg\neg\neg\varphi$. □

Note 9.9 The first item of Lemma 9.8 implies that if we take F to be an abbreviation of $\neg(P_1 \supset P_1)$ (say), then for every φ, $\vdash_{\mathbf{NB}} \mathsf{F} \supset \varphi$. It follows by the last item of Note 1.91 that we may assume that the language of **NB** is an extension of $\mathcal{L}_{CL}^{\mathsf{F}\neg}$ (Definition 1.61), and that **NB** is an extension of $\mathbf{CL}^{\mathsf{F}}$. Note also that since $(\mathsf{F} \supset \neg\mathsf{F}) \supset ((\mathsf{F} \vee \neg\mathsf{F}) \supset \neg\mathsf{F})$ is an instance of a positive classical tautology, axiom [t] ensures that $\vdash_{\mathbf{NB}} \neg\mathsf{F}$.

Our next goal is to provide a useful Gentzen-type system for **NB**.

Notation: $\neg S = \{\neg\varphi \mid \varphi \in S\}$.

Definition 9.10 (GNB) The system GNB is the system which is obtained from LK (Figures 1.2, 1.3) by replacing its rule $[\neg\Rightarrow]$ by the rule:

$$[\neg\Rightarrow]_B \quad \frac{\Gamma, \neg\Delta \Rightarrow \psi}{\neg\psi \Rightarrow \neg\Gamma, \Delta}.$$

9.2 The Modal Logic B as a Paraconsistent Logic

Note 9.11 As usual in paraconsistent logics, in *GNB* the standard classical rule $[\neg \Rightarrow]$ is replaced by some weaker form of it. Here this form, $[\neg \Rightarrow]_B$, is a variant of $[\neg \Rightarrow]$ in which the active formulas ($\neg\psi$ in the conclusion, ψ in the premise) are *displayed*. This is achieved by moving the context in the side where an active formula is displayed to the other side of $\Rightarrow$, with negations added to the formulas which are moved (so we have $\Gamma, \neg\Delta \Rightarrow \psi$ instead of $\Gamma \Rightarrow, \Delta, \psi$, and $\neg\psi \Rightarrow \neg\Gamma, \Delta$ instead of $\neg\psi, \Gamma \Rightarrow, \Delta$). Note that from the classical point of view, the premise and conclusion of $[\neg \Rightarrow]_B$ are equivalent (respectively) to the premise and conclusion of the classical $[\neg \Rightarrow]$.

Theorem 9.12 (Adequacy of *GNB*) $\mathcal{T} \vdash_{GNB} \varphi$ *iff* $\mathcal{T} \vdash_{NB} \varphi$.

Proof. We start by proving that if $\vdash_{NB} \varphi$ then $\vdash_{GNB} \Rightarrow \varphi$. By Note 9.5, it suffices to show that the set $Th(GKBT) = \{\varphi \mid \vdash_{GNB} \Rightarrow \varphi\}$ includes every axiom φ of HC_{min}, and is closed under [MP] and [CP]. If φ is an axiom of HCL^+ then obviously $\varphi \in Th(GKBT)$, since *GNB* includes LK^+, which is complete for $\mathbf{CL}^+$ (Theorem 1.134). For the other two axioms of HC_{min}, note that $\Rightarrow \varphi, \neg\varphi$ is derived in *GNB* from $\varphi \Rightarrow \varphi$ using $[\Rightarrow \neg]$, while $\neg\neg\varphi \Rightarrow \varphi$ is derived in *GNB* from $\neg\varphi \Rightarrow \neg\varphi$ using $[\neg \Rightarrow]_B$. With the help of rules of LK^+, this implies that $\neg\varphi \vee \varphi$ and $\neg\neg\varphi \supset \varphi$ are in $Th(GKBT)$ for every φ. That $Th(GKBT)$ is closed under [MP] follows (using cuts) from the fact that $\vdash_{LK^+} \varphi, \varphi \supset \psi \Rightarrow \psi$. Finally, suppose that $\varphi \supset \psi \in Th(GKBT)$. This implies (using cuts) that $\vdash_{GNB} \varphi \Rightarrow \psi$. Therefore by applying $[\neg \Rightarrow]_B$ we get that $\vdash_{GNB} \neg\psi \Rightarrow \neg\varphi$, implying that $\neg\psi \supset \neg\varphi \in Th(GKBT)$. It follows that $Th(GKBT)$ is closed under [CP].

Now, suppose that $\mathcal{T} \vdash_{NB} \varphi$. Since **NB** is an axiomatic extension of $HCL_\supset$, Theorem 1.94 implies that $\vdash_{NB} \psi_1 \supset (\ldots \supset (\psi_k \supset \varphi)\ldots)$ for some $\psi_1, \ldots, \psi_k \in \mathcal{T}$. It follows by what was proved above that $\vdash_{GNB} \Rightarrow \psi_1 \supset (\ldots \supset (\psi_k \supset \varphi)\ldots)$. Hence $\vdash_{GNB} \psi_1, \ldots, \psi_k \Rightarrow \varphi$, and so $\mathcal{T} \vdash_{GNB} \varphi$.

For the converse, define, as usual, the interpretation $I(s)$ of a sequent $s = \varphi_1, \ldots, \varphi_n \Rightarrow \psi_1, \ldots, \psi_k$ as $\bigwedge_{i=1}^n \varphi_i \supset \bigvee_{i=1}^k \psi_i$, where we take $\bigvee_{i=1}^k \psi_i$ to be F in case $k = 0$ (see Note 9.9). We first prove by induction on the structures of proofs in *GNB* that if $\vdash_{GNB} s$, then $I(s) \in Th(\mathbf{NB})$:

- The base case, where s is an axiom of *GNB*, is trivial, since $\varphi \supset \varphi$ is an instance of a positive classical tautology.

- Suppose that s is derived from s_1 and s_2 using a rule of *GNB* other than one of the two rules for $\neg$ (where we take $s_1 = s_2$ in case the rule has only one premise). It is easy to see that $I(s_1) \supset (I(s_2) \supset I(s))$ is then an instance of a classical tautology in $\mathcal{L}_{CL}^\mathsf{F}$. Hence it is in $Th(\mathbf{NB})$ by Note 9.9. Since by the induction hypothesis $I(s_1)$ and $I(s_2)$ are in $Th(\mathbf{NB})$ as well, it follows that so is $I(s)$.

- Suppose that $s = \Gamma \Rightarrow \Delta, \neg\varphi$ is derived from $s_1 = \varphi, \Gamma \Rightarrow \Delta$ using $[\Rightarrow \neg]$. It is again easy to see that this time $(\neg\varphi \vee \varphi) \supset (I(s_1) \supset I(s))$ is an instance of a classical tautology in $\mathcal{L}_{CL}^{\mathsf{F}}$. Hence, by Note 9.9, the induction hypothesis, and the fact that $\neg\varphi \vee \varphi \in Th(\mathbf{NB})$, together imply that $I(s) \in Th(\mathbf{NB})$.

- Suppose that $s = \neg\varphi \Rightarrow \neg\psi_1, \ldots, \neg\psi_n, \sigma_1, \ldots, \sigma_k$ is derived from $s_1 = \psi_1, \ldots, \psi_n, \neg\sigma_1, \ldots, \neg\sigma_k \Rightarrow \varphi$ using $[\neg \Rightarrow]_B$. If $n + k = 0$ then it follows from the induction hypothesis and the first item of Lemma 9.8 that $I(s) \in Th(\mathbf{NB})$. If $n + k \neq 0$, then by the induction hypothesis, $\psi_1 \wedge \ldots \wedge \psi_n \wedge \neg\sigma_1 \wedge \ldots \wedge \neg\sigma_k \supset \varphi \in Th(\mathbf{NB})$. By applying [CP] we get that $\neg\varphi \supset \neg(\psi_1 \wedge \ldots \wedge \psi_n \wedge \neg\sigma_1 \wedge \ldots \wedge \neg\sigma_k)$ is in $Th(\mathbf{NB})$. With the help of the second item of Lemma 9.8, and since $\neg\neg\sigma \supset \sigma \in Th(\mathbf{NB})$ for every σ, it easily follows that $I(s) \in Th(\mathbf{NB})$.

Suppose now that $\mathcal{T} \vdash_{GNB} \varphi$. Then $\vdash_{GNB} \psi_1, \ldots, \psi_n \Rightarrow \varphi$ for some $\psi_1, \ldots, \psi_n \in \mathcal{T}$. Hence $\psi_1 \wedge \ldots \wedge \psi_n \supset \varphi \in Th(\mathbf{NB})$, by what we have just proved. It follows that $\mathcal{T} \vdash_{\mathbf{NB}} \varphi$. □

The next proposition indicates a certain drawback of *GNB* in comparison to most other Gentzen-type systems studied in this book.

Proposition 9.13 *GNB does not admit cut-elimination.*

Proof. Obviously, $\vdash_{GNB} \neg(p \vee q), \neg(p \vee q) \supset r \Rightarrow r$. By applying $[\neg \Rightarrow]_B$ we get from this that $\vdash_{GNB} \neg r \Rightarrow \neg(\neg(p \vee q) \supset r), p \vee q$. Since also $\vdash_{GNB} p \vee q \Rightarrow p, q$, an application of the Cut rule yields that $\vdash_{GNB} \neg r \Rightarrow \neg(\neg(p \vee q) \supset r), p, q$. On the other hand, a straightforward (though tedious) search reveals that this sequent has no cut-free proof in *GNB*. □

Note 9.14 The fact that *GNB* does not admit cut-elimination is not such a serious drawback as it might seem. As we shall see in the next section, *GNB* does admit strong *analytic* cut-elimination (Definition 1.120), and this suffices here for making it a decidable system which has the crucial subformula property.[2] This is somewhat surprising, since the rule $[\neg \Rightarrow]_B$ does not enjoy (as a rule) the subformula property.

9.2.2 Semantics of NB

For providing adequate semantics for **NB**, we use the framework of modal Kripke frames (which is different from the framework of Kripke frames for intuitionistic logic which is given in Definition 1.66).

---
[2]See Notes 1.119 and 1.121 about the importance of analytic cut-elimination and the subformula property.

9.2 The Modal Logic B as a Paraconsistent Logic

Definition 9.15 (NB-frames) A triple $\langle W, R, \nu \rangle$ is called an **NB**-*frame* for $\mathcal{L}_{CL}$[3], if W is a nonempty (finite) set (of "worlds"), R is a reflexive and symmetric relation on W, and $\nu : W \times \mathcal{W}(\mathcal{L}_{CL}) \to \{t, f\}$ satisfies the following conditions:

- $\nu(w, \psi \wedge \varphi) = t$ iff $\nu(w, \psi) = t$ and $\nu(w, \varphi) = t$.
- $\nu(w, \psi \vee \varphi) = t$ iff $\nu(w, \psi) = t$ or $\nu(w, \varphi) = t$.
- $\nu(w, \psi \supset \varphi) = t$ iff $\nu(w, \psi) = f$ or $\nu(w, \varphi) = t$.
- $\nu(w, \neg\psi) = t$ iff there exists $w' \in W$ such that wRw', and $\nu(w', \psi) = f$.

Definition 9.16 (truth and validity) Let $\langle W, R, \nu \rangle$ be an **NB**-frame.

- A formula φ is *true* in a world $w \in W$ ($w \Vdash \varphi$) if $\nu(w, \varphi) = t$.
- A sequent $s = \Gamma \Rightarrow \Delta$ is *true* in a world $w \in W$ ($w \Vdash s$) if $\nu(w, \varphi) = f$ for some $\varphi \in \Gamma$, or $\nu(w, \varphi) = t$ for some $\varphi \in \Delta$. Equivalently, $w \Vdash s$ if $w \Vdash I(s)$, where $I(s)$ is the usual interpretation of s (as defined, e.g., in the proof of Theorem 9.12).
- A formula φ is *valid* in $\langle W, R, \nu \rangle$ ($\langle W, R, \nu \rangle \models \varphi$) if it is true in every world $w \in W$.
- A sequent s is *valid* in $\langle W, R, \nu \rangle$ ($\langle W, R, \nu \rangle \models s$) if it is true in every world $w \in W$.

Definition 9.17 (semantic consequence in NB)

- Let $\mathcal{T} \cup \{\varphi\}$ be a set of formulas in $\mathcal{L}_{CL}$. φ *semantically follows in* **NB** from $\mathcal{T}$ if for every **NB**-frame $\langle W, R, \nu \rangle$ and every $w \in W$: if $w \Vdash \psi$ for every $\psi \in \mathcal{T}$ then $w \Vdash \varphi$.
- Let $S \cup s$ be a set of sequents in $\mathcal{L}_{CL}$. s *semantically follows in* **NB** from S if for every **NB**-frame $\mathcal{W}$, if $\mathcal{W} \models s'$ for every $s' \in S$, then $\mathcal{W} \models s$. s is **NB**-*valid* if s semantically follows in **NB** from $\emptyset$ (that is, s is valid in every **NB**-frame).

Note 9.18 As is clear from the definitions, while the semantic consequence between formulas that we use is the local one, in defining our semantic consequence relation between sequents we employ the global approach. (See Note 9.6.) By this we follow the standard practice of dealing with formulas and sequents in classical first-order logic and in modal logics.

---
[3]Like in the case of our Definition 1.66, in the literature on modal logics one usually means by a "frame" just the pair $\langle W, R \rangle$, while we find it convenient to follow [225], and use this technical term a little bit differently, so that the valuation ν is a part of it.

Proposition 9.19 (strong soundness of NB)

1. If φ is a theorem of **NB** (that is, $\varphi \in Th(\mathbf{NB})$), then φ is valid in every **NB**-frame.

2. If $\mathcal{T} \vdash_{\mathbf{NB}} \varphi$ then φ semantically follows in **NB** from $\mathcal{T}$.

3. Let $S \cup s$ be a set of sequents. If $S \vdash_{GNB} s$ then s semantically follows in **NB** from S. In particular: if $\vdash_{GNB} s$ then s is **NB**-valid.

Proof.

1. Let $\mathcal{W} = \langle W, R, \nu \rangle$ be an **NB**-frame. We show that every theorem of **NB** is valid in $\mathcal{W}$.

   - Obviously, if φ is a positive classical tautology then $\mathcal{W} \models \varphi$, and if $\mathcal{W} \models \varphi$ and $\mathcal{W} \models \varphi \supset \psi$ them $\mathcal{W} \models \psi$.
   - Let $w \in W$. If $\nu(w, \varphi) = f$ then the reflexivity of R entails that there is $w' \in W$ ($w' = w$) such that wRw' and $\nu(w', \varphi) = f$. Hence $\nu(w, \neg\varphi) = t$. It follows that $\nu(w, \neg\varphi \vee \varphi) = t$ for every φ and $w \in W$, and so $\mathcal{W} \models \neg\varphi \vee \varphi$.
   - Let $w \in W$, and assume that $\nu(w, \varphi) = f$. Let w' be any element of W such that wRw'. Since R is symmetric, $w'Rw$ a well. It follows that there is $w'' \in W$ ($w'' = w$) such that $w'Rw''$, and $\nu(w'', \varphi) = f$. Hence, $\nu(w', \neg\varphi) = t$ for any such w'. This, in turn, implies that $\nu(w, \neg\neg\varphi) = f$. It follows that $\nu(w, \neg\neg\varphi \supset \varphi) = t$ for every φ and $w \in W$, and so $\mathcal{W} \models \neg\neg\varphi \supset \varphi$.
   - Suppose $\mathcal{W} \models \varphi \supset \psi$. Let $w \in W$, and assume that $\nu(w, \neg\psi) = t$. Then there is $w' \in W$ such that wRw' and $\nu(w', \psi) = f$. Since $\nu(w', \varphi \supset \psi) = t$ (because $\mathcal{W} \models \varphi \supset \psi$), this implies that $\nu(w', \varphi) = f$. Hence $\nu(w, \neg\varphi) = t$. It follows that $\nu(w, \neg\psi \supset \neg\varphi) = t$ for every φ, ψ, and $w \in W$, and so $\mathcal{W} \models \neg\psi \supset \neg\varphi$.

2. It is easy to see that ψ semantically follows in **NB** from $\{\varphi, \varphi \supset \psi\}$. Hence this item is an immediate corollary of the first one.

3. We need to prove that the axioms of *GNB* are valid in every **NB**-frame, and that the conclusion of every application of a rule of *GNB* semantically follows in **NB** from the premises of this application. We show this here for the case of $[\neg \Rightarrow]_B$, leaving all other (rather easy) cases for the reader. So let $\mathcal{W} = \langle W, R, \nu \rangle$ be an **NB**-frame, and suppose that $\mathcal{W} \models \Gamma, \neg\Delta \Rightarrow \psi$. We show that $\mathcal{W} \models \neg\psi \Rightarrow \neg\Gamma, \Delta$. So let $w \in W$, and assume that $\nu(w, \neg\psi) = t$. Then there is $w' \in W$ such that wRw' and $\nu(w', \psi) = f$. Since $\mathcal{W} \models \Gamma, \neg\Delta \Rightarrow \psi$, this implies

9.2 The Modal Logic B as a Paraconsistent Logic

that either $\nu(w', \varphi) = f$ for some $\varphi \in \Gamma$, or $\nu(w', \neg\varphi) = f$ for some $\varphi \in \Delta$. In the first case $\nu(w, \neg\varphi) = t$ for some $\varphi \in \Gamma$. In the second case there is $\varphi \in \Delta$ such that $\nu(w'', \varphi) = t$ for every w'' such that $w'Rw''$. But the symmetry of R implies that $w'Rw$. Hence, in the second case $\nu(w, \varphi) = t$ for some $\varphi \in \Delta$. In both cases we got that $\nu(w, \varphi) = t$ for some $\varphi \in \neg\Gamma \cup \Delta$. It follows that $w \Vdash \neg\psi \Rightarrow \neg\Gamma, \Delta$ for every $w \in W$, and so $\mathcal{W} \models \neg\psi \Rightarrow \neg\Gamma, \Delta$. □

Now we turn to prove the completeness of **NB** for its possible-worlds semantics, as well as the analyticity of *GNB*.

Theorem 9.20 *Let $S \cup s$ be a finite set of sequents in $\mathcal{L}_{CL}$. If s semantically follows in **NB** from S then s has an analytic proof in GNB from S.*

Proof. Suppose that s doesn't have an analytic proof in *GNB* from S. We construct an **NB**-frame in which the elements of S are valid, but s is not.

Denote by $\mathcal{F}$ the set of subformulas of formulas in $S \cup s$. Call a sequent $\Gamma \Rightarrow \Delta$ $\mathcal{F}$-*maximal* if the following conditions are satisfied:

(i) $\Gamma \cup \Delta = \mathcal{F}$.

(ii) $\Gamma \Rightarrow \Delta$ has no $\mathcal{F}$-analytic proof from S.

Lemma 1. *Suppose $\Gamma \cup \Delta \subseteq \mathcal{F}$, and $\Gamma \Rightarrow \Delta$ has no $\mathcal{F}$-analytic proof from S. Then $\Gamma \Rightarrow \Delta$ can be extended to an $\mathcal{F}$-maximal sequent $\Gamma' \Rightarrow \Delta'$ (that is, $\Gamma \subseteq \Gamma'$ and $\Delta \subseteq \Delta'$).*

Proof of Lemma 1. Let $\Gamma' \Rightarrow \Delta'$ be a maximal extension of $\Gamma \Rightarrow \Delta$ that consists of formulas in $\mathcal{F}$, and has no $\mathcal{F}$-analytic proof from S. (Such $\Gamma' \Rightarrow \Delta'$ exists, because $\mathcal{F}$ is finite.) To show that $\Gamma' \Rightarrow \Delta'$ is $\mathcal{F}$-maximal, assume for contradiction that there is $\varphi \in \mathcal{F}$ such that $\varphi \notin \Gamma' \cup \Delta'$. Then the maximality of $\Gamma' \Rightarrow \Delta'$ implies that both $\Gamma' \Rightarrow \Delta', \varphi$ and $\varphi, \Gamma' \Rightarrow \Delta'$ have $\mathcal{F}$-analytic proofs from S. But then we can get an $\mathcal{F}$-analytic proof of $\Gamma' \Rightarrow \Delta'$ from S using these two proofs together with an application of a cut on φ to their conclusions. (Note that since $\varphi \in \mathcal{F}$, the resulting proof of $\Gamma' \Rightarrow \Delta'$ is still $\mathcal{F}$-analytic.) This contradicts our assumption about $\Gamma' \Rightarrow \Delta'$.

Since s has no $\mathcal{F}$-analytic proof from S, it follows from Lemma 1 that s can be extended to an $\mathcal{F}$-maximal sequent $\Gamma^* \Rightarrow \Delta^*$.

Let W be the set of all $\mathcal{F}$-maximal sequents. Since $\mathcal{F}$ is finite, so is W. Since $(\Gamma^* \Rightarrow \Delta^*) \in W$, W is also nonempty. Define a relation R on W as follows: $(\Gamma_1 \Rightarrow \Delta_1) R (\Gamma_2 \Rightarrow \Delta_2)$ iff for every formula φ, if $\neg\varphi \in \Delta_1$ then $\varphi \in \Gamma_2$, and if $\neg\varphi \in \Delta_2$ then $\varphi \in \Gamma_1$. Obviously, R is symmetric. That it is also reflexive follows from the fact that if $\{\neg\varphi, \varphi\} \subseteq \Delta$ then $\Gamma \Rightarrow \Delta$ has a cut-free proof in GNB (since it can be derived from the axiom $\varphi \Rightarrow \varphi$ using $[\Rightarrow \neg]$ and weakenings). This fact and the $\mathcal{F}$-maximality of a sequent $\Gamma \Rightarrow \Delta$ in $\mathcal{W}$ imply that if $\neg\varphi \in \Delta$ then $\varphi \in \Gamma$, and so $(\Gamma \Rightarrow \Delta)R(\Gamma \Rightarrow \Delta)$. Next, let $\mathcal{W}$ be the **NB**-frame $\langle W, R, \nu\rangle$ in which W and R are as above, and ν is the unique valuation that satisfies the conditions in Definition 9.15, and for every propositional variable p, $\nu(\Gamma \Rightarrow \Delta, p) = t$ iff $p \in \Gamma$.

Lemma 2. Let $\Gamma \Rightarrow \Delta \in W$ and $\varphi \in \mathcal{F}$. Then $\nu(\Gamma \Rightarrow \Delta, \varphi) = t$ if $\varphi \in \Gamma$, and $\nu(\Gamma \Rightarrow \Delta, \varphi) = f$ if $\varphi \in \Delta$.

Proof of Lemma 2. By induction on the complexity of φ. The base case where φ is a propositional variable directly follows from the definition of ν and the fact that if $\varphi \in \Delta$ then $\varphi \notin \Gamma$ (because $\Gamma \Rightarrow \Delta$ has no $\mathcal{F}$-analytic proof). We show the induction step in the cases where the leading connective of φ is $\supset$ or $\neg$, leaving to the reader the cases of $\vee$ and $\wedge$.

- Suppose $\varphi = \psi \supset \sigma$, and $\varphi \in \Delta$. Since $\sigma \Rightarrow \varphi$ and $\Rightarrow \varphi, \psi$ have cut-free proofs in LK^+ in this case, the $\mathcal{F}$-maximality of $\Gamma \Rightarrow \Delta$ implies that $\sigma \in \Delta$ and $\psi \in \Gamma$. Therefore, the induction hypothesis implies that $\nu(\Gamma \Rightarrow \Delta, \psi) = t$ and $\nu(\Gamma \Rightarrow \Delta, \sigma) = f$. Hence $\nu(\Gamma \Rightarrow \Delta, \varphi) = f$.

- Suppose that $\varphi = \psi \supset \sigma$, and $\varphi \in \Gamma$. Since $\psi, \varphi \Rightarrow \sigma$ has a cut-free proof in LK^+ in this case, the $\mathcal{F}$-maximality of $\Gamma \Rightarrow \Delta$ implies that either $\psi \in \Delta$, or $\sigma \in \Gamma$. Therefore, the induction hypothesis implies that either $\nu(\Gamma \Rightarrow \Delta, \psi) = f$, or $\nu(\Gamma \Rightarrow \Delta, \sigma) = t$. In both cases $\nu(\Gamma \Rightarrow \Delta, \varphi) = t$.

- Suppose that $\varphi = \neg\psi$, and $\varphi \in \Delta$. By definition of R, this implies that if $(\Gamma \Rightarrow \Delta)R(\Gamma' \Rightarrow \Delta')$ then $\psi \in \Gamma'$. Therefore it follows from the induction hypothesis that $\nu(w, \psi) = t$ for every $w \in W$ such that $(\Gamma \Rightarrow \Delta)Rw$. Hence $\nu(\Gamma \Rightarrow \Delta, \varphi) = f$.

- Suppose that $\varphi = \neg\psi$, and $\varphi \in \Gamma$. Let $\Gamma_0 = \{\sigma \mid \neg\sigma \in \Delta\}$, $\Delta_o = \{\sigma \in \Delta \mid \neg\sigma \in \mathcal{F}\}$. Then $\Gamma_0, \neg\Delta_0 \Rightarrow \psi$ consists of formulas from $\mathcal{F}$, and it has no $\mathcal{F}$-analytic proof from S (because otherwise an application of $[\neg\Rightarrow]_B$ to this sequent, followed by weakenings, would provide an $\mathcal{F}$-analytic proof of $\Gamma \Rightarrow \Delta$ from S). It follows by Lemma 1 that $\Gamma_0, \neg\Delta_0 \Rightarrow \psi$ can be extended to an element $\Gamma' \Rightarrow \Delta'$ of W. We show that $(\Gamma \Rightarrow \Delta)R(\Gamma' \Rightarrow \Delta')$.

    - Suppose that $\neg\sigma \in \Delta$. Then $\sigma \in \Gamma_0$, and so $\sigma \in \Gamma'$.
    - Suppose that $\neg\sigma \in \Delta'$. Then $\neg\sigma \in \mathcal{F}$, and $\neg\sigma \notin \Gamma'$ (otherwise $\Gamma' \Rightarrow \Delta'$ would have an $\mathcal{F}$-analytic proof). Since $\neg\Delta_0 \subseteq \Gamma'$, this implies that $\sigma \notin \Delta$, and so $\sigma \in \Gamma$.

9.2 The Modal Logic B as a Paraconsistent Logic

Now the induction hypothesis implies that $\nu(\Gamma' \Rightarrow \Delta', \psi) = f$. Since $(\Gamma \Rightarrow \Delta)R(\Gamma' \Rightarrow \Delta')$, $\nu(\Gamma \Rightarrow \Delta, \varphi) = t$.

This ends the proof of Lemma 2.

Since $\Gamma^* \Rightarrow \Delta^*$ is an extension of s, it follows from Lemma 2 that if $s = \Gamma_s \Rightarrow \Delta_s$, then $\nu(\Gamma^* \Rightarrow \Delta^*, \varphi) = t$ for every $\varphi \in \Gamma_s$, while $\nu(\Gamma^* \Rightarrow \Delta^*, \varphi) = f$ for every $\varphi \in \Delta_s$. It follows that $\Gamma^* \Rightarrow \Delta^* \not\Vdash s$, and so $\mathcal{W} \not\Vdash s$.

Finally, let $s' = (\Gamma' \Rightarrow \Delta') \in S$, and let $w = (\Gamma \Rightarrow \Delta) \in W$. It is impossible that w is an extension of s', because w has no $\mathcal{F}$-analytic proof from S. It follows that either $\varphi \in \Delta$ for some $\varphi \in \Gamma'$, or $\varphi \in \Gamma$ for some $\varphi \in \Delta'$. By Lemma 2 this implies that either $\nu(w, \varphi) = f$ for some $\varphi \in \Gamma'$, or $\nu(w, \varphi) = t$ for some $\varphi \in \Delta'$. Hence $w \Vdash s'$ for every $w \in W$, and so $\mathcal{W} \Vdash s'$ for every $s' \in S$.

This ends our construction of the required **NB**-frame, and so the proof of the theorem. □

Next we present two important corollaries of Theorem 9.20.

Theorem 9.21 (analyticity of GNB) *GNB has the subformula property: if $S \vdash_{GNB} s$ then s has an analytic proof in GNB from S. In particular: if $\vdash_{GNB} s$ then s has an analytic proof in GNB.*[4]

Proof. Immediate from Proposition 9.19 and Theorem 9.20. □

Theorem 9.22 (weak completeness) *If Γ is finite then $\Gamma \vdash_{\mathbf{NB}} \varphi$ iff φ semantically follows in* **NB** *from Γ.*

Proof. This follows from Theorem 9.12, Proposition 9.19, and Theorem 9.20, where in the latter we take $S = \emptyset$ and $s = (\Gamma \Rightarrow \varphi)$. □

Theorem 9.22 can be strengthened to full completeness:

Theorem 9.23 (strong completeness) *For every theory $\mathcal{T}$, $\mathcal{T} \vdash_{\mathbf{NB}} \varphi$ iff φ semantically follows in* **NB** *from $\mathcal{T}$.*

Proof. (Outline) Call a pair of theories $\langle \mathcal{T}, \mathcal{S} \rangle$ *GNB-coherent* if there are no finite $\Gamma \subseteq \mathcal{T}$ and $\Delta \subseteq \mathcal{S}$ such that $\vdash_{GNB} \Gamma \Rightarrow \Delta$. Such a pair is called *GNB-maximal* if in addition $\mathcal{T} \cup \mathcal{S} = \mathcal{W}(\mathcal{L}_{CL})$. Using standard methods and the availability of the cut rule, it is not difficult to prove an analogue of Lemma 1 from the proof of Theorem 9.20: that every GNB-coherent pair

---

[4]This implies that if $\vdash_{GNB} s$ then s has a proof in GNE in which all cuts are analytic (that is, the cut formulas are subformulas of s).

of theories can be extended to a maximal pair. Let W be the set of all maximal pairs. Define the relation R on W and the valuation ν like in the proof of Theorem 9.20 (but with pairs $\langle \mathcal{T}, \mathcal{S} \rangle$ instead of sequents $\Gamma \Rightarrow \Delta$). Now it is easy to prove the obvious analogue of Lemma 2 from the proof of Theorem 9.20. Using these analogues of Lemmas 1 and 2, it is straightforward to adapt the proof of Theorem 9.20 to show that $\mathcal{W} = \langle W, R, \nu \rangle$ is a universal countermodel: If $\mathcal{T} \not\vdash_{\mathbf{NB}} \varphi$, then there is $w \in W$ such that $w \Vdash \psi$ for every $\psi \in \mathcal{T}$, while $w \not\Vdash \varphi$. From this the theorem immediately follows (again with the help of Theorem 9.12 and Proposition 9.19). □

9.2.3 Basic Properties of NB

From Proposition 2.14 and Note 2.6 it follows that a logic which enjoys the replacement property, as well as satisfies the properties provided in the desiderata list from Section 2.4, should contain the logic **NB**. Our next goal is to show that **NB** itself already satisfies all the required conditions, and so is the minimal logic of the type we are seeking. Now, in the previous parts of this section it was shown that **NB** has an analytic proof system which has the subformula property, and that it has also a useful, concrete denotational semantics. In this section we show that it is a strongly paraconsistent logic which satisfies all the other requirements in the desiderata list given in Section 2.4.

Theorem 9.24 (Decidability of NB) **NB** *is decidable.*

Proof. Given a formula φ (or a finite set of formulas $\Gamma \cup \{\varphi\}$), the number of sequents which consist only of subformulas of φ (or $\Gamma \Rightarrow \varphi$) is finite. Hence it easily follows from Theorem 9.21 that it is decidable whether $\vdash_{GNB} \varphi$ (or $\vdash_{GNB} \Gamma \Rightarrow \varphi$) or not. □

Note 9.25 Instead of using the Gentzen-type system *GNB*, one can use the semantics of **NB** in order to provide a decision procedure for the latter. This is due to the fact that from the proof of Theorem 9.20 it follows that a sequent s is **NB**-valid iff it is valid in every **NB**-frame in which the number of worlds is at most 2^n, where n is the number of subformulas of s.

Proposition 9.26

1. **NB** *is normal, with* $\wedge, \vee$ *and* $\supset$ *serving as its conjunction, disjunction, and implication (respectively).*

2. **NB** *is* $\neg$-*contained in classical logic.*

3. *The negation* $\neg$ *of* **NB** *is complete, contrapositive, and left-involutive.*

9.2 The Modal Logic B as a Paraconsistent Logic

Proof. The first item follows from the fact that **NB** is an axiomatic extension of **CL**$^+$ (see Note 9.6). The third item follows from the first one and Note 2.6. We leave the verification of the second item to the reader. □

Corollary 9.27 *NB is the minimal extension of* **CL**$^+$ *in* $\mathcal{L}_{CL}$ *in which* ¬ *is complete, contrapositive, and left-involutive.*

Proposition 9.28 *NB is self-extensional (that is, it has the replacement property).*

Proof. Immediate from Proposition 9.3 and the third item of Proposition 9.26. □

Proposition 9.29 *NB is strongly ¬-paraconsistent.*

Proof. It is easy to verify that no analytic proof of $p, \neg p \Rightarrow \neg q$ exists in GNB if p and q are distinct variables. Alternatively, construct an **NB**-frame $\mathcal{W} = \langle W, R, \nu \rangle$ in which $W = \{w_1, w_2\}$, $R = W \times W$, $\nu(w_1, p) = \nu(w_1, q) = \nu(w_2, q) = t$, while $\nu(w_2, p) = f$. Then $w_1 \Vdash \{p, \neg p\}$, while $w_1 \nVdash \neg q$. □

Corollary 9.30 *NB is a classical ¬-paraconsistent logic.*

Proof. Immediate from Propositions 9.29, 9.26, and Note 9.6. □

9.2.4 NB as an LFI

Proposition 9.31 *NB, as well as any of its classical ¬- paraconsistent extensions, is an LFI, and in fact a strong C-system.*

Proof. That any classical ¬- paraconsistent extension of **NB** is a strong C-system follows from Corollary 8.21, Note 9.9, Proposition 9.26, and the second item of Note 2.6. By Corollary 9.30, this applies to **NB** itself. □

Note 9.32 Note 9.9 entails that we may assume (as we do from now on) that the language of **NB** includes F, and that **NB** is an extension of **CL**$^\mathsf{F}$.[5] From Corollary 8.18 and the self-extensionality of **NB** it further follows that **NB** has a *unique* (up to *congruence*!) strong consistency operator ∘. Using the proof of Proposition 8.20, in the rest of this Chapter we assume that ∘ is defined in **NB** as $\circ\varphi =_{Df} \varphi \wedge \neg\varphi \supset \mathsf{F}$ (or as $\circ\varphi =_{Df} (\varphi \wedge \neg\varphi) \supset \neg(\varphi \supset \varphi)$).

---

[5] See the last items of Notes 1.91 and 1.131 (respectively) for Hilbert-type and Gentzen-type proof systems for **CL**$^\mathsf{F}$.

In this section we check which of the schemas studied in Chapter 8 is valid in **NB**, and which can be added to it without losing its paraconsistency and the contrapositivity of its negation. Unlike in Chapter 8, here it would be more convenient to consider for this task axiom [i] rather than the equivalent axiom [a$_\circ$]. (See Proposition 8.85.) What is more, it would be illuminating to split [i] into the following two axioms (to the conjunction of which it is equivalent already in **IL**$^+$):

[i$_1$] $\neg \circ \varphi \supset \varphi$,

[i$_2$] $\neg \circ \varphi \supset \neg \varphi$.

Proposition 9.33 *The following schemas are provable in* **NB**: [b], [k], [t], [c], [n$^l_\wedge$], [n$^r_\wedge$]$_1$, [n$^r_\wedge$]$_2$, [n$^l_\vee$], [n$^l_\supset$]$_2$, [a$_\neg$], [a$_\wedge$], *and* [a$_\vee$].

Proof. Let $\langle W, R, \nu \rangle$ be a **NB**-frame. From the definition of $\circ\varphi$ it follows that given $w \in W$, $w \Vdash \circ\varphi$ iff either $w \not\Vdash \varphi$, or $u \Vdash \varphi$ for every u such that wRu. This observation easily implies the validity of [a$_\neg$], [a$_\wedge$], and [a$_\vee$] in every **NB**-frame. Therefore they are all theorems of **NB** by Theorem 9.22.[6] That so is [n$^l_\wedge$] was shown in Lemma 9.8. [b] and [k] are theorems of **NB** by Proposition 9.31, while [t] and [c] are simply axioms of it. The remaining axioms are easily provable in HNB using [CP]. □

Note 9.34 The fact that [a$_\neg$], [a$_\wedge$], and [a$_\vee$] are all valid in **NB** means that **NB** is almost perfectly adequate to serve as an LFI according to da Costa's ideas. The only principle that it misses (as we show is the next theorem) is [a$_\supset$]. This is the price it pays for having both a contrapositive negation and the replacement property. However, this is not a high price, since the language of $\{\neg, \wedge, \vee\}$ suffices for classical reasoning (since its set of primitive connectives is functionally complete for two-valued matrices).

It is also remarkable that $\neg(\varphi \wedge \psi)$ is equivalent (and so congruent) in **NB** to $\neg\varphi \vee \neg\psi$. However, the next theorem shows that the other De Morgan rules are only partially valid in **NB**.

Our next goal is to show that **NB** is rather robust as a paraconsistent logic, since all the schemas studied in chapter 8 that are not already derivable in **NB** cannot even be added to it without losing its paraconsistency. For this we need first to clarify what we mean by 'adding a set of schemas X to **NB**' (and similarly defined systems). Note that by this we do *not* mean the axiomatic extension of **NB** by the axioms in X (Definition 1.9), or the logic induced by the immediate axiomatic extension of HNB by the

---

[6]Note that [a$_\neg$] is easily derivable already in $\mathbf{C}_{min}$, and was shown in Chapter 8 to be equivalent in **BK** to [c].

9.2 The Modal Logic B as a Paraconsistent Logic

axioms in X (see Definition 1.86). What we have in mind is a somewhat more complicated axiomatic extension of **NB**, in which [CP] still functions as a rule of proof (see Note 9.6). Below is the exact definition.

Definition 9.35 Let X be a set of schemas in $\mathcal{L}_{CL}$. We say that a logic **L** in $\mathcal{L}_{CL}$ is *obtained by adding to* **NB** *the axioms* in X iff the following hold:

- $Th(\mathbf{L})$, the set of valid formula of **L**, is the minimal set S of formulas in $\mathcal{L}_{CL}$ such that:
  1. S includes all axioms of $\mathbf{C}_{min}$ (Definition 7.37) as well as all instances of elements of X.
  2. S is closed under [MP] and [CP].

- HL, the system which is obtained by adding to HNB the axioms in X, is the Hilbert-type system whose set of axioms is $Th(\mathbf{L})$, and has [MP] for $\supset$ as its sole rule of inference.

- **L** is the logic induced by the Hilbert-type system HL which is defined in the previous item.

Proposition 9.36 *Suppose* **L** *is obtained by adding to* **NB** *some axioms.*

1. **L** *is normal, with* $\wedge, \vee$ *and* $\supset$ *serving as its conjunction, disjunction, and implication (respectively).*

2. *The negation* $\neg$ *of* **L** *is complete, contrapositive, and left-involutive.*

Proof. Similar to the proof of Proposition 9.26. □

Theorem 9.37 *Let* **L** *be obtained by adding to* **NB** *a non-empty subset of* $\{[\mathsf{e}], [\mathsf{n}^\mathsf{l}_\supset]_1, [\mathsf{n}^\mathsf{r}_\vee], [\mathsf{n}^\mathsf{r}_\supset], [\mathsf{a}_\supset], [\mathsf{a}_\circ], [\mathsf{i}_1], [\mathsf{l}], [\mathsf{d}]\} \cup \{[\mathsf{o}^\mathsf{i}_\#] \mid \sharp \in \{\wedge, \vee, \supset\}, i \in \{1,2\}\}$. *Then* **L** *is not* $\neg$-*paraconsistent.*

Proof. Recall in what follows that here $\circ\varphi =_{def} \varphi \wedge \neg\varphi \supset \mathsf{F}$ (Note 9.32).

- Suppose [e] is valid in **L**. Then from Propositions 2.14 and 9.36 it follows that $\vdash_\mathbf{L} \neg\varphi \supset (\varphi \supset \neg\neg\psi)$ for every φ, ψ. Since [c] is valid in **NB**, this implies that $\vdash_\mathbf{L} \neg\varphi \supset (\varphi \supset \psi)$.

- Suppose $[\mathsf{n}^\mathsf{l}_\supset]_1$ is valid in **L**. Then $\vdash_\mathbf{L} \neg(\varphi \supset \psi) \supset \varphi$. By applying [CP] we get that $\vdash_\mathbf{L} \neg\varphi \supset \neg\neg(\varphi \supset \psi)$. Hence $\vdash_\mathbf{L} \neg\varphi \supset (\varphi \supset \psi)$.

- Suppose $[\mathsf{n}^\mathsf{r}_\vee]$ is valid in **L**. Then $\vdash_\mathbf{L} \neg(\varphi \supset \psi) \wedge \neg\varphi \supset \neg((\varphi \supset \psi) \vee \varphi)$. But $\vdash_{\mathbf{CL}^+} (\varphi \supset \psi) \vee \varphi$, and so $\vdash_\mathbf{L} \neg((\varphi \supset \psi) \vee \varphi) \supset \mathsf{F}$ by Lemma 9.8. It follows that $\vdash_\mathbf{L} \neg(\varphi \supset \psi) \supset (\neg\varphi \supset \mathsf{F})$. Since it is easy to see that $\vdash_\mathbf{NB} (\neg\varphi \supset \mathsf{F}) \supset \varphi$, we get that $\vdash_\mathbf{L} \neg(\varphi \supset \psi) \supset \varphi$. Therefore it follows from the previous item that $\vdash_\mathbf{L} \neg\varphi \supset (\varphi \supset \psi)$.

- Suppose [$n_{\supset}^r$] is valid in **L**. In particular, $\vdash_\mathbf{L} \varphi \supset (\neg\varphi \supset \neg(\varphi \supset \varphi))$. Since $\vdash_{\mathbf{CL}^+} \varphi \supset \varphi$, it follows by Lemma 9.8 that $\vdash_\mathbf{L} \varphi \supset (\neg\varphi \supset \psi)$.

- Suppose [$a_\supset$] is valid in **L**. Then $\vdash_\mathbf{L} (\circ\varphi \wedge \circ\mathsf{F}) \supset \circ(\varphi \supset \mathsf{F})$. Since $\vdash_{\mathbf{CL}^+} \circ\mathsf{F}$, it follows that $\vdash_\mathbf{L} \circ\varphi \supset \circ(\varphi \supset \mathsf{F})$. But $\vdash_{\mathbf{CL}^+} (\varphi \supset \mathsf{F}) \supset \circ\varphi$. Hence $\vdash_\mathbf{L} (\varphi \supset \mathsf{F}) \supset \circ(\varphi \supset \mathsf{F})$. Using $\mathbf{CL}^+$ and the definition of $\circ$, this is equivalent to $\vdash_\mathbf{L} \neg(\varphi \supset \mathsf{F}) \supset ((\varphi \supset \mathsf{F}) \supset \mathsf{F})$. This in turn entails that $\vdash_\mathbf{L} \neg(\varphi \supset \mathsf{F}) \supset \varphi$, since $\vdash_{\mathbf{CL}^\mathsf{F}} ((\varphi \supset \mathsf{F}) \supset \mathsf{F}) \supset \varphi$. By applying [CP] and using the axiom [c] of *HNB*, we get that $\vdash_\mathbf{L} \neg\varphi \supset (\varphi \supset \mathsf{F})$, and so $\vdash_\mathbf{L} \neg\varphi \supset (\varphi \supset \psi)$.

- Suppose [i_1] is valid in **L**. Then $\vdash_\mathbf{L} \neg\circ\varphi \supset \varphi$. Using [CP] and [c], this entails that $\vdash_\mathbf{L} \neg\varphi \supset \circ\varphi$. It follows by the definition of $\circ$ that $\vdash_\mathbf{L} \neg\varphi \supset (\varphi \supset \mathsf{F})$, and so $\vdash_\mathbf{L} \neg\varphi \supset (\varphi \supset \psi)$.

- The case of [$a_\circ$] follows from that of [i_1] by Propositions 8.85 and 9.33.

- Suppose [l] is valid in **L**. Then $\vdash_\mathbf{L} \neg(\varphi \wedge \neg\varphi) \supset \circ\varphi$. Using [CP] and [c], this entails that $\vdash_\mathbf{L} \neg\circ\varphi \supset (\varphi \wedge \neg\varphi)$, and so that [$i_1$] is a theorem of **L**. Hence this item follows from the previous one.

- The proof in the case [d] is similar to that in the case of [l].

- Since both [$o_\supset^1$] and [$o_\supset^2$] entail [$a_\supset$] in $\mathbf{CL}^+$, the cases of [$o_\supset^1$] and [$o_\supset^2$] follow from the case of [$a_\supset$].

- Suppose e.g. that [$o_\vee^1$] is valid in **L**. (The proof in the case of [$o_\vee^2$] is similar.) Then in particular $\vdash_\mathbf{L} \circ\mathsf{F} \supset \circ(\mathsf{F} \vee \varphi)$. Since $\vdash_\mathbf{L} \circ\mathsf{F}$, this implies that $\vdash_\mathbf{L} \circ(\mathsf{F} \vee \varphi)$. But $\mathsf{F} \vee \varphi$ is equivalent in $\mathbf{CL}^\mathsf{F}$ to φ. It follows by the replacement property of **L** that $\vdash_\mathbf{L} \circ\varphi$. Hence $\vdash_\mathbf{L} \neg\varphi \supset (\varphi \supset \mathsf{F})$, and so $\vdash_\mathbf{L} \neg\varphi \supset (\varphi \supset \psi)$.

- Suppose, e.g., that [$o_\wedge^1$] is valid in **L**. (The proof in the case of [$o_\wedge^2$] is similar.) Then $\vdash_\mathbf{L} \circ(\neg\mathsf{F}) \supset \circ(\neg\mathsf{F} \wedge \varphi)$. Since $\vdash_\mathbf{L} \circ(\neg\mathsf{F})$ (using [c]), this implies that $\vdash_\mathbf{L} \circ(\neg\mathsf{F} \wedge \varphi)$. But since $\vdash_\mathbf{L} \neg\mathsf{F}$ (Note 9.9), $\neg\mathsf{F} \wedge \varphi$ is equivalent in **L** to φ. It follows by the replacement property of **L** that $\vdash_\mathbf{L} \circ\varphi$. Hence $\vdash_\mathbf{L} \neg\varphi \supset (\varphi \supset \mathsf{F})$, and so $\vdash_\mathbf{L} \neg\varphi \supset (\varphi \supset \psi)$. □

Note 9.38 After splitting [i], its part [i_2] remains the only negation axiom, among those which we have studied in the context of C-systems, that is not dealt with yet in Proposition 9.33 and Theorem 9.37. Now it is not difficult to show that [i_2] is not provable in **NB**. In the next section we show that it can nevertheless be added to **NB** without losing its paraconsistency, and that this addition leads to a very interesting logic.

9.2 The Modal Logic B as a Paraconsistent Logic

9.2.5 NB as a Version of the Modal Logic B

As hinted at the beginning of Section 9.2.2, the notion of '**NB**-frame' is very similar to the notion of a modal Kripke frame used in the study of modal logics. Indeed, **NB** is actually (equivalent to) the famous modal logic which is usually called **B** or **KTB** (see, e.g., [109]). The language of **B** is usually taken to be $\{\wedge, \vee, \supset, \mathsf{F}, \square\}$ (or $\{\wedge, \vee, \supset, \neg, \square\}$, where $\neg$ denotes the *classical* negation).[7] Its semantics is given by modal Kripke frames in which the accessibility relation R is again reflexive and symmetric, where the notion of a 'modal Kripke frame' is defined like that of an '**NB**-frame' (in Definition 9.15), except that instead of the clause there for $\neg$ we have the following clause for $\square$:

- $\nu(w, \square\psi) = t$ iff $\nu(w', \psi) = t$ for every $w' \in W$ such that wRw'.

Now, it is easy to see that with respect to modal Kripke frames, the language of our **NB** and the language of the modal logic **B** are equivalent in their expressive power. $\square$ is definable in the former by $\square\varphi =_{def} \sim\neg\varphi$, where $\sim\psi =_{def} \psi \supset \mathsf{F}$. On the other hand $\neg$ is definable in the language of **B** by $\neg\varphi =_{def} \sim\square\varphi$. It follows that the paraconsistent logic **NB** (whose language is just $\mathcal{L}_{CL}$) and the modal logic **B** are practically identical.

It is worth noting that the presentation of the modal **B** in the form **NB** is more concise (and in our opinion also clearer) than the usual one in two ways. First, **NB** really has only two basic connectives: $\supset$ and $\neg$. (F can be defined as $\neg(\varphi \supset \varphi)$, where φ is arbitrary, and $\vee$ and $\wedge$ can of course be defined in terms of $\supset$ and F.) The standard presentation of **B** needs three connectives: $\supset$, F, and $\square$. Second, the standard Hilbert-type proof system for **B** is more complicated than HNB. It is obtained from (the full) HCL by the addition of one rule of proof and three axioms. The rule is the necessitation rule (if $\vdash \varphi$ then $\vdash \square\varphi$). The three axioms are:

(K) $\square(\varphi \supset \psi) \supset (\square\varphi \supset \square\psi)$.

(T) $\square\varphi \supset \varphi$.

(B) $\varphi \supset \square\diamond\varphi$, where $\diamond\varphi =_{def} \sim\square\sim\varphi$.

In contrast, HNB is obtained from HCL^+ by the addition of one rule of proof (which admittedly is somewhat more complex than the necessitation rule), and just two extremely simple and natural axioms.

---

[7]In order not to confuse our paraconsistent $\neg$ with the classical negation, at the rest of this chapter we shall denote the latter by $\sim$ rather than by $\neg$.

9.3 S5 as a Paraconsistent Logic

9.3.1 The Logic NS5

Following Note 9.38, in this section we investigate the system that is obtained from **NB** by the addition of $[i_2]$. We start by presenting two schemas which are equivalent to $[i_2]$ over **NB**.

Lemma 9.39 *The following schemas are equivalent over logics which are obtained by adding schemas to* **NB**:

1. $[i_2]$ *(that is:* $\neg \circ \varphi \supset \neg \varphi$*).*

2. $\circ(\neg \varphi)$ *(that is:* $\neg \varphi \wedge \neg \neg \varphi \supset \mathsf{F}$*).*

3. $\neg \neg \varphi \supset (\neg \varphi \supset \psi)$.

Proof. That the second and third schemas are equivalent is obvious. We show the equivalence of the first two. So let **L** be obtained by adding schemas to **NB**.

- Suppose $\vdash_\mathbf{L} \neg \circ \varphi \supset \neg \varphi$. An application of [CP] to this, followed by a use of the axiom $\neg \neg \circ \varphi \supset \circ \varphi$, yields $\vdash_\mathbf{L} \neg \neg \varphi \supset \circ \varphi$, that is: $\vdash_\mathbf{L} \neg \neg \varphi \supset (\varphi \wedge \neg \varphi \supset \mathsf{F})$. Since $\vdash_\mathbf{L} \neg \neg \varphi \supset \varphi$, this entails that $\vdash_\mathbf{L} \neg \varphi \wedge \neg \neg \varphi \supset \mathsf{F}$.

- Suppose $\vdash_\mathbf{L} \neg \varphi \wedge \neg \neg \varphi \supset \mathsf{F}$. Using positive classical reasoning, this implies that $\vdash_\mathbf{L} \neg \neg \varphi \supset (\varphi \wedge \neg \varphi \supset \mathsf{F})$, that is: $\vdash_\mathbf{L} \neg \neg \varphi \supset \circ \varphi$. By applying [CP] we get from this that $\vdash_\mathbf{L} \neg \circ \varphi \supset \neg \neg \neg \varphi$. Therefore $\vdash_\mathbf{L} \neg \circ \varphi \supset \neg \varphi$ by axiom [c]. □

Definition 9.40 (NS5) Let *HNS5* be the Hilbert-type system which is obtained from *HNB* by adding to it as an axiom schema one of the three schemas which were proved equivalent in Lemma 9.39. **NS5** is the logic induced by *HNS5*. (See Definition 9.35.)

Note 9.41 The following observation leads to a simpler version of *HNS5*:

$$\neg \neg \varphi \supset (\neg \varphi \supset \varphi), \neg \varphi \vee \varphi \vdash_{\mathbf{IL}^+} \neg \neg \varphi \supset \varphi.$$

It follows that in order to axiomatize **NS5**, it suffices to add to HCL^+ the schemas $\neg \varphi \vee \varphi$, $\neg \neg \varphi \supset (\neg \varphi \supset \psi)$, and [CP] as a rule of proof (or to add to **CLuN** the schema $\neg \neg \varphi \supset (\neg \varphi \supset \psi)$, and [CP] as a rule of proof).

Like in the case of **NB** we start our investigation of **NS5** by providing a useful Gentzen-type system for it.

9.3 S5 as a Paraconsistent Logic

Definition 9.42 (GNS5) The system *GNS5* is the system which is obtained from *LK* (Figures 1.2, 1.3) by replacing its rule $[\neg \Rightarrow]$ by the rule:

$$[\neg \Rightarrow]_5 \quad \frac{\neg \Gamma \Rightarrow \psi, \neg \Delta}{\neg \Gamma, \neg \psi \Rightarrow \neg \Delta}.$$

Note 9.43 Again the difference between *GNS5* and *LK* is with respect to the classical rule $[\neg \Rightarrow]$. In *GNS5* its application is allowed only in case the context entirely consists of negated formulas.

Theorem 9.44 (Adequacy of GNS5) $\mathcal{T} \vdash_{GNS5} \varphi$ iff $\mathcal{T} \vdash_{\mathbf{NS5}} \varphi$.

Proof. The proof that if $\mathcal{T} \vdash_{\mathbf{NS5}} \varphi$ then $\mathcal{T} \vdash_{GNS5} \varphi$ is similar to the proof of the corresponding claim concerning **NB** and *GNB* (Theorem 9.12). By Note 9.41, we only to add a derivation of $\Rightarrow \neg\neg\varphi \supset (\neg\varphi \supset \psi)$ in *GNS5*. This is easy: an application of $[\neg \Rightarrow]_5$ to the axiom $\neg\varphi \Rightarrow \neg\varphi$ yields $\neg\neg\varphi, \neg\varphi \Rightarrow$. From this $\Rightarrow \neg\neg\varphi \supset (\neg\varphi \supset \psi)$ is derived using weakening and $[\Rightarrow \supset]$.

Like in the proof of Theorem 9.12, for the converse it again suffices to show that if $\vdash_{GNS5} s$, then $I(s) \in Th(\mathbf{NS5})$ (where $Th(\mathbf{NS5})$ is the set of theorems of **NS5**). The proof of this is almost identical to that of Theorem 9.12. The only change we have to make in the inductive proof given there is in the treatment of the rule $[\neg \Rightarrow]_5$. For this it is useful to first observe that $\vdash_{\mathbf{NS5}} (\neg\varphi \vee \psi) \supset (\neg\neg\varphi \supset \psi)$ (because $\vdash_{\mathbf{NS5}} \neg\varphi \supset (\neg\neg\varphi \supset \psi)$). Now suppose that $s = \neg\varphi, \neg\psi_1, \ldots, \neg\psi_n \Rightarrow \neg\sigma_1, \ldots, \neg\sigma_k$ is derived from $s_1 = \neg\psi_1, \ldots, \neg\psi_n \Rightarrow \neg\sigma_1, \ldots, \neg\sigma_k, \varphi$ using $[\neg \Rightarrow]_5$. Then the induction hypothesis concerning s_1 and repeated applications of the above observation imply that $\neg\psi_1 \wedge \ldots \wedge \neg\psi_n \wedge \neg\neg\sigma_1 \wedge \ldots \wedge \neg\neg\sigma_k \supset \varphi \in Th(\mathbf{NS5})$. By applying [CP] we get that $\neg\varphi \supset \neg(\neg\psi_1 \wedge \ldots \wedge \neg\psi_n \wedge \neg\neg\sigma_1 \wedge \ldots \wedge \neg\neg\sigma_k)$ is in $Th(\mathbf{NS5})$. Using the second item of Lemma 9.8, this implies that $\neg\varphi \supset (\neg\neg\psi_1 \vee \ldots \vee \neg\neg\psi_n \vee \neg\neg\neg\sigma_1 \vee \ldots \vee \neg\neg\neg\sigma_k)$ is in $Th(\mathbf{NS5})$. Hence it follows from the observation above that

$$\neg\varphi \wedge \neg\neg\neg\psi_1 \wedge \ldots \wedge \neg\neg\neg\psi_n \supset \neg\neg\neg\sigma_1 \vee \ldots \vee \neg\neg\neg\sigma_k \in Th(\mathbf{NS5}).$$

Finally, by the third item of Lemma 9.8 and Proposition 9.3, it follows from this that $I(s) \in Th(\mathbf{NS5})$. □

Note 9.45 The sequent $\neg\neg P_1 \Rightarrow P_1$ is derivable in *GNS5* by applying the cut rule to the sequents $\Rightarrow P_1, \neg P_1$ and $\neg\neg P_1, \neg P_1 \Rightarrow$ (both of which have very short cut-free proofs in *GNS5*). It is easy to see that this cut cannot be eliminated. Hence like in the case of *GNB*, the cut-elimination theorem fails for *GNS5*. Again we shall soon see that *GNS5* is analytic nevertheless. Then in the next section we show that by employing *hypersequents* rather than ordinary sequents, one can provide a proof system for **NS5** that *is* cut-free.

Our next subject is the semantics of **NS5**.

Definition 9.46 (NS5-frames)

- An **NB**-frame $\langle W, R, \nu \rangle$ is called an **NS5**-*frame* for $\mathcal{L}_{CL}$ if R is transitive (in addition to its being reflexive and symmetric).

- The notions of truth (in worlds) and validity in **NS5**-frames (of formulas and sequents) are defined like in Definition 9.16.

- Semantic consequence in **NS5** is defined like in Definition 9.17, using **NS5**-frames instead of **NB**-frames.

Note 9.47 In **NS5**-frames the accessibility relation R is an equivalence relation. Looking at each equivalence class as an independent frame, it is easy to see that we do note really need R here. We can take an **NS5**-frame to be just a *pair* $\langle W, \nu \rangle$, replacing the last item in Definition 9.15 by:

- $\nu(w, \neg \psi) = t$ iff there exists $w' \in W$ such that $\nu(w', \psi) = f$.

Note that with this notion of an **NS5**-frame, a formula of the form $\neg \varphi$ is either true in all worlds of an **NS5**-frame, or false in all of them. Note also that the latter is the case iff φ is true in all the worlds of the frame.

Proposition 9.48 (strong soundness of NS5)

1. If φ is a theorem of **NS5** then φ is valid in every **NS5**-*frame*.

2. If $\mathcal{T} \vdash_{\mathbf{NS5}} \varphi$ then φ semantically follows in **NS5** from $\mathcal{T}$.

3. Let $S \cup s$ be a set of sequents. If $S \vdash_{GNS5} s$ then s semantically follows in **NS5** from S. In particular: if $\vdash_{GNS5} s$ then s is **NS5**-*valid*.

Proof. From Note 9.47 it immediately follows that the additional axiom of *HNS*5, $\neg\neg\varphi \supset (\neg\varphi \supset \psi)$, is valid in every **NS5**-frame. The rest of the proof is similar to that of Proposition 9.19, and is left to the reader. □

Now we turn to prove the completeness of **NS5** for its possible-worlds semantics, as well as the analyticity of *GNS*5.

Theorem 9.49 *Let $S \cup s$ be a finite set of sequents in $\mathcal{L}_{CL}$. If s semantically follows in **NS5** from S then s has an analytic proof in GNS5 from S.*

Proof. We follow the proof of Theorem 9.20. We define $\mathcal{F}$, $\mathcal{F}$-maximality, and $\Gamma^* \Rightarrow \Delta^*$, and prove the corresponding version of Lemma 1 exactly as it is done there (but relative to *GNS*5 rather than *GNB*). Now let $\Gamma' =$

9.3 S5 as a Paraconsistent Logic

$\{\varphi \mid \neg\varphi \in \Gamma^*\}$ and $\Delta' = \{\varphi \mid \neg\varphi \in \Delta^*\}$. Obviously, $\neg\Gamma' \cap \neg\Delta' = \emptyset$, while $\neg\Gamma' \cup \neg\Delta'$ is the set of all elements of $\mathcal{F}$ which begins with $\neg$. Let W be the set of all $\mathcal{F}$-maximal sequents that extend $\neg\Gamma' \Rightarrow \neg\Delta'$. The two observations above about $\neg\Gamma'$ and $\neg\Delta'$ imply that for every $\Gamma \Rightarrow \Delta \in W$, $\neg\varphi \in \Gamma$ iff $\neg\varphi \in \Gamma'$, and $\neg\varphi \in \Delta$ iff $\neg\varphi \in \Delta'$. Now let $R = W^2$, and define ν as in the proof of Theorem 9.20 (that is: $\nu(\Gamma \Rightarrow \Delta, p) = t$ iff $p \in \Gamma$, in case p is a propositional variable). We show that Lemma 2 from the proof of Theorem 9.20 holds here as well. The proof is again by induction on the complexity of φ. We present the arguments in the only two cases in which they are different from those used in the proof of Theorem 9.20. Note that in both of them φ is of the form $\neg\psi$.

- Suppose $\varphi = \neg\psi$, and $\varphi \in \Delta$. Let $\Gamma_1 \Rightarrow \Delta_1 \in W$. By definition of W, $\varphi \in \Delta_1$. Hence $\psi \in \Gamma_1$ (because of the maximality of $\Gamma_1 \Rightarrow \Delta_1$, and the presence of $[\Rightarrow \neg]$). It follows by the induction hypothesis that $\nu(\Gamma_1 \Rightarrow \Delta_1, \psi) = t$ whenever $\Gamma_1 \Rightarrow \Delta_1 \in W$. Hence $\nu(\Gamma \Rightarrow \Delta, \varphi) = f$.

- Suppose $\varphi = \neg\psi$, and $\varphi \in \Gamma$. Then $\neg\Gamma' \Rightarrow \neg\Delta', \psi$ has no $\mathcal{F}$-analytic proof from S (because otherwise an application of $[\neg\Rightarrow]_5$ to this sequent, followed by weakenings, would provide an $\mathcal{F}$-analytic proof of $\Gamma \Rightarrow \Delta$ from S). It follows by Lemma 1 that $\neg\Gamma' \Rightarrow \neg\Delta', \psi$ can be extended to an element $\Gamma_1 \Rightarrow \Delta_1$ of W. By the induction hypothesis, $\nu(\Gamma_1 \Rightarrow \Delta_1, \psi) = f$. Hence $\nu(\Gamma \Rightarrow \Delta, \varphi) = t$.

Once we have Lemma 2, the rest of the proof is identical to what is done in the proof of Theorem 9.20. Details are left to the reader. □

Theorem 9.49 has for **NS5** the same important corollaries as Theorem 9.20 has for **NB**. We list them below, leaving their proofs (which are almost identical to their counterparts in the case of **NB**) to the reader.

Theorem 9.50 (analyticity of GNS5) *GNS5 has the subformula property: if $S \vdash_{GNS}$ s then s has an analytic proof in GNS5 from S. In particular: if $\vdash_{GNS5}$ s then s has an analytic proof in GNS5.*

Theorem 9.51 (weak completeness of NS5) *For finite Γ, $\Gamma \vdash_{NS5} \varphi$ iff φ semantically follows in **NS5** from Γ.*

Theorem 9.52 (Decidability of NS5) *NS5 is decidable.*

Proposition 9.53

1. **NS5** *is normal, with $\wedge, \vee$ and $\supset$ serving as its conjunction, disjunction, and implication (respectively).*

2. **NS5** is $\neg$-contained in classical logic.

3. The negation $\neg$ of **NS5** is complete, contrapositive, and left-involutive.

Proposition 9.54 **NS5** *is self-extensional (that is, it has the replacement property).*

Proposition 9.55 **NS5** *is strongly $\neg$-paraconsistent.*

Corollary 9.56 **NS5** *is a classical $\neg$-paraconsistent logic.*

Proposition 9.57 **NS5** *is a C-system in which all the schemas listed in Proposition 9.33 are valid, as well as* $[i_2]$ *and* $\circ\neg\varphi$.

Theorem 9.58 (Equivalence to S5) **NS5** *is equivalent to the famous modal logic* **S5** *(also known as* **KT5** *or* **KT45***).*

Note 9.59 The modal logic **S5** is the logic induced by the class of modal Kripke frames in which the accessibility relation is an equivalence relation. Theorem 9.58 follows from this characterization of **S5**. Note that the standard Hilbert-type system for **S5** is obtained from that of **B** by replacing the axiom (B) by the following axiom:

(5) $\Diamond\varphi \supset \Box\Diamond\varphi$.

It is worth noting also that in **NS5** $\Box\varphi$ (that is: $\sim\neg\varphi$) is equivalent to $\neg\neg\varphi$. (This can easily be shown by using the semantics, or by using $GNS5$.)

9.3.2 A Cut-free Hypersequential System for NS5

As pointed out in Note 9.45, the cut-elimination theorem fails for $GNS5$. However, by employing the more complicated data structure of hypersequents (Section 1.3.4), it is possible to provide an alternative Gentzen-type proof system for **NS5** that does not have this drawback. Such a system, $GNS5^h$, is presented in Figure 9.1 below. With the exceptions of the two standard external structural rules (External Contraction [EC] and External Weakening [EC]) and the two negation rules, all of the rules of $GNS5^h$ are just the obvious hypersequential versions of the rules of LK^+.[8] In contrast, the rule $[\neg \Rightarrow]$ of $GNS5^h$ is just a special case of the standard hypersequential extension of the corresponding classical rule (that is: from $G \mid \Gamma \Rightarrow \Delta, \varphi$ infer $G \mid \neg\varphi, \Gamma \Rightarrow \Delta$). On the other hand, the rule $[\Rightarrow \neg]$ of $GNS5^h$ is stronger than the standard hypersequential extension of the corresponding classical rule (that is: from $G \mid \varphi, \Gamma \Rightarrow \Delta$ infer $G \mid \Gamma \Rightarrow \Delta, \neg\varphi$), since the latter can be derived from the former with the aid of [IW] and [EC].

---

[8]It is not difficult to show that if '|' is interpreted as $\vee$, then the system obtained from $GNS5^h$ by deleting its two rules for $\neg$ is sound and complete for positive classical logic.

9.3 S5 as a Paraconsistent Logic

Axioms: $\varphi \Rightarrow \varphi$

Logical rules:

$$[\wedge \Rightarrow] \quad \frac{G|\Gamma, \varphi, \psi \Rightarrow \Delta}{G|\Gamma, \varphi \wedge \psi \Rightarrow \Delta} \qquad [\Rightarrow \wedge] \quad \frac{G|\Gamma \Rightarrow \Delta, \varphi \quad G|\Gamma \Rightarrow \Delta, \psi}{G|\Gamma \Rightarrow \Delta, \varphi \wedge \psi}$$

$$[\vee \Rightarrow] \quad \frac{G|\Gamma, \varphi \Rightarrow \Delta \quad G|\Gamma, \psi \Rightarrow \Delta}{G|\Gamma, \varphi \vee \psi \Rightarrow \Delta} \qquad [\Rightarrow \vee] \quad \frac{G|\Gamma \Rightarrow \Delta, \varphi, \psi}{G|\Gamma \Rightarrow \Delta, \varphi \vee \psi}$$

$$[\supset \Rightarrow] \quad \frac{G|\Gamma \Rightarrow \Delta, \varphi \quad G|\Gamma, \psi \Rightarrow \Delta}{G|\Gamma, \varphi \supset \psi \Rightarrow \Delta} \qquad [\Rightarrow \supset] \quad \frac{G|\Gamma, \varphi \Rightarrow \Delta, \psi}{G|\Gamma \Rightarrow \Delta, \varphi \supset \psi}$$

$$[\neg \Rightarrow] \quad \frac{G| \Rightarrow \varphi}{G|\neg \varphi \Rightarrow} \qquad [\Rightarrow \neg] \quad \frac{G|\Gamma, \varphi \Rightarrow \Delta}{G|\Gamma \Rightarrow \Delta| \Rightarrow \neg \varphi}$$

Structural rules:

$$[\text{Cut}] \quad \frac{G \mid \Gamma_1 \Rightarrow \Delta_1, \varphi \quad G \mid \varphi, \Gamma_2 \Rightarrow \Delta_2}{G \mid \Gamma_1, \Gamma_2 \Rightarrow \Delta_1, \Delta_2} \qquad [\text{IW}] \quad \frac{G \mid \Gamma \Rightarrow \Delta}{G \mid \Gamma', \Gamma \Rightarrow \Delta, \Delta'}$$

$$[\text{EC}] \quad \frac{G \mid s \mid s}{G \mid s} \qquad [\text{EW}] \quad \frac{G}{G \mid s}$$

Figure 9.1: The proof system $GNS5^h$

Example 9.60 In Note 9.45 it was observed that the sequent $\neg\neg P_1 \Rightarrow P_1$ has no cut-free proof in $GNS5$. Here is a cut-free proof of it in $GNS5^h$:

$$\frac{\dfrac{\dfrac{\dfrac{P_1 \Rightarrow P_1}{\Rightarrow P_1 \mid \Rightarrow \neg P_1} [\Rightarrow \neg]}{\Rightarrow P_1 \mid \neg\neg P_1 \Rightarrow} [\neg \Rightarrow]}{\neg\neg P_1 \Rightarrow P_1 \mid \neg\neg P_1 \Rightarrow P_1} [IW]}{\neg\neg P_1 \Rightarrow P_1} [EC]$$

(Note that two applications of [IW] have been done here in parallel.)

Now we turn to the semantics of $GNS5^h$.

Definition 9.61 (semantics of $GNS5^h$)

- A hypersequent G is *valid* in an **NS5**-frame $\langle W, \nu \rangle$ (See Note 9.47), or $\langle W, \nu \rangle$ is a *model* of G ($\langle W, \nu \rangle \models G$), if one of the components of G is valid in $\langle W, \nu \rangle$. (See definition 9.46.)

- Let $\mathcal{H} \cup \{G\}$ be a set of hypersequents in $\mathcal{L}_{CL}$. G semantically follows from $\mathcal{H}$ in **NS5** ($\mathcal{H} \vdash_{\mathbf{NS5}} G$) if every model of $\mathcal{H}$ is also a model of G.

Proposition 9.62 (strong soundness of $GNS5^h$**)** *Let* $\mathcal{H} \cup \{G\}$ *be a set of hypersequents. If* $\mathcal{H} \vdash_{GNS5^h} G$ *then* $\mathcal{H} \vdash_{\mathbf{NS5}} G$.

Proof. It is easy to see that the axioms of $GNS5^h$, its structural rules, and its logical rules for the positive connectives, all preserve validity of hypersequents in frames. (It is trivial with respect to the two external structural rules, and immediate from the third item of Proposition 9.48 for the rest.) Now we show that the same applies to the two negation rules.

- Suppose that $G \mid \Rightarrow \varphi$ is valid in $\langle W, \nu \rangle$. Then either one of the components of G is valid in $\langle W, \nu \rangle$, or $\Rightarrow \varphi$ is. The second case obtains iff φ is valid in $\langle W, \nu \rangle$, implying that $\neg \varphi \Rightarrow$ is valid in $\langle W, \nu \rangle$. Hence in both cases one of the components of $G \mid \neg \varphi \Rightarrow$ is valid in $\langle W, \nu \rangle$.

- Suppose that $G \mid \varphi, \Gamma \Rightarrow \Delta$ is valid in $\langle W, \nu \rangle$. Then either one of the components of G is valid in $\langle W, \nu \rangle$, or $\varphi, \Gamma \Rightarrow \Delta$ is. If the first case holds then obviously $G \mid \Gamma \Rightarrow \Delta \mid \Rightarrow \neg \varphi$ is valid in $\langle W, \nu \rangle$. So assume that $\varphi, \Gamma \Rightarrow \Delta$ is valid there. Then either $\nu(w, \varphi) = f$ for some $w \in W$, or $\Gamma \Rightarrow \Delta$ is valid in $\langle W, \nu \rangle$. In the first case $\Rightarrow \neg \varphi$ is valid in $\langle W, \nu \rangle$. Hence in either case $G \mid \Gamma \Rightarrow \Delta \mid \Rightarrow \neg \varphi$ is valid in $\langle W, \nu \rangle$. □

Now we turn to the strong completeness of $GNS5^h$ and to the cut-elimination theorem for it.

Notation. $\mathcal{H} \vdash^{cf}_{GNS5^h} G$ means that there is a proof in $GNS5^h$ of the hypersequent G from the set of hypersequents $\mathcal{H}$ in which each cut is on a formula φ such that $\varphi \in \Gamma \cup \Delta$ for some component $\Gamma \Rightarrow \Delta$ of some hypersequent in $\mathcal{H}$.

Proposition 9.63 (strong completeness of $GNS5^h$**)** *Let* $\mathcal{H} \cup \{G\}$ *be a finite set of hypersequents. If* $\mathcal{H} \vdash_{\mathbf{NS5}} G$ *then* $\mathcal{H} \vdash^{cf}_{GNS5^h} G$.

Proof. Suppose that $\mathcal{H} \nvdash^{cf}_{GNS5^h} G$. We construct a model of $\mathcal{H}$ which is not a model of G.

Let $\mathcal{F}$ be the set of subformulas of formulas in $\mathcal{H} \cup \{G\}$. We call a hypersequent G^* an $\mathcal{F}$-*hypersequent* if it has the following properties:

- Every formula which occurs in G^* belongs to $\mathcal{F}$.

- $\mathcal{H} \nvdash^{cf}_{GNS5^h} G^*$.

9.3 S5 as a Paraconsistent Logic

- If $\Gamma \cup \Delta \subseteq \mathcal{F}$ then either $\Gamma \Rightarrow \Delta \in G^*$, or $\mathcal{H} \vdash^{cf}_{GNS5^h} G^* \mid \Gamma \Rightarrow \Delta$.

Let $s_1, \ldots, s_n$ be an enumeration of all the sequents $\Gamma \Rightarrow \Delta$ such that $\Gamma \cup \Delta \subseteq \mathcal{F}$. ($n$ is finite because $\mathcal{H}$ is finite, and so $\mathcal{F}$ is finite.) Let $G_0 = G$. Define a sequence $G_1, \ldots, G_n$ of hypersequents by letting $G_i = G_{i-1} \mid s_i$ in case $\mathcal{H} \nvdash^{cf}_{GNS5^h} G_{i-1} \mid s_i$, and $G_i = G_{i-1}$ otherwise. Let $G^* = G_n$. Then G^* is an $\mathcal{F}$-hypersequent such that $G \subseteq G^*$. call a component $\Gamma^* \Rightarrow \Delta^*$ of G^* *maximal* if it has no proper extension in G^* (i.e. if $\Gamma' \Rightarrow \Delta' \in G^*$, $\Gamma^* \subseteq \Gamma'$ and $\Delta^* \subseteq \Delta'$, then $\Gamma^* = \Gamma'$ and $\Delta^* = \Delta'$). Let W be the set of all maximal components of G^*. Define a valuation $\nu : W \times \mathsf{Var} \to \{t, f\}$ by letting $\nu(\Gamma^* \Rightarrow \Delta^*, p) = t$ iff $p \in \Gamma^*$. We show by induction on the structure of formulas that the following hold for every $\varphi \in \mathcal{F}$ and every maximal component $\Gamma^* \Rightarrow \Delta^*$ of G^*:

1. If $\varphi \in \Gamma^*$ then $\nu(\Gamma^* \Rightarrow \Delta^*, \varphi) = t$.

2. If $\varphi \in \Delta^*$ then $\nu(\Gamma^* \Rightarrow \Delta^*, \varphi) = f$.

- The case where $\varphi \in \mathsf{Var}$ is immediate from the definition of ν, and the fact that if $\varphi \in \Delta^*$ then $\varphi \notin \Gamma^*$ (because $\mathcal{H} \nvdash^{cf}_{GNS5^h} G^*$).

- Suppose that $\varphi = \varphi_1 \wedge \varphi_2$.

    - Suppose $\varphi \in \Gamma^*$. Assume for contradiction that $\{\varphi_1, \varphi_2\} \nsubseteq \Gamma^*$. Then the maximality of $\Gamma^* \Rightarrow \Delta^*$ implies that $\varphi_1, \varphi_2, \Gamma^* \Rightarrow \Delta^*$ is not a component of G^*. Since G^* is an $\mathcal{F}$-hypersequent, it follows that $\mathcal{H} \vdash^{cf}_{GNS5^h} G^* \mid \varphi_1, \varphi_2, \Gamma^* \Rightarrow \Delta^*$. By applying $[\wedge \Rightarrow]$ to this hypersequent we get that $\mathcal{H} \vdash^{cf}_{GNS5^h} G^* \mid \Gamma^* \Rightarrow \Delta^*$, and so that $\mathcal{H} \vdash^{cf}_{GNS5^h} G^*$ (using [EC]). A contradiction. Hence $\{\varphi_1, \varphi_2\} \subseteq \Gamma^*$, and so $\nu(\Gamma^* \Rightarrow \Delta^*, \varphi_1) = \nu(\Gamma^* \Rightarrow \Delta^*, \varphi_2) = t$ by the induction hypothesis for φ_1, φ_2. Hence $\nu(\Gamma^* \Rightarrow \Delta^*, \varphi) = t$.

    - Suppose $\varphi \in \Delta^*$. Assume for contradiction that $\{\varphi_1, \varphi_2\} \cap \Delta^* = \emptyset$. Then the maximality of $\Gamma^* \Rightarrow \Delta^*$ and the fact that G^* is an $\mathcal{F}$-hypersequent imply that both $\mathcal{H} \vdash^{cf}_{GNS5^h} G^* \mid \Gamma^* \Rightarrow \Delta^*, \varphi_1$ and $\mathcal{H} \vdash^{cf}_{GNS5^h} G^* \mid \Gamma^* \Rightarrow \Delta^*, \varphi_2$. By applying $[\Rightarrow \wedge]$ to these two hypersequents (followed by [EC]) we get that $\mathcal{H} \vdash^{cf}_{GNS5^h} G^*$. A contradiction. Hence either $\varphi_1 \in \Delta^*$ or $\varphi_2 \in \Delta^*$. It follows by the induction hypothesis for φ_1 and φ_2 that either $\nu(\Gamma^* \Rightarrow \Delta^*, \varphi_1) = f$ or $\nu(\Gamma^* \Rightarrow \Delta^*, \varphi_2) = f$. In both cases $\nu(\Gamma^* \Rightarrow \Delta^*, \varphi) = f$.

- The cases $\varphi = \varphi_1 \vee \varphi_2$ and $\varphi = \varphi_1 \supset \varphi_2$ are similar to the case $\varphi = \varphi_1 \wedge \varphi_2$, and are left for the reader.

- Suppose $\varphi = \neg \psi$.

- Suppose $\varphi \in \Gamma^*$. Assume for contradiction that $\Rightarrow \psi \notin G^*$. Then $\mathcal{H} \vdash^{cf}_{GNS5^h} G^* \mid \Rightarrow \psi$, because G^* is an $\mathcal{F}$-hypersequent. By applying $[\neg \Rightarrow]$ to this hypersequent followed by internal weakenings, we get that $\mathcal{H} \vdash^{cf}_{GNS5^h} G^* \mid \Gamma^* \Rightarrow \Delta^*$, and so that $\mathcal{H} \vdash^{cf}_{GNS5^h} G^*$. A contradiction. Hence $\Rightarrow \psi \in G^*$. It follows that there is a maximal component $\Gamma^\# \Rightarrow \Delta^\#$ of G^* that extends it, i.e. $\psi \in \Delta^\#$. Therefore the induction hypothesis for ψ implies that $\nu(\Gamma^\# \Rightarrow \Delta^\#, \psi) = f$. Hence, $\nu(\Gamma^* \Rightarrow \Delta^*, \varphi) = t$.

- Suppose $\varphi \in \Delta^*$. Let $w = \Gamma^\# \Rightarrow \Delta^\# \in W$. We show that $\psi \in \Gamma^\#$. Assume otherwise. Then $\mathcal{H} \vdash^{cf}_{GNS5^h} G^* \mid \psi, \Gamma^\# \Rightarrow \Delta^\#$ (by the maximality of $\Gamma^\# \Rightarrow \Delta^\#$ and the fact that G^* is an $\mathcal{F}$-hypersequent). By applying $[\Rightarrow \neg]$ to $G^* \mid \psi, \Gamma^\# \Rightarrow \Delta^\#$, we get that $\mathcal{H} \vdash^{cf}_{GNS5^h} G^* \mid \Gamma^\# \Rightarrow \Delta^\# \mid \Rightarrow \varphi$, and so (using [EC]) $\mathcal{H} \vdash^{cf}_{GNS5^h} G^* \mid \Rightarrow \varphi$. Since $\varphi \in \Delta^*$, by applying [IW] to $G^* \mid \Rightarrow \varphi$ we get $\mathcal{H} \vdash^{cf}_{GNS5^h} G^* \mid \Gamma^* \Rightarrow \Delta^*$, and so $\mathcal{H} \vdash^{cf}_{GNS5^h} G^*$. A contradiction. It follows by the induction hypothesis for ψ that $\nu(w, \psi) = t$ for every $w \in W$. Hence $\nu(\Gamma^* \Rightarrow \Delta^*, \varphi) = f$.

Suppose now that $\Gamma \Rightarrow \Delta$ is some component of G. Since $G \subseteq G^*$, there is a maximal component $\Gamma^* \Rightarrow \Delta^*$ of G^* such that $\Gamma \subseteq \Gamma^*$ and $\Delta \subseteq \Delta^*$. Therefore, Properties 1 and 2 above imply that if $\varphi \in \Gamma$ then $\nu(\Gamma^* \Rightarrow \Delta^*, \varphi) = t$, while if $\varphi \in \Delta$ then $\nu(\Gamma^* \Rightarrow \Delta^*, \varphi) = f$. It follows that $\Gamma \Rightarrow \Delta$ is not true in the world $\Gamma^* \Rightarrow \Delta^*$ of W, and so $\Gamma \Rightarrow \Delta$ is not valid in $\langle W, \nu \rangle$. Hence $\langle W, \nu \rangle$ is not a model of G.

Finally, we prove that $\langle W, \nu \rangle$ is a model of $\mathcal{H}$. So let $G' \in \mathcal{H}$. It is impossible that every component of G' is a subsequent of some component of G^*, because otherwise G^* can be derived from G' (and so from $\mathcal{H}$) using just internal and external weakenings ([IW] and [EW]), and this contradicts the fact that $\mathcal{H} \not\vdash^{cf}_{GNS5^h} G^*$. Thus, there is a component $\Gamma \Rightarrow \Delta$ of G' which is not a subsequent of any component of G^*. We show that $\Gamma \Rightarrow \Delta$ is valid in $\langle W, \nu \rangle$. So let $\Gamma^* \Rightarrow \Delta^* \in W$. Then either $\Gamma \not\subseteq \Gamma^*$, or $\Delta \not\subseteq \Delta^*$. Assume e.g. the former. (The proof in the second case is similar). Then $\varphi \notin \Gamma^*$ for some $\varphi \in \Gamma$. Hence $\mathcal{H} \vdash^{cf}_{GNS5^h} G^* \mid \varphi, \Gamma^* \Rightarrow \Delta^*$ (because $\varphi \in \mathcal{F}$, $\Gamma^* \Rightarrow \Delta^*$ is maximal, and G^* is an $\mathcal{F}$-hypersequent). Assume for contradiction that $\varphi \notin \Delta^*$. Then $\mathcal{H} \vdash^{cf}_{GNS5^h} G^* \mid \Gamma^* \Rightarrow \Delta^*, \varphi$ as well. By applying a cut on φ to these two hypersequents, we get that $\mathcal{H} \vdash^{cf}_{GNS5^h} G^* \mid \Gamma^* \Rightarrow \Delta^*$, and so that $\mathcal{H} \vdash^{cf}_{GNS5^h} G^*$. A contradiction. It follows that $\varphi \in \Delta^*$, and so $\nu(\Gamma^* \Rightarrow \Delta^*, \varphi) = f$ by Property 2 above of maximal components of G^*. Since $\varphi \in \Gamma$, this means that $\Gamma \Rightarrow \Delta$ is true in the world $\Gamma^* \Rightarrow \Delta^*$. It follows that $\Gamma \Rightarrow \Delta$ is valid in $\langle W, \nu \rangle$, and so G' is valid $\langle W, \nu \rangle$. □

9.4 A Fully Maximal Extension of NS5

Theorem 9.64 (strong soundness and completeness) *Let $\mathcal{H} \cup \{G\}$ be a finite set of hypersequents. Then $\mathcal{H} \vdash_{\mathbf{NS5}} G$ iff $\mathcal{H} \vdash_{GNS5^h} G$.*

Proof. Immediate from Propositions 9.62 and 9.63. □

Note 9.65 Theorem 9.64 can be extended to the case in which $\mathcal{H}$ is an arbitrary set of hypersequents (not necessarily finite). We omit the proof.

Theorem 9.66 (cut-elimination) *$GNS5^h$ admits strong cut-elimination: If $\mathcal{H} \vdash_{GNS5^h} G$ then $\mathcal{H} \vdash_{GNS5^h}^{cf} G$. In particular: $I_*^f \vdash_{GNS5^h} G$ then G has a cut-free proof in $GNS5^h$.*

Proof. Immediate from Propositions 9.62 and 9.63. □

9.4 A Fully Maximal Extension of NS5

The modal logic **S5** is known to have the Scroggs' property ([264]): It has no finite-valued *weakly* characteristic matrix, but every proper extension of it does. The sets of theorems of those extensions form a sequence $\{\mathbf{S5}_n\}_{n=1}^{\infty}$, which is linearly ordered by inclusion. For each n, $\mathbf{S5}_n$ is the logic which is induced by the **S5**-frames $\langle W, \nu \rangle$ (see Note 9.47) in which W has exactly n elements. In turn, the use of such frames is equivalent to the use of a certain matrix with 2^n elements (corresponding to the set of subsets of W). We omit the general details, and concentrate on $\mathbf{S5}_2$, or rather on the equivalent logic $\mathbf{NS5}_2$, which is the maximal extension of **NS5** which is still paraconsistent. ($\mathbf{NS5}_1$, the counterpart of $\mathbf{S5}_1$, is just **CL**.) Like in the cases of **NB** and **NS5**, we start with a corresponding Hilbert-type representation, followed by a useful corresponding Gentzen-type system.

Definition 9.67 ($\mathbf{NS5}_2$) Let $HNS5_2$ be the Hilbert-type system which is obtained from HNB by adding to it as an axiom the following Schema:

[5$_2$] $\varphi \wedge \psi \wedge \neg\varphi \wedge \neg\psi \supset \neg(\varphi \vee \psi)$

$\mathbf{NS5}_2$ is the logic induced by $HNS5_2$.

Proposition 9.68 *$\mathbf{NS5}_2$ is an extension of* **NS5**.

Proof. By substituting $\neg\varphi$ for ψ in [5$_2$], and using $\mathbf{CL}^+$ and axiom [c], we get that $\vdash_{HNS5_2} \neg\neg\varphi \supset (\neg\varphi \supset \neg(\varphi \vee \neg\varphi))$. Since $\vdash_{\mathbf{NB}} \varphi \vee \neg\varphi$, this, Item 1 of Lemma 9.8, and the fact that $\supset$ is implication for every axiomatic extension of **NB**, together imply that $\vdash_{\mathbf{NS5}_2} \neg\neg\varphi \supset (\neg\varphi \supset \psi)$ for every φ, ψ. It follows by definition of **NS5** that $\mathbf{NS5}_2$ is extension of it. □

The following Lemma generalizes [5$_2$] to arbitrary numbers of formulas.

Lemma 9.69 *Let Σ be a non-empty finite set of formulas. Then $\vdash_{HNS5_2}$ $(\bigwedge \Sigma \wedge \bigwedge \neg \Sigma) \supset \neg(\bigvee \Sigma)$.*

Proof. The proof is by an induction on the cardinality n of Σ. The case $n = 1$ is trivial, while the case $n = 2$ is given by the axiom ([5$_2$]) of $HNS5_2$. For the induction step, suppose that the claim holds for every Σ such that $|\Sigma| = n \geq 1$. Let $|\Sigma| = n + 1$. Then $\Sigma = \Sigma' \cup \{\psi\}$ for some Σ' such that $|\Sigma'| = n$ and ψ. Then the induction hypothesis implies that $\vdash_{HNS5_2}$ $(\bigwedge \Sigma' \wedge \bigwedge \neg \Sigma') \supset \neg(\bigvee \Sigma')$. Since $n \geq 1$, also $\vdash_{HNS5_2} (\bigwedge \Sigma' \wedge \bigwedge \neg \Sigma') \supset \bigvee \Sigma'$. It follows that $\vdash_{HNS5_2} (\bigwedge \Sigma \wedge \bigwedge \neg \Sigma) \supset (\bigvee \Sigma' \wedge \neg(\bigvee \Sigma') \wedge \psi \wedge \neg \psi)$. Hence by using ([5$_2$]) with $\varphi = \bigvee \Sigma'$ we get that $\vdash_{HNS5_2} (\bigwedge \Sigma \wedge \bigwedge \neg \Sigma) \supset \neg(\bigvee \Sigma)$. $\square$

Definition 9.70 ($GNS5_2$) The system $GNS5_2$ is the system which is obtained from LK (Figures 1.2, 1.3) by replacing its rule $[\neg \Rightarrow]$ by the rule:

$$[\neg \Rightarrow]_{5_2} \quad \frac{\neg \Gamma \Rightarrow \Sigma, \neg \Delta}{\neg \Gamma, \neg \Sigma, \Sigma \Rightarrow \neg \Delta}.$$

Note 9.71 In the special case in which $\Sigma = \{\psi\}$, the rule $[\neg \Rightarrow]_{5_2}$ allows to infer $\neg \Gamma, \neg \psi, \psi \Rightarrow \neg \Delta$ from the premise $\neg \Gamma \Rightarrow \psi, \neg \Delta$. A cut on ψ of the conclusion of this instance of the rule with its premise yields $\neg \Gamma, \neg \psi \Rightarrow \neg \Delta$. It follows that the rule $[\neg \Rightarrow]_5$ of $GNS5$ is derivable in $GNS5_2$.

Theorem 9.72 (Adequacy of $GNS5_2$) $\mathcal{T} \vdash_{GNS5_2} \varphi$ *iff* $\mathcal{T} \vdash_{NS5_2} \varphi$.

Proof. The proof that if $\mathcal{T} \vdash_{NS5_2} \varphi$ then $\mathcal{T} \vdash_{GNS5_2} \varphi$ is similar to the proof of the corresponding claim concerning **NS5** and $GNS5$ (Theorem 9.44). We only have to provide a derivation of $\Rightarrow \varphi \wedge \psi \wedge \neg \varphi \wedge \neg \psi \supset \neg(\varphi \vee \psi)$ in $GNS5_2$. Now $\Rightarrow \neg(\varphi \vee \psi), \varphi, \psi$ is easily derivable in $GNS5_2$ using $[\vee \Rightarrow]$ and $[\Rightarrow \neg]$. By applying $[\neg \Rightarrow]_{5_2}$ to this sequent we get $\varphi, \psi, \neg \varphi, \neg \psi \Rightarrow \neg(\varphi \vee \psi)$. From that the desired sequent easily follows using standard rules of LK^+.

For the converse it again suffices to show that if $\vdash_{GNS5_2} s$, then $I(s)$ is a theorem of $HNS5_2$. The proof of this is almost identical to that of Theorem 9.44. The only change we have to make in the inductive proof given there is in the treatment of the rule $[\neg \Rightarrow]_{5_2}$. So, suppose that $s = \neg \Gamma, \neg \Sigma, \Sigma \Rightarrow \neg \Delta$ is derived from $s_1 = \neg \Gamma \Rightarrow \Sigma, \neg \Delta$. Then, the induction hypothesis concerning s_1 implies that $\vdash_{HNS5_2} (\bigwedge \neg \Gamma) \supset (\bigvee \Sigma \vee \bigvee \neg \Delta)$. Like in the proof of Theorem 9.44, it follows from this (using Proposition 9.68) that $\vdash_{HNS5_2} (\neg \bigvee \Sigma \wedge \bigwedge \neg \Gamma) \supset (\bigvee \neg \Delta)$. Hence Lemma 9.69 implies (using **CL**$^+$) that $\vdash_{HNS5_2} I(s)$. $\square$

Note 9.73 Again, it is easy to see that although $\neg \neg \varphi \Rightarrow \varphi$ is derivable in $GNS5_2$, it cannot have a cut-free proof there in case φ is a propositional variable. Hence like in the case of $GNS5$, the cut-elimination theorem fails for $GNS5_2$. Again, we shall soon see that $GNS5_2$ is analytic nevertheless.

9.4 A Fully Maximal Extension of NS5

Now we turn to the semantics of **NS5**$_2$.

Definition 9.74 (NS5$_2$-frames)

- An **NS5$_2$**-*frame* (for $\mathcal{L}_{CL}$) is an **NS5**-frame in which there are at most two worlds.

- The notions of truth (in worlds) and validity in **NS5$_2$**-frames (of formulas and sequents) are defined like in Definition 9.16.

- Semantic consequence in **NS5$_2$** is defined like in Definition 9.17, using **NS5$_2$**-frames instead of **NB**-frames.

Proposition 9.75 (strong soundness of NS5$_2$)

1. If φ is a theorem of **NS5$_2$** then φ is valid in every **NS5$_2$**-frame.

2. If $\mathcal{T} \vdash_{\mathbf{NS5}_2} \varphi$ then φ semantically follows in **NS5$_2$** from $\mathcal{T}$.

3. Let $S \cup s$ be a set of sequents. If $S \vdash_{GNS5_2} s$ then s semantically follows in **NS5$_2$** from S. In particular: if $\vdash_{GNS5_2} s$ then s is **NS5$_2$**-valid.

Proof. The proof is similar to that of Proposition 9.48. We only have to add to it a proof that the new axiom [5$_2$] is valid in every **NS5$_2$**-frame. So let $\langle \{w_0, w_1\}, \nu \rangle$ be such a frame (see Note 9.47). We show e.g. that $w_0 \Vdash (\mathbf{5}_2)$. So suppose that $w_0 \Vdash \varphi \wedge \psi \wedge \neg\varphi \wedge \neg\psi$. Then $w_0 \Vdash \varphi$, and $w_0 \Vdash \neg\varphi$. The latter fact implies that either $w_0 \nVdash \varphi$, or $w_1 \nVdash \varphi$. Hence the former fact implies that $w_1 \nVdash \varphi$. That $w_1 \nVdash \psi$ is proved similarly. It follows that $w_1 \nVdash \varphi \vee \psi$, and so $w_0 \Vdash \neg(\varphi \vee \psi)$. □

Theorem 9.76 *Let $S \cup s$ be a finite set of sequents in $\mathcal{L}_{CL}$. If s semantically follows in **NS5$_2$** from S, then s has an analytic proof in $GNS5_2$ from S.*

Proof. Again we follow the proof of Theorem 9.20. We define $\mathcal{F}$, $\mathcal{F}$-maximality, and $\Gamma^* \Rightarrow \Delta^*$, and prove the corresponding version of Lemma 1 exactly as it is done there (but this time relative to $GNS5_2$). Now let $\Gamma'_1 = \{\varphi \mid \neg\varphi \in \Gamma^*, \varphi \in \Gamma^*\}$, $\Gamma'_2 = \{\varphi \mid \neg\varphi \in \Gamma^*, \varphi \in \Delta^*\}$, and $\Delta' = \{\varphi \mid \neg\varphi \in \Delta^*\}$. Obviously, Γ'_1, Γ'_2 and Δ' are pairwise disjoint, their union is the set $\{\psi \mid \neg\psi \in \mathcal{F}\}$, and $\Gamma^* \Rightarrow \Delta^*$ is an extension of $\neg\Gamma'_2, \neg\Gamma'_1, \Gamma'_1 \Rightarrow \neg\Delta'$. Hence the $\mathcal{F}$-maximality of $\Gamma^* \Rightarrow \Delta^*$ implies that $\neg\Gamma'_2, \neg\Gamma'_1 \Rightarrow \Gamma'_1, \neg\Delta'$ has no $\mathcal{F}$-analytic proof from S (because otherwise an application of $[\neg\Rightarrow]_{5_2}$ to this sequent, followed by weakenings, would provide an $\mathcal{F}$-analytic proof of $\Gamma^* \Rightarrow \Delta^*$ from S). It follows by the above-mentioned Lemma 1 that $\neg\Gamma'_2, \neg\Gamma'_1 \Rightarrow \Gamma'_1, \neg\Delta'$ can be extended to an $\mathcal{F}$-maximal sequent $\Gamma^\# \Rightarrow \Delta^\#$. Let $w^* = \Gamma^* \Rightarrow \Delta^*$, $w^\# = \Gamma^\# \Rightarrow \Delta^\#$, $W = \{w^*, w^\#\}$, and define ν as in the proof of Theorem 9.20 (that is: $\nu(\Gamma \Rightarrow \Delta, p) = t$ iff $p \in \Gamma$, in case p is a propositional variable). We show that Lemma 2 from the proof of Theorem 9.20 (and Theorem 9.49) holds here as well. The proof is again by induction on the complexity of φ, and the only case which is different from the proof of Theorem 9.49 is when $\varphi = \neg\psi$, and $\varphi \in \Gamma^*$. In this case $\varphi \in \Gamma^\#$ as well (because φ begins with $\neg$), and either $\psi \in \Gamma'_1$, or $\psi \in \Gamma'_2$. In the latter case $\psi \in \Delta^*$, and so $\nu(w^*, \psi) = f$ by the induction hypothesis. In the former case $\psi \in \Delta^\#$, and so $\nu(w^\#, \psi) = f$. Therefore in both cases $\nu(w^*, \varphi) = \nu(w^\#, \varphi) = t$.

Again, once we have Lemma 2, the rest of the proof is identical to what is done in the proof of Theorem 9.20. □

Theorem 9.76 has for **NS5$_2$** the same important corollaries as Theorem 9.49 has for **NS5**. Again, we just list them below, leaving their proofs (which are almost identical to their counterparts in the cases of **NB** and **NS5**) to the reader.

Theorem 9.77 (analyticity of $GNS5_2$**)** $GNS5_2$ *has the subformula property: if* $S \vdash_{GNS5_2} s$ *then* s *has an analytic proof in* $GNS5_2$ *from* S. *In particular: if* $\vdash_{GNS5_2} s$ *then* s *has an analytic proof in* $GNS5_2$.

Theorem 9.78 (weak completeness of NS5$_2$) *For finite* Γ, $\Gamma \vdash_{\mathbf{NS5_2}} \varphi$ *iff* φ *semantically follows in* **NS5$_2$** *from* Γ.

Note 9.79 Like in the case of **NB** and **NS5**, here too Theorem 9.78 is true even if Γ is infinite.

Theorem 9.80 (Decidability of NS5$_2$) **NS5$_2$** *is decidable.*

Proposition 9.81

1. **NS5$_2$** *is normal, with* $\wedge, \vee$ *and* $\supset$ *serving as its conjunction, disjunction, and implication (respectively).*

2. **NS5$_2$** *is* $\neg$*-contained in classical logic.*

3. *The negation* $\neg$ *of* **NS5$_2$** *is complete, contrapositive, and left-involutive.*

Proposition 9.82 **NS5$_2$** *is self-extensional (that is, it has the replacement property).*

Proposition 9.83 **NS5$_2$** *is strongly* $\neg$*-paraconsistent.*

Corollary 9.84 **NS5$_2$** *is a classical* $\neg$*-paraconsistent logic.*

9.4 A Fully Maximal Extension of NS5

Proposition 9.85 $\mathbf{NS5_2}$ *is a C-system in which all the schemas listed in Proposition 9.33 are valid, as well as* $[i_2]$ *and* $\circ\neg\varphi$.

For the main results of this section, it would be useful to turn the possible worlds semantics of $\mathbf{NS5_2}$ into a weakly characteristic finite-valued matrix. It is well-known (see [264]) that this can actually be done for all axiomatic proper extensions of $\mathbf{S5}$, and so also all axiomatic proper extensions of the equivalent $\mathbf{NS5}$. The idea is that any finite $\mathbf{S5}$-frame $\langle W, \nu \rangle$ induces a finite matrix $\langle \mathcal{V}, \mathcal{D}, \mathcal{O} \rangle$ and a valuation ν^* in it as follows: $\mathcal{V} = 2^W$ (where 2^W is the powerset of W), $\mathcal{D} = \{W\}$, $\tilde{\wedge} = \cap$, $\tilde{\vee} = \cup$, $\tilde{\supset}(A, B) = (W - A) \cup B$, $\tilde{\neg}(A)$ is $\emptyset$ if $A = W$ and W otherwise, and $\nu^*(\varphi) = \{w \in W \mid w \Vdash \varphi\}$. Now in the case of $\mathbf{NS5_2}$ we may assume that $W = \{0, 1\}$. Denoting $t = \{0, 1\}$, $f = \emptyset$, $I_0 = \{0\}$, and $I_1 = \{1\}$, we are led to the following matrix:

Definition 9.86 ($\mathcal{MS}5_2$) $\mathcal{MS}5_2 = \langle \{t, f, I_0, I_1\}, \{t\}, \mathcal{O} \rangle$ *is the matrix for* $\mathcal{L}_{CL}$ *in which the interpretations of* $\neg, \vee, \wedge$, *and the defined connective* $\sim$ ($\sim\varphi =_{Df} \varphi \supset \neg(\neg\varphi \vee \varphi)$) *are given by the tables below, and* $a\tilde{\supset}b = \tilde{\sim}a\tilde{\vee}b$.

$\tilde{\vee}$	t	f	I_0	I_1
t	t	t	t	t
f	t	f	I_0	I_1
I_0	t	I_0	I_0	t
I_1	t	I_1	t	I_1

$\tilde{\wedge}$	t	f	I_0	I_1
t	t	f	I_0	I_1
f	f	f	f	f
I_0	I_0	f	I_0	f
I_1	I_1	f	f	I_1

$\tilde{\neg}$	
t	f
f	t
I_0	t
I_1	t

$\tilde{\sim}$	
t	f
f	t
I_0	I_1
I_1	I_0

Note 9.87 Let $\leq_t$ be the partial order on $\{t, f, I_0, I_1\}$ in which t is the maximal element, f is the minimal one, and I_0, I_1 are (incomparable) intermediate elements. (In view of what the names represent, $\leq_t$ is here just the $\subseteq$ relation on the subsets of $\{0, 1\}$.) Then the conjunction and disjunction are interpreted here, respectively, as the meet and the join of $\leq_t$, $\langle \{t, f, I_0, I_1\}, \leq_t \rangle$ is a distributive lattice (which is isomorphic to the lattice $\langle FOUR, \leq_t \rangle$ used in Chapter 5), and $\tilde{\sim}$ is one of the two possible involutions on this lattice. (The other one is the interpretation of $\neg$ used in $\mathcal{FOUR}$ and in the other four-valued matrices considered in Chapter 5.)

Theorem 9.88 $\mathcal{MS}5_2$ *is weakly characteristic for* $\mathbf{NS5_2}$, *that is:* $\vdash_{\mathbf{NS5_2}} \varphi$ *iff* φ *is valid in* $\mathcal{MS}5_2$.

Proof. It is straightforward to directly verify that $\mathbf{NS5_2}$ is sound for $\mathcal{MS}5_2$. Alternatively, let $\vdash_{\mathbf{NS5_2}} \varphi$, and suppose that ν is a valuation in $\mathcal{MS}5_2$. Let $\langle \{0, 1\}, \nu^* \rangle$ be the $\mathbf{NS5_2}$-frame in which for $w \in \{0, 1\}$ and p a propositional variable, $\nu^*(w, p) = t$ iff $I_w \leq_t \nu(p)$ (see Note 9.87). Since $\vdash_{\mathbf{NS5_2}} \varphi$, Theorem 9.78 implies that $\nu^*(0, \varphi) = \nu^*(1, \varphi) = t$. Now it is easy to check that for every formula ψ and $w \in \{0,1\}$, $\nu^*(w, \psi) = t$ iff $I_w \leq_t \nu(\psi)$. It follows that $I_0 \leq_t \nu(\varphi)$, and $I_1 \leq_t \nu(\varphi)$. Hence $\nu(\varphi) = t$.

For the converse, assume that φ is valid in $\mathcal{MS}5_2$. To show that $\vdash_{\mathbf{NS5}_2} \varphi$, it suffices by Theorem 9.78 to prove that φ is valid in any $\mathbf{NS5}_2$-frame. Let $\langle W, \nu \rangle$ be such a frame. Without a loss in generality, we may assume that $W = \{0, 1\}$. Define a valuation $\nu^{\#}$ in $\mathcal{MS}5_2$ as follows:

$$\nu^{\#}(\psi) = \begin{cases} t & \text{if } \nu(0, \psi) = t \text{ and } \nu(1, \psi) = t, \\ I_0 & \text{if } \nu(0, \psi) = t \text{ and } \nu(1, \psi) = f, \\ I_1 & \text{if } \nu(0, \psi) = f \text{ and } \nu(1, \psi) = t, \\ f & \text{if } \nu(0, \psi) = f \text{ and } \nu(1, \psi) = f. \end{cases}$$

It is not difficult to check that $\nu^{\#}$ is indeed an $\mathcal{MS}5_2$-valuation. Since φ is valid in $\mathcal{MS}5_2$, it follows that $\nu^{\#}(\varphi) = t$. By definition of $\nu^{\#}$, this means that $\nu(0, \varphi) = t$ and $\nu(1, \varphi) = t$. Hence φ is valid in $\langle W, \nu \rangle$. □

Note 9.89 It should be emphasized that the last theorem is only about *weak* completeness. This is why it does not contradict Corollary 3.57. In fact, $\mathcal{MS}5_2$ is *not* a paraconsistent matrix, and the (explosive!) logic it induces is different from $\mathbf{NS5}_2$. On the other hand, it follows from the first item of Proposition 9.81, Theorem 3.17, and the first item of Note 1.25, that $\mathcal{MS}5_2$ can still be used for characterizing the consequence relation of $\mathbf{NS5}_2$, using the method described in Note 1.25. (Footnote 6 and the last paragraph of Chapter 3 are relevant here.).

The next proposition is what makes $\mathbf{NS5}_2$ important:

Proposition 9.90 $\mathbf{NS5}_2$ *is a strongly maximal paraconsistent logic.*

For the proof of Proposition 9.90 we need the following definition and two lemmas.

Definition 9.91

- The operation $-$ on $\{t, f, I_0, I_1\}$ is defined by: $-t = t$, $-f = f$, $-I_0 = I_1$, $-I_1 = I_0$.

- The operation $-$ on valuations in $\mathcal{MS}5_2$ is defined by: $(-\nu)(p) = -\nu(p)$ (p is a propositional variable).

Lemma 9.92 *Let ν be a valuation in $\mathcal{MS}5_2$. Then $(-\nu)(\varphi) = -\nu(\varphi)$ for every formula φ.*

9.4 A Fully Maximal Extension of NS5

Proof. Induction on the structure of φ. The base case follows from the definition of $-\nu$. For the induction steps it obviously suffices to verify that $-$ commutes with the interpretations of the connectives (for example: that $-(a\tilde{\wedge}b) = -a\tilde{\wedge} - b$ for every $a, b \in \{t, f, I_0, I_1\}$). This is easy. □

Lemma 9.93 *Let ν be a valuation in $\mathcal{MS}5_2$ such that $\nu(P_1) \in \{I_0, I_1\}$. Then $\nu(\varphi[\sim P_1/P_1]) = (-\nu)(\varphi)$ for every φ such that $Fv(\varphi) = \{P_1\}$.*

Proof. Similar to the proof of Lemma 9.92, except that in the base case (that is: the case where $\varphi = P_1$) we use the fact that $\tilde{\sim}I_0 = -I_0$ and $\tilde{\sim}I_1 = -I_1$. □

Proof of Proposition 9.90. Let $\mathbf{L}$ be a proper extension of $\mathbf{NS5}_2$. Suppose that $\varphi_1, \ldots, \varphi_n \vdash_{\mathbf{L}} \sigma$, while $\varphi_1, \ldots, \varphi_n \nvdash_{\mathbf{NS5}_2} \sigma$. Then by the definition of $\mathbf{NS5}_2$ and Theorem 9.88, there is a valuation ν_0 in $\mathcal{MS}5_2$ such that $\nu_0(\varphi_1 \wedge \ldots \wedge \varphi_n \supset \sigma) \neq t$. Define a substitution θ as follows:

$$\theta(p) = \begin{cases} P_1 \supset P_1 & \nu_0(p) = t, \\ \neg(P_1 \supset P_1) & \nu_0(p) = f, \\ P_1 & \nu_0(p) = I_0, \\ \sim P_1 & \nu_0(p) = I_1. \end{cases}$$

Let ν be some valuation such that $\nu(P_1) = I_0$. It is easy to show by induction on the structure of τ that $\nu(\theta(\tau)) = \nu_0(\tau)$ for every formula τ. It follows that if $\varphi = \theta(\varphi_1 \wedge \ldots \wedge \varphi_n)$ (where we take $\varphi_1 \wedge \ldots \wedge \varphi_n$ to be $P_1 \supset P_1$ in case $n = 0$), and $\psi = \theta(\sigma)$, then $Fv(\varphi) = Fv(\psi) = \{P_1\}$; $\varphi \vdash_{\mathbf{L}} \psi$; and $\nu_0(\varphi \supset \psi) \neq t$. Next we consider all the cases which the last fact allows.

1. Let $\nu_0(\varphi) \in \{t, I_0\}$ and $\nu_0(\psi) \in \{f, I_1\}$. We show that this implies:

   (A) $\vdash_{\mathbf{NS5}_2} (P_1 \wedge \neg P_1) \supset \varphi$.
   (B) $\vdash_{\mathbf{NS5}_2} (P_1 \wedge \neg P_1 \wedge \psi) \supset \mathsf{F}$.

   Denote the formulas in (A) and (B) τ_A and τ_B (respectively). To prove (A) and (B) it suffices by Theorem 9.88 to show that τ_A and τ_B are valid in $\mathcal{MS}5_2$. So let ν be some valuation there. If $\nu(P_1) \in \{t, f\}$ then obviously $\nu(\tau_A) = \nu(\tau_B) = t$. If $\nu(P_1) = I_0 = \nu_0(P_1)$ then $\nu(\varphi) = \nu_0(\varphi) \in \{t, I_0\}$, while $\nu(\psi) = \nu_0(\psi) \in \{f, I_1\}$. The former fact implies that $\nu(\tau_A) = t$, while the latter implies that $\nu(P_1 \wedge \neg P_1 \wedge \psi) = f$, and so $= \nu(\tau_B) = t$. Finally, assume that $\nu(P_1) = I_1 = -\nu_0(P_1)$. Then Lemma 9.92 implies that in this case $\nu(\varphi) = -\nu_0(\varphi) \in \{t, I_1\}$, while $\nu(\psi) = -\nu_0(\psi) \in \{f, I_0\}$. These facts again easily imply that $\nu(\tau_A) = \nu(\tau_B) = t$.

Now, (A) implies that $P_1, \neg P_1 \vdash_{\mathbf{L}} \varphi$. Since $\varphi \vdash_{\mathbf{L}} \psi$, we get that $P_1, \neg P_1 \vdash_{\mathbf{L}} \psi$. But (B) implies that $\psi, P_1, \neg P_1 \vdash_{\mathbf{L}} \mathsf{F}$. It follows that $P_1, \neg P_1 \vdash_{\mathbf{L}} \mathsf{F}$, and so $\mathbf{L}$ is not paraconsistent.

2. Assume $\nu_0(\varphi) \in \{t, I_1\}$ and $\nu_0(\psi) = I_0$, or $\nu_0(\varphi) = I_1$ and $\nu_0(\psi) = f$. Since $Fv(\varphi) = Fv(\psi) = \{P_1\}$, this implies that $\nu(P_1) \in \{I_0, I_1\}$. Let $\varphi' = \varphi[\sim P_1/P_1]$ and $\psi' = \psi[\sim P_1/P_1]$. Then $Fv(\varphi') = Fv(\psi') = \{P_1\}$. Hence Lemma 9.93 implies that $\nu_0(\varphi') = -\nu_0(\varphi) \in \{t, I_0\}$ while $\nu_0(\psi') = -\nu_0(\psi) \in \{f, I_1\}$. Since $\vdash_{\mathbf{L}}$ is structural, we also have that $\varphi' \vdash_{\mathbf{L}} \psi'$. It follows that φ' and ψ' satisfy all the conditions that φ and ψ satisfy in case 1. Hence case 2 follows from case 1. □

Corollary 9.94 $\mathbf{NS5}_2$ *is maximally paraconsistent relative to* **CL**.

Proof. $\mathbf{NS5}_2$ is ¬-contained in classical logic and paraconsistent. Let σ be a classical tautology (in $\mathcal{L}_{CL}$) such that $\nvdash_{\mathbf{NS5}_2} \sigma$. By Proposition 9.90, the logic $\mathbf{L}$ which is obtained by adding φ to $\mathbf{L}$ as a new axiom is not paraconsistent. Hence $\neg\varphi, \varphi \vdash_{\mathbf{L}} \psi$ for every φ, ψ. Since $\supset$ is an implication for $\mathbf{NS5}_2$, and $\mathbf{L}$ is a finitary axiomatic extension of $\mathbf{NS5}_2$, $\supset$ is an implication for $\mathbf{L}$ too (Proposition 1.31). It follows that (every instance of) the axiom $[\neg \supset]$ of *HCL* (Figure 1.1) is provable in $\mathbf{L}$. Since all other axioms of *HCL* are provable already in $\mathbf{NS5}_2$, and *MP* is valid in $\mathbf{L}$, it follows that $\vdash_{HCL} \subseteq \vdash_{\mathbf{L}}$. Obviously $\vdash_{\mathbf{L}} \subseteq \vdash_{\mathbf{CL}}$ too. Therefore Theorem 1.90 entails that $\mathbf{L} = \mathbf{CL}$. □

Theorem 9.95 (full maximality) $\mathbf{NS5}_2$ *is a fully maximal paraconsistent logic*.

Proof. Immediate from Proposition 9.90 and Corollary 9.94. □

9.5 Using Other Standard Modal Logics

In the previous sections two famous modal logics, **B** and **S5**, were turned into strongly paraconsistent C-systems by letting $\neg\varphi =_{def} \sim \Box\varphi$ (where $\sim$ is the classical negation, that is $\sim \varphi =_{def} \varphi \supset \mathsf{F}$). Obviously, the same procedure can be applied to any other logic in the family known as 'normal modal logics'.[9] This family includes every logic in the language $\{\wedge, \vee, \supset, \mathsf{F}, \Box\}$ (or $\{\wedge, \vee, \supset, \sim, \Box\}$) which extends $\mathbf{CL}^+$, the necessitation rule [NEC] (if $\vdash \varphi$ then $\vdash \Box\varphi$) is valid in it, and every instance of the schema (K) (Section 9.2.5) is a theorem of it. For the sake of readers who have some

---

[9]Note again that the notion of 'normal modal logic' used in the literature on modal logics has nothing to do with the notion of 'normal logic' we use throughout this book, even though every 'normal modal logic' is normal also according to Definition 1.32.

9.5 Using Other Standard Modal Logics

knowledge of modal logics, in this section we briefly describe the C-systems which are obtained from the other three most famous normal modal logics.[10] However, from the paraconsistency point of view the logics considered in this section seem to be less interesting than **B** and **S5** investigated above.

9.5.1 K as a Paraconsistent Logic

The minimal normal modal logic is called **K**. Let **NK** be the corresponding logic in $\mathcal{L}_{CL}$. Now the most standard axiomatization of **K** is obtained from HCL (with $\sim$ replacing $\neg$) or HCL^F by the addition of [NEC] as a rule of proof, and of (K) as an axiom-schema. A well known alternative formulation (whose equivalence with the standard one can easily be proved), better suited for our needs, is obtained from HCL^F by adding to it the following rule and axioms:

- [MO] $\dfrac{\vdash \varphi \supset \psi}{\vdash \Box \varphi \supset \Box \psi}$

- $(K_\wedge)$ $\Box \varphi \wedge \Box \psi \supset \Box(\varphi \wedge \psi)$

- (NC) $\Box(\varphi \supset \varphi)$

Obviously, the rule [MO] is equivalent (above $\mathbf{CL^F}$) to the rule [CP] used in our axiomatizations of **NB** and **NS5**. The axioms $(K_\wedge)$ and (NC), in turn, are respectively equivalent to the following axioms:

- $(K_\neg)$ $\neg(\varphi \wedge \psi) \supset (\neg\varphi \vee \neg\psi)$

- $(NC_\neg)$ $\neg(\varphi \supset \varphi) \supset \varphi$

To see that $(K_\wedge)$ is indeed equivalent to $(K_\neg)$ is trivial. To show the second equivalence, note that $\Box(\varphi \supset \varphi)$ is obviously equivalent to $\neg(\varphi \supset \varphi) \supset \mathsf{F}$. Hence (NC) implies $(NC_\neg)$. For the converse, apply [CP] to the classical tautology $(\psi \supset \psi) \supset (\varphi \supset \varphi)$ and get $\neg(\varphi \supset \varphi) \supset \neg(\psi \supset \psi)$. Using $(NC_\neg)$, for every ψ this implies $\neg(\varphi \supset \varphi) \supset \psi$, and in particular $\neg(\varphi \supset \varphi) \supset \mathsf{F}$.

The above discussion leads to the following proposition.

Proposition 9.96 *A Hilbert-type system HNK in the language $\mathcal{L}_{CL}$ which induces the logic* **NK** *is obtained from HCL^+ by the addition of* [CP] *as a rule of proof, and the axioms $(K_\neg)$ and $(NC_\neg)$.*

---
[10] For more information, as well as for treating other modal logics, see [207].

Proof. A careful check of the above discussion reveals that for every φ and ψ, $\neg(\varphi \supset \varphi) \supset \psi$ is derivable in the system which is obtained from HCL^+ by the addition of [CP] as a rule of proof, and the axiom $(NC_\neg)$. Hence like in the case of HNB (see Note 9.9), we may assume that the language of HNK is an extension of $\mathcal{L}_{CL}^F$, and that every instance of a classical tautology in $\mathcal{L}_{CL}^F$ is provable in HNK. Therefore the discussion that precedes this proposition shows that HNK induces **K**. □

Finding a corresponding Gentzen-type system for **K** is also easy. All we have to do is to translate the standard one (see [297]) to the language $\mathcal{L}_{CL}$, using the simple connections between □ and ¬. Now GK, the standard system for **K**, has just one rule that involves □: from $\Gamma \Rightarrow \varphi$ infer $\Box\Gamma \Rightarrow \Box\varphi$. The translation of this rule into $\mathcal{L}_{CL}$ is:

$$[\neg \Rightarrow]_K \quad \frac{\Gamma \Rightarrow \psi}{\neg\psi \Rightarrow \neg\Gamma}.$$

It easily follows that by adding $[\neg\Rightarrow]_K$ To LK^+ we get a system GNK which is sound and complete for **NK**. Moreover, the fact that GK admits strong cut-elimination entails that so does GNK. In particular: GNK is analytic.

The semantics of **K** is given by arbitrary modal Kripke frames, in which there are no conditions on the accessibility relation R. Accordingly, the semantics of **NK** is given by **NK**-frames for $\mathcal{L}_{CL}$, where an **NK**-frame is defined like an **NB**-frame (Definition 9.15), but without demanding that R should be reflexive and symmetric. It is easy now to show that **NK** has all the nice properties of **NB** (and **NS5**) listed in Section 9.2.3, except that its negation ¬ is neither complete, nor left-involutive. Instead of Corollary 9.27 we have here that **NK** is the minimal extension of $\mathbf{CL}^+$ in $\mathcal{L}_{CL}$ in which ¬ is contrapositive.

Finally, we note that **NK** is again a C-system, and that the following schemas are provable in **NK** (in addition to [b] and [k]): $[n_\wedge^l]$, $[n_\wedge^r]$, $[n_\vee^l]$, $[n_\supset^l]_2$, $[a_\wedge]$, and $[a_\vee]$. Note that unlike **NB**, the schemas [t], [c], and $[a_\neg]$ are not valid in **NK**.

9.5.2 T as a Paraconsistent Logic

The modal logic **T** (also known as **KT**) is induced by the Hilbert-type system HTK, which is obtained from the one for **K** by adding to it as an axiom the schema (T) $\Box\varphi \supset \varphi$. The translation of this schema into $\mathcal{L}_{CL}$ is exactly the schema [t], that is: the law of excluded middle. If we denote (as usual) by **NT** the logic in $\mathcal{L}_{CL}$ that corresponds to **T** we get therefore:.

Proposition 9.97 *A Hilbert-type system HNT which induces the logic* **NT** *is obtained from HNK by the addition of the axiom* [t].

9.5 Using Other Standard Modal Logics

The semantics of **T** is given by the modal Kripke frames in which the accessibility relation R is reflexive. Accordingly, the semantics of **NT** is also given by **NK**-frames in which the accessibility relation is reflexive.

The standard Gentzen-type system GT for **T** is obtained from GK by the addition of the rule $[\Box \Rightarrow]$, which allows the inference of $\Box\varphi, \Gamma \Rightarrow \Delta$ from $\varphi, \Gamma \Rightarrow \Delta$. The reformulation of this rule using $\neg$ instead of $\Box$ is exactly the classical rule $[\Rightarrow \neg]$. (This is hardly surprising, since the latter is equivalent to axiom [t].) Hence by adding $[\Rightarrow \neg]$ to GK we obtain a Gentzen-type system GNT which induces **NT**. Again the well-known fact that GT admits strong cut-elimination easily entails that so does GNT.

It is again easy to show that **NT** has all the nice properties of **NB** (and **NS5**) listed in Section 9.2.3, except that its negation $\neg$ is not left-involutive. Instead of Corollary 9.27 we have now that **NT** is the minimal extension of $\mathbf{CL}^+$ in $\mathcal{L}_{CL}$ in which $\neg$ is contrapositive and *complete*.

Finally, **NT** is again a C-system, and from the schemas studied in chapter 8 it proves those that **NK** does, as well as [t]. Note that unlike **NB**, the schemas [c] and [a$_\neg$] are still not valid in **NT**.

9.5.3 S4 as a Paraconsistent Logic

The modal logic **S4** (also known as **KT4**) is induced by the Hilbert-type system $HS4$ which is obtained from the one for **T** by adding to it as an axiom the schema (4) $\Box\varphi \supset \Box\Box\varphi$. The translation of this schema into $\mathcal{L}_{CL}$ is $\neg\neg\sim\neg\varphi \supset \neg\varphi$. Hence a Hilbert-type system $HNS4$ which induces **NS4** is obtained from HNT by the addition of $\neg\neg\sim\neg\varphi \supset \neg\varphi$ as an axiom.

The semantics of **S4** is given by the modal Kripke frames in which the accessibility relation R is reflexive and transitive. Accordingly, the semantics of **NS4** is also given by **NK**-frames in which R is reflexive and transitive.

A Gentzen-type system $GNS4$ which induces **NS4** is obtained from LK^+ by adding to it the rule $[\Rightarrow \neg]$ and the following weaker version of $[\neg\Rightarrow]_5$:

$$[\neg\Rightarrow]_4 \quad \frac{\Rightarrow \psi, \neg\Delta}{\neg\psi \Rightarrow \neg\Delta}$$

As in the previous cases, $[\neg\Rightarrow]_4$ is just a reformulation (using $\neg$ rather than $\Box$) of the $[\Box \Rightarrow]$ rule, which is used in the standard Gentzen-type system $GS4$ for $S4$ (which allows the inference of $\Box\Gamma \Rightarrow \Box\varphi$ from $\Box\Gamma \Rightarrow \varphi$). Again, the well-known fact that $GS4$ admits strong cut-elimination easily entails that so does $GNS4$.

Finally, concerning properties as a paraconsistent system in general, and as a C-system in particular, there is no significant difference between **NS4** and **NT**, and what has been said here about **NT** applies to **NS4** as well.

9.6 Bibliographical Notes and Further Reading

For good introductions to modal logics we refer the reader to [109] and [117]. A good review of Gentzen-type systems for modal logics (on which the systems presented in this chapter are based) can be found in [297].

The idea of using modal logics for developing paraconsistent logics is an old one. It was in fact the main idea that was used in the first attempts to construct paraconsistent systems: the *discussive logics* of Jaśkowski [185, 186]. In this chapter we explored Béziau's [90, 91, 92] and Baten's [81] approaches, which in our opinion are better implementations of this idea. Both introduced what is called here **NS5**. (**NS5** was called ℤ by Béziau, and **A** by Batens.) Further study of this system was done by Osorio, Carballido, Zepeda, and Castellanos in [235] and [236].

Sections 9.1, 9.2, and 9.3.1 in this chapter are based on [73, 70]. We note that the same simplified axiomatization of **NS5** was independently discovered by Omori and Waragai in [233]. (Actually, our axiomatization *HNB* of the modal logic **KTB** is implicitly given there as well.) It should also be noted that the above analytic Gentzen-type system *GNB* for **NB** is a version of the Gentzen-type system for **B** given by Takano in [278] (see also [297]). Unlike the proof above, the analyticity of the assumptions-free fragment of that system is proved in [278] by syntactic means.

Hypersequential systems for the modal logic **S5** of the type described in Section 9.3.2 were first (independently) introduced in [215] and [243]. Since then several such systems for **S5** have been suggested in the literature. (See [100], [241], and [297] for further examples and references.) The system for **NS5** presented in Section 9.3.2 is an improved version, without any new structural rules, of the one given in [36].

Section 9.4 is mainly based on [66]. Further information about the logic **NS5**$_2$ studied there can be found in [236] (where it is called $\mathcal{P}\text{-}\mathcal{FOUR}$).

The realization that by applying the method that was applied to **S5** to other modal logics one gets other interesting paraconsistent logics was pursued independently by Marcos and Dodó in [139, 206, 207] and by Mruczek-Nasieniewska and Nasieniewski in [219, 220, 221]. These papers include much of the content of Section 9.5, as well as further relevant material.

Another investigation of paraconsistent logics from a modal viewpoint, including the study of analytic and cut-free sequent calculi for such logics, has been done by Lahav, Marcos, and Zohar in [196].

Chapter 10

Constructivity and Intuitionism

All the logics in the previous chapters were extensions of the positive classical logic $\mathbf{CL}^+$. However, $\mathbf{IL}^+$ (Section 1.2.3), the intuitionistic variant of $\mathbf{CL}^+$, might be a better starting point for investigating negations, and it is certainly the natural basis for investigating *constructive* negations. The main reason is that $\mathbf{IL}^+$ is the minimal normal logic (Proposition 1.70). As a result, all the valid sentences of $\mathbf{IL}^+$ are intuitively justified. In contrast, $\mathbf{CL}^+$ includes tautologies that may be considered counterintuitive, like:

$$(\psi \wedge \varphi \supset \tau) \supset ((\psi \supset \tau) \vee (\varphi \supset \tau)),$$
$$\psi \vee (\psi \supset \varphi),$$
$$(\varphi \supset \psi) \vee (\psi \supset \varphi).$$

Another consequence of the minimality of $\mathbf{IL}^+$ is that it (and not $\mathbf{CL}^+$) is the logic which is defined by the classical natural deduction rules for the positive connectives ($\wedge, \vee,$ and $\supset$). Therefore, it is only with the aid of the classical rules for its negation that one can prove in classical logic the counterintuitive positive tautologies mentioned above. Indeed, from the completeness of *HCL* for $\mathbf{CL}$ and Note 2.6 it follows that it is impossible to conservatively add to the intuitionistic positive logic a negation $\neg$ that will be both explosive and complete (Definition 2.5) in the resulting logic. With such an addition we get classical logic. The intuitionists indeed reject LEM (Proposition 1.78), and so give-up the completeness of their negation (where $\neg\varphi$ is usually defined by them as $\varphi \supset \mathsf{F}$), while they retain its explosive nature (see [180, 181]). At least for the purposes of this book, this is of course the wrong choice. In this chapter we shall determine and investigate all the (strongly) paraconsistent logics that are obtained by *conservatively* extending $\mathbf{IL}^+$ with a negation $\neg$ which has some of the *other* basic properties of classical negation, including completeness. We believe that these are the main logics which might be called "constructive paraconsistent logics".

Like in Chapter 7, the properties of negation (other than explosion) which we take as basic are those which are expressed by the axioms in A_{PAC} (Definition 7.35 and Figure 4.7). Accordingly, we define:

Definition 10.1 ($HIL^+(S)$, $HIL(S)$) For $S \subseteq A_{PAC}$, the Hilbert-type system $HIL^+(S)$ ($HIL(S)$) is the axiomatic extension in $\mathcal{L}_{CL}$ ($\mathcal{L}_{CL} \cup \{\mathsf{F}\}$) of HIL^+ (HIL) with the axioms in S.

Proposition 10.2 *Let $S \subseteq A_{PAC}$. $\mathbf{L}_{HIL(S)}$ and $\mathbf{L}_{HIL^+(S)}$ are normal and strongly paraconsistent for every $S \subseteq A_{PAC}$. Moreover, $\mathbf{L}_{HIL^+(S)}$ is boldly paraconsistent.*

Proof. The proof is practically identical to that of Proposition 7.36. □

Next, we show that the 2^{13} proof systems introduced in Definition 10.1 induce 2^{13} different paraconsistent logics.

Theorem 10.3 *Let S_1 and S_2 be two different subsets of A_{PAC}. Then $\mathbf{L}_{HIL^+(S_1)} \neq \mathbf{L}_{HIL^+(S_2)}$, $\mathbf{L}_{HIL^+(S_1)} \neq \mathbf{L}_{HIL(S_2)}$, and $\mathbf{L}_{HIL(S_1)} \neq \mathbf{L}_{HIL(S_2)}$.*

Proof. In the proof of Theorem 7.56 it was shown that if $(ax) \in A_{PAC} - S$, then (ax) is not provable in $HCL^+(S)$, and so it is not provable in $HIL^+(S)$. That it is not derivable in $HIL(S)$ is proved similarly. This fact and the bold paraconsistency of $HIL(S)$ (Proposition 10.2) easily imply the theorem. □

The following proposition shows that unlike the case of $\mathbf{CL}^+$, some care is required when trying to add to $\mathbf{IL}^+$ axioms from A_{PAC}, because $\mathbf{IL}^+$ might not remain the positive fragment of every logic of the form $\mathbf{L}_{HIL^+(S)}$.

Proposition 10.4 *If $\{[\mathsf{t}], [\mathsf{n}^!_\supset]_1\} \subseteq S \subseteq A_{PAC}$, then $\mathbf{L}_{HIL(S)} = \mathbf{L}_{HCL^\mathsf{F}(S)}$ and $\mathbf{L}_{HIL^+(S)} = \mathbf{L}_{HCL^+(S)}$. Hence $\mathbf{L}_{HIL^+(S)}$ and $\mathbf{L}_{HIL(S)}$ are not conservative extensions of $\mathbf{IL}^+$ and $\mathbf{IL}$ (respectively) in this case.*

Proof. It is easy to see that $\tau \vee (\varphi \supset \psi), \tau \supset \varphi \vdash_{HIL^+} ((\varphi \supset \psi) \supset \varphi) \supset \varphi$. By substituting $\neg(\varphi \supset \psi)$ for τ, we get:

$$\neg(\varphi \supset \psi) \vee (\varphi \supset \psi), \neg(\varphi \supset \psi) \supset \varphi \vdash_{HIL^+(\emptyset)} ((\varphi \supset \psi) \supset \varphi) \supset \varphi.$$

Since the two premises here are instances of [t] and $[\mathsf{n}^!_\supset]_1$ (respectively), we get that $\vdash_{HIL^+(S)} ((\varphi \supset \psi) \supset \varphi) \supset \varphi$. □

In view of Proposition 10.4, one of our main challenges in this chapter is to find out which of the logics in the families $\{\mathbf{L}_{HIL^+(S)} \mid S \subseteq A_{PAC}\}$ and $\{\mathbf{L}_{HIL(S)} \mid S \subseteq A_{PAC}\}$ are conservative extensions of $\mathbf{IL}^+$ and $\mathbf{IL}$ (respectively), and so are constructive logics. The semantic framework which is needed for solving this problem is introduced next.

10.1 The Semantic Framework

In Chapter 7 we provided semantics in a modular way to all the classical paraconsistent logics in $\mathcal{L}_{CL}$ which are obtained from $\mathbf{CL}^+$ by extending it with some subset of $\mathbf{A}_{PAC}$. This was achieved by replacing the use of the standard two-valued deterministic matrix with the use of four-valued (or three-valued) non-deterministic matrices. Similarly, in order to do the same for $\mathbf{IL}^+$ we should replace its standard two-valued deterministic possible worlds semantics with the use of four-valued (or three-valued) non-deterministic possible worlds semantics. For this, we first generalize the notion of a Kripke frame given in Definition 1.66.

Definition 10.5 ($\mathcal{M}$-(Kripke) frames, $\mathsf{IL}(\mathcal{M})$) Let $\mathcal{L}$ have $\supset$ as one of its binary connectives, and let $\mathcal{M} = \langle \mathcal{V}, \mathcal{D}, \mathcal{O} \rangle$ be an Nmatrix for $\mathcal{L}$.

- An $\mathcal{M}$-Kripke-frame for $\mathcal{L}$ (an $\mathcal{M}$-frame, for short) is a triple $\langle W, \leq, \nu \rangle$ in which $\langle W, \leq \rangle$ is a nonempty partially ordered set (of "worlds"), and $\nu : W \times \mathcal{W}(\mathcal{L}) \to \mathcal{V}$ satisfies the following conditions:
  - The persistence condition: for every $w, w' \in W$, if $w' \geq w$ and $\nu(w, \psi) \in \mathcal{D}$ then $\nu(w', \psi) \in \mathcal{D}$.
  - For every $w \in W$, $\lambda \psi.\nu(w, \psi)$ is an $\mathcal{M}$-valuation.
  - For every $w \in W$, $\nu(w, \psi \supset \varphi) \in \mathcal{D}$ iff $\nu(w', \varphi) \in \mathcal{D}$ for every $w' \geq w$ such that $\nu(w', \varphi) \in \mathcal{D}$.

- $\mathsf{IL}(\mathcal{M})$, the Kripke denotational semantics (see Definition 1.37) for $\mathcal{L}$ induced by $\mathcal{M}$, is the pair $\langle S, \models \rangle$ in which S is the class[1] of $\mathcal{M}$-frames, and the satisfiability relation $\models$ is defined by:

  $$\langle W, \leq, \nu \rangle \models \psi \text{ if } \nu(x, \psi) \in \mathcal{D} \text{ for every } x \in W.$$

  In this case we also say that ψ is *valid* in $\langle W, \leq, \nu \rangle$.

- Let $\mathbf{IL}(\mathcal{M}) = \langle \mathcal{L}, \vdash_{IL(\mathcal{M})} \rangle$, where $\vdash_{IL(\mathcal{M})}$ is the consequence relation that is induced by $\mathsf{IL}(\mathcal{M})$. In other words: $\mathcal{T} \vdash_{IL(\mathcal{M})} \psi$ iff ψ is valid in any $\mathcal{M}$-frame in which all the elements of $\mathcal{T}$ are valid.

An equivalent and useful alternative definition of $\vdash_{IL(\mathcal{M})}$ is given next.

Proposition 10.6 *Let $\mathcal{M} = \langle \mathcal{V}, \mathcal{D}, \mathcal{O} \rangle$ be an Nmatrix for $\mathcal{L}$. $\mathcal{T} \vdash_{IL(\mathcal{M})} \psi$ iff for every $\mathcal{M}$-frame $\langle W, \leq, \nu \rangle$ and for every $w \in W$ it holds that if $\nu(w, \varphi) \in \mathcal{D}$ for every $\varphi \in \mathcal{T}$, then $\nu(w, \psi) \in \mathcal{D}$ as well.*

---
[1] Footnote 22 in Chapter 1 is relevant here too.

Proof. Obviously, $\mathcal{T} \vdash_{IL(\mathcal{M})} \psi$ if the condition is satisfied. For the converse, assume that there are an $\mathcal{M}$-frame $\langle W, \leq, \nu \rangle$ and $w \in W$ such that $\nu(w, \varphi) \in \mathcal{D}$ for every $\varphi \in \mathcal{T}$, but $\nu(w, \psi) \notin \mathcal{D}$. Let $W_w = \{x \in W \mid w \leq x\}$, and define $\leq_w$ on W_w and ν_w by: $x \leq_w y$ iff $x \leq y$, $\nu_w(x, \varphi) = \nu(x, \varphi)$. It is easy to check that $\mathcal{M}_w = \langle W_w, \leq_w, \nu_w \rangle$ is an $\mathcal{M}$-frame. Now the assumption that $\nu(w, \varphi) \in \mathcal{D}$ for every $\varphi \in \mathcal{T}$ and the persistent condition imply that all the elements of $\mathcal{T}$ are valid in $\mathcal{M}_w$. On the other hand the fact that $\nu_w(w, \psi) \notin \mathcal{D}$ means that ψ is not. It follows that $\mathcal{T} \nvdash_{IL(\mathcal{M})} \psi$. □

Proposition 10.7 *If $\mathcal{M}_1$ is a refinement of $\mathcal{M}_2$, then any $\mathcal{M}_1$-frame is also an $\mathcal{M}_2$-frame.*

Proof. Obvious from the definitions. □

Example 10.8 Let $\mathcal{M}_2^I = (\{t, f\}, \{t\}, \mathcal{O})$ be the two-valued Nmatrix for $\mathcal{L}_{CL}^{\mathsf{F}}$ which is identical to the classical two-valued matrix for $\mathcal{L}_{CL}^{\mathsf{F}}$ (viewed as a deterministic Nmatrix), except that $f \tilde{\supset} f = \{t, f\}$ rather than $f \tilde{\supset} f = \{t\}$. It is easy to see that the $\mathcal{M}_2^I$-Kripke frames are exactly the ordinary Kripke frames for $\mathcal{L}_{CL}^{\mathsf{F}}$ given Definition 1.73.

It should be noted that although $\mathcal{M}_2^I$ is properly non-deterministic, the semantics of $\mathcal{M}_2^I$-frames is still deterministic in the sense that the truth-values which are assigned to a formula ψ are completely determined by the truth values which are assigned to the elements of $\mathsf{Var}(\psi)$. This is precisely defined in our next definition.

Definition 10.9 (deterministic $\mathcal{M}$-frames) Let $\mathcal{M} = \langle \mathcal{V}, \mathcal{D}, \mathcal{O} \rangle$ be an Nmatrix for $\mathcal{L}$. An $\mathcal{M}$-frame $\langle W, \leq, \nu \rangle$ is called *deterministic* if $\nu^* = \nu$ whenever $\langle W, \leq, \nu^* \rangle$ is another $\mathcal{M}$-frame (which has the same underlying poset $\langle W, \leq \rangle$), and $\nu^*(w, p) = \nu(w, p)$ for every $p \in \mathsf{Var}(\psi)$ and $w \in W$. The semantics $\mathsf{IL}(\mathcal{M})$ is called *deterministic* if every $\mathcal{M}$-frame is deterministic.

Next, we introduce the class of Nmatrices $\mathcal{M}$ such that (positive) intuitionistic logic is sound for the semantics $\mathsf{IL}(\mathcal{M})$.

Definition 10.10 (suitability for $\mathbb{IL}^+$ and $\mathbb{IL}$) Let $\mathcal{M} = \langle \mathcal{V}, \mathcal{D}, \mathcal{O} \rangle$ be an Nmatrix for a language $\mathcal{L}$ such that $\mathcal{L}_{CL}^+ \subseteq \mathcal{L}$.

- $\mathcal{M}$ is *suitable for* $\mathbb{IL}^+$ if the following conditions are satisfied:
  - $\wedge$ and $\vee$ are $\mathcal{M}$-conjunction and $\mathcal{M}$-disjunction, respectively.
  - If $b \in \mathcal{D}$ then $a \tilde{\supset} b \subseteq \mathcal{D}$,
  - If $a \in \mathcal{D}$ and $b \notin \mathcal{D}$ then $a \tilde{\supset} b \subseteq \overline{\mathcal{D}}$.

10.1 The Semantic Framework

- Suppose $\mathcal{L}_{CL}^F \subseteq \mathcal{L}$. Then $\mathcal{M}$ is *suitable for* $\mathbb{IL}$ if it is suitable for $\mathbb{IL}^+$, and $\tilde{\mathsf{F}} \subseteq \overline{\mathcal{D}}$.

Proposition 10.11 *If* $\mathcal{M} = \langle \mathcal{V}, \mathcal{D}, \mathcal{O} \rangle$ *is suitable for* $\mathbb{IL}^+$ *then* $\supset$ *is an implication for* $\vdash_{IL(\mathcal{M})}$.

Proof. From the definition of an $\mathcal{M}$-frame it follows that if $\mathcal{T} \vdash_{IL(\mathcal{M})} \varphi \supset \psi$ then $\mathcal{T}, \varphi \vdash_{IL(\mathcal{M})} \psi$. To show the converse, assume that $\mathcal{T} \nvdash_{IL(\mathcal{M})} \varphi \supset \psi$. By Proposition 10.6, there are $\mathcal{M}$-frame $\langle W, \leq, \nu \rangle$ and $w \in W$ such that $\nu(w, \tau) \in \mathcal{D}$ for every $\tau \in \mathcal{T}$, but $\nu(w, \varphi \supset \psi) \notin \mathcal{D}$. This implies that there is $w' \geq w$ such that $\nu(w', \varphi) \in \mathcal{D}$, while $\nu(w', \psi) \notin \mathcal{D}$. By the persistence condition, $\nu(w', \tau) \in \mathcal{D}$ for every $\tau \in \mathcal{T}$. It follows that $\nu(w', \tau) \in \mathcal{D}$ for every $\tau \in \mathcal{T} \cup \{\varphi\}$, while $\nu(w', \psi) \notin \mathcal{D}$. Therefore $\mathcal{T}, \varphi \nvdash_{IL(\mathcal{M})} \psi$ (by Proposition 10.6 again). $\square$

Theorem 10.12 (soundness of $\mathbb{IL}^+$ and $\mathbb{IL}$) *If* $\mathcal{M} = \langle \mathcal{V}, \mathcal{D}, \mathcal{O} \rangle$ *is suitable for* $\mathbb{IL}^+$ ($\mathbb{IL}$) *then* $\vdash_{IL^+} \subseteq \vdash_{IL(\mathcal{M})}$ ($\vdash_{IL} \subseteq \vdash_{IL(\mathcal{M})}$).

Proof. Suppose first that $\mathcal{M}$ is suitable for $\mathbb{IL}^+$. Then Proposition 10.6 and the first condition in Definition 10.10 easily imply that $\wedge$ and $\vee$ are conjunction and disjunction (respectively) for $\vdash_{IL(\mathcal{M})}$. By Proposition 10.11 $\supset$ is an implication for $\vdash_{IL(\mathcal{M})}$. Hence $\vdash_{IL^+} \subseteq \vdash_{IL(\mathcal{M})}$ by Proposition 1.70.

Now, assume that $\mathcal{M}$ is suitable for $\mathbb{IL}$. Then the condition in the second part of Definition 10.10 implies that $\vdash_{IL(\mathcal{M})} \mathsf{F} \supset \psi$ for every ψ. Hence, it follows from the fact that $\vdash_{IL^+} \subseteq \vdash_{IL(\mathcal{M})}$ and the third item of Theorem 1.90 that $\vdash_{IL} \subseteq \vdash_{IL(\mathcal{M})}$. $\square$

Next, we introduce the Nmatrix which serves as the basis for the semantics of the logics in $\{\mathbf{L}_{HIL^+(S)} \mid S \subseteq \mathsf{A}_{PAC}\}$.

Definition 10.13 ($\mathcal{M}_4^{Ib}$, $\mathcal{M}_{4F}^{Ib}$) The Nmatrix $\mathcal{M}_4^{Ib} = \langle \mathcal{V}, \mathcal{D}, \mathcal{O} \rangle$ for $\mathcal{L}_{CL}$ is defined like $\mathcal{M}_4^b$ (Definition 7.29), except that $a \tilde{\supset} b = \mathcal{V}$ in case both a and b are in $\overline{\mathcal{D}}$.[2] In other words:

- $\mathcal{V} = \{t, f, \top, \bot\}$, $\mathcal{D} = \{t, \top\}$

- $a \tilde{\supset} b = \begin{cases} \mathcal{D} & b \in \mathcal{D}, \\ \overline{\mathcal{D}} & a \in \mathcal{D} \text{ and } b \in \overline{\mathcal{D}}, \\ \mathcal{V} & a, b \in \overline{\mathcal{D}}. \end{cases}$

- $a \tilde{\vee} b = \begin{cases} \mathcal{D} & a \in \mathcal{D} \text{ or } b \in \mathcal{D}, \\ \overline{\mathcal{D}} & \text{otherwise.} \end{cases}$

---

[2] In [44] $\mathcal{M}_4^{Ib}$ is called $\mathcal{M}_{IP}$.

$$a \tilde{\wedge} b = \begin{cases} \mathcal{D} & a,b \in \mathcal{D}, \\ \overline{\mathcal{D}} & \text{otherwise.} \end{cases}$$

$$\tilde{\neg} a = \begin{cases} \mathcal{D} & a \in \{f, \top\}, \\ \overline{\mathcal{D}} & a \in \{t, \bot\}. \end{cases}$$

$\mathcal{M}_{4F}^{Ib}$ is the Nmatrix for the language $\mathcal{L}_{CL}^{F\neg} = \{\neg, \wedge, \vee, \supset, F\}$ in which $\tilde{\supset}$, $\tilde{\vee}$, $\tilde{\wedge}$, and $\tilde{\neg}$ are defined like $\mathcal{M}_4^{Ib}$, and in addition $\tilde{F} = \overline{\mathcal{D}}$.

Proposition 10.14 *Let $\mathcal{M}$ be a refinement of $\mathcal{M}_4^{Ib}$ (of $\mathcal{M}_{4F}^{Ib}$). Then HIL$^+$ (HIL) is sound for every $\mathcal{M}$-frame.*

Proof. Every refinement of $\mathcal{M}_4^{Ib}$ ($\mathcal{M}_{4F}^{Ib}$) is suitable for $\mathbf{IL}^+$ ($\mathbf{IL}$). Hence the claim follows from Proposition 10.7, and Theorems 10.12, 1.90. □

For refinements of $\mathcal{M}_4^{Ib}$ and $\mathcal{M}_{4F}^{Ib}$ it is possible to provide the following useful alternative to the persistence condition.

Proposition 10.15 *Let $\mathcal{M}$ be a refinement of $\mathcal{M}_4^{Ib}$ or of $\mathcal{M}_{4F}^{Ib}$. Then the persistence condition for $\mathcal{M}$-frames is equivalent to the following* monotonicity condition for $\mathcal{M}$-frames:

- If $x \leq y$ then $\nu(x, \varphi) \leq_k \nu(y, \varphi)$, where $\leq_k$ is the partial order on $\{t, f, \top, \bot\}$ which was used in Chapter 5 to define the bilattice $\mathcal{FOUR}$ (see Figure 5.1).

Proof. Let $\langle W, \leq, \nu \rangle$ be an $\mathcal{M}$-frame, and let $x \leq y$. We show that $\nu(x, \varphi) \leq_k \nu(y, \varphi)$. There are four cases to consider:

$\nu(x, \varphi) = \bot$: This case is trivial.

$\nu(x, \varphi) = t$: Then $\nu(y, \varphi) \in \{t, \top\}$ by persistence, whence $\nu(x, \varphi) \leq_k \nu(y, \varphi)$.

$\nu(x, \varphi) = f$: Then $\nu(x, \neg\varphi) \in \mathcal{D}$, whence $\nu(y, \neg\varphi) \in \mathcal{D}$ by persistence. This is possible only if $\nu(y, \varphi) \in \{f, \top\}$, and so again $\nu(x, \varphi) \leq_k \nu(y, \varphi)$.

$\nu(x, \varphi) = \top$: Then both $\nu(x, \varphi)$ and $\nu(x, \neg\varphi)$ are in $\mathcal{D}$. By persistence, this implies that both $\nu(y, \varphi)$ and $\nu(y, \neg\varphi)$ are in $\mathcal{D}$. This is possible only if $\nu(y, \varphi) = \top$ as well.

The converse — that the monotonicity condition implies the persistence condition — is trivial. □

Now we turn to the effects of the various negation rules in the context of our semantics for *HIL$^+$* and its extensions. The conditions we associated with these conditions in Chapter 7 lead this time to refinements of $\mathcal{M}_4^{Ib}$ (or $\mathcal{M}_{4F}^{Ib}$) on which the corresponding frames are based.

10.1 The Semantic Framework

Definition 10.16 ($\mathcal{M}_4^{Ib}(S)$, $\mathcal{M}_{4F}^{Ib}(S)$) For $S \subseteq A_{PAC}$, $\mathcal{M}_4^{Ib}(S)$ and $\mathcal{M}_{4F}^{Ib}(S)$ are the weakest refinements (Definition 7.43) of $\mathcal{M}_4^{Ib}$ and $\mathcal{M}_{4F}^{Ib}$ (respectively) in which the conditions in $C(S)$ (Definition 7.44) are satisfied.

Proposition 10.17 $\mathcal{M}_4^{Ib}(S)$ and $\mathcal{M}_{4F}^{Ib}(S)$ ($S \subseteq A_{PAC}$) are well-defined.

Proof. Similar to the proof of Proposition 7.45. □

Theorem 10.18 (soundness and completeness) Let $S \subseteq A_{PAC}$.

1. $\vdash_{HIL^+(S)} = \vdash_{IL(\mathcal{M}_4^{Ib}(S))}$.
2. $\vdash_{HIL(S)} = \vdash_{IL(\mathcal{M}_{4F}^{Ib}(S))}$.

Proof. We show the first item; The proof for $HIL(S)$ is similar.

To prove soundness (that is, that $\vdash_{HIL^+(S)} \subseteq \vdash_{IL(\mathcal{M}_4^{Ib}(S))}$) it suffices by Proposition 10.14 (and the second condition in Definition 10.5 on valuations in $\mathcal{M}$-frames) to show that for any axiom $ax \in A_{PAC}$, ax is valid in any refinement of $\mathcal{M}_4^{Ib}$ in which the condition $C(ax)$ is satisfied. The proof of this is done exactly as in the proof of Theorem 7.50.

To prove completeness of $HIL^+(S)$ for $\mathcal{M}_4^{Ib}(S)$-frames, define a canonical $\mathcal{M}_4^{Ib}(S)$-frame $\mathcal{F} = \langle W, \leq, \nu \rangle$ as follows:

- W is the set of all theories $\mathcal{T}$ in $\mathcal{L}_{CL}$ having the following properties:

  1. $\psi \in \mathcal{T}$ iff $\mathcal{T} \vdash_{HIL^+(S)} \psi$.
  2. $\mathcal{T}$ is not trivial (i.e. $\mathcal{T} \neq \mathcal{W}(\mathcal{L}_{CL})$).
  3. $\varphi \lor \psi \in \mathcal{T}$ iff either $\varphi \in \mathcal{T}$ or $\psi \in \mathcal{T}$.
  4. $\varphi \land \psi \in \mathcal{T}$ iff both $\varphi \in \mathcal{T}$ and $\psi \in \mathcal{T}$.
  5. $\varphi \supset \psi \in \mathcal{T}$ iff $\mathcal{T}, \varphi \vdash_{HIL^+(S)} \psi$.

- $\leq$ is the subset relation ($\subseteq$).

- ν is defined as follows:

$$\nu(\mathcal{T}, \psi) = \begin{cases} \bot & \psi \notin \mathcal{T}, \neg\psi \notin \mathcal{T} \\ f & \psi \notin \mathcal{T}, \neg\psi \in \mathcal{T} \\ t & \psi \in \mathcal{T}, \neg\psi \notin \mathcal{T} \\ \top & \psi \in \mathcal{T}, \neg\psi \in \mathcal{T} \end{cases}$$

Now we show that $\mathcal{F}$ is indeed an $\mathcal{M}_4^{Ib}(S)$-frame.

Obviously, $\langle W, \leq \rangle$ is a partially ordered set, and $\nu(\mathcal{T}, \psi) \in \mathcal{D}$ iff $\psi \in \mathcal{T}$. The latter fact immediately implies that ν satisfies the persistence condition.

Showing that for every $\mathcal{T} \in W$, $\lambda\varphi.\nu(\mathcal{T},\varphi)$ is an $\mathcal{M}_4^{Ib}(S)$-valuation is practically identical to the way it was shown in the proof of Theorem 7.50 that the valuation ν defined there is an $\mathcal{M}_4^b(S)$-valuation. We show as an example that if $[\mathsf{n}_\supset^r] \in S$, then $\lambda\varphi.\nu(\mathcal{T},\varphi)$ satisfies $C([\mathsf{n}_\supset^r])$. So assume that $\nu(\varphi) \in \{t, \top\}$ and $\nu(\psi) \in \{f, \top\}$. Then $\varphi \in \mathcal{T}$ and $\neg\psi \in \mathcal{T}$. Since $[\mathsf{n}_\supset^r] \in S$, this and the fact that $\mathcal{T} \in W$ easily imply that $\neg(\varphi \supset \psi) \in \mathcal{T}$. Now, if $\nu(\psi) = f$ then $\psi \notin \mathcal{T}$, and so $\varphi \supset \psi \notin \mathcal{T}$ (because $\varphi \in \mathcal{T}$ and property 1 of $\mathcal{T}$). Hence, also $\nu(\varphi \supset \psi) = f$ in this case (by definition of ν). On the other hand, if $\nu(\psi) = \top$, then $\psi \in \mathcal{T}$, and so $\varphi \supset \psi \in \mathcal{T}$ (because $\psi \vdash_{HIL^+} \varphi \supset \psi$). Hence, also $\nu(\varphi \supset \psi) = \top$ in this case (again by the definition of ν). It follows that in both cases $\nu(\varphi \supset \psi) = \nu(\psi)$, as required.

Next, we show that ν satisfies the frame condition concerning $\supset$.

- Suppose first that $\nu(\mathcal{T}, \varphi \supset \psi) \in \mathcal{D}$. Then $\varphi \supset \psi \in \mathcal{T}$. Let $\mathcal{T}'$ be an element in W such that $\mathcal{T} \leq \mathcal{T}'$ and $\nu(\mathcal{T}', \varphi) \in \mathcal{D}$. Then $\varphi \in \mathcal{T}'$, and since $\mathcal{T} \leq \mathcal{T}'$, also $\varphi \supset \psi \in \mathcal{T}'$. By Property 1 of $\mathcal{T}'$, it follows that $\psi \in \mathcal{T}'$, and so $\nu(\mathcal{T}', \psi) \in \mathcal{D}$ as well.

- Now suppose that $\nu(\mathcal{T}, \varphi \supset \psi) \notin \mathcal{D}$. Then $\varphi \supset \psi \notin \mathcal{T}$. Hence $\mathcal{T}, \varphi \nvdash_{HIL^+(S)} \psi$ (by Property 5 of $\mathcal{T}$). Therefore Theorems 1.97 and 1.94 imply that there is $\mathcal{T}' \in W$ such that $\mathcal{T} \cup \{\varphi\} \subseteq \mathcal{T}'$ while $\psi \notin \mathcal{T}'$. It follows that there is $\mathcal{T}' \in W$ such that $\mathcal{T} \leq \mathcal{T}'$, $\nu(\mathcal{T}', \varphi) \in \mathcal{D}$, and $\nu(\mathcal{T}', \psi) \notin \mathcal{D}$.

This completes the proof that W is an $\mathcal{M}_4^{Ib}(S)$-frame. Assume now that $\mathcal{T} \nvdash_{HIL^+(S)} \psi_0$. Then by Theorems 1.97 and 1.94 there exists $\mathcal{T}^* \in W$ such that $\mathcal{T} \subseteq \mathcal{T}^*$ and $\psi_0 \notin \mathcal{T}^*$. Hence $\nu(\mathcal{T}^*, \varphi) \in \mathcal{D}$ for every $\varphi \in \mathcal{T}$, while $\nu(\mathcal{T}^*, \psi_0) \notin \mathcal{D}$. By Proposition 10.6, it follows that $\mathcal{T} \nvdash_{IL(\mathcal{M}_4^{Ib}(S))} \psi$. □

Example 10.19 $HIL^+(\{[\mathsf{t}], [\mathsf{c}]\})$ is the system C_ω of da Costa ([127]). Theorem 10.18 provides illuminating semantics for it which can be used for a decision procedure (see the next section).[3] Here is a compact description of this semantics: A frame for C_ω is a triple $\langle W, \leq, \nu \rangle$ such that $\langle W, \leq \rangle$ is a nonempty partially ordered set, and $\nu : W \times \mathcal{W}(\mathcal{L}_{CL}) \to \{t, f, \top\}$ is a valuation which satisfies the following conditions:

- Suppose $x \leq y$. If $\nu(y, \varphi) \in \{t, f\}$ then $\nu(x, \varphi) = \nu(y, \varphi)$.

- $\nu(x, \varphi \wedge \psi) = f$ iff $\nu(x, \varphi) = f$ or $\nu(x, \psi) = f$

- $\nu(x, \varphi \vee \psi) = f$ iff $\nu(x, \varphi) = f$ and $\nu(x, \psi) = f$

---

[3] A more complicated Kripke-type semantics for C_ω was given by Baaz [76], while Loparić [199] provides bivaluations semantics for it.

10.1 The Semantic Framework

- $\nu(x, \varphi \supset \psi) = f$ iff for some $y \geq x$, $\nu(y, \varphi) \neq f$ while $\nu(y, \psi) = f$
- $\nu(x, \neg \varphi) = f$ iff $\nu(x, \varphi) = t$
- If $\nu(x, \varphi) = f$ then $\nu(x, \neg\varphi) = t$

A frame for C_ω is a model of a formula φ if $\nu(x, \varphi) \neq f$ for every $x \in W$.

A frame for $HIL^+(\{[t]\})$ and a model of a formula in it are defined similarly, but without the last condition.

We end this section by presenting the semantics introduced in it for $HIL^+(S)$ and $HIL(S)$ ($S \subseteq \mathsf{A}_{PAC}$) in an equivalent useful form, that follows the presentation given in Note 7.3 (and used in Footnote 4 in Chapter 7).

Definition 10.20 (double Kripke frames) Let $\mathcal{L} = \mathcal{L}_{CL}$ or $\mathcal{L} = \mathcal{L}_{CL}^{F\neg}$. A *double Kripke frame* for $\mathcal{L}$ (*d-frame*, for short) is a quadruple of the form $\langle W, \leq, \nu^+, \nu^- \rangle$ such that:

- $\langle W, \leq \rangle$ is a nonempty partially ordered set (of "worlds").
- ν^+, ν^- are functions from $W \times \mathcal{W}(\mathcal{L})$ to $\{1,0\}$ that satisfy the persistence condition: If $w \leq w'$ then $\nu^*(w, \psi) \leq \nu^*(w', \psi)$ ($* \in \{+,-\}$).
- $\nu^+(w, \neg\psi) = \nu^-(w, \psi)$.
- $\langle W, \leq, \nu^+ \rangle$ is a Kripke frame (Definitions 1.66, 1.73) for $\mathcal{L} - \{\neg\}$, where 1 and 0 are used instead of t and f (respectively). That is:
    - $\nu^+(w, \psi \wedge \varphi) = 1$ iff $\nu^+(w, \psi) = 1$ and $\nu^+(w, \varphi) = 1$.
    - $\nu^+(w, \psi \vee \varphi) = 1$ iff $\nu^+(w, \psi) = 1$ or $\nu^-(w, \varphi) = 1$
    - $\nu^+(w, \psi \supset \varphi) = 1$ iff $\nu^+(w', \varphi) = 1$ for every $w' \geq w$ such that $\nu^+(w', \psi) = 1$.
    - If $\mathsf{F} \in \mathcal{L}$ then $v^+(w, \mathsf{F}) = 0$.

We say that a formula ψ is *valid* in a d-frame $\mathcal{W} = \langle W, \leq, \nu^+, \nu^- \rangle$ ($\mathcal{W} \models \psi$) if $\nu^+(w, \psi) = 1$ for every $w \in W$.

Note 10.21 The definition of a d-frame is fully deterministic in the case of ν^+ (in the sense that the values assigned by ν^+ to a formula φ are completely determined by the values that ν^+ and ν^- assign to its immediate subformulas). In contrast, the only constraint imposed on the values of ν^- is the persistence condition. On the other hand, when translated into the framework of d-frames, the semantic conditions that correspond to the axioms in A_{PAC} become constraints on ν^- (that can very easily be read from the axioms). The list of these constraints is included in the next definition.

Definition 10.22 (*S*-d-frames) Let $\mathcal{L} \in \{\mathcal{L}_{CL}, \mathcal{L}_{CL}^{F-}\}$, and let $S \subseteq A_{PAC}$. An *S-d-frame* is a d-frame $\langle W, \leq, \nu^+, \nu^- \rangle$ in which ν^- satisfies all the conditions in $\{C^d(ax) \mid ax \in S\}$, where $C^d(ax)$ is given in the following list:

$C^d([\mathsf{t}])$: $\nu^-(w, \varphi) = 1$ if $\nu^+(w, \varphi) = 0$.

$C^d([\mathsf{c}])$: $\nu^-(w, \neg\varphi) = 0$ if $\nu^+(w, \varphi) = 0$.

$C^d([\mathsf{e}])$: $\nu^-(w, \neg\varphi) = 1$ if $\nu^+(w, \varphi) = 1$.

$C^d([\mathsf{n}_\supset^l]_1)$: $\nu^-(w, \varphi \supset \psi) = 0$ if $\nu^+(w, \varphi) = 0$.

$C^d([\mathsf{n}_\supset^l]_2)$: $\nu^-(w, \varphi \supset \psi) = 0$ if $\nu^-(w, \psi) = 0$.

$C^d([\mathsf{n}_\supset^r])$: $\nu^-(w, \varphi \supset \psi) = 1$ if $\nu^+(w, \varphi) = 1$ and $\nu^-(w, \psi) = 1$.

$C^d([\mathsf{n}_\vee^l]_1)$: $\nu^-(w, \varphi \vee \psi) = 0$ if $\nu^-(w, \varphi) = 0$.

$C^d([\mathsf{n}_\vee^l]_2)$: $\nu^-(w, \varphi \vee \psi) = 0$ if $\nu^-(w, \psi) = 0$.

$C^d([\mathsf{n}_\vee^r])$: $\nu^-(w, \varphi \vee \psi) = 1$ if $\nu^-(w, \varphi) = 1$ and $\nu^-(w, \psi) = 1$.

$C^d([\mathsf{n}_\wedge^l])$: $\nu^-(w, \varphi \wedge \psi) = 0$ if $\nu^-(w, \varphi) = 0$ and $\nu^-(w, \varphi) = 0$.

$C^d([\mathsf{n}_\wedge^r]_1)$: $\nu^-(w, \varphi \wedge \psi) = 1$ if $\nu^-(w, \varphi) = 1$.

$C^d([\mathsf{n}_\wedge^r]_2)$: $\nu^-(w, \varphi \wedge \psi) = 1$ if $\nu^-(w, \psi) = 1$.

The connections between the two representations of the semantics are described in the next proposition, in which we again make the following identifications: $t = \langle 1, 0 \rangle$, $f = \langle 0, 1 \rangle$, $\top = \langle 1, 1 \rangle$, and $\bot = \langle 0, 0 \rangle$.

Proposition 10.23 *Let $S \subseteq A_{PAC}$.*

1. *Let $\mathcal{W} = \langle W, \leq, \nu \rangle$ be an $\mathcal{M}_4^{Ib}(S)$-frame or $\mathcal{M}_{4F}^{Ib}(S)$-frame. Define $\nu^+(w, \varphi) = \pi_1(\nu(w, \varphi))$, $\nu^-(w, \varphi) = \pi_2(\nu(w, \varphi))$ (where π_1 and π_2 are the projection functions on pairs). Then the structure $\mathcal{W}^d = \langle W, \leq, \nu^+, \nu^- \rangle$ is an S-d-frame, and for every φ: $\mathcal{W} \models \varphi$ iff $\mathcal{W}^d \models \varphi$.*

2. *Conversely, let $\mathcal{W}^d = \langle W, \leq, \nu^+, \nu^- \rangle$ be an S-d-frame, and define $\nu(w, \varphi) = \langle \nu^+(w, \varphi), \nu^-(w, \varphi) \rangle$. Then (depending on whether F is in the language or not) $\mathcal{W} = \langle W, \leq, \nu \rangle$ is an $\mathcal{M}_4^{Ib}(S)$-frame or $\mathcal{M}_{4F}^{Ib}(S)$-frame, and for every φ: $\mathcal{W} \models \varphi$ iff $\mathcal{W}^d \models \varphi$.*

3. *$\mathcal{T} \vdash_{HIL^+(S)} \varphi$ (or $\mathcal{T} \vdash_{HIL(S)} \varphi$) iff φ is valid in every S-d-frame in which all the elements of $\mathcal{T}$ are valid.*

Proof. Using the constraints on ν^+ and ν^- given in Definition 10.20, the proof is just a straightforward exercise. Details are left to the reader. □

10.2 Analyticity, Conservativity, and Decidability

Theorem 10.18 does not have much value in itself. Indeed, it does not guarantee that $HIL^+(S)$ is conservative over HIL^+, and neither does it provide a decision procedure for $HIL^+(S)$. The reason for this is that the current semantic framework (of Nmatrices combined with intuitionistic frames) is not necessarily *analytic*. Next, we provide a definition of this notion which is suitable for the present context.

Definition 10.24 ($\mathcal{M}$-$\mathcal{F}$-semiframes, $\mathcal{F}$-analyticity) Let $\mathcal{M}$ be either $\mathcal{M}_4^{Ib}(S)$ or $\mathcal{M}_{4F}^{Ib}(S)$, where $S \subseteq A_{PAC}$, and let $\mathcal{F}$ be a set of formulas in the language of $\mathcal{M}$ which is closed under subformulas. Suppose that $\mathcal{M} = \langle \mathcal{V}, \mathcal{D}, \mathcal{O} \rangle$.

- An $\mathcal{M}$-$\mathcal{F}$-semiframe is a triple $\mathcal{W} = \langle W, \leq, \nu \rangle$ such that:
  - $\langle W, \leq \rangle$ is a nonempty partially ordered set.
  - $\nu : W \times \mathcal{F} \to \mathcal{V}$ is a partial valuation such that:
    * ν satisfies the monotonicity condition: if $y \geq x$ and $\varphi \in \mathcal{F}$, then $\nu(x, \varphi) \leq_k \nu(y, \varphi)$.
    * ν respects $\mathcal{M}$: If $\diamond(\psi_1, \ldots, \psi_n) \in \mathcal{F}$, then $\nu(x, \diamond(\psi_1, \ldots, \psi_n))$ is in $\tilde{\diamond}(\nu(x, \psi_1), \ldots, \nu(x, \psi_n))$.
    * If $\varphi \supset \psi \in \mathcal{F}$ then $\nu(x, \varphi \supset \psi) \in \mathcal{D}$ iff $\nu(y, \psi) \in \mathcal{D}$ for every $y \geq x$ such that $\nu(y, \varphi) \in \mathcal{D}$.

- If $\mathcal{W} = \langle W, \leq, \nu \rangle$ is an $\mathcal{M}$-$\mathcal{F}$-semiframe, and $\psi \in \mathcal{F}$, then $\mathcal{W}$ is a *model* of ψ (in symbols: $\mathcal{W} \models \psi$) if $\nu(x, \psi) \in \mathcal{D}$ for every $x \in W$.

- $\mathcal{M}$ is $\mathcal{F}$-analytic if for every $\mathcal{M}$-$\mathcal{F}$-semiframe $\langle W, \leq, \nu \rangle$ there exists an $\mathcal{M}$-frame $\langle W, \leq, \nu' \rangle$ such that ν' extends ν.

Note 10.25 By Proposition 10.15, an $\mathcal{M}$-frame is an $\mathcal{M}$-$\mathcal{W}(\mathcal{L})$-semiframe, where $\mathcal{L}$ is the language of $\mathcal{M}$.

Note 10.26 Proposition 10.15 entails that the persistence condition and the monotonicity condition are equivalent for *full* valuations. However, this is *not* necessarily true for partial ones. As a result, the fact that the definition of a semiframe uses the *monotonicity condition* is critical, since no $\mathcal{M}$ would have been analytic had we used the persistence condition instead. Thus, if we take $W = \{a, b\}$ with $a < b$, and define $\nu(a, p) = f, \nu(b, p) = t$, then the persistence condition is met by ν, but there is no way to extend ν to an appropriate ν'. Indeed, for any full extension ν' of ν we should have $\nu'(a, \neg p) \in \mathcal{D}$, $\nu'(b, \neg p) \notin \mathcal{D}$, contradicting the persistence condition. Theorem 10.29 below shows that the situation is completely different when semiframes are defined as above.

Note 10.27 For $S \subseteq \mathsf{A}_{PAC}$, the notion of S-$\mathcal{F}$-d-frame can be defined in a way which is completely analogous to Definition 10.24, and an obvious counterpart of Proposition 10.23 then holds. We leave the details to the reader.

Definition 10.28 ($\mathsf{N}_4, \mathsf{N}_3^\top$) In what follows, we denote: $\mathsf{N}_4 = \mathsf{A}_{PAC} - \{[\mathsf{t}]\}$ and $\mathsf{N}_3^\top = \mathsf{A}_{PAC} - \{[\mathsf{n}_\supset^!]_1\}$.

Theorem 10.29 Let $\mathcal{M} = \mathcal{M}_4^{Ib}(S)$ or $\mathcal{M} = \mathcal{M}_{4\mathsf{F}}^{Ib}(S)$ where $S \subseteq \mathsf{A}_{PAC}$. $\mathcal{M}$ is analytic iff $\{[\mathsf{t}], [\mathsf{n}_\supset^!]_1\} \not\subseteq S$ (i.e., iff either $S \subseteq \mathsf{N}_4$ or $S \subseteq \mathsf{N}_3^\top$).

Proof. To show that the condition is necessary, take $W = \{a, b\}$ with $a < b$, and define $\nu(a, P_1) = \nu(a, P_2) = \nu(b, P_2) = f$, $\nu(b, P_1) = \top$. Then ν respects the monotonicity condition, but if $\{[\mathsf{t}], [\mathsf{n}_\supset^!]_1\} \subseteq S$ then there is no extension ν' of ν such that $\langle W, \leq, \nu' \rangle$ is an $\mathcal{M}$-frame: $\nu'(a, P_1 \supset P_2)$ should on the one hand be f according to the definition of an $\mathcal{M}$-frame (because the presence of $[\mathsf{t}]$ implies that $\bot$ is not available), while according to $C([\mathsf{n}_\supset^!]_1)$ it should be t (again, because $\bot$ is not available).

Assume next that either $[\mathsf{t}] \notin S$, or $[\mathsf{n}_\supset^!]_1 \notin S$. We show that $\mathcal{M}$ is analytic. So let $\mathcal{M} = \langle \mathcal{V}, \mathcal{D}, \mathcal{O} \rangle$, and let $\langle W, \leq, \nu \rangle$ (where $\nu : W \times \mathcal{F} \to \mathcal{V}$) be an $\mathcal{M}$-$\mathcal{F}$-semiframe. We extend it to an $\mathcal{M}$-frame $\langle W, \leq, \nu' \rangle$ by defining ν' inductively as follows:

- $\nu'(x, \psi) = \nu(x, \psi)$ if $\psi \in \mathcal{F}$.

- $\nu'(x, \psi) = f$ if ψ is a variable, and $\psi \notin \mathcal{F}$.

- $\nu'(x, \neg \psi) = \tilde{\neg}_{\mathcal{FOUR}} \nu'(x, \psi)$ if $\neg \psi \notin \mathcal{F}$, where $\tilde{\neg}_{\mathcal{FOUR}}$ is the interpretation of $\neg$ in $\mathcal{FOUR}$ (Definition 5.10).

- $\nu'(x, \psi_1 \vee \psi_2) = sup_t(\nu'(x, \psi_1), \nu'(x, \psi_2))$ if $\psi_1 \vee \psi_2 \notin \mathcal{F}$.

- $\nu'(x, \psi_1 \wedge \psi_2) = inf_t(\nu'(x, \psi_1), \nu'(x, \psi_2))$ if $\psi_1 \wedge \psi_2 \notin \mathcal{F}$.

- If $\psi_1 \supset \psi_2 \notin \mathcal{F}$ then there are two cases:

    - If $[\mathsf{t}] \notin S$ then
    $$\nu'(x, \psi_1 \supset \psi_2) = \begin{cases} v & \nu'(x, \psi_1) \in \mathcal{D} \text{ and } \nu'(x, \psi_2) = v, \\ \bot & \nu'(x, \psi_1) \notin \mathcal{D} \text{ and} \\ & \exists y \geq x. \nu'(y, \psi_1) \in \mathcal{D} \wedge \nu'(y, \psi_2) \notin \mathcal{D}, \\ t & \text{otherwise.} \end{cases}$$

10.2 Analyticity, Conservativity, and Decidability 343

– If $[t] \in S$ but $[n_\supset^!]_1 \notin S$ then

$$\nu'(x, \psi_1 \supset \psi_2) = \begin{cases} t & \nu'(x, \psi_2) = t, \\ f & \exists y \geq x. \nu'(y, \psi_1) \in \mathcal{D} \wedge \nu'(y, \psi_2) \notin \mathcal{D}, \\ \top & \text{otherwise.} \end{cases}$$

We prove now by induction on the complexity of ψ that $\nu'(x, \psi)$ is well-defined for every $x \in W$, and that $\lambda x.\nu'(x, \psi)$ is monotonic. This follows from our assumption on ν if $\psi \in \mathcal{F}$, and is trivial if ψ is a variable. The cases where ψ is of one of the forms $\neg \psi_1$, $\psi_1 \vee \psi_2$, or $\psi_1 \wedge \psi_2$ follow easily from the induction hypothesis concerning ψ_1, ψ_2, and the monotonicity of the interpretations of $\neg, \vee$ and $\wedge$ in $\mathcal{FOUR}$. It remains to prove the case where $\psi = \psi_1 \supset \psi_2$, and $\psi \notin \mathcal{F}$. Now, a problem with the coherence of the definition of $\nu'(x, \psi)$ may occur in this case only if $[t] \in S$ and $\nu'(x, \psi_2) = t$. However, by induction hypothesis for ψ_2, if $\nu'(x, \psi_2) = t$ then $\nu'(z, \psi_2) \in \mathcal{D}$ for all $z \geq x$, and so only the first clause in the definition of $\nu'(x, \psi_1 \supset \psi_2)$ is applicable, implying that $\nu'(x, \psi)$ is well-defined in this case too. We show now that under the same assumptions concerning ψ, $\nu'(x, \psi) \leq_k \nu'(y, \psi)$ if $y \geq x$. There are two cases to consider:

- Assume that $[t] \notin S$.
  If $\nu'(x, \psi_1) \in \mathcal{D}$ then by the induction hypothesis also $\nu'(y, \psi_1) \in \mathcal{D}$ and $\nu'(x, \psi_2) \leq_k \nu'(y, \psi_2)$. Hence in this case $\nu'(x, \psi) = \nu'(x, \psi_2) \leq_k \nu'(y, \psi_2) = \nu'(y, \psi)$. If $\nu'(x, \psi_1) \notin \mathcal{D}$ and $\nu'(x, \psi) = \bot$ then trivially $\nu'(x, \psi) \leq_k \nu'(y, \psi)$. Finally, if $\nu'(x, \psi_1) \notin \mathcal{D}$ and $\nu'(x, \psi) = t$, then the fact that $y \geq x$ and the induction hypotheses concerning ψ_1 and ψ_2 together imply (using an analysis of the three cases in the definition of $\nu'(y, \psi)$) that $\nu'(y, \psi) \in \mathcal{D}$, and so again $\nu'(x, \psi) \leq_k \nu'(y, \psi)$.

- Assume that $[t] \in S$.
  In this case $\nu'(z, \varphi) \neq \bot$ for every $z \in W$ and every $\varphi \in \mathcal{F}$. Hence the definition of ν' implies that $\nu'(z, \varphi) \neq \bot$ for every $z \in W$ and every φ. It follows that $\nu'(z, \varphi) \notin \mathcal{D}$ iff $\nu'(z, \varphi) = f$, and $\nu'(z, \varphi) \neq t$ iff $\nu'(z, \varphi) \in \{f, \top\}$. Accordingly, the induction hypothesis implies that if $\nu'(z, \psi_2) \neq t$ for some z, then $\nu'(w, \psi_2) \neq t$ for every $w \geq z$. Hence if $\nu'(x, \psi) = f$ then $\nu'(y, \psi) \neq t$, and so $\nu'(x, \psi) \leq_k \nu'(y, \psi)$. If $\nu'(x, \psi) = t$ then $\nu'(x, \psi_2) = t$, and so by the induction hypothesis $\nu'(y, \psi_2) \in \mathcal{D}$, and also $\nu'(z, \psi_2) \in \mathcal{D}$ for every $z \geq y$. Thus in this case $\nu'(y, \psi) \in \mathcal{D}$ by the definition of ν', and so $\nu'(x, \psi) = t \leq_k \nu'(y, \psi)$. Finally, assume that $\nu'(x, \psi) = \top$. Therefore $\nu'(x, \psi_2) \neq t$ and $\forall z \geq x. \nu'(z, \psi_1) = f \vee \nu'(z, \psi_2) \in \mathcal{D}$. Since $y \geq x$, the first fact implies that also $\nu'(y, \psi_2) \neq t$, whence $\nu'(y, \psi) \neq t$. The second fact implies that

also $\forall z \geq y. \nu'(z, \psi_1) = f \vee \nu'(z, \psi_2) \in \mathcal{D}$, whence $\nu'(y, \psi) \neq f$. In consequence $\nu'(y, \psi) = \top$, and so $\nu'(x, \psi) \leq_k \nu'(y, \psi)$.

The monotonicity of ν', together with its definition and our assumption about ν', easily imply that $\nu'(x, \psi_1 \supset \psi_2) \notin \mathcal{D}$ iff $\exists y \geq x. \nu'(y, \psi_1) \in \mathcal{D} \wedge \nu'(y, \psi_2) \notin \mathcal{D}$, and that $\mathcal{W} = \langle W, \leq, \nu' \rangle$ is an $\mathcal{M}_4^{BI}$-frame (or $\mathcal{M}_{4F}^{BI}$-frame in case F is in the language). The assumption about ν' and the definition of ν' also easily imply that every condition in $C(S)$ concerning $\tilde{\neg}$, $\tilde{\vee}$, and $\tilde{\wedge}$ is satisfied in $\mathcal{W}$. We show that this is the case also with respect to the other four conditions:

- Assume that $[\mathsf{t}] \in S$. Then $\nu'(z, \varphi) \neq \bot$ for every $z \in W$ and every $\varphi \in \mathcal{F}$. Hence by induction on the structure of φ we have $\nu'(z, \varphi) \neq \bot$ for every $z \in W$ and every φ.

- Assume that $[\mathsf{n}^l_\supset]_1 \in S$. Then $[\mathsf{t}] \notin S$. Hence our assumption concerning ν and the definition of ν' trivially imply in this case that if $\nu'(\psi_1) \notin \mathcal{D}$ then $\nu'(x, \psi_1 \supset \psi_2) \in \{t, \bot\}$.

- Assume that $[\mathsf{n}^l_\supset]_2 \in S$. We show that $\nu'(x, \psi_1 \supset \psi_2) \in \{t, \bot\}$ in case $\nu'(\psi_2) \in \{t, \bot\}$. This is trivial from our assumption concerning ν' if $\psi_1 \supset \psi_2 \in \mathcal{F}$, and from the definition of $\nu'(x, \psi_1 \supset \psi_2)$ in case $\psi_1 \supset \psi_2 \notin \mathcal{F}$. (Note that if $[\mathsf{t}] \in S$ then $C([\mathsf{n}^l_\supset]_2)$ means that $\nu'(x, \psi_1 \supset \psi_2) = t$ if $\nu'(\psi_2) = t$.)

- Assume that $[\mathsf{n}^r_\supset] \in S$. We show that if $\nu'(x, \psi_1) \in \mathcal{D}$ and $\nu'(x, \psi_2) \in \{f, \top\}$ then $\nu'(x, \psi_1 \supset \psi_2) = \nu'(x, \psi_2)$. This is obvious in the cases where $\psi_1 \supset \psi_2 \in \mathcal{F}$, or $[\mathsf{t}] \notin S$. If $[\mathsf{t}] \in S$ and $\psi_1 \supset \psi_2 \notin \mathcal{F}$, then the definition of ν' implies first of all that $\nu'(x, \psi_1 \supset \psi_2) \in \{f, \top\}$ if $\nu'(x, \psi_2) \in \{f, \top\}$. If in addition $\nu'(x, \psi_1) \in \mathcal{D}$ then this fact, together with the definition of ν' and its monotonicity, implies that indeed $\nu'(x, \psi_1 \supset \psi_2) = \nu'(x, \psi_2)$.

It follows that $\mathcal{W}$ is an $\mathcal{M}$-frame as required. □

Corollary 10.30 *If $S \subseteq \mathsf{N}_4$ or $S \subseteq \mathsf{N}_3^\top$ then every S-$\mathcal{F}$-d-semiframe (see Note 10.27) can be extended to a full S-d-frame.*

Proof. This easily follows from Theorem 10.29 and Proposition 10.23. □

Note 10.31 The problem with the combination of $[\mathsf{t}]$ and $[\mathsf{n}^l_\supset]_1$ is that the condition imposed by the latter is not consistent with the condition of $\leq_k$-monotonicity in case $\bot$ is not available (that is: in the presence of $[\mathsf{t}]$).

10.2 Analyticity, Conservativity, and Decidability

As a first application, we use Theorem 10.29 to determine for which $S \subseteq A_{PAC}$ the system $HIL^+(S)$ is conservative over HIL^+. This is done in the next theorem, the proof of which nicely demonstrates how our semantic framework can be used, as well as the crucial role of the analyticity property.

Theorem 10.32 *Assume that $S \subseteq A_{PAC}$. Then $HIL^+(S)$ is a conservative extension of HIL^+ iff $\{[\mathsf{t}], [\mathsf{n}_\supset^!]_1\} \not\subseteq S$ (i.e., iff either $S \subseteq \mathsf{N}_4$ or $S \subseteq \mathsf{N}_3^\top$). Similarly, $HIL(S)$ is a conservative extension of HIL iff $\{[\mathsf{t}], [\mathsf{n}_\supset^!]_1\} \not\subseteq S$.*

Proof. The condition is necessary by Proposition 10.4.

We show that the condition is also sufficient in the case of HIL^+. (The case of HIL is similar.) It suffices for that to show that both $HIL^+(\mathsf{N}_4)$ and $HIL^+(\mathsf{N}_3^\top)$ are conservative over HIL^+. So let $\mathcal{F} = \mathcal{W}(\mathcal{L}_{CL}^+)$, and suppose that $\mathcal{T} \not\vdash_{HIL^+} \psi$, where $\mathcal{T} \cup \{\psi\} \subseteq \mathcal{F}$. We show that ψ is provable from $\mathcal{T}$ in neither $HIL^+(\mathsf{N}_4)$ nor $HIL^+(\mathsf{N}_3^\top)$. Since $\mathcal{T} \not\vdash_{HIL^+} \psi$, there is an *ordinary* two-valued Kripke frame $\langle W, \leq, \nu \rangle$ (where $\nu : W \times \mathcal{F} \to \{t, f\}$) which is a model of $\mathcal{T}$, but not of ψ (i.e. $\nu(x_0, \psi) = f$ for some $x_0 \in W$). Now we define corresponding semiframes for $HIL^+(\mathsf{N}_4)$ and $HIL^+(\mathsf{N}_3^\top)$.

$HIL^+(\mathsf{N}_4)$: Define ν_N on $W \times \mathcal{F}$ by:

$$\nu_N(x, \varphi) = \begin{cases} t & \text{if } \nu(x, \varphi) = t, \\ \bot & \text{if } \nu(x, \varphi) = f. \end{cases}$$

It is easy to check that $\langle W, \leq, \nu_N \rangle$ is an $\mathcal{M}_4^{Ib}(\mathsf{N}_4)$-$\mathcal{F}$-semiframe. (Note that any condition concerning $\tilde{\neg}$ is vacuously satisfied, since there is no sentence of the form $\neg \varphi$ in $\mathcal{F}$.)

$HIL^+(\mathsf{N}_3^\top)$: Define ν_P on $W \times \mathcal{F}$ by:

$$\nu_P(x, \varphi) = \begin{cases} \top & \text{if } \nu(x, \varphi) = t, \\ f & \text{if } \nu(x, \varphi) = f. \end{cases}$$

Again, it easy to check that $\langle W, \leq, \nu_P \rangle$ is an $\mathcal{M}_4^{Ib}(\mathsf{N}_3^\top)$-$\mathcal{F}$-semiframe.

By Theorem 10.29, $\langle W, \leq, \nu_N \rangle$ and $\langle W, \leq, \nu_F \rangle$ can respectively be extended to an $\mathcal{M}_4^{Ib}(\mathsf{N}_4)$-frame $\langle W, \leq, \nu_N' \rangle$ and an $\mathcal{M}_4^{Ib}(\mathsf{N}_3^\top)$-frame $\langle W, \leq, \nu_P' \rangle$. Obviously, both $\langle W, \leq, \nu_N' \rangle$ and $\langle W, \leq, \nu_P' \rangle$ are models of $\mathcal{T}$ in which ψ is not valid. (The latter because $\nu_N'(x_0, \psi) = \nu_N(x_0, \psi) = \bot$, and $\nu_P'(x_0, \psi) = \nu_P(x_0, \psi) = f$.) Hence, Theorem 10.18 implies that ψ is not provable from $\mathcal{T}$ in either $HIL^+(\mathsf{N}_4)$ or $HIL^+(\mathsf{N}_3^\top)$. □

Next, we turn to show that all the systems studied in this chapter are decidable. For this we need first two results concerning completeness with respect to semi-frames.

Theorem 10.33 (soundness and completeness 2) *Let $S \subseteq \mathsf{N}_4$ or $S \subseteq \mathsf{N}_3^\top$. If $\mathcal{F} = \mathsf{SF}(\mathcal{T} \cup \{\varphi\})$, then $\mathcal{T} \vdash_{HIL+(S)} \varphi$ iff $\mathcal{W} \models \varphi$ for every $\mathcal{M}_4^{Ib}(S)$-$\mathcal{F}$-semiframe $\mathcal{W}$ such that $\mathcal{W} \models \mathcal{T}$. A similar result holds for $HIL(S)$.*

Proof. Suppose $\mathcal{T} \vdash_{HIL+(S)} \varphi$, and let $\mathcal{W} = \langle W, \leq, \nu \rangle$ be an $\mathcal{M}_4^{Ib}(S)$-$\mathcal{F}$-semiframe such that $\mathcal{W} \models \mathcal{T}$. By Theorem 10.29 there is an $\mathcal{M}_4^{Ib}(S)$-frame $\mathcal{W}' = \langle W, \leq, \nu' \rangle$ such that ν' extends ν. Since $\mathcal{T} \subseteq \mathcal{F}$, this implies that $\mathcal{W}' \models \mathcal{T}$. Hence $\mathcal{W}' \models \varphi$ (because $\mathcal{T} \vdash_{HIL+(S)} \varphi$ and Theorem 10.18). Since $\varphi \in \mathcal{F}$, it follows that $\mathcal{W} \models \varphi$ as well.

For the converse, assume that $\mathcal{T} \not\vdash_{HIL+(S)} \varphi$. By Theorem 10.18 there is an $\mathcal{M}_4^{Ib}(S)$-frame $\mathcal{W}' = \langle W, \leq, \nu' \rangle$ such that $\mathcal{W}' \models \mathcal{T}$, $\mathcal{W}' \not\models \varphi$. Obviously, $\mathcal{W} = \langle W, \leq, \nu' \restriction \mathcal{F} \rangle$ is an $\mathcal{M}_4^{Ib}(S)$-$\mathcal{F}$-semiframe. Since $\mathcal{T} \cup \{\varphi\} \subseteq \mathcal{F}$, this implies that $\mathcal{W} \models \mathcal{T}$, while $\mathcal{W} \not\models \varphi$.

The proof in the case of $HIL(S)$ is identical. □

Theorem 10.34 (soundness and completeness 3) *Let S and $\mathcal{F}$ be like in Theorem 10.33. If $\mathcal{T}$ is finite, then $\mathcal{T} \vdash_{HIL+(S)} \varphi$ iff $\mathcal{W} \models \varphi$ for every $\mathcal{M}_4^{Ib}(S)$-$\mathcal{F}$-semiframe $\mathcal{W} = \langle W, \leq, \nu \rangle$ such that $\mathcal{W} \models \mathcal{T}$, and $|W| \leq 4^{|\mathcal{F}|}$. A similar result holds for $HIL(S)$.*

Proof. Again, we prove only the case of $HIL^+(S)$.

The soundness direction is immediate from Theorem 10.33.

For completeness, assume that $\mathcal{T} \not\vdash_{HIL+(S)} \varphi$. We construct an $\mathcal{M}_4^{Ib}(S)$-$\mathcal{F}$-semiframe $\mathcal{W} = \langle W, \leq, \nu \rangle$ such that $\mathcal{W} \models \mathcal{T}$, $\mathcal{W} \not\models \varphi$, and $|W| \leq 4^{|\mathcal{F}|}$. By Theorem 10.33, there is an $\mathcal{M}_4^{Ib}(S)$-$\mathcal{F}$-semiframe $\mathcal{W}^* = \langle W^*, \leq^*, \nu^* \rangle$ such that $\mathcal{W}^* \models \mathcal{T}$, $\mathcal{W}^* \not\models \varphi$. Let

$$\sim \; = \{\langle x, y \rangle \in (W^*)^2 \mid \forall \varphi \in \mathcal{F}. \nu^*(x, \varphi) = \nu^*(y, \varphi)\}.$$

Obviously, $\sim$ is an equivalence relation on W^*. Let $W = \{[x] \mid x \in W^*\}$, where $[x]$ is the equivalence class of x relative to $\sim$. Define $\leq$ on W by: $[x] \leq [y]$ iff $\nu^*(x, \psi) \leq_k \nu^*(y, \psi)$ for every $\psi \in \mathcal{F}$. Finally, define $\nu : W \times \mathcal{F} \to \{t, f, \top, \bot\}$ by: $\nu([x], \psi) = \nu(x, \psi)$. It is easy to see that $\leq$ and ν are well-defined, and that $\leq$ is a partial order on W. Now we show that ν has all the properties required for making $\mathcal{W} = \langle W, \leq, \nu \rangle$ an $\mathcal{M}_4^{Ib}(S)$-$\mathcal{F}$-semiframe. That ν satisfies the monotonicity condition and respects $\mathcal{M}_4^{Ib}(S)$ easily follows from the definitions of $\leq$ and ν, and that $\mathcal{W}^*$ respects $\mathcal{M}_4^{Ib}(S)$. We show that it satisfies also the condition concerning $\supset$.

- Suppose $\nu([x], \varphi \supset \psi) \in \mathcal{D}$, and so $\nu^*(x, \varphi \supset \psi) \in \mathcal{D}$. Let w be an element of W such that $[x] \leq w$ and $\nu(w, \varphi) \in \mathcal{D}$, and let $w = [y]$. That $[x] \leq [y]$ implies that $\nu^*(x, \varphi \supset \psi) \leq_k \nu^*(y, \varphi \supset \psi)$, and so

$\nu^*(y, \varphi \supset \psi) \in \mathcal{D}$. That $\nu([y], \varphi) \in \mathcal{D}$ implies that $\nu^*(y, \varphi) \in \mathcal{D}$. Since $\mathcal{W}^*$ respects $\mathcal{M}_4^{Ib}(S)$, the last two facts about ν^* entail that $\nu^*(y, \psi) \in \mathcal{D}$, and so $\nu(w, \psi) \in \mathcal{D}$.

- Suppose $\nu([x], \varphi \supset \psi) \notin \mathcal{D}$, and so $\nu^*(x, \varphi \supset \psi) \notin \mathcal{D}$. Since $\mathcal{W}^*$ is an $\mathcal{M}_4^{Ib}(S)$-$\mathcal{F}$-semiframe, this implies that there is $y \in W^*$ such that $x \leq^* y$, $\nu^*(y, \varphi) \in \mathcal{D}$, and $\nu^*(y, \psi) \notin \mathcal{D}$. It follows that $\nu([y], \varphi) \in \mathcal{D}$, while $\nu([y], \psi) \notin \mathcal{D}$. In turn, the facts that $x \leq^* y$ and that $\mathcal{W}^*$ satisfies the monotonicity condition imply that $[x] \leq [y]$. It follows that there is an element $w = [y]$ of W such that $[x] \leq w$, $\nu(w, \varphi) \in \mathcal{D}$, and $\nu(w, \psi) \notin \mathcal{D}$.

Now, since $\mathcal{W}^* \models \mathcal{T}$ and $\mathcal{W}^* \not\models \varphi$, the definition of ν immediately implies that $\mathcal{W}^* \models \mathcal{T}$ while $\mathcal{W}^* \not\models \varphi$. Finally, the definition of $\sim$ ensures that $\lambda \varphi \in \mathcal{F}.\nu(x, \varphi) \neq \lambda \varphi \in \mathcal{F}.\nu(y, \varphi)$ in case $[x] \neq [y]$. It follows that $|W|$ is at most $4^{|\mathcal{F}|}$, the number of functions from $\mathcal{F}$ to $\{t, f, \top, \bot\}$. Hence $\mathcal{W}$ is an $\mathcal{M}_4^{Ib}(S)$-$\mathcal{F}$-semiframe, as required. □

The soundness and completeness theorems can now be used to prove the following very important property of the constructive paraconsistent logics we study.

Theorem 10.35 (decidability) *The systems* $HIL^+(S)$ *and* $HIL(S)$ *are decidable for every* $S \subseteq \mathsf{A}_{PAC}$.

Proof. If $S \subseteq \mathsf{N}_4$ or $S \subseteq \mathsf{N}_3^\top$, then this follows from Theorem 10.34, since if $\mathcal{F}$ is finite, then so is the number of non-isomorphic $\mathcal{M}_4^{Ib}(S)$-$\mathcal{F}$-semiframes which have at most $4^{|\mathcal{F}|}$ worlds.

On the other hand, if $\{[\mathsf{t}], [\mathsf{n}_\supset^l]_1\} \subseteq S$, then Proposition 10.4 entails that $HIL^+(S) = HCL^+(S)$, and $HIL(S) = HCL^\mathsf{F}(S)$. Therefore the decidability in this case follows from Corollary 7.52. □

10.3 N4 and Other Important Systems

The constructive logics studied in this chapter split into two families:

The $\mathsf{N}_3^\top$ family : These are the logics which are induced by $HIL^+(S)$ or $HIL(S)$ for some $S \subseteq \mathsf{N}_3^\top$.

The N_4 family : These are the logics which are induced by $HIL^+(S)$ or $HIL(S)$ for some $S \subseteq \mathsf{N}_4$.

The most important advantage of the logics in the $N_3^\top$ family is that their negation $\neg$ is *complete* — a particularly important property for a paraconsistent negation, while that of the logics in the other family is not. Also, the semantics of the logics in the $N_3^\top$ family is simpler than that of the other family, because it is based on three-valued (non-deterministic) Kripke-frames, while the other uses four values. On the other hand, in the next section we show that while the cut-elimination theorem obtains for the Gentzen-types systems of the logics in the N_4 family, it fails for the logics in the $N_3^\top$ family. Instead a certain form of analytic cut-elimination does hold for logics in that family, and this form suffices for a decision procedure and other applications of cut-elimination.

Among the logics in the $N_3^\top$ family the most important are those which are induced by the following systems:

HIL$^+$({[t]}): The induced logic is the minimal (conservative) extension of positive intuitionistic logic which has a complete negation $\neg$.

HIL({[t]}): This is the minimal LFI (Chapter 8) with respect to $\neg$ which is based on positive intuitionistic logic (or full intuitionistic logic). The proof of this is very similar to the proof of Proposition 8.20: if we define $\circ\varphi = (\varphi \wedge \neg\varphi) \supset F$, then $\circ$ is a consistency operator for $\neg$ in every extension of *HIL*({[t]}). Note that here this $\circ$ is *not* a strong consistency operator, since [k] is not derivable in any system of the form *HIL*(S) (where $S \subseteq N_3^\top$). Indeed, suppose that $S \subseteq N_3^\top$. Let $W = \{w_1, w_2\}$ with $w_1 \leq w_2$, and let $\nu(w_1, P_1) = f$, $\nu(w_2, P_1) = \top$. Then $\langle W, \leq, \nu \rangle$ is an $\mathcal{M}$-$\{P_1\}$-semiframe for every $\mathcal{M}$ of the form $\mathcal{M}_{4F}^{Ib}(S)$ such that $S \subseteq N_3^\top$. By Theorem 10.29, $\langle W, \leq, \nu \rangle$ can be extended to an $\mathcal{M}$-frame $\mathcal{W} = \langle W, \leq, \nu' \rangle$, and it is easy to check that $\nu'(w_1, \circ P_1 \vee (P_1 \wedge \neg P_1)) \notin \{t, \top\}$.

HIL$^+$({[t], [c]}): As mentioned in Example 10.19, this is the system C_ω of da Costa ([127]). The induced logic is the minimal extension of **IL**$^+$ which has a complete and right involutive negation $\neg$.

HIL$^+$({[t], [c], [e]}): The induced logic is the minimal extension of **IL**$^+$ which has a complete and involutive negation $\neg$.

HIL$^+$($N_3^\top$): The induced logic is the strongest logic in $\mathcal{L}_{CL}$ of the $N_3^\top$ family, and as such it is a a very attractive system for constructive negation: It is a boldly paraconsistent logic which satisfies all the properties in the desiderata list given in Section 2.4. What is more, it is non-exploding (since its extension **PAC** is), and in addition to being complete and involutive, its negation satisfies de-Morgan rules as well as the most

10.3 N4 and Other Important Systems

intuitive connections between negation and implication. In fact, its defining proof system is obtained from $H_{\mathbf{PAC}}$ (the Hilbert-type system for the classical paraconsistent logic **PAC** which was studied in Chapter 4) by deleting its two unintuitive (at least from a constructive point of view) axioms: Pierce's axiom and $[\mathsf{n}^!_\supset]_1$. With respect to the latter it worth noting that it is a close relative of the principle which is rejected by paraconsistent logics, since $\neg\varphi \supset (\varphi \supset \psi)$ is obtained from $[\mathsf{n}^!_\supset]_1$ by one of the standard classical contra-position principles, and the converse is true as well. Another significant fact about $HIL^+(\mathsf{N}_3^\top)$ is that although it is strongly (and even boldly) paraconsistent, the basic law of contradiction $\neg(\varphi \wedge \neg\varphi)$ *is* valid in it.

$HIL(\mathsf{N}_3^\top)$: This is the strongest logic of the $\mathsf{N}_3^\top$ family. Most of what was said about $HIL^+(\mathsf{N}_3^\top)$ applies to it as well, except that it is neither non-exploding, nor boldly paraconsistent. On the other hand, it is an LFI which is based on $\mathbf{IL}^+$ (and of course the strongest such logic among those studied in this chapter).

Turning to the N_4 family, the most important logic in $\mathcal{L}_{CL}$ which belongs to it is the strongest one, $HIL^+(\mathsf{N}_4)$. This logic is better known as Nelson's paraconsistent (four-valued) logic **N4** (sometimes called $\mathbf{N}^-$), and so from now on we shall use this name for it. **N4** is a paraconsistent variant of the well-known Nelson's constructive (three-valued) logic **N3** (also known as **N**) [223, 294]. It is also the most extensively studied constructive paraconsistent logic[4], and so we concentrate on it in the rest of this section. (More information with references is given in Section 10.5.)

Our first observation concerning **N4** is that in terms of d-frames, its semantics can be described in a rather concise form.

Definition 10.36 (N4-d-frame) A d-frames $\langle W, \leq, \nu^+, \nu^- \rangle$ for $\mathcal{L}_{CL}$ is an N4-*d-frame*, if ν^- satisfies the following conditions:

- $\nu^-(w, \neg\psi) = 1$ iff $\nu^+(w, \psi) = 1$.
- $\nu^-(w, \psi \wedge \varphi) = 1$ iff $\nu^-(w, \psi) = 1$ or $\nu^-(w, \varphi) = 1$.
- $\nu^-(w, \psi \vee \varphi) = 1$ iff $\nu^-(w, \psi) = 1$ and $\nu^-(w, \varphi) = 1$.
- $\nu^-(w, \psi \supset \varphi) = 1$ iff $\nu^+(w, \psi) = 1$ and $\nu^-(w, \varphi) = 1$.

Proposition 10.37 **N4** *is sound and complete with respect to the class of* **N4**-*d-frames.*

---
[4]Thus the whole of [192] is devoted to **N4**-related paraconsistent logics.

Proof. It can easily be checked that an **N4**-d-frame is just an **N4**-d-frame for $\mathcal{L}_{CL}$. Therefore the proposition follows from Proposition 10.23. □

Corollary 10.38 *The semantics of* **N4** *that has been given in this chapter is* deterministic *(in the sense of Definition 10.9).*

Proof. Definition 10.36 implies that the values that ν^- assigns to a formula φ in an **N4**-d-frame $\langle W, \leq, \nu^+, \nu^- \rangle$ are completely determined by the values it assigns to its immediate subformulas. Using this fact, the corollary follows from Note 10.21 and Proposition 10.23. □

Note 10.39 It is straightforward to check that **N4** is actually the *only* logic of the form $\mathbf{L}_{HIL^+(S)}$ or $\mathbf{L}_{HIL(S)}$ (where $S \subseteq \mathsf{N}_4$ or $S \subseteq \mathsf{N}_3^\top$) whose semantics (as given in this chapter) is deterministic. Note that even the semantics of $\mathbf{L}_{HIL(\mathsf{N}_4)}$ is not deterministic, since in the corresponding d-frames $\nu^-(F)$ can be either 0 or 1.

Another important property of **N4** is that it can be reduced to $\mathbf{IL}^+$ by the following syntactical embedding (from [191]):

Definition 10.40 (m) Given some bijection e between the set Var and the set $\mathsf{Var} \cup \{\neg p \mid p \in \mathsf{Var}\}$ (of literals) (e.g. $e(P_{2n}) = P_n$ and $e(P_{2n+1}) = \neg P_n$ for every n), define $\mathsf{m} : \mathcal{W}(\mathcal{L}_{CL}) \to \mathcal{W}(\mathcal{L}_{CL}^+)$ inductively as follows:

- $\mathsf{m}(\psi) = e^{-1}(\psi)$ in case $\psi \in \mathsf{Var} \cup \{\neg p \mid p \in \mathsf{Var}\}$.
- For $\diamond \in \{\vee, \wedge, \supset\}$, $\mathsf{m}(\psi \diamond \varphi) = \mathsf{m}(\psi) \diamond \mathsf{m}(\varphi)$.
- $\mathsf{m}(\neg\neg\psi) = \mathsf{m}(\psi)$,
  $\mathsf{m}(\neg(\psi \wedge \varphi)) = \mathsf{m}(\neg\psi) \vee \mathsf{m}(\neg\varphi)$,
  $\mathsf{m}(\neg(\psi \vee \varphi)) = \mathsf{m}(\neg\psi) \wedge \mathsf{m}(\neg\varphi)$,
  $\mathsf{m}(\neg(\psi \supset \varphi)) = \mathsf{m}(\psi) \wedge \mathsf{m}(\neg\varphi)$.

In order to show that m can be used to reduce **N4** to $\mathbf{IL}^+$ we need the following connections between their semantics:

Proposition 10.41 *Let e and m be like in Definition 10.40.*

1. *For each* **N4**-*d-frame* $\langle W, \leq, \nu^+, \nu^- \rangle$ *there is an ordinary Kripke frame (Definition 1.66)* $\langle W, \leq, \nu \rangle$ *such that for every formula* $\psi \in \mathcal{W}(\mathcal{L}_{CL})$ *and* $w \in W$, $\nu^+(w, \psi) = 1$ *iff* $\nu(w, \mathsf{m}(\psi)) = t$ *and* $\nu^-(w, \psi) = 1$ *iff* $\nu(w, \mathsf{m}(\neg\psi)) = t$.

2. *For each ordinary Kripke frame* $\langle W, \leq, \nu \rangle$ *there is an* **N4**-*d-frame* $\langle W, \leq, \nu^+, \nu^- \rangle$ *such that for every* $\psi \in \mathcal{W}(\mathcal{L}_{CL})$ *and every* $w \in W$, $\nu(w, \mathsf{m}(\psi)) = t$ *iff* $\nu^+(w, \psi) = 1$; $\nu(w, \mathsf{m}(\neg\psi)) = t$ *iff* $\nu^-(w, \psi) = 1$.

10.3 N4 and Other Important Systems

Proof. Define a bijection $H: \{1,0\} \to \{t,f\}$ by: $H(1) = t$, $H(0) = f$.

To show the first item, let $\langle W, \leq, \nu^+, \nu^- \rangle$ be an **N4**-d-frame. Define a Kripke frame $\langle W, \leq, \nu \rangle$ by letting $\nu(w,p) = H(\nu^-(w, e(p)))$ for a variable p. Since ν^+ satisfies the persistence condition, so does ν. We show by induction on the complexity of ψ that ν has the required connections with ν^+ and ν^-.

- Suppose that ψ is a variable. Then:

$$\begin{aligned}
\nu(w, \mathsf{m}(\psi)) &= \nu(w, e^{-1}(\psi)) \\
&= H(\nu^+(w, e(e^{-1}(\psi)))) \\
&= H(\nu^+(w, \psi)), \\
\nu(w, \mathsf{m}(\neg\psi)) &= \nu(w, e^{-1}(\neg\psi)) \\
&= H(\nu^+(w, e(e^{-1}(\neg\psi)))) \\
&= H(\nu^+(w, \neg\psi)) \\
&= H(\nu^-(w, \psi)).
\end{aligned}$$

  The above equations easily imply the required connections.

- Suppose that $\psi = \neg\varphi$. Then:
  - $\nu(w, \mathsf{m}(\psi)) = t$ iff $\nu(w, \mathsf{m}(\neg\varphi)) = t$, iff (by induction hypothesis on φ) $\nu^-(w, \varphi) = 1$, iff $\nu^+(w, \neg\varphi) = 1$, iff $\nu^+(w, \psi) = 1$.
  - $\nu(w, \mathsf{m}(\neg\psi)) = t$ iff $\nu(w, \mathsf{m}(\neg\neg\varphi)) = t$, iff $\nu(w, \mathsf{m}(\varphi)) = t$, iff (by induction hypothesis on φ) $\nu^+(w, \varphi) = 1$, iff $\nu^-(w, \psi) = 1$.

- Suppose that $\psi = \tau \wedge \sigma$. Then:
  - $\nu(w, \mathsf{m}(\psi)) = t$ iff $\nu(w, \mathsf{m}(\tau) \wedge \mathsf{m}(\sigma)) = t$, iff $\nu(w, \mathsf{m}(\tau)) = t$ and $\nu(w, \mathsf{m}(\sigma)) = t$, iff (by induction hypothesis) $\nu^+(w, \tau) = 1$ and $\nu^+(w, \sigma) = 1$, iff $\nu^+(w, \psi) = 1$.
  - $\nu(w, \mathsf{m}(\neg\psi)) = t$ iff $\nu(w, \mathsf{m}(\neg\tau) \vee \mathsf{m}(\neg\sigma)) = t$, iff $\nu(w, \mathsf{m}(\neg\tau)) = t$ or $\nu(w, \mathsf{m}(\neg\sigma)) = t$, iff (by induction hypothesis) $\nu^-(w, \tau) = 1$ or $\nu^-(w, \sigma) = 1$, iff $\nu^-(w, \psi) = 1$.

- The case where $\psi = \tau \vee \sigma$ is similar to the previous one.

- Suppose that $\psi = \tau \supset \sigma$. Then:
  - $\nu(w, \mathsf{m}(\psi)) = t$ iff $\nu(w, \mathsf{m}(\tau) \supset \mathsf{m}(\sigma)) = t$, iff $\nu(w', \mathsf{m}(\tau)) = f$ or $\nu(w', \mathsf{m}(\sigma)) = t$ for every $w' \geq w$, iff (by induction hypothesis) $\nu^+(w', \tau) = 0$ or $\nu^+(w', \sigma) = 1$ in case $w' \geq w$, iff $\nu^+(w', \psi) = 1$.

- $\nu(w, \mathsf{m}(\neg\psi)) = t$ iff $\nu(w, \mathsf{m}(\tau) \wedge \mathsf{m}(\neg\sigma)) = t$, iff $\nu(w, \mathsf{m}(\tau)) = t$ and $\nu(w, \mathsf{m}(\neg\sigma)) = t$, iff (by induction hypothesis) $\nu^+(w, \tau) = 1$ and $\nu^-(w, \sigma) = 1$, iff $\nu^-(w, \psi) = 1$.

To show the second item, let $\langle W, \leq, \nu \rangle$ be an ordinary Kripke frame. For $w \in W$ and $\psi \in \mathcal{W}(\mathcal{L}_{CL})$ define: $\nu^+(w, \psi) = H^{-1}(\nu(w, \mathsf{m}(\psi)))$, $\nu^-(w, \psi) = H^{-1}(\nu(w, \mathsf{m}(\neg\psi)))$. Then the fact that $\langle W, \leq, \nu \rangle$ is Kripke frame and the definition of m easily imply that $\langle W, \leq, \nu^+, \nu^- \rangle$ is an **N4**-d-frame. (We leave the rather straightforward, but tedious, details to the reader.) That this **N4**-d-frame has the required connections with $\langle W, \leq, \nu \rangle$ follows from its very definition. □

We are ready to present the reduction of **N4** to **IL**$^+$.

Proposition 10.42 $\mathcal{T} \vdash_{\mathbf{N4}} \psi$ iff $\{\mathsf{m}(\varphi) \mid \varphi \in \mathcal{T}\} \vdash_{\mathbf{IL}^+} \mathsf{m}(\psi)$.

Proof. It is very easy to check that $\mathsf{m}(\varphi)$ is a theorem of **IL**$^-$ for every axiom φ of $HIL^+(\mathbf{N_4})$. Obviously, also $\mathsf{m}(\varphi), \mathsf{m}(\varphi \supset \psi) \vdash_{\mathbf{IL}^+} \mathsf{m}(\psi)$ for every φ and ψ. These two facts imply (using an induction on proofs in $HIL^+(\mathbf{N_4})$) that if $\mathcal{T} \vdash_{\mathbf{N4}} \psi$ then $\{\mathsf{m}(\varphi) \mid \varphi \in \mathcal{T}\} \vdash_{\mathbf{IL}^+} \mathsf{m}(\psi)$.

To show the converse, assume that $\mathcal{T} \nvdash_{\mathbf{N4}} \psi$. By Proposition 10.37 there is an **N4**-d-frame $\langle W, \leq, \nu^+, \nu^- \rangle$ which is model of $\mathcal{T}$, but $\nu^+(w, \psi) = 0$ for some $w \in W$. Hence the first part of Proposition 10.41 implies that there is ν such that $\langle W, \leq, \nu \rangle$ is a Kripke frame which is a model of $\{\mathsf{m}(\varphi) \mid \varphi \in \mathcal{T}\}$, but $\nu(w, \mathsf{m}(\psi)) = f$. It follows that $\{\mathsf{m}(\varphi) \mid \varphi \in \mathcal{T}\} \nvdash_{\mathbf{IL}^+} \mathsf{m}(\psi)$. □

Note 10.43 Using the Gentzen-type system for **N4** which is studied in the next section, it is possible to provide an alternative, purely syntactic, proof also of the 'if' part of Proposition 10.42. (This is indeed the method that is used in [191].) In turn, the 'only if' part of that proposition can be given a purely semantic proof, very similar to to that given above to the 'if' part, using the second part of Proposition 10.41.

10.4 Gentzen-type Systems

In this section we provide quasi-canonical Gentzen-type systems for all the constructive logics studied in this chapter.

10.4.1 The N$_4$ family

With logics from the N$_4$ family it is natural to associate single-conclusion Gentzen-type systems that extend Gentzen's LJ^+ or LJ (depending whether F is in the language or not).

10.4 Gentzen-type Systems

Definition 10.44 ($LJ^+(S)$, $LJ(S)$) *For $S \subseteq \mathbf{N_4}$, $LJ^+(S)$ and $LJ(S)$ are the single-conclusion Gentzen-type systems which are obtained from LJ^+ and LJ (respectively) by adding to them the single-conclusion versions of the Gentzen rules which are induced by the axioms in S (Figures 7.1, 7.2).*

Example 10.45 The Gentzen-type single-conclusion system $LJ^+(\mathbf{N_4})$ for $\mathbf{N_4}$ is presented in Figure 10.1. Note that the rules acronyms in this figure are subscripted by "I" to emphasize that these are intuitionistic, single-conclusion rules. Obviously, for every $S \subseteq \mathbf{N_4}$ the system $LJ^+(S)$ for $\mathbf{L}_{HIL^+(S)}$ is obtained from $LJ^+(\mathbf{N_4})$ by deleting the irrelevant rules (i.e., those that are neither rules of LJ^+ nor induced by the axioms in S).

Theorem 10.46 (soundness and completeness 4) *If S is a subset of $\mathbf{N_4}$ then $\vdash_{LJ^+(S)} \,=\, \vdash_{IL(\mathcal{M}_4^{\mathit{Ib}}(S))}$ and $\vdash_{LJ(S)} \,=\, \vdash_{IL(\mathcal{M}_{4F}^{\mathit{Ib}}(S))}$.*

Proof. A proof which is similar to the proof of Theorem 4.100 (but simpler) shows that $\mathcal{T} \vdash_{LJ^+(S)} \psi$ iff $\mathcal{T} \vdash_{HIL^+(S)} \psi$, and $\mathcal{T} \vdash_{LJ(S)} \psi$ iff $\mathcal{T} \vdash_{HIL(S)} \psi$. Hence the claim follows from Theorem 10.18. (An alternative direct proof is included within the proof of Theorem 10.47 below.) □

Theorem 10.47 (cut-elimination for the $\mathbf{N_4}$ family) *If $S \subseteq \mathbf{N_4}$ then $LJ^+(S)$ and $LJ(S)$ admit cut-elimination.*

Proof. The theorem can be proved using an easy adaptation of the original proof of Gentzen for LJ (that is: by a double induction on the complexity of the cut formula and the height of the proof). In what follows we present an alternative semantic proof, which is at the same time a direct proof of the completeness of $LJ^+(S)$ and $LJ(S)$ for the semantics of S-d-frames. We do the case of $LJ^+(S)$. (The case of $LJ(S)$ is practically identical.)

First we note that it is a routine matter to check that $LJ^+(S)$ (*with the cut rule*) is strongly sound for the semantics of S-d-frames, in the sense that if $s_1, \ldots, s_n \vdash_{LJ^+(S)} s$ and $\langle W, \leq, \nu^+, \nu^- \rangle$ is an S-d-frame, then s is true in every $w \in W$ in which $s_1, \ldots, s_n$ are all true (where $\Gamma \Rightarrow \varphi$ is true in w if either $\nu^+(w, \psi) = 0$ for some $\psi \in \Gamma$, or $\nu^+(w, \varphi) = 1$).[5] Therefore, to prove the theorem it suffices to show that if $\Gamma \Rightarrow \psi$ is true in every world of any S-d-frame, then there is a cut-free proof of $\Gamma \Rightarrow \psi$ in $LJ^+(S)$ (in symbols: $\vdash_{LJ^+(S)}^{\mathsf{cf}} \Gamma \Rightarrow \psi$). Accordingly, in the rest of the proof we assume that $\nvdash_{LJ^+(S)}^{\mathsf{cf}} \Gamma \Rightarrow \psi$, and construct an appropriate countermodel.

Let $\mathcal{F}$ be the set of all formulas φ such that φ is either a subformula of some formula in $\Gamma \cup \{\psi\}$, or a negation of such subformula. Obviously,

---

[5] This easily follows also from the equivalence between $LJ^+(S)$ and $HIL^+(S)$, the fact that a sequent $\varphi_1, \ldots, \varphi_n \Rightarrow \psi$ is valid in an S-d-frame iff the formula $\varphi_1 \wedge \ldots \wedge \varphi_n \supset \psi$ is valid there, Proposition 10.6, and Proposition 10.23.

Axioms: $\psi \Rightarrow \psi$

Structural Rules:

$$[\text{Cut}_I] \quad \frac{\Gamma \Rightarrow \psi \quad \Sigma, \psi \Rightarrow \varphi}{\Gamma, \Sigma \Rightarrow \varphi} \qquad [\text{Weakening}_I] \quad \frac{\Gamma \Rightarrow \varphi}{\Gamma, \psi \Rightarrow \varphi}$$

Logical Rules:

$$[\neg\neg\Rightarrow_I] \quad \frac{\Gamma, \psi \Rightarrow \varphi}{\Gamma, \neg\neg\psi \Rightarrow \varphi} \qquad [\Rightarrow\neg\neg_I] \quad \frac{\Gamma \Rightarrow \varphi}{\Gamma \Rightarrow \neg\neg\varphi}$$

$$[\wedge\Rightarrow_I] \quad \frac{\Gamma, \psi, \varphi \Rightarrow \tau}{\Gamma, \psi \wedge \varphi \Rightarrow \tau} \qquad [\Rightarrow\wedge_I] \quad \frac{\Gamma \Rightarrow \psi \quad \Gamma \Rightarrow \varphi}{\Gamma \Rightarrow \psi \wedge \varphi}$$

$$[\neg\wedge\Rightarrow_I] \quad \frac{\Gamma, \neg\psi \Rightarrow \tau \quad \Gamma, \neg\varphi \Rightarrow \tau}{\Gamma, \neg(\psi \wedge \varphi) \Rightarrow \tau}$$

$$[\Rightarrow\neg\wedge_I]_1 \quad \frac{\Gamma \Rightarrow \neg\psi}{\Gamma \Rightarrow \neg(\psi \wedge \varphi)} \qquad [\Rightarrow\neg\wedge_I]_2 \quad \frac{\Gamma \Rightarrow \neg\varphi}{\Gamma \Rightarrow \neg(\psi \wedge \varphi)}$$

$$[\vee\Rightarrow_I] \quad \frac{\Gamma, \psi \Rightarrow \tau \quad \Gamma, \varphi \Rightarrow \tau}{\Gamma, \psi \vee \varphi \Rightarrow \tau}$$

$$[\Rightarrow\vee_I]_1 \quad \frac{\Gamma \Rightarrow \psi}{\Gamma \Rightarrow \psi \vee \varphi} \qquad [\Rightarrow\vee_I]_2 \quad \frac{\Gamma \Rightarrow \varphi}{\Gamma \Rightarrow \psi \vee \varphi}$$

$$[\neg\vee\Rightarrow_I]_1 \quad \frac{\Gamma, \neg\psi, \Rightarrow \tau}{\Gamma, \neg(\psi \vee \varphi) \Rightarrow \tau} \qquad [\neg\vee\Rightarrow_I]_2 \quad \frac{\Gamma, \neg\varphi \Rightarrow \tau}{\Gamma, \neg(\psi \vee \varphi) \Rightarrow \tau}$$

$$[\Rightarrow\neg\vee_I] \quad \frac{\Gamma \Rightarrow \neg\psi \quad \Gamma \Rightarrow \neg\varphi}{\Gamma \Rightarrow \neg(\psi \vee \varphi)}$$

$$[\supset\Rightarrow_I] \quad \frac{\Gamma \Rightarrow \psi \quad \Gamma, \varphi \Rightarrow \tau}{\Gamma, \psi \supset \varphi \Rightarrow \tau} \qquad [\Rightarrow\supset_I] \quad \frac{\Gamma, \psi \Rightarrow \varphi}{\Gamma \Rightarrow \psi \supset \varphi}$$

$$[\neg\supset\Rightarrow_I]_1 \quad \frac{\Gamma, \psi \Rightarrow \tau}{\Gamma, \neg(\psi \supset \varphi) \Rightarrow \tau} \qquad [\neg\supset\Rightarrow_I]_2 \quad \frac{\Gamma, \neg\varphi \Rightarrow \tau}{\Gamma, \neg(\psi \supset \varphi) \Rightarrow \tau}$$

$$[\Rightarrow\neg\supset_I] \quad \frac{\Gamma \Rightarrow \psi \quad \Gamma \Rightarrow \neg\varphi}{\Gamma \Rightarrow \neg(\psi \supset \varphi)}$$

Figure 10.1: $LJ^+(\mathbf{N}_4)$: A proof system for $\mathbf{N4}$

10.4 Gentzen-type Systems

$\mathcal{F}$ is closed under subformulas. Given $\varphi \in \mathcal{F}$, we call $\Delta \subseteq \mathcal{F}$ φ-*maximal* if $\vdash^{cf}_{LJ+(S)} \Delta \Rightarrow \varphi$, but $\nvdash^{cf}_{LJ+(S)} \tau, \Delta \Rightarrow \varphi$ for every $\tau \in \mathcal{F} - \Delta$. Clearly, if $\Delta \cup \{\varphi\} \subseteq \mathcal{F}$ and $\Delta \nvdash^{cf}_{LJ+(S)} \varphi$ then there is Δ' such that $\Delta \subseteq \Delta' \subseteq \mathcal{F}$ and Δ' is φ-maximal. Let W be the set of all subsets of $\mathcal{F}$ that are φ-maximal for some $\varphi \in \mathcal{F}$. Obviously, $\langle W, \subseteq \rangle$ is a partially ordered set. In order to turn it into an S-$\mathcal{F}$-d-semiframe, we simultaneously define $\nu^+, \nu^- : W \times \mathcal{F} \to \{0, 1\}$ by the following recursive clauses:

- If $\tau \in \mathcal{F}$ is a variable and $\Delta \in W$, we let $\nu^-(\Delta, \tau) = 1$ iff $\tau \in \Delta$. If $\tau \in \mathcal{F}$ is not a variable, then $\nu^+(\Delta, \tau)$ is determined by the conditions on ν^+ that are given in Definition 10.20 (see Note 10.21).

- If $\tau \in \mathcal{F}$ we let $\nu^-(\Delta, \tau) = 1$ iff either $\neg \tau \in \Delta$, or if this is dictated by $C^d(ax)$ for some $ax \in S \cap \{[\mathsf{e}], [\mathsf{n}^r_\supset], [\mathsf{n}^r_\vee], [\mathsf{n}^r_\wedge]_1, [\mathsf{n}^r_\wedge]_2\}$.

First of all we show that $\mathcal{W} = \langle W, \subseteq, \nu^+, \nu^- \rangle$ is a $\emptyset - \mathcal{F}$-d-semiframe. Most of the conditions directly follow from the very definition of ν^+. The only exception is that both ν^+ and ν^- satisfy the persistence condition. So suppose that $\Delta_1 \subseteq \Delta_2$. We show by induction on the structure of τ that if $\nu^+(\Delta_1, \tau) = 1$ then $\nu^+(\Delta_2, \tau) = 1$, and if $\nu^-(\Delta_1, \tau) = 1$ then $\nu^-(\Delta_2, \tau) = 1$.

- Suppose that $\nu^+(\Delta_1, \tau) = 1$ because τ is a variable and $\tau \in \Delta_1$. Since $\Delta_1 \subseteq \Delta_2$, $\tau \in \Delta_2$ as well, and so $\nu^+(\Delta_2, \tau) = 1$.

- Suppose that $\nu^-(\Delta_1, \tau) = 1$ because $\neg \tau \in \Delta_1$. Then $\neg \tau \in \Delta_2$ as well, and so $\nu^-(\Delta_2, \tau) = 1$.

- Suppose that $\nu^+(\Delta_1, \tau) = 1$ because $\tau = \neg \sigma$, and $\nu^-(\Delta_1, \sigma) = 1$. By induction hypothesis for σ, $\nu^-(\Delta_2, \sigma) = 1$ too, and so $\nu^+(\Delta_2, \tau) = 1$.

- Suppose that $\tau = \sigma_1 \diamond \sigma_2$ where $\diamond \in \{\wedge, \vee, \supset\}$, and that $\nu^+(\Delta_1, \tau) = 1$ because this is dictated by the relevant condition on ν^+. It is easy to see that because $\Delta_1 \leq \Delta_2$, the induction hypothesis for σ_1 and σ_2 entails that the same condition forces that $\nu^+(\Delta_2, \tau) = 1$. (Details are just like in the case of Kripke frames, and are left to the reader.)

- Suppose that $\nu^-(\Delta_1, \tau) = 1$ because this is dictated by $C^d(ax)$ for some $ax \in S \cap \{[\mathsf{e}], [\mathsf{n}^r_\supset], [\mathsf{n}^r_\vee], [\mathsf{n}^r_\wedge]_1, [\mathsf{n}^r_\wedge]_2\}$. Then the induction hypothesis for the subformulas of τ easily implies that $C^d(ax)$ dictates also that $\nu^-(\Delta_2, \tau) = 1$. For example: suppose that $\tau = \sigma_1 \supset \sigma_2$, and $\nu^-(\Delta_1, \tau) = 1$ because $\nu^+(\Delta_1, \sigma_1) = 1$, $\nu^-(\Delta_1, \sigma_1) = 1$, and $[\mathsf{n}^r_\supset] \in S$. Then by the induction hypothesis for σ_1 and σ_2, $\nu^+(\Delta_2, \sigma_1) = 1$ and $\nu^-(\Delta_2, \sigma_1) = 1$. Hence $C^d([\mathsf{n}^r_\supset])$ dictates that $\nu^-(\Delta_2, \tau) = 1$ too.

Next we show by induction on the length of φ that every $\Delta \in W$ and $\varphi \in \mathcal{F}$ satisfy the following conditions:

(L) If $\varphi \in \Delta$ then $\nu^+(\Delta, \varphi) = 1$.

(R) If $\nvdash^{cf}_{LJ+(S)} \Delta \Rightarrow \varphi$ then $\nu^+(\Delta, \varphi) = 0$.

- Suppose that φ is a variable.

  (L) If $\varphi \in \Delta$ then $\nu^+(\Delta, \varphi) = 1$ by definition of ν^+.

  (R) If $\nvdash^{cf}_{LJ+(S)} \Delta \Rightarrow \varphi$ then $\varphi \notin \Delta$, and so $\nu^+(\Delta, \varphi) = 0$.

- Suppose that $\varphi = \varphi_1 \wedge \varphi_2$.

  (L) Suppose $\varphi \in \Delta$. Since $\Delta \in W$, there is $\tau \in \mathcal{F}$ such that Δ is τ-maximal. Hence $\nvdash^{cf}_{LJ+(S)} \Delta \Rightarrow \tau$. Since $\Delta \Rightarrow \tau$ can be derived from $\varphi_1, \Delta \Rightarrow \tau$ using weakening and $[\wedge \Rightarrow_l]$, also $\nvdash^{cf}_{LJ+(S)} \varphi_1, \Delta \Rightarrow \tau$. Therefore the τ-maximality of Δ implies that $\varphi_1 \in \Delta$. Hence $\nu^+(\Delta, \varphi_1) = 1$ by the induction hypothesis on φ_1. That $\nu^+(\Delta, \varphi_2) = 1$ as well is shown similarly. It follows from these two facts and the definition of ν^+ that $\nu^+(\Delta, \varphi) = 1$.

  (R) Suppose $\nvdash^{cf}_{LJ+(S)} \Delta \Rightarrow \varphi$. Since $\Delta \Rightarrow \varphi$ can be derived by an application of $[\Rightarrow \wedge_I]$ from $\Delta \Rightarrow \varphi_1$ and $\Delta \Rightarrow \varphi_2$, either $\nvdash^{cf}_{LJ+(S)} \Delta \Rightarrow \varphi_1$ or $\nvdash^{cf}_{LJ+(S)} \Delta \Rightarrow \varphi_2$. It follows by the induction hypothesis on φ_1 and φ_2 that either $\nu^+(\Delta, \varphi_1) = 0$ or $\nu^+(\Delta, \varphi_2) = 0$. In both cases $\nu^+(\Delta, \varphi) = 0$.

- The case where $\varphi = \varphi_1 \vee \varphi_2$ is left to the reader.

- Suppose that $\varphi = \varphi_1 \supset \varphi_2$.

  (L) Suppose $\varphi \in \Delta$. Let $\Delta \leq \Delta'$ (i.e. $\Delta \subseteq \Delta'$), where $\Delta' \in W$. Then there is $\tau' \in \mathcal{F}$ such that Δ' is τ'-maximal. Since $\Delta \subseteq \Delta'$ and $\varphi \in \Delta$, $\varphi \in \Delta'$ as well. Therefore $\Delta' \Rightarrow \tau'$ can be derived from $\Delta' \Rightarrow \varphi_1$ and $\varphi_2, \Delta' \Rightarrow \tau'$ using $[\supset \Rightarrow_l]$. Since $\nvdash^{cf}_{LJ+(S)} \Delta' \Rightarrow \tau'$ (because Δ' is τ'-maximal), this implies that either $\nvdash^{cf}_{LJ+(S)} \Delta' \Rightarrow \varphi_1$, or $\nvdash^{cf}_{LJ+(S)} \varphi_2, \Delta' \Rightarrow \tau'$. In the first case the induction hypothesis for φ_1 implies that $\nu^+(\Delta', \varphi_1) = 0$. In the second case the τ'-maximality of Δ' implies that $\varphi_2 \in \Delta'$, and so $\nu^+(\Delta', \varphi_2) = 1$ by the induction hypothesis for φ_2. Since this holds for every $\Delta' \in W$ such that $\Delta \leq \Delta'$, $\nu^+(\Delta, \varphi) = 1$.

  (R) Suppose $\nvdash^{cf}_{LJ+(S)} \Delta \Rightarrow \varphi$. Since $\Delta \Rightarrow \varphi$ can be derived by an application of $[\Rightarrow \supset_I]$ from $\Delta, \varphi_1 \Rightarrow \varphi_2$, $\nvdash^{cf}_{LJ+(S)} \Delta, \varphi_1 \Rightarrow \varphi_2$.

Let Δ' be an extension of $\Delta \cup \{\varphi_1\}$ which is φ_2-maximal. This implies that $\Delta' \in W$, and so $\nu^+(\Delta', \varphi_1) = 1$ and $\nu^+(\Delta', \varphi_2) = 0$ by the induction hypothesis on φ_1 and φ_2. Since $\Delta \subseteq \Delta'$ (and so $\Delta \leq \Delta'$), this implies that $\nu^+(\Delta, \varphi) = 0$.

- Suppose that $\varphi = \neg \sigma$. Then $\nu^+(\Delta, \varphi) = \nu^-(\Delta, \sigma)$.

  (L) Suppose that $\varphi \in \Delta$. Then $\nu^-(\Delta, \sigma) = 1$, and so $\nu^+(\Delta, \varphi) = 1$.

  (R) Suppose that $\vdash^{\text{cf}}_{LJ+(S)} \Delta \Rightarrow \varphi$. We show that $\nu^-(\Delta, \sigma) = 0$ (and so $\nu^+(\Delta, \varphi) = 0$) by showing that none of the cases in which $\nu^-(\Delta, \sigma) = 1$ is applicable here. Since $\vdash^{\text{cf}}_{LJ+(S)} \Delta \Rightarrow \varphi$, $\neg \sigma \notin \Delta$. It remains to show that no condition of the form $C^d(ax)$ where $ax \in S \cap \{[e], [n^r_{\supset}], [n^r_{\vee}], [n^r_{\wedge}]_1, [n^r_{\wedge}]_2\}$ can force $\nu^-(\Delta, \sigma)$ to be 1.

  – Suppose that $\sigma = \neg \sigma'$, and $[e] \in S$. Then $\varphi = \neg\neg\sigma'$. Since $\Delta \Rightarrow \varphi$ is derivable from $\Delta \Rightarrow \sigma'$ by $[\Rightarrow \neg\neg_I]$ (the counterpart of $[e]$), and $[e] \in S$, the fact that $\vdash^{\text{cf}}_{LJ+(S)} \Delta \Rightarrow \varphi$ implies that $\vdash^{\text{cf}}_{LJ+(S)} \Delta \Rightarrow \sigma'$. Hence $\nu^+(\Delta, \sigma') = 0$ by the induction hypothesis for σ', and so $C^d([e])$ is not applicable here.

  – Suppose that $\sigma = \sigma_1 \supset \sigma_2$, and $[n^r_{\supset}] \in S$. Then $\Delta \Rightarrow \varphi$ can be derived from $\Delta \Rightarrow \sigma_1$ and $\Delta \Rightarrow \neg \sigma_2$ using $[\Rightarrow \neg \supset_I]$ (the rule that corresponds here to $[n^r_{\supset}]$). Since $\vdash^{\text{cf}}_{LJ+(S)} \Delta \Rightarrow \varphi$, it follows that either $\vdash^{\text{cf}}_{LJ+(S)} \Delta \Rightarrow \sigma_1$, or $\vdash^{\text{cf}}_{LJ+(S)} \Delta \Rightarrow \neg \sigma_2$. By the induction hypothesis for σ_1 and σ_2 (which are shorter than φ), either $\nu^+(\Delta, \sigma_1) = 0$, or $\nu^+(\Delta, \neg \sigma_2) = 0$, and so $\nu^-(\Delta, \sigma_2) = 0$. In both cases $C^d([n^r_{\supset}])$ is not applicable.

  – The cases of $[n^r_{\vee}]$, $[n^r_{\wedge}]_1$, and $[n^r_{\wedge}]_2\}$ are left to the reader.

This completes the proof of **(L)** and **(R)**. Now we show that W is an S-$\mathcal{F}$-d-semiframe by showing that ν^- satisfies $C^d(ax)$ for every $ax \in S$. This follows from the definition of ν^- in case ax is a right introduction rule (i.e., if $ax \in \{[e], [n^r_{\supset}], [n^r_{\vee}], [n^r_{\wedge}]_1, [n^r_{\wedge}]_2\}$). Next we show two cases in which ax is a left introduction rule. The other four cases can be handled similarly. So let $\Delta \in W$, and τ – a formula such that Δ is τ-maximal. Then $\vdash^{\text{cf}}_{LJ+(S)} \Delta \Rightarrow \tau$.

- Suppose that $[c] \in S$, and $\nu^+(\Delta, \varphi) = 0$. Then $C^d([e])$ is not applicable. To show that $\nu^-(\Delta, \neg\varphi) = 0$, it suffices therefore to show that $\neg\neg\varphi \notin \Delta$. Assume otherwise. Since $\Delta \Rightarrow \tau$ is derivable by $[\neg\neg \Rightarrow]_I$ (the rule of $LJ^+(S)$ that corresponds to $[c]$) from $\varphi, \Delta \Rightarrow \tau$ in case $\neg\neg\varphi \in \Delta$, also $\vdash^{\text{cf}}_{LJ+(S)} \varphi, \Delta \Rightarrow \tau$. Since Δ is τ-maximal, this implies that $\varphi \in \Delta$. Hence $\nu^+(\Delta, \varphi) = 1$ by (L). This contradicts the assumption that $\nu^+(\Delta, \varphi) = 0$.

- Suppose that $[\mathsf{n}_\supset^l]_2 \in S$, and $\nu^-(\Delta, \psi) = 0$. Then $C^d([\mathsf{n}_\supset^r])$ is not applicable. To show that $\nu^-(\Delta, \varphi \supset \psi) = 0$, it suffices therefore to show that $\neg(\varphi \supset \psi) \notin \Delta$. Assume otherwise. Since $\Delta \Rightarrow \tau$ is derivable by $[\neg \supset \Rightarrow]_2$ (the rule of $LJ^+(S)$ that corresponds to $[\mathsf{n}_\supset^l]_2$) from $\neg\psi, \Delta \Rightarrow \tau$ in case $\neg(\varphi \supset \psi) \in \Delta$, also $\vdash^{cf}_{LJ^+(S)} \neg\psi, \Delta \Rightarrow \tau$. Since Δ is τ-maximal, this implies that $\neg\psi \in \Delta$. Hence $\nu^-(\Delta, \psi) = 1$ (by definition of ν^-). This contradicts the assumption that $\nu^-(\Delta, \psi) = 0$.

It is now simple to prove that $\Gamma \Rightarrow \psi$ is refuted in the S-$\mathcal{F}$-d-frame $\mathcal{W}$. Since $\nvdash^{cf}_{LJ^+(S)} \Gamma \Rightarrow \psi$, Γ can be extended to a ψ-maximal subset Γ^* of $\mathcal{F}$. Since $\Gamma^* \in W$ (by definition of W), it has properties (L) and (R). Now (L) implies that $\nu^+(\Gamma^*, \varphi) = 1$ for $\varphi \in \Gamma$, while (R) implies that $\nu^+(\Gamma^*, \psi) = 0$.

Finally, by Corollary 10.30 the S-$\mathcal{F}$-d-frame $\langle W, \leq, \nu^+, \nu^- \rangle$ can be extended to a full S-d-frame $\langle W, \leq, \nu_0^+, \nu_0^- \rangle$. This S-d-frame has a world (Γ^*) in which the sequent $\Gamma \Rightarrow \psi$ is not true. □

Corollary 10.48 $LJ^+(\mathsf{N}_4)$ and $LJ(\mathsf{N}_4)$ admit cut-elimination.

10.4.2 The Constructive Logics with Excluded Middle

The single-conclusion framework used for the N_4 family is not adequate for logics which have a complete negation. Thus, for such logics we use instead extensions of the multiple-conclusion systems LJ_m^+ and LJ_m (Definition 1.138). (This can be done for *all* the logics studied in this chapter.)

Definition 10.49 ($LJ_m^+(S)$, $LJ_m(S)$) Let $S \subseteq \mathsf{N}_4 \cup \mathsf{N}_3^\top$. The Gentzen-type systems $LJ_m^+(S)$ and $LJ_m(S)$ are obtained from LJ_m^+ and LJ_m (respectively) by adding to them the Gentzen-type rules which are induced by the axioms in S (according to Figures 7.1 and 7.2).

Theorem 10.50 (soundness and completeness 5) *Let* $S \subseteq \mathsf{N}_4 \cup \mathsf{N}_3^\top$. *Then* $\vdash_{LJ_m^+(S)} = \vdash_{IL(\mathcal{M}_4^{lb}(S))}$ *and* $\vdash_{LJ_m(S)} = \vdash_{IL(\mathcal{M}_{4F}^{lb}(S))}$.

Proof. Again a proof which is similar to the proof of Theorem 4.100 shows that $\mathcal{T} \vdash_{LJ_m^+(S)} \psi$ iff $\mathcal{T} \vdash_{HIL^+(S)} \psi$, and $\mathcal{T} \vdash_{LJ_m(S)} \psi$ iff $\mathcal{T} \vdash_{HIL(S)} \psi$. Hence the claim follows from Theorem 10.18. (Again, an alternative direct proof is included within the proof of Theorem 10.52 below.) □

When it comes to the issue of cut-elimination things become more complicated here than they are in the single-conclusion case.

Proposition 10.51 *Suppose that* $[\mathsf{t}] \in S \subseteq \mathsf{A}_{PAC}$. *Then* $LJ_m^+(S)$ *and* $LJ_m(S)$ *do not admit cut-elimination.*

10.4 Gentzen-type Systems

Proof. Using [t] ($= [\Rightarrow \neg]$), $[\supset \Rightarrow]$, and $[\Rightarrow \supset]$ one can derive $\Rightarrow p, \neg p$ and $\neg p \Rightarrow (\neg p \supset q) \supset q$, where p and q are distinct propositional variables. Using a cut, we derive $\Rightarrow p, (\neg p \supset q) \supset q$. However, it is obvious that no cut-free proof of this sequent exists even in $LJ_m(\mathsf{N}_3^\top)$. □

The next theorem provides a quite good substitute (which still suffices for a decision procedure) for the systems dealt with in Proposition 10.51.[6]

Theorem 10.52 (quasi-analytic cut-elimination) *Let $S \subseteq \mathsf{N}_3^\top$. A sequent s is provable in $LJ_m^+(S)$ or $LJ_m(S)$ iff it has there a proof in which every cut is quasi-analytic in the sense that the cut formula is either a subformula of some formula of s, or a negation of such a subformula.*

Proof. As usual, we show only the case of $LJ_m^+(S)$.

Again, it is a routine matter to check that $LJ_m^+(S)$ (with the cut rule) is strongly sound for the semantics of $\mathcal{M}_4^{lb}(S)$-frames. Therefore to prove the theorem it suffices to show that if a sequent s is true in every world of any $\mathcal{M}_4^{lb}(S)$-frame, then there is a quasi-analytic proof of s in $LJ_m^+(S)$. So assume that s is a sequent which has no quasi-analytic proof in $LJ_m^+(S)$. Let $\mathcal{F}'$ be the set of subformulas of s. By Theorem 10.29, it suffices to construct an $\mathcal{M}_4^{lb}(S)$-$\mathcal{F}'$-semiframe in which s is not valid. Let $\mathcal{F}'' = \mathcal{F}' \cup \{\neg \psi \mid \psi \in \mathcal{F}'\}$, and call a proof in $LJ_m^+(S)$ an *s-proof* if every cut in it is on some formula in $\mathcal{F}''$. Let W be the set of all sequents which do not have s-proofs in $LJ_m^+(s)$, and the union of their two sides is $\mathcal{F}''$. Obviously, if $\Gamma \cup \Delta \subseteq \mathcal{F}''$, $\Gamma \Rightarrow \Delta$ does not have an s-proof in $LJ_m^+(s)$, and $\psi \in \mathcal{F}''$, then either $\psi, \Gamma \Rightarrow \Delta$ or $\Gamma \Rightarrow \Delta, \psi$ does not have an s-proof in $LJ_m^+(s)$. It follows that any sequent which consists of elements of $\mathcal{F}''$ and has no s-proof in $LJ_m^+(s)$ can be extended to an element of W. In particular, s itself is a subsequent of some sequent $\Gamma_0 \Rightarrow \Delta_0 \in W$. Define now a partial order $\leq$ on W as follows: $\Gamma_1 \Rightarrow \Delta_1 \leq \Gamma_2 \Rightarrow \Delta_2$ if $\Gamma_1 \subseteq \Gamma_2$ (iff $\Delta_2 \subseteq \Delta_1$, since $\Gamma_1 \cup \Delta_1 = \Gamma_2 \cup \Delta_2 = \mathcal{F}''$). Finally, define $\nu : W \times \mathcal{F}' \to \mathcal{T}$ by:

$$\nu(\Gamma \Rightarrow \Delta, \psi) = \begin{cases} f & \psi \in \Delta, \\ t & \psi \in \Gamma, \neg \psi \in \Delta, \\ \top & \psi \in \Gamma, \neg \psi \in \Gamma. \end{cases}$$

ν is well-defined, because if $\Gamma \Rightarrow \Delta \in W$ and $\psi \in \mathcal{F}'$ then $\Gamma \cap \Delta = \emptyset$ and $\Gamma \cup \Delta = \mathcal{F}''$ (and so $\{\psi, \neg\psi\} \subseteq \Gamma \cup \Delta$). That ν is persistent is immediate from the definitions. As examples of the satisfaction of the other conditions needed to make $\langle W, \leq, \nu \rangle$ an $\mathcal{M}_4^{lb}(S)$-$\mathcal{F}'$-semiframe, we prove here the four conditions concerning $\supset$. (Note that $[\mathsf{n}_\supset^l]_1 \notin S$, since $S \subseteq \mathsf{N}_3^\top$.)

---

[6]Unfortunately, it was shown in [45] that even *analytic* cut-elimination (Definition 1.120 and Note 1.121) fails in $LJ_m^+(S)$ and $LJ_m(S)$ for some $S \subseteq \mathsf{N}_3^\top$ (including $\mathsf{N}_3^\top$ itself).

- Suppose that $\varphi \supset \psi \in \mathcal{F}'$, and $\nu(\Gamma \Rightarrow \Delta, \varphi \supset \psi) \in \mathcal{D}$. Then $\varphi \supset \psi \in \Gamma$. Let $\Gamma \Rightarrow \Delta \leq \Gamma' \Rightarrow \Delta'$, and suppose $\nu(\Gamma' \Rightarrow \Delta', \varphi) \in \mathcal{D}$. Then $\varphi \in \Gamma'$. Since also $\varphi \supset \psi \in \Gamma'$ (as $\Gamma \subseteq \Gamma'$), it is not possible that $\psi \in \Delta'$ (because otherwise $\Gamma' \Rightarrow \Delta'$ would have a cut-free proof, and so an s-proof). Hence $\psi \in \Gamma'$, and so $\nu(\Gamma' \Rightarrow \Delta', \psi) \in \mathcal{D}$ as well.

- Suppose that $\varphi \supset \psi \in \mathcal{F}'$, and $\nu(w, \varphi \supset \psi) \notin \mathcal{D}$, where $w = \Gamma \Rightarrow \Delta$. Then $\varphi \supset \psi \in \Delta$. Thus $\Gamma \Rightarrow \varphi \supset \psi$ does not have an s-proof, whence $\Gamma, \varphi \Rightarrow \psi$ does not have an s-proof either. Since $\Gamma \cup \{\varphi, \psi\} \subseteq \mathcal{F}''$, this implies that there is $w' = \Gamma' \Rightarrow \Delta' \in W$ such that $\Gamma \cup \{\varphi\} \subseteq \Gamma', \psi \in \Delta'$. Hence $w \leq w'$, and $\nu(w', \varphi) \in \mathcal{D}$, while $\nu(w', \psi) \notin \mathcal{D}$.

- Suppose $[n^r_{\supset}] \in S$. To show that $C([n^r_{\supset}])$ is satisfied, let $\varphi \supset \psi \in \mathcal{F}'$, and assume that $\nu(\Gamma \Rightarrow \Delta, \varphi) \in \mathcal{D}$, $\nu(\Gamma \Rightarrow \Delta, \psi) \in \{f, \top\}$. The first assumption implies that $\varphi \in \Gamma$, while the second implies that $\neg\psi \in \Gamma$, because $\{\psi, \neg\psi\} \not\subseteq \Delta$ (since $\Rightarrow \psi, \neg\psi$ has a cut-free proof in $LJ^+_m(S)$ in case $[t] \in S$). Since $\varphi, \neg\psi \Rightarrow \neg(\varphi \supset \psi)$ has a cut-free proof in case $[n^l_{\supset}] \in S$, $\neg(\varphi \supset \psi)$ cannot be in Δ. As $\neg(\varphi \supset \psi) \in \mathcal{F}''$, we get $\neg(\varphi \supset \psi) \in \Gamma$, and so $\nu(\Gamma \Rightarrow \Delta, \varphi \supset \psi) \in \{f, \top\}$. Given the global conditions concerning ν and $\supset$ proved above, this implies $C([n^r_{\supset}])$.

- Suppose $[n^l_{\supset}]_2 \in S$. Let $\varphi \supset \psi \in \mathcal{F}'$, and assume that $\nu(\Gamma \Rightarrow \Delta, \psi) = t$. Then $\neg\psi \in \Delta$ (by the definition of ν). Since $\neg(\varphi \supset \psi) \Rightarrow \neg\psi$ has a cut-free proof in case $[n^l_{\supset}]_2 \in S$, $\neg(\varphi \supset \psi) \notin \Gamma$. Hence $\neg(\varphi \supset \psi) \in \Delta$ (because $\neg(\varphi \supset \psi) \in \mathcal{F}''$). It follows that $\nu(\Gamma \Rightarrow \Delta, \varphi \supset \psi) \neq \top$ (by definition of ν), and $\nu(\Gamma \Rightarrow \Delta, \varphi \supset \psi) \neq f$ (because $\tau \notin \Delta$ in case $\neg\tau \in \Delta$). Hence $\nu(\Gamma \Rightarrow \Delta, \varphi \supset \psi) = t$. This implies $C([n^l_{\supset}]_2)$.

Now, by the definitions of ν and $\Gamma_0 \Rightarrow \Delta_0$, ν refutes s in the world $\Gamma_0 \Rightarrow \Delta_0$ of W. Hence, $\langle W, \leq, \nu \rangle$ is a finite countermodel of s as required. □

10.5 Bibliographical Notes and Further Reading

A general introduction to intuitionism and constructive mathematics is given, e.g., in [85, 180, 181, 284]. The logic **N4** was introduced in [2]. It serves as a basis of various useful paraconsistent logics (see, e.g., [149, 191, 228, 269, 300]), and a number of applications in computer science, such as logic programming [6, 295] and default logic [240]. A comprehensive study of this logic and related ones, with many more references, is given in Kamide and Wansing's book [192] (see also [191]). The main sources of the results on the other systems studied in this chapter are [44] and [45].

Part V

Relevance

Chapter 11

What is a Relevant Logic?

There is a common belief that in valid inferences the assumptions should somehow be *relevant* to the conclusion. This belief is perhaps the most intuitive reason for rejecting the principle which allows the inference of any proposition q from the contradicting assumptions p and $\neg p$. From this point of view the problem of developing appropriate paraconsistent logics is a part of a more general problem: that of developing appropriate logics in which the factor of relevance is taken into account. Such logics are known as *relevant logics*. Obviously, any such logic should be paraconsistent (though not vice versa). However, paraconsistency is not the end here, but a welcome by-product of the more general principle of relevance.

It is not the purpose of this book to provide solutions to difficult philosophical problems, or to try to settle deep philosophical debates. Accordingly, in what follows we shall not try to provide a characterization of relevant logics which is based on the search for "the one true logic" (as Anderson and Belnap tried to do in [4]), or even just the one which best reflects our intuitions about relevance. Instead, we shall start this part with an attempt to present basic formal criteria for relevance that seem to be adequate. The family of logics which satisfy these criteria includes the main logics that have been suggested in the literature (see e.g., [151, 210]) as relevant logics. The choice which logics in this family to study here in detail is then made according to the usual criteria that have served us throughout this book: the availability of effective semantics (on which a decision procedure can be based), the availability of decent proof theory (again, on which a decision procedure can be based), and maximality properties. Moreover: since the book is about paraconsistency, we shall mainly concentrate on relevant logics with negation (even though in the literature on relevant logics pure implicational calculi are already of great interest).

11.1 General Considerations

We begin with some considerations about relevance that are independent of any language and are not connected with any particular connective.

11.1.1 The Basic Relevance Criterion

Our starting point is the notion of *uniformity*, introduced in Definition 3.23. Being uniform seems to be a very natural demand from a logic. This is because if a theory $\mathcal{T}_2$ shares no content with $\mathcal{T}_1 \cup \{\psi\}$ then it should not be relevant to the question whether $\mathcal{T}_1 \vdash \psi$ or not. However, the definition of uniformity does allow one exceptional case: when $\mathcal{T}_2$ is inconsistent in the strongest possible meaning of the term, i.e., when it implies every formula. Now, the possibility of this exception is not compatible with the idea of relevance. Accordingly, the most basic criterion for relevance seems to us to be the following:

Definition 11.1 (basic relevance criterion) A logic $\mathbf{L} = \langle \mathcal{L}, \vdash_\mathbf{L} \rangle$ satisfies the *basic relevance criterion* if its consequence relation $\vdash_\mathbf{L}$ is uniform (Definition 3.23) and non-exploding (Definition 2.17).

The following proposition provides an equivalent formulation of the basic relevance criterion, which presents it as our first version of the fundamental relevant principle of *variable-sharing*:

Proposition 11.2 *A logic* $\mathbf{L} = \langle \mathcal{L}, \vdash_\mathbf{L} \rangle$ *satisfies the* basic relevance criterion *iff for every two theories* $\mathcal{T}_1, \mathcal{T}_2$ *and formula* ψ*, we have that* $\mathcal{T}_1 \vdash_\mathbf{L} \psi$ *whenever* $\mathcal{T}_1, \mathcal{T}_2 \vdash_\mathbf{L} \psi$ *and* $\mathcal{T}_2$ *has no propositional variables in common with* $\mathcal{T}_1 \cup \{\psi\}$.

Proof. Suppose that $\mathbf{L}$ satisfies the basic relevance criterion, and let $\mathcal{T}_1, \mathcal{T}_2 \vdash_\mathbf{L} \psi$, where $\mathcal{T}_2$ has no propositional variables in common with $\mathcal{T}_1 \cup \{\psi\}$. The latter assumption implies that $\mathsf{Var}(\mathcal{T}_2) \neq \mathsf{Var}(\mathcal{L})$. Since $\mathbf{L}$ is non-exploding, this and the uniformity of $\mathbf{L}$ imply that $\mathcal{T}_1 \vdash_\mathbf{L} \psi$.

For the converse, assume that $\mathbf{L}$ satisfies the condition given in the proposition. This immediately implies that $\mathbf{L}$ is uniform. To show that it is also non-exploding, suppose that $\mathsf{Var}(\mathcal{T}) \neq \mathsf{Var}(\mathcal{L})$. Let q be a variable that does not occur in any formula of $\mathcal{T}$, and assume for contradiction that $\mathcal{T} \vdash_\mathbf{L} q$. Then by substituting $\mathcal{T}_1 = \emptyset$ and $\mathcal{T}_2 = \mathcal{T}$ in the above condition we get that $\vdash_\mathbf{L} q$. This contradicts the non-triviality of $\mathbf{L}$. □

Corollary 11.3 *Suppose that* $\mathbf{L} = \langle \mathcal{L}, \vdash_\mathbf{L} \rangle$ *is a logic that satisfies the basic relevance criterion. Then:*

11.1 General Considerations

1. *If $\mathcal{T} \vdash_\mathbf{L} \psi$ then either $\vdash_\mathbf{L} \psi$, or $\mathcal{T}$ and ψ share a propositional variable.*

2. *$\mathcal{T} \nvdash_\mathbf{L} q$ if q is a variable that does not occur in any formula of $\mathcal{T}$.*

Corollary 11.4 *Suppose that $\mathbf{L} = \langle \mathcal{L}, \vdash_\mathbf{L} \rangle$ is a logic that satisfies the basic relevance criterion, and $\neg$ is a unary connective of $\mathcal{L}$. Then:*

1. *$\mathbf{L}$ is pre-$\neg$-paraconsistent with respect to $\neg$.*

2. *If $\neg$ is a negation for $\mathbf{L}$ then $\mathbf{L}$ is boldly (so strongly) $\neg$-paraconsistent, and not controllably $\neg$-explosive in contact with any σ.*

Example 11.5 No LFI (Chapter 8) satisfies the basic relevance criterion, since in all of them $p, \neg p, \circ p \vdash q$.

Note 11.6 For some of the most well-known relevant logics, it is not absolutely clear what is the intended consequence relation (see [35] for a discussion). Still, it seems that almost all of them satisfy the basic relevance criterion according to any of the sensible candidates for the associated consequence relation (even though this is not always easy to prove!). The only obvious exception are those 'relevant logics' which include in their language a primitive or definable propositional constant F (sometimes denoted by $\bot$ or 0) with the property that $\mathsf{F} \vdash \psi$ is valid for every ψ. In our opinion the inclusion of such a constant is indeed incoherent with the idea of relevance.

Proposition 11.7 *Suppose that $\mathbf{L} = \langle \mathcal{L}, \vdash_\mathbf{L} \rangle$ is a finitary logic that satisfies the basic relevance criterion. Then $\mathbf{L}$ has a characteristic matrix.*

Proof. This follows from Theorem 3.24. □

Example 11.8

1. Since q follows from $\{p, \neg p\}$ in classical logic and in intuitionistic logic, these logics do not satisfy the basic relevance criterion. However, their *positive fragments* are easily seen to satisfy it. Thus, in the intuitionistic case, if $\mathcal{T}_1 \nvdash_\mathbf{L} \psi$ then there is Kripke model $\mathcal{K}$ of $\mathcal{T}_1$ which is not a model of ψ. Now, if $\mathcal{T}_2$ has no propositional variable in common with $\mathcal{T}_1 \cup \{\psi\}$, then by assigning in every world t to every variable occurring in $\mathcal{T}_2$, $\mathcal{K}$ is turned into a model of $\mathcal{T}_1, \mathcal{T}_2$ which is not a model of ψ. Hence $\mathcal{T}_1, \mathcal{T}_2 \nvdash_\mathbf{L} \psi$ in such a case.

2. Let $\mathcal{M} = \langle \{t, \top, f\}, \{t, \top\}, \mathcal{O} \rangle$ be a three-valued paraconsistent logic (see Proposition 4.4). Assume that all operations of $\mathcal{O}$ are $\{\top\}$-closed (i.e., that $\tilde{\diamond}(\top, \top, \ldots, \top) = \top$ for every connective $\diamond$ of the language). Then $\mathbf{L}_\mathcal{M}$ satisfies the basic relevance criterion: if $\mathsf{Var}(\mathcal{T}_2) \cap \mathsf{Var}(\mathcal{T}_1 \cup$

$\{\psi\}) = \emptyset$, then by assigning $\top$ to any p in $\mathsf{Var}(\mathcal{T}_2)$ we can turn any countermodel of $\mathcal{T}_1 \vdash_\mathcal{M} \psi$ into a countermodel of $\mathcal{T}_1, \mathcal{T}_2 \vdash_\mathcal{M} \psi$. On the other hand, if the language of $\mathcal{M}$ includes an operation $\diamond$ which is not $\{\top\}$-closed, then $\mathcal{M}$ does *not* satisfy the basic relevance criterion. For example, if $\tilde{\diamond}(\top, \top, \ldots, \top) = t$ then by Proposition 4.4 $\mathcal{T} = \{p, \neg p, \neg \diamond (p, p, \ldots, p)\}$ has no model in $\mathcal{M}$, and so $p, \neg p, \neg \diamond (p, p, \ldots, p) \vdash_\mathcal{M} q$.

Example 11.8 shows that we cannot be satisfied with the basic relevance criterion. Thus both of the positive logics mentioned in its first item have $q \supset (p \supset q)$ as a valid formula, while the rejection of this 'paradox of material implication' has been one of the main motivations for developing relevant logics.

11.1.2 A Plausible Semantic Criterion

Proposition 11.7 suggests a natural direction for going beyond the basic relevance criterion: to impose appropriate constraints on the characteristic matrix whose existence is guaranteed by that proposition for every logic which satisfies that criterion.

Let **L** be one of the three-valued logics mentioned in Example 11.8. Then any two paradoxical formulas are necessarily congruent in **L** (formally this is reflected by the fact that $p, \neg p, q, \neg q, \psi[p/r] \vdash_\mathbf{L} \psi[q/r]$ for every p, q, r and ψ). Intuitively, this state of affairs is in a direct conflict with principles of relevance. More generally, if a logic is induced by a finite matrix with n elements, then in any state of affairs any set of $n+1$ formulas necessarily includes two different formulas which are absolutely congruent in that state of affairs (formally: if ν is a valuation, and $\psi_1, \ldots \psi_{n+1}$ are formulas, then there are $1 \leq i < j \leq n+1$ such that for any formula φ and any propositional variable p: $\nu(\varphi[\psi_i/p]) = \nu(\varphi[\psi_j/p])$). This again is in conflict with the idea of relevance. It seems counterintuitive that there is an a-priori, logically dictated, fixed finite bound on the number of distinct propositions (or even just distinct paradoxical propositions). According to this intuition, any characteristic matrix for a relevant logic should necessarily be infinite. Actually, it seems reasonable to make a little bit stronger demand:

Definition 11.9 (minimal semantic relevance criterion) A logic satisfies the *minimal semantic relevance criterion* if it does not have a finite weakly characteristic matrix.

Note 11.10 The main reason that the minimal semantic relevance criterion forbids a relevant logic **L** to have even a finite *weakly* characteristic matrix is that for the crucial Definition 11.30 below (in which the relevant entailment

11.1 General Considerations

induced by a reasonable relevant logic is defined), only the set of valid formulas of **L** is relevant. Another reason is that the existence of a finite weakly characteristic matrix is often reflected in the validity of counter-intuitive (from a relevance point of view) formulas. Thus, the existence of a 3-valued weakly characteristic matrix is frequently reflected by a formula of the form:

$$(p_1 \leftrightarrow p_2) \vee (p_1 \leftrightarrow p_3) \vee (p_1 \leftrightarrow p_4) \vee (p_2 \leftrightarrow p_3) \vee (p_2 \leftrightarrow p_4) \vee (p_3 \leftrightarrow p_4),$$

where $\leftrightarrow$ and $\vee$ are appropriate equivalence and disjunction (respectively) available in the logic.

Note 11.11 To the best of our knowledge, our minimal semantic relevance criterion has never been suggested before as a criterion for relevance.[1] Nevertheless, all the main systems that have been designed in the literature to be relevant logics do satisfy it. This includes all the main systems studied in this part of the book. (See in particular Theorem 12.20 and Theorem 13.33 below.)

Definition 11.12 (semi-relevant logics) A logic **L** which satisfies both the basic relevance criterion and the minimal semantic relevance criterion is called *semi-relevant*.

Note 11.13 The term 'semi-relevant' has frequently been used in the literature in connection with the logic **RM** (to be studied in Chapter 15), but it has never been precisely defined. Now, **RM** is indeed semi-relevant according to Definition 11.12, but not relevant according to the definition given later in this chapter. Thus, Definition 11.12 is coherent with the relevant literature.

Note 11.14 By Corollary 11.4, every semi-relevant logic is necessarily pre-$\neg$-paraconsistent (for every unary connective $\neg$ of the language).

Note 11.15 By Proposition 11.7, every semi-relevant logic has a characteristic matrix, but any such matrix is necessarily infinite.

Example 11.16 It is easy to see that Batens' paraconsistent logic **CLuN** satisfies the basic relevance criterion. Hence Proposition 7.12 implies that **CLuN** is a semi-relevant logic. In Chapter 15 we show that so is the logic **RM** (a logic which is extensively studied in [4, 151]). Nevertheless, since $\vdash_{\mathbf{CLuN}} p \supset (q \supset q)$ and $\vdash_{\mathbf{RM}} \neg(p \to p) \to (q \to q)$, neither of these logics has ever been considered to be a relevant logic.

---

[1] Not even in [52], on which this chapter is based.

Example 11.16 shows that the two general criteria suggested above should be taken only as *necessary* conditions for relevance, but not as sufficient. In the next section we consider a more radical approach to relevance.

11.2   Relevant Entailments

To get a better characterization of relevance it might seem intuitive to demand a stronger form of the first item of Corollary 11.3: that $\mathcal{T} \not\vdash_\mathbf{L} \psi$ if there is no variable which occurs both in ψ and $\mathcal{T}$. However, if $\mathbf{L}$ has a formula ψ such that $\vdash_\mathbf{L} \psi$, then this strong version is in a direct conflict with the monotonicity property of $\mathbf{L}$. Hence such a strong version is compatible only with a *nonstandard* notion of "consequence relation", in which the condition of monotonicity is abandoned. This is the approach that was adopted (somewhat implicitly) by the relevance logicians. Next we present the notion they seem to use. The main idea behind it is to replace the condition of monotonicity by an appropriate version of the variable-sharing principle, which is stronger than that reflected by the basic relevance criterion.

Note: *This section should best be read in parallel to Section 1.1 above.*

Definition 11.17 (relevant entailment) Let $\mathcal{L}$ be a propositional language.

- A *pre-entailment* on $\mathcal{L}$ is a relation $\mathrel{\vdash}$ between finite *multisets* of formulas of $\mathcal{L}$ and formulas of $\mathcal{L}$ that has the following properties:

    Reflexivity: $\varphi \mathrel{\vdash} \varphi$ for every φ.
    Transitivity: If $\Gamma_1 \mathrel{\vdash} \varphi$, and $\Gamma_2, \varphi \mathrel{\vdash} \psi$, then $\Gamma_1, \Gamma_2 \mathrel{\vdash} \psi$.

- An *entailment* on $\mathcal{L}$ is a pre-entailment on $\mathcal{L}$ which satisfies the following two further conditions:

    Structurality: Suppose that θ is an $\mathcal{L}$-substitution, and that $\Gamma \mathrel{\vdash} \varphi$. Then $\theta(\Gamma) \mathrel{\vdash} \theta(\varphi)$.
    Weak consistency: $\not\mathrel{\vdash} p$ (i.e., $[\,] \not\mathrel{\vdash} p$) in case p is a variable.

- A (pre-) entailment on $\mathcal{L}$ is *relevant* if it has the following properties:

    Strong relevance criterion: If Γ_2 is not empty, and $\Gamma_1, \Gamma_2 \mathrel{\vdash} \varphi$, then Γ_2 and Γ_1, φ share some variable. In particular: if Γ is not empty, and $\Gamma \mathrel{\vdash} \varphi$, then Γ and φ share some variable.
    Replacement property: Whenever φ and ψ are equivalent in $\mathrel{\vdash}$ (i.e., $\varphi \mathrel{\vdash} \psi$ and $\psi \mathrel{\vdash} \varphi$), then they are congruent (or "indistinguishable") in $\mathrel{\vdash}$ (i.e., $\sigma[\varphi/p] \mathrel{\vdash} \sigma[\psi/p]$ and $\sigma[\psi/p] \mathrel{\vdash} \sigma[\varphi/p]$ for every formula σ and variable p).

11.2 Relevant Entailments

Note 11.18 From the strong relevance criterion it follows that the condition of non-triviality used in Definition 1.3 (i.e., $p \not\vdash q$) holds for any relevant pre-entailment. However, since monotonicity is not demanded for such relations, this fact does not entail the above condition of weak consistency.

Note 11.19 Having the replacement property intuitively means that if each of φ and ψ relevantly entails the other in **L**, then the two formulas express *the same proposition*, and so each of them can replace the other in any context. As we saw in previous chapters, most of the paraconsistent logics do *not* have this property. However, we do expect it from a *relevant* entailment.

Note 11.20 It might seem strange that relevant entailment is a relation between *multisets* of formulas and formulas, rather than between *sets* of formulas and formulas (as consequence relations are usually and most naturally taken to be). Unfortunately, the relevant entailments induced by the most famous relevant logics studied in the literature (and described in the next chapters) cannot be viewed as relations between sets and formulas. To see exactly why, we need concepts that will be introduced later in this chapter (in Section 11.3.1). However, since this is an important point, let us give an explanation already at this stage. So in most relevant logics the underlying relation of relevant entailment is reflected within the language by a connective $\rightarrow$ such that $\Gamma, \varphi \mathrel{\mid\!\sim} \psi$ iff $\Gamma \mathrel{\mid\!\sim} \varphi \rightarrow \psi$. In particular: $\varphi \rightarrow (\varphi \rightarrow \varphi)$ should be taken as valid iff $\varphi, \varphi \mathrel{\mid\!\sim} \varphi$. Now, if by the left hand side of the latter we mean the *set* $\{\varphi, \varphi\}$ (which is identical to $\{\varphi\}$) then the reflexivity of $\mathrel{\mid\!\sim}$ implies that $\varphi \rightarrow (\varphi \rightarrow \varphi)$ should be valid. However, actually this schema is *not* valid in most relevant logics. This can be justified only if the left hand side of '$\varphi, \varphi \mathrel{\mid\!\sim} \varphi$' is understood as the *multiset* $[\varphi, \varphi]$. Still, we do believe that seeing assumptions as coming in sets is more intuitive and natural then seeing them as coming in multisets. This belief motivates our next definition.

Definition 11.21 (s-entailment) Let $\mathcal{L}$ be a propositional language. An *s-entailment* on $\mathcal{L}$ is a relation $\mathrel{\mid\!\sim}$ between finite *sets* of formulas of $\mathcal{L}$ and formulas of $\mathcal{L}$ that satisfies all the conditions listed in the first two items of Definition 11.17 (understood as conditions involving sets rather than multisets). A *relevant s-entailment* on $\mathcal{L}$ is an s-entailment which has the replacement property, and satisfies the strong relevance criterion.

Note 11.22 Obviously, *every relevant s-entailment can be viewed as a relevant entailment*. The converse is not true. To see this, it should be noted that although the conditions in Definitions 11.17 and 11.21 are identically formulated, the meaning of what is written is different. The crucial difference concerns the transitivity condition. In the multiset version used in

Definition 11.17, by "Γ, φ" we mean the multiset union of Γ and $[\varphi]$ (i.e., the multiset obtained from Γ by adding one occurrence of φ). It is completely insignificant for it whether φ already occurs in Γ or not. Similarly, by Γ_1, Γ_2 we mean the multiset union of Γ_1 and Γ_2, and again different occurrences of formulas are practically treated as different formulas. In contrast, the exact meaning of the transitivity condition for Definition 11.21 is: If $\Gamma_1 \mathrel{\mid\!\sim} \varphi$, and $\Gamma_2 \cup \{\varphi\} \mathrel{\mid\!\sim} \psi$, then $\Gamma_1 \cup \Gamma_2 \mathrel{\mid\!\sim} \psi$. Here we may assume, if we wish, that $\varphi \in \Gamma_2$ (because if $\Gamma_2 \cup \{\varphi\} \mathrel{\mid\!\sim} \psi$, then also $(\Gamma_2 \cup \{\varphi\}) \cup \{\varphi\} \mathrel{\mid\!\sim} \psi$). Hence we get that the following rule is derivable: if $\Gamma_1 \mathrel{\mid\!\sim} \varphi$, and $\Gamma_2, \varphi \mathrel{\mid\!\sim} \psi$, then $\Gamma_1, \Gamma_2, \varphi \mathrel{\mid\!\sim} \psi$. However, this rule is not necessarily valid for relevant entailments (where what is written is understood in terms of multisets).

Every relevant entailment induces a logic in a natural way:

Definition 11.23 (L($\mathrel{\mid\!\sim}$)) Let $\mathrel{\mid\!\sim}$ be a relevant entailment on $\mathcal{L}$. **L**($\mathrel{\mid\!\sim}$), the *logic induced by* $\mathrel{\mid\!\sim}$, is the pair $\langle \mathcal{L}, \vdash_{\mathrel{\mid\!\sim}} \rangle$, where $\mathcal{T} \vdash_{\mathrel{\mid\!\sim}} \varphi$ iff there exists a finite multiset Γ such that each element of Γ is an element of $\mathcal{T}$, and $\Gamma \mathrel{\mid\!\sim} \varphi$.

Proposition 11.24 *If $\mathrel{\mid\!\sim}$ is a relevant entailment on $\mathcal{L}$ then* **L**($\mathrel{\mid\!\sim}$) *is indeed a logic. This logic is finitary, and it satisfies the basic relevance criterion.*

Proof. The proofs of the various conditions are straightforward. We only show here that **L**($\mathrel{\mid\!\sim}$) is non-trivial, i.e., $p \not\vdash_{\mathrel{\mid\!\sim}} q$, where p and q are different variables. Assume otherwise. Then there is a multiset $\Gamma = [p, p, \ldots, p]$ such that $\Gamma \mathrel{\mid\!\sim} q$. Now Γ cannot be empty because of the weak consistency of $\mathrel{\mid\!\sim}$, and it cannot be non-empty because of the strong relevance criterion that $\mathrel{\mid\!\sim}$ satisfies. A contradiction. □

Like in the case of ordinary consequence relations, it is useful to introduce multiple-conclusioned counterparts of relevant entailments and the related notions. This is what we do next.

Definition 11.25 (Scott entailment) Let $\mathcal{L}$ be a propositional language.

- A *Scott pre-entailment* on $\mathcal{L}$ is a binary relation $\mathrel{\mid\!\sim}$ between multisets of formulas of $\mathcal{L}$ that satisfies the following conditions:

  Reflexivity: $\varphi \mathrel{\mid\!\sim} \varphi$ for every φ.
  Transitivity: If $\Gamma_1 \mathrel{\mid\!\sim} \Delta_1, \varphi$ and $\varphi, \Gamma_2 \mathrel{\mid\!\sim} \Delta_2$, then $\Gamma_1, \Gamma_2 \mathrel{\mid\!\sim} \Delta_1, \Delta_2$.

- A *Scott entailment* on $\mathcal{L}$ is a Scott pre-entailment on $\mathcal{L}$ which satisfies the following two further conditions:

  Structurality: Suppose that θ is an $\mathcal{L}$-substitution, and that $\Gamma \mathrel{\mid\!\sim} \Delta$. Then $\theta(\Gamma) \mathrel{\mid\!\sim} \theta(\Delta)$.

Weak consistency: $[\,]\not\hspace{-2pt}\sim [\,]$.

- A Scott (pre-)entailment on $\mathcal{L}$ is called *relevant* iff it has the replacement property, and it satisfies the following version of the strong relevance criterion: If Γ_1, Δ_1 and Γ_2, Δ_2 are not empty, and $\Gamma_1, \Gamma_2 \sim \Delta_1, \Delta_2$, then Γ_1, Δ_1 and Γ_2, Δ_2 share some variable. In particular: if Γ and Δ are not empty, and $\Gamma \sim \Delta$, then Γ and Δ share some variable.

- A *Scott relevant s-entailment* on $\mathcal{L}$ is a relation $\sim$ between *sets* of formulas of $\mathcal{L}$ and formulas of $\mathcal{L}$ that satisfies all the conditions listed in the previous items (understood as conditions involving sets rather than multisets).

11.3 Basic Relevant Connectives

It seems safe to say that the general feeling is that a relevant logic that deserves this name should naturally define a relevant entailment. However, Example 11.8 shows that the question what connectives are available in a given logic (and what are their properties) has crucial importance for classifying it as a relevant logic. Indeed, it is usually expected that such a logic should contain in its language certain *relevant connectives*, which provide appropriate substitutes for the central connectives of classical logic (as well as many other logics). In this section we describe the relevant counterparts of the classical implication, conjunction, and negation. (Note that any classical connective is definable using these connectives, and the first two even suffice for defining any positive classical connective.)

11.3.1 Weak Relevant Implication

We start by what was called in [4] 'the heart' of (a relevant) logic:

Definition 11.26 (internal implication) Let $\mathcal{L}$ be a propositional language, and let $\sim$ be a (Scott) pre-entailment on $\mathcal{L}$. A binary connective $\to$ of $\mathcal{L}$ is called an *internal implication* for $\sim$ if for every Γ, $(\Delta,)$ φ and ψ: $\Gamma, \varphi \sim \psi$ iff $\Gamma \sim \varphi \to \psi$ ($\Gamma, \varphi \sim \psi, \Delta$ iff $\Gamma \sim \varphi \to \psi, \Delta$).[2]

Obviously, if $\to$ is an internal implication for a pre-entailment $\sim$ then for every $\varphi_1, \ldots, \varphi_n$ and ψ we have:

$$[\varphi_1, \ldots, \varphi_n] \sim \psi \quad \text{iff} \quad \sim \varphi_1 \to (\varphi_2 \to \cdots (\varphi_n \to \psi) \cdots).$$

---
[2] In this part of the book we follow the tradition in relevant logic and use $\to$ to denote a relevant implication or any related "intensional" implication. The notation $\supset$ will be used for the implications introduced in previous parts (especially the classical implication).

It follows that in such a case the pre-entailment is completely determined by $TH(\mathrel{\vdash\mkern-9mu\sim})$, the set of valid formulas of $\mathrel{\vdash\mkern-9mu\sim}$ (i.e., formulas φ such that $\mathrel{\vdash\mkern-9mu\sim} \varphi$). Next we describe the minimal logic whose set of valid formulas should be contained in $TH(\mathrel{\vdash\mkern-9mu\sim})$ (and also naturally induces, in a sense to be described below, a relevant entailment with an internal implication $\to$).

Axioms:

[Id] $\varphi \to \varphi$ (Identity)

[Tr] $(\varphi \to \psi) \to ((\psi \to \sigma) \to (\varphi \to \sigma))$ (Transitivity)

[Pe] $(\varphi \to (\psi \to \sigma)) \to (\psi \to (\varphi \to \sigma))$ (Permutation)

Rule of inference:

[MP] $\dfrac{\varphi \quad \varphi \to \psi}{\psi}$

Figure 11.1: The proof system $HLL_\to$

Proposition 11.27 *Let $\mathcal{L}$ be a language, and let $\mathrel{\vdash\mkern-9mu\sim}$ be a pre-entailment on $\mathcal{L}$, for which $\to$ is an internal implication. Then the set of formulas φ such that $\mathrel{\vdash\mkern-9mu\sim} \varphi$ includes every $\mathcal{L}$-instance of any theorem of the system $HLL_\to$ presented in Figure 11.1, and it is closed under the rule of inference of that system (i.e., if $\mathrel{\vdash\mkern-9mu\sim} \varphi$ and $\mathrel{\vdash\mkern-9mu\sim} \varphi \to \psi$ then $\mathrel{\vdash\mkern-9mu\sim} \psi$). Moreover: $\varphi \to \psi, \varphi \mathrel{\vdash\mkern-9mu\sim} \psi$.*

Proof. Since $\varphi \to \psi \mathrel{\vdash\mkern-9mu\sim} \varphi \to \psi$, and $\to$ is an internal implication, we have that $\varphi \to \psi, \varphi \mathrel{\vdash\mkern-9mu\sim} \psi$. By the transitivity condition, this entails that if $\mathrel{\vdash\mkern-9mu\sim} \varphi$ and $\mathrel{\vdash\mkern-9mu\sim} \varphi \to \psi$ then $\mathrel{\vdash\mkern-9mu\sim} \psi$. Therefore to finish the proof it suffices to show that $\mathrel{\vdash\mkern-9mu\sim} \varphi$ for every axiom φ of $HLL_\to$:

(Id) Immediate from the fact that $\varphi \mathrel{\vdash\mkern-9mu\sim} \varphi$.

(Tr) Since $\varphi, \varphi \to \psi \mathrel{\vdash\mkern-9mu\sim} \psi$ for every φ and ψ (see above), $\varphi, \varphi \to \sigma \mathrel{\vdash\mkern-9mu\sim} \sigma$, and $\sigma, \sigma \to \psi \mathrel{\vdash\mkern-9mu\sim} \psi$. Therefore that $\varphi \to \sigma, \sigma \to \psi, \varphi \mathrel{\vdash\mkern-9mu\sim} \psi$ follows from the transitivity of $\mathrel{\vdash\mkern-9mu\sim}$. Since $\to$ is an internal implication for $\mathrel{\vdash\mkern-9mu\sim}$, this entails that $\mathrel{\vdash\mkern-9mu\sim} (\varphi \to \sigma) \to ((\sigma \to \psi) \to (\varphi \to \psi))$.

(Pe) From the fact that $\varphi, \varphi \to \psi \mathrel{\vdash\mkern-9mu\sim} \psi$ for every φ and ψ, it follows in particular that $\varphi \to (\psi \to \sigma), \varphi \mathrel{\vdash\mkern-9mu\sim} \psi \to \sigma$, and $\psi \to \sigma, \psi \mathrel{\vdash\mkern-9mu\sim} \sigma$. By the transitivity of $\mathrel{\vdash\mkern-9mu\sim}$, it follows that $\varphi \to (\psi \to \sigma), \varphi, \psi \mathrel{\vdash\mkern-9mu\sim} \sigma$, and so $\varphi \to (\psi \to \sigma), \psi, \varphi \mathrel{\vdash\mkern-9mu\sim} \sigma$, and $\mathrel{\vdash\mkern-9mu\sim} (\varphi \to (\psi \to \sigma)) \to (\psi \to (\varphi \to \sigma))$. □

11.3 Basic Relevant Connectives

Proposition 11.28 *Let $HRM0_\to$ be the system obtained from $HLL_\to$ by adding to it the following axioms:*

$$[Ct] \quad (\varphi \to (\varphi \to \psi)) \to (\varphi \to \psi) \quad \text{(Contraction)}$$

$$[Mi] \quad \varphi \to (\varphi \to \varphi) \quad \text{(Mingle)}$$

If $\vdash$ is a relevant s-entailment on $\mathcal{L}$ for which $\to$ is an internal implication, then the set of formulas φ such that $\vdash \varphi$ includes every $\mathcal{L}$-instance of any theorem of the system $HRM0_\to$, and it is closed under the rule of inference of that system. [3]

Proof. By Proposition 11.27, it suffices to prove the validity in $\vdash$ of the two extra axioms of $HRM0_\to$.

(Ct) From Proposition 11.27 it follows that $\{\varphi \to (\varphi \to \psi), \varphi\} \vdash \varphi \to \psi$, and $\{\varphi \to \psi, \varphi\} \vdash \psi$. Hence $\{\varphi \to (\varphi \to \psi), \varphi\} \vdash \psi$ by transitivity, implying that $\vdash (\varphi \to (\varphi \to \psi)) \to (\varphi \to \psi)$ (since $\to$ is an internal implication for $\vdash$).

(Mi) Since $\{\varphi\} \cup \{\varphi\} \vdash \varphi$, the fact that $\to$ is an internal implication for $\vdash$ implies that $\varphi \vdash \varphi \to \varphi$, and so $\vdash \varphi \to (\varphi \to \varphi)$. □

Note 11.29 Let $\mathbf{LL}_\to$ and $\mathbf{RM0}_\to$ be the logics in the language of $\{\to\}$ which are induced by the Hilbert-type systems $HLL_\to$ and $HRM0_\to$ (respectively). It is known (see, e.g., [28]) that $\mathbf{LL}_\to$ is the pure implicational fragment of Linear logic ([174]). On the other hand, although $HRM0_\to$ consists of the pure implicational axioms of the standard Hilbert-type formulation of the system **RM** mentioned above, it was shown by Meyer that $\mathbf{RM0}_\to$ is *not* the pure implicational fragment of that system (see [4, p.98]).

Using an internal implication is the standard method of defining relevant entailments. The way this is done is given in the next definition and the propositions that follow it. It is again based on the crucial fact that $[\varphi_1, \ldots, \varphi_n] \vdash \psi$ iff $\vdash \varphi_1 \to (\varphi_2 \to \cdots (\varphi_n \to \psi) \cdots)$.

Definition 11.30 ($\vdash_\mathbf{L}^\to$) Let $\mathbf{L} = \langle \mathcal{L}, \vdash_\mathbf{L} \rangle$ be a logic, and let $\to$ be a (primitive or defined) binary connective of $\mathcal{L}$.

- The relation $\vdash_\mathbf{L}^\to$ between finite *lists* of formulas and formulas which is induced by $\to$ is defined by:

$$\varphi_1, \ldots, \varphi_n \vdash_\mathbf{L}^\to \psi \quad \text{iff} \quad \vdash_\mathbf{L} \varphi_1 \to (\varphi_2 \to \cdots (\varphi_n \to \psi) \cdots).$$

---
[3] Actually, the only conditions needed here are reflexivity and transitivity. Hence the proposition holds for what we might have called pre-s-entailments.

- Suppose $S_1 \mathrel{\vdash_{\mathbf{L}}^{\rightarrow}} \varphi$ iff $S_2 \mathrel{\vdash_{\mathbf{L}}^{\rightarrow}} \varphi$ in case the multisets which correspond to the lists S_1 and S_2 are identical. Then $\mathrel{\vdash_{\mathbf{L}}^{\rightarrow}}$ is extended to a relation between finite *multisets* of formulas and formulas by defining:

$$[\varphi_1, \ldots, \varphi_n] \mathrel{\vdash_{\mathbf{L}}^{\rightarrow}} \psi \text{ iff } \varphi_1, \ldots, \varphi_n \mathrel{\vdash_{\mathbf{L}}^{\rightarrow}} \psi.$$

- Suppose $S_1 \mathrel{\vdash_{\mathbf{L}}^{\rightarrow}} \varphi$ iff $S_2 \mathrel{\vdash_{\mathbf{L}}^{\rightarrow}} \varphi$ whenever the sets which correspond to S_1 and S_2 are identical. Then $\mathrel{\vdash_{\mathbf{L}}^{\rightarrow}}$ is extended to be also a relation between finite *sets* of formulas and formulas by defining:

$$\{\varphi_1, \ldots, \varphi_n\} \mathrel{\vdash_{\mathbf{L}}^{\rightarrow}} \psi \text{ iff } \varphi_1, \ldots, \varphi_n \mathrel{\vdash_{\mathbf{L}}^{\rightarrow}} \psi.$$

Proposition 11.31 *Let* **L** *and* $\rightarrow$ *be like in Definition 11.30, and let the logics* **LL**$_\rightarrow$ *and* **RM0**$_\rightarrow$ *be like in Note 11.29.*

1. $\mathrel{\vdash_{\mathbf{L}}^{\rightarrow}}$ *is a well-defined entailment (for which* $\rightarrow$ *is an internal implication) iff the set of formulas* φ *such that* $\vdash_{\mathbf{L}} \varphi$ *includes every* $\mathcal{L}$*-instance of any theorem of* **LL**$_\rightarrow$, *and is closed under MP.*

2. $\mathrel{\vdash_{\mathbf{L}}^{\rightarrow}}$ *is a well-defined s-entailment (for which* $\rightarrow$ *is an internal implication) iff the set of formulas* φ *such that* $\vdash_{\mathbf{L}} \varphi$ *includes every* $\mathcal{L}$*-instance of any theorem of* **RM0**$_\rightarrow$, *and is closed under MP.*

Proof. Obviously, if $\mathrel{\vdash_{\mathbf{L}}^{\rightarrow}}$ is a well-defined pre-entailment then $\rightarrow$ is an internal implication for it. Therefore it is immediate from the definitions and Propositions 11.27 and 11.28 that in all parts of the proposition the specified conditions are necessary. For the converse, assume first that $\vdash_{\mathbf{L}} \varphi$ for every $\mathcal{L}$-instance φ of a theorem of **LL**$_\rightarrow$, and that the set of theorems of **L** is closed under MP. Then the validity of [Tr] and [Pe] easily entails (using the closure of the set of theorems of **L** under MP) that $\vdash_{\mathbf{L}} (\sigma \rightarrow \varphi) \rightarrow (\sigma \rightarrow \psi)$ whenever $\vdash_{\mathbf{L}} \varphi \rightarrow \psi$. This fact and the validity of [Pe] ensure that $\mathrel{\vdash_{\mathbf{L}}^{\rightarrow}}$ is well-defined as a relation between finite multisets of formulas and formulas. The validity of [Id] and [Tr], together with the fact that **L** is a *logic*, entail then that $\mathrel{\vdash_{\mathbf{L}}^{\rightarrow}}$ is an entailment. This proves the first part of the proposition. For the second part it suffices now to show that if in addition [Ct] and [Mi] are valid too, then $\mathrel{\vdash_{\mathbf{L}}^{\rightarrow}}$ is well-defined also as a relation between finite *sets* of formulas and formulas. This is easy. □

Definition 11.32 (weak relevant implication) A connective $\rightarrow$ is called a *weak relevant implication* of a logic **L** iff $\mathrel{\vdash_{\mathbf{L}}^{\rightarrow}}$ is a well-defined *relevant* (pre-)entailment such that **L** is an extension of $\mathbf{L}(\mathrel{\vdash_{\mathbf{L}}^{\rightarrow}})$ (i.e., if $\Gamma \mathrel{\vdash_{\mathbf{L}}^{\rightarrow}} \psi$ then $\Gamma \vdash_{\mathbf{L}} \psi$).

11.3 Basic Relevant Connectives

Note 11.33 Since **L** is a logic, $\mathrel{\vdash\mkern-9mu\sim}_{\mathbf{L}}^{\rightarrow}$ is a well-defined (relevant) entailment iff it is a well-defined (relevant) pre-entailment. Therefore, Definition 11.32 can be read either with the 'pre-' or without it. It is also easily seen that if $\mathrel{\vdash\mkern-9mu\sim}_{\mathbf{L}}^{\rightarrow}$ is indeed a well-defined (pre-)entailment then:

- $\rightarrow$ is an internal implication for $\mathrel{\vdash\mkern-9mu\sim}_{\mathbf{L}}^{\rightarrow}$.

- The condition $\mathrel{\vdash\mkern-9mu\sim}_{\mathbf{L}}^{\rightarrow} \subseteq \vdash_{\mathbf{L}}$ from Definition 11.32 is equivalent to:

$$\varphi, \varphi \rightarrow \psi \vdash_{\mathbf{L}} \psi.$$

Therefore, by Proposition 11.31 $\rightarrow$ is a weak relevant implication of **L** iff **L** contains $\mathbf{LL}_{\rightarrow}$, and $\mathrel{\vdash\mkern-9mu\sim}_{\mathbf{L}}^{\rightarrow}$ satisfies the strong relevance criterion and has the replacement property. Next we provide a condition on $\rightarrow$ that is obviously necessary for the former (and by Proposition 11.63 below it is usually also sufficient), and an easy-to-check condition which is equivalent to the latter.

Definition 11.34 (variable sharing property, VSP) Let **L** and $\rightarrow$ be like in Definition 11.30. $\rightarrow$ has in **L** (or **L** has with respect to $\rightarrow$) the *variable sharing property* (VSP) if $\mathsf{Var}(\varphi) \cap \mathsf{Var}(\psi) \neq \emptyset$ whenever $\vdash_{\mathbf{L}} \varphi \rightarrow \psi$.

Definition 11.35 (replacement property for $\rightarrow$) $\rightarrow$ has in a logic **L** the *replacement property* if $\vdash_{\mathbf{L}} \sigma\{\varphi/p_i\} \rightarrow \sigma\{\psi/p_i\}$ ($1 \leq i \leq n$) whenever $\vdash_{\mathbf{L}} \varphi \rightarrow \psi$, $\vdash_{\mathbf{L}} \psi \rightarrow \varphi$, and $\sigma = \diamond(p_1, \ldots, p_n)$ (where $\diamond$ is an n-ary connective and $p_1, \ldots, p_n$ are n distinct variables).

Proposition 11.36 *If $\rightarrow$ has the replacement property in **L** then $\mathrel{\vdash\mkern-9mu\sim}_{\mathbf{L}}^{\rightarrow}$ has the replacement property.*

Proof. Assume that $\rightarrow$ has the replacement property in **L**, and suppose that $\vdash_{\mathbf{L}} \varphi \rightarrow \psi$ and $\vdash_{\mathbf{L}} \psi \rightarrow \varphi$ (i.e., $\varphi \mathrel{\vdash\mkern-9mu\sim}_{\mathbf{L}}^{\rightarrow} \psi$ and $\psi \mathrel{\vdash\mkern-9mu\sim}_{\mathbf{L}}^{\rightarrow} \varphi$). We prove by an induction on the structure of σ that for *every* σ and variable p, $\vdash_{\mathbf{L}} \sigma[\varphi/p] \rightarrow \sigma[\psi/p]$. We leave the standard details to the reader. □

Note 11.37 Recall that the demand from $\rightarrow$ to have the replacement property intuitively means that if each of φ and ψ relevantly implies the other in **L**, then each of the two formulas can replace the other in any context. Note that most of the implication connectives used in the previous chapters do *not* have this property. The reason is that so far the validity of $\varphi \supset \psi$ and $\psi \supset \varphi$ (for some implication connective $\supset$) intuitively only meant that ψ is true whenever φ is, and vice versa. This does not imply that the two formulas have the same content. In contrast, if $\rightarrow$ is a weak relevant implication of **L**, and both $\varphi \rightarrow \psi$ and $\psi \rightarrow \varphi$ are theorems of **L**, then φ and ψ should be congruent in **L** (even though in general **L** might not have the full replacement property).

11.3.2 (Strong) Relevant Implication

Satisfying the conditions listed in Definition 11.32 is a minimal demand from a relevant implication of a logic **L**. Intuitively, we would expect to have a much stronger connection between such a connective (or in fact *any* "implication connective" of **L**) and the consequence relation of **L**. Note however that no connective that satisfies VSP can be an implication for **L** in the sense of Definition 1.32, because it cannot satisfy the standard deduction theorem (Indeed, since $\vdash_\mathbf{L} p \to p$, also $q \vdash_\mathbf{L} p \to p$. Therefore had $\to$ satisfied that theorem, we would have that $\vdash_\mathbf{L} q \to (p \to p)$, in contradiction to VSP). Still, we would expect to have $\mathcal{T} \vdash_\mathbf{L} \varphi \to \psi$ whenever $\mathcal{T}, \varphi \vdash_\mathbf{L} \psi$, and φ is *relevant* to ψ. Intuitively, the latter should necessarily be true at least in the case where $\mathcal{T}, \varphi \vdash_\mathbf{L} \psi$, but $\mathcal{T} \nvdash_\mathbf{L} \psi$. This observation leads to Definition 1.26 above, which we now repeat for the reader's convenience:

Definition 11.38 (RDP) Let **L** be a logic for a language $\mathcal{L}$, and let $\to$ be a (primitive or defined) connective of $\mathcal{L}$. The connective $\to$ has in **L** (or **L** has with respect to $\to$) the *relevant deduction property* (RDP), if:

$$\mathcal{T}, \varphi \vdash_\mathbf{L} \psi \text{ iff either } \mathcal{T} \vdash_\mathbf{L} \psi \text{ or } \mathcal{T} \vdash_\mathbf{L} \varphi \to \psi.$$

Note 11.39 The main idea behind some basic relevant logics (like the logic $\mathbf{R}_\to$ described below) has been the following version of RDP: $\mathcal{T} \vdash \varphi \to \psi$ iff there is a proof of ψ from $\mathcal{T} \cup \{\varphi\}$ in which φ is actually *used*. This version applies only to a logic which is defined by a proof system, and it depends on the choice of the proof system. Moreover: this version requires an exact definition of "a use of a formula in a given proof".[4] In contrast, our RDP is meaningful for every logic.[5]

The naturalness of RDP leads to the following definition:

Definition 11.40 (relevant implication) Let **L** be a logic for a language $\mathcal{L}$, and let $\to$ be a (primitive or defined) connective of $\mathcal{L}$. $\to$ is called a *relevant implication* of **L** if it is a weak relevant implication for **L** which has the RDP.

In what follows we investigate the basic properties of the class of logics which have the RDP with respect to a connective $\to$, as well as its subclass consisting of the logics that have also the VSP (with respect to $\to$) and its subclass consisting of the logics in which $\to$ is a relevant implication.

---

[4] See [4, 151] for attempts to define this notion and [35] for a critical discussion.
[5] It should be noted that RDP is indeed the way in which A. Church stated his deduction theorem in [118] (the paper in which the system $HR_\to$ described in Definition 11.44 below was introduced).

11.3 Basic Relevant Connectives

Proposition 11.41 *Suppose $\to$ has in $\mathbf{L}$ the RDP. Then:*

1. *$\mathbf{L}$ contains $\mathbf{LL}_\to$.*

2. *$\vdash_{\mathbf{L}}^{\to}$ is a well-defined entailment.*

3. *$\mathbf{L} = \mathbf{L}(\vdash_{\mathbf{L}}^{\to})$.*

Proof. Since $\mathbf{L}$ is a logic, and so structural, it suffices for the first part to show the validity of the axioms and inference rule of $HLL_\to$ for the case where φ, ψ and σ are different variables p, q, r (respectively).

(Id) Since $p \vdash_{\mathbf{L}} p$, but $\nvdash_{\mathbf{L}} p$, the RDP implies that $\vdash_{\mathbf{L}} p \to p$.

(MP) Since $p \to q \vdash_{\mathbf{L}} p \to q$, the RDP entails that $p, p \to q \vdash_{\mathbf{L}} q$.

(Tr) From the validity of [MP] it follows that $p, p \to q, q \to r \vdash_{\mathbf{L}} r$. It is impossible that $p \to q, q \to r \vdash_{\mathbf{L}} r$, since otherwise we would get $\vdash_{\mathbf{L}} p$ by substituting p for q and r, and using the provability of [Id]. Hence $p \to q, q \to r \vdash_{\mathbf{L}} p \to r$ by the RDP. Now, it is impossible that $p \to q \vdash_{\mathbf{L}} p \to r$, since otherwise by substituting p for q we would get $\vdash_{\mathbf{L}} p \to r$, although $p \nvdash_{\mathbf{L}} r$. Hence, $p \to q \vdash_{\mathbf{L}} (q \to r) \to (p \to r)$ by the RDP. Again, it is impossible that $\vdash_{\mathbf{L}} (q \to r) \to (p \to r)$, since otherwise we would get $\vdash_{\mathbf{L}} (p \to q)$ by substituting q for r. Hence, the RDP entails that $\vdash_{\mathbf{L}} (p \to q) \to ((q \to r) \to (p \to r))$.

(Pe) The validity of [MP] entails that $p \to (q \to r), p, q \vdash_{\mathbf{L}} r$. On the other hand, $p \to (q \to r), q \nvdash_{\mathbf{L}} r$, because otherwise we would get $q \vdash_{\mathbf{L}} r$ by substituting $q \to r$ for p. Thus, it follows by the RDP that $p \to (q \to r), q \vdash_{\mathbf{L}} p \to r$. Now, $p \to (q \to r) \nvdash_{\mathbf{L}} p \to r$, because otherwise by substituting $q \to q$ for p and q for r, we would get $\vdash_{\mathbf{L}} q$ (using previous items). Hence, $p \to (q \to r) \vdash_{\mathbf{L}} q \to (p \to r)$ by the RDP. Finally, $\nvdash_{\mathbf{L}} q \to (p \to r)$, since otherwise we would get $\vdash_{\mathbf{L}} r$ by substituting $q \to q$ for p and q. Hence, the RDP implies that $\vdash_{\mathbf{L}} (p \to (q \to r)) \to (q \to (p \to r))$.

The second part follows from the first by Proposition 11.31. We leave the proof of the third part to the reader. □

Corollary 11.42 *If $\to$ has in $\mathbf{L}$ the RDP then it is a relevant implication of $\mathbf{L}$ iff it has in $\mathbf{L}$ the replacement property (Definition 11.35), and $\vdash_{\mathbf{L}}^{\to}$ satisfies the strong relevance criterion (implying that $\mathbf{L}$ has the VSP with respect to $\to$).*

Proof. Immediate from Proposition 11.41 and Note 11.33. □

Corollary 11.43 *Let* **L** *be a logic for a language* $\mathcal{L}$, *and let* $\to$ *be a connective of* $\mathcal{L}$ *such that* **L** *has both the* VSP *and the* RDP *with respect to* $\to$. *Then* $\to$ *is a semi-implication for* **L** *(Definition 1.27). In particular: if* $\to$ *is a relevant implication of* **L** *then* $\to$ *is a semi-implication for* **L**.

Proof. From $\vdash_{LL_\to} (p \to q) \to (p \to q)$ we get $\vdash_{LL_\to} p \to ((p \to q) \to q)$ by [Pe]. By substituting $p \to p$ for p, $q \to q$ for q, and using [Id] we get: $\vdash_{LL_\to} ((p \to p) \to (q \to q)) \to (q \to q)$. Now, assume that $\to$ is not a semi-implication for **L** even though it has in **L** the RDP. Then, by the definition of a semi-implication and Proposition 11.41, the above implies that $\vdash_{\mathbf{L}} (q \to q) \to ((p \to p) \to (q \to q))$, and so $\vdash_{\mathbf{L}} (p \to p) \to (q \to q)$. This contradicts the VSP. □

Next we describe the minimal logic that is included in any logic which has a relevant implication $\to$, or even a semi-implication $\to$.

Definition 11.44 (The logic $R_\to$) The proof system $HR_\to$ is the system obtained from $HLL_\to$ (Figure 11.1) by adding to it the axiom [Ct] (from Proposition 11.28). $R_\to$ is the logic in $\{\to\}$ induced by $HR_\to$.

Theorem 11.45 *Let* **L** *be a logic for a language* $\mathcal{L}$, *and let* $\to$ *be a connective of* $\mathcal{L}$ *such that* **L** *has the* RDP *with respect to* $\to$. *Then either* **L** *contains* $\mathbf{R}_\to$, *or* $\vdash_{\mathbf{L}} (\varphi \to \psi) \to (\psi \to \varphi)$ *for every* φ *and* ψ.

Proof. By Proposition 11.41, [MP] for $\to$ is valid in **L**. This entails that $p \to (p \to q), p \vdash_{\mathbf{L}} q$. Hence, by the RDP, either $p \to (p \to q) \vdash_{\mathbf{L}} p \to q$, or $p \to (p \to q) \vdash_{\mathbf{L}} q$. Since $\not\vdash_{\mathbf{L}} p \to q$ (because $p \not\vdash_{\mathbf{L}} q$), in the first case we get that $\vdash_{\mathbf{L}} (p \to (p \to q)) \to (p \to q)$, and so [Ct] is valid in **L**. It follows by Proposition 11.41 that in this case **L** contains $\mathbf{R}_\to$. Since $\not\vdash_{\mathbf{L}} q$, in the second case the RDP implies that $\vdash_{\mathbf{L}} (p \to (p \to q)) \to q$. Substituting in this $\sigma = \psi \to \varphi$ for p and $\tau = (\varphi \to \psi) \to (\psi \to \varphi)$ for q we get:

$$(*) \quad \vdash_{\mathbf{L}} (\sigma \to (\sigma \to \tau)) \to \tau.$$

But it is easy to see that $\vdash_{LL_\to} \sigma \to (\sigma \to \tau)$ (for these particular σ and τ). Hence, (*) and Proposition 11.41 imply that in the second case $\vdash_{\mathbf{L}} \tau$, i.e., $\vdash_{\mathbf{L}} (\varphi \to \psi) \to (\psi \to \varphi)$ for every φ and ψ. □

As an immediate corollary of Theorem 11.45 we get:

Theorem 11.46 *If* $\to$ *is a semi-implication for* **L**, *then* **L** *contains* $\mathbf{R}_\to$.

Here is another easy corollary of Theorem 11.45:

11.3 Basic Relevant Connectives

Theorem 11.47 *Let $\mathbf{L}$ be a logic for a language $\mathcal{L}$, and let $\to$ be a connective of $\mathcal{L}$ such that $\mathbf{L}$ has both the VSP and the RDP with respect to $\to$. Then $\mathbf{L}$ contains $\mathbf{R}_\to$. In particular, if $\to$ is a relevant implication of $\mathbf{L}$ then $\mathbf{L}$ includes $\mathbf{R}_\to$.*

Proof. Immediate from Corollaries 11.46 and 11.43. □

The next corollary establishes an important connection between semi-implications (and so relevant implications) and the classical implication:

Corollary 11.48 *If $\to$ is a semi-implication for a logic $\mathbf{L}$ (in particular: if $\to$ is a relevant implication for a logic $\mathbf{L}$), and $\mathbf{L}$ is $\mathbf{F}$-contained in classical logic, then $\mathbf{F}(\to)$ is the classical (material) implication.*

Proof. Denote $\mathbf{F}(\to)$ by $\widetilde{\supset}$. Since $\vdash_\mathbf{L} p \to p$, $\widetilde{\supset}(t,t) = \widetilde{\supset}(f,f) = t$. From $p, p \to q \vdash_\mathbf{L} q$ it easily follows that $\widetilde{\supset}(t,f) = f$. Finally, by Theorem 11.46 we have that $\vdash_\mathbf{L} (p \to (p \to p)) \to (p \to p)$. This entails that $\widetilde{\supset}(f,t) = t$, since had $\widetilde{\supset}(f,t) = f$, we would have $(f \widetilde{\supset} (f \widetilde{\supset} f)) \widetilde{\supset} (f \widetilde{\supset} f) = f$. □

Next, we describe some other important properties of logics which have a relevant implication.

Corollary 11.49 *Let $\mathbf{L}$ be a logic for a language $\mathcal{L}$, and let $\to$ be a connective of $\mathcal{L}$ such that $\mathbf{L}$ has both the VSP and the RDP with respect to $\to$. Then neither $q \to (p \to q)$ nor $\neg p \to (p \to q)$ is a theorem of $\mathbf{L}$. In particular, this is true if $\to$ is a relevant implication of $\mathbf{L}$.*

Proof. We prove that $\neg p \to (p \to q)$ is not a theorem of $\mathbf{L}$, leaving the easier $q \to (p \to q)$ to the reader. So suppose that $\vdash_\mathbf{L} \neg p \to (p \to q)$. Substituting $p \to p$ for p, and using [Id] and [Pe] (which are valid in $\mathbf{L}$ by Proposition 11.41), we get that $\vdash_\mathbf{L} \neg (p \to p) \to q$. This contradicts the variable-sharing property of $\mathbf{L}$. □

Note 11.50 The last corollary implies that logics in which $\to$ has the RDP and VSP (in particular, logics in which $\to$ is a relevant implication) avoid (with respect to $\to$) *all* of the so-called "paradoxes of material implication" (according to which if p is false then it implies everything, while if q is true then it is implied by everything). However, Theorem 11.53 and Corollary 11.54 below show that there is a price to pay for this.

Lemma 11.51 *Suppose $\to$ has in $\mathbf{L}$ the RDP. Then $\to$ is an implication for $\mathbf{L}$ iff $\vdash_\mathbf{L} \psi \to (\varphi \to \psi)$ for every φ and ψ, iff $q \vdash_\mathbf{L} p \to q$ for some (any) two distinct variables p, q.*

Proof. Suppose $\to$ has in $\mathbf{L}$ the RDP. By Proposition 11.41, $\varphi, \varphi \to \psi \vdash_\mathbf{L} \psi$ for every φ and ψ, and so if $\mathcal{T} \vdash_\mathbf{L} \varphi \to \psi$ then $\mathcal{T}, \varphi \vdash_\mathbf{L} \psi$. It follows that $\to$ is an implication for $\mathbf{L}$ iff the converse holds as well, i.e., if that $\mathcal{T}, \varphi \vdash_\mathbf{L} \psi$ implies that $\mathcal{T} \vdash_\mathbf{L} \varphi \to \psi$. Next we show that this is equivalent to $\psi \vdash_\mathbf{L} \varphi \to \psi$ for every φ and ψ. The latter is obviously a necessary condition for the former. To show that it is also sufficient, assume that $\mathcal{T}, \varphi \vdash_\mathbf{L} \psi$. By the RDP, either $\mathcal{T} \vdash_\mathbf{L} \varphi \to \psi$ or $\mathcal{T} \vdash_\mathbf{L} \psi$. In the first case we are done, while the assumption that $\psi \vdash_\mathbf{L} \varphi \to \psi$ implies this in the second case too.

Now, since $\mathbf{L}$ is structural, that $\psi \vdash_\mathbf{L} \varphi \to \psi$ for every φ and ψ is equivalent to $q \vdash_\mathbf{L} p \to q$ for every distinct variables p, q. By the RDP, this in turn is equivalent to $\vdash_\mathbf{L} q \to (p \to q)$ (otherwise we would have had $\vdash_\mathbf{L} p \to q$. Hence, $p \vdash_\mathbf{L} q$), and so to $\vdash_\mathbf{L} \psi \to (\varphi \to \psi)$ for every φ, ψ. □

Proposition 11.52 *If $\mathbf{L}$ has a conjunction or a disjunction, and $\to$ has in $\mathbf{L}$ the RDP, then $\to$ is an implication for $\mathbf{L}$. In particular: every semi-implication for $\mathbf{L}$ is in such a case also an implication for $\mathbf{L}$.*

Proof. Suppose that $\to$ has in $\mathbf{L}$ the RDP.

- Assume that $\wedge$ is a conjunction for $\mathbf{L}$. From the three properties of $\wedge$ listed in Note 1.25 it easily follows that if p and q are two distinct variables, then $p, q \vdash_\mathbf{L} p \wedge q$, while $p \not\vdash_\mathbf{L} p \wedge q$ (otherwise we would have got that $p \vdash_\mathbf{L} q$). Since $\to$ has in $\mathbf{L}$ the RDP, this implies that $p \vdash_\mathbf{L} q \to (p \wedge q)$. Similarly, these properties of $\wedge$ implies that $\vdash_\mathbf{L} (p \wedge q) \to p$. Since by Proposition 11.41 $\varphi \to \psi, \psi \to \tau \vdash_\mathbf{L} \varphi \to \tau$, it follows that $p \vdash_\mathbf{L} q \to p$, and so $\to$ is an implication for $\mathbf{L}$ by Lemma 11.51.

- Assume that $\vee$ is a disjunction for $\mathbf{L}$. Since $p \vee q \vdash_\mathbf{L} p \vee q$, we have (i) $p \vdash_\mathbf{L} p \vee q$ and (ii) $q \vdash_\mathbf{L} p \vee q$. Now, Proposition 11.41 implies that $p, p \to q \vdash_\mathbf{L} q$. Obviously, also $q, p \to q \vdash_\mathbf{L} q$. As $\vee$ is a disjunction for $\mathbf{L}$, the last two facts imply that $p \vee q, p \to q \vdash_\mathbf{L} q$. Since $p \vee q \not\vdash_\mathbf{L} q$ (otherwise we would get $p \vdash_\mathbf{L} q$ by (i)), this implies that $p \vee q \vdash_\mathbf{L} (p \to q) \to q$, and so by (ii) that $q \vdash_\mathbf{L} (p \to q) \to q$. By substituting $p \to q$ for p in the last fact we get (*) $q \vdash_\mathbf{L} ((p \to q) \to q) \to q$. But it is easy to see that $\vdash_{\mathbf{LL}_\to} (((p \to q) \to q) \to q) \to (p \to q)$. Hence, Proposition 11.41 and (*) together imply that $q \vdash_\mathbf{L} p \to q$. It follows that $\to$ is an implication for $\mathbf{L}$ by Lemma 11.51. □

Theorem 11.53 *Let $\mathbf{L}$ be a logic for a language $\mathcal{L}$, and let $\to$ be a connective of $\mathcal{L}$ such that $\mathbf{L}$ has both the VSP and the RDP with respect to $\to$. Then $\mathbf{L}$ has no implication, no conjunction, and no disjunction (all in the sense of Definition 1.23).*

11.3 Basic Relevant Connectives

Proof. Assume for contradiction that **L** has an implication $\supset$. Let p and q be two distinct variables. Then $p, p \supset q \vdash_{\mathbf{L}} q$. Since $p \supset q \not\vdash_{\mathbf{L}} q$ (because otherwise we would have $q \supset q \vdash_{\mathbf{L}} q$, and so that $\vdash_{\mathbf{L}} q$), this implies that $p \supset q \vdash_{\mathbf{L}} p \to q$. But $\vdash_{\mathbf{L}} p \supset (q \to q)$ (since $p \vdash_{\mathbf{L}} q \to q$, and $\supset$ is an implication for **L**). It follows that $\vdash_{\mathbf{L}} p \to (q \to q)$. This contradicts VSP for $\to$. Hence **L** has no implication. That it does not have a conjunction or disjunction either follows from this and from Proposition 11.52. □

Corollary 11.54 *If a logic* **L** *has a relevant implication then it has no implication, no conjunction, and no disjunction. In particular: it is not normal.*

Note 11.55 While the availability of a relevant implication compensates for the lack of an implication, the lack of a conjunction is definitely a drawback. The desire to have one is the main motivation for introducing relevant logics in which the implication is only weakly relevant.

A partial converse to Theorem 11.47 is given in the next Proposition:[6]

Proposition 11.56 *Let $\mathcal{H}$ be an axiomatic extension of $HR_\to$. Then the logic induced by $\mathcal{H}$ has the* RDP.

Proof. The "if" direction is trivial. The converse is proved by induction on the length of the proof of ψ from $\mathcal{T} \cup \{\varphi\}$: If ψ is an axiom of $\mathcal{H}$ or $\psi \in \mathcal{T}$ then $\mathcal{T} \vdash_{\mathcal{H}} \psi$. If $\psi = \varphi$ then $\mathcal{T} \vdash_{\mathcal{H}} \varphi \to \psi$ by axiom [Id]. Finally, if ψ was inferred from σ and $\sigma \to \psi$ then by the induction hypothesis there are four cases to consider:

1. If $\mathcal{T} \vdash_{\mathcal{H}} \sigma$ and $\mathcal{T} \vdash_{\mathcal{H}} \sigma \to \psi$ then $\mathcal{T} \vdash_{\mathcal{H}} \psi$.

2. If $\mathcal{T} \vdash_{\mathcal{H}} \sigma$ and $\mathcal{T} \vdash_{\mathcal{H}} \varphi \to (\sigma \to \psi)$ then $\mathcal{T} \vdash_{\mathcal{H}} \varphi \to \psi$ using axiom [Pe].

3. If $\mathcal{T} \vdash_{\mathcal{H}} \varphi \to \sigma$ and $\mathcal{T} \vdash_{\mathcal{H}} \sigma \to \psi$ then $\mathcal{T} \vdash_{\mathcal{H}} \varphi \to \psi$ using axiom [Tr].

4. Suppose $\mathcal{T} \vdash_{\mathcal{H}} \varphi \to \sigma$ and $\mathcal{T} \vdash_{\mathcal{H}} \varphi \to (\sigma \to \psi)$. Then using axioms [Pe] and [Tr] we get that $\mathcal{T} \vdash_{\mathcal{H}} \varphi \to (\varphi \to \psi)$. Hence $\mathcal{T} \vdash_{\mathcal{H}} \varphi \to \psi$ by axiom [Ct]. □

Corollary 11.57 *The connective $\to$ is a relevant implication of an axiomatic extension* **L** *of* $\mathbf{R}_\to$ *iff $\to$ has in* **L** *the replacement property, and $\vdash_{\mathbf{L}}^{\to}$ satisfies the strong relevance criterion.*[7]

---

[6]This proposition is due to [118]. It is a 'use'-free variant of the relevant deduction theorem of $HR_\to$ from [4, 151]. The proof is essentially identical to the proof of the theorem given there.

[7]This proposition can be used for showing that in particular, $\to$ is a relevant implication of $\mathbf{R}_\to$ itself. The proof of this is similar to the proof of Theorem 12.21 below.

Proof. Immediate from Proposition 11.56 and Corollary 11.42. □

We end this section by showing that a relevant implication (if such exists) is necessarily unique (and so from now on we shall speak in case it exists on *the* relevant implication of **L**).

Proposition 11.58 *A logic* **L** *can have at most one relevant implication. More precisely: if both* $\to$ *and* $\supset$ *are relevant implications of* **L**, *then* $\varphi \to \psi$ *and* $\varphi \supset \psi$ *are congruent in* **L** *for any* φ, ψ.

Proof. It was shown in the proof of Proposition 11.41 that if p and q are distinct variables, then $p, p \to q \vdash_\mathbf{L} q$, while $p \to q \nvdash_\mathbf{L} q$. Since $\supset$ is a relevant implication, this implies that $p \to q \vdash_\mathbf{L} p \supset q$. Since $\nvdash_\mathbf{L} p \supset q$ (because $p \nvdash_\mathbf{L} q$), this in turn implies that $\vdash_\mathbf{L} (p \to q) \to (p \supset q)$. That $\vdash_\mathbf{L} (p \supset q) \to (p \to q)$ is proved similarly. By the replacement property of $\to$, this implies that $p \to q$ and $p \supset q$ are congruent in **L**. Since **L** is structural, this is true for any φ, ψ. □

11.3.3 Relevant Conjunction

Relevant implication is the most important connective in relevant logics. Next, we introduce another connective that is in general convenient to have, and is usually available in such logics.

Definition 11.59 (internal conjunction) Let $\mathcal{L}$ be a propositional language, and let $\mid\!\sim$ be a (Scott) pre-entailment on $\mathcal{L}$. A binary connective $\otimes$ of $\mathcal{L}$ is called an *internal conjunction* for $\mid\!\sim$ if the following condition is satisfied:

$$\Gamma, \varphi, \psi \mid\!\sim \sigma \text{ iff } \Gamma, \varphi \otimes \psi \mid\!\sim \sigma \quad (\Gamma, \varphi, \psi \mid\!\sim \Delta \text{ iff } \Gamma, \varphi \otimes \psi \mid\!\sim \Delta)$$

for every Γ, φ, ψ and σ (Δ).[8]

Proposition 11.60 *Let* $\mid\!\sim$ *be a pre-entailment.*

1. *Suppose* $\otimes$ *is an internal conjunction for* $\mid\!\sim$. *Then* $[\varphi_1, \ldots, \varphi_n] \mid\!\sim \psi$ *iff* $\varphi_1 \otimes \varphi_2 \otimes \cdots \otimes \varphi_n \mid\!\sim \psi$, *where* $\varphi_1 \otimes \cdots \otimes \varphi_n$ *is any formula obtained from* $[\varphi_1, \ldots, \varphi_n]$ *by applications of* $\otimes$, *provided that for each* $1 \leq i \leq n$, φ_i *is used exactly once.*

---

[8] This connective (also known as "intensional conjunction", "multiplicative conjunction" or "fusion") was first introduced by Dunn [146]. In works on relevant logic it is usually denoted by ∘. In works on the more general subject of substructural logics it is nowadays more common to denote it by $\otimes$, which is Girard's notation for it in linear logic (see [174]). Note that this connective $\otimes$ should not be confused with the one denoted by the same symbol in Chapter 5 in connection with $\mathcal{FOUR}$ and bilattices.

11.3 Basic Relevant Connectives

2. *Suppose $\to$ is an internal implication, and $\otimes$ is an internal conjunction for $\mathrel{\vert\!\sim}$. Then $[\varphi_1, \ldots, \varphi_n] \mathrel{\vert\!\sim} \psi$ iff $\mathrel{\vert\!\sim} \varphi_1 \otimes \cdots \otimes \varphi_n \to \psi$ (where $\varphi_1 \otimes \cdots \otimes \varphi_n \to \psi$ is ψ in case $n = 0$).*

Proof. The proof immediately follows from the definitions. □

Note 11.61 It follows from the first part of Proposition 11.60 that if $\otimes$ is an internal conjunction for a pre-entailment $\mathrel{\vert\!\sim}$, then any two formulas that represent $\varphi_1 \otimes \cdots \otimes \varphi_n$ (i.e., formulas obtained from $[\varphi_1, \ldots, \varphi_n]$ by applications of $\otimes$, provided that for each $1 \leq i \leq n$, φ_i is used exactly once) are equivalent in $\mathrel{\vert\!\sim}$. It follows that if $\mathrel{\vert\!\sim}$ has in addition the replacement property (which is the case in all relevant logics ever studied), then any two formulas of this sort are congruent in it. Henceforth we shall use therefore the notation $\varphi_1 \otimes \cdots \otimes \varphi_n$ freely, without caring what is the exact formula this expression denotes.

Definition 11.62 (associated conjunction) Let $\mathbf{L}$ be a logic with a connective $\to$ such that $\mathrel{\vert\!\sim}_{\mathbf{L}}^{\to}$ is well defined as a relation between multisets of formulas and formulas. A connective $\otimes$ is called *a conjunction associated with $\to$ in $\mathbf{L}$* if it is an internal conjunction for $\mathrel{\vert\!\sim}_{\mathbf{L}}^{\to}$.

Proposition 11.63 *Let $\mathbf{L}$ and $\to$ be like in Definition 11.62, and let $\otimes$ be a conjunction associated with $\to$ in $\mathbf{L}$.*

1. $[\varphi_1, \ldots, \varphi_n] \mathrel{\vert\!\sim}_{\mathbf{L}}^{\to} \psi$ iff $\vdash_{\mathbf{L}} \varphi_1 \otimes \cdots \otimes \varphi_n \to \psi$.

2. $\vdash_{\mathbf{L}} \varphi_1 \to (\varphi_2 \to \cdots (\varphi_n \to \psi) \cdots)$ iff $\vdash_{\mathbf{L}} \varphi_1 \otimes \cdots \otimes \varphi_n \to \psi$.

3. $\mathrel{\vert\!\sim}_{\mathbf{L}}^{\to}$ *satisfies the strong relevance criterion iff $\mathbf{L}$ has the variable sharing property* (VSP).[9]

Proof. The first part is immediate from the second part of Proposition 11.60. The second part is a reformulation of the first.

For the third part, assume first that $\mathrel{\vert\!\sim}_{\mathbf{L}}^{\to}$ satisfies the strong relevance criterion. Suppose that $\vdash_{\mathbf{L}} \varphi \to \psi$. Then $\varphi \mathrel{\vert\!\sim}_{\mathbf{L}}^{\to} \psi$. Hence the strong relevance criterion for $\mathrel{\vert\!\sim}_{\mathbf{L}}^{\to}$ implies that $\mathsf{Var}(\varphi) \cap \mathsf{Var}(\psi) \neq \emptyset$.

For the converse, assume that $\mathbf{L}$ has the VSP. Suppose that $\Gamma_2 \neq \emptyset$, and $\Gamma_1, \Gamma_2 \mathrel{\vert\!\sim}_{\mathbf{L}}^{\to} \varphi$. Let $\Gamma_2 = \{\psi_1, \ldots, \psi_n\}$ (where $n > 0$), and $\Gamma_1 = \{\varphi_1, \ldots, \varphi_k\}$. Then by definition, $[\psi_1, \ldots, \psi_n] \mathrel{\vert\!\sim}_{\mathbf{L}}^{\to} \varphi_1 \to (\varphi_2 \to \cdots (\varphi_k \to \varphi) \cdots)$. Hence, Item 1 implies that $\vdash_{\mathbf{L}} (\psi_1 \otimes \cdots \otimes \psi_n) \to (\varphi_1 \to (\varphi_2 \to \cdots (\varphi_k \to \varphi) \cdots))$. Therefore VSP implies that $\mathsf{Var}(\psi_1 \otimes \cdots \otimes \psi_n) \cap \mathsf{Var}((\varphi_1 \to (\varphi_2 \to \cdots (\varphi_k \to \varphi) \cdots))$ is not empty, i.e., Γ_2 and Γ_1, φ share some variable. □

---

[9] Henceforth we shall usually omit the reference to $\to$ when we talk about VSP, and rely on the context to determine what connective we have in mind.

Corollary 11.64 *If $\to$ has in $\mathbf{L}$ the RDP and an associated conjunction, then $\to$ is a relevant implication of $\mathbf{L}$ iff $\to$ has in $\mathbf{L}$ the VSP and the replacement property.*

Proof. Immediate from Corollaries 11.63 and 11.42. □

Note 11.65 The strong relevance criterion for $\mathrel{\vdash}\!\!\!\!\!\to_{\mathbf{L}}$ is the natural substitute for the relevance criterion for $\vdash_{\mathbf{L}}$ that was discussed and rejected at Section 11.2. Therefore, together with Note 11.50, the last proposition and corollary might justify the fact that in the literature on relevant logics VSP is usually taken as the most basic criterion for relevance.

Proposition 11.66 *Let $\mathbf{L}$ and $\to$ be like in Definition 11.62. $\otimes$ is a conjunction associated with $\to$ in $\mathbf{L}$ iff the following residuation axioms are provable in $\mathbf{L}$ (for every φ, ψ, and σ):*

$$[\text{R1}] \quad (\varphi \to (\psi \to \sigma)) \to (\varphi \otimes \psi \to \sigma),$$
$$[\text{R2}] \quad (\varphi \otimes \psi \to \sigma) \to (\varphi \to (\psi \to \sigma)).$$

Proof. Let $\otimes$ be a conjunction associated with $\to$. Since we have that $\varphi \to (\psi \to \sigma), \varphi, \psi \mathrel{\vdash}\!\!\!\!\!\to_{\mathbf{L}} \sigma$, this implies that $\varphi \to (\psi \to \sigma), \varphi \otimes \psi \mathrel{\vdash}\!\!\!\!\!\to_{\mathbf{L}} \sigma$, and so, by definition of $\mathrel{\vdash}\!\!\!\!\!\to_{\mathbf{L}}$, [R1] is provable. For [R2], note that $\varphi, \psi \mathrel{\vdash}\!\!\!\!\!\to_{\mathbf{L}} \varphi \otimes \psi$ follows from $\varphi \otimes \psi \mathrel{\vdash}\!\!\!\!\!\to_{\mathbf{L}} \varphi \otimes \psi$, and our assumption on $\otimes$. This and the fact that $\varphi \otimes \psi \to \sigma, \varphi \otimes \psi \mathrel{\vdash}\!\!\!\!\!\to_{\mathbf{L}} \sigma$ imply that $\varphi \otimes \psi \to \sigma, \varphi, \psi \mathrel{\vdash}\!\!\!\!\!\to_{\mathbf{L}} \sigma$ (using transitivity). This implies [R2].

For the converse, assume that both [R1] and [R2] are valid in $\mathbf{L}$. Then $\varphi \otimes \psi \to \sigma \mathrel{\vdash}\!\!\!\!\!\to_{\mathbf{L}} \varphi \to (\psi \to \sigma)$, and $\varphi \to (\psi \to \sigma) \mathrel{\vdash}\!\!\!\!\!\to_{\mathbf{L}} \varphi \otimes \psi \to \sigma$. Now $\Gamma, \varphi, \psi \mathrel{\vdash}\!\!\!\!\!\to_{\mathbf{L}} \sigma$ iff $\Gamma \mathrel{\vdash}\!\!\!\!\!\to_{\mathbf{L}} \varphi \to (\psi \to \sigma)$, iff (by the above two facts) $\Gamma \mathrel{\vdash}\!\!\!\!\!\to_{\mathbf{L}} \varphi \otimes \psi \to \sigma$, iff $\Gamma, \varphi \otimes \psi \mathrel{\vdash}\!\!\!\!\!\to_{\mathbf{L}} \sigma$. It follows that $\otimes$ is an internal conjunction for $\mathrel{\vdash}\!\!\!\!\!\to_{\mathbf{L}}$. □

Note 11.67 Some of the axiom schemas of $HLL_\to$ and $HRM0_\to$ have shorter and more transparent formulations in the presence of an internal conjunction $\otimes$. Thus it is easy to see that in a logic in which every instance of [Id], [Tr], [R1], and [R2] is provable, and the set of theorems is closed under (MP), the axioms of Permutation, Contraction, and Mingle are respectively equivalent to the following axioms:

$$[\text{Pe}] \quad \varphi \otimes \psi \to \psi \otimes \varphi,$$
$$[\text{Ct}] \quad \varphi \to \varphi \otimes \varphi,$$
$$[\text{Mi}] \quad \varphi \otimes \varphi \to \varphi.$$

In other words: the permutation axiom is equivalent to the commutativity of $\otimes$, while together the contraction and mingle axioms are equivalent to the idempotency of $\otimes$.

11.3 Basic Relevant Connectives

It is also worth noting that in a logic which weakly contains $\mathbf{LL}_\to$, the axiom schema [R2] is equivalent to the schema $\varphi \to (\psi \to \varphi \otimes \psi)$.

We end this section with another important connection between the relevant connectives and the corresponding classical ones (see Corollary 11.48).

Proposition 11.68 *Suppose that $\to$ is a relevant implication for a logic* $\mathbf{L}$ *which is* $\mathbf{F}$*-contained in classical logic, and $\otimes$ is an internal conjunction associated with $\to$. Then $\mathbf{F}(\otimes)$ is the classical conjunction.*

Proof. For convenience, we denote $\mathbf{F}(\to)$ by $\tilde{\supset}$ and $\mathbf{F}(\otimes)$ by $\tilde{\&}$. Since $\vdash_\mathbf{L} (p \to (q \to q)) \to (p \otimes q \to q)$, by assigning f to q we easily get (using Corollary 11.48) that $\tilde{\&}(t,f) = \tilde{\&}(f,f) = f$. Hence the validity in $\mathbf{L}$ of $p \otimes q \to q \otimes p$ entails that $\tilde{\&}(f,t) = f$. That $\tilde{\&}(t,t) = t$ is immediate from the validity in $\mathbf{L}$ of $p \to (q \to p \otimes q)$ (and Corollary 11.48). □

11.3.4 Relevant Negation

Important as implication is, negation is the focus of this book, and we believe that there is not much point to investigate relevant logics which lack an appropriate negation connective. (In fact, already the first item of Example 11.8 indicates its crucial role here.)

So what is relevant negation? The point of view which was (implicitly) adopted by the relevant logicians about this question follows that of classical logic: *the main function of a relevant negation is to make it possible to conservatively extend in a natural way a given relevant entailment into a Scott (multiple-conclusioned) relevant entailment.*

Definition 11.69 (internal negation for Scott pre-entailment) Let $\mathcal{L}$ be a propositional language, and let $\mathrel{\vdash\mkern-10mu\sim}$ be a Scott pre-entailment on $\mathcal{L}$. A unary connective $\neg$ of $\mathcal{L}$ is called an *internal negation* for $\mathrel{\vdash\mkern-10mu\sim}$ if the following conditions are satisfied for every Γ, Δ and φ:

- $\Gamma, \varphi \mathrel{\vdash\mkern-10mu\sim} \Delta$ iff $\Gamma \mathrel{\vdash\mkern-10mu\sim} \Delta, \neg\varphi$, and
- $\Gamma \mathrel{\vdash\mkern-10mu\sim} \Delta, \varphi$ iff $\Gamma, \neg\varphi \mathrel{\vdash\mkern-10mu\sim} \Delta$.[10]

Lemma 11.70 *Let $\mathcal{L}$ be a propositional language, and let $\mathrel{\vdash\mkern-10mu\sim}$ be a Scott pre-entailment on $\mathcal{L}$. A unary connective $\neg$ of $\mathcal{L}$ is an internal negation for $\mathrel{\vdash\mkern-10mu\sim}$ if either of the two conditions specified in Definition 11.69 is satisfied.*

---

[10] In the literature on relevant logic the relevant negation is usually denoted by $\sim$. Here we use the more standard notation, which is used throughout this book.

Proof. Suppose, e.g., that the first condition is satisfied, i.e., $\Gamma, \varphi \mathrel{\mid\!\sim} \Delta$ iff $\Gamma \mathrel{\mid\!\sim} \Delta, \neg\varphi$. Since $\varphi \mathrel{\mid\!\sim} \varphi$, this implies that $\mathrel{\mid\!\sim} \varphi, \neg\varphi$, and since $\neg\varphi \mathrel{\mid\!\sim} \neg\varphi$, also $\neg\varphi, \varphi \mathrel{\mid\!\sim}$. Assume now that $\Gamma \mathrel{\mid\!\sim} \Delta, \varphi$. Then a cut on φ of this and $\neg\varphi, \varphi \mathrel{\mid\!\sim}$ yields $\Gamma, \neg\varphi \mathrel{\mid\!\sim} \Delta$. For the converse, assume that $\Gamma, \neg\varphi \mathrel{\mid\!\sim} \Delta$. Then a cut on $\neg\varphi$ of this and $\mathrel{\mid\!\sim} \varphi, \neg\varphi$ yields $\Gamma \mathrel{\mid\!\sim} \Delta, \varphi$.

The proof that the second condition implies the first is similar. □

Definition 11.71 ($\mathrel{\mid\!\sim}^1$) Let $\mathrel{\mid\!\sim}$ be a Scott pre-entailment on $\mathcal{L}$. The pre-entailment *induced* by $\mathrel{\mid\!\sim}$ is the relation $\mathrel{\mid\!\sim}^1$ defined by: $\Gamma \mathrel{\mid\!\sim}^1 \varphi$ if $\Gamma \mathrel{\mid\!\sim} \varphi$.

Lemma 11.72 *If $\mathrel{\mid\!\sim}$ is a Scott (relevant) (pre-)entailment on $\mathcal{L}$ which has an internal negation $\neg$, then $\mathrel{\mid\!\sim}^1$ is a (relevant) (pre)-entailment. Moreover: if $\to$ ($\otimes$) is an internal implication (conjunction) for $\mathrel{\mid\!\sim}$ then it is also an internal implication (conjunction) for $\mathrel{\mid\!\sim}^1$.*

Proof. With the possible exception of strong consistency, all the conditions are immediate from the definitions. For strong consistency, assume that $\mathrel{\mid\!\sim}^1 p$. Then $\mathrel{\mid\!\sim} p$, and so also $\mathrel{\mid\!\sim} \neg p$ (because of structurality). Since $\neg$ is an internal negation for $\mathrel{\mid\!\sim}$, this implies that $p \mathrel{\mid\!\sim}$. But $\mathrel{\mid\!\sim} p$ and $p \mathrel{\mid\!\sim}$ imply (by transitivity) $[\,] \mathrel{\mid\!\sim} [\,]$, contradicting the strong consistency of $\mathrel{\mid\!\sim}$. □

Definition 11.73 (internal negation) Let $\mathrel{\mid\!\sim}$ be a pre-entailment on $\mathcal{L}$, and let $\neg$ be a unary connective of $\mathcal{L}$. $\neg$ is an *internal negation* for $\mathrel{\mid\!\sim}$ if $\mathrel{\mid\!\sim}$ is induced by some Scott pre-entailment for which $\neg$ is an internal negation.

The next proposition shows that given a (relevant) entailment $\mathrel{\mid\!\sim}$ for $\mathcal{L}$ and a unary connective $\neg$ of $\mathcal{L}$, there is at most one way to extend $\mathrel{\mid\!\sim}$ to a Scott (relevant) entailment for which $\neg$ is an internal negation. It also provides necessary and sufficient conditions for such an extension to exist.

Proposition 11.74 *Let $\mathrel{\mid\!\sim}$ be an entailment on $\mathcal{L}$, and let $\neg$ be a unary connective of $\mathcal{L}$.*

1. *There is at most one Scott entailment which induces $\mathrel{\mid\!\sim}$, and for which $\neg$ is an internal negation.*

2. *$\neg$ is an internal negation for $\mathrel{\mid\!\sim}$ iff $\neg$ satisfies the following conditions:*

   - *If $\Gamma, \varphi \mathrel{\mid\!\sim} \neg\psi$ then $\Gamma, \psi \mathrel{\mid\!\sim} \neg\varphi$.*
   - *$\neg\neg\varphi \mathrel{\mid\!\sim} \varphi$.*
   - *There is no formula φ such that both $\mathrel{\mid\!\sim} \varphi$ and $\mathrel{\mid\!\sim} \neg\varphi$.*

Proof. Suppose that $\mathrel{\mid\!\sim}$ and $\neg$ are as in the proposition.

11.3 Basic Relevant Connectives

1. Suppose $\mathrel{\mid\!\sim}^*$ is a Scott entailment which induces $\mathrel{\mid\!\sim}$, and for which $\neg$ is an internal negation. Then in case $n > 0$ we have that $[\varphi_1, \ldots, \varphi_n] \mathrel{\mid\!\sim}^*$ iff $[\varphi_1, \ldots, \varphi_{n-1}] \mathrel{\mid\!\sim} \neg\varphi_n$, while $[\varphi_1, \ldots, \varphi_n] \mathrel{\mid\!\sim}^* [\psi_1, \ldots, \psi_k]$ in case $k > 0$ iff $[\varphi_1, \ldots, \varphi_n, \neg\psi_1, \ldots, \neg\psi_{k-1}] \mathrel{\mid\!\sim} \psi_k$. It follows that $\mathrel{\mid\!\sim}^*$ is completely determined by $\mathrel{\mid\!\sim}$.

2. It is easy to see that the definition of an internal negation implies the necessity of the first two conditions for the existence of a Scott entailment $\mathrel{\mid\!\sim}^*$ which induces $\mathrel{\mid\!\sim}$, and for which $\neg$ is an internal negation. As for the third, assume that there is a formula φ such that both $\mathrel{\mid\!\sim} \varphi$ and $\mathrel{\mid\!\sim} \neg\varphi$. Then both $\mathrel{\mid\!\sim}^* \varphi$ and $\varphi \mathrel{\mid\!\sim}^*$. Hence (by transitivity) $[\,] \mathrel{\mid\!\sim}^* [\,]$, contradicting the strong consistency of $\mathrel{\mid\!\sim}^*$.

For the converse, assume that the three conditions are satisfied. Since $\neg\psi \mathrel{\mid\!\sim} \neg\psi$, the first condition implies that $\psi \mathrel{\mid\!\sim} \neg\neg\psi$. Together with the first two conditions this in turn implies that if $\Gamma, \neg\varphi \mathrel{\mid\!\sim} \psi$ then $\Gamma, \neg\psi \mathrel{\mid\!\sim} \varphi$. (If $\Gamma, \neg\varphi \mathrel{\mid\!\sim} \psi$ then $\Gamma, \neg\varphi \mathrel{\mid\!\sim} \neg\neg\psi$, and so by the first condition $\Gamma, \neg\psi \mathrel{\mid\!\sim} \neg\neg\varphi$. This and the second condition entail that $\Gamma, \neg\psi \mathrel{\mid\!\sim} \varphi$.) Now define $\mathrel{\mid\!\sim}^\neg$ as in the proof of the first part, i.e.,

$$[\varphi_1, \ldots, \varphi_n] \mathrel{\mid\!\sim}^\neg [\psi_1, \ldots, \psi_k] \text{ iff } [\varphi_1, \ldots, \varphi_n, \neg\psi_1, \ldots, \neg\psi_{k-1}] \mathrel{\mid\!\sim} \psi_k$$

$$[\varphi_1, \ldots, \varphi_n] \mathrel{\mid\!\sim}^\neg \text{ iff } [\varphi_1, \ldots, \varphi_{n-1}] \mathrel{\mid\!\sim} \neg\varphi_n$$

$$[\,] \mathrel{\not\mid\!\sim}^\neg [\,]$$

It is a routine matter to check that the first two conditions, together with the extra two we have deduced from them, ensure that $\mathrel{\mid\!\sim}^\neg$ is well-defined. It is also straightforward to show that it satisfies all the requirements. Here we do it for the transitivity condition. So assume that $\Gamma_1 \mathrel{\mid\!\sim}^\neg \Delta_1, \varphi$, and $\varphi, \Gamma_2 \mathrel{\mid\!\sim}^\neg \Delta_2$. Then $\Gamma_1, \neg\Delta_1 \mathrel{\mid\!\sim} \varphi$. Suppose first that $\Gamma_2 \neq \emptyset$, so $\Gamma_2 = \Gamma_2', \psi$ for some ψ. Then $\varphi, \Gamma_2', \neg\Delta_2 \mathrel{\mid\!\sim} \neg\psi$. Hence the transitivity of $\mathrel{\mid\!\sim}$ implies that $\Gamma_1, \neg\Delta_1, \Gamma_2', \neg\Delta_2 \mathrel{\mid\!\sim} \neg\psi$, and so $\Gamma_1, \Gamma_2 \mathrel{\mid\!\sim}^\neg \Delta_1, \Delta_2$. The case where $\Delta_2 \neq [\,]$ is similar. Next assume that both Γ_2 and Δ_2 are empty. Then $\varphi, \Gamma_2 \mathrel{\mid\!\sim}^\neg \Delta_2$ means that $\mathrel{\mid\!\sim} \neg\varphi$. Because of our third condition, this implies that $\mathrel{\not\mid\!\sim} \varphi$. It follows that $\Gamma_1 \cup \Delta_1 \neq \emptyset$. Assume e.g. that $\Gamma_1 = \Gamma_1', \psi$.(The case where Δ_1 is not empty is similar.) Then $\Gamma_1 \mathrel{\mid\!\sim}^\neg \Delta_1, \varphi$ implies that $\neg\varphi, \neg\Delta_1, \Gamma_1' \mathrel{\mid\!\sim} \neg\psi$. This and $\mathrel{\mid\!\sim} \neg\varphi$ together imply that $\neg\Delta_1, \Gamma_1' \mathrel{\mid\!\sim} \neg\psi$, and so again $\Gamma_1, \Gamma_2 \mathrel{\mid\!\sim}^\neg \Delta_1, \Delta_2$. □

Definition 11.75 (associated negation; $\neg$-consistency) Suppose that **L** is a logic with a connective $\to$ such that $\mathrel{\mid\!\sim}^{\to}_{\mathbf{L}}$ is well defined. Let $\neg$ be a unary connective of **L**.

- ¬ is a *negation associated with* → if ¬ is an internal negation for $\vdash^{\rightarrow}_{\mathbf{L}}$.
- **L** is ¬-*consistent* if there is no φ such that both $\vdash_{\mathbf{L}} \varphi$ and $\vdash_{\mathbf{L}} \neg\varphi$.

Proposition 11.76 *Let* **L** *and* → *be like in Definition 11.75.* ¬ *is a negation associated with* → *iff the following conditions are satisfied:*

- *The following* negation axioms *are provable in* **L** *(for every φ and ψ):*

  [N1] $\quad (\varphi \to \neg\psi) \to (\psi \to \neg\varphi),$
  [N2] $\quad \neg\neg\varphi \to \varphi.$

- **L** *is* ¬-*consistent.*

Proof. This easily follows from the second part of Proposition 11.74, the definition of $\vdash^{\rightarrow}_{\mathbf{L}}$, and the first part of Proposition 11.31. □

Proposition 11.77 *Let* **L** *and* → *be like in Definition 11.75, and suppose that* ¬ *is a negation associated with* →. *If* **L** *is* **F**-*contained in classical logic then* $\mathbf{F}(\neg)$ *is the classical negation.*

Proof. Denote $\mathbf{F}(\neg)$ by $\widetilde{\neg}$. By assigning in [N1] f to φ and t to ψ we easily get (by Corollary 11.48) that $\widetilde{\neg} f = t$. Therefore, by assigning in [N2] f to φ we get (Corollary 11.48 again) that $\widetilde{\neg} t = \widetilde{\neg}\widetilde{\neg} f = f$. □

Next we introduce the minimal logic which has a weak relevant implication and an associated negation, as well as the minimal logic which has a relevant implication together with an associated negation. For this we use the axioms which were introduced in Proposition 11.76.

Definition 11.78 ($HLL_{\overrightarrow{\neg}}$, $HR_{\overrightarrow{\neg}}$, $\mathbf{LL}_{\overrightarrow{\neg}}$, $\mathbf{R}_{\overrightarrow{\neg}}$)

- $HLL_{\overrightarrow{\neg}}$ and $HR_{\overrightarrow{\neg}}$ are the Hilbert-type proof systems which are obtained from $HLL_{\rightarrow}$ and $HR_{\rightarrow}$ (respectively) by adding ¬ to their language, and the axioms [N1] and [N2] to their lists of axioms.

- $\mathbf{LL}_{\overrightarrow{\neg}}$ and $\mathbf{R}_{\overrightarrow{\neg}}$ are the logics in the language of $\{\to, \neg\}$ which are induced by the systems $HLL_{\overrightarrow{\neg}}$ and $HR_{\overrightarrow{\neg}}$ (respectively).[11]

Corollary 11.79

1. *If* → *is a weak relevant implication of a logic* **L**, *and* ¬ *is a negation associated with* →, *then* **L** *contains* $\mathbf{LL}_{\overrightarrow{\neg}}$.

---

[11]$\mathbf{LL}_{\overrightarrow{\neg}}$ is equivalent to the multiplicative fragment without the propositional constants of Linear Logic (see [28, 174]). The logic $\mathbf{R}_{\overrightarrow{\neg}}$ is equivalent to the corresponding fragment of the relevant logic **R**. It will be discussed in detail in the next chapter.

11.3 Basic Relevant Connectives

2. If $\to$ is a relevant implication of a logic **L**, and $\neg$ is a negation associated with $\to$, then **L** contains $\mathbf{R}_{\to}^{\neg}$.

Proof. Immediate from Propositions 11.31, 11.47, and 11.76. □

Next, we show that a weak relevant implication which has an associated negation also has an associated internal conjunction.

Proposition 11.80 *Let* **L** *and* $\to$ *be like in Definition 11.75, and let* $\neg$ *be a negation associated with* $\to$. *Define:* $\varphi \otimes \psi =_{Df} \neg(\varphi \to \neg\psi)$. *Then* $\otimes$ *is an internal conjunction associated with* $\to$.

Proof. Immediate from the definitions of internal connectives. □

Example 11.81 The connective $\to$ is a weak relevant implication of $\mathbf{LL}_{\to}^{\neg}$, and the connective $\neg$ is a negation associated with it. Indeed, most of the relevant conditions are easily proved using Propositions 11.31, 11.76, and Note 11.33. The only exception is the strong relevance criterion. To show that $\mathbf{LL}_{\to}$ satisfies this criterion it suffices by Propositions 11.80 and 11.63 to show that it has the VSP. This is done in Corollary 13.22 below for a system that is stronger than $\mathbf{LL}_{\to}^{\neg}$ (and even $\mathbf{R}_{\to}^{\neg}$).

Example 11.82 Let $\mathbf{LL}_{mm}$ be the logic obtained from $\mathbf{LL}_{\to}$ by adding to its language two new unary connectives $\neg$ and $\sim$, and to its proof system $HLL_{\to}$ the four axioms that are needed according to Proposition 11.76 in order to make both $\neg$ and $\sim$ negations associated with $\to$ (two axioms for $\neg$, two axioms for $\sim$). It is not too difficult to show that $\to$ is a weak relevant implication in $\mathbf{LL}_{mm}$, and that both $\neg$ and $\sim$ are negations associated with it. We show that $\nvdash_{\mathbf{LL}_{mm}} \neg\varphi \to \sim\varphi$. For this we use the matrix $\langle Z, \{0\}, O \rangle$ where Z is the set of integers, and O interprets the connective as follows: $\widetilde{\to}(a,b) = b - a$, $\widetilde{\neg}(a) = -a$, and $\widetilde{\sim}(a) = 1 - a$. It is not difficult to show that $\mathbf{LL}_{mm}$ is strongly sound with respect to this matrix. In particular: $\nu(\psi) = 0$ for every valuation ν in it and for every theorem ψ of $\mathbf{LL}_{mm}$. This is not true for $\neg\varphi \to \sim \varphi$. (Actually, the latter has a model in this matrix for no φ!) It follows that a weak relevant implication may have more than one negation associated with it.

Things are different when it comes to (strong) relevant implications. Situations of the type described in Example 11.82 are then impossible:

Theorem 11.83 *Let* $\to$ *be a relevant implication for a logic* **L**. *Then* $\to$ *has in* **L** *at most one associated negation: if both* $\neg$ *and* $\sim$ *are negations associated with* $\to$, *then* $\neg\varphi$ *and* $\sim\varphi$ *are congruent in* **L** *for every* φ.

Proof. Suppose both $\neg$ and $\sim$ are negations associated with $\to$. Then the following are theorems of **L** for every variables p, q and formula φ in the language $\{\neg, \sim, \to\}$:

(1) $(p \to p) \to (\varphi \to \varphi)$ if $p \in \mathsf{Var}(\varphi)$,

(2) $((p \to p) \to (q \to q)) \to (((q \to \neg(p \to p)) \to \neg(p \to p)) \to q)$,

(3) $\neg p \to (p \to \neg(p \to p))$,

(4) $(\sim \varphi \to \varphi) \to \varphi$,

(5) $(p \to \sim(p \to p)) \to \sim p$.

Item (1) is proved by an easy induction on the structure of φ, using the second part of Corollary 11.79. By the same corollary, to prove (2)-(5) it suffices to show that they are theorems of $\mathbf{R}_{\to}^{\neg}$. Items (2) and (3) can most easily be proved using the Gentzen-type system $GR_{\to}^{\neg}$ for $\mathbf{R}_{\to}^{\neg}$ which is described in the next chapter.[12] Items (4) and (5) are proved similarly, by using $GR_{\to}^{\neg}$ with $\neg$ replaced by $\sim$.

Let $\varphi = \neg(p \to p)$. By (1), $\vdash_{\mathbf{L}} (p \to p) \to \sim(\sim \varphi \to \sim \varphi)$. Hence by substituting $\sim \varphi$ for q in (2) we get with the help of (4) that $\vdash_{\mathbf{L}} \sim \varphi$. This and and the fact we inferred above from (1) about $\sim \varphi$ easily imply that $\vdash_{\mathbf{L}} (p \to p) \to \sim \varphi$. Using the axioms for $\sim$ this in turn entails that $\vdash_{\mathbf{L}} \varphi \to \sim(p \to p)$, i.e., that $\vdash_{\mathbf{L}} \neg(p \to p) \to \sim(p \to p)$. By (3), this implies that $\vdash_{\mathbf{L}} \neg p \to (p \to \sim(p \to p))$, and so $\vdash_{\mathbf{L}} \neg p \to \sim p$ follows by (5). The converse implication is proved similarly. Since **L** is structural, and $\to$ has in it the replacement property, this entails the congruence of $\neg \varphi$ and $\sim \varphi$ in **L** for every formula φ. □

11.4 Relevant Logics

We are ready at last to present our two definitions of the notion of a "relevant logic".[13] We start with the stronger one.

---

[12] If we replace in $GR_{\to}^{\neg}$ $\neg$ by $\sim$ then we get of course a system every theorem of which is also a theorem of **L**. Moreover: by combining these two Gentzen-type systems into one, in which $\neg$ and $\sim$ have the same rules, we get a system in which the congruence of $\neg \varphi$ and $\sim \varphi$ is very easily shown. However, the soundness of this extended system with respect to **L** is equivalent to Theorem 11.83, and so we cannot use it in the proof of the latter.

[13] The name *"classical relevant logic"* might be more appropriate here, since we demand the availability of a relevant counterpart of the classical negation. However, this name has already been used in the literature on relevant logics for something different, and we shall not study any other type of "relevant logics" in this book, so the shorter terminology would be more convenient for our purposes.

11.4 Relevant Logics

Definition 11.84 (Strongly Relevant Logic) A logic **L** is a *strongly relevant logic* if the following conditions are satisfied:

- **L** is finitary.

- **L** has a relevant implication having an associated negation.

- **L** satisfies the minimal semantic relevance criterion.

Note 11.85 In [52], from which most of the material of this section (as well as Section 12.1 in the next chapter) is taken, the last condition in Definition 11.84 was not included in the definitions of 'strongly relevant logic' and of 'relevant logic'. Indeed, all the propositions of the present section and of Section 12.1 remain true if we delete the minimal semantic relevance criterion from these definitions. However, we think that this criterion is very natural, and it *is* used in Chapter 13 for determining *maximal* (strongly) relevant logics.

Proposition 11.86 *If* **L** *is a strongly relevant logic then* $\mathbf{L} = \mathbf{L}(\mid\!\sim_{\mathbf{L}}^{\rightarrow})$.

Proof. Immediate from Proposition 11.41. □

Proposition 11.87 *Every strongly relevant logic* **L** *satisfies the basic relevance criterion.*

Proof. Let $\rightarrow$ be a relevant implication for **L**, and let $\otimes$ be the internal conjunction associated with $\rightarrow$ (which exists by Proposition 11.80). Suppose that $\mathcal{T}_1, \mathcal{T}_2$ are theories and ψ is a formula such that $\mathcal{T}_1, \mathcal{T}_2 \vdash_{\mathbf{L}} \psi$ and $\mathcal{T}_2$ has no variable in common with $\mathcal{T}_1 \cup \{\psi\}$. Since **L** is finitary, and $\rightarrow$ has in it the RDP, this easily implies that there are elements $\varphi_1, \ldots, \varphi_n$ of $\mathcal{T}_2$ ($n \geq 0$), and elements $\sigma_1, \ldots, \sigma_k$ of $\mathcal{T}_1$ ($k \geq 0$), such that

$$\vdash_{\mathbf{L}} \varphi_1 \rightarrow \cdots \rightarrow \varphi_n \rightarrow \sigma_1 \rightarrow \cdots \rightarrow \sigma_k \rightarrow \psi.$$

Assume that $n > 0$. Then $\vdash_{\mathbf{L}} \varphi_1 \otimes \cdots \otimes \varphi_n \rightarrow \sigma_1 \rightarrow \cdots \rightarrow \sigma_k \rightarrow \psi$. Now, the VSP implies that $\varphi_1 \otimes \cdots \otimes \varphi_n$ and $\sigma_1 \rightarrow \cdots \rightarrow \sigma_k \rightarrow \psi$ have a variable in common. This contradicts our assumption about $\mathcal{T}_2$ and $\mathcal{T}_1 \cup \{\psi\}$. It follows that $n = 0$, and so $\vdash_{\mathbf{L}} \sigma_1 \rightarrow \cdots \rightarrow \sigma_k \rightarrow \psi$. Hence $\mathcal{T}_1 \vdash_{\mathbf{L}} \psi$. □

Proposition 11.88 *A strongly relevant logic has exactly one relevant implication* $\rightarrow$, *and this* $\rightarrow$ *has exactly one associated negation* $\neg$ *and exactly one associated conjunction* $\otimes$, *given by* $\varphi \otimes \psi =_{Df} \neg(\varphi \rightarrow \neg\psi)$.

Proof. Immediate from Propositions 11.58, 11.80, and Theorem 11.83. (We leave the uniqueness of $\otimes$ as an exercise for the reader.) □

Next, we note an intimate connection which exists between strongly relevant logics and classical logic.

Definition 11.89 (F_0) The interpretation F_0 of the language $\{\rightarrow, \neg\}$ is the one which assigns the classical bivalent interpretations of implication and negation to $\rightarrow$ and $\neg$ (respectively).

Theorem 11.90 *Let L be a strongly relevant logic, where $\rightarrow$ is the the relevant implication of L, and $\neg$ is its associated negation. Let $L_{\rightarrow}^{\neg}$ be the $\{\rightarrow, \neg\}$-fragment of L. Then any simple extension of $L_{\rightarrow}^{\neg}$ is F_0-contained in classical logic. Moreover, for each such extension, F_0 is the only bivalent interpretation of $\{\rightarrow, \neg\}$ which has this property.*

Proof. That F_0 has this property follows from Theorems 12.14 and 12.9 of Chapter 12 (whose proofs do not use the present theorem). That it is the only one follows from Corollary 11.48, and Propositions 11.68, 11.77. □

Example 11.91 The matrix described in Example 11.82 is a model of the extension of $LL_{\rightarrow}^{\neg}$ by the axiom $((\varphi \rightarrow \psi) \rightarrow \psi) \rightarrow \varphi$, which is not a classical tautology (if we interpret $\rightarrow$ as the classical implication). Hence, Theorem 11.90 is not valid for $LL_{\rightarrow}^{\neg}$.

Corollary 11.92 *Every strongly relevant logic L is boldly paraconsistent.*

Proof. The $\{\neg, \rightarrow\}$-fragment of L is semi-normal (since $\rightarrow$ is a semi-implication for it), and it is $\neg$-contained in classical logic by Theorem 11.90. Hence L is $\neg$-coherent with classical logic. From Proposition 11.87 and Corollary 11.4 it follows that L is also boldly pre-paraconsistent. □

The next proposition shows that in contrast, a strongly relevant logic is never paracomplete.

Proposition 11.93 *The negation $\neg$ of a strongly relevant logic L is complete for L.*

Proof. For every φ and ψ, $\vdash_{R_{\rightarrow}^{\neg}} (\neg\varphi \rightarrow \psi) \rightarrow ((\varphi \rightarrow \psi) \rightarrow \psi)$. (This will be particularly easy to verify using the Gentzen-type system $GR_{\rightarrow}^{\neg}$ presented in Chapter 12.) It follows by Corollary 11.79 that $\{\neg\varphi \rightarrow \psi, \varphi \rightarrow \psi\} \vdash_L \psi$. Hence the RDP for $\rightarrow$ implies that $\neg$ is complete for L. □

The notion of a strongly relevant logic is in our opinion the one that most faithfully reflects the main intuitions and motivations of the relevantists as

11.4 Relevant Logics

they are described, e.g., in [4, 151]. Unfortunately, the corresponding class of logics does not include the major relevant logics studied in the literature (at least according to their natural interpretations as *logics* in the sense of Definition 1.3). Next, we introduce an alternative weaker (but still quite reasonable) notion that does not have this drawback.[14]

Definition 11.94 (relevant logic) A logic **L** is a *relevant logic* if the following conditions are satisfied:

- **L** is finitary.
- **L** has a *weak* relevant implication $\to$ having an associated negation.
- $\mathbf{L}(\vdash_\mathbf{L}^{\to})$ is a strongly relevant logic.
- **L** satisfies the basic relevance criterion.

Note 11.95 Since **L** and $\mathbf{L}(\vdash_\mathbf{L}^{\to})$ have the same set of valid formulas, **L** satisfies the minimal semantic relevance criterion iff $\mathbf{L}(\vdash_\mathbf{L}^{\to})$ does. Hence, every relevant logic satisfies the minimal semantic relevance criterion.

Corollary 11.96 *Every strongly relevant logic is a relevant logic.*

Proof. Immediate from Propositions 11.41 and 11.87.

Note 11.97 Unlike in the case of strongly relevant logics, the satisfaction of the basic relevance criterion should explicitly be demanded for relevant logics. Indeed, if **L** is such a logic, and we add explosion (from φ and $\neg\varphi$ infer ψ) as a new rule of inference, then the resulting logic still satisfies the first three conditions in the definition of relevant logics (because it has the same set of valid formulas as **L**), but by Corollary 11.4 it does not satisfy the basic relevance criterion. Actually, it might not be easy to show that a given logic which is not a strongly relevant logic in our sense satisfies what we take here as the basic relevance criterion. Surprisingly, the logic **R** (the major relevant logic of Anderson and Belnap's school — see Chapter 14) is a case in point. In fact, for **R** the only proof we know uses the strong tool of Routley and Meyer's relational semantic (see e.g. [151]).

Note 11.98 By definition, if **L** is a relevant logic that is not a strongly relevant logic, then $\mathbf{L} \neq \mathbf{L}(\vdash_\mathbf{L}^{\to})$. This means that the "implication" connective $\to$ of **L** does not faithfully reflect the consequence relation of **L**. Usually **L**

---

[14]This weaker notion still does not apply to systems like **E** (Anderson and Belnap's favorite system) or **T**. However, these systems are not motivated solely by considerations of relevance (even though they have been introduced by Anderson and Belnap's school).

is obtained in such a case from $\mathbf{L}(\vdash_{\mathbf{L}}^{\rightarrow})$ by adding to it some inference rules that do not involve $\rightarrow$, and are not directly reflected by valid implications of $\mathbf{L}$. See Chapters 14 and 15 for examples of this sort (like $\mathbf{R}$ and $\mathbf{RM}$).

It should again be emphasized that $\mathbf{L}$ and $\mathbf{L}(\vdash_{\mathbf{L}}^{\rightarrow})$ always have the same valid formulas. However, they may differ in their consequence relations.

Note 11.99 Unlike strongly relevant logic (Corollary 11.92), there seems to be no guarantee that every relevant logic $\mathbf{L}$ is necessarily paraconsistent (according to our definition of paraconsistency). Obviously, such $\mathbf{L}$ is always boldly pre-paraconsistent by Corollary 11.4. However, $\rightarrow$ may not be a semi-implication for its $\{\neg, \rightarrow\}$-fragment, and so it may have no semi-normal fragment containing $\neg$. Still, all relevant logics studied in the literature are also paraconsistent (and so boldly paraconsistent).

11.5 Bibliographical Notes and Further Reading

Relevant logics were introduced by Anderson and Belnap in [3]. (Though their work has its roots in Church's paper [118], where $\mathbf{R}_{\rightarrow}$ and the RDP were first introduced.) With the exception of linear logic and its fragments (which were first introduced in [174]), all the logics which are mentioned in this chapter have been developed within the work of their school. (See Chapters 12 and 14 for more information on this work, and [4, 5, 99, 151, 210, 252] for some extensive books, surveys, and further references.) Nowadays both relevant logics and linear logic are taken to belong to the bigger family of *substructural logics*. The two current main books on this family are [238] and [254].

What is called here 'the basic relevance criterion' is implicit in RM87, on Page 418 of [4], and almost explicit in the discussion that follows it. It is argued there that it is stronger than what has usually been taken as the main criterion for relevance: the variable-sharing property.

Most of this chapter is based (with several improvements) on [52]. (See Note 11.85 for one main difference.) The main direction of Theorem 11.90 has essentially first been proved in [49] (though the result here is a littlebit stronger). The general results on semi-implications are taken from [53].

Chapter 12

The Minimal Relevant Logic

This chapter is mainly devoted to the study of the logic $\mathbf{R}_{\neg\to}$, which was introduced in the previous chapter using a Hilbert-type system. In the first section we describe a natural proof-theoretical characterization of it in terms of Gentzen-type Systems. In the second section we show that $\mathbf{R}_{\neg\to}$ can be characterized as the minimal relevant logic. In the third section we briefly discuss the advantages and drawbacks of $\mathbf{R}_{\neg\to}$ as a paraconsistent logic. (The main drawback is that no effective semantics for it is known at the time this book is written.)

12.1 The Logic $\mathbf{R}_{\neg\to}$ and Its Proof Systems

The logic $\mathbf{R}_{\neg\to}$ was introduced in Definition 11.78 using the Hilbert-type system $HR_{\neg\to}$. For the convenience of the reader, we present that system in Figure 12.1, and repeat the definition of $\mathbf{R}_{\neg\to}$ in the next definition.

Definition 12.1 ($\mathbf{R}_{\neg\to}$) Let $\mathcal{IL} = \{\neg, \to\}$. We denote by $\mathbf{R}_{\neg\to}$ the logic $\langle \mathcal{IL}, \vdash_{HR_{\neg\to}} \rangle$.

Following the notations introduced in the previous chapter, in what follows $\varphi \otimes \psi$ abbreviates $\neg(\varphi \to \neg\psi)$ and $\varphi + \psi$ abbreviates $\neg\varphi \to \psi$. We shall usually treat $\otimes$ as if it is a primitive connective of the language of $\mathbf{R}_{\neg\to}$.

Turning to the point of view of Gentzen-type systems, it is clear that the weakening rule is what allows the introduction of irrelevant elements into proofs. In fact, only two rules are involved in the classical (or intuitionistic) derivation of the sequent $\neg p, p \Rightarrow q$ from the standard axiom $p \Rightarrow p$: $[\neg \Rightarrow]$ and weakening. This observation leaves two possible choices. The one taken in the paraconsistent logics which were presented in the previous chapters was the rejection of $[\neg \Rightarrow]$. In contrast, relevant logics are essentially logics

---

Axioms:

[Id] $\varphi \to \varphi$ (Identity)

[Tr] $(\varphi \to \psi) \to ((\psi \to \sigma) \to (\varphi \to \sigma))$ (Transitivity)

[Pe] $(\varphi \to (\psi \to \sigma)) \to (\psi \to (\varphi \to \sigma))$ (Permutation)

[Ct] $(\varphi \to (\varphi \to \psi)) \to (\varphi \to \psi)$ (Contraction)

[N1] $(\varphi \to \neg\psi) \to (\psi \to \neg\varphi)$ (Contraposition)

[N2] $\neg\neg\varphi \to \varphi$ (Double negation)

Rule of inference:

[MP] $\dfrac{\varphi \quad \varphi \to \psi}{\psi}$

Figure 12.1: The proof system $HR_\to^\neg$

that reject weakening. In particular, $\mathbf{R}_\to^\neg$ can be characterized (see Proposition 12.4 below) as the logic in the language of $\{\to, \neg\}$, which is induced by the following weakening-free Gentzen-type system:

Definition 12.2 ($GR_\to^\neg$) We denote by $GR_\to^\neg$ the system in $\mathcal{IL}$ which is obtained from the standard purely multiplicative version of LK (Figure 1.3) by replacing $\supset$ by $\to$, and deleting the weakening rule.

The following notes are needed in order to understand this definition of $GR_\to^\neg$, and the characterization of $\mathbf{R}_\to^\neg$ in terms of it:

- Deleting weakening does *not* mean that the induced logic $\mathbf{R}_\to^\neg$ is not monotonic, because the monotonicity of the two Tarskian consequence relations that are associated with a Gentzen-type system G (see Definition 1.107 and Note 1.116) does not depend on the availability of weakening. Neither does the fact that these two consequence relations are identical in case G is *pure* (see Note 1.116). Since it is easy to see that $GR_\to^\neg$ is indeed pure, this system naturally induces exactly one Tarskian consequence relation.[1]

- A choice that becomes crucial when weakening is not present is whether to formulate the logical rules in an additive way (like the logical rules of LK in Chapter 1), or in a multiplicative way — see Note 1.115. In $GR_\to^\neg$ they

---

[1] Note that the relevantists have mainly concentrated in their work on the sets of *theorems* of their systems, not the corresponding consequence relations. See Note 11.6.

12.1 The Logic $R_\to$ and Its Proof Systems

are taken as *multiplicative* rules. Thus the logical rules of $GR_\to$ for negation and implication, and the derived rules for $\otimes$ and $+$ are:

$$[\neg \Rightarrow] \quad \frac{\Gamma \Rightarrow \Delta, \varphi}{\neg \varphi, \Gamma \Rightarrow \Delta} \qquad \frac{\varphi, \Gamma \Rightarrow \Delta}{\Gamma \Rightarrow \Delta, \neg \varphi} \quad [\Rightarrow \neg]$$

$$[\to \Rightarrow] \quad \frac{\Gamma_1 \Rightarrow \Delta_1, \varphi \quad \psi, \Gamma_2 \Rightarrow \Delta_2}{\Gamma_1, \Gamma_2, \varphi \to \psi \Rightarrow \Delta_1, \Delta_2} \qquad \frac{\Gamma, \varphi \Rightarrow \Delta, \psi}{\Gamma \Rightarrow \Delta, \varphi \to \psi} \quad [\Rightarrow \to]$$

$$[\otimes \Rightarrow] \quad \frac{\Gamma, \varphi, \psi \Rightarrow \Delta}{\Gamma, \varphi \otimes \psi \Rightarrow \Delta} \qquad \frac{\Gamma_1 \Rightarrow \Delta_1, \varphi \quad \Gamma_2 \Rightarrow \Delta_2, \psi}{\Gamma_1, \Gamma_2 \Rightarrow \Delta_1, \Delta_2, \varphi \otimes \psi} \quad [\Rightarrow \otimes]$$

$$[+ \Rightarrow] \quad \frac{\varphi, \Gamma_1 \Rightarrow \Delta_1 \quad \psi, \Gamma_2 \Rightarrow \Delta_2}{\Gamma_1, \Gamma_2, \varphi + \psi \Rightarrow \Delta_1, \Delta_2} \qquad \frac{\Gamma \Rightarrow \Delta, \varphi, \psi}{\Gamma \Rightarrow \Delta, \varphi + \psi} \quad [\Rightarrow +]$$

- Another crucial choice that should be made once a structural rule like weakening is deleted, is what data structure to use on both sides of a "sequent". Using finite *sets* (as we usually do in this book) is the most natural choice. However, in Gentzen's original formulation the two sides of a sequent consist of finite *sequences*. This does not matter much in the case of classical logic, since Gentzen adopted the structural rules of exchange, contraction, and weakening, and the latter includes expansion (Note 1.111) as a (very) special case. However, things are different when one of these rules is missing. Now Anderson and Belnap (implicitly) chose to delete weakening from the multiplicative version of *Gentzen's original formulation* of classical logic, and retain (the multiplicative versions of) all other rules of *that* system (without adding anything new). According to this choice, the structural rules of $GR_\to$ are exchange, contraction and cut (where cut can be shown to be eliminable). This means e.g. that although $p \Rightarrow p$ is derivable in $GR_\to$ (and so $p \to p$ is logically valid in $\mathbf{R}_\to$), the sequents $q, p \Rightarrow p$, $p, q \Rightarrow p$, and even $p, p \Rightarrow p$ are not derivable in $GR_\to$ (and so $q \to (p \to p)$, $p \to (q \to p)$, and even $p \to (p \to p)$ are not theorems of $\mathbf{R}_\to$). A more convenient presentation of $GR_\to$ (which we adopt here) uses *multisets* as the data structure, so the exchange rule is built-in. This leaves contraction and cut as the only structural rules of this system.

Note 12.3 Instead of taking as axioms all sequents of the form $\varphi \Rightarrow \varphi$, it suffices to take as axioms the sequents of the form $p \Rightarrow p$ where p is a variable. It is then easy to show that every sequent of the form $\varphi \Rightarrow \varphi$ is derivable in the resulting system.

Proposition 12.4

1. $\vdash_{GR_\to} = \vdash_{\mathbf{R}_\to}$.

2. $\vdash_{GR_\neg} \varphi_1, \ldots, \varphi_n \Rightarrow \psi$ iff $\vdash_{R_\neg} \varphi_1 \otimes \varphi_2 \otimes \ldots \otimes \varphi_n \to \psi$.

Proof. By definition of $\mathbf{R}_\neg$, it suffices to show the claims for $HR_\neg$ rather than for $\mathbf{R}_\neg$ itself.

1. For one direction, we prove by induction on length of proofs in $HR_\neg$ that if $\mathcal{T} \vdash_{HR_\neg} \varphi$ then $\mathcal{T} \vdash_{GR_\neg} \varphi$. (Recall that the latter means that there exists a finite $\Gamma \subseteq \mathcal{T}$ such that $\vdash_{GR_\neg} \Gamma \Rightarrow \varphi$.) For this it suffices to show (since cut is one of the rules of $GR_\neg$) that $\vdash_{GR_\neg} \Rightarrow \varphi$ for every axiom φ of $HR_\neg$, and that $\vdash_{GR_\neg} \varphi, \varphi \to \psi \Rightarrow \psi$. This is easy.

   For the converse, let $s = \varphi_1, \ldots, \varphi_n \Rightarrow \psi_1, \ldots, \psi_k$ be a non-empty sequent (so $k + n > 0$). Define a *translation* of s to be any formula which has one of the following forms:

   - $(\varphi_1 \otimes \cdots \otimes \varphi_n \otimes \neg\psi_1 \otimes \cdots \otimes \neg\psi_{j-1} \otimes \neg\psi_{j+1} \otimes \cdots \otimes \neg\psi_k) \to \psi_j$ for some $1 \leq j \leq k$.
   - $(\varphi_1 \otimes \cdots \otimes \varphi_{i-1} \otimes \varphi_{i+1} \otimes \cdots \otimes \varphi_n \otimes \neg\psi_1 \otimes \cdots \otimes \neg\psi_k) \to \neg\varphi_i$ for some $1 \leq i \leq n$.

   It is not difficult to show by induction on length of proofs in $GR_\neg$, that if a sequent is provable in $GR_\neg$ then each of its translations is a theorem of $HR_\neg$. (The tedious details are routine, and we omit them.) It follows that if $\vdash_{GR_\neg} \varphi_1, \ldots, \varphi_n \Rightarrow \psi$ then $\vdash_{HR_\neg} \varphi_1 \otimes \cdots \otimes \varphi_n \to \psi$. By the inference rule of $HR_\neg$ and its (derived) residuation axioms (Proposition 11.66), this entails that $\vdash_{GR_\neg} \subseteq \vdash_{HR_\neg}$.

2. This has already been shown in the proof of the first part. □

Note 12.5 Let $\varphi_{\Gamma \Rightarrow \Delta}$ be any of the translations of $\Gamma \Rightarrow \Delta$ described in the proof of Proposition 12.4. It is very easy to show that $\varphi_{\Gamma \Rightarrow \Delta}$ has the following properties:

1. $\vdash_{GR_\neg} \varphi_{\Gamma \Rightarrow \Delta}, \Gamma \Rightarrow \Delta$.

2. For every Γ^*, Δ^*, there is a cut-free proof in $GR_\neg$ of $\Gamma^* \Rightarrow \Delta^*, \varphi_{\Gamma \Rightarrow \Delta}$ from $\Gamma, \Gamma^* \Rightarrow \Delta^*, \Delta$. In particular, $\Gamma \Rightarrow \Delta \vdash_{GR_\neg} \Rightarrow \varphi_{\Gamma \Rightarrow \Delta}$.

It immediately follows from these two facts that if $\varphi_{\Gamma \Rightarrow \Delta}$ and $\varphi^*_{\Gamma \Rightarrow \Delta}$ are two translations of $\Gamma \Rightarrow \Delta$ as above then $\vdash_{GR_\neg} \varphi_{\Gamma \Rightarrow \Delta} \Rightarrow \varphi^*_{\Gamma \Rightarrow \Delta}$.

Proposition 12.6 (cut elimination for $GR_\neg$) *If both $\Gamma_1 \Rightarrow \Delta_1, \varphi$ and $\varphi, \Gamma_2 \Rightarrow \Delta_2$ have cut-free proofs in $GR_\neg$, then so does $\Gamma_1, \Gamma_2 \Rightarrow \Delta_1, \Delta_2$.*

Proof. We prove something stronger: that if both $\vdash_{GR_\rightarrow\neg} \Gamma_1 \Rightarrow \Delta_1, \varphi, \ldots, \varphi$ and $\vdash_{GR_\rightarrow\neg} \varphi, \ldots, \varphi, \Gamma_2 \Rightarrow \Delta_2$, then $\vdash_{GR_\rightarrow\neg} \Gamma_1, \Gamma_2 \Rightarrow \Delta_1, \Delta_2$. This is done by the usual double induction (see [170]) on the complexity of φ and on the sum of the lengths of the proofs of the two premises of the cut. □

Theorem 12.7 (decidability of $R_{\rightarrow\neg}$) *The logic $R_{\rightarrow\neg}$ is decidable.*

Proof. This is a consequence of the cut-elimination theorem for the Gentzen-type system for $R_{\rightarrow\neg}$ (Proposition 12.6). However, the argument is not simple, and we omit it (see [151] for details). □

12.2 Minimality and Other Properties of $R_{\rightarrow\neg}$

The importance of $R_{\rightarrow\neg}$ is due to the following crucial fact:

Proposition 12.8 *Every relevant logic is an extension of $R_{\rightarrow\neg}$.*

Proof. That $R_{\rightarrow\neg}$ is contained in any strongly relevant logic is immediate from Corollary 11.79. Since always $L(\vdash_L^{\rightarrow}) \subseteq L$, this and the definition of a relevant logic imply that $R_{\rightarrow\neg}$ is contained also in any relevant logic. □

Another important property that $R_{\rightarrow\neg}$ and its simple extensions have is given in the next Theorem.

Theorem 12.9 *Every simple extension of $R_{\rightarrow\neg}$ (including $R_{\rightarrow\neg}$ itself) is F_0-contained in classical logic (where F_0 is the bivalent interpretation given in Definition 11.89).*

Proof. We begin by introducing some notations.

- Let φ be a sentence in the language $\{\neg, \rightarrow\}$. φ^c, the *classical translation* of φ, is the sentence in $\mathcal{L}_{CL}$ obtained from φ by replacing every occurrence of $\rightarrow$ by $\supset$.

- Let $\Gamma = \varphi_1, \ldots, \varphi_k$ be a list of sentences in the language $\{\neg, \rightarrow\}$. Then Γ^c denotes the list $\varphi_1^c, \ldots, \varphi_k^c$.

Obviously, $\Gamma \vdash_{2F_0} \psi$ iff $\Gamma^c \vdash_{CL} \psi^c$.

Next, we prove two useful lemmas.

Lemma 12.10 *Suppose Γ and Δ are lists of sentences in $\{\neg, \rightarrow\}$ such that $Fv(\Gamma) \subseteq \{p\}$, $Fv(\Delta) \subseteq \{p\}$, and $\vdash_{LK} \Gamma^c \Rightarrow \Delta^c$. Then*

$$\vdash_{GR_\rightarrow\neg} p \rightarrow (p \rightarrow (p+p)), \Gamma \Rightarrow \Delta$$

(where $GR_{\rightarrow\neg}$ is the Gentzen-type system for $R_{\rightarrow\neg}$ given in Section 12.1).

Proof. By induction on the number n of connectives in $\Gamma \Rightarrow \Delta$.

Suppose that $n = 0$ (the base case). Then Γ consists of $k > 0$ copies of p, while Δ consists of $l > 0$ copies of p. Because of the contraction rule, it suffices to prove the claim in the case where $k = l$. We do this by induction on k. The cases $k = 2$ and $k = 1$ are easy (with the second following from the first using contractions). For the induction step, assume that $\vdash_{GR_{\rightarrow}} p \rightarrow (p \rightarrow (p+p)), \Gamma' \Rightarrow \Gamma'$, where Γ' consists of $k > 0$ copies of p. A cut on p of this sequent with the sequent $p \rightarrow (p \rightarrow (p+p)), p, p \Rightarrow p, p$ (the case $k = 2$) yields (using contractions) $p \rightarrow (p \rightarrow (p+p)), \Gamma', p \Rightarrow \Gamma', p$, i.e. $p \rightarrow (p \rightarrow (p+p)), \Gamma \Rightarrow \Gamma$, where Γ consists of $k+1$ copies of p.

The induction step is split into four cases:

- $\Delta = \Delta_1, \varphi$, where $\varphi = \varphi_1 \rightarrow \varphi_2$. Then $\vdash_{LK} \Gamma^c \Rightarrow \Delta_1^c, \varphi_1^c \rightarrow \varphi_2^c$, and so $\vdash_{LK} \Gamma^c, \varphi_1^c \Rightarrow \Delta_1^c, \varphi_2^c$. By induction hypothesis we get that $\vdash_{GR_{\rightarrow}} p \rightarrow (p \rightarrow (p+p)), \Gamma, \varphi_1 \Rightarrow \Delta_1, \varphi_2$, and so $\vdash_{GR_{\rightarrow}} p \rightarrow (p \rightarrow (p+p)), \Gamma \Rightarrow \Delta_1, \varphi_1 \rightarrow \varphi_2$, i.e. $\vdash_{GR_{\rightarrow}} p \rightarrow (p \rightarrow (p+p)), \Gamma \Rightarrow \Delta$.

- $\Gamma = \Gamma_1, \varphi$, where $\varphi = \varphi_1 \rightarrow \varphi_2$. Then $\vdash_{LK} \Gamma_1^c, \varphi_1^c \rightarrow \varphi_2^c \Rightarrow \Delta^c$, and so $\vdash_{LK} \Gamma_1^c, \varphi_2^c \Rightarrow \Delta^c$, and $\vdash_{LK} \Gamma_1^c \Rightarrow \Delta^c, \varphi_1^c$. It follows by induction hypothesis that $\vdash_{GR_{\rightarrow}} p \rightarrow (p \rightarrow (p+p)), \Gamma_1, \varphi_2 \Rightarrow \Delta$, and $\vdash_{GR_{\rightarrow}} p \rightarrow (p \rightarrow (p+p)), \Gamma_1 \Rightarrow \Delta, \varphi_1$. Hence $\vdash_{GR_{\rightarrow}} p \rightarrow (p \rightarrow (p+p)), \Gamma_1, \varphi \Rightarrow \Delta$, i.e. $\vdash_{GR_{\rightarrow}} p \rightarrow (p \rightarrow (p+p)), \Gamma \Rightarrow \Delta$.

We leave the remaining two cases to the reader. □

Lemma 12.11 *Let $\mathbf{L}$ be a simple extension of $\mathbf{R}_{\rightarrow}$, and let p be a variable. If there exist sentences $\varphi_1, \ldots, \varphi_n$ and ψ such that $\varphi_1, \ldots, \varphi_n \vdash_{\mathbf{L}} \psi$ but $\varphi_1^c, \ldots, \varphi_n^c \nvdash_{CL} \psi^c$, then there exist such sentences which have the further properties that $Fv(\varphi_i) = \{p\}$ for $1 \leq i \leq n$, $Fv(\psi) = \{p\}$, φ_i^c is a classical tautology for $1 \leq i \leq n$, and $(\psi \rightarrow p)^c = \psi^c \supset p$ is a classical tautology too.*

Proof. Let v_0 be a classical valuation such that $v_0(\psi^c) = f$, while $v_0(\varphi_i^c) = t$ for $1 \leq i \leq n$. Let $\varphi_1', \ldots, \varphi_n'$ and ψ' be the sentences obtained from $\varphi_1, \ldots, \varphi_n$ and ψ (respectively) by substituting $\neg(p \rightarrow p)$ for every variable q such that $v_0(q) = f$, and $p \rightarrow p$ for every variable q such that $v_0(q) = t$. Obviously, $\varphi_1', \ldots, \varphi_n'$ and ψ' satisfy all the required conditions. □

Proof of Theorem 12.9. Assume for contradiction that there are sentences $\varphi_1, \ldots, \varphi_n$ and ψ such that $\varphi_1, \ldots, \varphi_n \vdash_{\mathbf{L}} \psi$ but $\varphi_1^c, \ldots, \varphi_n^c \nvdash_{CL} \psi^c$. By Lemma 12.11 we may assume that $Fv(\varphi_i) = \{p\}$ for $1 \leq i \leq n$, $Fv(\psi) = \{p\}$, φ_i^c is a classical tautology for $1 \leq i \leq n$, and $(\psi \rightarrow p)^c$ is a classical tautology. Therefore, it follows from Lemma 12.10 that $p \rightarrow (p \rightarrow (p+p)) \vdash_{\mathbf{L}} \varphi_i$

12.2 Minimality and Other Properties of $\mathbf{R}_{\rightarrow\neg}$

for $1 \leq i \leq n$, and that $p \rightarrow (p \rightarrow (p+p)), \psi \vdash_\mathbf{L} p$. These two facts and the assumption that $\varphi_1, \ldots, \varphi_n \vdash_\mathbf{L} \psi$ together imply that

$$(*) \quad p \rightarrow (p \rightarrow (p+p)) \vdash_\mathbf{L} p.$$

Denote $p \rightarrow (p \rightarrow (p+p))$ by ψ. Substituting ψ for p in (*), we get:

$$(**) \quad (\psi \rightarrow (\psi \rightarrow (\psi + \psi))) \vdash_\mathbf{L} \psi.$$

Now, $(\psi \rightarrow (\psi + \psi))^c$ is a classical tautology, and $Fv(\psi \rightarrow (\psi + \psi)) = \{p\}$. Hence, Lemma 12.10 implies that $\vdash_\mathbf{L} (p \rightarrow (p \rightarrow (p+p))) \rightarrow (\psi \rightarrow (\psi + \psi))$. By definition of ψ, this means that $\vdash_\mathbf{L} \psi \rightarrow (\psi \rightarrow (\psi + \psi))$. Hence, (**) implies that $\vdash_\mathbf{L} \psi$. But by (*) and the definition of ψ, also $\psi \vdash_\mathbf{L} p$. From the last two facts it follows that $\vdash_\mathbf{L} p$. A contradiction. □

Next, we introduce a particularly important type of extensions of $\mathbf{R}_{\rightarrow\neg}$.

Proposition 12.12 *Every axiomatic extension of $\mathbf{R}_{\rightarrow\neg}$ has the RDP.*

Proof. Immediate from Proposition 11.56. □

Corollary 12.13 *The $\{\neg, \rightarrow\}$-fragment of any axiomatic extension of $\mathbf{R}_{\rightarrow\neg}$ is a semi-normal logic which is $\neg$-contained in classical logic.*

Proof. This follows from Theorem 12.9 and Proposition 12.12. □

Theorem 12.14

1. *Every strongly relevant logic is an axiomatic extension of $\mathbf{R}_{\rightarrow\neg}$.*

2. *An axiomatic extension of $\mathbf{R}_{\rightarrow\neg}$ is a strongly relevant logic iff it is $\neg$-consistent, it satisfies the minimal semantic criterion, and $\rightarrow$ has in it the VSP and the replacement property.*

Proof.

1. Let $\mathbf{L}$ be a strongly relevant logic. As $\mathbf{L}$ is finitary, the RDP of $\rightarrow$ entails that $\mathcal{T} \vdash_L \varphi$ iff $\vdash_L \varphi_1 \rightarrow (\varphi_2 \rightarrow (\cdots \rightarrow (\varphi_n \rightarrow \varphi)\cdots))$ for some $\varphi_1, \ldots, \varphi_n \in \mathcal{T}$. It follows that $\mathbf{L}$ has a strongly sound and complete Hilbert-type system HL which has MP for $\rightarrow$ as the sole rule of inferences, and all valid formulas of $\mathbf{L}$ as axioms. By Proposition 12.8 this implies that HL can be obtained from $HR_{\rightarrow\neg}$ by adding to it a set of axiom schemas. Hence $\mathbf{L}$ is an axiomatic extension of $\mathbf{R}_{\rightarrow\neg}$.

2. Let **L** be an axiomatic extension of $\mathbf{R}_\to$. By Proposition 11.76, **L** is ¬-consistent, while by the first part of Proposition 11.31, $\vdash_{\mathbf{R}_\to}$ is a well-defined entailment relation, for which $\to$ is an internal implication. It follows by Proposition 11.80 that $\to$ has in $\mathbf{R}_\to$ an associated internal conjunction. Hence, Proposition 12.12 and Corollary 11.64 imply that **L** is a strongly relevant logic iff it satisfies the minimal semantic criterion, and $\to$ has in it the VSP and the replacement property. □

Corollary 12.15 *Let **L** be a ¬-consistent extension of $\mathbf{R}_\to$ which satisfies the minimal semantic criterion, and in which $\to$ has the VSP and the replacement property. Then $\mathbf{L}(\vdash_{\mathbf{L}})$ is a strongly relevant logic.*

Proposition 12.16 *The connective $\to$ has the replacement property in every axiomatic simple extension of $\mathbf{R}_\to$.*

Proof. This is easily shown using the axioms [Tr], [N1], and [N2]. □

Corollary 12.17 *An axiomatic simple extension of $\mathbf{R}_\to$ is strongly relevant if it satisfies the minimal semantic criterion, and $\to$ has in it the VSP.*

Proof. From Proposition 12.9 it easily follows that every non-trivial simple extension of $\mathbf{R}_\to$ is necessarily ¬-consistent. Hence the claims follows from Proposition 12.16 and Corollary 12.15. □

Note 12.18 Proposition 12.16 can be strengthened by noting that in $\mathbf{R}_\to$ one can define an equivalence operation by $\varphi \leftrightarrow \psi =_{Df} (\varphi \to \psi) \otimes (\psi \to \varphi)$. It is not difficult then to show that $\leftrightarrow$ has in $\mathbf{R}_\to$ the following properties:

1. $\vdash_{\mathbf{R}_\to} (\varphi \leftrightarrow \psi) \to (\varphi \to \psi)$,

2. $\vdash_{\mathbf{R}_\to} (\varphi \leftrightarrow \psi) \to (\psi \to \varphi)$,

3. $\vdash_{\mathbf{R}_\to} (\varphi \to \psi) \to ((\psi \to \varphi) \to (\varphi \leftrightarrow \psi))$,

4. $\varphi \leftrightarrow \psi \vdash_{\mathbf{R}_\to} \theta[\varphi/p] \leftrightarrow \theta[\psi/p]$ for every sentence θ and variable p.

Our next goal is to show that $\mathbf{R}_\to$ is a strongly relevant logic. For this we prove first the following:

Theorem 12.19 *The logic $\mathbf{R}_\to$ has no finite weakly characteristic Nmatrix.*

12.2 Minimality and Other Properties of $\mathbf{R}_{\rightarrow}$

Proof. Suppose for contradiction that $\mathbf{R}_{\rightarrow}$ has a weakly characteristic Nmatrix $\mathcal{M} = \langle \mathcal{V}, \mathcal{D}, \mathcal{O} \rangle$, where $\mathcal{V}$ has n elements. Let $p^k \to \varphi$ be defined inductively for $k \geq 1$ by: $p^1 \to \varphi = p \to \varphi$, $p^{k+1} \to \varphi = p \to (p^k \to \varphi)$. Let p and q be distinct variables. Define ψ_0 to be the sentence $(p^n \to q) \to (p^{n+1} \to ((q \to q) \to q))$. An easy induction on the structure of cut-free proofs in $GR_{\rightarrow}$ shows that sequents of the form $p^{n_1} \to q, \ldots, p^{n_k} \to q \Rightarrow p^m \to q$ and $p^{n_1} \to q, \ldots, p^{n_k} \to q \Rightarrow p^m \to ((q \to q) \to q)$ are provable in $GR_{\rightarrow}$ iff $k = 1$ and $n_1 \geq m$. This easily implies that $\nvdash_{\mathbf{R}_{\rightarrow}} \psi_0$. Hence, there is a valuation ν_0 in $\mathcal{M}$ which refutes ψ_0. Since $\mathcal{V}$ has n elements, there are two possibilities:

1. There is $1 \leq k \leq n$ such that $\nu_0(p^{n+1} \to ((q \to q) \to q)) = \nu_0(p^k \to q)$.

   Let $\psi_1 = (p^n \to q) \to (p^k \to q)$. We define a partial valuation ν_1 on the set of subformulas of ψ_1 as follows:

   - $\nu_1(\varphi) = \nu_0(\varphi)$ for every subformula φ of $p^n \to q$ (and so also of $p^k \to q$).
   - $\nu_1((p^n \to q) \to (p^k \to q)) = \nu_0(\psi_0)$.

   Then ν_1 refutes ψ_1, even though ψ_1 is a theorem of $\mathbf{R}_{\rightarrow}$ (since $1 \leq k \leq n$). This contradicts the fact that $\mathbf{R}_{\rightarrow}$ is weakly sound for $\mathcal{M}$.

2. There are $1 \leq i < j \leq n$ such that $\nu_0(p^j \to q) = \nu_0(p^i \to q)$.

   For this case we first introduce the following notations:

   $l = j - i$ (note that $l > 0$).

   m is the smallest integer such that $m \geq n+1$ and $m \equiv j \pmod{l}$.

   $n_1 = m + (n - j)$ (note that $n_1 \geq n+1$).

   $\psi_2 = (p^{n_1} \to q) \to (p^{n+1} \to ((q \to q) \to q))$.

   Define a partial valuation ν_2 on the set of subformulas of ψ_2 as follows:

   - $\nu_2(\varphi) = \nu_0(\varphi)$ for every subformula φ of either $p^j \to q$ or $p^{n+1} \to ((q \to q) \to q)$. (Note that p and q are the only subformulas of both.)
   - For $j < k \leq m$, $\nu_2(p^k \to q) = \nu_0(p^{i+((k-j) \bmod l)} \to q)$. Note that because $\nu_0(p^j \to q) = \nu_0(p^i \to q)$, $\nu_2(p^m \to q) = \nu_0(p^j \to q)$.
   - For $m < k \leq n_1$, $\nu_2(p^k \to q) = \nu_0(p^{j+(k-m)} \to q)$. In particular: $\nu_2(p^{n_1} \to q) = \nu_0(p^n \to q)$.
   - $\nu_2(\psi_2) = \nu_0(\psi_0)$.

It is easy to see that ν_2 is indeed a partial $\mathcal{M}$-valuation. Obviously, it refutes ψ_2, which is a theorem of $\mathbf{R}_{\overrightarrow{\neg}}$ (since $n_1 \geq n+1$). This again contradicts the fact that $\mathbf{R}_{\overrightarrow{\neg}}$ is weakly sound for $\mathcal{M}$. □

Corollary 12.20 *The logic* $\mathbf{R}_{\overrightarrow{\neg}}$ *satisfies the minimal semantic criterion.*

Now we are ready to prove the most important characterization of $\mathbf{R}_{\overrightarrow{\neg}}$:

Theorem 12.21 *The logic* $\mathbf{R}_{\overrightarrow{\neg}}$ *is the minimal (strongly) relevant logic: it is a strongly relevant logic, every relevant logic is an extension of it, and every strongly relevant logic is an axiomatic extension of it.*

Proof. By Proposition 12.8 and Theorem 12.14, it suffices to show that $\mathbf{R}_{\overrightarrow{\neg}}$ is itself a strongly relevant logic. In turn, to show this it suffices, by Corollary 12.20 and Corollary 12.17, to show that $\to$ has in $\mathbf{R}_{\overrightarrow{\neg}}$ the VSP. This is proved in Corollary 13.22 below for a stronger system. □

Corollary 12.22 *The logic* $\mathbf{R}_{\overrightarrow{\neg}}$ *has all the properties of a strongly relevant logic proved in the previous chapter. In particular:*

1. $\mathbf{R}_{\overrightarrow{\neg}}$ *is boldly paraconsistent.*

2. $\mathbf{R}_{\overrightarrow{\neg}}$ *has exactly one relevant implication:* $\to$, *which has exactly one associated negation:* $\neg$, *and exactly one associated conjunction:* $\otimes$.

3. $\neg$ *is complete for* $\mathbf{R}_{\overrightarrow{\neg}}$.

4. $\mathbf{R}_{\overrightarrow{\neg}} = \mathbf{L}(\vdash_{\overrightarrow{\mathbf{R}_{\overrightarrow{\neg}}}})$.

Proof. Immediate from Theorem 12.21 and Corollary 11.92, Proposition 11.88, Proposition 11.93, and Proposition 11.86. □

We end with another characterization of the class of relevant logics:

Proposition 12.23 *A finitary logic* $\mathbf{L}$ *is relevant logic iff the following conditions are satisfied:*

- $\mathbf{L}$ *contains* $\mathbf{R}_{\overrightarrow{\neg}}$,

- $\mathbf{L}$ *is* $\neg$-*consistent.*

- $\to$ *has in* $\mathbf{L}$ *the* VSP *and the replacement property.*

- $\mathbf{L}$ *satisfies the basic relevance criterion.*

- $\mathbf{L}$ *satisfies the minimal semantic criterion.*

Proof. This follows from Propositions 11.31 and 11.76, and from Corollary 12.15. □

12.3 Advantages and Drawbacks of $\mathbf{R}_{\neg\to}$

The results of the previous sections show that $\mathbf{R}_{\neg\to}$ is quite an appealing system: it is decidable, has a nice proof theory, and the fact that it is the minimal (strongly) relevant logic makes it a particularly natural paraconsistent logic. Nevertheless, from the point of view of this book it currently has one serious drawback: even though it is decidable, to the best of our knowledge no effective semantics is known for it. Nevertheless, we believe that the virtues of $\mathbf{R}_{\neg\to}$ still justify its inclusion in this book. (Note that it is the only exception of this type that we are making in it!) However, it should be noted that as a paraconsistent logic $\mathbf{R}_{\neg\to}$ has several other drawbacks:

1. In the Gentzen-type formulation of $\mathbf{R}_{\neg\to}$ one is forced to work with multisets of formulas rather than with sets. In our view this is less natural and less convenient. What is more: the use of multisets make the contraction rule very difficult to control. As a result, the decision procedure provided by the Gentzen-type system for $\mathbf{R}_{\neg\to}$ (which is currently the only known one) is very complex and impractical.

2. By Proposition 11.53, $\mathbf{R}_{\neg\to}$ has neither a conjunction nor a disjunction, and so it is not normal. (This is a drawback of *every* strongly relevant logic, and it can be overcome only by giving up strong relevance).

3. The best (and the only natural) substitute for conjunction that $\mathbf{R}_{\neg\to}$ can offer, its "relevant conjunction" $\otimes$, can hardly be regarded as such. First of all, it is not even idempotent: although $\varphi \to \varphi \otimes \varphi$ is valid in $\mathbf{R}_{\neg\to}$, the converse implication is not. Even worse: as the next theorem shows, the elimination rule for $\otimes$ (from $\varphi \otimes \psi$ infer φ) is not even admissible in $\mathbf{R}_{\neg\to}$ (i.e. the set of theorems of $\mathbf{R}_{\neg\to}$ is not closed under this rule). In our opinion, this (somewhat surprising) fact indicates that something is missing in $\mathbf{R}_{\neg\to}$. (Note that it is very easy to see that this strange phenomenon does not occur, e.g., in $\mathbf{LL}_{\neg\to}$.)

Theorem 12.24 *There are φ and ψ such that $\vdash_{\mathbf{R}_{\neg\to}} \varphi \otimes \psi$, but $\nvdash_{\mathbf{R}_{\neg\to}} \varphi$.*

Proof. Let φ be some sentence in the language of $GR_{\neg\to}$, and let p and q be two distinct variables. It is easy to prove that the sequents $p, p \to \varphi \Rightarrow \varphi$; $\Rightarrow \neg q \to \neg q \otimes \neg q, q$; and $\Rightarrow \neg q \to \neg q \otimes \neg q$ are derivable in $GR_{\neg\to}$. (Note that the contraction rule is needed for deriving the third one.) Using just the logical rules of $GR_{\neg\to}$, it is not difficult to derive the following sequent from these sequents and from $\varphi \Rightarrow \varphi$:

$$p \to \varphi, (p \to q) \to \varphi \Rightarrow \varphi \otimes (\neg q \to \neg q \otimes \neg q), p \otimes (\neg q \to \neg q \otimes \neg q).$$

Using contraction again, this immediately implies:

$$(*) \quad p \to \varphi, (p \to q) \to \varphi \vdash_{\mathbf{R}_{\to}} \varphi \otimes (\neg q \to \neg q \otimes \neg q).$$

Now it can easily be seen that for $\varphi = \big(((p \to q) \to p) \to p \otimes ((p \to q) \to p)\big)$ we have: $\vdash_{\mathbf{R}_{\to}} p \to \varphi$ and $\vdash_{\mathbf{R}_{\to}} (p \to q) \to \varphi$. Thus, (*) implies that $\vdash_{\mathbf{R}_{\to}} \varphi \otimes (\neg q \to \neg q \otimes \neg q)$. It remains to show that $\nvdash_{\mathbf{R}_{\to}} \varphi$. Assume otherwise. Since φ is in the language of $\{\to, \otimes\}$, it should be provable in the $\{\to, \otimes\}$-fragment of $GR_{\to}$. This fragment, however, is contained in LJ^+ (interpreting $\to$ as $\supset$ and $\otimes$ as $\wedge$). We conclude therefore that $((p \to q) \to p) \to [p \wedge ((p \to q) \to p)]$ is valid intuitionistically. This, in turn, implies that $((p \to q) \to p) \to p$ is valid intuitionistically, contradicting Proposition 1.78. □

12.4 Bibliographical Notes and Further Reading

More information on $\mathbf{R}_{\to}$ can be found in [4] and [151]. These sources also contain (sometimes implicitly) the following results of this chapter (and their proofs): Proposition 12.4, Proposition 12.6, Theorem 12.7, Proposition 12.12, and Proposition 12.16.

The Gentzen-type formulation of $\mathbf{R}_{\to}$ and the decidability of this system are due to Kripke ([195]).

Theorem 12.9 was first prove in [49]. Theorem 12.24 is taken from [39].

Theorem 12.19 is due to [57]. It considerably strengthens a result of Phai (stated in [237]), according to which $\mathbf{R}_{\to}$ has no finite weakly characteristic matrix. (That result is also mentioned, with a wrong outline of proof, in Section 8.17 of Chapter II of [4].)

Chapter 13

Two Maximal Relevant Logics

While the previous chapter was devoted to the *minimal* (strongly) relevant logic, in this Chapter we turn to *maximal* ones in the same basic language. In the first section we investigate a very natural maximal strongly relevant logic, while in the second — a closely related maximal *normal* relevant logic.

13.1 A Maximal Strong Relevant Logic

Interesting as $\mathbf{R}_\to$ is, it has the serious drawbacks noted in the previous chapter. In this section we study another strongly relevant logic in the same language, $\mathbf{RMI}_\to$, in which almost all these drawbacks disappear. (The only exception is the lack of conjunction and disjunction, but this cannot be avoided because of Theorem 11.53.) Moreover: $\mathbf{RMI}_\to$ is a *maximal* strongly relevant logic, and (unlike $\mathbf{R}_\to$) it has a useful, concrete semantics. Accordingly, $\mathbf{RMI}_\to$ can be viewed as an "ideal" strongly relevant logic.

13.1.1 The Logic $\mathbf{RMI}_\to$ and Its Proof Systems

$\mathbf{RMI}_\to$ is obtained from $\mathbf{R}_\to$ by extending either of its two formulations.

Definition 13.1 ($HRMI_\to$) $HRMI_\to$ is the system obtained from $HR_\to$ (Figure 12.1) by the addition of the mingle axiom from Proposition 11.28:

$$[\text{Mi}] \quad \varphi \to (\varphi \to \varphi)$$

Note 13.2 It is easy to see that axiom [Id] is redundant in the formulation of $HRMI_\to$. We note also that axiom [Mi] is obviously equivalent to in $HR_\to$ to $\varphi \otimes \varphi \to \varphi$. Since the converse implication is provable in $HR_\to$, the addition of the mingle axiom is equivalent to making $\otimes$ idempotent.

Definition 13.3 ($GRMI_\to$) $GRMI_\to$ is the system obtained from $GR_\to$ by adding to it the structural rule of Expansion (Note 1.116).

Note 13.4 Alternatively, $GRMI_\to$ may be defined as the system which in addition to the cut rule (as the sole structural rule) has exactly the same axioms and logical rules as $GR_\to$, but its sequents consist of *finite sets of formulas* on both side of $\Rightarrow$.[1] Note also that as in the case of $GR_\to$, instead of taking $\varphi \Rightarrow \varphi$ as an axiom for every φ, it would suffice to take as axioms only sequents of the form $p \Rightarrow p$, where p is a variable.

Proposition 13.5 $\vdash_{GRMI_\to} = \vdash_{HRMI_\to}$.

Proof. Similar to the proof in the case of $\mathbf{R}_\to$ (Proposition 12.4). □

Definition 13.6 ($RMI_\to$) $RMI_\to$ is the logic $\langle \mathcal{IL}, \vdash_{HRMI_\to} \rangle$ (or the logic $\langle \mathcal{IL}, \vdash_{GRMI_\to} \rangle$, which is identical to $= \langle \mathcal{IL}, \vdash_{HRMI_\to} \rangle$ by Proposition 13.5).

Next, we describe some important properties of the proof system $GRMI_\to$.

Definition 13.7 (full relevance) A formula φ is *fully relevant* to a formula ψ (a sequent s) if $\mathsf{Var}(\varphi) \subseteq \mathsf{Var}(\psi)$ ($\mathsf{Var}(\varphi) \subseteq \mathsf{Var}(s)$).

Proposition 13.8

1. $GRMI_\to$ *is closed under* relevant weakening: *If* $\vdash_{GRMI_\to} \Gamma \Rightarrow \Delta$, *then* $\vdash_{GRMI_\to} \varphi, \Gamma \Rightarrow \Delta$ *and* $\vdash_{GRMI_\to} \Gamma \Rightarrow \Delta, \varphi$ *whenever* φ *is fully relevant to* $\Gamma \Rightarrow \Delta$.

2. $GRMI_\to$ *is closed under the following* strong relevant mingle *rule: If* $\mathsf{Var}(\Gamma_1 \Rightarrow \Delta_1) \cap \mathsf{Var}(\Gamma_2 \Rightarrow \Delta_2) \neq \emptyset$, *and both* $\vdash_{GRMI_\to} \Gamma_1 \Rightarrow \Delta_1$ *and* $\vdash_{GRMI_\to} \Gamma_2 \Rightarrow \Delta_2$, *then* $\vdash_{GRMI_\to} \Gamma_1, \Gamma_2 \Rightarrow \Delta_1, \Delta_2$.

Proof. The first part is proved by induction on the complexity of φ. The base case (where φ is a variable) is done by an inner induction on the length of the proof of $\Gamma \Rightarrow \Delta$.

For the second part, assume that that p occurs in both $\Gamma_1 \Rightarrow \Delta_1$ and $\Gamma_2 \Rightarrow \Delta_2$, and that the two sequents are provable in $GRMI_\to$. Then by the first part, both $\Gamma_1 \Rightarrow \Delta_1, p$ and $p, \Gamma_2 \Rightarrow \Delta_2$ are provable in $GRMI_\to$. Hence so is $\Gamma_1, \Gamma_2 \Rightarrow \Delta_1, \Delta_2$ (by an application of a cut). □

Note 13.9 As long as the use of $GRMI_\to$ is restricted to the language $\mathcal{IL}$, the two rules described in Proposition 13.8 are actually *derivable*, not only admissible. Indeed, suppose first that p is a variable which occurs in $\Gamma \Rightarrow \Delta$.

---

[1] Note again that this means that in the formulation of the rules, notation of the form "Γ, φ" denotes $\Gamma \cup \{\varphi\}$, where φ may be an element of Γ. Thus, $\varphi \Rightarrow \varphi \to \varphi$ and $\varphi \otimes \varphi \Rightarrow \varphi$ may be derived by ($\Rightarrow \to$) and ($\otimes \Rightarrow$) (respectively) from $\varphi \Rightarrow \varphi$.

13.1 A Maximal Strong Relevant Logic

Then $p \in \mathsf{Var}(\psi)$ for some $\psi \in \Gamma \cup \Delta$. Therefore it follows from Proposition 13.8 that $\vdash_{GRMI_{\rightarrow}} \psi \Rightarrow \psi, p$. Hence $\Gamma \Rightarrow \Delta, p$ is derivable in $GRMI_{\rightarrow}$ from $\Gamma \Rightarrow \Delta$ using a cut of the latter with $\psi \Rightarrow \psi, p$. That $p, \Gamma \Rightarrow \Delta$ is derivable in $GRMI_{\rightarrow}$ from $\Gamma \Rightarrow \Delta$ is proved similarly. From this point the proof of the derivability of the two rules exactly follows the proof of Proposition 13.8.

Proposition 13.10 (cut elimination for $GRMI_{\rightarrow}$) *Suppose $\Gamma_1 \Rightarrow \Delta_1, \varphi$ and $\varphi, \Gamma_2 \Rightarrow \Delta_2$ have cut-free proofs in $GRMI_{\rightarrow}$. Then $\Gamma_1, \Gamma_2 \Rightarrow \Delta_1, \Delta_2$ has such a proof too.*

Proof. By the usual double induction (see [170]) on the complexity of φ and on the sum of the lengths of the proofs of the premises of the cut. □

Corollary 13.11 (decidability of $\mathbf{RMI}_{\rightarrow}$) *$\mathbf{RMI}_{\rightarrow}$ is decidable.*

Proof. It is easy to see that any cut-free proof in $GRMI_{\rightarrow}$ of a sequent $\Gamma \Rightarrow \Delta$ can include only sequents which consist of subformulas of formulas in $\Gamma \cup \Delta$. Assuming that we take both sides of a sequent as *sets* of formulas (see Note 13.4), there are only finitely many sequents that may occur in a proof of a sequent $\Gamma \Rightarrow \Delta$, and so only finitely many possible lists of sequents that may serve as a cut-free proof of it. Hence the decidability of $\mathbf{RMI}_{\rightarrow}$ follows from Proposition 13.10. □

Note 13.12 In the next section we provide an alternative, *semantic* proof of the decidability of $\mathbf{RMI}_{\rightarrow}$. (See Corollary 13.21.)

We end this section with the following deduction theorems:

Proposition 13.13 (Deduction theorems for $\mathbf{RMI}_{\rightarrow}$) *Let $\mathbf{L}$ be an axiomatic extension of $\mathbf{RMI}_{\rightarrow}$.*

1. *$\mathbf{L}$ has the RDP: $\mathcal{T}, \varphi \vdash_{\mathbf{L}} \psi$ iff $\mathcal{T} \vdash_{\mathbf{L}} \psi$ or $\mathcal{T} \vdash_{\mathbf{L}} \varphi \rightarrow \psi$.*

2. *If φ is fully relevant to ψ then $\mathcal{T}, \varphi \vdash_{\mathbf{L}} \psi$ iff $\mathcal{T} \vdash_{\mathbf{L}} \varphi \rightarrow \psi$.*

Proof. The first part is a corollary of Proposition 12.12.
The second part immediately follows from the first, once we prove that if φ is fully relevant to ψ then $\vdash_{\mathbf{RMI}_{\rightarrow}} \psi \rightarrow (\varphi \rightarrow \psi)$. The latter easily follows from Proposition 13.5 and the first part of Proposition 13.8. □

13.1.2 Weakly Characteristic Matrix for $\mathbf{RMI}_{\to}$

Our starting point in developing appropriate semantics for $\mathbf{RMI}_{\to}$ is classical logic. This logic maybe viewed as based on the following two principles:

(T) Whatever is not *absolutely* true is false.

(F) Whatever is not *absolutely* false is true.

These two intuitive principles might look fuzzy, but they can be translated into completely precise ones using the semantic framework of matrices. Recall that the conditions in Definition 3.1 concerning the set $\mathcal{D}$ imply that any matrix $\mathcal{M}$ contains at least two different truth values, t and f, so that $t \in \mathcal{D}$ while $f \notin \mathcal{D}$. We may take these two elements as respectively denoting absolute truth and absolute falsehood. This leads to the following interpretations of the terms used in the formulation of (T) and (F) above:

- "φ is true" means $\nu(\varphi) \in \mathcal{D}$,
- "φ is absolutely true" means $\nu(\varphi) = t$,
- "φ is false" means $\nu(\varphi) \notin \mathcal{D}$,
- "φ is absolutely false" means $\nu(\varphi) = f$.

Note that together with the condition $f \neq t$ (which we henceforth assume) each of the two principles above already implies that $t \in \mathcal{D}$ while $f \notin \mathcal{D}$, and that whatever is absolutely true is true, and whatever is absolutely false is false. Together, the two principles imply that the set of truth values is $\{t, f\}$, and we get the classical, bivalent semantics. t and f may indeed be identified with the classical truth values, and in this section we shall make this identification. Accordingly, on the classical truth values $\neg$ behaves in this section exactly like the classical negation.

After formulating the two classical principles in precise terms we immediately see that paraconsistency is in a direct conflict with Principle (T). Indeed, we have seen in Chapter 3 that any many-valued paraconsistent logic should be based on a matrix in which there exists at least one designated "paradoxical" element I such that both I and $\neg I$ are in $\mathcal{D}$. The simplest many-valued structures with this property are the three-valued logics that were considered in Chapter 4 (like $\mathbf{J_3}$). It should be noted that although these logics reject Principle (T), they still adhere to Principle (F). However, from the point of view of this chapter, the three-valued paraconsistent logics have the drawback that they do not take relevance into consideration: any two paradoxical propositions are equivalent according to them. In order to avoid this, but still keep at least one of the two classical principles, we

13.1 A Maximal Strong Relevant Logic

should allow for more than one paradoxical truth-value. The most natural alternative is to have a potentially infinite number of them. Paradoxical propositions that get different paradoxical truth-value should then be considered as irrelevant to each other, and neither of them should imply the other. These considerations, together with the desire to validate $\mathbf{RMI}_{\neg\to}$'s theorems, have led to the following structure:

Definition 13.14 ($\mathcal{A}_\omega$)

1. The matrix $\mathcal{A}_\omega = \langle A_\omega, \mathcal{D}_\omega, \mathcal{O} \rangle$ for $\mathcal{IL}$ is defined as follows:
   - $A_\omega = \{t, f, I_1, I_2, I_3, \ldots\}$,
   - $\mathcal{D}_\omega = A_\omega - \{f\} = \{t, I_1, I_2, I_3, \ldots\}$,
   - The operations in $\mathcal{O}$ are the following:

   $$\widetilde{\neg} t = f, \quad \widetilde{\neg} f = t, \quad \widetilde{\neg} I_k = I_k \quad (k = 1, 2, \ldots),$$

   $$a \widetilde{\to} b = \begin{cases} t & a = f \text{ or } b = t, \\ I_k & a = b = I_k, \\ f & \text{otherwise.} \end{cases}$$

2. $\mathcal{A}_n$ ($n \geq 0$) is the submatrix of $\mathcal{A}_\omega$ that consists of $\{t, f, I_1, \ldots, I_n\}$.

Note 13.15 Like in $\mathbf{R}_{\neg\to}$, it is useful to introduce in the language the connective $\otimes$, defined by $\varphi \otimes \psi = \neg(\varphi \to \neg\psi)$, and the connective $+$, defined by $\varphi + \psi = \neg(\neg\varphi \otimes \neg\psi) = \neg\varphi \to \psi$. Obviously, the corresponding operations on A_ω are given by:

$$a \widetilde{\otimes} b = \begin{cases} f & a = f \text{ or } b = f, \\ I_k & a = b = I_k, \\ t & \text{otherwise.} \end{cases} \qquad a \widetilde{+} b = \begin{cases} t & a = t \text{ or } b = t, \\ I_k & a = b = I_k, \\ f & \text{otherwise.} \end{cases}$$

Note 13.16 To understand the definition of $\mathcal{A}_\omega$, note that t and f are intended to be the possible truth-values of "normal" propositions, while the neutral values $I_1, I_2, \ldots$ are the possible truth-values of "paradoxical" (or "abnormal") propositions. This intuition dictates the truth-table of $\neg$. Now, in the intended semantics of $\mathcal{A}_\omega$ two propositions are irrelevant to each other iff they are both abnormal, but they are assigned different neutral values. Accordingly, the fact that $I_i \widetilde{\to} I_j = f$ in case $i \neq j$ reflects the intuition that relevance is a necessary condition for entailment. This explains the truth-table of $\to$. The truth-tables of $\otimes$ and $+$ are then dictated by their definitions in terms of $\neg$ and $\to$. Note that according to the resulting truth-table for $+$ (described in the previous note), $\varphi + \psi$ is true iff either φ or ψ is true, and φ and ψ are relevant to each other. Hence $+$ behaves exactly as we would expect from a *relevant disjunction*. Thus the disjunctive syllogism is indeed valid for $+$, while the inference of $\varphi + \psi$ from φ (or from ψ) is not.

The next definition and proposition are useful for understanding the structure of $\mathcal{A}_\omega$, and the interpretations of the connectives in it.

Definition 13.17 The relations $\preceq^1_\otimes$ and $\preceq^1$ on A_ω are defined as follows:

- $a \preceq^1_\otimes b$ if either $a = b$, or $a = f$, or $a = t$ and $b = I_i$ for some i.

- $a \preceq^1 b$ if either $a = b$, or $a = f$, or $b = t$.

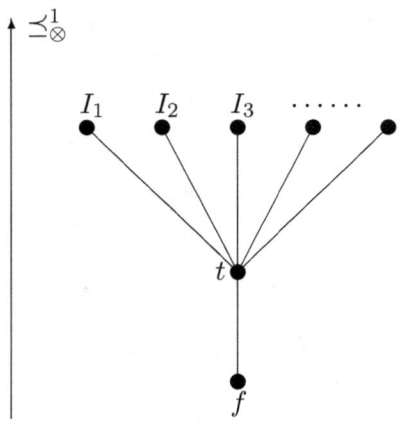

Figure 13.1: The lower-semilattice $\langle A_\omega, \preceq^1_\otimes \rangle$

Proposition 13.18

1. $\langle A_\omega, \preceq^1_\otimes \rangle$ is a lower semilattice, and $a \widetilde{\otimes} b = \inf_{\preceq^1_\otimes} \{a, b\}$. (See Figure 13.1.)

2. $\langle A_\omega, \preceq^1 \rangle$ is a bounded lattice, in which t is the maximal element, f is the minimal one, and the structure obtained by deleting t is a flat ordered set.

3. $\widetilde{\neg}$ is an involution on $\langle A_\omega, \preceq^1 \rangle$, that is: for every $a, b \in A_\omega$: $\widetilde{\neg}\widetilde{\neg}a = a$ and $\widetilde{\neg}b \preceq^1 \widetilde{\neg}a$ in case $a \preceq^1 b$.

4. Let $a, b \in A_\omega$. Then $a \widetilde{\to} b \in \mathcal{D}_\omega$ iff $a \preceq^1 b$.

Proof. Left for the reader. □

13.1 A Maximal Strong Relevant Logic

Theorem 13.19 (weak completeness of $\mathbf{RMI}_{\to}$) $\vdash_{\mathbf{RMI}_{\to}} \varphi$ iff $\vdash_{\mathcal{A}_\omega} \varphi$.

Proof. For the soundness part we prove by induction on length of derivations in $GRMI_{\to}$ that if $\vdash_{GRMI_{\to}} \Gamma \Rightarrow \Delta$ then for every valuation ν in $\mathcal{A}_\omega$, either $\nu(\varphi) = f$ for some $\varphi \in \Gamma$, or $\nu(\varphi) = t$ for some $\varphi \in \Delta$, or there exists i such that $\nu(\varphi) = I_i$ for every $\varphi \in \Gamma \cup \Delta$. In particular: if $\vdash_{\mathbf{RMI}_{\to}} \varphi$ (i.e., $\vdash_{GRMI_{\to}} \Rightarrow \varphi$) then $\nu(\varphi) \neq f$ for every ν in $\mathcal{A}_\omega$.

For completeness, assume $\not\vdash \varphi$ (here and in the rest of the proof we write just "$\vdash$" instead of "$\vdash_{\mathbf{RMI}_{\to}}$"). We construct a valuation ν in $\mathcal{A}_\omega$ such that $\nu(\varphi) = f$. Let $\mathcal{T} = \emptyset$. Then $\mathcal{T}$ is a set of sentences which are fully relevant to φ, and $\mathcal{T} \not\vdash \varphi$. Extend $\mathcal{T}$ to a maximal theory $\mathcal{T}^*$ which has these two properties. Then for ψ such that $\mathsf{Var}(\psi) \subseteq \mathsf{Var}(\varphi)$, $\mathcal{T}^* \vdash \psi$ iff $\psi \in \mathcal{T}^*$, and $\psi \notin \mathcal{T}^*$ iff $\mathcal{T}^* \cup \{\psi\} \vdash \varphi$. By the deduction theorem of $\mathbf{RMI}_{\to}$ (Proposition 13.13) we get:

(1) If $\mathsf{Var}(\psi) \subseteq \mathsf{Var}(\varphi)$ then $\psi \notin \mathcal{T}^*$ iff $\mathcal{T}^* \vdash \psi \to \varphi$.

Now using $GRMI_{\to}$ it is easy to show that $\vdash ((\varphi \to \psi) \to \varphi) \to \varphi$ whenever $\mathsf{Var}(\psi) \subseteq \mathsf{Var}(\varphi)$ (from $\varphi \Rightarrow \varphi$ infer $\varphi \Rightarrow \varphi, \psi$ using relevant weakening. From this infer $\Rightarrow \varphi, \varphi \to \psi$. Finally, from the last sequent and $\varphi \Rightarrow \varphi$ infer $(\varphi \to \psi) \to \varphi \Rightarrow \varphi$, and so also $\Rightarrow ((\varphi \to \psi) \to \varphi) \to \varphi)$. This and two applications of (1) immediately entail:

(2) $\varphi \to \psi \in \mathcal{T}^*$ for every ψ such that $\mathsf{Var}(\psi) \subseteq \mathsf{Var}(\varphi)$.

Since $\vdash (\varphi \to \neg\psi) \to (\psi \to \neg\varphi)$, (2) implies:

(3) $\psi \to \neg\varphi \in \mathcal{T}^*$ for every ψ such that $\mathsf{Var}(\psi) \subseteq \mathsf{Var}(\varphi)$.

By taking $\psi = (\varphi \to \varphi)$ we get:

(4) $\neg\varphi \in \mathcal{T}^*$.

Now, Item (2) in Proposition 13.8 implies that $\neg\psi, \tau \vdash \psi \to \tau$ in case $\mathsf{Var}(\psi) \cap \mathsf{Var}(\tau) \neq \emptyset$. Hence:

(5) If $\mathsf{Var}(\psi) \cap \mathsf{Var}(\tau) \neq \emptyset$, $\neg\psi \in \mathcal{T}^*$, and $\tau \in \mathcal{T}^*$, then $\psi \to \tau \in \mathcal{T}^*$.

Now, we construct the Lindenbaum algebra of $\mathcal{T}^*$ on the set of sentences which are fully relevant to φ. For this let $\psi \equiv \tau$ iff $\mathcal{T}^* \vdash \psi \to \tau$ and $\mathcal{T}^* \vdash \tau \to \psi$. This is obviously a congruence relation. Define the operations on the corresponding set of equivalence classes in the obvious way (e.g. $[\psi] \widetilde{\to} [\tau] = [\psi \to \tau]$ etc., where $[\psi]$ is the equivalence class of ψ). We shall denote $[\varphi]$ by f, $[\neg\varphi]$ by t. Obviously, $t \neq f$, since $\mathcal{T}^* \vdash \neg\varphi$ while $\mathcal{T}^* \not\vdash \varphi$. Moreover: for ψ which is fully relevant to φ we have:

(6) $[\psi] \notin \{t, f\}$ iff both $\mathcal{T}^* \vdash \psi$ and $\mathcal{T}^* \vdash \neg\psi$.

To see this, suppose $[\psi] \notin \{t, f\}$. Then $[\psi] \neq f = [\varphi]$ and so by (2) $\mathcal{T}^* \nvdash \psi \to \varphi$. This means, by (1), that $\mathcal{T}^* \vdash \psi$. Similarly, $[\psi] \neq t = [\neg\varphi]$, and so, by (3), $\mathcal{T}^* \nvdash \neg\varphi \to \psi$. It follows that $\mathcal{T}^* \nvdash \neg\psi \to \varphi$, and so $\mathcal{T}^* \vdash \neg\psi$. For the converse, assume both ψ and $\neg\psi$ are in $\mathcal{T}^*$. Then $\psi \not\equiv \varphi$ since $\mathcal{T}^* \nvdash \varphi$, and $\psi \not\equiv \neg\varphi$, since $\mathcal{T}^* \nvdash \neg\neg\varphi$. Hence $[\psi] \neq f$ and $[\psi] \neq t$.

Now, (5) and (6) together imply:

(7) If $\{[\psi], [\tau]\} \cap \{t, f\} = \emptyset$, and $\mathsf{Var}(\psi) \cap \mathsf{Var}(\tau) \neq \emptyset$, then $[\psi] = [\tau]$.

Let n be the number of variables in $\mathsf{Var}(\varphi)$. From (7) it follows that the number of equivalence classes which are different from t and f is at most n. Denote these equivalence classes by $I_1, I_2, \ldots, I_k$ (where $k \leq n$). We show that the Lindenbaum algebra of $\mathcal{T}^*$ is isomorphic to $\mathcal{A}_k$. For this it suffices to show that the operations of this algebra are identical to those of $\mathcal{A}_\omega$.

The case of $\neg$: Obviously $\widetilde{\neg} f = t$ and $\widetilde{\neg} t = f$. Now let $I_j = [\psi]$. By (6), $\mathcal{T}^* \vdash \psi$ and $\mathcal{T}^* \vdash \neg\psi$. But both $\psi \to (\neg\psi \to (\psi \to \neg\psi))$ and $\psi \to (\neg\psi \to (\neg\psi \to \psi))$ are theorems of $\mathbf{RMI}_{\to}$. Hence $\psi \equiv \neg\psi$ and $\widetilde{\neg} I_j = [\neg\psi] = [\psi]$.

The case of $\otimes$: Suppose ψ is fully relevant to φ. Using relevant weakening, it is easy to derive $\varphi \otimes \psi \to \varphi$ in $GRMI_{\to}$. It follows that $\mathcal{T}^* \nvdash \varphi \otimes \psi$, and so $[\varphi \otimes \psi] = f$, by (1) and (2). Hence $f \widetilde{\otimes} [\psi] = f$. Similarly, $[\psi] \widetilde{\otimes} f = f$. If $[\psi] = [\tau]$ then $[\psi] \widetilde{\otimes} [\tau] = [\psi] \widetilde{\otimes} [\psi] = [\psi \otimes \psi] = [\psi]$, since $\psi \to \psi \otimes \psi$ and $\psi \otimes \psi \to \psi$ are theorems of $\mathbf{RMI}_{\to}$. Finally, suppose $[\psi] \neq f$, $[\tau] \neq f$ and $[\psi] \neq [\tau]$. Then $\mathcal{T}^* \vdash \psi$ and $\mathcal{T}^* \vdash \tau$. Assume for contradiction that $[\neg\varphi] = t \neq [\psi] \widetilde{\otimes} [\tau] (= [\psi \otimes \tau])$. Then by (3) $\mathcal{T}^* \nvdash \neg\varphi \to \psi \otimes \tau$. Hence $\mathcal{T}^* \nvdash \neg(\psi \otimes \tau) \to \varphi$, and so by (1) $\mathcal{T}^* \vdash \neg(\psi \otimes \tau)$. However, both $\vdash_{RMI_{\to}} \psi \to (\tau \to (\neg(\psi \otimes \tau) \to (\psi \to \tau)))$ and $\vdash_{RMI_{\to}} \psi \to (\tau \to (\neg(\psi \otimes \tau) \to (\tau \to \psi)))$. Hence $\psi \equiv \tau$, contradicting $[\psi] \neq [\tau]$.

The case of $\to$: Since $\psi \to \tau$ and $\neg(\psi \otimes \neg\tau)$ are equivalent in $\mathbf{RMI}_{\to}$, and the corresponding operations are identical on $\mathcal{A}_\omega$, this case follows from the previous two.

Now, let $\nu(\psi) = [\psi]$. This is easily seen to be a valuation (the canonical one) in the Lindenbaum Algebra of $\mathcal{T}^*$, and so in $\mathcal{A}_k$, and so in $\mathcal{A}_\omega$. Obviously $\nu(\varphi) = f$. □

Here are some easy corollaries of the last theorem:

Corollary 13.20 $\mathcal{T} \vdash_{\mathbf{RMI}_{\to}} \psi$ *iff there exists a subset* $\{\varphi_1, \ldots, \varphi_n\}$ *of* $\mathcal{T}$ $(n \geq 0)$ *such that* $\varphi_1 \to (\varphi_2 \to \cdots \to (\varphi_n \to \psi) \cdots)$ *is valid in* $\mathcal{A}_\omega$.

13.1 A Maximal Strong Relevant Logic

Proof. Immediate from Proposition 13.13 and the last theorem. □

Corollary 13.21 (decidability) *For a finite theory Γ and a formula φ it is decidable whether $\Gamma \vdash_{\mathbf{RMI}_\to} \varphi$ or not.*

Proof. It is easy to check that if $\nu \in \Lambda_{\mathcal{A}_\omega}$, and $\nu(p) \in \{t, f, I_{i_1}, \ldots, I_{i_k}\}$ for every p in $\mathsf{Var}(\psi)$, then $\nu(\psi) \in \{t, f, I_{i_1}, \ldots, I_{i_k}\}$. It follows that if the cardinality of $\mathsf{Var}(\psi)$ is n, then $\vdash_{\mathbf{RMI}_\to} \psi$ iff ψ is valid in the finite matrix $\mathcal{A}_n$. Therefore the corollary follows from the previous one. □

Corollary 13.22 (variable-sharing property) *If $\vdash_{\mathbf{RMI}_\to} \varphi \to \psi$ then φ and ψ share a variable.*

Proof. Suppose that φ and ψ do not share a variable. Let ν be a valuation in $\mathcal{A}_\omega$ that assigns I_1 to all formulas in $\mathsf{Var}(\varphi)$ and I_2 to all formulas in $\mathsf{Var}(\psi)$. Then $\nu(\varphi) = I_1$, $\nu(\psi) = I_2$, and $\nu(\varphi \to \psi) = f$. Hence $\varphi \to \psi$ is not a theorem of $\mathbf{RMI}_\to$. □

Corollary 13.23 *If $\vdash_{\mathbf{RMI}_\to} \varphi \otimes \psi$ then $\vdash_{\mathbf{RMI}_\to} \varphi$ and $\vdash_{\mathbf{RMI}_\to} \psi$.*

Proof. Suppose that $\not\vdash_{\mathbf{RMI}_\to} \varphi$. Then there is a valuation ν in $\mathcal{A}_\omega$ such that $\nu(\varphi) = f$. It follows that $\nu(\varphi \otimes \psi) = f$ as well. Hence $\varphi \otimes \psi$ is not valid in $\mathcal{A}_\omega$, and so it is not provable in $\mathbf{RMI}_\to$. □

Note 13.24 Recall that by Theorem 12.24, $\mathbf{R}_\to$ does *not* have the property of $\mathbf{RMI}_\to$ described in Corollary 13.23. This again indicates that the connective $\otimes$ of $\mathbf{R}_\to$ cannot be taken as a "conjunction".

Note 13.25 From the proof of Corollary 13.23 it follows that $\varphi \otimes \psi \vdash_{\mathcal{A}_\omega} \varphi$. On the other hand, Corollary 13.20 implies that $P_1 \otimes P_2 \not\vdash_{\mathbf{RMI}_\to} P_1$. It follows that $\mathcal{A}_\omega$ is only *weakly* characteristic for $\mathbf{RMI}_\to$, but not (strongly) characteristic for it. Still, Corollary 13.20 means that $\mathcal{A}_\omega$ suffices for characterizing (in an indirect way) the consequence relation of $\mathbf{RMI}_\to$. In Section 13.1.6 we present a more complex matrix which does strongly characterize $\mathbf{RMI}_\to$.

13.1.3 Maximal Strong Relevance of $\mathbf{RMI}_\to$

The logic $\mathbf{RMI}_\to$ is of course an axiomatic extensions of $\mathbf{R}_\to$. Hence it has all the properties proved in Chapter 12 of such extensions. Thus, we have:

Theorem 13.26 *Let $\mathbf{L}$ be an axiomatic simple extension of $\mathbf{RMI}_\to$ (including $\mathbf{RMI}_\to$ itself).*

1. **L** is a semi-normal logic which is $\neg$-contained in classical logic.

2. **L** is $\mathbf{F}_0$-contained in classical logic (where $\mathbf{F}_0$ is the bivalent interpretation given in Definition 11.89).

Proof. Immediate from Corollary 12.13 and Theorem 12.9. □

Our next goal is to show that $\mathbf{RMI}_\to$ is a strongly relevant logic.

Theorem 13.27 $\mathbf{RMI}_\to$ has no finite weakly characteristic Nmatrix.

Proof. Suppose $\mathbf{RMI}_\to$ has a finite weakly characteristic Nmatrix $\mathcal{M}$ having n truth values. Let $\mathbf{t}_n = (P_1 \to P_1) \otimes (P_2 \to P_2) \otimes \cdots \otimes (P_n \to P_n)$, and let φ_n be the $+$-disjunction of $\mathbf{t}_{n+1} \otimes (P_k \to P_l)$ for $1 \leq k < l \leq n+1$. Then $\nvdash_{\mathbf{RMI}_\to} \varphi_n$, since by assigning I_k to P_k ($k = 1, \ldots, n+1$) we get a countermodel in $\mathcal{A}_\omega$. Hence there is a valuation ν in $\mathcal{M}$ which assigns φ_n a non-designated value. Since $\mathcal{M}$ has only n truth values, there exist $1 \leq i < j \leq n+1$ such that $\nu(P_i) = \nu(P_j)$. For every subformula ψ of φ_n let ψ' be the sentence obtained from ψ by replacing every subformula of the form $P_i \to P_j$ by $P_i \to P_i$. (Note that every subformula ψ of φ_n contains at most one occurrence of $P_i \to P_j$.) Define a partial valuation ν' on the set of subformulas of φ'_n by letting $\nu'(\psi') = \nu(\psi)$ for every subformula ψ of φ_n. It is easy to see that ν' is a partial valuation in $\mathcal{M}$. (That $\nu'(P_i \to P_j) \in \widetilde{\to}(\nu'(P_i), \nu'(P_j))$, where $\widetilde{\to}$ is the interpretation of $\to$ in $\mathcal{M}$, follows from the the fact that $\nu(P_i) = \nu(P_j)$. Legality of $\nu'(\psi')$ in any other case is now proved by a straightforward induction on the structure of ψ'.) Now, $\nu'(\varphi'_n) = \nu(\varphi_n)$, and so φ'_n is not valid in $\mathcal{M}$. This contradicts the assumption that $\mathcal{M}$ is weakly characteristic for $\mathbf{RMI}_\to$, because φ'_n is valid in $\mathcal{A}_\omega$, and so it is provable in $\mathbf{RMI}_\to$. □

Corollary 13.28 $\mathbf{RMI}_\to$ satisfies the minimal semantic criterion.

Theorem 13.29 $\mathbf{RMI}_\to$ is a strongly relevant logic.

Proof. This follows from Corollaries 12.17, 13.22, and 13.28. □

Corollary 13.30 $\mathbf{RMI}_\to$ is boldly paraconsistent, and $\neg$ is complete for it.

Proof. By Theorem 13.29, Corollary 11.92, and Proposition 11.93. □

Now we turn to the main goal of this section: to show that $\mathbf{R}_\to$ is a *maximal* strongly relevant logic.

13.1 A Maximal Strong Relevant Logic

Theorem 13.31 *If* **L** *is a strongly proper, non-trivial simple extension of* **RMI**$_\rightarrow$ *then there exists $l \geq 0$ such that $\mathcal{A}_l$ is weakly characteristic for* **L** *(i.e., for all φ, $\vdash_\mathbf{L} \varphi$ iff φ is valid in $\mathcal{A}_l$).*

Proof. Let HL be the Hilbert-type system obtained from $HRMI_\rightarrow$ by adding the valid formulas of **L** as axioms. Obviously, $\vdash_\mathbf{L} \varphi$ iff $\vdash_{HL} \varphi$. Using HL instead of $HRMI_\rightarrow$, the proof of Theorem 13.19 can easily be adapted to show the following generalization:

(*) If φ is a sentence of $\mathcal{IL}$ having exactly n variables, and $\nvdash_\mathbf{L} \varphi$, then there exists some $k \leq n$ such that any theorem of **L** is valid in $\mathcal{A}_k$, but φ is not.[2]

On the other hand, it is obvious that $\mathcal{A}_i$ is a submatrix of $\mathcal{A}_j$ if $i < j$. Hence, Proposition 3.14 entails:

(**) If φ is not valid in $\mathcal{A}_i$, and $i < j$, then φ is not valid in $\mathcal{A}_j$.

Now, since **L** is not trivial, (*) entails that there is $k \in N$ such that any theorem of **L** is valid in $\mathcal{A}_k$. Since **L** is a *strongly proper* extension of **RMI**$_\rightarrow$, (*) and (**) together entail that there is a maximal l such that any theorem of **L** is valid in $\mathcal{A}_l$ (and so also in $\mathcal{A}_i$ for every $i < l$). Finally, it easily follows from (*) and (**) that $\mathcal{A}_l$ is in fact weakly characteristic for **L**. □

Note 13.32 There are proper simple extensions of **RMI**$_\rightarrow$ which are not strongly proper. $\mathbf{L}_{\mathcal{A}_\omega}$ (investigated in the next section) provides an example.

Theorem 13.33 *The logic* **RMI**$_\rightarrow$ *has the Scroggs' property: it does not have a finite weakly characteristic matrix, but every strongly proper simple extension of it does.*[3]

Proof. This follows from Theorems 13.31 and Corollary 13.28. □

Finally, we are ready to prove:

Theorem 13.34 **RMI**$_\rightarrow$ *is a maximal strongly relevant logic in $\mathcal{IL}$: it is a strongly relevant logic, but every proper simple extension of it is not.*

Proof. By Theorem 13.33 no axiomatic proper simple extension of **RMI**$_\rightarrow$ satisfies the minimal semantic criterion. Therefore the claim follows from Theorem 13.29 and the first part of Theorem 12.14. □

---
[2]This is true for every simple extension of **RMI**$_\rightarrow$, not only strongly proper ones.
[3]The name *'Scroggs' property* is because a similar result was first shown for $S5$ in [264].

13.1.4 Axiomatic Simple Extensions of $\mathbf{RMI}_\rightarrow$

The main goal of this section is to provide Hilbert-type formulations for all axiomatic simple extensions of $\mathbf{RMI}_\rightarrow$.

Definition 13.35 ($Th(\mathcal{A}_k)$) $Th(\mathcal{A}_k)$ is the set of formulas in $\mathcal{IL}$ which are valid in $\mathcal{A}_k$. (This is a particular case of Definition 3.11.)

Theorem 13.36 *The sequence $\{Th(\mathcal{A}_k)\}_{k=0}^\infty$ is strictly decreasing, and contains the sets of valid formulas of all possible strongly proper simple extensions of $\mathbf{RMI}_\rightarrow$. Moreover: the set of theorems of $\mathbf{RMI}_\rightarrow$ is the intersection of $\{Th(\mathcal{A}_k)\}_{k=0}^\infty$.*

Proof. Let $\mathbf{t}_n = (P_1 \to P_1) \otimes (P_2 \to P_2) \otimes \cdots \otimes (P_n \to P_n)$, and $R^n(P_i, P_j) = \mathbf{t}_n \otimes ((P_i \to P_i) + (P_j \to P_j))$. If $\nu \in \Lambda_{\mathcal{A}_\omega}$, and $1 \leq i, j \leq n$, then:

$$\nu(R^n(P_i, P_j)) = \begin{cases} f & \text{if } \nu(P_i) = I_m,\ \nu(P_j) = I_l, \text{ and } m \neq l, \\ I_k & \text{if } \nu(P_1) = \nu(P_2) = \ldots = \nu(P_n) = I_k, \\ t & \text{otherwise.} \end{cases}$$

Now, let $\theta_0 = P_2 \to (P_1 \to P_2)$, and for $n > 0$ let θ_n be the +-disjunction of $R^{n+1}(P_i, P_j)$ for $1 \leq i < j \leq n+1$. Using the pigeonhole principle, it is easy to see that θ_n is valid in $\mathcal{A}_n$, but not in $\mathcal{A}_{n+1}$. This fact and fact (**) from the proof of Theorem 13.31 together entail that $\{Th(\mathcal{A}_k)\}_{k=0}^\infty$ is strictly decreasing. The other parts of the proposition are immediate corollaries of Theorem 13.31 and its proof. □

Definition 13.37 ($\mathbf{RMI}_\rightarrow^n$) $\mathbf{RMI}_\rightarrow^n$ is the axiomatic simple extension of $\mathbf{RMI}_\rightarrow$ which is induced by the Hilbert-type system, obtained by adding to $HRMI_\rightarrow$ (as new axioms) all the substitution instances of the sentence θ_n defined in the proof of Theorem 13.36 (i.e., by turning θ_n into an axiom-schema and adding the result to $HRMI_\rightarrow$).

Proposition 13.38 (weak completeness of $\mathbf{RMI}_\rightarrow^n$) *A formula φ is in $Th(\mathcal{A}_n)$ iff $\vdash_{\mathbf{RMI}_\rightarrow^n} \varphi$.*

Proof. From the proof of Theorem 13.31 it follows that for every n, $\mathbf{RMI}_\rightarrow^n$ can be axiomatized by adding to $HRMI_\rightarrow$ any schema which is valid in $\mathcal{A}_n$, but not in $\mathcal{A}_{n+1}$. In particular: by adding the schematic counterpart of θ_n as an extra axiom to $HRMI_\rightarrow$, we get an axiomatization of $Th(\mathcal{A}_n)$. □

Note 13.39 It follows from Theorem 13.31 that each possible strongly proper simple extension of $\mathbf{RMI}_\rightarrow$ just limits the potential number of different (equivalence classes of) contradictory propositions to some finite number n. From the point of view of relevance, any particular choice of n would obviously be artificial.

13.1 A Maximal Strong Relevant Logic

Three of the proper simple extensions of $\mathbf{RMI}_{\to}$ deserve to be mentioned. They corresponds to the cases $k = 0, 1, 2$ in Definition 13.37:

- Since $\mathcal{A}_0$ is exactly the classical two-valued matrix (in the language of negation, implication, and conjunction), by Proposition 13.38 $\mathbf{RMI}_{\to}^0$ is just classical logic.

- $\mathbf{RMI}_{\to}^1 = \mathbf{RM}_{\to}$ is equivalent (by Proposition 13.38) to any system which is weakly characteristic for the three-valued matrix $\mathcal{A}_1$ (which was discussed already in Section 4.4.2) and has MP for $\to$ as its sole rule of inference. In particular, the system introduced by Sobociński in [273] has been shown there to have these properties, and so $\mathbf{RMI}_{\to}^1$ is equivalent to it.[4] That system, in turn, is known (see [4]) to be equivalent to the purely intensional (or "multiplicative") fragment of the semi-relevant system $\mathbf{RM}$ (a proof of this result is given in Chapter 15 below). This is the reason for the name $\mathbf{RM}_{\to}$ that we shall use for this system from now on. Now θ_1 (from Definition 13.37) is equivalent in $\mathbf{RMI}_{\to}$ to $\neg(P_1 \to P_1) \to (P_2 \to P_2)$. It follows by Proposition 13.38 that $\mathbf{RM}_{\to}$ can be axiomatized by adding to $HRMI_{\to}$ the axiom schema $\neg(\varphi \to \varphi) \to (\psi \to \psi)$. Hence, the variable-sharing property fails for $\mathbf{RM}_{\to}$. On the other hand, $\mathbf{RM}_{\to}$ is still boldly paraconsistent (this easily follows from its soundness with respect to $\mathcal{A}_1$ and the RDP). Since classical logic is its only strongly proper simple extension, $\mathbf{RM}_{\to}$ is a *weakly maximal* paraconsistent logic (which is also maximal relative to classical logic). However, as mentioned in Example 4.43, it is not a (strongly) maximal paraconsistent logic. (This is an immediate corollary of Proposition 13.68 below.)

  A cut-free Gentzen-type formulation $GRM_{\to}$ of $\mathbf{RM}_{\to}$ is obtained from $GRMI_{\to}$ (or directly from $GR_{\to}$) by the addition of the *(full) mingle* rule (also called "mix" in [174]):

  $$\frac{\Gamma_1 \Rightarrow \Delta_1 \quad \Gamma_2 \Rightarrow \Delta_2}{\Gamma_1, \Gamma_2 \Rightarrow \Delta_1, \Delta_2}$$

  The proofs that $GRM_{\to}$ is equivalent to $\mathbf{RM}_{\to}$, and that the cut-elimination theorem obtains for it, are similar to the proofs of the corresponding propositions about $GRMI_{\to}$, and so we omit them.

- $\mathbf{RMI}_{\to}^2$ is interesting because while it has the VSP (the proof of this is like that for $\mathbf{RMI}_{\to}$, using $\mathcal{A}_2$ instead of $\mathcal{A}_\omega$), its only two simple

---

[4] As noted in Section 4.4.2, $\mathcal{A}_1$ was first introduced in [273], and so it is known as "Sobociński's three-valued matrix".

axiomatic extensions do not. In fact, it is easy to see that $\mathbf{RMI}^2_{\to}$ would have been considered a *maximal strongly relevant logic* had we not demanded in this book (unlike in [52]) such logics to satisfy the minimal semantic relevance criterion.

13.1.5 Expressive Power of the Language of $\mathbf{RMI}_{\to}$

What logical connectives (in addition to negation) should be used in a logic which is based on $\mathcal{A}_\omega$ (and $\mathcal{A}_k$ where $k < \omega$), given the intuitions that underly it? We suggest two main criteria. The main one (which we think is absolutely necessary) is *symmetry*: there should be no way to distinguish between two different basic paradoxical values on a *logical* basis. The other is *isolation of contradictions*: a formula may be assigned a paradoxical value like I_k only if all its constituents are assigned that paradoxical value. These intuitive criteria are formalized in our next definition. The theorem that follows it shows that the connectives which are definable in $\mathcal{IL}$ are exactly those that meet the strongest form of the two criteria mentioned above.

Definition 13.40 Let F be an n-ary operation on A_ω (A_m).

- F satisfies the *symmetry* condition if
$$F(h(x_1), \ldots, h(x_n)) = h(F(x_1, \ldots, x_n))$$
for all $x_1, \ldots, x_n \in A_\omega$ (A_m) and for every injective function h from A_ω to A_ω (A_m to A_m) such that $h(t) = t$ and $h(f) = f$.

- F satisfies the *strong isolation* condition if there exists a subset $D(F)$ of $1, \ldots, n$ such that:
  - If $\vec{x}, \vec{y} \in A_\omega^n$, and $x_i = y_i$ for every $i \in D(F)$, then $F(\vec{x}) = F(\vec{y})$;
  - For every $k < \omega$, $F(\vec{x}) = I_k$ iff $x_i = I_k$ for every $i \in D(F)$.

Definition 13.41 Let ψ be a formula in $\mathcal{IL}$ such that $\mathsf{Var}(\psi) \subseteq \{P_1, \ldots, P_n\}$ (where $P_1, \ldots, P_n$ are the first n variables). g_ψ^n, the function from A_ω^n to A_ω that is induced by ψ, is defined as follows: if $\vec{x} = \langle x_1, \ldots, x_n \rangle$ then $g_\psi^n(\vec{x}) = \nu_{\vec{x}}(\psi)$, where $\nu_{\vec{x}}$ is a valuation in A_ω such that $\nu_{\vec{x}}(P_i) = x_i$ for $1 \leq i \leq n$.

Note 13.42 Obviously, $g_{P_i}^n(\vec{x}) = x_i$ ($1 \leq i \leq n$), while $g_{\neg\psi}^n(\vec{x}) = \widetilde{\neg} g_\psi^n(\vec{x})$ and $g_{\varphi \to \psi}^n(\vec{x}) = g_\varphi^n(\vec{x}) \widetilde{\to} g_\psi^n(\vec{x})$.

Theorem 13.43 *An n-ary operation F on A_ω (A_m) is induced by some formula ψ of $\mathcal{IL}$ such that $\mathsf{Var}(\psi) \subseteq \{P_1, \ldots, P_n\}$ iff it satisfies both the symmetry condition and the strong isolation condition.*

13.1 A Maximal Strong Relevant Logic

Proof. We do the case of A_ω. (The proof in the case of A_m where $m < \omega$ is almost identical.)

For the 'only if' direction, let ψ be a formula of $\mathcal{IL}$ such that $\mathsf{Var}(\psi) \subseteq \{P_1, \ldots, P_n\}$. We show that g_ψ^n satisfies the two conditions.

- To prove that g_ψ^n satisfies the symmetry condition, let h be an injective function from A_ω to A_ω such that $h(t) = t$ and $h(f) = f$. It is easy to see that h is actually a homomorphism, that is: $h(\tilde{\neg} a) = \tilde{\neg} h(a)$ and $h(a\tilde{\to} b) = h(a)\tilde{\to} h(b)$ for every $a, b \in A_\omega$. Using these two facts and Note 13.42, it is straightforward to show by an induction on the structure of ψ that $g_\psi^n(h(x_1), \ldots, h(x_n)) = h(g_\psi^n(x_1, \ldots, x_n))$ for every $\psi \in \mathcal{W}(\mathcal{IL})$ and $x_1, \ldots, x_n \in A_\omega$. (Thus, for example, if $\psi = \neg\varphi$ then $g_\psi^n(h(x_1), \ldots, h(x_n)) = \tilde{\neg} g_\varphi^n(h(x_1), \ldots, h(x_n)) = \tilde{\neg} h(g_\varphi^n(x_1, \ldots, x_n)) = h(\tilde{\neg} g_\varphi^n(x_1, \ldots, x_n)) = h(g_\psi^n(x_1, \ldots, x_n))$.)

- Let $D(\psi) = \{i \mid P_i \in \mathsf{Var}(\psi)\}$. We prove that g_ψ^n satisfies the strong isolation condition with $D(g_\psi^n) = D(\psi)$. The proof is again by an induction on the structure of ψ, using Note 13.42. The claim is obvious in case $\psi = P_i$ for some $1 \le i \le n$. We do the induction step in case $\psi = \psi_1 \to \psi_2$, leaving the case where $\psi = \neg\psi_1$ to the reader. Since in this case $\mathsf{Var}(\psi) = \mathsf{Var}(\psi_1) \cup \mathsf{Var}(\psi_2)$, $D(\psi) = D(\psi_1) \cup D(\psi_2)$. Suppose first that $x_i = y_i$ for every $i \in D(\psi)$. By induction hypothesis for ψ_1 and ψ_2, this implies that $g_{\psi_1}^n(\vec{x}) = g_{\psi_1}^n(\vec{y})$ and $g_{\psi_2}^n(\vec{x}) = g_{\psi_2}^n(\vec{y})$. Hence $g_\psi^n(\vec{x}) = g_\psi^n(\vec{y})$ (by Note 13.42). Finally, since in A_ω $a\tilde{\to} b = I_k$ iff $a = b = I_k$, we have that $g_\psi^n(\vec{x}) = I_k$ iff $g_{\psi_1}^n(\vec{x}) = I_k$ and $g_{\psi_2}^n(\vec{x}) = I_k$, iff (by induction hypothesis for ψ_1 and ψ_2) $x_i = I_k$ for every $i \in D(\psi_1)$ and every $i \in D(\psi_2)$, iff $x_i = I_k$ for every $i \in D(\psi)$.

For the 'if' direction we introduce first some notations:

- Let $\mathsf{Var}(\psi) = \{P_1, \ldots, P_n\}$. $S[\psi]$, the subset of A_ω^n which is *characterized* by ψ, is:

$$S[\psi] = \{\langle a_1, \ldots, a_n\rangle \in A_\omega^n \mid g_\psi(a_1, \ldots, a_n) \ne f\}.$$

- Denote by I_k^n the n-tuple $\langle I_k, I_k, \ldots, I_k\rangle$. Let $I(n) = \{I_k^n \mid k < \omega\}$.

The facts about g_ψ^n proved above and the definitions of the interpretation of $\otimes$ and $+$ in A_ω easily entail:

1. If $\mathsf{Var}(\psi) = \{P_1, \ldots, P_n\}$, and $\vec{x} \notin I(n)$, then $g_\psi(\vec{x}) \in \{t, f\}$.

2. If $\mathsf{Var}(\psi) = \{P_1, \ldots, P_n\}$, then $I(n) \subseteq S[\psi]$.

3. If $\mathsf{Var}(\psi_i) = \{P_1, \ldots, P_n\}$ for $1 \leq i \leq k$ then:

$$\bigcap_{i=1}^{k} S[\psi_i] = S[\psi_1 \otimes \psi_2 \otimes \cdots \otimes \psi_k],$$

$$\bigcup_{i=1}^{k} S[\psi_i] = S[\psi_1 + \psi_2 + \cdots + \psi_k].$$

Next, define $\langle b_1, \ldots, b_n \rangle \in A_\omega^n$ to be *similar* to $\vec{a} = \langle a_1, \ldots, a_n \rangle \in A_\omega^n$ if there exists an injective function h from A_ω to A_ω such that $h(t) = t$, $h(f) = f$, and $h(a_i) = b_i$ for $1 \leq i \leq n$. Obviously, similarity of tuples is an equivalence relation, and two vectors $\langle a_1, \ldots, a_n \rangle$ and $\langle b_1, \ldots, b_n \rangle$ are similar iff the following conditions are satisfied for all $1 \leq i, j \leq n$:

(i) If $a_i = t$ then $b_i = t$,

(ii) If $a_i = f$ then $b_i = f$,

(iii) If $a_i \in I(1)$ then $b_i \in I(1)$,

(iv) If $a_i = a_j$ then $b_i = b_j$,

(v) If $a_i \neq a_j$ then $b_i \neq b_j$.

It follows that every $\vec{b} \in A_\omega^n$ is similar to some $\vec{a} \in A_n^n$.

Now, define a subset $C \subseteq A_\omega^n$ to be *strictly characterizable* in IL if $C = S[\psi]$ for some formula ψ of IL such that $\mathsf{Var}(\psi) = \{P_1, \ldots, P_n\}$. We show:

Lemma. Let $C \subseteq A_\omega^n$. If C satisfies the following two conditions:

- $I(n) \subseteq C$.

- If $\vec{a} \in C$ and $\vec{b}$ is similar to $\vec{a}$ then $\vec{b} \in C$.

Then C is strictly characterizable in IL.

To prove the lemma, assume that $C \subseteq A_\omega^n$ satisfies the two conditions. Then the basic properties of similarity entail that $C = \bigcup_{\vec{a} \in A_n^n \cap C} S^{\vec{a}}$, where $S^{\vec{a}}$ is the union of $I(n)$ and the set of tuples which are similar to $\vec{a}$. In turn, $S^{\vec{a}} = (\bigcap_{1 \leq i \leq n} S_i^{\vec{a}}) \cap (\bigcap_{1 \leq i,j \leq n} S_{i,j}^{\vec{a}})$ by (i)-(v) above, where:

$$S_i^{\vec{a}} = \begin{cases} I(n) \cup \{\vec{x} \in A_\omega^n \mid x_i = t\} & a_i = t, \\ I(n) \cup \{\vec{x} \in A_\omega^n \mid x_i = f\} & a_i = f, \\ \{\vec{x} \in A_\omega^n \mid x_i \in I(1)\} & a_i \in I(1). \end{cases}$$

13.1 A Maximal Strong Relevant Logic

$$S_{i,j}^{\vec{a}} = \begin{cases} I(n) \cup \{\vec{x} \in A_\omega^n \mid x_i = x_j\} & a_i = a_j, \\ I(n) \cup \{\vec{x} \in A_\omega^n \mid x_i \neq x_j\} & a_i \neq a_j. \end{cases}$$

By (3) above it remains therefore to prove that these $S_i^{\vec{a}}$ and $S_{i,j}^{\vec{a}}$ are strictly characterizable in $\mathcal{IL}$ for all i and j. For this define:

$$\mathbf{t}_n = (P_1 \to P_1) \otimes (P_2 \to P_2) \otimes \cdots \otimes (P_n \to P_n).$$

Then $\mathbf{t}_n$ is a sentence such that

$$g_{\mathbf{t}_n}(\vec{x}) = \begin{cases} I_k & \vec{x} = I_k^n, \\ t & \vec{x} \notin I(n). \end{cases}$$

That $S_i^{\vec{a}}$ and $S_{i,j}^{\vec{a}}$ are strictly characterizable in $\mathcal{IL}$ now follows from the following easily established equations (where $\leftrightarrow$ is defined as in Note 12.18):

$$I(n) \cup \{\vec{x} \in A_\omega^n \mid x_i = t\} = S[\mathbf{t}_n \to P_i],$$
$$I(n) \cup \{\vec{x} \in A_\omega^n \mid x_i = f\} = S[\mathbf{t}_n \to \neg P_i],$$
$$\{\vec{x} \in A_\omega^n \mid x_i \in I(1)\} = S[\mathbf{t}_n \otimes P_i \otimes \neg P_i],$$
$$I(n) \cup \{\vec{x} \in A_\omega^n \mid x_i = x_j\} = S[\mathbf{t}_n \otimes (P_i \leftrightarrow P_j)],$$
$$I(n) \cup \{\vec{x} \in A_\omega^n \mid x_i \neq x_j\} = S[\mathbf{t}_n \to \neg(P_i \leftrightarrow P_j)].$$

This end the proof of the lemma. □

Now we turn at last to prove the 'if' direction. So let $F : A_\omega^n \to A_\omega$ satisfy the conditions of symmetry and strong isolation.

Assume first that $D(F) = \{1, \ldots, n\}$. This assumption and the conditions of symmetry and strong isolation together imply that the set $E = \{\vec{x} \in A_\omega^n \mid F(\vec{x}) \neq f\}$ satisfies the two conditions given in the lemma. Hence $E = S[\psi_F]$ for some formula ψ_F in $\mathcal{IL}$ such that $\mathsf{Var}(\psi_F) = \{P_1, \ldots, P_n\}$. Now we show that $F = g_{\psi_F}$. We consider three cases:

- If $F(\vec{x}) = I_k$ for some k then (by isolation) $\vec{x} = I_k^n$, and since ψ_F is in $\mathcal{IL}$ it follows that $g_{\psi_F}(\vec{x}) = I_k = F(\vec{x})$.

- If $F(\vec{x}) = f$ then $\vec{x} \notin E$. Hence $\vec{x} \notin S[\psi_F]$, and so $g_{\psi_F}(\vec{x}) = f = F(\vec{x})$

- If $F(\vec{x}) = t$ then $\vec{x} \in E = S[\psi_F]$, and (by strong isolation) $\vec{x} \notin I(n)$. Hence $g_{\psi_F}(\vec{x}) \neq f$, and (by isolation) $g_{\psi_F}(\vec{x}) \notin I(1)$. It follows that $g_{\psi_F}(\vec{x}) = t = F(\vec{x})$.

Finally, assume that $D(F) = \{i_1, \ldots, i_k\}$. Define $F^* : A_\omega^k \to A_\omega$ by: $F^*(x_1, \ldots, x_k) = F(y_1, \ldots, y_n)$, where $y_j = x_l$ if $j = i_l$ for some $l \leq k$, and $y_j = f$ otherwise. Then F^* satisfies the conditions of symmetry and strong isolation, with $D(F^*) = \{1,\ldots,k\}$. It follows by the case we have already proved that there is a formula ψ_{F^*} such that $F^* = g_{\psi_{F^*}}$, and $\mathsf{Var}(\psi_{F^*}) = \{P_1,\ldots,P_k\}$. Let ψ_F be obtained from ψ_{F^*} by simultaneously substituting P_{i_j} for P_j $(j = 1,\ldots,k)$. It is not difficult to see that $F = g_{\psi_F}$. □

Corollary 13.44 *An n-ary operation F on $\mathcal{A}_1$ is definable in in the language of $\mathcal{A}_1$ by a formula ψ such that $\mathsf{Var}(\psi) \subseteq \{P_1,\ldots,P_k\}$ iff it satisfies the strong isolation condition.*

13.1.6 Strongly Characteristic Matrix for $\mathbf{RMI}_{\to}$

As its name suggests, Theorem 13.19 implies that $\mathcal{A}_\omega$ is *weakly* characteristic for $\mathbf{RMI}_{\to}$. On the other hand, it was shown in Note 13.25 that $\mathcal{A}_\omega$ is *not* (strongly) characteristic for $\mathbf{RMI}_{\to}$. In this section we present a more complex matrix which does strongly characterize $\mathbf{RMI}_{\to}$. It is based on adding to the ideas of relevance and of paraconsistency also the idea of *"degrees of truth"* (to be explained and further exploited in Section 15.5).

Definition 13.45 ($\mathcal{SA}_m$) The matrix $\mathcal{SA}_m$ for $\mathcal{IL}$ is $\langle SA, \mathcal{D}_{SA}, \mathcal{O} \rangle$

- $SA = [0,1] \times A_\omega$.
  If $v = \langle x, a \rangle \in A$ then x is called the *degree* of v and is denoted by $deg(v)$, while a is called the *value* of v and is denoted by $val(v)$.

- $\mathcal{D}_{SA} = [0,1] \times \mathcal{D}_\omega$. (Recall that $\mathcal{D}_\omega = \{t, I_1, I_2, I_3, \ldots\}$.)

- The operations in $\mathcal{O}$ are defined as follows:

$$\tilde{\neg}\langle x, t\rangle = \langle x, f\rangle \quad \tilde{\neg}\langle x, f\rangle = \langle x, t\rangle \quad \tilde{\neg}\langle x, I_k\rangle = \langle x, I_k\rangle$$

$$deg(u \tilde{\to} v) = min\{deg(u), deg(v)\},$$

$$val(u \tilde{\to} v) = \begin{cases} I_k & u = v \text{ and } val(u) = I_k, \\ t & deg(u) \leq deg(v) \text{ and } val(u) = f, \\ t & deg(v) \leq deg(u) \text{ and } val(v) = t, \\ f & \text{otherwise.} \end{cases}$$

The next proposition provides useful characterizations of the operation $\tilde{\neg}$ of $\mathcal{SA}_m$, and of its set $\mathcal{D}_{SA}$ of designated elements. We leave its easy proof to the reader.

Proposition 13.46

1. $\mathcal{D}_{SA} = \{a \mid val(a) \neq f\}$.

13.1 A Maximal Strong Relevant Logic

2. The operation $\tilde{\neg}$ can be characterized as follows:

$$deg(\tilde{\neg}u) = deg(u),$$

$$val(\tilde{\neg}u) = \begin{cases} I_k & val(u) = I_k, \\ t & val(u) = f, \\ f & val(u) = t. \end{cases}$$

Theorem 13.47 (strong completeness 1) $\mathcal{T} \vdash_{\mathbf{RMI}_\rightarrow} \varphi$ iff $\mathcal{T} \vDash_{\mathcal{SA}_m} \varphi$.

Proof. To show soundness, we prove first by induction on length of proofs that if a sequent is provable in $GRMI_\rightarrow$ then it is not empty, and all its translations (see the proof of Proposition 12.4) are valid in $\mathcal{SA}_m$. It follows that every theorem of $\mathbf{RMI}_\rightarrow$ is valid in $\mathcal{SA}_m$, and so a sentence φ is valid in $\mathcal{SA}_m$ iff it is valid in $\mathcal{A}_\omega$. Since (MP) for $\rightarrow$ is obviously valid for $\vDash_{\mathcal{SA}_m}$, from this the soundness of $\mathbf{RMI}_\rightarrow$ with respect to $\mathcal{SA}_m$ easily follows with the help of Corollary 13.20.

For completeness, assume $\mathcal{T} \nvdash \varphi$ (again we write here just "$\vdash$" instead of "$\vdash_{\mathbf{RMI}_\rightarrow}$"). Extend $\mathcal{T}$ to a maximal theory $\mathcal{T}^*$ such that $\mathcal{T}^* \nvdash \varphi$. Then by Proposition 13.13, $\sigma \notin \mathcal{T}^*$ iff $\mathcal{T}^* \vdash \sigma \rightarrow \varphi$ (for every sentence σ). Hence $\mathcal{T}^*$ is complete, since the fact that $\vdash (\sigma \rightarrow \varphi) \rightarrow ((\neg\sigma \rightarrow \varphi) \rightarrow \varphi)$ implies that for every σ, either $\sigma \in \mathcal{T}^*$, or $\neg\sigma \in \mathcal{T}^*$. Using Proposition 12.16, define a congruence relation $\equiv$ on the set of sentences by: $\sigma \equiv \tau$ iff $\mathcal{T}^* \vdash \sigma \rightarrow \tau$ and $\mathcal{T}^* \vdash \tau \rightarrow \sigma$, and denote by $[\sigma]$ the equivalence class of σ with respect to $\equiv$. Let $\mathcal{F} = \{[\sigma] \mid \sigma \notin \mathcal{T}^*\}$, and define a relation $\sqsubseteq$ on $\mathcal{F}$ by letting $[\sigma] \sqsubseteq [\tau]$ iff $\sigma \rightarrow \tau \in \mathcal{T}^*$. It is easy to see that $\sqsubseteq$ is a well-defined partial order relation on $\mathcal{F}$ with a maximal element $[\varphi]$. That it is actually *linear* follows from the fact[5] that $\vdash_{GRMI_\rightarrow} \sigma \rightarrow \varphi, \tau \rightarrow \varphi, (\sigma \rightarrow \tau) \rightarrow \varphi, (\tau \rightarrow \sigma) \rightarrow \varphi \Rightarrow \varphi$ (and so there are no σ, τ such that $[\sigma] \in \mathcal{F}, [\tau] \in \mathcal{F}, [\sigma] \not\sqsubseteq [\tau]$, and $[\tau] \not\sqsubseteq [\sigma]$). It follows that there is an injective homomorphism g from $\langle \mathcal{F}, \sqsubseteq \rangle$ to $\langle [0,1], \leq \rangle$. Now since $\vdash ((\sigma \rightarrow \sigma) \rightarrow \varphi) \rightarrow \varphi$, we have that $[(\sigma \rightarrow \sigma) \rightarrow \varphi] \in \mathcal{F}$ for every formula σ, and so $g([(\sigma \rightarrow \sigma) \rightarrow \varphi])$ is defined for every σ. Denote this number by $d(\sigma)$. Next, let $\mathbf{I}_1, \mathbf{I}_2, \ldots$ be an enumeration of all the equivalence classes of formulas σ such that both $\sigma \in \mathcal{T}^*$ and $\neg\sigma \in \mathcal{T}^*$. Define $V(\sigma)$ to be f if $\sigma \notin \mathcal{T}^*$ (i.e. $\sigma \rightarrow \varphi \in \mathcal{T}^*$), t if $\neg\sigma \notin \mathcal{T}^*$ (i.e. $\neg\sigma \rightarrow \varphi \in \mathcal{T}^*$), and I_k if $\sigma \in \mathbf{I}_k$. (Note that the completeness of $\mathcal{T}^*$ implies that $V(\sigma)$ is well-defined for every formula σ.) Finally, define $\nu(\sigma) = \langle d(\sigma), V(\sigma) \rangle$. This means that $deg(\nu(\sigma)) = d(\sigma)$ and $val(\nu(\sigma)) = V(\sigma)$. We show that ν is a model of $\mathcal{T}^*$ (and so of $\mathcal{T}$) in $\mathcal{SA}_m$ which is not a model of φ. It is obvious

---

[5]This fact, as all other facts about $\vdash_{GRMI_\rightarrow}$ which are used below, can easily be shown with the help of Proposition 13.8. Alternatively, such facts can be shown by checking validity in $\mathcal{A}_\omega$.

that ν assigns to every formula σ an element of SA, and Proposition 13.46 implies that $\nu(\sigma) \in \mathcal{D}_{SA}$ for every $\sigma \in \mathcal{T}^*$, while $\nu(\varphi) \notin \mathcal{D}_{SA}$. It is also immediate that ν respects the interpretation of $\neg$ in $\mathcal{SA}_m$. It remains to show that it respects the interpretation of $\to$ as well.

To show that $deg(\nu(\sigma \to \tau)) = min\{deg(\nu(\sigma)), deg(\nu(\tau))\}$, we prove that $[((\sigma \to \tau) \to (\sigma \to \tau)) \to \varphi] = min_\sqsubseteq\{[(\sigma \to \sigma) \to \varphi], [(\tau \to \tau) \to \varphi]\}$. Suppose first that $min_\sqsubseteq\{[(\sigma \to \sigma) \to \varphi], [(\tau \to \tau) \to \varphi]\} = [(\sigma \to \sigma) \to \varphi]$. Then the formula $\sigma \leq \tau =_{Df} ((\sigma \to \sigma) \to \varphi) \to ((\tau \to \tau) \to \varphi)$ is in $\mathcal{T}^*$. That in this case $[((\sigma \to \tau) \to (\sigma \to \tau)) \to \varphi] = [(\sigma \to \sigma) \to \varphi]$ follows therefore from the following two theorems of $\mathbf{RMI}_{\to}$:

$$(((\sigma \to \tau) \to (\sigma \to \tau)) \to \varphi) \to ((\sigma \to \sigma) \to \varphi),$$

$$(\sigma \leq \tau) \to (((\sigma \to \sigma) \to \varphi) \to (((\sigma \to \tau) \to (\sigma \to \tau)) \to \varphi)).$$

If $min_\sqsubseteq\{[(\sigma \to \sigma) \to \varphi], [(\tau \to \tau) \to \varphi]\} = [(\tau \to \tau) \to \varphi]$, the proof of the required identity is similar.

Finally, we show that $val(\nu(\sigma \to \tau))$ (i.e. $V(\sigma \to \tau)$) is always what it should be according to Definition 13.45. We divide the proof of this into cases, according to that Definition.

- Suppose that $\nu(\sigma) = \nu(\tau)$ and $val(\nu(\sigma)) = I_k$. Then $V(\sigma) = I_k$, and so $V(\tau) = I_k$ too. Hence $[\sigma] = [\tau] = I_k$. It follows that $\sigma, \sigma \to \tau$ and $\tau \to \sigma$ are all in $\mathcal{T}^*$. But both $\vdash_{GRMI_{\to}} \sigma, \tau \to \sigma \Rightarrow (\sigma \to \tau) \to \sigma$ and $\vdash_{GRMI_{\to}} \sigma \to \tau \Rightarrow \sigma \to (\sigma \to \tau)$. It follows that $(\sigma \to \tau) \to \sigma$ and $\sigma \to (\sigma \to \tau)$ are both in $\mathcal{T}^*$. Hence $[\sigma \to \tau] = [\sigma] = I_k$, implying that $V(\sigma \to \tau) = I_k$.

- Suppose that $deg(\nu(\sigma)) \leq deg(\nu(\tau))$, and $val(\nu(\sigma)) = f$. Then $\sigma \to \varphi$ and $((\sigma \to \sigma) \to \varphi) \to ((\tau \to \tau) \to \varphi)$ are both in $\mathcal{T}^*$. Since $\{((\sigma \to \sigma) \to \varphi) \to ((\tau \to \tau) \to \varphi), \sigma \to \varphi\} \vdash_{RMI_{\to}} \neg(\sigma \to \tau) \to \varphi$, we get that $\neg(\sigma \to \tau) \to \varphi \in \mathcal{T}^*$, and so $V(\sigma \to \tau) = t$.

- Suppose that $deg(\nu(\tau)) \leq deg(\nu(\sigma))$, and $val(\nu(\tau)) = t$. Then $\neg\tau \to \varphi$ and $((\tau \to \tau) \to \varphi) \to ((\sigma \to \sigma) \to \varphi)$ are both in $\mathcal{T}^*$. But like in the previous case, $\neg(\sigma \to \tau) \to \varphi$ follows in $RMI_{\to}$ from these two sentences. Hence again, we get that $V(\sigma \to \tau) = t$.

- Suppose that none of the three cases above obtains. We show that $val(\nu(\sigma \to \tau)) = V(\sigma \to \tau) = f$. We consider three subcases:

    - Suppose that $d(\sigma) = d(\tau)$. Since the second and the third cases above do not obtain, it follows that $V(\sigma) \neq f$ and $V(\tau) \neq t$. Hence $\sigma \in \mathcal{T}^*$ and $\neg\tau \in \mathcal{T}^*$. Suppose for contradiction that

13.1 A Maximal Strong Relevant Logic

$V(\sigma \to \tau) \neq f$. Then $\sigma \to \tau \in \mathcal{T}^*$. Since both $\vdash_{GRMI_{\to}} \sigma, \neg\tau, \sigma \to \tau \Rightarrow \neg\sigma$ and $\vdash_{GRMI_{\to}} \sigma, \neg\tau, \sigma \to \tau \Rightarrow \tau \to \sigma$, we get that $\tau \to \sigma$ and $\neg\sigma$ are also in $\mathcal{T}^*$. Hence $\sigma \equiv \tau$ and $V(\sigma) = I_k$ for some k. This contradicts our assumption that the first case does not obtain.

- Suppose that $d(\sigma) < d(\tau)$. Since the second case above does not obtain, $V(\sigma) \neq f$, and so $\sigma \in \mathcal{T}^*$. Since $d(\tau) \not\leq d(\sigma)$, $((\tau \to \tau) \to \varphi) \to ((\sigma \to \sigma) \to \varphi) \notin \mathcal{T}^*$, and so the formula $(((\tau \to \tau) \to \varphi) \to ((\sigma \to \sigma) \to \varphi)) \to \varphi$ is in $\mathcal{T}^*$. But the sequent $(((\tau \to \tau) \to \varphi) \to ((\sigma \to \sigma) \to \varphi)) \to \varphi, \sigma, \sigma \to \tau \Rightarrow \varphi$ is derivable in $GRMI_{\to}$. It follows that $\sigma \to \tau \notin \mathcal{T}^*$, and so $V(\sigma \to \tau) = f$.

- Suppose that $d(\tau) < d(\sigma)$. Since the third case above does not obtain, $V(\tau) \neq t$, and so $\neg\tau \in \mathcal{T}^*$. Since $d(\sigma) \not\leq d(\tau)$, the formula $((\sigma \to \sigma) \to \varphi) \to ((\tau \to \tau) \to \varphi)$ is not in $\mathcal{T}^*$, and so $(((\sigma \to \sigma) \to \varphi) \to ((\tau \to \tau) \to \varphi)) \to \varphi \in \mathcal{T}^*$. But the sequent $(((\sigma \to \sigma) \to \varphi) \to ((\tau \to \tau) \to \varphi)) \to \varphi, \neg\tau, \sigma \to \tau \Rightarrow \varphi$ is derivable in $GRMI_{\to}$. It follows that $\sigma \to \tau \notin \mathcal{T}^*$, and so $V(\sigma \to \tau) = f$. □

Despite the fact that $\mathcal{SA}_m$ is much more complex than $\mathcal{A}_\omega$, it shares with the latter the property that up to isomorphism, the number of different submatrices of $\mathcal{SA}_m$ which have exactly k elements ($k < \omega$) is finite. Our next theorem shows that this fact can be used to provide a direct decision procedure for the consequence relation of $\mathbf{RMI}_{\to}$:

Theorem 13.48 *Suppose that $\Gamma \nvdash_{\mathbf{RMI}_{\to}} \varphi$, and that $\Gamma \cup \{\varphi\}$ involves at most n different variables. Then there is a submatrix $\mathcal{SA}_m(\Gamma, \varphi)$ of $\mathcal{SA}_m$ such that $\mathcal{SA}_m(\Gamma, \varphi)$ has at most $3n$ elements, and there is a valuation ν in it which is a model of Γ, but not a model of φ.*

Proof. Since $\Gamma \nvdash_{\mathbf{RMI}_{\to}} \varphi$, Theorem 13.47 entails that there is valuation ν in $\mathcal{SA}_m$ which is a model of Γ but not of φ. Now suppose that all the variables in $\Gamma \cup \{\varphi\}$ are among $\{p_1, \ldots, p_n\}$. Let $X = \{\nu(p_1), \ldots, \nu(p_n)\}$. From Definition 13.45 and Proposition 13.46 it follows that $\overline{X} = X \cup \{\langle deg(u), t\rangle \mid u \in X\} \cup \{\langle deg(u), f\rangle \mid u \in X\}$ is closed under the operations of $\mathcal{AS}$. This fact easily entails and $\mathcal{M}_X = \langle \overline{X}, \mathcal{D}_X, \mathcal{O}_X \rangle$ is a sub-matrix of $\mathcal{AS}$, where $\mathcal{D}_X = \mathcal{D}_{SA} \cap \overline{X}$, and the operations in $\mathcal{O}_X$ are the reductions to $\overline{X}$ of the operations in $\mathcal{SA}_m$. Let ν^* be any valuation in $\mathcal{M}_X$ such that $\nu^*(p_i) = \nu(p_i)$ for every $1 \leq i \leq n$. Obviously $\nu^*(\psi) = \nu(\psi)$ for every $\psi \in \Gamma \cup \{\varphi\}$. It follows that $\mathcal{M}_X$ can serve as $\mathcal{SA}_m(\Gamma, \varphi)$. □

It is not too difficult to show that the bound $3n$ given in Theorem 13.48 can be replaced by $3n - 1$. However, the next proposition shows that the latter bound cannot be further improved.

Proposition 13.49 *For $n > 0$ there is a theory T_n in $P_1, \ldots, P_n$ such that T_n has n elements, $T_n \not\vdash_{\mathbf{RMI}_\to} P_1$, but any model of T_n in $\mathcal{SA}_m$ which is not a model of P_1 involves at least n different degrees, and at least $3n - 1$ different elements of SA.*

Proof. Let $T_1 = \emptyset$, $T_{n+1} = T_n \cup \{\neg(P_{n+1} \to P_{n+1}) \otimes \neg(P_1 \otimes P_2 \otimes \ldots \otimes P_n \to P_1 \otimes P_2 \otimes \ldots \otimes P_n)\}$. First we show by an induction on n that if ν is model of T_n in $\mathcal{SA}_m$ which is not a model of P_1, then for every $i < n$, $deg(\nu(P_{i+1})) < deg(\nu(P_i))$, and $val(\nu(P_i)) \notin \{t, f\}$ in case $i > 1$. The case $n = 1$ is trivial. Suppose the claim holds for T_n, and let ν be a model of T_{n+1} which is not a model of P_1. Then $val(\nu(P_1)) = f$, and ν is a model of T_n as well as of $\neg(P_{n+1} \to P_{n+1}) \otimes \psi_n$, where $\psi_n = \neg(P_1 \otimes P_2 \otimes \ldots \otimes P_n \to P_1 \otimes P_2 \otimes \ldots \otimes P_n)$. From the induction hypothesis and the definition of $\otimes$ in $\mathcal{SA}_m$ it easily follows that $\nu(\psi_n) = \langle deg(\nu(P_n)), f\rangle$. Hence ν can be a model of $\neg(P_{n+1} \to P_{n+1}) \otimes \psi_n$ only if $deg(\nu(P_{n+1})) < deg(\nu(P_n))$, and $val(\nu(P_{n+1})) \notin \{t, f\}$.

Now from what we have just proved it easily follows that if ν is a model of T_n in $\mathcal{SA}_m$ which is not a model of P_1 then $deg(\nu(P_1)), \ldots, deg(\nu(P_n))$ are all different, and that ν uses at least three elements of SA of degree $deg(\nu(P_i))$ for every $i > 1$ ($\langle deg(\nu(P_i)), t\rangle, \langle deg(\nu(P_i)), f\rangle$, and $\nu(P_i)$ itself). Since $val(\nu(P_1)) = f$, ν also uses at least two elements of SA of degree $deg(\nu(P_1))$. Hence ν uses at least $3n - 1$ different elements of SA. □

13.1.7 More General Useful Semantics for $\mathbf{RMI}_\to$

The semantics for $\mathbf{RMI}_\to$ given by $\mathcal{SA}_m$ is a special case of a general family of matrices for which $\mathbf{RMI}_\to$ is strongly sound and (by Theorem 13.47) strongly complete. The basic intuitive ideas behind this family are:

- Propositions are divided into "domains of discourse".

- The domains are partially ordered according to certain 'degrees of priority' or 'degrees of dependency'. This partial order induces a hierarchical, tree-like structure, in which the root has the highest 'degree of priority', and a leaf — a lowest one.

- Each domain has its own truth-values. Usually it has two, corresponding to the classical truth-values, and classical logic is valid within it. However, domains which are leaves of the tree of domains may be degenerate, having just one truth-value I, with the intended meaning

13.1 A Maximal Strong Relevant Logic

of "inconsistent", or "both true and false". (See Figure 13.2 for an example of such a finite tree of domains with their truth-values.)

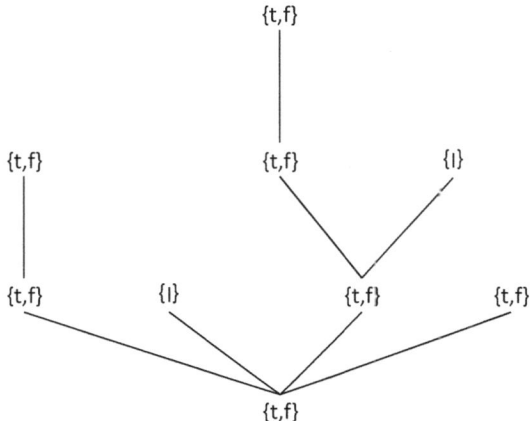

Figure 13.2: An example of an $RMI_\to$-tree

- Two propositions are relevant to each other iff their associated domains are related by the grading relation. A logical operator demands relevance if it produces a false proposition whenever it is applied to propositions which are not relevant to each other. $\to$ and $+$ should be operations of this sort.

Now we translate the above intuitions into formal definitions.

Definition 13.50 (tree) A *tree* is a lower semi-lattice $\langle S, \preceq \rangle$ [6] which has a minimal element, and for every $a \in S$ the set $\{b \in S \mid b \preceq a\}$ is linearly ordered by $\preceq$.[7]

Definition 13.51 ($\text{RMI}_\to$-matrix) An $\text{RMI}_\to$-*matrix* is a matrix $\mathcal{M} = \langle \mathcal{V}, \mathcal{D}, \mathcal{O} \rangle$ for $\mathcal{IL}$ such that:

- There is a tree $\langle S, \preceq \rangle$ such that $\mathcal{V}$ is a subset of $S \times \{t, f, I\}$ which satisfies the following conditions for every $a \in S$: $\langle a, t \rangle \in \mathcal{V}$ iff $\langle a, f \rangle \in$

---

[6] This means that $\preceq$ partially orders S, and for every $a, b \in S$ there is an element $inf_\preceq\{a, b\}$ (frequently denoted by $a \wedge b$) such that for every $x \in S$, $x \preceq inf_\preceq\{a, b\}$ iff $x \preceq a$ and $x \preceq b$.

[7] This is a slight generalization of the standard definition of a tree in mathematics, in which the set $\{b \in S \mid b \preceq a\}$ is usually demanded to be *well-ordered*, not just linearly ordered. In the finite case the two definitions are equivalent. Note also that the demand of having a minimal element is not really needed in what follows, and it anyway holds for any *finite* lower semi-lattice.

$\mathcal{V}$ iff $\langle a, I \rangle \notin \mathcal{V}$, and $\langle a, I \rangle \in \mathcal{V}$ only if a is a leaf (i.e. a $\preceq$-maximal element) of S.

- $\langle a, x \rangle \in \mathcal{D}$ iff $x \neq f$ (i.e., $x \in \{t, I\}$).

- The operations in $\mathcal{O}$ are the following:

  $\widetilde{\neg}\langle a, t\rangle = \langle a, f\rangle$, $\widetilde{\neg}\langle a, f\rangle = \langle a, t\rangle$, $\widetilde{\neg}\langle a, I\rangle = \langle a, I\rangle$.

  $u \widetilde{\otimes} v = \inf_{\preceq_\otimes}\{u, v\}$, where $\langle x, a\rangle \preceq_\otimes \langle y, b\rangle$ if either $x \prec y$ (i.e. $x \neq y$ and $x \preceq y$), or $x = y$ and $a \preceq^1 b$.

  $u \widetilde{\to} v = \widetilde{\neg}(u \widetilde{\otimes} \widetilde{\neg} v)$.

We shall refer to $\langle S, \preceq \rangle$ as the *tree on which $\mathcal{M}$ is based*. If $\nu = \langle a, x \rangle \in \mathcal{V}$ then we denote a by $dom(\nu)$ and call it the *domain* of ν, and we denote x by $bval(\nu)$ and call it the *basic value* of ν. Two elements a, b of S are *relevant* (in $\mathcal{M}$) if either $a \preceq b$ or $b \preceq a$. Two elements of $\mathcal{V}$ are *relevant* (in $\mathcal{M}$) if their domains are relevant.

Note 13.52 Every submatrix of either $\mathcal{A}_\omega$ or $\mathcal{SA}_m$ (including $\mathcal{A}_\omega$ and $\mathcal{SA}_m$) is isomorphic to an $\mathbf{RMI}_{\to}$-matrix. Thus $\mathcal{A}_\omega$ itself is isomorphic to an $\mathbf{RMI}_{\to}$-matrix which is based on the flat tree $\langle N \cup \{\bot\}, \preceq \rangle$, where N is the set of natural numbers, and $a \preceq b$ if either $a = b$ or $a = \bot$. In turn, $\mathcal{SA}_m$ is isomorphic to an $\mathbf{RMI}_{\to}$-matrix which is based on the tree $\langle [0, 1] \times (N \cup \{\bot\}), \preceq \rangle$, where $\langle a, x \rangle \preceq \langle b, y \rangle$ if either $a < b$ and $x = \bot$, or $a = b$ and either $x = y$ or $x = \bot$. In both cases $\langle a, I \rangle$ is a truth-value if and only if a is a leaf of the underlying tree (that is, $a \preceq b$ iff $b = a$).

Theorem 13.53 (strong completeness 2)

1. $\mathcal{T} \vdash_{\mathbf{RMI}_{\to}} \varphi$ iff $\mathcal{T} \vdash_{\mathcal{F}} \varphi$ (Definition 3.18), where $\mathcal{F}$ is the class of $\mathbf{RMI}_{\to}$-matrices.

2. If Γ is finite then $\Gamma \vdash_{\mathbf{RMI}_{\to}} \varphi$ iff $\mathcal{T} \vdash_{\mathcal{F}_f} \varphi$, where $\mathcal{F}_f$ is the class of finite $\mathbf{RMI}_{\to}$-matrices.

Proof. The proof of the soundness of $\mathbf{RMI}_{\to}$ for $\mathbf{RMI}_{\to}$-matrices is similar to the corresponding part of the proof of Theorem 13.47. Details are left for the reader. The rest of the theorem follows from Theorem 13.47, Theorem 13.48, and Note 13.52. □

Note 13.54 Proofs which are very similar to those of Theorems 13.47 and 13.53 show that:

1. A general semantics for $\mathbf{RMI}^1_{\to} = \mathbf{RM}_{\to}$ is given by $\mathbf{RMI}_{\to}$-matrices in which the underlying tree is linearly ordered.

13.1 A Maximal Strong Relevant Logic   431

2. A strongly characteristic matrix for this logic is given by the submatrix of $\mathcal{SA}_m$ whose set of truth-values is $([0,1) \times \{t,f\}) \cup \{\langle 1, I_1 \rangle\}$ (which is an $\mathbf{RMI}_\to$-matrix if we identify I_1 with I).[8]

Now we turn to discuss the expressive power of $\mathcal{IL}$ with respect to the general semantics of $\mathbf{RMI}_\to$-matrices. Here the most important observation is that the relation of relevance between formula is definable:

Definition 13.55 ($\mathcal{R}$) The binary connective $\mathcal{R}$ is defined as follows:

$$\varphi \mathcal{R} \psi =_{Df} (\varphi \to \varphi) + (\psi \to \psi).$$

Note 13.56 It is easy to verify that $\varphi \mathcal{R} \psi$ is equivalent in $\mathbf{RMI}_\to$ (or $\mathbf{R}_\to$) to each of the following formulas: $(\varphi \to \psi) + (\psi \to \varphi)$; $\varphi \to (\psi \to (\varphi + \psi))$; $\varphi \to (\psi \to (\neg \psi \to \varphi))$; $\neg \varphi + \varphi + \neg \psi + \psi$. Note that by the axiomatization given in Section 13.1.4, the addition of any of these formulas (including $\varphi \mathcal{R} \psi$ itself) to $\mathbf{RMI}_\to$ leads to $\mathbf{RM}_\to$.

The proof of the following proposition is straightforward:

Proposition 13.57 Let $\mathcal{M}$ be an $\mathbf{RMI}_\to$-matrix, and let ν be a valuation in $\mathcal{M}$. Then $\nu \models_\mathcal{M} \varphi \mathcal{R} \psi$ iff $\nu(\varphi)$ and $\nu(\psi)$ are relevant in $\mathcal{M}$.

Similarly, we have:

Proposition 13.58 Let $\mathcal{M}$ be an $\mathbf{RMI}_\to$-matrix which is based on the tree $\langle S, \preceq \rangle$, and let ν be a valuation in $\mathcal{M}$.

1. $dom(\nu(\varphi)) \preceq dom(\nu(\psi))$ iff $\nu \models_\mathcal{M} (\varphi \to \varphi) \to (\psi \to \psi)$.

2. $dom(\nu(\varphi)) = dom(\nu(\psi))$ iff $\nu \models_\mathcal{M} ((\varphi \to \varphi) \to (\psi \to \psi)) \otimes ((\psi \to \psi) \to (\varphi \to \varphi))$.

Now suppose that a sentence σ of $\mathcal{IL}$ is a propositional combination of two sentences φ and ψ. Propositions 13.57 and 13.58 imply that all the semantical facts that are needed in order to determine the truth or falsehood of σ (as well as other important facts about it) are expressible in $\mathcal{IL}$. This fact has important consequences. Suppose, for example, that we feel sufficiently safe to believe that $\varphi_1, \ldots, \varphi_m$ belong all to the same domain of relevance. In such a case we can include in the theory we use sentences that express this belief. Applying then the resulting theory to propositional

---

[8] It is not difficult to see that this matrix is isomorphic to the multiplicative Sugihara matrix $\mathcal{M}_m([0,1])$ which will be shown to be strongly characteristic for $\mathbf{RM}_\to$ in Theorem 15.15 below.

combinations of $\varphi_1, \ldots, \varphi_n$ is equivalent to applying classical logic to them. On the other hand, if we want to distinguish between, say, simple arithmetical facts and propositions concerning proper classes, then we just need to include in our theory sentences that express the view that propositions of the first kind precede those of the second kind in the partial order of priority. A contradiction that might be discovered concerning proper classes would not affect then our knowledge concerning arithmetical facts.

13.2 A Maximal Normal Relevant Logic

The logic $\mathbf{RMI}_\to$ studied in the previous section has the same big drawback that all strongly relevant logics have according to Theorem 11.53: it is not normal (and in particular has no conjunction). In this section we show that it has a single simple extension which is both normal and relevant (though not strongly relevant, of course): $\mathbf{L}_{\mathcal{A}_\omega}$[9] (where $\mathcal{A}_\omega$ is the matrix described in the previous section). This implies of course that this logic is maximal among those which have these two properties.

13.2.1 The Normal Simple Extensions of $\mathbf{RMI}_\to$

Our first goal is to characterize all the normal simple extensions of $\mathbf{RMI}_\to$. For this we need the following definition:

Definition 13.59 $\varphi \supset \psi =_{Df} \varphi \to \psi \otimes \varphi$

Theorem 13.60 *For every* $0 \leq n \leq \omega$, $\mathbf{L}_{\mathcal{A}_n}$ *is normal. In particular:* $\otimes$ *is a conjunction for* $\mathbf{L}_{\mathcal{A}_n}$, *and* $\supset$ *is an implication for it.*

Proof. It is easy to check that for every $0 \leq n \leq \omega$, $\otimes$ is an $\mathcal{A}_n$-conjunction, while $\supset$ is an $\mathcal{A}_n$-implication. Hence the theorem follows from Proposition 3.32 and Note 3.30. □

The next theorem shows that Theorem 13.60 provides *all* the normal simple extensions of $\mathbf{RMI}_\to$.

Theorem 13.61 *Let* $\mathbf{L}$ *be a normal simple finitary extension of* $\mathbf{L}_{\mathcal{A}_\omega}$. *Then there exists* $0 \leq n \leq \omega$ *such that* $\mathbf{L} = \mathbf{L}_{\mathcal{A}_n}$.

Proof. Let $Th(\mathbf{L})$ be the set of theorems of $\mathbf{L}$. By Theorems 13.19 and 13.31, there exists $0 \leq n \leq \omega$ such that $\varphi \in Th(\mathbf{L})$ iff $\vdash_{\mathcal{A}_n} \varphi$. We show that $\mathbf{L} = \mathbf{L}_{\mathcal{A}_n}$, that is: for every theory $\mathcal{T}$ and formula φ, $\mathcal{T} \vdash_{\mathcal{A}_n} \varphi$ iff $\mathcal{T} \vdash_{\mathbf{L}} \varphi$.

---
[9]This logic was called $\mathbf{SRMI}_\to$ in [43].

13.2 A Maximal Normal Relevant Logic

Since **L** is normal, it has an implication $\leadsto$.[10] Let $\tilde{\leadsto}$ be the interpretation of $\leadsto$ in $\mathcal{A}_n$. $\tilde{\leadsto}$ has the following properties:

(a) $x\tilde{\leadsto}y = I_i$ iff $x = y = I_i$.

  This is true in $\mathcal{A}_n$ for every definable connective of $\mathcal{IL}$.

(b) $t\tilde{\leadsto}t = f\tilde{\leadsto}f = t$.

  Since $\leadsto$ is an implication for **L**, $\vdash_{\mathbf{L}} p \leadsto p$. Hence also $\vdash_{\mathcal{A}_n} p \leadsto p$. This implies that $t\tilde{\leadsto}t \neq f$ and $f\tilde{\leadsto}f \neq f$, and so (b) follows from (a).

(c) $x\tilde{\leadsto}t = t$.

  Since $\leadsto$ is an implication for **L**, $\vdash_{\mathbf{L}} q \leadsto (p \leadsto p)$. Hence also $\vdash_{\mathcal{A}_n} q \leadsto (p \leadsto p)$. It follows that $x\tilde{\leadsto}(t\tilde{\leadsto}t) \neq f$, and so (a) and (b) imply that $x\tilde{\leadsto}t = t$.

(d) If $x \neq I_i$ then $x\tilde{\leadsto}I_i = t$.

  From the fact shown above that $\vdash_{\mathcal{A}_n} q \leadsto (p \leadsto p)$ it follows also that $x\tilde{\leadsto}(I_i\tilde{\leadsto}I_i) \neq f$. Hence, (a) and (b) imply that if $x \neq I_i$ then $x\tilde{\leadsto}I_i = t$.

(e) If $x \neq f$ then $x\tilde{\leadsto}f = f$.

  From (c) and (a) it follows that $(t\tilde{\leadsto}t)\tilde{\leadsto}t \neq f$ and $(I_i\tilde{\leadsto}I_i) \to I_i \neq f$. Therefore if $(f\tilde{\leadsto}f)\tilde{\leadsto}f \neq f$ as well, then $\vdash_{\mathcal{A}_n} (p \leadsto p) \leadsto p$. This in turn implies that $\vdash_{\mathbf{L}} (p \leadsto p) \leadsto p$. Since $\leadsto$ is an implication for **L**, this entails that $\vdash_{\mathbf{L}} p$, and so **L** is not a logic. A contradiction. It follows that $(f\tilde{\leadsto}f)\tilde{\leadsto}f = f$, and so $t\tilde{\leadsto}f = f$ by (b).

  To show that also $I_i\tilde{\leadsto}f = f$, observe that from the fact that $\leadsto$ is an implication for **L** it easily follows that $\vdash_{\mathbf{L}} ((q \leadsto q) \leadsto p) \leadsto ((p \leadsto r) \leadsto r)$. Hence $\vdash_{\mathcal{A}_n} ((q \leadsto q) \leadsto p) \leadsto ((p \leadsto r) \leadsto r)$. Using (a), this implies that $((t\tilde{\leadsto}t)\tilde{\leadsto}I_i)\tilde{\leadsto}((I_i\tilde{\leadsto}f)\tilde{\leadsto}f) = t$, and so $t\tilde{\leadsto}((I_i\tilde{\leadsto}f)\tilde{\leadsto}f) = t$ by (b) and (d). Therefore, had $I_i\tilde{\leadsto}f$ been equal to t, we would have got that $t\tilde{\leadsto}(t\tilde{\leadsto}f) = t$ which is impossible, since we have shown that $t\tilde{\leadsto}f = f$. Hence $I_i\tilde{\leadsto}f \neq t$, and so by (a) $I_i\tilde{\leadsto}f = f$.

Taken together, (a)–(e) mean that $\tilde{\leadsto} = \tilde{\supset}$, where $\supset$ is the implication connective of $L_{\mathcal{A}_n}$ given in Definition 13.59 and Proposition 13.60. It follows that $\vdash_{\mathcal{A}_n} (\varphi \leadsto \psi) \leftrightarrow (\varphi \supset \psi)$, and so also $\vdash_{\mathbf{L}} (\varphi \leadsto \psi) \leftrightarrow (\varphi \supset \psi)$. Since **L** is a simple extension of $\mathbf{RMI}_{\to}$, this implies that $\varphi \leadsto \psi$ and $\varphi \supset \psi$ are equivalent in **L**, and so $\supset$ is an implication for **L**.

---

[10] Recall that this means that $p \leadsto q$ is an abbreviation for a formula θ of $\mathcal{IL}$ such that $Fv(\theta) = \{p,q\}$, and for every $\mathcal{T}, \varphi, \psi$ it holds that $\mathcal{T}, \varphi \vdash_{\mathbf{L}} \psi$ iff $\mathcal{T} \vdash_{\mathbf{L}} \theta\{\varphi/p, \psi/q\}$.

Now, assume that $\mathcal{T} \vdash_{\mathcal{A}_n} \varphi$. Then by Theorem 3.17 and Theorem 13.73 below (whose proof does not depend on the present theorem), there exist $\psi_1, \ldots, \psi_\ell$ in $\mathcal{T}$ s.t. $\psi_1, \ldots, \psi_\ell \vdash_{\mathcal{A}_n} \varphi$. Hence, by Theorem 13.60, the formula $\psi_1 \supset (\psi_2 \supset (\cdots \supset (\psi_\ell \supset \varphi) \cdots))$ is valid in $\mathcal{A}_n$, and so it is also in $Th(\mathbf{L})$. On the other hand, the fact shown above that $\supset$ is an implication for $\mathbf{L}$ implies that $\psi_1, \ldots, \psi_\ell, \psi_1 \supset (\psi_2 \supset (\cdots \supset (\psi_\ell \supset \varphi) \cdots)) \vdash_\mathbf{L} \varphi$. It follows that $\psi_1, \ldots, \psi_\ell \vdash_\mathbf{L} \varphi$ and so $\mathcal{T} \vdash_\mathbf{L} \varphi$. The proof that if $\mathcal{T} \vdash_\mathbf{L} \varphi$ then $\mathcal{T} \vdash_{\mathcal{A}_n} \varphi$ is similar (since it is assumed that $\mathbf{L}$ is finitary). □

Corollary 13.62 *The only normal simple extension of* $\mathbf{RMI}_\rightarrow$ *which satisfies the minimal semantic relevance criterion is* $\mathbf{L}_{\mathcal{A}_\omega}$.

13.2.2 Properties of $\mathbf{L}_{\mathcal{A}_\omega}$

Corollary 13.62 means that the only candidate for being a simple extension of $\mathbf{RMI}_\rightarrow$ which is both relevant and normal is $\mathbf{L}_{\mathcal{A}_\omega}$. In this section we show that $\mathbf{L}_{\mathcal{A}_\omega}$ is indeed a relevant logic, and that it is also a classical paraconsistent logic. We also prove some other basic properties of it (like decidability), and provide a sound and complete Hilbert-type system for it.

Proposition 13.63

1. $\vdash_{\mathbf{RMI}_\rightarrow} \varphi$ iff $\vdash_{\mathcal{A}_\omega} \varphi$, i.e., $Th(\mathbf{RMI}_\rightarrow) = Th(\mathbf{L}_{\mathcal{A}_\omega})$.

2. $\mathbf{L}(\vdash^\rightarrow_{\mathbf{L}_{\mathcal{A}_\omega}}) = \mathbf{RMI}_\rightarrow$.

3. $\mathbf{L}(\vdash^\rightarrow_{\mathbf{L}_{\mathcal{A}_\omega}})$ is a strongly relevant logic.

4. $\rightarrow$ is a weak relevant implication for $\mathbf{L}_{\mathcal{A}_\omega}$, $\neg$ is an associated negation for $\rightarrow$, and $\otimes$ is an associated conjunction for it.

Proof. 1. is just Theorem 13.19. The second item follows from the first together with Corollary 13.20. The last two items follow from the second together with Theorem 13.29 (and its proof). □

Corollary 13.64 (decidability) *For a finite theory* Γ *and a formula* φ *it is decidable whether* $\Gamma \vdash_{\mathbf{L}_{\mathcal{A}_\omega}} \varphi$ *or not.*

Proof. Similar to that of Corollary 13.21. It is easy to check that if ν is a valuation in $\Lambda_{\mathcal{A}_\omega}$, and $\nu(p) \in \{t, f, I_{i_1}, \ldots, I_{i_k}\}$ for every p in $\mathsf{Var}(\Gamma \cup \{\varphi\})$, then $\nu(\psi) \in \{t, f, I_{i_1}, \ldots, I_{i_k}\}$ for every $\psi \in \Gamma \cup \{\varphi\}$. It follows that if the cardinality of $\mathsf{Var}(\Gamma \cup \{\varphi\})$ is n, then $\Gamma \vdash_{\mathbf{L}_{\mathcal{A}_\omega}} \varphi$ iff $\Gamma \vdash_{\mathbf{L}_{\mathcal{A}_n}} \varphi$. The latter is decidable, since $\mathcal{A}_n$ is finite. □

Proposition 13.65 $\mathbf{L}_{\mathcal{A}_\omega}$ *satisfies the basic relevance criterion.*

13.2 A Maximal Normal Relevant Logic

Proof. Suppose $\mathcal{T}_1, \mathcal{T}_2 \vdash_{\mathbf{L}_{\mathcal{A}_\omega}} \psi$, and $\mathcal{T}_2$ has no variables in common with $\mathcal{T}_1 \cup \{\psi\}$. Let ν be a model of $\mathcal{T}_1$ in $\mathcal{A}_\omega$. Define a new valuation ν' in $\mathcal{A}_\omega$ by letting $\nu'(p) = I_1$ if p occurs in $\mathcal{T}_2$, and $\nu'(p) = \nu(p)$ otherwise. Then $\nu'(\varphi) = I_1$ for every $\varphi \in \mathcal{T}_2$, and $\nu'(\varphi) = \nu(\varphi)$ for every $\varphi \in \mathcal{T}_1 \cup \{\psi\}$ (because $\mathcal{T}_2$ has no variables in common with $\mathcal{T}_1 \cup \{\psi\}$). It follows that ν' is a model of $\mathcal{T}_1 \cup \mathcal{T}_2$, and so it is a model of ψ. Since $\nu(\psi) = \nu'(\psi)$, ν is a model of ψ. It follows that $\mathcal{T}_1 \vdash_{\mathbf{L}_{\mathcal{A}_\omega}} \psi$. □

Corollary 13.66 $\mathbf{L}_{\mathcal{A}_\omega}$ *is boldly paraconsistent.*

Proof. This follows from Theorem 12.9, Theorem 13.60, Proposition 13.65, and Corollary 11.4. □

Next, we provide a Hilbert-type system for $\mathbf{L}_{\mathcal{A}_\omega}$, and use it to show that $\mathbf{L}_{\mathcal{A}_\omega}$ is finitary.

Definition 13.67 ($SRMI_\to$) $SRMI_\to$ is the system obtained from $HRMI_\to$ by adding the following elimination rule for $\otimes$ as a rule of inference:

$$[\otimes E] \quad \frac{\varphi \otimes \psi}{\varphi}.$$

Proposition 13.68 *The rule* $[\otimes E]$ *is valid in* $\mathcal{A}_\omega$, *but it is not derivable even in* $RM_\to$.

Proof. The validity of $[\otimes E]$ immeiately follows from the definition of $\widetilde{\otimes}$. On the other hand, the formula $P_1 \otimes P_2 \to P_1$ is not valid in $\mathcal{A}_1$. (Take $\nu(P_1) = I_1$, $\nu(P_2) = I_2$.) Hence this formula is not a theorem of $RM_\to$. Therefore, it easily follows from the relevant deduction theorem for $RM_\to$ that $[\otimes E]$ is not derivable in it. □

The validity of $[\otimes E]$ for $\vdash_{\mathcal{A}_\omega}$ was already observed at the beginning of Section 13.1.6. This immediately implies the soundness of $SRMI_\to$ for $\vdash_{\mathcal{A}_\omega}$. To prove that it is also strongly complete, we first show that $\supset$ is an implication for the logic induced by $SRMI_\to$.

Theorem 13.69 (deduction theorem for $SRMI_\to$) $\mathcal{T}, \varphi \vdash_{SRMI_\to} \psi$ iff $\mathcal{T} \vdash_{SRMI_\to} \varphi \supset \psi$. *The same is true for any axiomatic extension of* $SRMI_\to$.

Proof. Obviously, $\varphi, \varphi \supset \psi \vdash_{SRMI_\to} \psi$. Hence the "if" part.

For the converse, let $\mathbf{L}$ be any axiomatic extension of $SRMI_\to$. We prove by induction on the length of a derivation in $\mathbf{L}$ of ψ from $\mathcal{T} \cup \{\varphi\}$, that $\mathcal{T} \vdash_{\mathbf{L}} \varphi \supset \psi$. If ψ is an axiom of $\mathbf{L}$ or $\psi \in \mathcal{T}$ then we use the fact that $\vdash_{R_\to} \psi \to (\varphi \supset \psi)$. If $\psi = \varphi$ we use the fact that $\vdash_{R_\to} \varphi \supset \varphi$. In the induction step we have two cases. Suppose, first, that ψ was inferred from θ and $\theta \to \psi$. By induction hypothesis, $\mathcal{T} \vdash_{\mathbf{L}} \varphi \supset \theta$ and $\mathcal{T} \vdash_{\mathbf{L}} \varphi \supset (\theta \to \psi)$. But $\vdash_{RMI_{\to}} (\varphi \supset \theta) \to ((\varphi \supset (\theta \to \psi)) \to (\varphi \supset \psi))$. (Use $GRMI_{\to}$ or $\mathcal{A}_\omega$.) It follows that $\mathcal{T} \vdash_{\mathbf{L}} \varphi \supset \psi$ in this case. Finally, assume that ψ was inferred from $\psi \otimes \theta$. By induction hypothesis, $\mathcal{T} \vdash_{\mathbf{L}} \varphi \supset \psi \otimes \theta$. Since Theorem 13.60 easily implies that $\vdash_{\mathcal{A}_\omega} (\varphi \supset \psi \otimes \theta) \supset (\varphi \supset \psi)$, the weak completeness of $HRMI_{\to}$ for $\mathcal{A}_\omega$ entails that $\mathcal{T} \vdash_{\mathbf{L}} \varphi \supset \psi$. □

Theorem 13.70 (completeness of $SRMI_{\to}$) $\mathcal{T} \vdash_{\mathcal{A}_\omega} \varphi$ iff $\mathcal{T} \vdash_{SRMI_{\to}} \varphi$.

Proof. The soundness part is easy.

For completeness, assume $\mathcal{T} \not\vdash \varphi$ (we shall write just "$\vdash$" instead of "$\vdash_{SRMI_{\to}}$" until the end of this proof). We construct a valuation ν which shows that $\mathcal{T} \not\vdash_{\mathcal{A}_\omega} \varphi$. For this extend first $\mathcal{T}$ to a maximal theory $\mathcal{T}^*$ such that $\mathcal{T}^* \not\vdash \varphi$. By the deduction theorem for $SRMI_{\to}$ (Theorem 13.69), $\mathcal{T}^* \not\vdash \sigma$ iff $\mathcal{T} \vdash \sigma \supset \varphi$.

Fact 1. $\mathcal{T}^* \vdash \varphi \to \sigma$ for all σ.

Proof. Otherwise $\mathcal{T}^* \vdash (\varphi \to \sigma) \supset \varphi$ for some σ. But

$$\vdash_{\mathcal{A}_\omega} ((\varphi \to \sigma) \supset \varphi) \to \varphi \otimes (\sigma \to \sigma).$$

Thus, it follows from Theorem 13.19 that $\mathcal{T}^* \vdash \varphi$. A contradiction.

Fact 2. $\mathcal{T}^* \not\vdash \sigma$ iff $\mathcal{T}^* \vdash \sigma \to \varphi$.

Proof. The "if" part is obvious. For the converse, suppose $\mathcal{T} \not\vdash \sigma$. Then $\mathcal{T}^* \vdash \sigma \supset \varphi$. By definition of $\supset$, this means that $\mathcal{T}^* \vdash \sigma \to \varphi \otimes \sigma$. But $\mathcal{T}^* \vdash \varphi \to (\sigma \to \varphi)$ by Fact 1, and so $\mathcal{T}^* \vdash \varphi \otimes \sigma \to \varphi$. It follows that $\mathcal{T}^* \vdash \sigma \to \varphi$, by transitivity.

Fact 3. $\mathcal{T}^* \vdash \sigma \to \neg\varphi$, for every σ.

Proof. Immediate from Fact 1, by contraposition and double negation.

Fact 4. $\mathcal{T}^* \vdash \neg\varphi$.

Proof. By Fact 3, taking $\sigma = (\varphi \to \varphi)$.

Fact 5. $\mathcal{T}^* \vdash \neg\varphi \to \sigma$ iff ($\mathcal{T}^* \vdash \sigma$ and $\mathcal{T}^* \not\vdash \neg\sigma$) iff $\mathcal{T}^* \not\vdash \neg\sigma$.

Proof. $\mathcal{T}^* \vdash \neg\varphi \to \sigma$ iff $\mathcal{T}^* \vdash \neg\sigma \to \varphi$ iff $\mathcal{T}^* \not\vdash \neg\sigma$ (by Fact 2). In addition, Fact 4 implies that if $\mathcal{T}^* \vdash \neg\varphi \to \sigma$ then $\mathcal{T}^* \vdash \sigma$.

Construct now the Lindenbaum algebra of $\mathcal{T}^*$ by defining $\sigma \equiv \tau$ iff $\mathcal{T}^* \vdash \sigma \to \tau$ and $\mathcal{T}^* \vdash \tau \to \sigma$. This is obviously a congruence relation. Define then the operations on the set of equivalence classes in the obvious way (e.g. $[\sigma] \to [\tau] = [\sigma \to \tau]$ etc., where $[\sigma]$ is the equivalence class of σ).

13.2 A Maximal Normal Relevant Logic

We shall denote $[\varphi]$ by f, $[\neg\varphi]$ by t. Obviously, $t \neq f$, since $\mathcal{T}^* \vdash \neg\varphi$ while $\mathcal{T}^* \not\vdash \varphi$. We shall denote the other equivalence classes by $I_1, I_2, \ldots$ (The set of these classes is obviously enumerable.)

Fact 6. $[\sigma] = I_n$ for some n iff both $\mathcal{T}^* \not\vdash \sigma$ and $\mathcal{T}^* \not\vdash \neg\sigma$.

Proof. Suppose that $[\sigma] = I_n$. Then $[\sigma] \neq f = [\varphi]$ and so $\mathcal{T}^* \not\vdash \sigma \to \varphi$, by Fact 1. This means, by Fact 2, that $\mathcal{T} \not\vdash \sigma$. Similarly, $[\sigma] \neq t = [\neg\varphi]$, and so, by Fact 3, $\mathcal{T} \not\vdash \neg\varphi \to \sigma$. This entails, by Fact 5, that $\mathcal{T}^* \not\vdash \neg\sigma$.

For the converse, assume both σ and $\neg\sigma$ are provable. Then $\sigma \not\equiv \varphi$ since $\mathcal{T}^* \not\vdash \varphi$, and $\sigma \not\equiv \neg\varphi$, since $\mathcal{T}^* \not\vdash \neg\neg\varphi$ by the double negation axiom. Hence $[\sigma] \neq f$ and $[\sigma] \neq t$, and so $[\sigma] = I_k$ for some k.

Next, we show that the Lindenbaum algebra of $\mathcal{T}^*$ is exactly $\mathcal{A}_\omega$ by showing that the operations are identical to those of $\mathcal{A}_\omega$.

- *The case of $\neg$:* Obviously $\neg f = t$ and $\neg t = f$. Now let $I_n = [\sigma]$. By Fact 6, $\mathcal{T}^* \vdash \sigma$ and $\mathcal{T}^* \vdash \neg\sigma$. But both $\sigma \to (\neg\sigma \to (\sigma \to \neg\sigma))$ and $\sigma \to (\neg\sigma \to (\neg\sigma \to \sigma))$ are theorems of $\mathbf{RMI}_{\vec{\to}}$. Hence $\sigma \equiv \neg\sigma$ and $\neg I_n = [\neg\sigma] = [\sigma]$.

- *The case of $\otimes$:* Since $\varphi \otimes \sigma \vdash \varphi$, $\mathcal{T}^* \not\vdash \varphi \otimes \sigma$ and so $[\varphi \otimes \sigma] = f$ by the first two facts. Hence $f \otimes [\sigma] = f$ for all σ. Similarly, $[\sigma] \otimes f = f$. If $[\sigma] = [\tau]$ then $[\sigma] \otimes [\tau] = [\sigma] \otimes [\sigma] = [\sigma \otimes \sigma] = [\sigma]$, since $\sigma \to \sigma \otimes \sigma$ and $\sigma \otimes \sigma \to \sigma$ are theorems of $\mathbf{RMI}_{\vec{\to}}$. Finally, suppose $[\sigma] \neq f$, $[\tau] \neq f$ and $[\sigma] \neq [\tau]$. Then $\mathcal{T}^* \vdash \sigma$ and $\mathcal{T}^* \vdash \tau$. Assume for contradiction that $t \neq [\sigma] \otimes [\tau] (= [\sigma \otimes \tau])$. Then $\mathcal{T}^* \vdash \neg(\sigma \otimes \tau)$, by Facts 3 and 5. But $\vdash_{RMI_{\vec{\to}}} \sigma \to (\tau \to \neg(\sigma \otimes \tau) \to (\sigma \to \tau))$ and $\vdash_{RMI_{\vec{\to}}} \sigma \to (\tau \to \neg(\sigma \otimes \tau) \to (\tau \to \sigma))$. Hence, $\sigma \equiv \tau$, contradicting $[\sigma] \neq [\tau]$.

- *The case of $\to$:* Since $\sigma \to \tau$ and $\neg(\sigma \otimes \neg\tau)$ are equivalent in $\mathbf{RMI}_{\vec{\to}}$, and the corresponding operations are identical on $\mathcal{A}_\omega$, this case follows from the previous two.

Define now $\nu(\sigma) = [\sigma]$. This is easily seen to be a valuation (the canonical one). Obviously $\nu(\sigma) \neq f$ for $\sigma \in \mathcal{T}$, while $\nu(\varphi) = f$. Hence $\mathcal{T} \not\vdash_{\mathcal{A}_\omega} \varphi$. □

Corollary 13.71 *Let $SRMI_{\vec{\to}}^n$ be obtained from $HRMI_{\vec{\to}}^n$ by adding $[\otimes E]$ as a second rule of inference. Then $SRMI_{\vec{\to}}^n$ is a strongly sound and complete axiomatization of $\mathbf{L}_{\mathcal{A}_n}$.*

Proof. $SRMI_{\vec{\to}}^n$ is an axiomatic simple extension of $SRMI_{\vec{\to}}$. The logic induced by the latter is normal by Theorems 13.70 and 13.60. Hence so is the logic induced by $SRMI_\to^n$. Therefore, the corollary easily follows from Proposition 13.38, and the proof of Theorem 13.61. □

Note 13.72 $\mathbf{L}_{\mathcal{A}_1}$ is identical to the logic $\mathbf{SRM}_\to$ investigated in Chapter 4 (see Sections 4.4.2 and 4.5.1). The last corollary provides the Hilbert-type system for it promised in Section 4.5.2 of that chapter. By the results of Section 13.1.4, it is obtained from $HRMI_\to$ by adding to it the axiom schema $\neg(\varphi \to \varphi) \to (\psi \to \psi)$ and the rule $[\otimes E]$.

Theorem 13.73 (compactness theorem) $\mathcal{T} \vdash_{\mathcal{A}_\omega} \psi$ iff there exists a finite $\Gamma \subseteq \mathcal{T}$ such that $\Gamma \vdash_{\mathcal{A}_\omega} \psi$.

Proof. Immediate from Theorem 13.70. □

Corollary 13.74 $\mathcal{T} \vdash_{\mathcal{A}_\omega} \varphi$ iff $\mathcal{T} \vdash_{\mathbf{RMI}_\to} \varphi \otimes \psi$ for some formula ψ.

Proof. Obviously, if such ψ exists then $\mathcal{T} \vdash_{SRMI_\to} \varphi$, and so $\mathcal{T} \vdash_{\mathcal{A}_\omega} \varphi$. Conversely, if $\mathcal{T} \vdash_{\mathcal{A}_\omega} \varphi$ then by Theorem 13.73 there exists $\Gamma = \{\psi_1, \ldots, \psi_n\} \subseteq \mathcal{T}$ such that $\Gamma \vdash_{\mathcal{A}_\omega} \varphi$. By Theorems 13.60 this implies that $\vdash_{\mathcal{A}_\omega} \psi \to \varphi \otimes \psi$, where $\psi = \psi_1 \otimes \psi_2 \otimes \cdots \otimes \psi_n$. It follows by Theorem 13.19 that $\vdash_{HRMI_\to} \psi \to \varphi \otimes \psi$, and so $\psi_1, \ldots, \psi_n \vdash_{HRMI_\to} \varphi \otimes \psi$. Hence $\mathcal{T} \vdash_{\mathbf{RMI}_\to} \varphi \otimes \psi$. □

Note 13.75 We have shown, in fact, a stronger result: If $\varphi_1, \ldots, \varphi_n \vdash_{\mathcal{A}_\omega} \psi$ then there is a proof of this fact which is almost entirely in $HRMI_\to$, except for one application of $[\otimes E]$ at the very end.

Theorem 13.76 $\mathbf{L}_{\mathcal{A}_\omega}$ is a maximal normal relevant logic.

Proof. That $\mathbf{L}_{\mathcal{A}_\omega}$ is relevant follows from Theorem 13.73 and Propositions 13.63, 13.65. That it is normal follows from Theorem 13.60, and that it is maximal with respect to these properties follows from Corollary 13.62. □

We end this section by showing that although $\mathbf{L}_{\mathcal{A}_\omega}$ is a relevant logic, it can also be viewed as a classical paraconsistent logic.

Proposition 13.77 $\mathbf{L}_{\mathcal{A}_\omega}$ is $\neg$-contained in classical logic, and $\neg$ is complete for it.

Proof. This follows from Proposition 3.47 and Corollary 3.44, since $\mathcal{A}_\omega$ is obviously a $\neg$-subclassical f-matrix.[11] □

Proposition 13.78 Let $\varphi \vee \psi = (\varphi \supset \psi) \supset \psi$. Then $\mathbf{L}_{\mathcal{A}_\omega}$ is a classical $\neg$-paraconsistent logic. Moreover, if $\varphi \vee \psi = (\varphi \supset \psi) \supset \psi$, then the $\{\supset, \vee, \otimes\}$-fragment of $\mathbf{L}_{\mathcal{A}_\omega}$ is identical to $\mathbf{CL}^+$ (where $\wedge$ is replaced by $\otimes$).

Proof. This follows from Proposition 3.54, using Corollary 13.66, Proposition 13.77, Theorem 13.60 (and its proof), and Theorem 13.73. □

---

[11] Note that since $\mathcal{A}_\omega$ is a $\neg$-subclassical f-matrix, it also has all the other properties that were proved in Chapter 3 for such matrices, like Propositions 3.36 and 3.49.

13.2.3 Gentzen-type System for $\mathbf{L}_{\mathcal{A}_\omega}$

As we have seen, the system $GRMI_{\rightarrow}$ is only *weakly* complete with respect to $\mathcal{A}_\omega$. In order to construct a cut-free Gentzen-type system which is *strongly* complete we need to use the richer data structure of hypersequents, which is described in Section 1.3.4. (See Definition 1.143.)

In Figure 13.3 we present a hypersequential Gentzen-type proof system $GSRMI_{\rightarrow}$ for $\mathbf{L}_{\mathcal{A}_\omega}$. (Note that like in Note 13.4, Γ, Δ denote here sets.)

Axioms: $\psi \Rightarrow \psi$

Logical rules:

$$[\neg \Rightarrow] \ \frac{G \mid \Gamma \Rightarrow \Delta, \varphi}{G \mid \neg\varphi, \Gamma \Rightarrow \Delta} \qquad [\Rightarrow \neg] \ \frac{G \mid \varphi, \Gamma \Rightarrow \Delta}{G \mid \Gamma \Rightarrow \Delta, \neg\varphi}$$

$$[\otimes \Rightarrow] \ \frac{G \mid \Gamma, \varphi, \psi \Rightarrow \Delta}{G \mid \Gamma, \varphi \otimes \psi \Rightarrow \Delta} \qquad [\Rightarrow \otimes] \ \frac{G \mid \Gamma_1 \Rightarrow \Delta_1, \varphi \quad G \mid \Gamma_2 \Rightarrow \Delta_2, \psi}{G \mid \Gamma_1, \Gamma_2 \Rightarrow \Delta_1, \Delta_2, \varphi \otimes \psi}$$

$$[\rightarrow \Rightarrow] \ \frac{G \mid \Gamma_1 \Rightarrow \Delta_1, \varphi \quad G \mid \psi, \Gamma_2 \Rightarrow \Delta_2}{G \mid \Gamma_1, \Gamma_2, \varphi \rightarrow \psi \Rightarrow \Delta_1, \Delta_2} \qquad [\Rightarrow \rightarrow] \ \frac{G \mid \Gamma, \varphi \Rightarrow \Delta, \psi}{G \mid \Gamma \Rightarrow \Delta, \varphi \rightarrow \psi}$$

Structural rules:

$$\frac{G \mid s \mid s}{G \mid s}$$

$$\frac{G \mid \Gamma_1 \Rightarrow \Delta_1, \varphi \quad G \mid \varphi, \Gamma_2 \Rightarrow \Delta_2}{G \mid \Gamma_1, \Gamma_2 \Rightarrow \Delta_1, \Delta_2}$$

$$\frac{G \mid \Gamma_1, \Gamma_2 \Rightarrow \Delta_1, \Delta_2}{G \mid \Gamma_1 \Rightarrow \Delta_1 \mid \Gamma_2, \Gamma' \Rightarrow \Delta_2, \Delta'}$$

(external contraction [EC], cut, and strong splitting [SS], respectively).

Figure 13.3: The proof system $GSRMI_{\rightarrow}$

Note 13.79 The rule [SS] is another example of how the use of hypersequents allows to introduce completely new types of useful structural rules.

Note 13.80 As noted in Section 1.3.4, in hypersequential calculi we have a choice between treating side components in a multiplicative manner or in an additive one. For $\mathbf{L}_{\mathcal{A}_\omega}$ the choice of the additive version has proved to be better, since it makes the problematic rule of external contraction superfluous (Theorem 13.89).

Example 13.81 The following is a proof in $GSRMI_{\to}$ of the hypersequent $\Rightarrow \varphi \mid (\varphi \to \psi) \to \varphi \Rightarrow \varphi$:[12]

$$\dfrac{\dfrac{\dfrac{\varphi \Rightarrow \varphi}{\Rightarrow \varphi \mid \varphi \Rightarrow \psi} \; [SS]}{\Rightarrow \varphi \mid \Rightarrow \varphi \to \psi} \; [\Rightarrow\to] \qquad \varphi \Rightarrow \varphi}{\Rightarrow \varphi \mid (\varphi \to \psi) \to \varphi \Rightarrow \varphi} \; [\to\Rightarrow]$$

Example 13.82 Below is a proof in $GSRMI_{\to}$ of a single-conclusion sequent, which is valid in $\mathcal{A}_\omega$, and by Corollary 13.74 indirectly says that $(\varphi \to \psi) \to \varphi \vdash_{\mathcal{A}_\omega} \varphi$. Note that the proof here uses proper hypersequents, but unlike the proofs of this sequent in $GRMI_{\to}$, they have only single-conclusion components. (Note that in this example the two sides of a sequent are treated as multisets of formulas rather than sets.)

$$\dfrac{\dfrac{\dfrac{\dfrac{\dfrac{\dfrac{\dfrac{\dfrac{\dfrac{\varphi \Rightarrow \varphi}{\varphi,\varphi \Rightarrow \varphi} \quad \psi \Rightarrow \psi}{\varphi,\varphi,\varphi \to \psi \Rightarrow \psi}}{\varphi \to \psi \Rightarrow \varphi \to \psi \quad (\varphi \to \psi) \to \varphi, \; \varphi \to \psi, \; \varphi \to \psi, \; \varphi \Rightarrow \psi}}{(\varphi \to \psi) \to \varphi, \; \varphi \to \psi, \; \varphi \Rightarrow \psi}}{\dfrac{(\varphi \to \psi) \to \varphi, \; \varphi \to \psi \Rightarrow \varphi \to \psi}{(\varphi \to \psi) \to \varphi \Rightarrow (\varphi \to \psi) \to (\varphi \to \psi)}}}{\dfrac{(\varphi \to \psi) \to \varphi \Rightarrow \varphi \otimes \big((\varphi \to \psi) \to (\varphi \to \psi)\big)}{\cdots}}}{(\varphi \to \psi) \to \varphi \Rightarrow \varphi \otimes \big((\varphi \to \psi) \to (\varphi \to \psi)\big)}$$

Next we turn to the semantics of $GSRMI_{\to}$.

Definition 13.83

1. A valuation ν is a *model* in $\mathcal{A}_\omega$ of a sequent $\Gamma \Rightarrow \Delta$ if either $\nu(\varphi) = f$ for some $\varphi \in \Gamma$, or $\nu(\psi) = t$ for some $\psi \in \Delta$, or $\Gamma \Rightarrow \Delta$ is not empty and there exists k such that $\nu(\varphi) = I_k$ for all $\varphi \in \Gamma \cup \Delta$, or $\Gamma \Rightarrow \Delta$ is empty and there exists k such that $\nu(P) = I_k$ for some variable P (i.e., ν is a model of the empty sequent iff it is not a classical valuation).

2. A valuation ν is a *model* of a hypersequent G in $\mathcal{A}_\omega$ ($\nu \models_{\mathcal{A}_\omega} G$) if ν is a model in $\mathcal{A}_\omega$ of at least one of the components of G.

3. A hypersequent G is valid in $\mathcal{A}_\omega$ ($\vdash_{\mathcal{A}_\omega} G$) if every valuation in $\mathcal{A}_\omega$ is a model of G.

---

[12]This is a direct proof in $GSRMI_{\to}$ that $(\varphi \to \psi) \to \varphi \vdash_{\mathcal{A}_\omega} \varphi$. See Theorem 13.86.

13.2 A Maximal Normal Relevant Logic

Note 13.84 It is easy to see that if s is an ordinary non-empty sequent, then $\nu \models_{\mathcal{A}_\omega} s$ iff $\nu(\psi_s) \neq f$, where ψ_s is any translation of s into a formula (see the proof of Proposition 13.5). This and the weak soundness and completeness of $\mathbf{RMI}_\to$ easily imply that $\vdash_{GRMI_\to} s$ iff $\vdash_{\mathcal{A}_\omega} s$. The next theorem shows that the situation with respect to $GSRMI_\to$ and hypersequents is similar.

Proposition 13.85 *If* $\vdash_{GSRMI_\to} G$ *then* $\vdash_{\mathcal{A}_\omega} G$.

Proof. For the proof, it suffices to show that the axioms of $GSRMI_\to$ are valid in $\mathcal{A}_\omega$, and all its rules are *truth* preserving: every model of all the premises of a rule is also a model of its conclusion. This is straightforward. We only note that the cut rule is non-trivially sound here even in case the resulting component is empty. Indeed, ν can be a model of both $\Rightarrow \varphi$ and $\varphi \Rightarrow$ only if $\nu(\varphi) = I_k$ for some k. Such ν is not a classical valuation, and so it is a model of $\Rightarrow$ as well. □

Next we use hypersequents and $GSRMI_\to$ to characterize the consequence relation $\vdash_{\mathcal{A}_\omega}$ between *formulas*.

Theorem 13.86 *Let* $\Gamma = \{\psi_1, \ldots, \psi_n\}$. *The following are equivalent:*

1. $\vdash_{GSRMI_\to} \Rightarrow \varphi \mid \Gamma \Rightarrow \varphi$.

2. $\vdash_{GSRMI_\to} \Rightarrow \varphi \mid \psi_1 \Rightarrow \varphi \mid \cdots \mid \psi_n \Rightarrow \varphi$.

3. *There are* $\Gamma_1, \ldots, \Gamma_k \subseteq \Gamma$ *such that* $\vdash_{GSRMI_\to} \Gamma_1 \Rightarrow \varphi \mid \cdots \mid \Gamma_k \Rightarrow \varphi$.

4. $\Gamma \vdash_{\mathcal{A}_\omega} \varphi$.

Proof. Item 2 follows from Item 1 by a repeated use of strong splitting ($[SS]$). That Item 2 implies Item 3 is trivial, and Item 4 easily follows from Item 3 using Proposition 13.85.

Finally, to show that Item 4 implies Item 1, assume that $\Gamma \vdash_{\mathcal{A}_\omega} \varphi$. By Corollary 13.74, this implies that there is $\Delta \subseteq \Gamma$ and ψ such that $\vdash_{GRMI_\to} \Delta \Rightarrow \varphi \otimes \psi$. A cut on $\varphi \otimes \psi$ of this with $\varphi \otimes \psi \Rightarrow \varphi \mid \Rightarrow \varphi$ (which is easily derivable from $\varphi \Rightarrow \varphi$ using $[SS]$ followed by $[\otimes \Rightarrow]$) yields $\Delta \Rightarrow \varphi \mid \Rightarrow \varphi$. From this, $\Rightarrow \varphi \mid \Gamma \Rightarrow \varphi$ follows by applications of $[SS]$ and $[EC]$ (external contraction). □

Example 13.87 The last theorem implies that Example 13.81 contains a direct proof in $GSRMI_\to$ that $(\varphi \to \psi) \to \varphi \vdash_{\mathcal{A}_\omega} \varphi$. (Recall that Example 13.82 indirectly shows the same.)

Corollary 13.88 *Let $\mathcal{T}$ be a theory and φ a formula. The following conditions on $\mathcal{T}$ and φ are equivalent:*

1. *There is a finite $\Gamma \subseteq \mathcal{T}$ such that $\vdash_{GSRMI_{\to}} \Rightarrow \varphi \mid \Gamma \Rightarrow \varphi$.*

2. *$\vdash_{GSRMI_{\to}} \Rightarrow \varphi \mid \psi_1 \Rightarrow \varphi \mid \cdots \mid \psi_n \Rightarrow \varphi$ for some $\psi_1, \ldots, \psi_n \in \mathcal{T}$.*

3. *There are $\Gamma_1, \ldots, \Gamma_k \subseteq \mathcal{T}$ such that $\vdash_{GSRMI_{\to}} \Gamma_1 \Rightarrow \varphi \mid \cdots \mid \Gamma_k \Rightarrow \varphi$.*

4. *$\mathcal{T} \vdash_{\mathcal{A}_\omega} \varphi$.*

Proof. Immediate from Theorem 13.86 and Theorem 13.73. □

Our next goals are to show the completeness of $GSRMI_{\to}$, and that both cut and external contraction are superfluous in it.

Theorem 13.89 *If a hypersequent G is valid in $\mathcal{A}_\omega$ then G has a proof in $GSRMI_{\to}$ in which there are no cuts or external contractions.*

Proof. We start with some notations and definitions. First, in this proof $\vdash G$ means that G has a proof in $GSRMI_{\to}$ in which cut and external contraction are not used. Second, If $G = \Gamma_1 \Rightarrow \Delta_1 \mid \ldots \mid \Gamma_n \Rightarrow \Delta_n$ then we denote by G_i the hypersequent which is obtained from G by deleting $\Gamma_i \Rightarrow \Delta_i$. (Hence, $G = \Gamma_i \Rightarrow \Delta_i \mid G_i$ up to the order of components. Note that G_i may be empty.) Finally, we say that a hypersequent $\Gamma'_1 \Rightarrow \Delta'_1 \mid \ldots \mid \Gamma'_n \Rightarrow \Delta'_n$ *relevantly extends* the hypersequent $\Gamma_1 \Rightarrow \Delta_1 \mid \ldots \mid \Gamma_n \Rightarrow \Delta_n$ if for all $1 \leq i \leq n$ we have that $\Gamma_i \subseteq \Gamma'_i$, $\Delta_i \subseteq \Delta'_i$, and every formula in $\Gamma'_i \Rightarrow \Delta'_i$ is a subformula of some formula in $\Gamma_i \Rightarrow \Delta_i$. It is easy to see that relevant extension is a transitive relation: if G_1 relevantly extends G_2, and G_2 relevantly extends G_3, then G_1 relevantly extends G_3.

Lemma 13.89A. *A model of a hypersequent G is also a model of every relevant extension of G.*

Proof. Let ν be a model of G and let G' be a relevant extension of G. Then ν is a model of some component $\Gamma_i \Rightarrow \Delta_i$ of G. If $\nu(\psi) = f$ for some $\psi \in \Gamma_i$, or $\nu(\psi) = t$ for some $\psi \in \Delta_i$, then the same is true for the corresponding component $\Gamma'_i \Rightarrow \Delta'_i$ of G'. If $\nu(\psi) = I_k$ for all formulas of $\Gamma_i \Rightarrow \Delta_i$ (including the case in which $\Gamma_i \Rightarrow \Delta_i$ is empty), then the same is again true for $\Gamma'_i \Rightarrow \Delta'_i$, since it consists only of subformulas of formulas in $\Gamma_i \Rightarrow \Delta_i$. In either case ν is also a model of $\Gamma'_i \Rightarrow \Delta'_i$ and so of G'. □

Next, call a hypersequent G such that $\not\vdash G$ *saturated* if every component $\Gamma_i \Rightarrow \Delta_i$ of G satisfies the following conditions:

13.2 A Maximal Normal Relevant Logic

(i) If $\neg\varphi \in \Gamma_i$ then $\varphi \in \Delta_i$.

(ii) If $\neg\varphi \in \Delta_i$ then $\varphi \in \Gamma_i$.

(iii) If $\varphi \otimes \psi \in \Gamma_i$ then $\varphi \in \Gamma_i$ and $\psi \in \Gamma_i$.

(iv) If $\varphi \otimes \psi \in \Delta_i$ and $\not\vdash \Gamma_i \Rightarrow \Delta_i, \varphi \mid G_i$ then $\varphi \in \Delta_i$.

(v) If $\varphi \otimes \psi \in \Delta_i$ and $\not\vdash \Gamma \Rightarrow \Delta_i, \psi \mid G_i$ then $\psi \in \Delta_i$.

Lemma 13.89B. *If $\not\vdash G$ then there is a saturated relevant extension G' of G such that $\not\vdash G'$.*

Proof. If $\not\vdash G$ and G is not saturated, then it is possible to properly and relevantly extend G to a hypersequent G^* such that $\not\vdash G^*$. (This is obvious and standard if one of the conditions (i)–(iii) is violated by some component $\Gamma \Rightarrow \Delta$ of G, and is trivial in the special cases (iv)–(v).) Since G has only finitely many subformulas, this process must stop (because of the transitivity of the relation of relevant extension) with a saturated sequent G' such that G' relevantly extends G, and $\not\vdash G'$. □

Lemma 13.89C. *Every unprovable saturated hypersequent has a countermodel in $\mathcal{A}_\omega$.*

Proof. Let $G = \Gamma_1 \Rightarrow \Delta_1 \mid \cdots \mid \Gamma_n \Rightarrow \Delta_n$ be an unprovable saturated sequent. Define:

$$\Gamma = \bigcup_{i=1}^n \Gamma_i, \quad \Delta = \bigcup_{i=1}^n \Delta_i$$
$$I(G) = \{p \in \mathsf{Var}(G) \mid p \in \Gamma \cap \Delta\},$$
$$R = \{\langle p, q\rangle \in I(G)^2 \mid \exists \Gamma' \subseteq \Gamma \exists \Delta' \subseteq \Delta. \vdash_{GRMI_{\vec{\neg}}} \Gamma' \Rightarrow \Delta' \text{ and } \{p,q\} \subseteq \mathsf{Var}(\Gamma' \Rightarrow \Delta')\}.$$

We first show that **R** is an equivalence relation. That **R** is reflexive follows immediately from the definition of $I(G)$, and the symmetry of **R** is trivial. It remains to show that **R** is transitive. So assume that pRq and qRr. Then there exist $\Gamma', \Delta', \Gamma'', \Delta''$ such that $\Gamma', \Gamma'' \subseteq \Gamma$, $\Delta', \Delta'' \subseteq \Delta$, $\vdash_{GRMI_{\vec{\neg}}} \Gamma' \Rightarrow \Delta'$, $\vdash_{GRMI_{\vec{\neg}}} \Gamma'' \Rightarrow \Delta''$, $\{p,q\} \subseteq \mathsf{Var}(\Gamma' \Rightarrow \Delta')$, and $\{q,r\} \subseteq \mathsf{Var}(\Gamma'' \Rightarrow \Delta'')$. Since q belongs to both $\mathsf{Var}(\Gamma' \Rightarrow \Delta')$ and $\mathsf{Var}(\Gamma'' \Rightarrow \Delta'')$, Proposition 13.8 entails that $\vdash_{GRMI_{\vec{\neg}}} \Gamma', \Gamma'' \Rightarrow \Delta', \Delta''$. But $\{p,r\} \subseteq \{p,q,r\} \subseteq \mathsf{Var}(\Gamma', \Gamma'' \Rightarrow \Delta', \Delta'')$. Hence pRr.

Let $C_1, \ldots, C_\ell$ be the equivalence classes of **R** (in some order). Obviously, ℓ has at most the cardinality of $\mathsf{Var}(G)$. From the proof of the transitivity of **R** it also easily follows that for every $1 \leq i \leq \ell$ there exist $\Gamma_i \subseteq \Gamma, \Delta_i \subseteq \Delta$ such that $\vdash_{GRMI_{\vec{\neg}}} \Gamma_i \Rightarrow \Delta_i$ and $C_i \subseteq \mathsf{Var}(\Gamma_i \Rightarrow \Delta_i)$.

We now define a countermodel ν of G in $\mathcal{A}_\ell$ (and so in $\mathcal{A}_\omega$) as follows:

$$\nu(p) = \begin{cases} I_i & p \in C_i, \\ t & p \in \Gamma, p \notin \Delta, \\ f & p \notin \Gamma. \end{cases}$$

To show that ν indeed refutes G, we first show by induction on the complexity of φ that if $\varphi \in \Gamma$ then $\nu(\varphi) \neq f$, and if $\varphi \in \Delta$ then $\nu(\varphi) \neq t$. This is obvious in case φ is a variable. In case $\varphi = \neg\psi$ the claim follows easily from the induction hypothesis and conditions (i)–(ii) in the definition of a saturated sequent. If $\varphi = \psi_1 \otimes \psi_2$ and $\varphi \in \Gamma$ then the claim follows from the induction hypothesis concerning ψ_1 and ψ_2 and condition (iii) of the definition of a saturated hypersequent. Finally assume that $\varphi = \psi_1 \otimes \psi_2$ and $\varphi \in \Delta$. So $\varphi \in \Delta_i$ for some i. Had both $\Gamma_i \Rightarrow \Delta_i, \psi_1 \mid G_i$ and $\Gamma_i \Rightarrow \Delta_i, \psi_2 \mid G_i$ been provable, so would have been G (using our externally additive version of $(\Rightarrow \otimes)$, and the fact that $\varphi \in \Delta_i$). Hence one of these sequents is unprovable. Assume, e.g., that $\not\vdash \Gamma_i \Rightarrow \Delta_i, \psi_1 \mid G_i$. Then $\psi_1 \in \Delta_i$ by condition (iv) of the definition of a saturated hypersequent. Hence $\nu(\psi_1) \neq t$ by induction hypothesis. If $\nu(\psi_1) = f$ then $\nu(\varphi) = f \neq t$. Assume, therefore, that $\nu(\psi_1) = I_k$ for some k. Then $\nu(P) = I_k$ for every $P \in \mathsf{Var}(\psi_1)$. Hence $\mathsf{Var}(\psi_1) \subseteq C_k$. By the observation above concerning C_k there exist $\Gamma'_j \subseteq \Gamma_j, \Delta'_j \subseteq \Delta_j$ $(j = 1, \ldots, n)$ such that $\vdash_{GRMI_\to} \Gamma' \Rightarrow \Delta'$ and $\mathsf{Var}(\psi_1) \subseteq C_k \subseteq \mathsf{Var}(\Gamma' \Rightarrow \Delta')$, where $\Gamma' = \bigcup_{j=1}^n \Gamma'_j, \Delta' = \bigcup_{j=1}^n \Delta'_j$. Hence $\vdash_{GRMI_\to} \Gamma' \Rightarrow \Delta', \psi_1$ by Proposition 13.8. Using the strong splitting rule of $GSRMI_\to$, this implies that $\vdash \Gamma'_i \Rightarrow \Delta'_i, \psi_1 \mid G_i$. It is not possible therefore that $\vdash \Gamma_i \Rightarrow \Delta_i, \psi_2 \mid G_i$, since otherwise we would have (using again $(\Rightarrow \otimes)$, and the facts that $\Gamma'_i \subseteq \Gamma_i, \Delta'_i \subseteq \Delta_i$, and $\varphi \in \Delta_i$) that $\vdash G$. It follows by condition (v) of the definition of a saturated hypersequent that $\psi_2 \in \Delta_i$, and so $\nu(\psi_2) \neq t$ by induction hypothesis. If $\nu(\psi_2) = f$ then again $\nu(\varphi) = f \neq t$. Assume therefore that $\nu(\psi_2) = I_m$ for some m. Then again, $\mathsf{Var}(\psi_2) \subseteq C_m$, and there exist $\Gamma''_j \subseteq \Gamma_j, \Delta''_j \subseteq \Delta_j$ $(j = 1, \ldots, n)$ such that $\vdash_{GRMI_\to} \Gamma'' \Rightarrow \Delta''$ and $\mathsf{Var}(\psi_2) \subseteq C_m \subseteq \mathsf{Var}(\Gamma'' \Rightarrow \Delta'')$, where $\Gamma'' = \bigcup_{j=1}^n \Gamma''_j, \Delta'' = \bigcup_{j=1}^n \Delta''_j$. Hence $\vdash_{GRMI_\to} \Gamma'' \Rightarrow \Delta'', \psi_2$ by Proposition 13.8. This and the fact that $\vdash_{GRMI_\to} \Gamma' \Rightarrow \Delta', \psi_1$ entail that $\vdash_{GRMI_\to} \Gamma', \Gamma'' \Rightarrow \Delta', \Delta'', \varphi$. Since $\varphi \in \Delta$, this fact entails that pRq for every $p, q \in \mathsf{Var}(\varphi)(= \mathsf{Var}(\psi_1) \cup \mathsf{Var}(\psi_2))$. It follows that $C_k = C_m$ and so $I_k = I_m$ and $\nu(\varphi) = I_k \otimes I_k = I_k \neq f$.

Next, we observe that if $p \in I(G)$ then $G \mid \Rightarrow$ is derivable from the axiom $p \Rightarrow p$ using strong splittings.[13] It follows that if the empty sequent is a component of G than $I(G)$ is empty, and so $\nu(p) \in \{t, f\}$ for all p. Hence ν refutes the empty sequent in case it is a component of G.

---

[13] Note that the usual rule of external weakening is a special case of strong splitting.

13.2 A Maximal Normal Relevant Logic

To show that ν is a countermodel of G, it remains now only to eliminate the possibility that there exist $1 \leq i \leq n$ and $1 \leq k \leq \ell$ such that $\nu(\varphi) = I_k$ for all $\varphi \in \Gamma_i \cup \Delta_i$. Well, had there been such i and k we would have that $\nu(p) = I_k$ for all $p \in \mathsf{Var}(\Gamma_i \Rightarrow \Delta_i)$, and so $\mathsf{Var}(\Gamma_i \Rightarrow \Delta_i) \subseteq C_k$. Hence there would have been $\Gamma' \subseteq \Gamma$, $\Delta' \subseteq \Delta$ such that $\vdash_{GRMI_{\rightarrow}} \Gamma' \Rightarrow \Delta'$ and $\mathsf{Var}(\Gamma_i \Rightarrow \Delta_i) \subseteq C_k \subseteq \mathsf{Var}(\Gamma' \Rightarrow \Delta')$. Proposition 13.8 would have implied then that $\vdash_{GRMI_{\rightarrow}} \Gamma_i, \Gamma' \Rightarrow \Delta_i, \Delta'$. From this it is possible to derive G using strong splittings (since $\Gamma_i \Rightarrow \Delta_i$ is a component of G and $\Gamma' \subseteq \Gamma$, $\Delta' \subseteq \Delta$). A contradiction. □

Theorem 13.89 follows from Lemmas 13.89A, 13.89B, and 13.89C. □

Theorem 13.90 *(soundness & completeness, cut-elimination)*

1. $\vdash_{GSRMI_{\rightarrow}} G$ iff $\vdash_{\mathcal{A}_\omega} G$.

2. *The cut elimination theorem holds for $GSRMI_{\rightarrow}$. What is more: if a hypersequent is provable in it then it has a proof there in which the cut rule and the external contraction rule are not used.*

3. *$GSRMI_{\rightarrow}$ is a conservative extension of $GRMI_{\rightarrow}$.*

Proof. Immediate from Proposition 13.85, Theorem 13.89, and Theorem 13.19. □

Corollary 13.90 can be generalized. For this we first need a definition.

Definition 13.91 The n-part of a hypersequential calculus is the fragment in which only hypersequents with at most n components are allowed.

Corollary 13.92 *If G has n components then $\vdash_{\mathcal{A}_\omega} G$ iff G is provable in the (external contraction-free and cut-free) n-part of $GSRMI_{\rightarrow}$.*

A close examination of the proof of the completeness theorem reveals that Corollary 13.92 can be strengthened as follows:

Corollary 13.93 *Let $GSRMI_{\rightarrow}(n)$ be the system for hypersequents with n components which has as rules the logical rules of $GSRMI_{\rightarrow}$ (not including cut!) and as axioms the hypersequents with n components which can be derived from a theorem of $GRMI_{\rightarrow}$ using strong splittings. Then a hypersequent G with n components is valid in $\mathcal{A}_\omega$ iff it has a proof in $GSRMI_{\rightarrow}(n)$.*

Corollary 13.94 *$\mathcal{T} \vdash_{\mathcal{A}_\omega} \varphi$ iff there exists a finite subset Γ of $\mathcal{T}$ such that $\Gamma \Rightarrow \varphi \mid \Rightarrow \varphi$ has a cut-free (and external contraction-free) proof in the 2–part of $GSRMI_{\rightarrow}$.*

Proof. Immediate from Corollary 13.88 and Corollary 13.92. □

Another important characterization of $\vdash_{\mathcal{A}_\omega}$ in terms of $GSRMI_{\to}$ is given in the following corollary:

Corollary 13.95 $\mathcal{T} \vdash_{\mathcal{A}_\omega} \varphi$ *iff the sequent* $\Rightarrow \varphi$ *is derivable in the 2-part of* $GSRMI_{\to}$ *from the set of sequents* $\{\Rightarrow \psi \mid \psi \in \mathcal{T}\}$.

Proof. The "if" part follows from the soundness of the rules of $GSRMI_{\to}$. The "only if" part follows from Corollary 13.94, since $\Rightarrow \varphi$ is derivable from $\Gamma \Rightarrow \varphi \mid \Rightarrow \varphi$ and the set $\{\Rightarrow \psi \mid \psi \in \Gamma\}$ using n cuts followed by an external contraction. □

Both of the last two corollaries mean that for characterizing the consequence relation induced by $\mathcal{A}_\omega$, only the 2–part of $GSRMI_{\to}$ is needed.

Note 13.96 From the semantic definitions (especially the definition of a model of the empty sequent) it immediately follows that a sequent s is classically valid iff $s^* \mid \Rightarrow$ is valid in $\mathcal{A}_\omega$, where s^* is obtained from s by replacing each $\supset$ by $\to$, each $\wedge$ by $\otimes$, and each $\vee$ by $+$. Therefore, it follows from Corollary 13.93 that s is classically valid iff $s^* \mid \Rightarrow$ has a cut-free (and external contraction-free) proof in the 2–part of $GSRMI_{\supset}$.

13.3 Bibliographical Notes and Further Reading

That **RMI**$_{\to}$ has the variable-sharing property was observed by Parks already in [239] (See also [4, Page 148]). The rest of the content of the first four subsections of Section 13.1 is mainly taken from [24]. (Theorem 13.27 is due to [57].) Section 13.1.5 (in particular: Theorem 13.43) is a corrected version of material taken from [43]. Section 13.1.6 is mostly new, though it is based on work described in [29, 30, 35]. These three papers are also the sources of Section 13.1.7, and we refer the reader to these three papers for more information about the general semantics for **RMI**$_{\to}$ described there, and for its motivation. Other sources of Section 13.1, which contain further information about its subject, are [38, 39, 41].

Section 13.2 is mainly based on [38] and [43].

Chapter 14

Relevance in Richer Languages

So far we have considered only relevant logics which are in the basic language for relevant logics: the language $\mathcal{IL}$ of $\to$ and $\neg$ (which includes also $\otimes$ and $+$). This chapter is devoted to what we believe to be the most important (among those that have been studied in the literature) relevant extensions of these logics to richer languages.

14.1 Adding Propositional Constants

The simplest way to extend the logics discussed in the previous chapters to stronger language is to add to their language the following standard propositional constants (which are in common use in the literature on substructural logics — see e.g. [238, 254]), together with their standard defining axioms and rules. Following the relevance literature, we denote here these constants by t, f, T, and F. (In [174] they are denoted by 1, $\bot$, $\top$ and 0, respectively.) The associated axioms and rules are the following:

t: The axioms t and $t \to (\varphi \to \varphi)$ in the Hilbert-type case, and the axiom $\Rightarrow t$ together with the rule $\dfrac{\Gamma \Rightarrow \Delta}{t, \Gamma \Rightarrow \Delta}$ in the Gentzen-type case.

f: The axioms $\neg f$ and $\neg \varphi \to (\varphi \to f)$ in the Hilbert-type case, and the axiom $f \Rightarrow$ together with the rule $\dfrac{\Gamma \Rightarrow \Delta}{\Gamma \Rightarrow \Delta, f}$ in the Gentzen-type case.

T: The axiom $\varphi \to T$ in the Hilbert-type case, and the axiom $\Gamma \Rightarrow \Delta, T$ in the Gentzen-type case.

F: The axiom $F \to \varphi$ in the Hilbert-type case, and the axiom $F, \Gamma \Rightarrow \Delta$ in the Gentzen-type case.

Definition 14.1 Let $\mathcal{L}$ be a propositional language such that $\{\neg, \to\} \subseteq \mathcal{L}$ and $\mathcal{L} \cap \{\mathsf{t}, \mathsf{f}, \mathsf{T}, \mathsf{F}\} = \emptyset$, and let X be a nonempty subset of $\{\mathsf{t}, \mathsf{f}, \mathsf{T}, \mathsf{F}\}$.

- $\mathcal{L}^X = \mathcal{L} \cup X$.

- Let **L** be an extension in $\mathcal{L}$ of $\mathbf{R}_\to$. $\mathbf{L}^X$ is the axiomatic extension of **L** by the axioms which were associated above with the elements of X.

- Let H be a Hilbert-type system in $\mathcal{L}$. H^X is the Hilbert-type system in $\mathcal{L}^X$ which is obtained by adding to H the axioms which were associated above with the elements of X.

- Let G be a Gentzen-type system in $\mathcal{L}$. G^X is the Gentzen-type system in $\mathcal{L}^X$ which is obtained by adding to G the Gentzen-type rules and axioms which were associated above with the elements of X.

Notation 14.2 We shall usually omit the curly brackets from X, and write e.g. just $\mathbf{L}^{\mathsf{t},\mathsf{F}}$ instead of $\mathbf{L}^{\{\mathsf{t},\mathsf{F}\}}$.

We leave the easy proofs of the following two propositions to the reader:

Proposition 14.3 *Let $\mathcal{L}$, X, and **L** be like in Definition 14.1.*

1. *If HL is a Hilbert-type system which is (strongly) sound and complete for **L**, then HL^X is (strongly) sound and complete for $\mathbf{L}^X$.*

2. *If GL is a Gentzen-type system which is (strongly) sound and complete for **L**, then GL^X is (strongly) sound and complete for $\mathbf{L}^X$.*

Proposition 14.4 *Let $\mathcal{L}$, X, and **L** be like in Definition 14.1.*

1. *If $\mathsf{f} \in X$ then $\neg \varphi \leftrightarrow (\varphi \to \mathsf{f})$ is a theorem of $\mathbf{L}^X$*

2. *If $\{\mathsf{t},\mathsf{f}\} \subseteq X$ then the following equivalences are theorems of $\mathbf{L}^X$:*

    (a) $\mathsf{t} \leftrightarrow \neg \mathsf{f}$ and $\mathsf{f} \leftrightarrow \neg \mathsf{t}$.

    (b) $\mathsf{t} \leftrightarrow (\mathsf{f} \to \mathsf{f})$.

3. *If $\{\mathsf{T},\mathsf{F}\} \subseteq X$ then the following equivalences are theorems of $\mathbf{L}^X$:*

    (a) $\mathsf{T} \leftrightarrow \neg \mathsf{F}$ and $\mathsf{F} \leftrightarrow \neg \mathsf{T}$.

    (b) $\mathsf{T} \leftrightarrow (\mathsf{F} \to \mathsf{F})$.

Proposition 14.4 implies that the addition of some of the four constants to an extension **L** of $\mathbf{R}_\to$ can practically lead to just three new extensions of **L** (rather than seven). Therefore we mainly concentrate in what follows on the effects of adding t and/or F to such **L**, assuming that f and T are then defined using one of the options given in Proposition 14.4.

14.1 Adding Propositional Constants

Theorem 14.5 $\mathbf{R}_{\to}^{t,F}$ *is a conservative extension of* $\mathbf{R}_{\to}$

Proof. It is easy to extend the proof of the cut-elimination theorem for $GR_{\to}$ (Proposition 12.6) to $GR_{\to}^{t,F}$. This implies that if Γ and Δ are in $\mathcal{IL}$ and $\Gamma \Rightarrow \Delta$ is provable in $GR_{\to}^{t,F}$, then it is provable in $GR_{\to}$ too. Hence $\mathbf{R}_{\to}^{t,F}$ is a conservative extension of $\mathbf{R}_{\to}$. □

The situation is completely different when it comes to adding t and f to $\mathbf{RMI}_{\to}$. This is due to the following important property of $\mathbf{RMI}_{\to}$:

Proposition 14.6 *Let* $\mathbf{L}$ *be an extension of* $\mathbf{RMI}_{\to}$.

1. *Suppose that* $\mathcal{T}$ *is a theory, and* φ, ψ, *and* σ *are sentences in the language of* $\mathbf{L}$ *such that* $\mathcal{T} \vdash_{\mathbf{L}} \sigma$, $\mathcal{T} \vdash_{\mathbf{L}} \sigma \to \varphi$, *and* $\mathcal{T} \vdash_{\mathbf{L}} \sigma \to \psi$. *Then* $\mathcal{T} \vdash_{\mathbf{L}} \varphi + \psi$ (*i.e.* $\mathcal{T} \vdash_{\mathbf{L}} \neg \varphi \to \psi$).

2. *Suppose that for every two sentences* φ *and* ψ *in the language of* $\mathbf{L}$ *there is* σ *such that* $\vdash_{\mathbf{L}} \sigma$, $\vdash_{\mathbf{L}} \sigma \to (\varphi \to \varphi)$, *and* $\vdash_{\mathbf{L}} \sigma \to (\psi \to \psi)$. *Then* $\vdash_{\mathbf{L}} \varphi \mathcal{R} \psi$ (*Definition 13.55*), *and* $\mathbf{L}$ *is an extension of* $\mathbf{RM}_{\to}$.

Proof. The first part is immediate from the fact that the relevant mingle rule (Proposition 13.8) easily entails that $\vdash_{GRMI_{\to}} \sigma, \sigma \to \varphi, \sigma \to \psi \Rightarrow \varphi, \psi$. For the second, assume that $\vdash_{\mathbf{L}} \sigma$, $\vdash_{\mathbf{L}} \sigma \to (\varphi \to \varphi)$, and $\vdash_{\mathbf{L}} \sigma \to (\psi \to \psi)$. Then the first part entails that $\vdash_{\mathbf{L}} \varphi \mathcal{R} \psi$. Hence $\mathbf{L}$ is an extension of $\mathbf{RM}_{\to}$ by the axiomatization of it given in Section 13.1.4 (see Note 13.56). □

Corollary 14.7

1. $\mathbf{RMI}_{\to}^{t} = \mathbf{RM}_{\to}^{t}$.

2. $\mathbf{RMI}_{\to}^{t}$ *is not weakly conservative over* $\mathbf{RMI}_{\to}$, *and it lacks the VSP.*

Proof. The second part of Proposition 14.6 entails that $\mathbf{RMI}_{\to}^{t}$ is an extension of $\mathbf{RM}_{\to}^{t}$. The converse is obvious. Hence the first part. The second one follows from the first, and the fact that $\neg(P_1 \to P_1) \to (P_2 \to P_2)$ is a theorem of $\mathbf{RM}_{\to}$ that violates the VSP, and is not provable in $\mathbf{RMI}_{\to}$. □

Note 14.8 It follows from Proposition 14.6 that one cannot conservatively add to the basic relevant logic $\mathbf{R}_{\to}$ both the mingle axiom $\varphi \to (\varphi \to \varphi)$ and the t-axioms $\mathsf{t} \to (\varphi \to \varphi)$ and t. Moreover: by doing so the VSP is lost.[1] Another interesting observation is that Proposition 14.6 means that from the point of view of **RMI**$_\to$, two true sentences are relevant to each other iff there is a true sentence that implies them both.

Next we show that in contrast to the addition of f or t, the addition of T and F to **RMI**$_\to$ (or its extensions) is not problematic at all.

Definition 14.9 ($\mathcal{A}_n^\mathsf{F}$) For $0 \leq n \leq \omega$, $\mathcal{A}_n^\mathsf{F}$ is the extension of $\mathcal{A}_n$ to the language $\{\neg, \to, \mathsf{F}\}$ in which F is interpreted as the truth-value f.

Theorem 14.10 *Let* **L** *be some simple finitary extension of* **RMI**$_\to$.

1. *There exists $0 \leq n \leq \omega$ such that* $\mathbf{L}^\mathsf{F}$ *is weakly sound and complete for* $\mathcal{A}_n^\mathsf{F}$ *(i.e.* $\vdash_{\mathbf{L}^\mathsf{F}} \varphi$ *iff φ is valid in* $\mathcal{A}_n^\mathsf{F}$*).*

2. *If* **L** *is normal then there exists $0 \leq n \leq \omega$ such that* $\mathbf{L}^\mathsf{F}$ *is strongly sound and complete for* $\mathcal{A}_n^\mathsf{F}$.

Proof.

1. Similar to the proofs of Theorems 13.19 and 13.36.

2. Similar to the proof of Theorem 13.61. □

Corollary 14.11 *Let* **L** *be a simple finitary extension of* **RMI**$_\to$ *(including* **RMI**$_\to$ *itself) which is either axiomatic or normal. Then* $\mathbf{L}^\mathsf{F}$ *is a conservative extension of* **L**.

Proof. The case where **L** is normal follows from Theorem 13.61 and the second part of Theorem 14.10. The case where **L** is an axiomatic extension of **RMI**$_\to$ follows from Theorems 13.19 and 13.36, the first part of Theorem 14.10, and the first part of Proposition 13.13. □

Note 14.12 Most of the properties of **RMI**$_\to$ and $\mathbf{L}_{\mathcal{A}_\omega}$ described in Chapter 13 are shared (with similar proofs) by **RMI**$_\to^\mathsf{F}$ and $\mathbf{L}_{\mathcal{A}_\omega}^\mathsf{F}$ (respectively). In particular: both of them are decidable and strongly paraconsistent; $\mathcal{A}_\omega^\mathsf{F}$ is weakly characteristic for **RMI**$_\to^\mathsf{F}$ and strongly characteristic for $\mathbf{L}_{\mathcal{A}_\omega}^\mathsf{F}$; $HRMI_\to^\mathsf{F}$ and $GRMI_\to^\mathsf{F}$ are strongly sound and complete for **RMI**$_\to^\mathsf{F}$, while $SRMI_\to^\mathsf{F}$ and $GSRMI_\to^\mathsf{F}$ are strongly sound and complete for $\mathbf{L}_{\mathcal{A}_\omega}^\mathsf{F}$. Note however that neither of these logics is boldly paraconsistent.

---

[1] Actually, one may anyway ask in what sense, if any, $\mathbf{R}_\to^\mathsf{t}$ does have the VSP. It seems that it has it only if we take t to "contain" every propositional variable p. This may be justified by the standard view of t as the "conjunction of all logically valid formulas". However, from the purely syntactic point of view this convention is questionable (and it means that every implication in which t occurs respects the VSP).

14.2 Anderson and Belnap's R

As already noted, the most famous (and extensively studied) approach to logical relevance is that of the family of logics initiated by Anderson and Belnap (see [3, 4, 5, 99, 151, 210]). Unfortunately, the most known systems in this family are undecidable, lack useful corresponding Gentzen-type systems, and their semantics is too complicated for providing decision procedures. As such, they are out of the limited scope of this book (see the preface). However, we feel that it would be very inappropriate to completely ignore them in a book on propositional paraconsistent logics. Therefore we now provide a short review of the logic **R**, which is the principal relevant logic in the family of Anderson and Belnap (see [151]), and is a conservative extension of the minimal relevant logic $\mathbf{R}_{\to}$ studied in Chapter 12.

Historically, **R** was developed on a purely proof-theoretical basis. It was obtained as the result of an attempt to combine two different fragments: the purely *intensional* fragment $\mathbf{R}_{\to}$ considered above, and the purely *extensional* fragment $\mathbf{R}_{fde}$. Each fragment separately is quite natural, but the attempt to combine them causes problems. Below we first present $\mathbf{R}_{fde}$. Then we describe how **R** is obtained from the two fragments, and review its main properties. Our presentation is mainly technical, and uses the notions and tools described in earlier chapters. It can be argued that it is also faithful to the ideas that have stood behind the historical development of **R** and its fragments, but we shall not discuss this here.

14.2.1 The Construction of R

In retrospect, from the list of drawbacks of $\mathbf{R}_{\to}$ given in Chapter 12, there seems to be just one drawback that Anderson and Belnap were not ready to accept: its lack of a conjunction. Of course, by Theorem 11.53 this is unavoidable in strongly relevant logics. So Anderson and Belnap made a compromise. Their full logics are relevant logics which are *not strongly* relevant logics, but they do have a conjunction. Unlike the method used in Chapter 13, this has been achieved by first adding to the language of $\mathbf{R}_{\to}$ a *new* connective $\wedge$, and then extending $\mathbf{R}_{\to}$ to a relevant logic (in the enriched language) in which this $\wedge$ serves as a conjunction. The strategy which Anderson and Belnap have used for the latter task involves two stages:

- Deciding what formulas of the form $\varphi \to \psi$ should be taken as valid in case φ and ψ are in the language of $\{\neg, \wedge, \vee\}$, where $\varphi \vee \psi =_{Df} \neg(\neg\varphi \wedge \neg\psi)$. In the literature on relevant logics such implications are called *first-degree entailments*, and the set of valid first-degree entailments is denoted by $\mathbf{R}_{fde}$.

- Combining $\mathbf{R}_{\to}$ and $\mathbf{R}_{fde}$ into one coherent whole, called $\mathbf{R}$.

Without entering into the historical development and motivations, the outcome of the first stage was that a first-degree entailment $\varphi \to \psi$ was considered valid by Anderson and Belnap iff the sequent $\varphi \Rightarrow \psi$ is derivable in the ($\{\neg, \vee, \wedge\}$-fragment of) the system $G_{\mathbf{4CL}}$ (Definition 5.61), i.e., iff $\varphi \Rightarrow \psi$ is valid in the four-valued matrix $\mathcal{FOUR}$ of Dunn and Belnap (Theorem 5.62). Accordingly, $\mathbf{R}_{fde}$ has both an effective semantics and an effective, cut-free Gentzen-type system.

Things become much more complicated when it comes to the second stage of combining $\mathbf{R}_{\to}$ and $\mathbf{R}_{fde}$ into one system $\mathbf{R}$. What is common to these two systems is that both can be characterized using cut-free Gentzen-type systems. However, one of them includes weakening among its structural rules, while the other rejects it. Therefore there seems to be no way of combining the two calculi into one coherent system. Accordingly, the system $\mathbf{R}$ is defined by using a Hilbert-type system. That system is essentially constructed by extending $HR_{\to}$ with axioms that are obtained by translating the rules of the pure $\{\vee, \wedge\}$-fragment of the system $G_{\mathbf{4CL}}$ into implications. (Recall that $G_{\mathbf{4CL}}$ characterizes $\mathbf{R}_{fde}$.) The idea is that a sequent $\varphi_1, \ldots, \varphi_n \Rightarrow \psi_1, \ldots, \psi_k$ is translated into the formula $\varphi_1 \wedge \ldots \wedge \varphi_n \to \psi_1 \vee \ldots \vee \psi_k$. Take as an example the rule $[\vee \Rightarrow]$ of $G_{\mathbf{4CL}}$. In order to get its effect using the above translation, the following inference rules should be derivable in $\mathbf{R}$:

- Infer $\psi \vee \varphi \to \sigma$ from $\psi \to \sigma$ and $\varphi \to \sigma$.

- Infer $\chi \wedge (\psi \vee \varphi) \to \sigma$ from $\chi \wedge \psi \to \sigma$. and $\chi \wedge \varphi \to \sigma$

(Here σ is the disjunction of the formulas in Δ, while χ is the conjunction of the formulas in Γ, provided that Γ is not empty).

Now, the second of these two inference rules can be derived from the first (using $HR_{\to}$) with the help of $\chi \wedge (\psi \vee \varphi) \to (\chi \wedge \psi) \vee (\chi \wedge \varphi)$. Since the corresponding sequent is provable in $G_{\mathbf{4CL}}$, this may be achieved by including this distributivity principle among the axioms of $\mathbf{R}$. As for the first principle above, it would have been nice had it been possible to translate it into the axiom $(\psi \to \sigma) \to (\varphi \to \sigma) \to (\psi \vee \varphi \to \sigma)$. Unfortunately, by adding this axiom to $\mathbf{R}_{\to}$ (together with the direct translations of simpler rules) we get classical logic. So, the first principle is translated instead into the axiom $(\psi \to \sigma) \wedge (\varphi \to \sigma) \to (\psi \vee \varphi \to \sigma)$. However, this would not suffice for inferring $\psi \vee \varphi \to \sigma$ from $\psi \to \sigma$ and $\varphi \to \sigma$, unless we can infer $(\psi \to \sigma) \wedge (\varphi \to \sigma)$ from these two premises. More generally: the translation process can work only if we can infer $\varphi \wedge \psi$ from φ and ψ. Again, it would have been nice had it been possible to achieve this by

14.2 Anderson and Belnap's R

including $\varphi \to \psi \to \varphi \wedge \psi$ among our axioms, but again doing so would lead to classical logic. Instead this principle is turned into an *extra rule of inference* ("Adjunction").

The final Hilbert-type system *HR* defining **R** which is obtained using the process outlined above is presented in Figure 14.1.

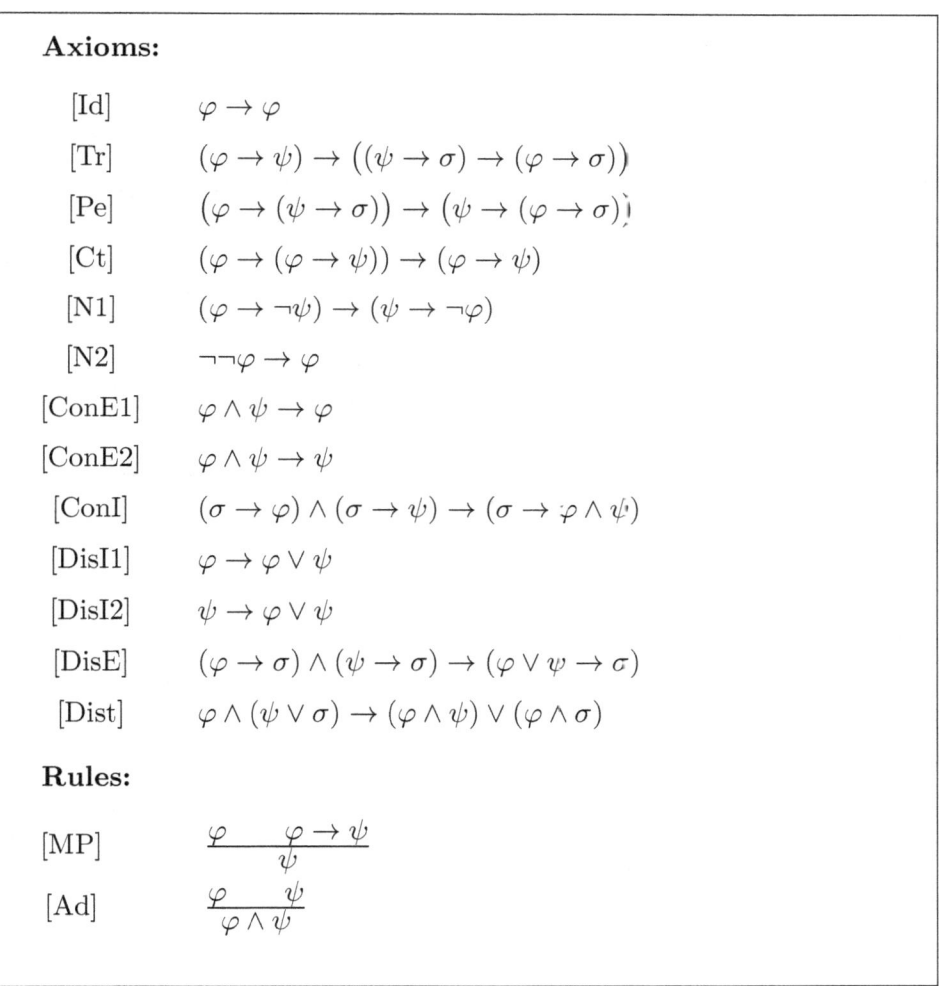

Axioms:

[Id] $\varphi \to \varphi$

[Tr] $(\varphi \to \psi) \to ((\psi \to \sigma) \to (\varphi \to \sigma))$

[Pe] $(\varphi \to (\psi \to \sigma)) \to (\psi \to (\varphi \to \sigma))$

[Ct] $(\varphi \to (\varphi \to \psi)) \to (\varphi \to \psi)$

[N1] $(\varphi \to \neg\psi) \to (\psi \to \neg\varphi)$

[N2] $\neg\neg\varphi \to \varphi$

[ConE1] $\varphi \wedge \psi \to \varphi$

[ConE2] $\varphi \wedge \psi \to \psi$

[ConI] $(\sigma \to \varphi) \wedge (\sigma \to \psi) \to (\sigma \to \varphi \wedge \psi)$

[DisI1] $\varphi \to \varphi \vee \psi$

[DisI2] $\psi \to \varphi \vee \psi$

[DisE] $(\varphi \to \sigma) \wedge (\psi \to \sigma) \to (\varphi \vee \psi \to \sigma)$

[Dist] $\varphi \wedge (\psi \vee \sigma) \to (\varphi \wedge \psi) \vee (\varphi \wedge \sigma)$

Rules:

[MP] $\dfrac{\varphi \quad \varphi \to \psi}{\psi}$

[Ad] $\dfrac{\varphi \quad \psi}{\varphi \wedge \psi}$

Figure 14.1: The proof system *HR*

Definition 14.13 (R)

- $\mathcal{L}_R = \{\neg, \to, \wedge, \vee\}$.[2]

- **R** is the logic in $\mathcal{L}_R$ which is induced by *HR*.

---

[2] It is convenient (but not essential) to take here both $\wedge$ and $\vee$ as primitives.

14.2.2 Some Useful Facts about R

In this section we prove some basic properties of **R** and its axiomatic extensions that will be used in the next section.

Lemma 14.14

1. $\varphi \to \psi \vdash_\mathbf{R} (\sigma \wedge \varphi) \to (\sigma \wedge \psi)$, $\quad \varphi \to \psi \vdash_\mathbf{R} (\varphi \wedge \sigma) \to (\psi \wedge \sigma)$.

2. $\varphi \to \psi \vdash_\mathbf{R} (\sigma \vee \varphi) \to (\sigma \vee \psi)$, $\quad \varphi \to \psi \vdash_\mathbf{R} (\varphi \vee \sigma) \to (\psi \vee \sigma)$.

Proof. We show that $\varphi \to \psi \vdash_\mathbf{R} \varphi \vee \sigma \to \psi \vee \sigma$ as an example. From $\vdash_\mathbf{R} \psi \to \psi \vee \sigma$ it follows that $\varphi \to \psi \vdash_\mathbf{R} \varphi \to \psi \vee \sigma$. Obviously $\vdash_\mathbf{R} \sigma \to \psi \vee \sigma$. Using Adjunction, we get that $\varphi \to \psi \vdash_\mathbf{R} (\varphi \to \psi \vee \sigma) \wedge (\sigma \to \psi \vee \sigma)$. Using [DisE] and [MP], we get from this $\varphi \to \psi \vdash_\mathbf{R} \varphi \vee \sigma \to \psi \vee \sigma$. □

Proposition 14.15 *Let* **L** *be an axiomatic extension of* **R**. *Then* **L** *has the replacement property. In particular:* **R** *has this property.*

Proof. Similar to the proof of Proposition 12.16. We only have to show that the addition of $\vee$ and $\wedge$ preserves the replacement property. This easily follows from Lemma 14.14. □

Next, we list the most useful equivalences which are provable in in **R**.

Lemma 14.16 *The following equivalences are theorems of* **R**.

1. $\neg\neg\varphi \leftrightarrow \varphi$,

2. $(\neg\varphi \to \neg\psi) \leftrightarrow (\psi \to \varphi)$,

3. $(\varphi \to (\psi \to \sigma)) \leftrightarrow (\varphi \otimes \psi \to \sigma)$,

4. $\varphi \vee \psi \leftrightarrow \neg(\neg\varphi \wedge \neg\psi)$ and $\varphi \wedge \psi \leftrightarrow \neg(\neg\varphi \vee \neg\psi)$,

5. $\neg(\varphi \vee \psi) \leftrightarrow \neg\varphi \wedge \neg\psi$ and $\neg(\varphi \wedge \psi) \leftrightarrow \neg\varphi \vee \neg\psi$,

6. $\varphi \vee \psi \leftrightarrow \psi \vee \varphi$ and $\varphi \wedge \psi \leftrightarrow \psi \wedge \varphi$,

7. $\varphi \vee (\psi \vee \sigma) \leftrightarrow (\varphi \vee \psi) \vee \sigma$ and $\varphi \wedge (\psi \wedge \sigma) \leftrightarrow (\varphi \wedge \psi) \wedge \sigma$,

8. $\varphi \vee \varphi \leftrightarrow \varphi$ and $\varphi \wedge \varphi \leftrightarrow \varphi$,

9. $\varphi \wedge (\psi \vee \sigma) \leftrightarrow (\varphi \wedge \psi) \vee (\varphi \wedge \sigma)$ and $\varphi \vee (\psi \wedge \sigma) \leftrightarrow (\varphi \vee \psi) \wedge (\varphi \vee \sigma)$.

14.2 Anderson and Belnap's R

Proof. The first three items are already theorems of $\mathbf{R}_\rightarrow$. The rest easily follow from the axioms of $\mathbf{R}$, sometimes using previous items in the list, Lemma 14.14, and Proposition 14.15. Details are left to the reader. □

Our next goal is to show that $\lor$ is a disjunction for $\mathbf{R}$.

Lemma 14.17 *The following equivalences are theorems of* $\mathbf{R}$.

1. $(\varphi \land (\varphi \to \psi)) \to \psi$,

2. $(\varphi + \psi) \to \varphi \lor \psi$.

Proof.

1. That $\vdash_\mathbf{R} (\varphi \land (\varphi \to \psi)) \to (\varphi \land (\varphi \to \psi)) \to \psi$ follows with the help of [ConE1] and [ConE2] from the fact that $\vdash_{\mathbf{R}_\rightarrow} \varphi \to ((\varphi \to \psi) \to \psi)$. Therefore, by the contraction axiom we get $\vdash_\mathbf{R} (\varphi \land (\varphi \to \psi)) \to \psi$.

2. Since $\vdash_{\mathbf{R}_\rightarrow} (\varphi \to \sigma) \to ((\psi \to \sigma) \to (\varphi + \psi \to \sigma))$, this follows with the help of [DisI1] and [DisI2]. □

Theorem 14.18 $\lor$ *is a disjunction for any axiomatic extension of* $\mathbf{R}$.

Proof. Let $\mathbf{L}$ be an axiomatic extension of $\mathbf{R}$. The axioms [DisI1] and [DisI2] of $\mathbf{R}$ ensure that if $\mathcal{T}, \varphi \lor \psi \vdash_\mathbf{L} \sigma$ then $\mathcal{T}, \varphi \vdash_\mathbf{L} \sigma$ and $\mathcal{T}, \psi \vdash_\mathbf{L} \sigma$. For the converse we first prove:

(*) If $\mathcal{T}, \varphi \vdash_\mathbf{L} \sigma$ then $\mathcal{T}, \varphi \lor \psi \vdash_\mathbf{L} \sigma \lor \psi$.

Let HL be some Hilbert-type system for $\mathbf{L}$ which is obtained from HR by the addition of some schemes, and let $\sigma_1, \ldots, \sigma_n = \sigma$ be a derivation in HL of σ from $\mathcal{T} \cup \{\varphi\}$. To prove (*), it suffices to show by induction on i that $\mathcal{T}, \varphi \lor \psi \vdash_\mathbf{L} \sigma_i \lor \psi$ for every $1 \le i \le n$. Because of [DisI1], this is obvious in the base cases where σ_i is an axiom of HL or $\sigma_i \in \mathcal{T} \cup \{\varphi\}$. For the induction step there are two cases to consider:

- Suppose that σ_i is inferred from σ_j and σ_k (where $j, k < i$) using [Ad]. Then $\sigma_i = \sigma_j \land \sigma_k$. By induction hypothesis, $\mathcal{T}, \varphi \lor \psi \vdash_\mathbf{L} \sigma_j \lor \psi$ and $\mathcal{T}, \varphi \lor \psi \vdash_\mathbf{L} \sigma_k \lor \psi$. By applying [Ad], we get:

$$\mathcal{T}, \varphi \lor \psi \vdash_\mathbf{L} (\sigma_j \lor \psi) \land (\sigma_k \lor \psi).$$

By Items 6 and 9 of Lemma 14.16, this implies that $\mathcal{T}, \varphi \lor \psi \vdash_\mathbf{L} \sigma_i \lor \psi$.

- Suppose that σ_i is inferred from σ_j and σ_k (where $j, k < i$) using [MP]. Then $\sigma_k = \sigma_j \to \sigma_i$. By induction hypothesis, $\mathcal{T}, \varphi \vee \psi \vdash_\mathbf{L} \sigma_j \vee \psi$ and $\mathcal{T}, \varphi \vee \psi \vdash_\mathbf{L} (\sigma_j \to \sigma_i) \vee \psi$. By applying [Ad], we get:

$$\mathcal{T}, \varphi \vee \psi \vdash_\mathbf{L} (\sigma_j \vee \psi) \wedge ((\sigma_j \to \sigma_i) \vee \psi).$$

This implies that $\mathcal{T}, \varphi \vee \psi \vdash_\mathbf{L} (\sigma_j \wedge (\sigma_j \to \sigma_i)) \vee \psi$ by Items 6 and 9 of Lemma 14.16. Using the first item of Lemma 14.17 and Lemma 14.14, this in turn easily entails that $\mathcal{T}, \varphi \vee \psi \vdash_\mathbf{L} \sigma_i \vee \psi$.

Returning to theorem's proof, suppose that $\mathcal{T}, \varphi \vdash_\mathbf{L} \sigma$ and $\mathcal{T}, \psi \vdash_\mathbf{L} \sigma$. By (*), the first assumption implies that $\mathcal{T}, \varphi \vee \psi \vdash_\mathbf{L} \sigma \vee \psi$, while the second assumption implies that $\mathcal{T}, \psi \vee \sigma \vdash_\mathbf{L} \sigma \vee \sigma$. These two facts and Items 6 and 8 of Lemma 14.16 entail that $\mathcal{T}, \varphi \vee \psi \vdash_\mathbf{L} \sigma$. □

14.2.3 Main Properties of R

We end this section with a list of the main properties of **R**. Unless we give other references, the proofs of these properties can be found in [4] or [151].

1. **R** *is a conservative extension of* $\mathbf{R}_\to$ *and* $\mathbf{R}_{fde}$*:*

    (a) If all sentences in $\mathcal{T} \cup \{\varphi\}$ are in the language of $\mathbf{R}_\to$, then $\mathcal{T} \vdash_\mathbf{R} \varphi$ iff $\mathcal{T} \vdash_{R_\to} \varphi$. In particular: **R** *is strongly pre-paraconsistent*.

    (b) If φ and ψ are in the extensional language (of $\{\vee, \wedge, \neg\}$) then $\vdash_\mathbf{R} \varphi \to \psi$ iff $\varphi \Rightarrow \psi$ is provable in the system $G_{\mathbf{4CL}}$.

2. If all formulas in $\mathcal{T} \cup \{\varphi\}$ are in the language of $\{\neg, \vee, \wedge\}$, then $\mathcal{T} \vdash_\mathbf{R} \varphi$ iff $\mathcal{T} \vdash_{G_{\mathbf{LP}}} \varphi$ (see Section 4.5.1), and so (by Theorem 4.95) $\mathcal{T} \vdash_\mathbf{R} \varphi$ iff $\mathcal{T} \vdash_{\mathbf{LP}} \varphi$. The purely extensional fragment of **R** is therefore identical with the standard paraconsistent 3-valued logic **LP**. In particular, if φ is in the language of $\{\vee, \wedge, \neg\}$ then $\vdash_\mathbf{R} \varphi$ iff φ is a classical tautology (Proposition 4.51).

3. $\mathbf{L}(\vdash_\mathbf{R}^\to)$ *is a strongly relevant logic*. To see this, note that from Corollary 12.15 all we need to show is that **R** satisfies the minimal semantic criterion, and $\to$ has in it the VSP and the replacement property. Now, from Corollary 12.20 and the first item of this list it immediately follows that **R** satisfies the minimal semantic criterion (another proof is given in Proposition 15.39 below). The replacement property for $\to$ was proved in Proposition 14.15, and the VSP for $\to$ is again proved in [4] and [151]. Note that it follows from the VSP that neither $\neg p \wedge p \to q$ nor $q \to (p \vee \neg p)$ is valid in **R**.

14.2 Anderson and Belnap's R

4. **R** *is a relevant logic, but it is not a strongly relevant logic.* To demonstrate the first part, it suffices by the previous item to show that **R** satisfies the basic relevance criterion. The proof of this requires the Routley-Meyer semantics of **R** (see the last item below), and so we omit it. The second part follows from the fact that **R** lacks the RDP: $p \not\vdash_{\mathbf{R}} q \to p \wedge q$, even though $p, q \vdash_{\mathbf{R}} p \wedge q$, and $p \not\vdash_{\mathbf{R}} p \wedge q$.[3]

5. *The disjunctive syllogism (DS) is not valid in **R**:* if p and q are distinct variables then $\neg p, p \vee q \not\vdash_{\mathbf{R}} q$. However, Meyer and Dunn have shown that DS is *admissible* for **R**: if $\vdash_{\mathbf{R}} \neg \varphi$ and $\vdash_{\mathbf{R}} \varphi \vee \psi$ then $\vdash_{\mathbf{R}} \psi$. [4]

6. *Deduction theorems:* As is noted in item 4, **R** does not have the RDP. Instead we have that if $\mathcal{T}$ is a set of formulas then $\mathcal{T} \vdash_{\mathbf{R}} \varphi$ iff there exists a sentence ψ of the form $\psi_1 \wedge \ldots \wedge \psi_n$ such that $\vdash_{\mathbf{R}} \psi \to \varphi$, and each conjunct of ψ is either an element of $\mathcal{T}$, or of the form $p \to p$ where p is a variable.

7. $\wedge$ is a conjunction for **R**, and by Theorem 14.18 $\vee$ is a disjunction for **R**. Hence **R** is semi-normal. However, at present it is not known whether **R** has an implication, and so whether it is normal.

8. It is possible to conservatively add to **R** the standard propositional constants mentioned in the first section of this chapter, together with their characteristic axioms. Doing so again leads to three new variations of **R**. The main advantage of adding t (or f) to **R** is that every axiomatic extension of $\mathbf{R}^{\mathsf{t}}$ (including $\mathbf{R}^{\mathsf{t}}$ itself and $\mathbf{R}^{\mathsf{t},\mathsf{F}}$) is normal, since in addition to the conjunction $\wedge$ and the disjunction $\vee$ it also has an implication, defined by: $\varphi \supset \psi =_{Df} \varphi \wedge \mathsf{t} \to \psi$.

9. Extend $\mathbf{F}_0$ from Definition 11.89 to a bivalent interpretation $\mathbf{F}_R$ of the language of **R** by letting $\mathbf{F}_R(\wedge)$ be the classical conjunction (and so $\mathbf{F}_R(\vee) = \mathbf{F}_R(+)$ is the classical disjunction, and $\mathbf{F}_R(\otimes)$ is the classical conjunction). In [49] it is proved that every simple extension of **R** is $\mathbf{F}_R$-contained in classical logic. In particular, **R** itself is $\neg$-contained in classical logic. This fact and items 1(a) and 7 of this list imply that **R** is boldly paraconsistent.

---

[3] A word of caution should be added here. What we say applies to our natural definition of **R**, as well as to any other definition of it according to which $p, q \vdash_{\mathbf{R}} p \wedge q$. Note however that the previous item provides an example of a strongly relevant logic which has the same valid formulas as **R**, but does not satisfy this condition.

[4] This implies that if $\mathbf{R}^{DS}$ is **R** extended with (DS), then $\mathbf{R}^{DS}$ has the same logical theorems as **R**. In particular: $\neg p \wedge p \to q$ is not provable in $\mathbf{R}^{DS}$. Nevertheless, by Lewis' argument (see first section of Chapter 2), $\mathbf{R}^{DS}$ is *not* paraconsistent! Indeed, $p \vee q$ follows from p in $\mathbf{R}^{DS}$ using the axiom [DisI1], and from this and $\neg p$ one can infer q using (DS). Hence $\neg p, p \vdash_{\mathbf{R}^{DS}} q$.

10. **R** *is undecidable* (This was shown by Urquhart in [289]).

11. Various types of semantics for **R** and its fragments are described in [151]. The most important and useful of which is the relational semantics of Routley and Meyer ([259]). This semantics is based on a possible-worlds approach which employs a *ternary* accessibility relation. Though this semantics is useful for proving some meta-theorems about **R** (like the admissibility of the disjunctive syllogism), it does not provide a decision procedure. In fact, the undecidability of **R** implies that it has no effective semantics.

Note 14.19 It should be noted that the family of logics which have been introduced by the school of Anderson and Belnap is rather big, and includes also many logics which are not relevant logics according to the characterization given in Chapter 11 (though they are still boldly paraconsistent). The two most famous examples of the kind just described are **E** and **T**, which are both strictly weaker than **R**. The first, in which in addition to relevance also a modal necessity is incorporated, was the favorite of Anderson and Belnap themselves and is the central logic in [4]. The second is the central logic in [99]. Neither of them is decidable[5], and they are more complicated than **R**. Hence they too are out of the scope of this book.[6]

14.3 RMI — A Purely Relevant Logic

14.3.1 Relevant Conjunction and RMI_{min}

As we saw in the previous section, the fact that $\mathbf{R}_\rightarrow$ lacks a conjunction was solved by Anderson and Belnap by adding to it a connective $\wedge$, for which $\varphi \wedge \psi \rightarrow \varphi$ and $\varphi \wedge \psi \rightarrow \psi$ are logically valid, but the inference of $\varphi \wedge \psi$ from $\{\varphi, \psi\}$ is achieved only via an additional rule of inference. The next proposition implies that this method is not applicable to $\mathbf{RMI}_\rightarrow$.

Proposition 14.20 *Let* **L** *be an extension of* $\mathbf{RMI}_\rightarrow$ *that has a connective* $\wedge$ *for which the axioms [ConE1], [ConE2], and the Adjunction rule [Ad] (see Figure 14.1) are all valid. Then the* $\{\neg, \rightarrow\}$*-fragment of* **L** *is an (not necessarily proper) extension of* $\mathbf{RM}_\rightarrow = \mathbf{RMI}^1_\rightarrow$*, and so it does not have the variable-sharing property.*

---

[5]This was again shown by Urquhart in [289].
[6]There are many contraction-free logics which are closely related to Anderson and Belnap's relevant logics, and *are* decidable (like **RW** [107] or the multiplicative-additive fragment of Girard's linear Logic). However, logics without contraction are not relevant logics according to our understanding of this notion (as described in Chapter 11).

14.3 RMI — A Purely Relevant Logic

Proof. Given φ and ψ, let $\sigma = (\varphi \to \varphi) \wedge (\psi \to \psi)$. Then the assumptions about $\wedge$ imply that $\vdash_\mathbf{L} \sigma$, $\vdash_\mathbf{L} \sigma \to (\varphi \to \varphi)$, and $\vdash_\mathbf{L} \sigma \to (\psi \to \psi)$. Hence $\mathbf{L}$ is an extension of $\mathbf{RM}_\to$ by item 2 of Proposition 14.6. □

Proposition 14.20 shows that $\mathbf{RMI}_\to$ is purely intensional in a very strong sense: It seems that any attempt to add extensional connectives to it which would interact with its basic connectives in a non-trivial way, would lead to a non-conservative extension, and the VSP would be lost. On the other hand, what *is* still in full agreement with the spirit of $\mathbf{RMI}_\to$ is to add to it a *relevant* conjunction. For such a connective $\wedge$, $\varphi \wedge \psi$ should be true if and only if both φ and ψ are true, *and* they are *relevant* to each other. As we shall see later in this section, the semantics of $\mathbf{RMI}_\to$-matrices provides a natural interpretation for such a connective. However, we start constructing our extension of $\mathbf{RMI}_\to$ with a relevant conjunction by assuming only some simple and intuitive *syntactic* properties that the corresponding relevance relation between formulas should satisfy:

Definition 14.21 The relation SR (syntactic relevance) between formulas of $\mathcal{L}_R$ is inductively defined as follows:

- $\varphi SR \varphi$.

- If $\varphi SR \psi$ then $\psi SR \varphi$.

- If $\varphi SR \psi$ then $\varphi SR \neg \psi$.

- If $\varphi SR \psi$ then $\varphi SR(\psi \to \sigma)$. and $\varphi SR(\sigma \to \psi)$

Lemma 14.22 *If $\varphi SR \psi$ then $(\varphi \to \sigma) SR(\psi \to \tau)$, $(\varphi \to \sigma) SR(\tau \to \psi)$.*

Proof. If $\varphi SR \psi$ then $\varphi SR(\psi \to \tau)$, and so $(\psi \to \tau) SR \varphi$, implying that $(\psi \to \tau) SR(\varphi \to \sigma)$, and so $(\varphi \to \sigma) SR(\psi \to \tau)$. The proof of the other claim is similar. □

Notation Let $\Gamma \cup \{\varphi\} \subseteq \mathcal{W}(\mathcal{L})$, where $\mathcal{IL} \subseteq \mathcal{L}$. By writing $\Gamma \vdash_{\mathbf{RMI}_\to} \varphi$ we mean that there are Γ^*, φ^* in $\mathcal{IL}$ and a substitution Θ such that $\Gamma = \Theta(\Gamma^*)$, $\varphi = \Theta(\varphi^*)$, and $\Gamma^* \vdash_{\mathbf{RMI}_\to} \varphi^*$. As before, whenever we claim that $\Gamma \vdash_{\mathbf{RMI}_\to} \varphi$, we usually leave the easy proof of this (using either $GRMI_\to$ or $\mathcal{A}_\omega$) to the reader.

Lemma 14.23 *If $\varphi SR \psi$ then $\vdash_{\mathbf{RMI}_\to} \varphi \mathcal{R} \psi$.* (See Definition 13.55.)

Proof. This is immediate from the fact that for every φ, ψ and σ we have: $\vdash_{\mathbf{RMI}_\to} \varphi \mathcal{R} \varphi$ and $\vdash_{\mathbf{RMI}_\to} \varphi \mathcal{R} \psi \to \psi \mathcal{R} \varphi$; $\vdash_{\mathbf{RMI}_\to} \varphi \mathcal{R} \psi \to \varphi \mathcal{R} \neg \psi$; $\vdash_{\mathbf{RMI}_\to} \varphi \mathcal{R} \psi \to \varphi \mathcal{R}(\psi \to \sigma)$ and $\vdash_{\mathbf{RMI}_\to} \varphi \mathcal{R} \psi \to \varphi \mathcal{R}(\sigma \to \psi)$. □

Definition 14.24 ($HRMI_{min}$, $\mathbf{RMI}_{min}$)

- The system $HRMI_{min}$ in $\mathcal{L}_R$ is obtained from $HRMI_\to$ by adding to it all the axioms of HR except [Dist] that mention $\land$ or $\lor$ (i.e. [ConE1], [ConE2], [ConI], [DisI1], [DisI2], [DisE]), and the rule [SRA] of *strict relevant adjunction*: If $\varphi SR\psi$ then from φ and ψ infer $\varphi \land \psi$.

- $\mathbf{RMI}_{min}$ is the logic induced by $HRMI_{min}$.

Definition 14.25 ($HRMI^*_{min}$, $HRMI^{**}_{min}$)

- $HRMI^*_{min}$ is obtained from $HRMI_\to$ by adding to it the axioms [ConE1], [ConE2], [DisI1], [DisI2], and the following inference rules:

$$[\to \land] \; \frac{\sigma \to \varphi \quad \sigma \to \psi}{\sigma \to \varphi \land \psi}, \qquad [\lor \to] \; \frac{\varphi \to \sigma \quad \psi \to \sigma}{\varphi \lor \psi \to \sigma}.$$

- $HRMI^{**}_{min}$ is obtained from $HRMI_\to$ by adding to it all the axioms of HR that mention $\land$ or $\lor$ except [Dist], as well as the following rule [RA] of relevant adjunction:

$$\frac{\varphi \quad \psi \quad \varphi \mathcal{R} \psi}{\varphi \land \psi}.$$

Proposition 14.26 *Both $HRMI^*_{min}$ and $HRMI^{**}_{min}$ induce $\mathbf{RMI}_{min}$.*

Proof. To show that $\vdash_{HRMI^*_{min}} \subseteq \vdash_{\mathbf{RMI}_{min}}$, it suffices to show that $\sigma \to \varphi, \sigma \to \psi \vdash_{\mathbf{RMI}_{min}} \sigma \to \varphi \land \psi$ and $\varphi \to \sigma, \psi \to \sigma \vdash_{\mathbf{RMI}_{min}} \varphi \lor \psi \to \sigma$. We show the latter as an example. Since by Lemma 14.22 $(\varphi \to \sigma)SR(\psi \to \sigma)$, we may infer $(\varphi \to \sigma) \land (\psi \to \sigma)$ in $\mathbf{RMI}_{min}$ from $\varphi \to \sigma$ and $\psi \to \sigma$ by applying [SRA]. From this $\varphi \lor \psi \to \sigma$ follows using [DisE].

That $\vdash_{\mathbf{RMI}_{min}} \subseteq \vdash_{HRMI^{**}_{min}}$ is immediate from Lemma 14.23.

Finally, to show that $\vdash_{HRMI^{**}_{min}} \subseteq \vdash_{HRMI^*_{min}}$ it suffices to derive [ConI] and [DisE] in $HRMI^*_{min}$, and to show that $\varphi, \psi, \varphi\mathcal{R}\psi \vdash_{HRMI^*_{min}} \varphi \land \psi$.

- By [ConE1], $\vdash_{HRMI^*_{min}} (\sigma \to \varphi) \land (\sigma \to \psi) \to (\sigma \to \varphi)$. Hence $((\sigma \to \varphi) \land (\sigma \to \psi) \otimes \sigma) \to \varphi$ is provable in $HRMI^*_{min}$ (using valid inferences of $\mathbf{RMI}_\to$). Similarly, $((\sigma \to \varphi) \land (\sigma \to \psi) \otimes \sigma) \to \psi$ is provable in $HRMI^*_{min}$. Now an application of $[\to \land]$ yields $\vdash_{HRMI^*_{min}} ((\sigma \to \varphi) \land (\sigma \to \psi) \otimes \sigma) \to (\varphi \land \psi)$, and so $\vdash_{HRMI^*_{min}} [ConI]$.

- By [DisI1], $(\varphi \to \sigma) \land (\psi \to \sigma) \to (\varphi \to \sigma)$ is provable in $HRMI^*_{min}$. Hence $\vdash_{HRMI^*_{min}} \varphi \to ((\varphi \to \sigma) \land (\psi \to \sigma) \to \sigma)$ (using valid inferences of $\mathbf{RMI}_\to$). Similarly, $\psi \to ((\varphi \to \sigma) \land (\psi \to \sigma) \to \sigma)$ is provable in $HRMI^*_{min}$. Now, an application of $[\lor \to]$ yields $\vdash_{HRMI^*_{min}} (\varphi \lor \psi) \to ((\varphi \to \sigma) \land (\psi \to \sigma) \to \sigma)$, and so $\vdash_{HRMI^*_{min}} [DisE]$.

14.3 RMI — A Purely Relevant Logic

- Let $\Gamma = \{\varphi, \psi, \varphi \mathcal{R}\psi\}$. Since $\Gamma \vdash_{\mathbf{RMI}_{\to}} \neg\varphi \to \psi$ and $\Gamma \vdash_{\mathbf{RMI}_{\to}} \neg\varphi \to \varphi$ (the latter because $\vdash_{\mathbf{RMI}_{\to}} \varphi \to (\neg\varphi \to \varphi)$), we have $\Gamma \vdash_{HRMI^*_{min}} \neg\varphi \to \varphi \wedge \psi$ (using $[\to \wedge]$). Therefore, $\Gamma \vdash_{HRMI^*_{min}} \neg(\varphi \wedge \psi) \to \varphi$. Similarly, $\Gamma \vdash_{HRMI^*_{min}} \neg(\varphi \wedge \psi) \to \psi$. Hence $\Gamma \vdash_{HRMI^*_{min}} \neg(\varphi \wedge \psi) \to (\varphi \wedge \psi)$, by another application of $[\to \wedge]$. Since $\neg\sigma \to \sigma \vdash_{\mathbf{RMI}_{\to}} \sigma$ for every σ, it follows that $\Gamma \vdash_{HRMI^*_{min}} \varphi \wedge \psi$. □

Corollary 14.27

1. $\varphi \to \psi \vdash_{\mathbf{RMI}_{min}} (\sigma \wedge \varphi) \to (\sigma \wedge \psi)$; $\varphi \to \psi \vdash_{\mathbf{RMI}_{min}} (\varphi \wedge \sigma) \to (\psi \wedge \sigma)$.
$\varphi \to \psi \vdash_{\mathbf{RMI}_{min}} (\sigma \vee \varphi) \to (\sigma \vee \psi)$; $\varphi \to \psi \vdash_{\mathbf{RMI}_{min}} (\varphi \vee \sigma) \to (\psi \vee \sigma)$.

2. Every simple extension of $\mathbf{RMI}_{min}$ enjoys the replacement property.

3. $\varphi \mathcal{R}\psi \vdash_{\mathbf{RMI}_{min}} (\sigma \wedge \varphi)\mathcal{R}(\sigma \wedge \psi)$; $\varphi \mathcal{R}\psi \vdash_{\mathbf{RMI}_{min}} (\sigma \vee \varphi)\mathcal{R}(\sigma \vee \psi)$.

Proof.

1. This is easily established using $HRMI^*_{min}$.

2. The proof is identical to that of Proposition 14.15 (using Item 1).

3. Let $\Gamma = \{\sigma \wedge \varphi \to \sigma, \sigma \wedge \varphi \to \varphi, \sigma \wedge \psi \to \sigma, \sigma \wedge \psi \to \psi\}$. Then:

    (a) $\Gamma \vdash_{\mathbf{RMI}_{\to}} \varphi\mathcal{R}\psi \to (\sigma \wedge \varphi \to (\sigma \wedge \psi \to (\neg\varphi \to \psi)))$,

    (b) $\Gamma \vdash_{\mathbf{RMI}_{\to}} \sigma \wedge \varphi \to (\sigma \wedge \psi \to (\neg\varphi \to \sigma))$,

    (c) $\Gamma \vdash_{\mathbf{RMI}_{\to}} \sigma \wedge \varphi \to (\sigma \wedge \psi \to (\neg(\sigma \wedge \psi) \to \sigma))$.

    We prove (b) as an example. For this we use the following fact:

    $\vdash_{GRMI_{\to}} P_3 \to P_1, P_3 \to P_2, P_4 \to P_2 \Rightarrow P_3 \to (P_4 \to (\neg P_1 \to P_2))$.

    Now, (b) follows from this fact by substituting $\sigma \wedge \varphi$ for P_3, φ for P_1, σ for P_2, and $\sigma \wedge \psi$ for P_4. The proofs of (ε) and (c) are similar.

    Since all elements of Γ are theorems of $HRMI^*_{min}$, (a) and (b) imply that $\varphi\mathcal{R}\psi \vdash_{HRMI^*_{min}} \sigma \wedge \varphi \to (\sigma \wedge \psi \to (\neg\varphi \to \sigma \wedge \psi))$, and so $\varphi\mathcal{R}\psi \vdash_{HRMI^*_{min}} \sigma \wedge \varphi \to (\sigma \wedge \psi \to (\neg(\sigma \wedge \psi) \to \varphi))$. In turn, this and (c) implies that $\varphi\mathcal{R}\psi \vdash_{HRMI^*_{min}} \sigma \wedge \varphi \to (\sigma \wedge \psi \to (\neg(\sigma \wedge \psi) \to \sigma \wedge \varphi))$, and so $\varphi\mathcal{R}\psi \vdash_{\mathbf{RMI}_{min}} (\sigma \wedge \varphi)\mathcal{R}(\sigma \wedge \psi)$ (by Note 13.56 and Proposition 14.26).

    The proof that $\varphi\mathcal{R}\psi \vdash_{\mathbf{RMI}_{min}} (\sigma \vee \varphi)\mathcal{R}(\sigma \vee \psi)$ is similar. □

Next, we provide a Gentzen-type system for $\mathbf{RMI}_{min}$, which is more useful for showing derivability in it than each of its three Hilbert-type representations given above.

Definition 14.28 (*GRMI$_{min}$*) The system *GRMI$_{min}$* in $\mathcal{L}_R$ is obtained from *GRMI$_\rightarrow$* (taken as in Note 13.4) by adding to it the following rules:

- *Relevant Mingle*:
$$\frac{\Gamma_1 \Rightarrow \Delta_1 \quad \Gamma_2 \Rightarrow \Delta_2}{\Gamma_1, \Gamma_2 \Rightarrow \Delta_1, \Delta_2}$$
provided $(\Gamma_1 \cup \Delta_1) \cap (\Gamma_2 \cup \Delta_2) \neq \emptyset$.

- The rules for $\wedge$ and $\vee$ given in Note 1.131:
$$\frac{\Gamma, \psi \Rightarrow \Delta}{\Gamma, \psi \wedge \varphi \Rightarrow \Delta}, \quad \frac{\Gamma, \varphi \Rightarrow \Delta}{\Gamma, \psi \wedge \varphi \Rightarrow \Delta}, \quad \frac{\Gamma \Rightarrow \Delta, \psi}{\Gamma \Rightarrow \Delta, \psi \vee \varphi}, \quad \frac{\Gamma \Rightarrow \Delta, \varphi}{\Gamma \Rightarrow \Delta, \psi \vee \varphi}.$$

- Relevant versions of the following two *additive* rules from Figure 1.2:
$$\frac{\Gamma \Rightarrow \Delta, \psi \quad \Gamma \Rightarrow \Delta, \varphi}{\Gamma \Rightarrow \Delta, \psi \wedge \varphi}, \quad \frac{\Gamma, \psi \Rightarrow \Delta \quad \Gamma, \varphi \Rightarrow \Delta}{\Gamma, \psi \vee \varphi \Rightarrow \Delta}.$$

In these relevant versions there is a side condition on the applicability of the rules: that the context (i.e. $\Gamma \cup \Delta$) should be *nonempty*.[7]

Note 14.29 The relevant mingle rule is actually derivable from the other rules of *GRMI$_{min}$*. For example, using cuts with $\varphi \Rightarrow \varphi, \varphi$ (which is obtained from the axiom $\varphi \Rightarrow \varphi$ using expansion), one can derive $\Gamma_1, \Gamma_2 \Rightarrow \Delta_1, \Delta_2$ from $\Gamma_1 \Rightarrow \Delta_1$ and $\Gamma_2 \Rightarrow \Delta_2$ in case $\varphi \in \Gamma_1 \cap \Gamma_2$. Similarly, if $\varphi \in \Delta_1 \cap \Gamma_2$, then $\Gamma_1 \Rightarrow \Delta_1, \varphi$ and $\varphi, \Gamma_2 \Rightarrow \Delta_2$ follow in *GRMI$_{min}$* from $\Gamma_1 \Rightarrow \Delta_1$ and $\Gamma_2 \Rightarrow \Delta_2$ (respectively), and then a cut on φ yields $\Gamma_1, \Gamma_2 \Rightarrow \Delta_1, \Delta_2$. (We leave the other two cases of relevant mingle to the reader.) However, without the relevant mingle rule *GRMI$_{min}$* would not admit cut-elimination. Thus, it is very easy to provide a cut-free proof in *GRMI$_{min}$* of $p \wedge q \Rightarrow p, q$, but it is not difficult to see that such a proof necessarily uses the relevant mingle rule.

A proof that the cut-elimination theorem does obtain for *GRMI$_{min}$* (when relevant mingle is taken as an official rule) can be found in [33].[8] It follows from it and from the next proposition that **RMI**$_{min}$ is decidable.

Proposition 14.30

1. For each non-empty sequent s let φ_s be one of the translations of s described in the proof of Proposition 12.4. Then for every set $S \cup \{s^*\}$ of non-empty sequents, $S \vdash_{GRMI_{min}} s^*$ iff $\{\varphi_s \mid s \in S\} \vdash_{\mathbf{RMI}_{min}} \varphi_{s^*}$.

---

[7]Thus, in *GRMI$_{min}$* one *cannot* in general derive $\Rightarrow \varphi \wedge \psi$ from $\Rightarrow \varphi$ and $\Rightarrow \psi$.

[8]What makes the proof here more complicated than in the case of *GRMI$_\rightarrow$* are the side conditions on the applications of relevance mingle, $[\Rightarrow \wedge]$, and $[\vee \Rightarrow]$: If the corresponding reductions are done carelessly, than these conditions might not be preserved after them.

14.3 RMI — A Purely Relevant Logic

2. $\mathcal{T} \vdash_{\mathbf{RMI}_{min}} \varphi$ iff $\mathcal{T} \Vdash_{GRMI_{min}} \varphi$ (i.e., $\{\Rightarrow \psi \mid \psi \in \mathcal{T}\} \vdash_{GRMI_{min}} \Rightarrow \varphi$ — see Note 1.116).

Proof. Suppose first that $S \cup \{s^*\}$ is a set of non-empty sequents such that $S \vdash_{GRMI_{min}} s^*$. Let $s_1, \ldots, s_k = s^*$ be a proof in $GRMI_{min}$ of s^* from S. Since the rules of $GRMI_{min}$ do not allow to infer anything from the empty sequent (except itself), we may assume that for every i, s_i is not empty. By Note 14.29 we may assume also that the proof does not use relevant mingle. Now we prove by induction that for every $1 \leq i \leq k$, $\mathcal{T}_S \vdash_{\mathbf{RMI}_{min}} \varphi_{s_i}$, where $\mathcal{T}_S = \{\varphi_s \mid s \in S\}$. The induction on i is straightforward. We do one case as an example. So, suppose that $s_i = \psi \vee \sigma, \Gamma \Rightarrow \Delta$, and it was inferred by $[\vee \Rightarrow]$ from $s_l = \psi, \Gamma \Rightarrow \Delta$ and $s_j = \sigma, \Gamma \Rightarrow \Delta$, where $l, j < i$. Since φ_{s_l} is equivalent in $\mathbf{R}_{\rightarrow}$ to $\psi \rightarrow \varphi_{\Gamma \Rightarrow \Delta}$, and φ_{s_j} is equivalent in $\mathbf{R}_{\rightarrow}$ to $\sigma \rightarrow \varphi_{\Gamma \Rightarrow \Delta}$, the induction hypothesis implies that $\mathcal{T}_S \vdash_{\mathbf{RMI}_{min}} \psi \rightarrow \varphi_{\Gamma \Rightarrow \Delta}$, and $\mathcal{T}_S \vdash_{\mathbf{RMI}_{min}} \sigma \rightarrow \varphi_{\Gamma \Rightarrow \Delta}$. Using an application of the relevant rule of $HRMI^*_{min}$, we get that $\mathcal{T}_S \vdash_{\mathbf{RMI}_{min}} \psi \vee \sigma \rightarrow \varphi_{\Gamma \Rightarrow \Delta}$. Hence $\mathcal{T}_S \vdash_{\mathbf{RMI}_{min}} \varphi_{s_i}$.

Since $\varphi_{\Rightarrow \psi} = \psi$, what we have just proved implies that if $\mathcal{T} \Vdash_{GRMI_{min}} \varphi$ then $\mathcal{T} \vdash_{\mathbf{RMI}_{min}} \varphi$. This shows one direction of the second part of the proposition. The other direction is proved by showing that if $\psi_1, \ldots, \psi_k$ is a proof in $HRMI^*_{min}$ from a theory $\mathcal{T}$, then for every $1 \leq i \leq k$, $\mathcal{T} \Vdash_{GRMI_{min}} \psi_i$. Proving this by induction on i is standard, and is left to the reader. (What makes the induction particularly easy here is the obvious correspondence between the axioms and rules for $\vee$ and $\wedge$ of $HRMI^*_{min}$ and the rules for $\vee$ and $\wedge$ of $GRMI_{min}$.)

Finally, let again $S \cup \{s^*\}$ be a set of non-empty sequents, where $s^* = \Gamma^* \Rightarrow \Delta^*$, and suppose that $\{\varphi_s \mid s \in S\} \vdash_{\mathbf{RMI}_{min}} \varphi_{s^*}$. Then $\{\Rightarrow \varphi_s \mid s \in S\} \vdash_{GRMI_{min}} \Rightarrow \varphi_{s^*}$, by what we have just proved. Hence there is a finite subset $S_0 = \{\Gamma_1 \Rightarrow \Delta_1, \ldots, \Gamma_n \Rightarrow \Delta_n\}$ such that $\{\Rightarrow \varphi_s \mid s \in S_0\} \vdash_{GRMI_{min}} \Rightarrow \varphi_{s^*}$. Now by Note 12.5, $\Gamma_i \Rightarrow \Delta_i \vdash_{GR_{\rightarrow}} \Rightarrow \varphi_{\Gamma_i \Rightarrow \Delta_i}$ for $1 \leq i \leq n$. Hence $S_0 \vdash_{GRMI_{min}} \Rightarrow \varphi_{s^*}$. Since Note 12.5 entails also that $\vdash_{GR_{\rightarrow}} \varphi_{s^*}, \Gamma^* \Rightarrow \Delta^*$, it follows that $S_0 \vdash_{GRMI_{min}} \Gamma^* \Rightarrow \Delta^*$, and so $S \vdash_{GRMI_{min}} s^*$. This completes the proof of also the first part of the proposition. □

Note 14.31 Due to the side conditions on some of the rules of $GRMI_{min}$, the standard consequence relation $\vdash_{GRMI_{min}}$ which is associated with it according to Definition 1.107 is *not* identical to $\vdash_{\mathbf{RMI}_{min}}$. Thus there is no subset Γ of $\{p, q, p\mathcal{R}q\}$ for which $\Gamma \vdash_{GRMI_{min}} p \wedge q$ (while $\{p, q, p\mathcal{R}q\} \vdash_{\mathbf{RMI}_{min}} p \wedge q$ by Proposition 14.26). Indeed, had there been such Γ, then we would have also that $\Gamma \vdash_{GRMI_{min}} p$ and $\Gamma \vdash_{GRMI_{min}} q$. This is impossible, since $GRMI_{min}$ is a conservative extension of $GRMI_{\rightarrow}$[9], and there is no subset Γ of $\{p, q, p\mathcal{R}q\}$ for which both $\Gamma \vdash_{GRMI_{\rightarrow}} p$ and $\Gamma \vdash_{GRMI_{\rightarrow}} q$.

---

[9]This is immediate from the cut-elimination theorem for $GRMI_{\rightarrow}$. In Corollary 14.51

Proposition 14.32

1. With the exception of the two equivalences concerning distribution which are given in its last item, all the other equivalences listed in Lemma 14.16 are theorems of $\mathbf{RMI}_{min}$.

2. The two schemes given in Lemma 14.17 are theorems of $\mathbf{RMI}_{min}$.

Proof. The proofs are practically identical to those of Lemma 14.16 and 14.17. Alternatively, all these schemas are easily proved in $GRMI_{min}$. □

Corollary 14.33 Let $\varphi \mathcal{R}^\wedge \psi =_{Df} (\varphi \to \varphi) \wedge (\psi \to \psi)$. Then $\varphi \mathcal{R}^\wedge \psi$ is equivalent in $\mathbf{RMI}_{min}$ to $\varphi \mathcal{R} \psi$ in the sense that the following hold:

1. $\varphi \mathcal{R}^\wedge \psi \vdash_{\mathbf{RMI}_{min}} \varphi \mathcal{R} \psi$.

2. $\varphi \mathcal{R} \psi \vdash_{\mathbf{RMI}_{min}} \varphi \mathcal{R}^\wedge \psi$.

Proof. The first item is immediate from the second item of Proposition 14.32. The second follows from the validity in $\mathbf{RMI}_{min}$ of the rule [RA] of $HRMI^{**}_{min}$, since $\varphi \mathcal{R} \psi \vdash_{\mathbf{RMI}_\to} (\varphi \to \varphi) \mathcal{R} (\psi \to \psi)$. □

14.3.2 The Logic RMI and Its Semantics

Interesting as $\mathbf{RMI}_{min}$ is, it has the drawback that no useful, concrete semantics is currently known for it. However, it is at least sound for a natural extension to $\mathcal{L}_R$ of the semantics for $RMI_\to$ given in Sections 13.1.6 and 13.1.7. The basis for that semantics is provided by the next proposition (the easy proof of which we leave to the reader):

Proposition 14.34 Let $\mathcal{M} = \langle \mathcal{V}, \mathcal{D}, \mathcal{O} \rangle$ be an $\mathbf{RMI}_\to$-matrix which is based on the tree $\langle S, \preceq \rangle$. Define a relation $\leq$ on $\mathcal{V}$ by letting $u \leq v$ iff $u \to v \in \mathcal{D}$. Then $\langle \mathcal{V}, \leq \rangle$ is a lattice, in which the following hold (where $\wedge$ and $\vee$ denote, as usual, the lattice operations):

1. $u \leq v$ if either $u = v$, or $dom(u) \preceq dom(v)$ and $bval(u) = f$, or $dom(v) \preceq dom(u)$ and $bval(v) = t$.

2. $u \wedge v$ is u if $u \leq v$, v if $v \leq u$, and $\langle inf_\preceq \{dom(u), dom(v)\}, f \rangle$ otherwise (i.e. if u and v are not relevant).

3. $u \vee v$ is v if $u \leq v$, u if $v \leq u$, and $\langle inf_\preceq \{dom(u), dom(v)\}, t \rangle$ otherwise (i.e. if u and v are not relevant).

---

below we show by semantic means something stronger: that even the full logic $\mathbf{RMI}$ is a conservative extension of $\mathbf{RMI}_\to$.

14.3 RMI — A Purely Relevant Logic

Definition 14.35 (RMI-matrix) An **RMI**-matrix is a matrix $\langle \mathcal{V}, \mathcal{D}, \mathcal{O} \rangle$ for $\mathcal{L}_R$ such that $\langle \mathcal{V}, \mathcal{D}, \mathcal{O} \restriction \mathcal{IL} \rangle$ is an $\mathbf{RMI}_{\rightarrow}$-matrix, and the interpretations of $\wedge$ and $\vee$ are according to Proposition 14.34 (that is, $\wedge$ and $\vee$ are interpreted as the lattice operations of the lattice $\langle \mathcal{V}, \leq \rangle$ defined in that proposition).

Proposition 14.36 *Every* $\mathbf{RMI}_{\rightarrow}$-*matrix can be extended to an* **RMI**-*matrix in a unique way.*

Proof. Immediate from the definitions and Proposition 14.34. □

Definition 14.37 $\mathcal{SA}$ is the unique extension of $\mathcal{SA}_m$ to an **RMI**-matrix.

Note 14.38 An alternative characterization of **RMI**-matrices can be given in terms of the induced structure $\langle \mathcal{V}, \leq, \neg \rangle$. It is easy to see that if $\langle \mathcal{V}, \mathcal{D}, \mathcal{O} \rangle$ is an **RMI**-matrix, then $\neg$ is an involution on the corresponding lattice $\langle \mathcal{V}, \leq \rangle$ (as defined in Proposition 14.34), and the following condition is satisfied: if uRv then $uR\neg v$, where uRv iff u and v are related by $\leq$ (i.e., either $u \leq v$, or $v \leq u$). Conversely, if $\langle \mathcal{V}, \leq, \neg \rangle$ is a structure with these properties, then there is a unique way to turn it into an **RMI**-matrix (by defining $\mathcal{D}$ and interpretations for $\rightarrow$, $\wedge$, and $\vee$). See [29] for details.

Lemma 14.39 $\mathbf{RMI}_{min}$ *is sound for the semantics of* **RMI**-*matrices.*

Proof. Since $HRMI_{\rightarrow}$ is sound for the semantics of $\mathbf{RMI}_{\rightarrow}$-matrices (Theorem 13.53), by Proposition 14.26 it suffices to check that the axioms and rules of $HRMI^*$ are sound for **RMI**-matrices. This easily follows from the definitions of $\leq$, $\wedge$, and $\vee$ in **RMI**-matrices. □

Unfortunately, $\mathbf{RMI}_{min}$ is not also complete for the semantics of **RMI**-matrices. Thus $\varphi \wedge (\psi \vee \neg \psi) \rightarrow (\varphi \wedge \psi) \vee (\varphi \wedge \neg \psi)$ is easily seen to be valid in any **RMI**-matrix, but in [30] it is shown that it is not a theorem of $\mathbf{RMI}_{min}$. (See the proof of Proposition B.8 there.) Next we extend $\mathbf{RMI}_{min}$ to a system which is both sound and complete for **RMI**-matrices.

Definition 14.40 (*HRMI*, **RMI**)

- *HRMI* is the system obtained from $HRMI^*_{min}$ by adding to it the following relevant version of the axiom [Dist] from Figure 14.1:

    [Dist]$_r$ $\quad \varphi \wedge (\psi \vee \sigma) \rightarrow (\varphi \wedge \psi) \vee (\varphi \wedge \sigma)$ provided that $\psi SR\sigma$.

- **RMI** is the logic induced by *HRMI*.

Note 14.41 By Proposition 14.26, we could have chosen to use in Definition 14.40 $HRMI^*_{min}$ or $HRMI^{**}_{min}$ rather than $HRMI_{min}$.

Proposition 14.42 **RMI** *is sound for the semantics of* **RMI**-*matrices.*

Proof. By Lemma 14.39, it suffices to prove that $[\text{Dist}]_r$ is valid in **RMI**-matrices. So suppose that $\psi SR\sigma$. Then $\vdash_{\mathbf{RMI}_{min}} \psi\mathcal{R}\sigma$ by Lemma 14.23. Therefore, Lemma 14.39 implies that $\psi\mathcal{R}\sigma$ is valid in any **RMI**-matrix. It follows from Proposition 13.57 that if $\mathcal{M}$ is an **RMI**-matrix, and ν is a valuation in $\mathcal{M}$, then either $\nu(\psi) \leq \nu(\sigma)$, or $\nu(\sigma) \leq \nu(\psi)$. Hence either $\nu(\psi \vee \sigma) = \nu(\psi)$ or $\nu(\psi \vee \sigma) = \nu(\sigma)$. This immediately entails that $\nu \models_{\mathcal{M}} \varphi \wedge (\psi \vee \sigma) \to (\varphi \wedge \psi) \vee (\varphi \wedge \sigma)$. □

Obviously, Corollary 14.27 and Proposition 14.32 are true for **RMI**. Next we turn to important syntactic properties of **RMI** which $\mathbf{RMI}_{min}$ lacks. For the first one we need the following simple lemma.

Lemma 14.43 *If* $\sigma SR\psi$ *then* $\vdash_{\mathbf{RMI}} (\sigma \vee \varphi) \wedge (\psi \vee \varphi) \to (\sigma \wedge \psi) \vee \varphi$.

Proof. From Proposition 14.32 and the second part of Corollary 14.27 it easily follows that $(\sigma \vee \varphi) \wedge (\psi \vee \varphi) \to (\sigma \wedge \psi) \vee \varphi$ is equivalent in $\mathbf{RMI}_{min}$ to $(\neg\sigma \vee \neg\psi) \wedge \neg\varphi \to (\neg\sigma \wedge \neg\varphi) \vee (\neg\psi \wedge \neg\varphi)$. The latter formula is equivalent to an axiom of $HRMI$ in case $\sigma SR\psi$, since if $\sigma SR\psi$ then $\neg\sigma SR\neg\psi$. □

Theorem 14.44 $\vee$ *is a disjunction for any axiomatic extension of* **RMI**.

Proof. The proof is very similar to that of Theorem 14.18. Because of Proposition 14.32, we practically only have to check that the two applications of [Adj] to $\sigma_j \vee \varphi$ and $\sigma_k \vee \varphi$ that are done in the induction step of that proof, as well as the two times that the provability of $(\sigma_j \vee \varphi) \wedge (\sigma_k \vee \varphi) \to (\sigma_j \wedge \sigma_k) \vee \varphi$ is used there, are permissible in $HRMI$. For this note that in both cases of the induction step we have that $\sigma_j SR\sigma_k$. (In the first one — because [Adj] can be applied in $HRMI$ to σ_j and σ_k only if $\sigma_j SR\sigma_k$; in the second — because $\sigma_k = \sigma_j \to \sigma_i$ in it.) By Lemma 14.43, this justifies the two applications of distribution. By the third item of Corollary 14.27, the fact that $\sigma_j SR\sigma_k$ entails also that $\vdash_{HRMI} (\sigma_j \vee \varphi)\mathcal{R}(\sigma_k \vee \varphi)$. Hence the application of [Adj] to these two formulas is an application of the rule [RA] of $HRMI^{**}_{min}$, and so it is justified by Proposition 14.26. □

Corollary 14.45 *Let* $HRMI^*$ *be obtained from HR by adding to it (or replacing the axiom* [Id] *by) the mingle axiom* [Mi]*, and respectively replacing the rule* [Ad] *and the axiom* [Dist] *by the following relevant versions:*

$$[RA] \; \frac{\varphi \quad \psi \quad \varphi\mathcal{R}\psi}{\varphi \wedge \psi}, \qquad [RD] \; \frac{\psi\mathcal{R}\sigma}{\varphi \wedge (\psi \vee \sigma) \to (\varphi \wedge \psi) \vee (\varphi \wedge \sigma)}.$$

Then $HRMI^*$ *induces* **RMI**.

14.3 RMI — A Purely Relevant Logic

Proof. That $\vdash_{\mathbf{RMI}} \subseteq \vdash_{HRMI^*}$ follows from Proposition 14.26 together with Lemma 14.23. For the converse it suffices by Proposition 14.26 to show that $\psi \mathcal{R} \sigma \vdash_{\mathbf{RMI}} \varphi \wedge (\psi \vee \sigma) \to (\varphi \wedge \psi) \vee (\varphi \wedge \sigma)$. Now Note 13.56 and Proposition 14.32 entail that $\psi \mathcal{R} \sigma \vdash_{\mathbf{RMI}} (\psi \to \sigma) \vee (\sigma \to \psi)$. It follows therefore from Theorem 14.44 that it would suffices to prove:

1. $\psi \to \sigma \vdash_{\mathbf{RMI}} \varphi \wedge (\psi \vee \sigma) \to (\varphi \wedge \psi) \vee (\varphi \wedge c)$,

2. $\sigma \to \psi \vdash_{\mathbf{RMI}} \varphi \wedge (\psi \vee \sigma) \to (\varphi \wedge \psi) \vee (\varphi \wedge c)$.

We prove the second claim. (The proof of the former is similar.) So from From $\psi \to \psi$ and $\sigma \to \psi$ one can infer $\psi \vee \sigma \to \psi$ by an application of $[\vee \to]$. Hence $\sigma \to \psi \vdash_{\mathbf{RMI}} \psi \vee \sigma \to \psi$. Therefore, the first item of Corollary 14.27 entails that $\sigma \to \psi \vdash_{\mathbf{RMI}} \varphi \wedge (\psi \vee \sigma) \to (\varphi \wedge \psi)$, and so $\sigma \to \psi \vdash_{\mathbf{RMI}} \varphi \wedge (\psi \vee \sigma) \to (\varphi \wedge \psi) \vee (\varphi \wedge \sigma)$. □

Next we show that **RMI** has an implication too.

Definition 14.46 $\varphi \supset \psi =_{Df} (\varphi \to \psi) \vee \psi$.

Proposition 14.47 $\supset$ *is an implication for* **RMI** *as well as for any axiomatic extension of it.*[10]

Proof. To show that if $\mathcal{T} \vdash_{\mathbf{RMI}} \varphi \supset \psi$ then $\mathcal{T}, \varphi \vdash_{\mathbf{RMI}} \psi$, it suffices to show that $\varphi, \varphi \supset \psi \vdash_{\mathbf{RMI}} \psi$. In turn, by definition of $\supset$ and Theorem 14.44, it suffices for this to show that both $\varphi, \psi \vdash_{\mathbf{RMI}} \psi$ and $\varphi, \varphi \to \psi \vdash_{\mathbf{RMI}} \psi$. Both are obvious.

As usual, for the converse we prove by induction on i that if $\psi_1, \ldots, \psi_n = \psi$ is a proof in *HRMI* of ψ from $\mathcal{T} \cup \{\varphi\}$, then for every $1 \leq i \leq n$ $\mathcal{T} \vdash_{\mathbf{RMI}} \varphi \supset \psi_i$. There are four cases to consider.

- If ψ_i is an axiom of *HRMI* or $\psi_i \in \mathcal{T}$ then it suffices to show that $\psi_i \vdash_{\mathbf{RMI}} \varphi \supset \psi_i$. This is obvious by definition of $\supset$.

- The case where $\psi_i = \varphi$ is immediate too, since obviously $\vdash_{\mathbf{RMI}} \varphi \supset \varphi$.

- Suppose $\psi_k = \psi_j \to \psi_i$, where $j, k < i$. By induction hypothesis, $\mathcal{T} \vdash_{\mathbf{RMI}} \varphi \supset \psi_j$ and $\mathcal{T} \vdash_{\mathbf{RMI}} \varphi \supset \psi_k$. Hence we need to prove that for every φ, σ and ψ, $\varphi \supset \sigma, \varphi \supset (\sigma \to \psi) \vdash_{\mathbf{RMI}} \varphi \supset \psi$. By Theorem 14.44 and the axioms [DisI1] and [DisI2], it suffices for this to show the following:

    1. $\sigma, \sigma \to \psi \vdash_{\mathbf{RMI}} \psi$;

---

[10] For **RMI** this proposition was proved in [30] using a semantic method. Here we provide a direct syntactic proof that works for all axiomatic extensions of **RMI**.

2. $\varphi \to \sigma, \sigma \to \psi \vdash_{\mathbf{RMI}} \varphi \to \psi$;
3. $\sigma, \varphi \to (\sigma \to \psi) \vdash_{\mathbf{RMI}} \varphi \to \psi$;
4. $\varphi \to \sigma, \varphi \to (\sigma \to \psi) \vdash_{\mathbf{RMI}} \varphi \to \psi$.

These can all be proved already in $\mathbf{R}_{\to}$.[11]

- Suppose that $\psi_i = \psi_j \wedge \psi_k$, where $j, k < i$ and $\psi_j SR \psi_k$. By induction hypothesis, $\mathcal{T} \vdash_{\mathbf{RMI}} \varphi \supset \psi_j$ and $\mathcal{T} \vdash_{\mathbf{RMI}} \varphi \supset \psi_k$. Hence we need to prove that for every φ, σ and ψ, if $\sigma SR\psi$ then $\varphi \supset \sigma, \varphi \supset \psi \vdash_{\mathbf{RMI}} \varphi \supset \sigma \wedge \psi$. By Theorem 14.44 and the axioms [DisI1] and [DisI2], it suffices for this to show the following:

  1. $\sigma, \psi \vdash_{\mathbf{RMI}} \sigma \wedge \psi$ in case $\sigma SR\psi$;
  2. $\sigma, \varphi \to \psi \vdash_{\mathbf{RMI}} \varphi \supset \sigma \wedge \psi$ in case $\sigma SR\psi$;
  3. $\varphi \to \sigma, \psi \vdash_{\mathbf{RMI}} \varphi \supset \sigma \wedge \psi$ in case $\sigma SR\psi$;
  4. $\varphi \to \sigma, \varphi \to \psi \vdash_{\mathbf{RMI}} \varphi \to \sigma \wedge \psi$ in case $\sigma SR\psi$.

The first and forth items of this list are trivial. We show the second. (The proof of the third is almost identical.) For this, we use the following obvious facts:

- $\varphi \to \psi, \psi \to \sigma \vdash_{\mathbf{RMI}} \varphi \to \sigma$. Hence, $\varphi \to \psi, \psi \to \sigma \vdash_{\mathbf{RMI}} \varphi \to \sigma \wedge \psi$, and so $\varphi \to \psi, \psi \to \sigma \vdash_{\mathbf{RMI}} \varphi \supset \sigma \wedge \psi$.
- $\sigma, \sigma \to \psi \vdash_{\mathbf{RMI}} \psi$. Therefore, if $\sigma SR\psi$ then $\sigma, \sigma \to \psi \vdash_{\mathbf{RMI}} \sigma \wedge \psi$, and so $\sigma, \sigma \to \psi \vdash_{\mathbf{RMI}} \varphi \supset \sigma \wedge \psi$.

From this two facts it follows (by Theorem 14.44) that:

$$\sigma, \varphi \to \psi, (\psi \to \sigma) \vee (\sigma \to \psi) \vdash_{\mathbf{RMI}} \varphi \supset \sigma \wedge \psi.$$

But by Lemma 14.23, Note 13.56, and Proposition 14.32 (see second item of Lemma 14.17), if $\sigma SR\psi$ then $\vdash_{\mathbf{RMI}} (\psi \to \sigma) \vee (\sigma \to \psi)$. Hence $\sigma, \varphi \to \psi \vdash_{\mathbf{RMI}} \varphi \supset \sigma \wedge \psi$. □

Now we are able to prove the completeness of **RMI** with respect to **RMI**-matrices in general and $\mathcal{SA}$ in particular.

Theorem 14.48 (strong completeness of RMI)

1. $\mathcal{T} \vdash_{\mathbf{RMI}} \varphi$ iff $\mathcal{T} \vdash_{\mathcal{SA}} \varphi$.

---

[11]Note the similarity of the proof of this case and that of Proposition 11.56. In fact, the intuition behind the definition of $\supset$ comes from Proposition 11.56.

14.3 RMI — A Purely Relevant Logic

2. $\mathcal{T} \vdash_{\mathbf{RMI}} \varphi$ iff $\mathcal{T} \vdash_{\mathcal{F}} \varphi$ (Definition 3.18), where $\mathcal{F}$ is the class of **RMI**-matrices.

Proof. The 'only if' direction follows in both parts from Proposition 14.42. Therefore, the second part follows from the first. It remains to show that if $\mathcal{T} \vdash_{\mathcal{S}\mathcal{A}} \varphi$ then $\mathcal{T} \vdash_{\mathbf{RMI}} \varphi$. So assume $\mathcal{T} \not\vdash \varphi$ (where we write just "$\vdash$" instead of "$\vdash_{\mathbf{RMI}}$"). Extend $\mathcal{T}$ to a maximal theory $\mathcal{T}^*$ such that $\mathcal{T}^* \not\vdash \varphi$. We show that for every sentence A, $A \notin \mathcal{T}^*$ iff $\mathcal{T}^* \vdash A \to \varphi$. So suppose for contradiction that there exists a sentence A such that $A \notin \mathcal{T}^*$ and $A \to \varphi \notin \mathcal{T}^*$. Then the maximality of $\mathcal{T}^*$ entails that $\mathcal{T}^*, A \vdash \varphi$ and $\mathcal{T}^*, A \to \varphi \vdash \varphi$. Since obviously $\mathcal{T}^*, \varphi \vdash \varphi$, the latter implies that $\mathcal{T}, A \supset \varphi \vdash \varphi$ by Theorem 14.44. On the other hand, that $\mathcal{T}^*, A \vdash \varphi$ implies that $\mathcal{T}^* \vdash A \supset \varphi$ by Proposition 14.47. Taken together, the two conclusions entail that $\mathcal{T}^* \vdash \varphi$. A contradiction.

Once we prove that for every sentence A, $A \notin \mathcal{T}^*$ iff $\mathcal{T}^* \vdash A \to \varphi$, the rest of the proof is almost identical to that of Theorem 13.47. The only differences are that instead of using Proposition 12.16 we need to use the second item of Corollary 14.27 (to ensure that the defined $\equiv$ is a congruence relation), and that we have to show that the valuation ν which is constructed there now respects also the interpretations of $\wedge$ and $\vee$. This is obvious from the definitions of those interpretations as the lattice operations, and the corresponding axioms and rules of $HRMI^*_{min}$. $\square$

Note 14.49 Unlike in the case of $\mathbf{RMI}_{\to}$, the unique extension of $\mathcal{A}_\omega$ to $\mathcal{L}_R$ (which we also denote by $\mathcal{A}_\omega$) is not sufficient even for weak characterization of **RMI**. Thus both $\varphi \vee (\varphi \to \psi)$ and $\varphi \vee (\varphi \supset \psi)$ are valid in that extension, but are not valid in $\mathcal{A}\mathcal{S}$. It is actually not difficult to show that in the presence of either of them the structure which is constructed in the proof of Theorem 14.48 is isomorphic to a submatrix of $\mathcal{A}_\omega$. This implies that by adding to $HRMI$ either of these schemes as a new axiom, we get a system which is strongly sound and complete with respect to $\mathcal{A}_\omega$.

14.3.3 Main Properties of RMI

In this section we use the semantics of **RMI** to investigate its properties. We start with the most important difference between **RMI** and **R**:

Theorem 14.50 (decidability of RMI) **RMI** *is decidable.*

Proof. First we show that if $\Gamma \not\vdash_{\mathbf{RMI}} \varphi$, and $\Gamma \cup \{\varphi\}$ involves at most n different variables, then again there is a submatrix $\mathcal{S}\mathcal{A}(\Gamma, \varphi)$ of $\mathcal{S}\mathcal{A}$ such that $\mathcal{S}\mathcal{A}(\Gamma, \varphi)$ has at most $3n$ elements, and there is a valuation ν in it which is a model of Γ, but not a model of φ. The proof of this is very similar to that of Theorem 13.48. We only have to add that the set $\overline{X}$ defined there is closed also under $\vee$ and $\wedge$. This easily follows from Proposition 14.34.

From this observation it follows that $\Gamma \vdash_{\mathbf{RMI}} \varphi$ iff $\Gamma \vdash_{\mathcal{M}} \varphi$ for every **RMI**-matrix $\mathcal{M}$ with at most $3n$ elements. Hence **RMI** is decidable. □

Next we present several important corollaries of the soundness of **RMI** with respect to **RMI**-matrices.

Proposition 14.51 **RMI** *is a conservative extension of* $\mathbf{RMI}_{\neg \rightarrow}$.

Proof. Suppose $\mathcal{T} \not\vdash_{\mathbf{RMI}_{\neg \rightarrow}} \varphi$. Then by Theorem 13.53 there is an $\mathbf{RMI}_{\neg \rightarrow}$-matrix in which all elements of $\mathcal{T}$ are valid, while φ is not. Hence it follows from Propositions 14.42 and 14.36 that $\mathcal{T} \not\vdash_{\mathbf{RMI}} \varphi$. □

Corollary 14.52 **RMI** *has no finite weakly characteristic Nmatrix. In particular: it satisfies the minimal semantic criterion.*

Proof. Immediate from Theorem 13.27 and Proposition 14.51. □

Proposition 14.53 **RMI** *has the VSP with respect to both* $\rightarrow$ *and* $\wedge$: *If either* $\vdash_{\mathbf{RMI}} \varphi \rightarrow \psi$ *or* $\vdash_{\mathbf{RMI}} \varphi \wedge \psi$ *then* $\mathsf{Var}(\varphi) \cap \mathsf{Var}(\psi) \neq \emptyset$.

Proof. The proof for both $\rightarrow$ and $\wedge$ is almost identical to that of Corollary 13.22, using the unique extension of $\mathcal{A}_\omega$ to an **RMI**-matrix (Proposition 14.36). □

Proposition 14.54 **RMI** *is semi-normal, but it has no conjunction, and so it is not normal.*

Proof. That **RMI** has a disjunction and an implication, and so is semi-normal, was shown in Theorem 14.44 and Proposition 14.47 (respectively). Now we show that it has no conjunction, and so it is not normal. So suppose for contradiction that φ is a formula such that $\mathsf{Var}(\varphi) = \{P_1, P_2\}$ and φ defines a conjunction for **RMI**. Consider the **RMI**-matrix $\mathcal{M} = \langle \mathcal{V}, \mathcal{D}, \mathcal{O} \rangle$ in which $\mathcal{V} = \{0, 1, 2\} \times \{t, f\}$ and $\mathcal{M}$ is based on the tree $\langle \{0, 1, 2\}, \preceq \rangle$, where $x \preceq y$ iff either $x = 0$ or $x = y$. Since φ defines a conjunction for **RMI**, it follows from the soundness of **RMI** for **RMI**-matrices that for every valuation ν in $\mathcal{M}$ it holds that $bval(\nu(\varphi)) = t$ iff $bval(\nu(P_1)) = bval(\nu(P_2)) = t$. To prove that this is impossible, it suffices to show that for every formula ψ such that $\mathsf{Var}(\psi) = \{P_1, P_2\}$ there is a function $g_\psi : \{t, f\} \rightarrow \{t, f\}$ such that exactly one of the following holds for all valuations ν in $\mathcal{M}$ for which $dom(\nu(P_1)) = 1$ and $dom(\nu(P_2)) = 2$:

1. $dom(\nu(\psi)) = 0$ and $bval(\nu(\psi)) = t$;

14.3 RMI — A Purely Relevant Logic

2. $dom(\nu(\psi)) = 0$ and $bval(\nu(\psi)) = f$;

3. $dom(\nu(\psi)) = 1$ and $bval(\nu(\psi)) = g(bval(\nu(P_1)))$;

4. $dom(\nu(\psi)) = 2$ and $bval(\nu(\psi)) = g(bval(\nu(P_2)))$.

We show this by induction on the complexity of ψ.

- If $\psi = P_1$ or $\psi = P_2$, we take g_ψ to be the identity function. Then ψ has Property 3 in the first case, and Property 4 in the other.

- If $\psi = \neg\psi_1$, we let $g_\psi = \lambda x.(\mathbf{F}_{CL}(\neg))(g_{\psi_1}(x))$ (see Definition 1.48). So if ψ_1 has Property 1 (Property 2) then ψ has Property 2 (Property 1). If ψ_1 has Property 3 or Property 4 then ψ has the same property.

- If $\psi = \psi_1 + \psi_2$, we let $g_\psi = \lambda x.(\mathbf{F}_{CL}(\vee))(g_{\psi_1}(x), g_{\psi_2}(x))$. Then if either ψ_1 or ψ_2 has Property 1, then so does ψ. If neither has Property 1, but one of them has Property 2, then so does ψ. If both have Property 3, or both have Property 4, then so does ψ. If one of them has Property 3, and the other has Property 4, then ψ has Property 2.

- If $\psi = \psi_1 \wedge \psi_2$, we let $g_\psi = \lambda x.(\mathbf{F}_{CL}(\wedge))(g_{\psi_1}(x), g_{\psi_2}(x))$. Then if either ψ_1 or ψ_2 has Property 2, then so does ψ. If one of them has Property 1, then ψ has the same property that the other has. The rest of the cases are like in the case of $+$. □

Proposition 14.55 *Let $\mathbf{F}_R$ be like in Item 9 of Section 14.2.3. Then* **RMI** *is $\mathbf{F}_R$-contained in classical logic, and so it is $\neg$-contained in classical logic.*

Proof. This is immediate from the fact that the matrix $\langle\{t, f\}, \{t\}, \mathbf{F}_R\rangle$ is an **RMI**-matrix which is isomorphic to the classical two-valued matrix. □

Note 14.56 In [49] it is proved that like in the case of **R**, actually *every* simple extension of $\mathbf{RMI}_{min}$ (not only **RMI**) is $\mathbf{F}_R$-contained in classical logic. Note also that $\mathbf{F}_R(\supset)$ is the classical implication.

Proposition 14.57 **RMI** *satisfies the basic relevance criterion.*

Proof. Suppose $\mathcal{T}_1, \mathcal{T}_2 \vdash_{\mathbf{RMI}} \psi$, and $\mathcal{T}_2$ has no variables in common with $\mathcal{T}_1 \cup \{\psi\}$. We show that $\mathcal{T}_1 \vdash_{\mathbf{L}} \psi$. Suppose otherwise. Then by Theorem 14.48 there is a valuation ν in $\mathcal{SA}$ which is a model of $\mathcal{T}_1$, but not of ψ. Since $\mathcal{T}_2$ has no variable in common with $\mathcal{T}_1 \cup \{\psi\}$, we may assume without loss in generality that $\nu(p) = \langle 1, I_1\rangle$ for every variable p which occurs in $\mathcal{T}_2$. But then $\nu(\varphi) = \langle 1, I_1\rangle$ for every $\varphi \in \mathcal{T}_2$, and so ν is a model in $\mathcal{SA}$ of $\mathcal{T}_1 \cup \mathcal{T}_2$, but not of ψ. By Theorem 14.48 again, this contradicts our assumption that $\mathcal{T}_1, \mathcal{T}_2 \vdash_{\mathbf{RMI}} \psi$. □

Proposition 14.58 **RMI** *is boldly paraconsistent.*

Proof. This follows from Proposition 14.55, Proposition 14.54, Proposition 14.57, and Corollary 11.4. □

Proposition 14.59 **RMI** *is a relevant logic.*

Proof. Obviously, **RMI** is finitary, and it contains $\mathbf{R}_\neg$. That it is $\neg$-consistent is a corollary of Proposition 14.55. Therefore its being a relevant logic follows from Proposition 12.23, Proposition 14.53, the second item of Corollary 14.27, Proposition 14.57, and Corollary 14.52. □

Note 14.60 Since **RMI** has a disjunction (as well as an implication), it follows from Theorem 11.53 that **RMI** is not a strongly relevant logic. The same Theorem and Proposition 14.53 imply that the RDP for $\to$ fails in **RMI**. A specific example of this failure is given by the fact that $p \to q \vdash_{\mathbf{RMI}} p \to p \wedge q$, but neither $\vdash_{\mathbf{RMI}} p \to p \wedge q$ nor $\vdash_{\mathbf{RMI}} (p \to q) \to (p \to p \wedge q)$. (The former would imply that $\vdash_{\mathbf{RMI}} p \to q$, while it is not difficult to show that by adding the latter to **RMI** we get classical logic.)

Now we turn to some interesting properties that $\supset$ has in **RMI**.

Proposition 14.61 (properties of $\supset$ in RMI)

1. $\varphi \to \psi$ *is equivalent in* **RMI** *to* $(\varphi \supset \psi) \wedge (\neg \psi \supset \neg \varphi)$. *Hence it is possible to choose* $\mathcal{L}_{CL}$ *(with the interpretations of its connectives in* **RMI**-*matrices) as the language of* **RMI**, *rather than* $\mathcal{L}_R$.

2. $\supset$ *has in* **RMI** *the following weak version of the VSP (which is still stronger than that of* **CL***): If* $\vdash_{\mathbf{RMI}} \varphi \supset \psi$ *then either* $\vdash_{\mathbf{RMI}} \psi$, *or* φ *and* ψ *share a variable.*[12]

3. *The* $\{\vee, \supset\}$-*fragment of* **RMI** *is identical to that of Gödel-Dummett logic* $\mathbf{G}_\infty$ *(see Definition 15.87 below).*

Proof.

1. We leave the proof to the reader.

2. By Proposition 14.47, $\vdash_{\mathbf{RMI}} \varphi \supset \psi$ iff $\varphi \vdash_{\mathbf{RMI}} \psi$. Hence the claim follows from Proposition 14.57.

---

[12]In [30] it is proved that **RMI** actually satisfies a strong form of Craig's interpolation theorem: if $\vdash_{\mathbf{RMI}} \varphi \supset \psi$ then either $\vdash_{\mathbf{RMI}} \psi$, or there is an interpolant σ such that $\vdash_{\mathbf{RMI}} \varphi \supset \sigma$, $\vdash_{\mathbf{RMI}} \sigma \supset \psi$, and σ contains only variables which are common to φ and ψ.

14.3 RMI — A Purely Relevant Logic

3. It is well-known (see [144]) that the $\{\vee, \supset\}$-fragment of $\mathbf{G}_\infty$ is obtained by adding to the corresponding fragment of *HIL* the axiom $(\varphi \supset \psi) \vee (\psi \supset \varphi)$. It is easy to check that the latter axiom is valid in any **RMI**-matrix, while Theorem 14.44 and Proposition 14.47 entail that the $\{\vee, \supset\}$-fragment of *HIL* is contained in **RMI**. Hence so does the $\{\vee, \supset\}$-fragment of $\mathbf{G}_\infty$. That the $\{\vee, \supset\}$-fragment of **RMI** is contained in $\mathbf{G}_\infty$ follows from Theorem 15.88 below, Since **RM** is an extension of **RMI**. □

Note 14.62 **RMI** has several interesting extensions, determined by certain subclasses of the class of **RMI**-matrices (or submatrices of $\mathcal{SA}$). The most important of them, **RM**, corresponds to linearly ordered **RMI**-matrices, and is described in the next chapter. Others are discussed in [26]. One particular extension which is worth mentioning is $\mathbf{L}(\mathcal{A}_\omega)$ (in the language $\mathcal{L}_R$). Proofs which are similar to those of Proposition 14.42 and Theorem 14.48 show that a strongly sound and complete axiomatization of it is obtained by adding to *HRMI* as a new axiom either $\varphi \vee (\varphi \to \psi)$ or $\varphi \vee (\varphi \supset \psi)$. Note that since the $\{\vee, \supset\}$-fragment of **CL** is obtained by adding to the $\{\vee, \supset\}$-fragment of *HIL* the latter axiom, the $\{\vee, \supset\}$-fragment of $\mathbf{L}(\mathcal{A}_\omega)$ is identical to that of classical logic.

Note 14.63 The effects of extending **RMI** with the standard propositional constants discussed in Section 14.1 is similar to their effects in the case of $\mathbf{RMI}_{\to}$. Thus Proposition 14.6 again implies that by adding f or t to **RMI** the scheme $\varphi \mathcal{R} \psi$ becomes derivable, and we get a system which is equivalent to $\mathbf{RM}^{\mathbf{t}}$ (see next chapter). On the other hand the constants T and F can be interpreted in $\mathcal{SA}$ as $\langle 0, t \rangle$ and $\langle 0, f \rangle$ (respectively). Hence $\mathbf{RMI}^\mathsf{F}$ is a conservative extension of **RMI**.

We end this section by noting two important rules which are *not* admissible in **RMI**.

Proposition 14.64

1. Unlike in $\mathbf{L}_{\mathcal{A}_\omega}$, the elimination rule $[\otimes E]$ for $\otimes$ (Definition 13.67) is not admissible in **RMI**.

2. Unlike in **R** (and **RM**), the disjunctive syllogism (DS) is not admissible in **RMI**.

Proof.

1. It suffices to check (using $\mathcal{SA}$) that $\vdash_{\mathbf{RMI}} (p \vee (p \to q)) \otimes (q \to q)$, while $\not\vdash_{\mathbf{RMI}} p \vee (p \to q)$.

2. Obviously, $\vdash_{\mathbf{RMI}} \neg(p \to p) \vee (\neg(q \to q) \vee p\mathcal{R}^\wedge q)$ and $\vdash_{\mathbf{RMI}} \neg\neg(p \to p)$. However, it is easy to see (using $\mathcal{SA}$) that $\nvdash_{\mathbf{RMI}} \neg(q \to q) \vee p\mathcal{R}^\wedge q$. □

Note 14.65 For a logic $\mathbf{L}$ in $\mathcal{L}_R$, denote by $\mathbf{L}^{DS}$ the logic obtained from $\mathbf{L}$ by adding the disjunctive syllogism (DS) as a rule of inference. Using the example given in the proof of the second item of Proposition 14.64, it is easy to see that $\vdash_{\mathbf{RMI}^{DS}} \varphi\mathcal{R}^\wedge\psi$. Now in the next chapter we show that by adding $\varphi\mathcal{R}^\wedge\psi$ to **RMI** as a new axiom-schema we get the semi-relevant system **RM**. Hence $\mathbf{RMI}^{DS} = \mathbf{RM}^{DS}$. Since the disjunctive syllogism is admissible in **RM** (Corollary 15.28 below), it follows that $\mathbf{RMI}^{DS}$ and **RM** have the same set of valid formulas.

14.3.4 Cut-free Gentzen-type System for RMI

Our next goal is to provide a decent Gentzen-type system *GRMI* for **RMI**. For doing this adequately, we again need to use hypersequents. (See Section 1.3.4, as well as Sections 9.3.2 and 13.2.3.) *GRMI* is essentially the natural hypersequential version of *GRMI$_{min}$* (Definition 14.28) extended by one special new structural rule called splitting ([Sp]). By this we mean (see Note 13.80) that all the rules of *GRMI$_{min}$* remain rules of *GRMI*, but now side components are allowed to be present in applications of them (in addition to the presence of side formulas), and in addition we add [Sp] and the external versions [EC] and [EW] of the structural rules of contraction and weakening (respectively).[13] The new extra rule, [Sp], allows to infer $G \mid \Gamma_1 \Rightarrow \Delta_1 \mid \Gamma_2 \Rightarrow \Delta_2$ from $G \mid \Gamma_1, \Gamma_2 \Rightarrow \Delta_1, \Delta_2$.

The resulting system, *GRMI*, is similar to the system *GRM* in Figure 15.1 below. The difference is that in *GRMI* the same side conditions are forced on applications of the rules [Mi], [$\Rightarrow \wedge$], and [$\vee \Rightarrow$] (as presented in Figure 15.1) as those which are forced in *GRMI$_{min}$* on their counterparts:

- [Mi] may be applied only if $(\Gamma_1 \cup \Delta_1) \cap (\Gamma_2 \cup \Delta_2) \neq \emptyset$.

- [$\Rightarrow \wedge$] and [$\vee \Rightarrow$] may be applied only if $\Gamma \cup \Delta \neq \emptyset$.

We turn now to the he connections between **RMI** and *GRMI*.

Definition 14.66 (translations of hypersequents) Let G be a hypersequent all components of which are non-empty. The translation σ_G of G is the $\vee$-disjunction of the translations of its components (where again the translation of a sequent s is one of the translations of s described in the proof of Proposition 12.4). In particular: $\sigma_{\Rightarrow\varphi} = \varphi$.

---
[13]Like in the case of *GSRMI$_{\Rightarrow}$* (Figure 13.3) it is actually possible to do without [EW].

Theorem 14.67 (adequacy of GRMI) $\mathcal{T} \vdash_{\mathbf{RMI}} \varphi$ iff $\mathcal{T} \Vdash_{GRMI} \varphi$ (see Proposition 14.30). In particular: $\vdash_{\mathbf{RMI}} \varphi$ iff $\vdash_{GRMI} \Rightarrow \varphi$.

Proof. Suppose first that $\mathcal{T} \vdash_{\mathbf{RMI}} \varphi$. Then there is proof $\psi_1, \ldots, \psi_k = \varphi$ in $HRMI^*$ (see Corollary 14.45) of φ from $\mathcal{T}$. We prove by induction on i that for every $1 \leq i \leq k$, $\mathcal{T} \Vdash_{GRMI} \psi_i$. This is easy if $\psi_i \in \mathcal{T}$, or ψ_i is an axiom of $HRMI^*$, or ψ_i is derived using [MP]. If ψ_i is derived using [RA], we use the fact that by Propositions 14.30 and 14.26, $\{\sigma, \tau, \sigma\mathcal{R}\tau\} \Vdash_{GRMI_{min}} \sigma \wedge \tau$. Finally, for the case where ψ_i is derived using [RD] we have to show that in general, $\Rightarrow \psi\mathcal{R}\sigma \vdash_{GRMI} \Rightarrow \mathcal{D}$, where $\mathcal{D} = \tau \wedge (\psi \vee \sigma) \to (\tau \wedge \psi) \vee (\tau \wedge \sigma)$. Now using cuts and [Sp], one can easily derive $\psi \Rightarrow \sigma \mid \sigma \Rightarrow \psi$ from $\Rightarrow \psi\mathcal{R}\sigma$ in $GRMI$. In turn, it is not difficult to see that $\psi \Rightarrow \sigma \vdash_{GRMI_{min}} \Rightarrow \mathcal{D}$, and $\sigma \Rightarrow \psi \vdash_{GRMI_{min}} \Rightarrow \mathcal{D}$. These facts imply that $\Rightarrow \psi\mathcal{R}\sigma \vdash_{GRMI} \Rightarrow \mathcal{D} \mid \Rightarrow \mathcal{D}$, and so $\Rightarrow \psi\mathcal{R}\sigma \vdash_{GRMI} \Rightarrow \mathcal{D}$ by [EC].

For the converse, we prove something more general. We show that if $\mathcal{S}$ is a set of hypersequents, G is a hypersequent that has no empty component, and $\mathcal{S} \vdash_{GRMI} G$, then $\{\sigma_H \mid H \in \mathcal{S}\} \vdash_{\mathbf{RMI}} \sigma_G$. As usual, the proof is by induction on the length of the derivation of G from $\mathcal{S}$. This induction is easy with the help of the fact that since $\vee$ is a disjunction for **RMI**, if $\psi_1, \ldots, \psi_n \vdash_{\mathbf{RMI}} \tau$ then $\sigma \vee \psi_1, \ldots \sigma \vee, \psi_n \vdash_{\mathbf{RMI}} \sigma \vee \tau$.[14] We leave details for the reader. We only note that for handling [Sp] we use this fact and the fact that $\vdash_{\mathbf{RMI}_{min}} \varphi + \psi \to \varphi \vee \psi$, and also that the assumption that G contains an empty component is needed for handling the cut rule. (Note that since an empty component cannot be eliminated, if G has a derivation from $\mathcal{S}$ at all, then it has a derivation in which no empty component occurs.) □

Theorem 14.68 *The cut-elimination theorem obtains for GRMI (that is, if $\vdash_{GRMI} G$ then G has in GRMI a cut-free proof). Hence, GRMI enjoys the subformula property.*

A syntactic proof of Theorem 14.68 has been given in [33]. It is very complicated, and so we omit it. We note that it is also possible to provide a semantic proof, similar in spirit to the (simpler!) one that is given in the next chapter for GRM (the hypersequential proof system of **RM**).

Note 14.69 It can be shown (we skip the details) that if G has no empty component, then $\vdash_{GRMI} G$ iff for every **RMI**-matrix $\mathcal{M}$ and for every valuation ν in $\mathcal{M}$, there is a component s of G such that $\nu \models \varphi_s$ (where φ_s is like in Proposition 14.30). More generally, $\vdash_{GRMI} G$ iff for every **RMI**-matrix $\mathcal{M}$, either G has an empty component and one of the truth-values of $\mathcal{M}$

---

[14] Actually, proving this is the key step in proving that $\vee$ is a disjunction for **RMI**. See the proof of Theorem 14.44.

is of the form $\langle a, I \rangle$, or for every valuation ν in $\mathcal{M}$ there is a non-empty component s of G such that $\nu \models \varphi_s$.

Note 14.70 It should be noted that in contrast to *HRMI*, the single **RMI**-matrix $\mathcal{SA}$ is *not* sufficient for characterizing *GRMI*. Let for example G be the hypersequent:

$$\Rightarrow p \mid \Rightarrow q \mid p, q \Rightarrow p, q.$$

Then, for every valuation ν in $\mathcal{SA}$ there is a component s of G such that $\nu \models \varphi_s$. However, this is not true for every **RMI**-matrix, and so G is not a theorem of *GRMI*. (This follows also from Theorem 14.68, since it is easy to see that G has no cut-free proof in *GRMI*.)

Note 14.71 While if $\mathcal{S} \vdash_{GRMI} G$ then $\{\sigma_H \mid H \in \mathcal{S}\} \vdash_{\mathbf{RMI}} \sigma_G$ (see the proof of Theorem 14.67), it can happen that $\vdash_{\mathbf{RMI}} \sigma_G$ even though $\nvdash_{GRMI} G$. Again, the hypersequent G mentioned in Note 14.70 (that is: $\Rightarrow p \mid \Rightarrow q \mid p, q \Rightarrow p, q$) is a good example. As shown in that note, This G is not provable in *GRMI*, but σ_G is the formula $p \vee q \vee p\mathcal{R}q$, which is valid in $\mathcal{SA}$ and so *is* a theorem of **RMI**.

14.4 Bibliographical Notes and Further Reading

The seminal paper in which the study of Anderson and Belnap's family of relevance logics began is [3]. The main sources of information about this family in general and the logic **R** in particular (with extensive lists of references) are the two volumes [4, 5], as well as [99, 151, 210]. Two other important sources which are devoted to the more general family of substructural logics are the books of Restall [254] and Paoli [238].

The logic **RMI** was introduced and studied in [29, 30, 33]. A motivating introduction to it appears in [35] (with a comparison to **R**). Further information about it and related systems can be found in [38] and [41].

Chapter 15

A Maximal Semi-Relevant Logic

In this chapter we investigate what happens if we do apply to $\mathbf{RMI}_{\to}$ exactly the same procedure by which $\mathbf{R}$ was obtained from $\mathbf{R}_{\to}$. As we shall see, we get by this a very interesting system.

15.1 The Logic RM and Its Semantics

Definition 15.1 (RM) *HRM* is the Hilbert-type system which is obtained from $HRMI_{\to}$ by adding to it the Adjunction rule [Ad] and the axioms that mention $\land$ and $\lor$ in Figure 14.1. Equivalently, *HRM* is the Hilbert-type system which is obtained from *HR* by replacing [Id] with the Mingle axiom [Mi]: $\varphi \to (\varphi \to \varphi)$. **RM** is the logic in $\mathcal{L}_R$ which is induced by *HRM*.

Definition 15.1 is the official definition of **RM** in the literature (see, e.g., [4, 151]), and it reflects the way **RM** was historically introduced. Now we present an equivalent definition which better reflects the connections between **RM** and the logic **RMI** from the previous chapter.

Proposition 15.2 **RM** *is an axiomatic extension of* **RMI**. *Specifically:* **RM** *is induced by the system* HRM^* *which is obtained from* $HRMI^*$ *(Corollary 14.45) by adding either* $\varphi\mathcal{R}\psi$ *or* $\varphi\mathcal{R}^{\land}\psi$ *as a new axiom-schema.*

Proof. Obviously, $\vdash_{\mathbf{RM}} \varphi\mathcal{R}^{\land}\psi$. Therefore, it follows from Corollary 14.33 (or from Proposition 14.20) that $\vdash_{\mathbf{RM}} \varphi\mathcal{R}\psi$. It follows that if $\mathcal{T} \vdash_{HRM^*} \varphi$ then $\mathcal{T} \vdash_{\mathbf{RM}} \varphi$. The converse is immediate from Corollary 14.45. □

Corollary 15.3 **RM** *does not have the VSP with respect to either* $\to$ *or* $\land$. *Hence it is a proper extension of* **RMI**.

Proof. Immediate from Proposition 15.2 and Proposition 14.53. □

Now we turn to the semantics of **RM**. In principle, we can follow Note 13.54, and show that a proof which is very similar to that of Theorem 14.48 establishes the following:

1. A general semantics for **RM** is given by **RMI**-matrices in which the underlying tree is linearly ordered. Such matrices are known in the literature as "Sugihara matrices".

2. A strongly characteristic matrix for **RM** is given by the sub-matrix of $\mathcal{SA}$ whose set of truth-values is $([0,1) \times \{t, f\}) \cup \{\langle 1, I_1 \rangle\}$ (which is an **RMI**-matrix if we identify I_1 with I). This matrix is isomorphic to the Sugihara matrix $\mathcal{M}([0,1])$ defined below.

Nevertheless, to make our investigation of **RM** independent (to a great extent) of that of the more complicated **RMI**, in this chapter we introduce Sugihara matrices the way they are usually defined in the literature, and directly prove the above theorems. We leave to the reader the straightforward verification that the definition of a Sugihara matrix given below is indeed equivalent to the characterization of that notion which is given above.

Definition 15.4 (Sugihara chains) A *Sugihara chain* is a triple $\langle \mathcal{V}, \leq, - \rangle$ such that $\mathcal{V}$ has at least two elements, $\leq$ is a linear order on $\mathcal{V}$, and $-$ is an involution for $\leq$ on $\mathcal{V}$. (The last condition means that for every $a, b \in \mathcal{V}$ we have that $--a = a$ and that $-b \leq -a$ whenever $a \leq b$.)

Example 15.5 There are plenty of examples of Sugihara chains in all areas of mathematics. The most important for our needs are the following:

- $S_{\mathbb{R}} = \langle \mathbb{R}, \leq, - \rangle$, $S_{\mathbb{Z}} = \langle \mathbb{Z}, \leq, - \rangle$ and $S_{\mathbb{Z}^*} = \langle \mathbb{Z} - \{0\}, \leq, - \rangle$, where $\mathbb{R}$ is the set of real numbers, $\mathbb{Z}$ is the set of integers, $\leq$ is the usual order relation on $\mathbb{R}$, and $-a$ is the usual additive inverse of a.

- The *finite* substructures $S_{\mathbb{Z}_n} = \langle \mathbb{Z}_n, \leq, - \rangle$ and $S_{\mathbb{Z}_n^*} = \langle \mathbb{Z}_n^*, \leq, - \rangle$ of $S_{\mathbb{Z}}$, where for $n > 0$: $\mathbb{Z}_n = \{z \in \mathbb{Z} \mid -n \leq z \leq n\}$, and $\mathbb{Z}_n^* = \mathbb{Z}_n - \{0\}$.

- $S_{[0,1]} = \langle [0,1], \leq, \lambda x.1 - x \rangle$, where $\leq$ is again the usual order relation. Note that the underlying ordered set is here *bounded and complete*.

The following two lemmas will be useful in the sequel:

Lemma 15.6 *Let $n > 0$ be a natural number. Every finite Sugihara chain which has $2n + 1$ elements is isomorphic to $S_{\mathbb{Z}_n}$, and every finite Sugihara chain which has $2n$ elements is isomorphic to $S_{\mathbb{Z}_n^*}$.*

Proof. By an easy induction on n. □

15.1 The Logic RM and Its Semantics

Lemma 15.7 *Every countable Sugihara chain can be embedded in $S_{[0,1]}$.*

Proof. It is well known that every countable linearly ordered set can be embedded in any closed interval $[a,b]$ of $\mathbb{R}$, so that a is assigned to the minimal element of the set (if such exists), and b is assigned to the maximal element of the set (if such exists). Now, let $\langle \mathcal{V}, \leq, - \rangle$ be a countable Sugihara chain, and let $D = \{a \in \mathcal{V} \mid -a \leq a\}$. Suppose first that there is $a \in \mathcal{V}$ such that $-a = a$. It is easy to prove that in such a case a is unique, and it is the minimal element of D. Let f be an embedding of D into $[1/2, 1]$ such that $f(a) = 1/2$, and extend f to the whole of $\mathcal{V}$ by letting $f(x) = -f(-x)$ in case $x \notin D$. (Note that if $x \notin D$ then $-x \in D$ because of the linearity of $\leq$ and the fact that $--x = x$.) If there is no $a \in \mathcal{V}$ such that $-a = a$ we let f be any embedding of D into $[2/3, 1]$ (say), and we again extend f to the whole of $\mathcal{V}$ by letting $f(x) = -f(-x)$ in case $x \notin D$. In both cases f is easily seen to be an embedding of $\langle \mathcal{V}, \leq, - \rangle$ into $[0,1]$. □

Definition 15.8 Let $S = \langle \mathcal{V}, \leq, - \rangle$ be a Sugihara chain, and let $a, b \in \mathcal{V}$.

- $a < b$ if $a \leq b$ and $a \neq b$.
- $|a| = max(-a, a)$.
- $a \preceq_+ b$ iff either $|a| < |b|$ or $|a| = |b|$ and $a < b$.

The following lemma is easily verified:

Lemma 15.9 *If $S = \langle \mathcal{V}, \leq, - \rangle$ is a Sugihara chain, then $\preceq_+$ is a linear order on $\mathcal{V}$.*

Definition 15.10 (Sugihara matrix) Let $S = \langle \mathcal{V}, \leq, - \rangle$ be a Sugihara chain.

- The *multiplicative Sugihara matrix* based on S is the matrix $\mathcal{M}_m(S) = \langle \mathcal{V}, \mathcal{D}, \mathcal{O} \rangle$ for $\{\neg, \to\}$ in which $\mathcal{D} = \{a \in \mathcal{V} \mid -a \leq a\}$ (equivalently, $\mathcal{D} = \{a \in \mathcal{V} \mid |a| = a\}$), $\neg a = -a$, and $a \tilde{\to} b = max_{\preceq_+}(-a, b)$.

- The *Sugihara matrix* $\mathcal{M}(S)$ based on S is the extension of $\mathcal{M}_m(S)$ to $\mathcal{L}_R$ in which $a \tilde{\wedge} b = min_{\leq}(a, b)$ and $a \tilde{\vee} b = max_{\leq}(a, b)$.

- A matrix $\mathcal{M}$ for $\mathcal{L}_R$ ($\mathcal{IL}$) is a *(multiplicative) Sugihara matrix* if for some Sugihara chain S, $\mathcal{M}$ is the (multiplicative) Sugihara matrix which is based on S.

Notation. For $A \in \{\mathbb{R}, \mathbb{Z}, \mathbb{Z}^*, [0,1], \mathbb{Z}_n, \mathbb{Z}_n^*\}$ we shall henceforth write just $\mathcal{M}(A)$ instead of $\mathcal{M}(S_A)$, and $\mathcal{M}_m(A)$ instead of $\mathcal{M}_m(S_A)$. (See Example 15.5.) We shall also abbreviate $min_{\leq}(a, b)$ and $max_{\leq}(a, b)$ to $min(a, b)$ and $max(a, b)$ (respectively).

Note 15.11 Obviously, we have that in a (multiplicative) Sugihara matrix $a \tilde{+} b = max_{\leq_+}(a,b)$. (Here, as usual, the connective $+$ is defined by: $\varphi + \psi =_{Df} \neg\varphi \to \psi$, and so $a \tilde{+} b = -a \tilde{\to} b$.) It is also easy to see that the above definition of $\tilde{\to}$ in Sugihara matrices is equivalent to the following original definition of Sugihara ([275]):

$$a \tilde{\to} b = \begin{cases} max(-a,b) & a \leq b, \\ min(-a,b) & a > b. \end{cases}$$

It easily follows that $a \tilde{\to} b \in \mathcal{D}$ iff $a \leq b$.

Note 15.12 It is easy to see that the set $\mathcal{D}$ is upward close in a Sugihara matrix $\mathcal{M}$: If $a \in \mathcal{D}$ and $a \leq b$ (where $\leq$ is the order relation of the Sugihara chain which underlies $\mathcal{M}$), then $b \in \mathcal{D}$.

Proposition 15.13 Let $S = \langle \mathcal{V}, \leq, - \rangle$ be a Sugihara chain, and suppose that $\mathcal{V}'$ is a subset of $\mathcal{V}$ which is closed under $-$, and has at least two elements. Then $S' = \langle \mathcal{V}', \leq, - \rangle$ is also a Sugihara chain, and $\mathcal{M}(S')$ ($\mathcal{M}_m(S')$) is a submatrix of $\mathcal{M}(S)$ ($\mathcal{M}_m(S)$).

Proof. The definitions of the operations immediately imply that if $\mathcal{V}'$ is closed under -, then it is closed under $\tilde{\to}$, $\tilde{\wedge}$, and $\tilde{\vee}$ as well. The proposition easily follows from this fact and from the way the set $\mathcal{D}$ of designated elements is defined in Sugihara matrices. □

Proposition 15.14 Let $S = \langle \mathcal{V}, \leq, - \rangle$ be a Sugihara chain, and let $a \in \mathcal{V}$. Suppose that ν_1 and ν_2 are valuations in $\mathcal{M}(S)$ such that for every variable p: if $max\{\nu_1(p), -\nu_1(p), \nu_2(p), -\nu_2(p)\} \geq a$, then $\nu_1(p) = \nu_2(p)$. Then for every φ: if $max\{\nu_1(\varphi), -\nu_1(\varphi), \nu_2(\varphi), -\nu_2(\varphi)\} \geq a$ then $\nu_1(\varphi) = \nu_2(\varphi)$.

Proof. By induction on the complexity of φ.

- The case where φ is a variable is what is assumed about ν_1 and ν_2.

- The case where $\varphi = \neg\psi$ is immediate from the induction hypothesis and the fact that $--x = x$ for every $x \in \mathcal{V}$.

- Suppose that $\varphi = \psi_1 \vee \psi_2$, and $max\{\nu_1(\varphi), -\nu_1(\varphi), \nu_2(\varphi), -\nu_2(\varphi)\} \geq a$. There are here four subcases to consider:

    - Suppose that $-\nu_1(\varphi) \geq a$. Since $-$ is an involution, this implies that $-\nu_1(\psi_1) \geq a$ and $-\nu_1(\psi_2) \geq a$. It follows by the induction hypothesis for ψ_1 and ψ_2 that $\nu_1(\psi_1) = \nu_2(\psi_1)$ and $\nu_1(\psi_2) = \nu_2(\psi_2)$. Hence $\nu_1(\varphi) = \nu_2(\varphi)$.

15.1 The Logic RM and Its Semantics

- The case where $-\nu_2(\varphi) \geq a$ is similar to the previous one.
- Suppose that $\nu_1(\varphi) \geq a$. Without loss in generality, we may assume that $\nu_1(\varphi) = \nu_1(\psi_1)$. Thus it follows by the induction hypothesis for ψ_1 that $\nu_1(\psi_1) = \nu_2(\psi_1) \geq a$. If also $\nu_2(\psi_2) \geq a$, then $\nu_1(\psi_2) = \nu_2(\psi_2)$ as well (by the induction hypothesis for ψ_2), and so $\nu_1(\varphi) = \nu_2(\varphi)$. If $\nu_2(\psi_2) < a$ then $\nu_2(\varphi) = \nu_2(\psi_1)$, and so again $\nu_1(\varphi) = \nu_2(\varphi)$.
- The case where $\nu_2(\varphi) \geq a$ is similar to the previous one.

- The case where $\varphi = \psi_1 \wedge \psi_2$ follows from the previous ones, since in lattices with an involution we have that $a \wedge b = -(-a \vee -b)$.

- Using Note 15.11, the proof in the case where $\varphi = \psi_1 \to \psi_2$ is similar to that in the case $\varphi = \psi_1 \vee \psi_2$. We leave details to the reader. □

Next, we prove a *strong* soundness and completeness theorem for $\mathbf{RM}_{\to}$ and for $\mathbf{RM}$.[1]

Theorem 15.15 (strong soundness and completeness of RM)

1. $\mathbf{RM}_{\to}$ is strongly sound and complete for the class of multiplicative Sugihara matrices.

2. $\mathbf{RM}_{\to}$ is strongly sound and complete for $\mathcal{M}_m([0,1])$.

3. $\mathbf{RM}$ is strongly sound and complete for the class of Sugihara matrices.

4. $\mathbf{RM}$ is strongly sound and complete for $\mathcal{M}([0,1])$.

Proof.

1. For the soundness part we prove by induction on length of cut-free derivations in $GRM_{\to}$ (Section 13.1.4) that if $\vdash_{GRM_{\to}} \varphi_1, \ldots, \varphi_k \Rightarrow \psi_1, \ldots, \psi_n$ then $\neg\varphi_1 + \cdots + \neg\varphi_k + \psi_1 + \cdots + \psi_n$ is valid in any Sugihara matrix. We leave the straightforward but tedious details to the reader.

   For completeness, assume $\mathcal{T} \nvdash_{\mathbf{RM}_{\to}} \varphi$. Extend $\mathcal{T}$ to a maximal theory $\mathcal{T}^*$ such that $\mathcal{T}^* \nvdash_{\mathbf{RM}_{\to}} \varphi$. Then the relevant deduction theorem of $\mathbf{RM}_{\to}$ (Proposition 13.13) implies that for every sentence ψ, $\psi \notin \mathcal{T}^*$ iff $\psi \to \varphi \in \mathcal{T}^*$. Since $\vdash_{\mathbf{RMI}_{\to}} (\psi \to \varphi) \to (\sigma \to \varphi) \to (\psi + \sigma \to \varphi)$, this in turn implies:

---

[1] Recall that in Chapter 13 we saw that $\mathbf{RM}_{\to}$ is *weakly* (but not strongly) sound and complete for the three-valued matrix $\mathcal{A}_1$ (which can easily be seen to be a multiplicative Sugihara matrix).

(1) If $\psi + \sigma \in \mathcal{T}^*$ then either $\psi \in \mathcal{T}^*$ or $\sigma \in \mathcal{T}^*$.

Now, (1) and the easily verified facts that $\vdash_{\mathbf{RM}_{\to}} (\psi \to \sigma) + (\sigma \to \psi)$ and $\vdash_{\mathbf{RM}_{\to}} \neg \psi + \psi$ imply:

(2) For every ψ, σ, either $\psi \to \sigma \in \mathcal{T}^*$ or $\sigma \to \psi \in \mathcal{T}^*$.

(3) For every sentence ψ, either $\psi \in \mathcal{T}^*$ or $\neg \psi \in \mathcal{T}^*$.

Now, construct the Lindenbaum algebra $\mathcal{M}_{\mathcal{T}^*}$ of $\mathcal{T}^*$ in the usual way: We define that $\psi \equiv \sigma$ iff both $\psi \to \sigma \in \mathcal{T}^*$ and $\sigma \to \psi \in \mathcal{T}^*$. By Proposition 12.16, this is obviously a congruence relation. Let $\mathcal{V}$ be the set of equivalence classes, and let $\mathcal{D} = \{[\psi] \mid \psi \in \mathcal{T}^*\}$. Define the operations $\tilde{\neg}$ and $\tilde{\to}$ on $\mathcal{V}$ by: $[\psi]\tilde{\to}[\sigma] = [\psi \to \sigma]$ and $\tilde{\neg}[\psi] = [\neg \psi]$. To show that the resulting matrix is a multiplicative Sugihara matrix we let $[\psi] \leq [\sigma]$ iff $\psi \to \sigma \in \mathcal{T}^*$. These are all legitimate definitions because $\equiv$ is congruence relation. It is a standard matter to show that $\leq$ is a partial order on $\mathcal{V}$ and that the negation axioms of $\mathbf{RMI}_{\to}$ ensure that $\tilde{\neg}$ is an involution on $\langle \mathcal{V}, \leq \rangle$. By (2) above $\leq$ is also linear. It follows that $S = \langle \mathcal{V}, \leq, \neg \rangle$ is a Sugihara chain. Next we show that $\mathcal{M}_{\mathcal{T}^*} = \mathcal{M}_m(S)$. That $[\psi] \in \mathcal{D}$ iff $\tilde{\neg}[\psi] \leq [\psi]$ easily follows from the definitions of $\mathcal{D}$ and $\leq$, and the fact that both $\psi \to (\neg \psi \to \psi)$ and $(\neg \psi \to \psi) \to \psi$ are theorems of $\mathbf{RMI}_{\to}$. It remains to show that the operation $\tilde{\to}$ of $\mathcal{M}_{\mathcal{T}^*}$ is identical to that of $\mathcal{M}_m(S)$. We use for that the characterization of the latter that is given in Note 15.11.

- Suppose that $[\psi] \leq [\sigma]$. Then $\psi \to \sigma \in \mathcal{T}^*$. Since in $\mathbf{RMI}_{\to}$ both $(\psi \to \sigma) \to (\neg \psi \to (\psi \to \sigma))$ and $(\psi \to \sigma) \to (\sigma \to (\psi \to \sigma))$ are provable, it follows that both $\neg \psi \to (\psi \to \sigma)$ and $\sigma \to (\psi \to \sigma)$ are in $\mathcal{T}^*$. Hence $[\psi]\tilde{\to}[\sigma] \geq max(\tilde{\neg}[\psi], [\sigma])$. To prove the converse, note that since $\leq$ is linear, $max(\tilde{\neg}[\psi], [\sigma])$ is either $[\sigma]$ or $[\neg \psi]$. In the first case $\tilde{\neg}[\psi] \leq [\sigma]$, and so $\neg \psi \to \sigma \in \mathcal{T}^*$. Since $\vdash_{\mathbf{RMI}_{\to}} (\neg \psi \to \sigma) \to ((\psi \to \sigma) \to \sigma)$, we get that in this case $[\psi \to \sigma] \leq [\sigma]$. In the second case $[\sigma] \leq \tilde{\neg}[\psi]$, and so $\sigma \to \neg \psi \in \mathcal{T}^*$. Since $\vdash_{\mathbf{RMI}_{\to}} (\sigma \to \neg \psi) \to ((\psi \to \sigma) \to \neg \psi)$, we get that in this case $[\psi \to \sigma] \leq \tilde{\neg}[\psi]$. In both cases we have that $[\psi]\tilde{\to}[\sigma] = [\psi \to \sigma] \leq max(\tilde{\neg}[\psi], [\sigma])$.

- Suppose that $[\psi] \not\leq [\sigma]$. Then $\psi \to \sigma \notin \mathcal{T}^*$. By (3) this implies that $\neg(\psi \to \sigma) \in \mathcal{T}^*$. Since $\neg(\psi \to \sigma) \to ((\psi \to \sigma) \to \sigma)$ and $\neg(\psi \to \sigma) \to ((\psi \to \sigma) \to \neg \psi)$ are theorems of $\mathbf{RMI}_{\to}$, it follows that both $(\psi \to \sigma) \to \sigma$ and $(\psi \to \sigma) \to \neg \psi$ are elements of $\mathcal{T}^*$. Hence $[\psi]\tilde{\to}[\sigma] \leq min(\tilde{\neg}[\psi], [\sigma])$. To prove the converse, note that since $\leq$ is linear, $min(\tilde{\neg}[\psi], [\sigma])$ is either $[\sigma]$

15.1 The Logic RM and Its Semantics

or $[\neg\psi]$. In the first case $\tilde{\neg}[\psi] \leq [\sigma]$, and so $\neg\psi \to \sigma \in \mathcal{T}^*$. Since $\vdash_{\mathbf{RMI}_\to} (\neg\psi \to \sigma) \to (\neg\psi \to (\psi \to \sigma))$, we get that $\tilde{\neg}[\psi] \leq [\psi \to \sigma]$ in this case. In the second case $[\sigma] \leq \tilde{\neg}[\psi]$, and so $\sigma \to \neg\psi \in \mathcal{T}^*$. Since $\vdash_{\mathbf{RMI}_\to} (\sigma \to \neg\psi) \to (\sigma \to (\psi \to \sigma))$, we get that $[\sigma] \leq [\psi \to \sigma]$ in this case. In both cases we have that $[\psi]\tilde{\to}[\sigma] = [\psi \to \sigma] \geq min(\tilde{\neg}[\psi], [\sigma])$.

The end of the proof is now standard. Let $\nu(\psi) = [\psi]$. This is easily seen to be a valuation (the canonical one) in $\mathcal{M}_{\mathcal{T}^*}$. Obviously, ν is a model ψ iff $\psi \in \mathcal{T}^*$. Hence ν is a model of $\mathcal{T}$ in the Sugihara matrix $\mathcal{M}_{\mathcal{T}^*}$ which is not a model of φ.

2. The multiplicative Sugihara matrix constructed in the proof of the first part is countable. Hence, the second part follows from the first (and its proof), Lemma 15.7 and Proposition 15.13.

3. Given the strong soundness of $\mathbf{RM}_\to$ for multiplicative Sugihara matrices, the proof of the strong soundness of $\mathbf{RM}$ for Sugihara matrices is straightforward, and is left to the reader.

   For completeness, assume that $\mathcal{T} \not\vdash_{\mathbf{RM}} \varphi$. Extend $\mathcal{T}$ to a maximal theory $\mathcal{T}^*$ such that $\mathcal{T}^* \not\vdash_{\mathbf{RM}} \varphi$. Then $\psi \notin \mathcal{T}^*$ iff $\mathcal{T}^*, \psi \vdash_{\mathbf{RM}} \varphi$. Hence Theorem 14.18 implies that $\mathcal{T}^*$ is *prime*, i.e.: if $\psi \vee \sigma \in \mathcal{T}^*$ then either $\psi \in \mathcal{T}^*$ or $\sigma \in \mathcal{T}^*$. Therefore it follows from the second item of Lemma 14.17 that (1) from the proof of the first part holds for $\mathcal{T}^*$. From this point the proof is almost identical to the proof of the first part, except that we show that $\mathcal{M}_{\mathcal{T}^*} = \mathcal{M}(S)$ (where S is defined like in that proof) rather than that $\mathcal{M}_{\mathcal{T}^*} = \mathcal{M}_m(S)$. For this, all we have to add to the proof of the first part is that $[\psi \wedge \sigma] = min([\psi], [\sigma])$ and $[\psi \vee \sigma] = max([\psi], [\sigma])$. This is obvious from the axioms concerning $\wedge$ and $\vee$ of $\mathbf{RM}$, and the linearity of $\leq$.

4. The proof is almost identical to that of the second part. □

As an easy corollary of the last theorem we have:

Theorem 15.16 *The logic $\mathbf{RM}$ is a conservative extension of $\mathbf{RM}_\to$: if $\mathcal{T} \cup \{\varphi\}$ is in the language $\{\neg, \to\}$ then $\mathcal{T} \vdash_{\mathbf{RM}} \varphi$ iff $\mathcal{T} \vdash_{\mathbf{RM}_\to} \varphi$.*

Proof. Immediate from Theorem 15.15, since every multiplicative Sugihara matrix can (uniquely) be extended to a Sugihara matrix. □

Next we show that for *weak* completeness of $\mathbf{RM}$ we do not need an uncountable matrix like $\mathcal{M}([0, 1])$. Instead, each of the countable Sugihara matrices $\mathcal{M}(\mathbb{Z})$ and $\mathcal{M}(\mathbb{Z}^*)$ suffices for this, and the same applies to the set of finite Sugihara matrices.

Definition 15.17 (The matrices $\mathcal{RM}_n$) For $k = 1, 2, \ldots$ we let $\mathcal{RM}_{2k} = \mathcal{M}(\mathbb{Z}_k^*)$ and $\mathcal{RM}_{2k+1} = \mathcal{M}(\mathbb{Z}_k)$.

Note 15.18 It is easy to see that $\mathcal{RM}_2$ is just the classical two-valued matrix for $\mathcal{L}_{CL}$, while $\mathcal{RM}_3$ is the three-valued extension of $\mathcal{A}_1$ which according to Note 4.60 is equivalent to the three-valued matrix PAC.

Proposition 15.19 *Every finite Sugihara matrix which has n elements is isomorphic to $\mathcal{RM}_n$. Hence, every such matrix is isomorphic to some finite submatrix of $\mathcal{M}(\mathbb{Z})$.*

Proof. This is an easy corollary of Lemma 15.6 and Proposition 15.13. □

Theorem 15.20 *Let $\mathcal{T}$ be a theory and φ a sentence such that $\mathsf{Var}(\mathcal{T} \cup \{\varphi\})$ is finite.*

1. *If n is the number of variables which occur in $\mathcal{T} \cup \{\varphi\}$, then $\mathcal{T} \vdash_{\mathbf{RM}} \varphi$ iff $\mathcal{T} \vdash_{\mathcal{RM}_k} \varphi$ for every $2 \leq k \leq 2n$.*

2. *$\mathcal{T} \vdash_{\mathbf{RM}} \varphi$ iff $\mathcal{T} \vdash_{\mathcal{M}(\mathbb{Z})} \varphi$.*

Proof. From the soundness of **RM** for Sugihara matrices it follows that if $\mathcal{T} \vdash_{\mathbf{RM}} \varphi$ then $\mathcal{T} \vdash_{\mathcal{M}(\mathbb{Z})} \varphi$, and $\mathcal{T} \vdash_{\mathcal{RM}_k} \varphi$ for every $k \geq 2$. For the converse, assume $\mathcal{T} \not\vdash_{\mathbf{RM}} \varphi$. By Theorem 15.15 there is a Sugihara chain $S = \langle \mathcal{V}, \leq, - \rangle$ and a valuation ν in $\mathcal{M}(S)$ which is a model of $\mathcal{T}$ but not of φ. Suppose $\mathsf{Var}(\mathcal{T} \cup \{\varphi\}) = \{p_1, \ldots, p_n\}$, and let $\mathcal{V}' = \{\nu(p_1), -\nu(p_1), \ldots, \nu(p_n), -\nu(p_n)\}$. An easy induction on the complexity of a sentence ψ shows that $\nu(\psi) \in \mathcal{V}'$ for every ψ such that $\mathsf{Var}(\psi) \subseteq \{p_1, \ldots, p_n\}$. Since ν is not a model of φ, this implies that $\mathcal{V}'$ has at least two elements (and of course not more than $2n$). Hence, Proposition 15.13 implies that $S' = \langle \mathcal{V}', \leq, - \rangle$ is also a Sugihara chain, and $\mathcal{M}(S')$ is a submatrix of $\mathcal{M}(S)$. Let ν' be any valuation in $\mathcal{M}(S')$ such that $\nu'(p_i) = \nu(p_i)$ for $1 \leq i \leq n$. Then $\nu'(\psi) = \nu(\psi)$ or every ψ such that $\mathsf{Var}(\psi) \subseteq \{p_1, \ldots, p_n\}$. It follows that ν' is a model of $\mathcal{T}$ in $\mathcal{M}(S')$ which is not a model of φ. Hence $\mathcal{T} \not\vdash_{\mathcal{M}(S')} \varphi$. By Proposition 15.19 this implies that $\mathcal{T} \not\vdash_{\mathcal{RM}_k} \varphi$ for some $2 \leq k \leq 2n$, and that $\mathcal{T} \not\vdash_{\mathcal{M}(\mathbb{Z})} \varphi$. □

Corollary 15.21 *If $\mathcal{T}$ is a finite theory then $\mathcal{T} \vdash_{\mathbf{RM}} \varphi$ iff $\mathcal{T} \vdash_{\mathcal{M}(\mathbb{Z})} \varphi$. In particular: $\mathcal{M}(\mathbb{Z})$ is weakly characteristic for **RM**.*

In contrast, we have:

Proposition 15.22 *$\mathcal{M}(\mathbb{Z})$ is not strongly characteristic for **RM**.*

15.1 The Logic RM and Its Semantics

Proof. Let $\mathcal{T} = \{P_i \mid i \geq 2\} \cup \{(P_i \to P_{i+1}) \to P_1 \mid i \geq 2\}$, and let $S = \langle \mathcal{V}, \leq, - \rangle$ be a Sugihara chain. It is not difficult to check that a valuation ν in $\mathcal{M}(S)$ can be a model of $\mathcal{T}$ which is not a model of P_1 iff $\nu(P_1) < -\nu(P_1)$, while for $i > 1$ $\nu(P_i) \geq -\nu(P_i)$ and $\nu(P_i) > \nu(P_{i+1})$. Such ν does not exist in $\mathcal{M}(\mathbb{Z})$, but it does in $\mathcal{M}([0,1])$. Hence $\mathcal{T} \vdash_{\mathcal{M}(\mathbb{Z})} P_1$, while $\mathcal{T} \nvdash_{\mathbf{RM}} P_1$. □

The characterization of **RM** in terms of finite matrices which is given in Theorem 15.20 can in fact be improved using the next proposition.

Proposition 15.23 *For every $n \geq 2$, if $\vdash_{\mathcal{RM}_{n+1}} \varphi$ then also $\vdash_{\mathcal{RM}_n} \varphi$.*

Proof. The claim is obvious in case n is even, since $\mathcal{RM}_{2k}$ is a submatrix of $\mathcal{RM}_{2k+1}$ for every $k \geq 1$.

Now suppose that $n = 2k+1$ for some $k \geq 1$, and that $\nvdash_{\mathcal{RM}_{2k+1}} \varphi$. We show that also $\nvdash_{\mathcal{RM}_{2k+2}} \varphi$. Let ν be a valuation in $\mathcal{RM}_{2k+1}$ such that $\nu(\varphi) < 0$. Define a valuation ν^* in $\mathcal{RM}_{2k+2}$ by letting $\nu^*(p) = \nu(p) + 1$ in case $\nu(p) \geq 0$, and $\nu^*(p) = \nu(p) - 1$ in case $\nu(p) < 0$. By induction on the complexity of ψ it is not difficult to show that for every sentence ψ we have:

- If $\nu(\psi) > 0$ then $\nu^*(\psi) = \nu(\psi) + 1$.
- If $\nu(\psi) = 0$ then $\nu^*(\psi) \in \{-1, 1\}$.
- If $\nu(\psi) < 0$ then $\nu^*(\psi) = \nu(\psi) - 1$.

It follows in particular that $\nu^*(\varphi) < 0$. Hence $\nvdash_{\mathcal{RM}_{2k+2}} \varphi$. □

Corollary 15.24 *For every $n \geq 2$, if $\vdash_{\mathcal{RM}_n} \varphi$ then $\vdash_{\mathcal{RM}_m} \varphi$ for every $2 \leq m \leq n$.*

Proposition 15.25 *Suppose that $|\mathrm{Var}(\varphi)| = n$. Then $\vdash_{\mathbf{RM}} \varphi$ iff $\vdash_{\mathcal{RM}_{2n}} \varphi$.*

Proof. Immediate from Proposition 15.20 and Corollary 15.24. □

Proposition 15.26 *$\mathcal{M}(\mathbb{Z}^*)$ is weakly characteristic for **RM**.*

Proof. That if $\vdash_{\mathbf{RM}} \varphi$ then $\vdash_{\mathcal{M}(\mathbb{Z}^*)} \varphi$ follows from the soundness of **RM** for Sugihara matrices. For the converse, assume that $\nvdash_{\mathbf{RM}} \varphi$. Then by Proposition 15.25, there is n such that $\nvdash_{\mathcal{RM}_{2n}} \varphi$. Since $\mathcal{RM}_{2n}$ is a submatrix of $\mathcal{M}(\mathbb{Z}^*)$, this implies that $\nvdash_{\mathcal{M}(\mathbb{Z}^*)} \varphi$. □

Corollary 15.27 *If Γ is a finite set of sentences, and $\Gamma \vdash_{\mathcal{M}(\mathbb{Z}^*)} \varphi$, then the rule Γ/φ is admissible in **RM**.*

Proof. Let θ be a substitution such that $\vdash_{\mathbf{RM}} \theta(\psi)$ for every $\psi \in \Gamma$. By Proposition 15.26, $\vdash_{\mathcal{M}(\mathbb{Z}^*)} \theta(\psi)$ for every $\psi \in \Gamma$. Since $\Gamma \vdash_{\mathcal{M}(\mathbb{Z}^*)} \varphi$, $\vdash_{\mathcal{M}(\mathbb{Z}^*)} \theta(\varphi)$ as well. Hence $\vdash_{\mathbf{RM}} \theta(\varphi)$ by Proposition 15.26 again. □

Corollary 15.28 *The disjunctive syllogism is admissible in* **RM**: $\vdash_{\mathbf{RM}} \psi$ *whenever both* $\vdash_{\mathbf{RM}} \neg\varphi$ *and* $\vdash_{\mathbf{RM}} \varphi \vee \psi$.

Proof. It is easy to see that $\neg p, p \vee q \vdash_{\mathcal{M}(\mathbb{Z}^*)} q$. Hence the claim follows from Corollary 15.27. □

Note 15.29 In contrast, $\neg p, p \vee q \nvdash_{\mathcal{M}(\mathbb{Z})} q$. By Theorem 15.20 this implies that $\neg p, p \vee q \nvdash_{\mathbf{RM}} q$. It follows that the analogue of Corollary 15.27 does *not* hold for $\mathcal{M}(\mathbb{Z}^*)$.

Note 15.30 Corollary 15.28 and the second part of Proposition 14.64 indicate another important difference between **RMI** and **RM**. The significance of this difference is reflected in the next proposition.

Proposition 15.31 *Let* $\mathbf{RMI}^{DS}$ *and* $\mathbf{RM}^{DS}$ *be the systems which are obtained from* **RMI** *and* **RM** *(respectively) by adding the disjunctive syllogism as an extra rule of inference.*

1. $\vdash_{\mathbf{RMI}^{DS}} \varphi$ *iff* $\vdash_{\mathbf{RM}^{DS}} \varphi$ *iff* $\vdash_{\mathbf{RM}} \varphi$.
2. $\mathbf{RMI}^{DS} = \mathbf{RM}^{DS}$.

Proof. Obviously, $\vdash_{\mathbf{RMI}} \neg(p \to p) \vee (\neg(q \to q) \vee p\mathcal{R}^\wedge q)$, $\vdash_{\mathbf{RMI}} \neg\neg(p \to p)$, and $\vdash_{\mathbf{RMI}} \neg\neg(q \to q)$. This implies that $\vdash_{\mathbf{RMI}^{DS}} p\mathcal{R}^\wedge q$. The two parts of the proposition easily follow from this fact using Proposition 15.2 and Corollary 15.28. □

15.2 The Nice Properties of RM

The logic **RM** has several nice properties. The first we present is one that the main logics developed by Anderson and Belnap's school lack.

Theorem 15.32 **RM** *is decidable.*

Proof. Immediate from Theorem 15.20. (See also Proposition 15.25 for the special case of theoremhood in **RM**). □

Proposition 15.33 **RM** *is normal.*

Proof. From Note 1.25 it follows that $\wedge$ is a conjunction for **RM**. That **RM** has also a disjunction and an implication follows from Theorem 14.18 and Proposition 14.47 (using Proposition 15.2). □

Proposition 15.34 *Let* $\mathbf{F}_R$ *be like in Item 9 of Section 14.2.3. Then* **RM** *is* $\mathbf{F}_R$-*contained in classical logic. Hence it is* $\neg$-*contained in classical logic.*

15.2 The Nice Properties of RM

Proof. Immediate from the fact that the matrix $\langle \{t, f\}, \{t\}, \mathbf{F}_R \rangle$ can easily be turned into a Sugihara matrix which is isomorphic to the classical two-valued matrix. □

Proposition 15.35 *RM satisfies the basic relevance criterion.*

Proof. Suppose that $\mathcal{T}_1, \mathcal{T}_2 \vdash_{\mathbf{RM}} \psi$ and $\mathcal{T}_2$ has no variable in common with $\mathcal{T}_1 \cup \{\psi\}$. We show that $\mathcal{T}_1 \vdash_{\mathbf{RM}} \psi$. Suppose otherwise. Then by Theorem 15.15 there is a valuation ν in $\mathcal{M}([0,1])$ such that $\nu(\varphi) \geq 1/2$ for every $\varphi \in \mathcal{T}_1$, while $\nu(\psi) < 1/2$. Since $\mathcal{T}_2$ has no variable in common with $\mathcal{T}_1 \cup \{\psi\}$, we may assume without loss in generality that $\nu(p) = 1/2$ for every variable p which occurs in $\mathcal{T}_2$. But then $\nu(\varphi) = 1/2$ for every $\varphi \in \mathcal{T}_2$, and so ν is a model in $\mathcal{M}([0,1])$ of $\mathcal{T}_1 \cup \mathcal{T}_2$ which is not model of ψ. By Theorem 15.15, this contradicts our assumption that $\mathcal{T}_1, \mathcal{T}_2 \vdash_{\mathbf{RM}} \psi$. □

Corollary 15.36 *RM is boldly paraconsistent.*

Proof. This follows from Proposition 15.34, Proposition 15.33, Proposition 15.35, and Corollary 11.4. □

Next we show that **RM** is a semi-relevant logic (Definition 11.12).

Definition 15.37 For $n \geq 2$ let φ_n^R be the formula $\bigvee_{1 \leq i < j \leq n+1} P_i \to P_j$.

Lemma 15.38 *Let $\mathcal{M} = \langle \mathcal{V}, \mathcal{D}, \mathcal{O} \rangle$ be a finite Nmatrix for $\mathcal{L}_R$. Suppose that $\mathbf{R}$ is weakly sound for $\mathcal{M}$, and that $\mathcal{V}$ has at most n elements. Then φ_n^R is valid in $\mathcal{M}$.*

Proof. Let ν be any valuation in $\mathcal{M}$. Since $\mathcal{M}$ has less than $n+1$ elements, there are necessarily $1 \leq i < j \leq n+1$ such that $\nu(P_i) = \nu(P_j)$. For every formula ψ let ψ^* be the formula which is obtained from ψ by replacing every subformula of ψ of the form $P_i \to P_i$ by $P_i \to P_j$. Define a valuation ν^* by letting $\nu^*(\psi) = \nu(\psi^*)$ for every ψ. It is easy to verify that ν^* is indeed a legitimate valuation in $\mathcal{M}$. Obviously, $\nu(\varphi_n^R) = \nu^*(\psi)$, where ψ is the formula which is obtained from φ_n^R by replacing the unique occurrence of $P_i \to P_j$ in φ_n^R by $P_i \to P_i$. But ψ has the form $\tau \vee (P_i \to P_i) \vee \sigma$ (where either τ or σ might be empty), and so it is a theorem of $\mathbf{R}$. Since $\mathbf{R}$ is weakly sound for $\mathcal{M}$, this implies that $\nu^*(\psi) \in \mathcal{D}$, and so $\nu(\varphi_n^R) \in \mathcal{D}$ too. It follows that φ_n^R is valid in $\mathcal{M}$. □

Proposition 15.39 *R and RM have no finite weakly characteristic Nmatrix. In particular: they satisfies the minimal semantic criterion.*

Proof. Using $\mathcal{M}(\mathbb{Z})$ and Corollary 15.21, it is easy to show that for no n is φ_n^R (Definition 15.37) a tautology of **RM** (and so of **R**). Since **R** and **RM** are simple extensions of **R**, Lemma 15.38 implies that neither of them has a finite weakly characteristic Nmatrix. □

Theorem 15.40 **RM** *is a normal semi-relevant logic.*

Proof. This follows from Propositions 15.33, 15.35, and 15.39. □

Another way in which the logic **RM** is "semi-relevant" is given in the next proposition:

Proposition 15.41

1. **RM** *does not have the VSP.*

2. *If* $\vdash_{\mathbf{RM}} \varphi \to \psi$ *then either* φ *and* ψ *share a variable, or both* $\neg\varphi$ *and* ψ *are provable in* **RM**.

Proof.

1. $\neg(p \to p) \to (q \to q)$ is a theorem of $\mathbf{RM}_{\to}$, and so also of **RM**.

2. Suppose that $\vdash_{\mathbf{RM}} \varphi \to \psi$, but φ and ψ share no variable. We show that both $\neg\varphi$ and ψ are provable in **RM**. Suppose e.g. that $\neg\varphi$ is not (the argument in the case where ψ is not valid in **RM** is similar). Then by Theorem 15.20 there is valuation ν in $\mathcal{M}(\mathbb{Z})$ such that $\nu(\neg\varphi) < 0$, and so $\nu(\varphi) > 0$. Without a loss in generality we may assume that $\nu(q) = 0$ for every variable $q \notin \mathsf{Var}(\varphi)$. Since φ and ψ share no variable, this implies that $\nu(q) = 0$ for every variable $q \in \mathsf{Var}(\psi)$. But then $\nu(\psi) = 0$. Since $\nu(\varphi) > 0$, this implies that $\nu(\varphi \to \psi) < 0$, contradicting the assumption that $\vdash_{\mathbf{RM}} \varphi \to \psi$. □

Note 15.42 In relevant logics, if $\varphi \to \psi$ is a tautology then φ and ψ share a variable. On the other hand, in classical logic there are in such a case two other possibilities: that $\neg\varphi$ is a tautology, and that ψ is a tautology. Proposition 15.41 shows that **RM** is intermediate in this respect between relevant logics and classical logic. This intuitively provides an additional justification for seeing **RM** as a 'semi-relevant' logic.

Next we show that **RM** has the Scroggs' property (Theorem 13.33) too.

Theorem 15.43 *Let* **L** *be a simple strongly proper extension of* **RM**. *Then there is a natural number* $n \geq 2$ *such that* $Th(\mathbf{L}) = Th(\mathcal{RM}_n)$, *i.e.* $\mathcal{RM}_n$ *is weakly characteristic for* **L**.

15.2 The Nice Properties of RM

Proof. First, we prove that all theorems of **L** are valid in $\mathcal{RM}_2$. Suppose for contradiction that there is a theorem φ of **L** which is not valid in $\mathcal{RM}_2$. Then there is a valuation ν_0 in $\mathcal{RM}_2$ such that $\nu_0(\varphi) = -1$. By substituting in φ $P_0 \to P_0$ for every variable p such that $\nu_0(p) = 1$ and $\neg(P_0 \to P_0)$ for every variable p such that $\nu_0(p) = -1$ we obtain a theorem ψ of **L** such that $\mathsf{Var}(\psi) = \{P_0\}$, and $\nu(\psi) = -1$ for any valuation ν in $\mathcal{RM}_2$. It follows that $\neg\psi$ is valid in $\mathcal{RM}_2$. Therefore, Proposition 15.25 implies that $\vdash_{\mathbf{RM}} \neg\psi$. Hence both ψ and $\neg\psi$ are theorems of **L**. But because $\mathsf{Var}(\psi) = \{P_0\}$, the first part of Theorem 15.20 implies that $\neg\psi, \psi \vdash_{\mathbf{RM}} P_0$. It follows that $\vdash_{\mathbf{L}} P_0$, contradicting the assumption that **L** is a logic.

Now, let A be the set of all natural numbers n such that all theorems of **L** are valid in $\mathcal{RM}_n$. By what we have just proved, $2 \in A$, and so A is not empty. On the other hand, the fact that **L** is a simple strongly proper extension of **RM** means that there is a sentence φ_0 of $\mathcal{L}_R$ such that $\vdash_{\mathbf{L}} \varphi_0$, but $\nvdash_{\mathbf{RM}} \varphi_0$. Thus, Proposition 15.25 implies that there is $n_0 \geq 2$ such that φ_0 is not valid in $\mathcal{RM}_{n_0}$, and so $n_0 \notin A$. It follows by Corollary 15.24 that A has a maximal element $k \geq 2$. Then by Corollary 15.24 again, every theorem of **L** is valid in $\mathcal{RM}_j$ for every $2 \leq j \leq k$, and there is a theorem of **L** which is not valid in $\mathcal{RM}_j$ for $j > k$.

We end the proof by showing that $\mathcal{RM}_k$ is weakly characteristic for **L**. Since $k \in A$, it suffices to show that if $\nvdash_{\mathbf{L}} \varphi$ then φ is not valid in $\mathcal{RM}_k$. So suppose that $\nvdash_{\mathbf{L}} \varphi$, and let $\mathsf{Var}(\varphi) = \{p_1, \ldots, p_n\}$. Define:

$$\mathcal{T} = \{\sigma \mid \mathsf{Var}(\sigma) \subseteq \{p_1, \ldots, p_n\} \text{ and } \vdash_{\mathbf{L}} \sigma\}.$$

Since $\nvdash_{\mathbf{L}} \varphi$, also $\mathcal{T} \nvdash_{\mathbf{RM}} \varphi$. Therefore, Theorem 15.20 and its proof imply that there is l and a valuation ν_0 in $\mathcal{RM}_l$ such that ν_0 is a model of $\mathcal{T}$ in $\mathcal{RM}_l$ which is not a model of φ, and for every element a of $\mathcal{RM}_l$ there is $1 \leq i \leq n$ such that either $a = \nu_0(p_i)$ or $a = -\nu_0(p_i) = \nu_0(\neg p_i)$. We show that $l \in A$. So let σ be a theorem of **L**, and let ν be a valuation in $\mathcal{RM}_l$. Let θ be a substitution that assigns to any variable q an element τ of $\{p_1, \neg p_1, \ldots, p_n, \neg p_n\}$ such that $\nu(q) = \nu_0(\tau)$. Then for any variable q, $\nu(q) = (\nu_0 \circ \theta)(q)$. Hence, Proposition 3.9 implies that $\nu = \nu_0 \circ \theta$, and so $\nu(\sigma) = \nu_0(\theta(\sigma))$. But since **L** is a logic, $\theta(\sigma)$ is also a theorem of **L**, and by definition of θ this implies that $\theta(\sigma) \in \mathcal{T}$. Since ν_0 is a model of $\mathcal{T}$, $\nu_0(\theta(\sigma))$ is designated, and so $\nu(\sigma)$ is designated. This was shown for every valuation ν in $\mathcal{RM}_l$ and any theorem σ of **L**, and so it follows that indeed $l \in A$. Hence $l \leq k$. Since φ is not valid in $\mathcal{RM}_l$ (because $\nu_0(\varphi)$ is not designated), φ is not valid in $\mathcal{RM}_k$ either. □

Theorem 15.44 *RM has the Scroggs' property: it does not have a finite weakly characteristic matrix, but every strongly proper extension of it does.*

Proof. This follows from Proposition 15.39 and Theorem 15.43. □

By Theorem 15.43, if **L** is a simple extension of **RM**, then $Th(\mathbf{L})$ belongs to the sequence $\{Th(\mathcal{RM}_n)\}_{n=2}^{\infty}$. Next we axiomatize each of the elements in this sequence, and show that they are all different from each other.

Definition 15.45 $\mathbf{RM}_n$ is the simple axiomatic extension of **RM** which is obtained by adding φ_n^R to *HRM* as an extra axiom-schema (i.e. by adding to *HRM* all substitution instances of φ_n^R as new axioms).

Note 15.46 From Note 15.18 it immediately follows that $\mathbf{RM}_2$ is just classical logic, while $\mathbf{RM}_3$ is equivalent to **PAC**. (See Section 4.4.4.)

Theorem 15.47

1. *For every $n \geq 2$ and $\varphi \in \mathcal{L}_R$, φ is valid in $\mathcal{RM}_n$ iff $\vdash_{\mathbf{RM}_n} \varphi$. (In other words: $Th(\mathcal{RM}_n) = Th(\mathbf{RM}_n)$ for every $n \geq 2$.)*

2. *The sequence $\{Th(\mathcal{RM}_n)\}_{n=2}^{\infty}$ is strictly decreasing, and includes every set of the form $Th(\mathbf{L})$ such that **L** is a simple strongly proper extension of **RM**.*

Proof. It is easy to check that for every $n \geq 2$, φ_n^R is valid in $\mathcal{RM}_n$ but not in $\mathcal{RM}_{n+1}$. Hence n is the maximal number k such that φ_n^R is valid in $\mathcal{RM}_k$. Therefore Theorem 15.43 and its proof imply the first part, and also that the sequence $\{Th(\mathcal{RM}_n)\}_{n=2}^{\infty}$ includes every set of the form $Th(\mathbf{L})$ such that **L** is a simple strongly proper extension of **RM**. Proposition 15.23 implies that this sequence is decreasing. That it is strictly decreasing again follows from the fact that φ_n^R is valid in $\mathcal{RM}_n$ but not in $\mathcal{RM}_{n+1}$. □

Now we turn to what is perhaps the most important property of **RM**.

Theorem 15.48 (maximality of RM)

1. **RM** *is a maximal finitary logic which is both normal and semi-relevant logic. In other words: every proper simple finitary extension of **RM** is either not normal or not semi-relevant.*

2. **RM** *is a maximal finitary paraconsistent logic which is normal and satisfies also the minimal semantic relevance criterion.*

Proof.

1. Let **L** be a simple finitary extension of **RM** which is both normal and semi-relevant. We show that $\mathbf{L} = \mathbf{RM}$. Now, by Theorem 15.43 no strongly proper extension of **RM** can be semi-relevant. It follows that

15.2 The Nice Properties of RM

$Th(\mathbf{L}) = Th(\mathbf{RM})$. Let $\Rightarrow$ be a defined connective of $\mathcal{L}_R$ which is an implication for $\mathbf{L}$. Then for every $\mathcal{T}, \varphi, \psi$: $\mathcal{T}, \varphi \vdash_\mathbf{L} \psi$ iff $\mathcal{T} \vdash_\mathbf{L} \varphi \Rightarrow \psi$. In the sequel we denote by $\tilde{\neg}, \tilde{\vee}, \tilde{\wedge}, \tilde{\rightarrow}$, and $\tilde{\Rightarrow}$ the interpretations in $\mathcal{M}(\mathbb{Z})$ of $\neg, \vee, \wedge, \rightarrow$, and $\Rightarrow$, respectively. In what follows we extensively use the following property of these operations:

> (*) If $f : \mathbb{Z}^n \to \mathbb{Z}$ is obtained from $\tilde{\neg}, \tilde{\vee}, \tilde{\wedge}, \tilde{\rightarrow}$, and $\tilde{\Rightarrow}$ using compositions, then for every $a_1, \ldots, a_n \in \mathbb{Z}$:
> $f(a_1, \ldots, a_n) \in \{a_1, -a_1, a_2, -a_2, \ldots, a_n, -a_n\}$.

Next, we prove some properties of $\tilde{\Rightarrow}$.

(a) For $n \geq 0$: $n\tilde{\Rightarrow}n = -n\tilde{\Rightarrow}n = -n\tilde{\Rightarrow}-n = n$, while $n\tilde{\Rightarrow}-n = -n$.

   Proof. Since $p \vdash_\mathbf{L} p$, $\vdash_\mathbf{L} p \Rightarrow p$. Hence $\vdash_\mathbf{RM} p \Rightarrow p$, and so $a\tilde{\Rightarrow}a \geq 0$ for every $a \in \mathbb{Z}$. By (*) this implies that $n\tilde{\Rightarrow}n = -n\tilde{\Rightarrow}-n = n$ for every $n \geq 0$. Now, the fact that $\vdash_\mathbf{L} p \Rightarrow p$ implies that $p \vdash_\mathbf{L} p \Rightarrow p$, and so $\vdash_\mathbf{L} p \Rightarrow (p \Rightarrow p)$. It follows that $\vdash_\mathbf{RM} p \Rightarrow (p \Rightarrow p)$. Hence, for every $n \geq 0$, $-n\tilde{\Rightarrow}(-n\tilde{\Rightarrow}-n) \geq 0$, and so $-n\tilde{\Rightarrow}n \geq 0$. By (*) this implies that $-n\tilde{\Rightarrow}n = n$. Next we note that since $\mathbf{L}$ is a logic (i.e. non-trivial), the fact that $\vdash_\mathbf{L} p \Rightarrow p$ implies that $p \Rightarrow p \not\vdash_\mathbf{L} p$, and so $\not\vdash_\mathbf{L} (p \Rightarrow p) \Rightarrow p$. It follows that there is some $a \in \mathbb{Z}$ such that $(a\tilde{\Rightarrow}a)\tilde{\Rightarrow}a < 0$. By what we have already shown, if $a \geq 0$ then $(a\tilde{\Rightarrow}a)\tilde{\Rightarrow}a = a \geq 0$. Hence, necessarily $a < 0$. So let $a = -k$ for some $k > 0$. Then $(-k\tilde{\Rightarrow}-k)\tilde{\Rightarrow}-k < 0$, and so $k\tilde{\Rightarrow}-k < 0$. By (*) this implies that $k\tilde{\Rightarrow}-k = -k$. Now by Lemma 15.6, for every $n > 0$ the submatrix of $\mathcal{M}(\mathbb{Z})$ induced by $\{-n, n\}$ is isomorphic to the submatrix of $\mathcal{M}(\mathbb{Z})$ induced by $\{-k, k\}$. It follows that $n\tilde{\Rightarrow}-n = -n$ for every $n > 0$, and (*) implies that $n\tilde{\Rightarrow}-n = -n$ also when $n = 0$.

(b) $a\tilde{\Rightarrow}k \in \{|a|, k\}$ for every $a \in \mathbb{Z}$ and $k \geq 0$.

   Proof. Since $\vdash_\mathbf{L} p \Rightarrow p$, also $q \vdash_\mathbf{L} p \Rightarrow p$. Hence $\vdash_\mathbf{L} q \Rightarrow (p \Rightarrow p)$, and so $\vdash_\mathbf{RM} q \Rightarrow (p \Rightarrow p)$. Hence $a\tilde{\Rightarrow}(k\tilde{\Rightarrow}k) \geq 0$ for every $a \in \mathbb{Z}$ and $k \geq 0$. By item 1 above this means that $a\tilde{\Rightarrow}k \geq 0$ for every $a \in \mathbb{Z}$ and $k \geq 0$. Hence (*) implies that $a\tilde{\Rightarrow}k \in \{|a|, k\}$ for every $a \in \mathbb{Z}$ and $k \geq 0$.

(c) For every $a \in \mathbb{Z}$ and $k \geq 0$, if $|a| \leq k$ then $-k\tilde{\Rightarrow}a \in \{|a|, k\}$.

   Proof. The fact that $\vdash_{\mathbf{RMI}_{\vec{\neg}}} \neg((p \to q) \to (p \to q)) \to p$ entails that $\neg((p \to q) \to (p \to q)) \vdash_\mathbf{L} p$. Hence $\vdash_\mathbf{L} \neg((p \to q) \to (p \to q)) \Rightarrow p$, and so $\vdash_\mathbf{RM} \neg((p \to q) \to (p \to q)) \Rightarrow p$. It follows that if $a \in \mathbb{Z}$ and $k \geq 0$ then $-((a\tilde{\to}k)\tilde{\to}(a\tilde{\to}k))\tilde{\Rightarrow}a \geq 0$. Now, if

$|a| \leq k$ then $-((a\tilde{\to}k)\tilde{\to}(a\tilde{\to}k)) = -k$, and so in such a case we get that $-k\tilde{\Rightarrow}a \geq 0$. By (*) this is equivalent to $-k\tilde{\Rightarrow}a \in \{|a|, k\}$.

(d) If $0 \leq k \leq n$ then $k\tilde{\Rightarrow}-n = -n$.

Proof. Since **L** is semi-relevant, $\neg(p \to p), (p \to p) \not\vdash_\mathbf{L} q$. Hence $\neg(p \to p) \not\vdash_\mathbf{L} (p \to p) \Rightarrow q$, and so $\neg(p \to p) \not\vdash_\mathbf{RM} (p \to p) \Rightarrow q$ as well. By Corollary 15.21 this implies that there is a valuation ν in $\mathcal{M}(\mathbb{Z})$ which is a model of $\neg(p \to p)$ but not of $(p \to p) \Rightarrow q$. The first fact implies that $\nu(p) = 0$, and so the second one implies that $0\tilde{\Rightarrow}\nu(q) < 0$. By (b) this is possible only if $\nu(q) = -n$ for some $n > 0$. But in such a case it easily follows from Proposition 15.19 that $0\tilde{\Rightarrow} -n < 0$ for *every* $n > 0$. By (*) and (a) it follows that $0\tilde{\Rightarrow} -n = -n$ for every n.

From the fact shown above that $\neg(p \to p) \not\vdash_\mathbf{L} (p \to p) \Rightarrow q$, it follows that $\not\vdash_\mathbf{L} \neg(p \to p) \Rightarrow ((p \to p) \Rightarrow q)$. Hence $\not\vdash_\mathbf{RM} \neg(p \to p) \Rightarrow ((p \to p) \Rightarrow q)$. Therefore Proposition 15.25 implies that there is a valuation ν in $\mathcal{RM}_4$ such that $\nu(\neg(p \to p) \Rightarrow ((p \to p) \Rightarrow q)) < 0$. By (c) it cannot be the case that $\nu(\neg(p \to p)) = -2$. Hence $|\nu(p)| = 1$, and we get that $-1\tilde{\Rightarrow}(1\tilde{\Rightarrow}\nu(q)) < 0$. By (a) and (b) this is impossible if $\nu(q) \in \{-1, 1, 2\}$. It follows that $\nu(q) = -2$, and so $-1\tilde{\Rightarrow}(1\tilde{\Rightarrow}-2) < 0$. This in turn implies (by (a) and (b) again) that $1\tilde{\Rightarrow}-2 = -2$. As usual, by Proposition 15.19 this means that $k\tilde{\Rightarrow} -n = -n$ in case $0 < k < n$. By (a) and what was shown above about $0\tilde{\Rightarrow} -n$, this equation holds also if $k = n$ or $k = 0$. Hence $k\tilde{\Rightarrow} -n = -n$ whenever $0 \leq k \leq n$.

(e) If $0 < n < k$ then $k\tilde{\Rightarrow}-n < 0$.

Proof. $(p \land \neg p) \lor (p \land \neg p \to q)$ is not a tautology of **RM** in case $p \neq q$ (take in $\mathcal{M}(\mathbb{Z})$ $\nu(p) = 1$, $\nu(q) = -2$). Hence it is not provable in **L** either, and so also $q \to q \not\vdash_\mathbf{L} (p \land \neg p) \lor (p \land \neg p \to q)$. It follows that $\not\vdash_\mathbf{L} (q \to q) \Rightarrow (p \land \neg p) \lor (p \land \neg p \to q)$, and so $\not\vdash_\mathbf{RM} (q \to q) \Rightarrow (p \land \neg p) \lor (p \land \neg p \to q)$. Therefore Proposition 15.25 implies that there is a valuation ν in $\mathcal{RM}_4$ such that $\nu((q \to q) \Rightarrow (p \land \neg p) \lor (p \land \neg p \to q)) < 0$. By (b) this is possible only if $\nu((p \land \neg p) \lor (p \land \neg p \to q)) < 0$. An easy check shows that this is the case only if $\nu(q) = -2$ and $|\nu(p)| = 1$. Hence the fact that $\nu((q \to q) \Rightarrow (p \land \neg p) \lor (p \land \neg p \to q)) < 0$ means that $2\tilde{\Rightarrow} - 1 < 0$. By Proposition 15.19 again, it follows that $k\tilde{\Rightarrow}-n < 0$ whenever $0 < n < k$.

Next we show that $[MP]$ for $\tilde{\Rightarrow}$ is valid in **RM**, i.e. $\varphi, \varphi \Rightarrow \psi \vdash_\mathbf{RM} \psi$ for every φ and ψ. Suppose otherwise. Then from corollary 15.21 it follows that there is a valuation ν in $\mathcal{M}(\mathbb{Z})$ such that $\nu(\varphi) \geq 0$,

$\nu(\psi) < 0$, and $\nu(\varphi \Rightarrow \psi) \geq 0$. But this is impossible because of (d) and (e) from the above list of properties of $\Rightarrow$.

Finally, we prove that $\mathbf{L} = \mathbf{RM}$. Since $\mathbf{L}$ is an extension of $\mathbf{RM}$, it suffices to show that if $\mathcal{T} \vdash_\mathbf{L} \varphi$ then $\mathcal{T} \vdash_\mathbf{RM} \varphi$. So assume that $\mathcal{T} \vdash_\mathbf{L} \varphi$. Since $\mathbf{L}$ is finitary, there are $\psi_1, \ldots, \psi_n \in \mathcal{T}$ such that $\{\psi_1, \ldots, \psi_n\} \vdash_\mathbf{L} \varphi$. It follows that $\vdash_\mathbf{L} \psi_1 \Rightarrow (\psi_2 \Rightarrow \cdots (\psi_n \Rightarrow \varphi) \cdots)$. This in turn implies that $\vdash_\mathbf{RM} \psi_1 \Rightarrow (\psi_2 \Rightarrow \cdots (\psi_n \Rightarrow \varphi) \cdots)$. But we have shown that $[MP]$ for $\Rightarrow$ is valid in $\mathbf{RM}$. Therefore $\{\psi_1, \ldots, \psi_n\} \vdash_\mathbf{RM} \varphi$, and so $\mathcal{T} \vdash_\mathbf{RM} \varphi$.

2. That $\mathbf{L}$ is semi-relevant was used in the proof of the first part only for deriving (d). Since $\neg p, p \vdash_\mathbf{RM} \neg(p \to p)$, while both $\neg(p \to p) \vdash_\mathbf{RM} p$ and $\neg(p \to p) \vdash_\mathbf{RM} \neg p$, practically the same proof shows that if $\mathbf{L}$ is a finitary proper simple extension of $\mathbf{RM}$ which is both normal and paraconsistent, then $\mathbf{L}$ has a finite weakly characteristic matrix. □

15.3 Cut-free Gentzen-type System for RM

In this section we turn to the proof theory of $\mathbf{RM}$. For doing this adequately, we again need to use hypersequents (See Section 1.3.4, as well as Sections 9.3.2, 13.2.3, and 14.3.4.) Figure 15.1 contains a hypersequential Gentzen-type proof system *GRM* for $\mathbf{RM}$. (Like in Note 13.4, Γ, Δ denote in it sets. The short names [EC], [EW], [Mi], and [Sp] stand for External Contraction, External Weakening, Mingle, and Splitting, respectively.)

Example 15.49 Below is a proof in *GRM* of $\Rightarrow (\varphi \to \psi) \vee (\psi \to \varphi)$. Note that two applications of $[\Rightarrow \to]$ and then of $[\Rightarrow \vee]$ are done in it in parallel:

$$\cfrac{\cfrac{\cfrac{\cfrac{\varphi \Rightarrow \varphi \qquad \psi \Rightarrow \psi}{\varphi, \psi \Rightarrow \varphi, \psi} [Mi]}{\varphi \Rightarrow \psi \mid \psi \Rightarrow \varphi} [Sp]}{\Rightarrow \varphi \to \psi \mid \Rightarrow \psi \to \varphi} [\Rightarrow \to]}{\cfrac{\Rightarrow (\varphi \to \psi) \vee (\psi \to \varphi) \mid \Rightarrow (\varphi \to \psi) \vee (\psi \to \varphi)}{\Rightarrow (\varphi \to \psi) \vee (\psi \to \varphi)} [EC]} [\Rightarrow \vee]$$

The next proposition provides several useful properties of *GRM*.

Proposition 15.50

1. $\vdash_{GRM} G \mid \neg\varphi, \Gamma \Rightarrow \Delta$ iff $\vdash_{GRM} G \mid \Gamma \Rightarrow \Delta, \varphi$.

2. $\vdash_{GRM} G \mid \Gamma \Rightarrow \Delta, \neg\varphi$ iff $\vdash_{GRM} G \mid \varphi, \Gamma \Rightarrow \Delta$.

Axioms: $G|\psi \Rightarrow \psi$

Logical rules:

$$[\neg \Rightarrow] \quad \frac{G \mid \Gamma \Rightarrow \Delta, \varphi}{G \mid \neg\varphi, \Gamma \Rightarrow \Delta} \qquad\qquad [\Rightarrow \neg] \quad \frac{G|\varphi, \Gamma \Rightarrow \Delta}{G|\Gamma \Rightarrow \Delta, \neg\varphi}$$

$$[\rightarrow \Rightarrow] \quad \frac{G|\Gamma_1 \Rightarrow \Delta_1, \varphi \quad G|\psi, \Gamma_2 \Rightarrow \Delta_2}{G|\Gamma_1, \Gamma_2, \varphi \rightarrow \psi \Rightarrow \Delta_1, \Delta_2} \qquad [\Rightarrow \rightarrow] \quad \frac{G|\Gamma, \varphi \Rightarrow \Delta, \psi}{G|\Gamma \Rightarrow \Delta, \varphi \rightarrow \psi}$$

$$[\wedge \Rightarrow] \quad \frac{G|\Gamma, \varphi \Rightarrow \Delta}{G|\Gamma, \varphi \wedge \psi \Rightarrow \Delta} \quad \frac{G|\Gamma, \psi \Rightarrow \Delta}{G|\Gamma, \varphi \wedge \psi \Rightarrow \Delta}$$

$$[\Rightarrow \wedge] \quad \frac{G|\Gamma \Rightarrow \Delta, \varphi \quad G|\Gamma \Rightarrow \Delta, \psi}{G|\Gamma \Rightarrow \Delta, \varphi \wedge \psi}$$

$$[\vee \Rightarrow] \quad \frac{G|\Gamma, \varphi \Rightarrow \Delta \quad G|\Gamma, \psi \Rightarrow \Delta}{G|\Gamma, \varphi \vee \psi \Rightarrow \Delta}$$

$$[\Rightarrow \vee] \quad \frac{G|\Gamma \Rightarrow \Delta, \varphi}{G|\Gamma \Rightarrow \Delta, \varphi \vee \psi} \quad \frac{G|\Gamma \Rightarrow \Delta, \psi}{G|\Gamma \Rightarrow \Delta, \varphi \vee \psi}$$

Structural rules:

$$[\text{EC}] \quad \frac{G \mid s \mid s}{G \mid s} \qquad\qquad [\text{EW}] \quad \frac{G}{G \mid s}$$

$$[\text{Sp}] \quad \frac{G \mid \Gamma_1, \Gamma_2 \Rightarrow \Delta_1, \Delta_2}{G \mid \Gamma_1 \Rightarrow \Delta_1 \mid \Gamma_2 \Rightarrow \Delta_2} \qquad [\text{Mi}] \quad \frac{G \mid \Gamma_1 \Rightarrow \Delta_1 \quad G \mid \Gamma_2 \Rightarrow \Delta_2}{G \mid \Gamma_1, \Gamma_2 \Rightarrow \Delta_1, \Delta_2}$$

$$[\text{Cut}] \quad \frac{G \mid \Gamma_1 \Rightarrow \Delta_1, \varphi \quad G \mid \varphi, \Gamma_2 \Rightarrow \Delta_2}{G \mid \Gamma_1, \Gamma_2 \Rightarrow \Delta_1, \Delta_2}$$

Figure 15.1: The proof system *GRM*

3. $\vdash_{GRM} G \mid \Gamma \Rightarrow \Delta, \varphi \rightarrow \psi$ iff $\vdash_{GRM} G \mid \varphi, \Gamma \Rightarrow \Delta, \psi$.

4. $\vdash_{GRM} G \mid \varphi \wedge \psi, \Gamma \Rightarrow \Delta$ iff $\vdash_{GRM} G \mid \varphi, \Gamma \Rightarrow \Delta \mid \psi, \Gamma \Rightarrow \Delta$.

5. $\vdash_{GRM} G \mid \Gamma \Rightarrow \Delta, \varphi \wedge \psi$ iff
   $\vdash_{GRM} G \mid \Gamma \Rightarrow \Delta, \varphi$ and $\vdash_{GRM} G \mid \Gamma \Rightarrow \Delta, \psi$.

6. $\vdash_{GRM} G \mid \varphi \vee \psi, \Gamma \Rightarrow \Delta$ iff
   $\vdash_{GRM} G \mid \varphi, \Gamma \Rightarrow \Delta$ and $\vdash_{GRM} G \mid \psi, \Gamma \Rightarrow \Delta$.

7. $\vdash_{GRM} G \mid \Gamma \Rightarrow \Delta, \varphi \vee \psi$ iff $\vdash_{GRM} G \mid \Gamma \Rightarrow \Delta, \varphi \mid \Gamma \Rightarrow \Delta, \psi$.

Proof. We show the last item as an example, leaving the others to the reader. One direction is easy: $G \mid \Gamma \Rightarrow \Delta, \varphi \vee \psi$ can be derived in *GRM*

15.3 Cut-free Gentzen-type System for RM

from $G \mid \Gamma \Rightarrow \Delta, \varphi \mid \Gamma \Rightarrow \Delta, \psi$ by using two applications of $[\Rightarrow \vee]$, followed by an application of $[EC]$. For the converse, note that $\varphi \vee \psi \Rightarrow \psi \mid \varphi \vee \psi \Rightarrow \varphi$ can be derived from $\varphi \Rightarrow \psi \mid \psi \Rightarrow \varphi$ (which is proved within the proof of $(\varphi \rightarrow \psi) \vee (\psi \rightarrow \varphi)$ given in Example 15.49), $\psi \Rightarrow \psi$, and $\varphi \Rightarrow \varphi$ by using two applications of $[\vee \Rightarrow]$. A cut on $\varphi \vee \psi$ of this sequent and $G \mid \Gamma \Rightarrow \Delta, \varphi \vee \psi$ yields $G \mid \varphi \vee \psi \Rightarrow \varphi \mid \Gamma \Rightarrow \Delta, \psi$. Another cut on $\varphi \vee \psi$ of this last sequent and $G \mid \Gamma \Rightarrow \Delta, \varphi \vee \psi$ yields $G \mid G \mid \Gamma \Rightarrow \Delta, \varphi \mid \Gamma \Rightarrow \Delta, \psi$. From this $G \mid \Gamma \Rightarrow \Delta, \varphi \mid \Gamma \Rightarrow \Delta, \psi$ can be derived using applications of $[EC]$. □

Note 15.51 Proposition 15.50 can be used for reducing the provability of a hypersequent G to the provability of a finite set of hypersequents, the components of each have only variables on their right-hand side, and only variables or implications on their left-hand side.

Next we turn to the semantics of the hypersequents of *GRM*.

Definition 15.52 Let ν be a valuation in $\mathcal{M}(\mathbb{Z})$.

- Let $\Gamma \Rightarrow \Delta$ be a non-empty sequent. Define:
  - $d_\nu(\Gamma \Rightarrow \Delta) =_{Df} max(\{|\nu(\varphi)| \mid \varphi \in \Gamma \cup \Delta\})$.
  - $\nu(\Gamma \Rightarrow \Delta) =_{Df} \begin{cases} d_\nu(\Gamma \Rightarrow \Delta) & \exists \varphi \in \Gamma.\nu(\varphi) = -d_\nu(\Gamma \Rightarrow \Delta), \\ d_\nu(\Gamma \Rightarrow \Delta) & \exists \psi \in \Delta.\nu(\psi) = d_\nu(\Gamma \Rightarrow \Delta), \\ -d_\nu(\Gamma \Rightarrow \Delta) & \text{otherwise}. \end{cases}$

- We say that ν is a *model* of a sequent $\Gamma \Rightarrow \Delta$ ($\nu \models \Gamma \Rightarrow \Delta$) if either $\Gamma \Rightarrow \Delta$ is empty and there is a variable p such that $\nu(p) = 0$, or $\Gamma \Rightarrow \Delta$ is not empty and $\nu(\Gamma \Rightarrow \Delta) \geq 0$ (i.e. there is $\varphi \in \Gamma$ such that $\nu(\varphi) = -d_\nu(\Gamma \Rightarrow \Delta)$, or $\psi \in \Delta$ such that $\nu(\psi) = d_\nu(\Gamma \Rightarrow \Delta)$).

- We say that ν is a *model* of a hypersequent G (in symbols: $\nu \models G$) if ν is a model of at least one component of G.

Definition 15.53 (RM-validity of hypersequents) A hypersequent G is **RM**-*valid* if every valuation in $\mathcal{M}(\mathbb{Z})$ is a model of G.

Note 15.54 Let the translation σ_s of $s = \varphi_1, \ldots, \varphi_n \Rightarrow \psi_1, \ldots, \psi_k$ (where $n+k > 0$) be the sentence $\neg\varphi_1 + \cdots \neg\varphi_n + \psi_1 + \cdots \psi_k$, and let the translation σ_G of a hypersequent G that has no empty component be the $\vee$-disjunction of the translations of its components. It is easy to check that $\nu \models G$ iff $\nu \models \sigma_G$. In particular: if φ is sentence then $\nu \models \varphi$ iff $\nu \models \Rightarrow \varphi$. Hence $\Rightarrow \varphi$ is **RM**-valid iff φ is valid in $\mathcal{M}(\mathbb{Z})$. By Corollary 15.21 this implies that $\Rightarrow \varphi$ is **RM**-valid iff $\vdash_{\mathbf{RM}} \varphi$.

Note 15.55 Instead of $\mathcal{M}(\mathbb{Z})$ we could have used $\mathcal{M}([0,1])$ in Definitions 15.52 and 15.53. This would not change much in what follows, and would in fact be a better choice, since the use of $\mathcal{M}([0,1])$ is good also for the natural first-order extension of **RM**, while $\mathcal{M}(\mathbb{Z})$ is practically useless for first-order **RM**. We have still used here $\mathcal{M}(\mathbb{Z})$ since it makes reading and understanding somewhat easier. We trust that the reader will be able to adapt the use of $\mathcal{M}(\mathbb{Z})$ which is done in this section to the use of $\mathcal{M}([0,1])$.

Proposition 15.56 (soundness of *GRM*) *Let G be a hypersequent in $\mathcal{L}_R$. If $\vdash_{GRM} G$ then G is* **RM**-*valid.*

Proof. Using an induction on length of proofs, we can prove something stronger: that if G follows in *GRM* from a set S of hypersequents, then every model ν (in $\mathcal{M}(\mathbb{Z})$) of all the elements of S is also a model of G. For this we need to show that all axioms of *GRM* are **RM**-valid, and that if a valuation ν in $\mathcal{M}(\mathbb{Z})$ is a model of all premises of some application of a rule of *GRM*, then it is also a model of the conclusion of that application. This is straightforward but tedious, so we omit the details. We only note one case that requires special attention: an application of the cut rule in the unusual case in which both of the contexts $\Gamma_1 \Rightarrow \Delta_1$ and $\Gamma_2 \Rightarrow \Delta_2$ are empty. In that case $G \mid \Rightarrow$ is derived from $G \mid \Rightarrow \varphi$ and $G \mid \varphi \Rightarrow$. Let ν be a model of the premises. Then either it is a model of some component of G, in which case it is a model of $G \mid \Rightarrow$ too, or it is a model of both $\Rightarrow \varphi$ and $\varphi \Rightarrow$. The latter is possible only if $\nu(\varphi) = 0$. A necessary condition for this is that $\nu(p) = 0$ for some variable $p \in \mathsf{Var}(\varphi)$. Hence, in that case ν is model of $\Rightarrow$, and so also of $G \mid \Rightarrow$. □

Our next (much harder!) goals are to show that *GRM* is also complete, and that the cut-elimination theorem holds for it.

Notation 15.57 $\vdash_{GRM}^{cf} G$ means that G has a cut-free proof in *GRM*.

Lemma 15.58 *If $\Sigma \neq \emptyset$ and $\mathsf{Var}(\Gamma \cup \Delta) \subseteq \Sigma$ then $\vdash_{GRM}^{cf} \Sigma, \Gamma \Rightarrow \Delta, \Sigma$.*

Proof. By induction on the complexity of $\Gamma \cup \Delta$. The basic case, where $\Gamma \cup \Delta$ consists of variables is proved using the *Mingle* rule. The induction steps are straightforward.

Definition 15.59 A hypersequent G is *maximal* if $\not\vdash_{GRM}^{cf} G$, but $\vdash_{GRM}^{cf} G \mid s$ whenever the following two conditions on s are satisfied:

- s consists only of subformulas of formulas in G.

- s is not a component of G.

15.3 Cut-free Gentzen-type System for RM

Notation 15.60 For finite A, $\sharp(A)$ denotes the number of elements in A.

Definition 15.61 Let ν be a partial function from Var to $\mathbb{Z}$. We call ν a *k-semivaluation* if the following two conditions are satisfied:

1. $|\nu(p)| \geq k$ for every $p \in Dom(\nu)$.
2. If $Dom(\nu) \neq \mathsf{Var}$ then $\sharp(\mathsf{Var} - Dom(\nu)) < k$.

Note 15.62 If ν is a 0-semivaluation or an 1-semivaluation then ν is a full valuation, i.e., $\nu : \mathsf{Var} \to \mathbb{Z}$.

Definition 15.63 A valuation $\nu^* : \mathsf{Var} \to \mathbb{Z}$ is a *k-completion* of a k-semivaluation ν if $\nu^*(p) = \nu(p)$ for $p \in Dom(\nu)$, while $|\nu^*(p)| < k$ otherwise.

Definition 15.64 Let ν be an l-semivaluation, and let $k \geq l$. A formula φ is a $\nu^l_{<k}$-*formula* (in symbols: $\varphi \in \nu^l_{<k}$) if $|\nu^*(\varphi)| < k$ for every l-completion ν^* of ν. φ is a $\nu^l_{\geq k}$-*formula* (in symbols: $\varphi \in \nu^l_{\geq k}$) if $|\nu^*(\varphi)| \geq k$ for every l-completion ν^* of ν. φ is a $\nu_{<l}$-*formula* (in symbols: $\varphi \in \nu_{<l}$) or a $\nu_{\geq l}$-*formula* (in symbols: $\varphi \in \nu_{\geq l}$) if it is a $\nu^l_{<l}$ or $\nu^l_{\geq l}$-formula (respectively).

Lemma 15.65 *Let ν be an l-semivaluation, and let $k \geq l$.*

1. *Every formula is either a $\nu^l_{<k}$-formula or a $\nu^l_{\geq k}$-formula.*
2. *φ is a $\nu^l_{\geq k}$-formula iff $|\nu^*(\varphi)| \geq k$ for some l-completion ν^* of ν. φ is a $\nu^l_{<k}$-formula if $|\nu^*(\varphi)| < k$ for some l-completion ν^* of ν.*
3. *If φ is an $\nu^l_{\geq k}$-formula, then $\nu_1(\varphi) = \nu_2(\varphi)$ whenever ν_1 and ν_2 are l-completions of ν.*

Proof. Immediate from Proposition 15.14. □

Definition 15.66 Let G be a hypersequent, and ν be a k-semivaluation. We say that ν is *k-adequate* for G if the following conditions are satisfied:

1. If $\Gamma \Rightarrow \Delta$ is a component of G, and $\Gamma \cup \Delta$ contains some $\nu_{\geq k}$-formula, then every k-completion of ν refutes $\Gamma \Rightarrow \Delta$.
2. If $k = 0$ and the empty sequent $\Rightarrow$ is a component of G, then $\nu(p) \neq 0$ for every $p \in \mathsf{Var}$.
3. Let Σ be the set of variables for which ν is not defined. Suppose $\Sigma \neq \emptyset$, and that $\Gamma \Rightarrow \Delta$ is a component of G which consists only of $\nu_{<k}$-formulas. Then $\vdash^{cf}_{GRM} G|\varphi, \Sigma \Rightarrow \Sigma$ for $\varphi \in \Gamma$, and $\vdash^{cf}_{GRM} G|\Sigma \Rightarrow \Sigma, \psi$ for $\psi \in \Delta$. (By using repeated applications of *Mingle*, this implies that $\vdash^{cf}_{GRM} G|\Sigma, \Gamma' \Rightarrow \Delta', \Sigma$ whenever $\Gamma' \subseteq \Gamma$ and $\Delta' \subseteq \Delta$.)

Lemma 15.67 *Assume that G is a maximal hypersequent, and let ν be a k-semivaluation which is k-adequate for G. Then ν has a k-completion which is not a model of G.*

Proof. By induction on k.

The case $k = 0$ is trivial, since if ν is a 0-semivaluation, then ν is a full valuation and every formula is $\nu_{\geq 0}$-formula. Hence, if ν is 0-adequate for G then (by definition) ν is a 0-completion of itself which is not a model of G.

Similarly, the case where $k > 0$ and $Dom(\nu) = \mathsf{Var}$ is trivial, since in this case ν is a k-completion of itself, every formula is $\nu_{\geq k}$-formula, and $\nu(p) \neq 0$ for every p. Hence, the k-adequacy of ν implies that ν itself is a k-completion as required. It follows that the lemma is true in the particular case where $k = 1$.

Now, assume that $k > 1$, and the claim is true for every $l < k$. Let ν be a k-semivaluation such that $Dom(\nu) \neq \mathsf{Var}$, and ν is k-adequate for G. Let $I = \{\Gamma_j \Rightarrow \Delta_j \mid 1 \leq j \leq m\}$ be the set of all components of G which consist only of $\nu_{<k}$-formulas.

Suppose first that I is empty (i.e., $m = 0$). Then every component of G contains some $\nu_{\geq k}$-formula, and so every k-completion of ν refutes G (by definition of k-adequacy).

Now, assume that I is not empty and let $\Gamma_0 = \bigcup_{1 \leq j \leq m} \Gamma_j$, $\Delta_0 = \bigcup_{1 \leq j \leq m} \Delta_j$. Suppose that $\Gamma_0 \Rightarrow \Delta_0$ is not a component of G. Then the maximality of G entails that there exists a cut-free proof in GRM of $\Gamma_0 \Rightarrow \Delta_0 | G$. By using the *Splitting* rule [Sp] we can get from such a proof a cut-free proof of $I | G$, and so of G (since $I \subseteq G$). A contradiction. It follows that $\Gamma_0 \Rightarrow \Delta_0 \in G$. Since this sequent consists only of $\nu_{<k}$-formulas, this entails that $\Gamma_0 \Rightarrow \Delta_0 \in I$, and so it is actually the maximal element of I. Let Σ be the set of variables for which ν is not defined. Since $\Sigma \neq \emptyset$ by our assumption, and ν is k-adequate for G, the third item of Definition 15.66 implies that $\vdash^{cf}_{GRM} \Sigma, \Gamma_0 \Rightarrow \Delta_0, \Sigma | G$. It follows that $\Sigma^{(=)} = \Sigma - \Gamma_0 \cap \Delta_0$ is not empty (because otherwise $\Sigma, \Gamma_0 \Rightarrow \Delta_0, \Sigma | G$ is identical to G). Let $\Sigma^{(<)} = \Sigma - \Sigma^{(=)} = \Sigma \cap \Gamma_0 \cap \Delta_0$, and let $l = 1 + \sharp(\Sigma^{(<)})$. Now, $0 < l < k$, since $\Sigma^{(<)}$ is a *proper* subset of Σ, while $\sharp(\Sigma) < k$ (because ν is a k-semivaluation). Extend ν to a semivaluation $\tilde{\nu}$ on $Dom(\nu) \cup \Sigma^{(=)}$ as follows:

$$\tilde{\nu}(p) = \begin{cases} \nu(p) & p \in Dom(\nu), \\ l & p \in \Sigma^{(=)} \cap \Gamma_0, \\ -l & p \in \Sigma^{(=)} - \Gamma_0. \end{cases}$$

Note that $\Sigma^{(<)}$ is the set of variables for which $\tilde{\nu}$ is not defined. It follows that $\tilde{\nu}$ is an l-semivaluation, and its definition implies that every l-completion of $\tilde{\nu}$ is a k-completion of ν. Moreover, for every l-completion ν^*

15.3 Cut-free Gentzen-type System for RM

of $\tilde{\nu}$ and every φ, either $|\nu^*(\varphi)| \geq k$, or $|\nu^*(\varphi)| \leq l$. Now we show that the following two claims are true for every $\varphi \in \Gamma_0 \cup \Delta_0$ and every l-completion ν^* of $\tilde{\nu}$:

(a) If $\varphi \in \Gamma_0$ then $\nu^*(\varphi) = l$, or $\varphi \in \tilde{\nu}_{<l}$ and $\vdash^{cf}_{GRM} \varphi, \Sigma^{(<)} \Rightarrow \Sigma^{(<)}|G$.

(b) If $\varphi \in \Delta_0$ then $\nu^*(\varphi) = -l$, or $\varphi \in \tilde{\nu}_{<l}$ and $\vdash^{cf}_{GRM} \Sigma^{(<)} \Rightarrow \Sigma^{(<)}, \varphi|G$.

What follows is a proof of these claims by an induction on the complexity of φ. We assume in it that ν^* is some l-completion of $\tilde{\nu}$, and so also a k-completion of ν. By Lemma 15.65, this implies that $|\nu^*(\varphi)| < l$ iff $\varphi \in \tilde{\nu}_{<l}$, and $|\nu^*(\varphi)| < k$ iff $\varphi \in \tilde{\nu}^l_{<k}$.

$\varphi \in \Gamma_0$, and φ is a variable.
Then $\varphi \in \Sigma$ (because ν is a k-semivaluation, and so a variable is a $\nu_{<k}$-formula iff ν is not defined for it). It follows that either $\varphi \in \Sigma^{(=)} \cap \Gamma_0$, or $\varphi \in \Sigma^{(<)}$. In the first case $\nu^*(\varphi) = l$ by definition of $\tilde{\nu}$. In the second $\varphi \in \tilde{\nu}_{<l}$ (since $\tilde{\nu}$ is an l-semivaluation, and $\Sigma^{(<)}$ is the set of variables for which $\tilde{\nu}$ is not defined), and $\varphi, \Sigma^{(<)} = \Sigma^{(<)}$. Hence $\vdash^{cf}_{GRM} \varphi, \Sigma^{(<)} \Rightarrow \Sigma^{(<)}|G$ by using [Mi] and [EW].

$\varphi \in \Delta_0$, and φ is a variable.
Then $\varphi \in \Sigma$, and so either $\varphi \in \Sigma^{(=)} - \Gamma_0$, or $\varphi \in \Sigma^{(<)}$. In the first case $\nu^*(\varphi) = -l$ by definition of $\tilde{\nu}$. In the second $\varphi \in \tilde{\nu}_{<l}$, and again we have $\vdash^{cf}_{GRM} \Sigma^{(<)} \Rightarrow \Sigma^{(<)}, \varphi|G$ by using [Mi] and [EW].

$\varphi \in \Gamma_0$, and $\varphi = \neg \psi$.
Then φ is a $\nu_{<k}$-formula, and so ψ is a $\nu_{<k}$-formula (since $|\nu^*(\neg\psi)| = |\nu^*(\psi)|$). It is impossible that $\vdash^{cf}_{GRM} \Gamma_0 \Rightarrow \Delta_0, \psi|G$, since otherwise we would have got that $\vdash^{cf}_{GRM} G$ by using $[\neg \Rightarrow]$. It follows that $\Gamma_0 \Rightarrow \Delta_0, \psi$ is in G (because G is maximal), and since it consists only of $\nu_{<k}$-formulas, necessarily $\psi \in \Delta_0$. Therefore we get by induction hypothesis on ψ that either $\nu^*(\psi) = -l$, or $\vdash^{cf}_{GRM} \Sigma^{(<)} \Rightarrow \Sigma^{(<)}, \psi|G$ and $\psi \in \tilde{\nu}_{<l}$. Hence, either $\nu^*(\varphi) = l$, or $\varphi \in \tilde{\nu}_{<l}$ and (using $[\neg \Rightarrow]$) $\vdash^{cf}_{GRM} \varphi, \Sigma^{(<)} \Rightarrow \Sigma^{(<)}|G$.

$\varphi \in \Delta_0$, and $\varphi = \neg \psi$.
This case is similar to the previous one.

$\varphi \in \Gamma_0$, and $\varphi = \sigma \to \psi$.
Then φ is a $\nu_{<k}$-formula, and so ψ and σ are $\nu_{<k}$-formulas (because $|\nu^*(\varphi)| = max\{|\nu^*(\psi)|, |\nu^*(\sigma)|\}$ in this case). It is impossible that both $\vdash^{cf}_{GRM} \psi, \Gamma_0 \Rightarrow \Delta_0|G$ and $\vdash^{cf}_{GRM} \Gamma_0 \Rightarrow \Delta_0, \sigma|G$, since otherwise we would have got that $\vdash^{cf}_{GRM} G$ by using $[\to \Rightarrow]$. Assume, for instance,

that $\nvdash^{cf}_{GRM} \psi, \Gamma_0 \Rightarrow \Delta_0|G$. Then $\psi, \Gamma_0 \Rightarrow \Delta_0$ is in G (because G is maximal), and since it consists only of $\nu_{<k}$-formulas, necessarily $\psi \in \Gamma_0$. It follows by induction hypothesis on ψ that either $\nu^*(\psi) = l$, or $\psi \in \tilde{\nu}_{<l}$ and $\vdash^{cf}_{GRM} \psi, \Sigma^{(<)} \Rightarrow \Sigma^{(<)}|G$. In the first case $\nu^*(\varphi) = l$ (since σ is a $\nu_{<k}$-formula, and so $|\nu^*(\sigma)| \leq l$). In the second we consider the following two subcases:

- Supposed that $\sigma \notin \Delta_0$. Then $\Gamma_0 \Rightarrow \Delta_0, \sigma \notin I$, and since this sequent consists only of $\nu_{<k}$-formulas, $\Gamma_0 \Rightarrow \Delta_0, \sigma \notin G$. Hence $\vdash^{cf}_{GRM} \Gamma_0 \Rightarrow \Delta_0, \sigma|G$. Using $[\rightarrow\Rightarrow]$, this and our assumption $\vdash^{cf}_{GRM} \psi, \Sigma^{(<)} \Rightarrow \Sigma^{(<)}|G$ imply $\vdash^{cf}_{GRM} \varphi, \Sigma^{(<)}, \Gamma_0 \Rightarrow \Delta_0, \Sigma^{(<)}|G$. Since $\Sigma^{(<)} \subseteq \Gamma_0 \cap \Delta_0$, and $\varphi \in \Gamma_0$, we get that $\vdash^{cf}_{GRM} G$. A contradiction. Hence this subcase is actually impossible.

- If $\sigma \in \Delta_0$, then by induction hypothesis either $\nu^*(\sigma) = -l$, or $\sigma \in \tilde{\nu}_{<l}$ and there is a cut-free proof of $\Sigma^{(<)} \Rightarrow \Sigma^{(<)}, \sigma|G$. In the first case again $\nu^*(\varphi) = l$. In the second $\varphi \in \tilde{\nu}_{<l}$ (because $\psi \in \tilde{\nu}_{<l}$ and $\sigma \in \tilde{\nu}_{<l}$), and we can get a cut-free proof of $\varphi, \Sigma^{(<)} \Rightarrow \Sigma^{(<)}|G$ by applying $(\rightarrow\Rightarrow)$ to $\psi, \Sigma^{(<)} \Rightarrow \Sigma^{(<)}|G$ and $\Sigma^{(<)} \Rightarrow \Sigma^{(<)}, \sigma|G$.

$\varphi \in \Delta_0$, **and** $\varphi = \sigma \rightarrow \psi$.
Then φ is a $\nu_{<k}$-formula, and so ψ and σ are $\nu_{<k}$-formulas. It is impossible that $\vdash^{cf}_{GRM} \Gamma_0, \sigma \Rightarrow \Delta_0, \psi|G$, since otherwise we would have got that $\vdash^{cf}_{GRM} G$ by using $[\Rightarrow\rightarrow]$. Hence $\Gamma_0, \sigma \Rightarrow \Delta_0, \psi$ is in G, and since it consists only of $\nu_{<k}$-formulas, necessarily $\psi \in \Delta_0$ and $\sigma \in \Gamma_0$. Hence by induction hypothesis **(b)** applies to ψ and **(a)** to σ. If $\nu^*(\psi) = -l$ and $\nu^*(\sigma) = l$, or $\nu^*(\psi) = -l$ and $|\nu^*(\sigma)| < l$, or $\nu^*(\sigma) = l$ and $|\nu^*(\psi)| < l$ then $\nu^*(\varphi) = -l$. Otherwise, $\psi \in \tilde{\nu}_{<l}$, $\sigma \in \tilde{\nu}_{<l}$, $\vdash^{cf}_{GRM} \Sigma^{(<)} \Rightarrow \Sigma^{(<)}, \psi|G$, and $\vdash^{cf}_{GRM} \sigma, \Sigma^{(<)} \Rightarrow \Sigma^{(<)}|G$. It follows that $\varphi \in \tilde{\nu}_{<l}$, and $\vdash^{cf}_{GRM} \sigma, \Sigma^{(<)} \Rightarrow \Sigma^{(<)}, \psi|G$ (using *Mingle*). By using $[\Rightarrow\rightarrow]$ we get $\vdash^{cf}_{GRM} \Sigma^{(<)} \Rightarrow \Sigma^{(<)}, \varphi|G$.

$\varphi \in \Gamma_0$, **and** $\varphi = \psi \vee \sigma$.
It is impossible that $\vdash^{cf}_{GRM} \psi, \Gamma_0 \Rightarrow \Delta_0|G$ and $\vdash^{cf}_{GRM} \sigma, \Gamma_0 \Rightarrow \Delta_0|G$ are both true, since this would have implied $\vdash^{cf}_{GRM} G$ by using $[\vee \Rightarrow]$. Assume e.g. that $\nvdash^{cf}_{GRM} \psi, \Gamma_0 \Rightarrow \Delta_0|G$. Then $\psi, \Gamma_0 \Rightarrow \Delta_0$ is in G. Assume for contradiction that $\psi \in \nu_{\geq k}$. Then ν^* refutes $\psi, \Gamma_0 \Rightarrow \Delta_0$ (since ν^* is a k-completion of ν, and ν is k-adequate). This is possible only if $\nu^*(\psi) \geq k$, because $\Gamma_0 \cap \Delta_0 \subseteq \nu_{<k}$, and so $\nu^*(\psi, \Gamma_0 \Rightarrow \Delta_0) = -\nu^*(\psi)$ under our assumptions. But in such a case also $\nu^*(\varphi) \geq k$, contradicting the fact that $\varphi \in \Gamma_0$, and so $\varphi \in \nu_{<k}$. It follows that $\psi \in \nu_{<k}$, hence $\psi, \Gamma_0 \Rightarrow \Delta_0 \in I$, and so $\psi \in \Gamma_0$. By applying the induction hypothesis to ψ we get that either $\nu^*(\psi) = l$, or $\psi \in \tilde{\nu}_{<l}$

15.3 Cut-free Gentzen-type System for RM

(and so $|\nu^*(\psi)| < l$) and $\vdash^{cf}_{GRM} \psi, \Sigma^{(<)} \Rightarrow \Sigma^{(<)}|G$. In the first case $\nu^*(\varphi) \geq l$, and since $\varphi \in \nu_{<k}$ (because $\varphi \in \Gamma_0$), $\nu^*(\varphi) = l$. In the second case we consider the following three subcases:

- Assume that $\sigma \in \nu_{\geq k}$. Then either $\nu^*(\sigma) \geq k$ or $\nu^*(\sigma) \leq -k$. In the first case also $\nu^*(\varphi) \geq k$ as well, contradicting the fact that $\varphi \in \nu_{<k}$. So $\nu^*(\sigma) \leq -k$. Since $|\nu^*(p)| < l$ for every $p \in \Sigma^{(<)}$, $\nu^*(\sigma, \Sigma^{(<)} \Rightarrow \Sigma^{(<)}) = -\nu^*(\sigma) \geq k$ in this case, and so ν^* is a model of the sequent $\sigma, \Sigma^{(<)} \Rightarrow \Sigma^{(<)}$. Hence $\sigma, \Sigma^{(<)} \Rightarrow \Sigma^{(<)}$ is not in G (since ν is k-adequate for G, ν^* is a k-completion of ν, and $\sigma, \Sigma^{(<)} \Rightarrow \Sigma^{(<)}$ contains the $\nu_{\geq k}$-formula σ). It followed that $\vdash^{cf}_{GRM} \sigma, \Sigma^{(<)} \Rightarrow \Sigma^{(<)}|G$. From this and our assumption that $\vdash^{cf}_{GRM} \psi, \Sigma^{(<)} \Rightarrow \Sigma^{(<)}|G$, it follows that $\vdash^{cf}_{GRM} \varphi, \Sigma^{(<)} \Rightarrow \Sigma^{(<)}|G$ (using $[\vee \Rightarrow]$). Moreover, $|\nu^*(\varphi)| < l$ because $\nu^*(\sigma) \leq -k < -l$ while $|\nu^*(\psi)| < l$. Hence $\varphi \in \tilde{\nu}_{<l}$.

- Assume that $\nu^*(\sigma) = l$. Then $\nu^*(\varphi) = l$ (since $|\nu^*(\psi)| < l$).

- Assume that $\sigma \in \nu_{<k}$ and $\nu^*(\sigma) \neq l$. Then $\nu^*(\sigma) = -l$ or $|\nu^*(\sigma)| < l$ (recall that $|\nu^*(\alpha)| \geq k$ or $|\nu^*(\alpha)| \leq l$ for every α). In the first case $\nu^*(\varphi) = \nu^*(\psi)$, and so $|\nu^*(\varphi)| < l$. In the second case our assumption on $\nu^*(\psi)$ again implies that $|\nu^*(\varphi)| < l$. It remains to prove that $\vdash^{cf}_{GRM} \varphi, \Sigma^{(<)} \Rightarrow \Sigma^{(<)}|G$. Since $\vdash^{cf}_{GRM} \psi, \Sigma^{(<)} \Rightarrow \Sigma^{(<)}|G$, it suffices to prove that $\vdash^{cf}_{GRM} \sigma, \Sigma^{(<)} \Rightarrow \Sigma^{(<)}|G$. If $\sigma \in \Gamma_0$ this would follows from our induction hypothesis applied to σ (since $\nu^*(\sigma) \neq l$). If $\sigma \notin \Gamma_0$ then $\sigma, \Sigma^{(<)} \Rightarrow \Sigma^{(<)}$ is not in I, and since we are assuming that $\sigma \in \nu_{<k}$, it follows that $\sigma, \Sigma^{(<)} \Rightarrow \Sigma^{(<)}$ is not in G. Hence the maximality of G implies that $\vdash^{cf}_{GRM} \sigma, \Sigma^{(<)} \Rightarrow \Sigma^{(<)}|G$.

$\varphi = \psi \vee \sigma$ **and** $\varphi \in \Delta_0$.

Without a loss of generality, we may assume that $\nu^*(\sigma) \leq \nu^*(\psi)$, and so $\nu^*(\varphi) = \nu^*(\psi)$. Now, $\varphi \in \Delta_0 \subseteq \nu_{<k}$, and so $|\nu^*(\varphi)| < k$. It follows that $|\nu^*(\psi)| < k$, and so $\psi \in \nu_{<k}$. Suppose that $\psi \notin \Delta_0$. Then $\Gamma_0 \Rightarrow \Delta_0, \psi$ is not in I, and since this sequent consists only of formulas in $\nu_{<k}$, $\Gamma_0 \Rightarrow \Delta_0, \psi$ is not in G either. Hence $\vdash^{cf}_{GRM} \Gamma_0 \Rightarrow \Delta_0, \psi|G$, and so $\vdash^{cf}_{GRM} G$ (by $[\Rightarrow \vee]$). A contradiction. It follows that $\psi \in \Delta_0$, and so by the induction hypothesis either $\nu^*(\psi) = -l$, or $\psi \in \tilde{\nu}_{<l}$ and $\vdash^{cf}_{GRM} \Sigma^{(<)} \Rightarrow \Sigma^{(<)}, \psi|G$. In the first case, $\nu^*(\varphi) = -l$. In the second, $|\nu^*(\varphi)| = |\nu^*(\psi)| < l$, and so $\varphi \in \tilde{\nu}_{<l}$. Using $(\Rightarrow \vee)$, that $\vdash^{cf}_{GRM} \Sigma^{(<)} \Rightarrow \Sigma^{(<)}, \varphi|G$ in this case follows from the assumption that $\vdash^{cf}_{GRM} \Sigma^{(<)} \Rightarrow \Sigma^{(<)}, \psi|G$.

$\varphi = \psi \wedge \sigma$ and $\varphi \in \Gamma_0$.
The proof in this case is similar to that in the previous one.

$\varphi = \psi \wedge \sigma$ and $\varphi \in \Delta_0$.
The proof is similar to the case where $\varphi = \psi \vee \sigma$ and $\varphi \in \Gamma_0$.

This concludes the proof of **(a)** and **(b)**. Next, we show that $\tilde{\nu}$ is l-adequate for G. Since $l < k$, this will end the proof of the lemma by the induction hypothesis for l (and the fact that any l-completion of $\tilde{\nu}$ is a k-completion of ν).

Now, the second condition in Definition 15.66 is vacuously satisfied (since $l > 0$), while the third is immediate from **(a)** and **(b)**, since if $\Gamma \Rightarrow \Delta$ is a component of G which consists only of $\tilde{\nu}_{<l}$-formulas, then $\Gamma \subseteq \Gamma_0$ and $\Delta \subseteq \Delta_0$ (This follows from the definitions of I, Γ_0, and Δ_0, and the fact that $\tilde{\nu}_{<l} \subset \tilde{\nu}_{<k} = \nu_{<k}$).

To show the first condition, assume that ν^* is an l-completion of $\tilde{\nu}$, $\Gamma \Rightarrow \Delta$ is a component of G, and $\Gamma \cup \Delta$ contains some $\tilde{\nu}_{\geq l}$-formula. If $\Gamma \Rightarrow \Delta \notin I$, then ν^* refutes $\Gamma \Rightarrow \Delta$, because ν is k-adequate, and ν^* is a k-completion of ν. If $\Gamma \Rightarrow \Delta \in I$, then $\Gamma \subseteq \Gamma_0$ and $\Delta \subseteq \Delta_0$, and so **(a)** and **(b)** imply that $\nu^*(\varphi) = l$ for every $\tilde{\nu}_{\geq l}$-formula $\varphi \in \Gamma$, and $\nu^*(\varphi) = -l$ for every $\tilde{\nu}_{\geq l}$-formula $\varphi \in \Delta$. Now, since $\Gamma \cup \Delta$ contains some $\tilde{\nu}_{\geq l}$-formula, only such formulas determine whether ν^* is a model of $\Gamma \Rightarrow \Delta$ or not. It follows that ν^* refutes $\Gamma \Rightarrow \Delta$. □

Theorem 15.68 *Let G be a hypersequent in $\mathcal{L}_R$. Then either $\vdash_{GRM}^{cf} G$, or G is not **RM**-valid.*

Proof. Suppose that $\nvdash_{GRM}^{cf} G$. Let Σ be the set of variables which occur in G, and assume that the number of elements in Σ is n. Let G^* be a maximal hypersequent which extends G. (Such an extension exists, because the number of hypersequents which consist only of subformulas of formulas in G is finite.) Define $\nu(p) = n+1$ for every $p \in \mathsf{Var} - \Sigma$. Obviously, ν is an $n+1$-semivaluation, and if φ occurs in G^* then φ is a $\nu_{<n+1}$-formula such that $\mathsf{Var}(\varphi) \subseteq \Sigma$. These facts and Lemma 15.58 easily imply (using External Weakenings) that ν is $n+1$-adequate for G^*. Therefore it follows from Lemma 15.67 that ν has an $n+1$-completion ν^* which is not a model of G^*, and so not a model of G either. Hence G is not **RM**-valid. □

Theorem 15.69 (adequacy of GRM) $\vdash_{GRM} G$ *iff G is **RM**-valid.*

Proof. Immediate from Proposition 15.56 and Theorem 15.68. □

Corollary 15.70 $\vdash_{GRM} \Rightarrow \varphi$ *iff* $\vdash_{\mathbf{RM}} \varphi$.

15.3 Cut-free Gentzen-type System for RM

Proof. Immediate from Theorem 15.69 and Note 15.54. □

Corollary 15.70 shows how GRM can be used for characterizing validity of formulas in **RM**. The next theorem generalizes it by showing how GRM can be used to characterize the consequence relation of **RM**. Since the latter is finitary, it suffices to treat the case of inferences from finite theories.

Theorem 15.71 Let $\mathcal{T} = \{\varphi_1, \ldots, \varphi_n\}$. Then $\mathcal{T} \vdash_{\mathbf{RM}} \psi$ iff the hypersequent $\varphi_1 \Rightarrow \psi \mid \cdots \mid \varphi_n \Rightarrow \psi \mid \Rightarrow \psi$ is provable in GRM.

Proof. Since $\supset$ (Definition 14.46) is an implication for **RM** (Proposition 14.47), and $\wedge$ is a conjunction for it, $\mathcal{T} \vdash_{\mathbf{RM}} \psi$ iff $\vdash_{\mathbf{RM}} \varphi_1 \wedge \cdots \wedge \varphi_n \supset \psi$. By Corollary 15.70 this is equivalent to $\vdash_{GRM} \Rightarrow \varphi_1 \wedge \cdots \wedge \varphi_n \supset \psi$, which in turn is equivalent to $\vdash_{GRM} \varphi_1 \Rightarrow \psi \mid \cdots \mid \varphi_n \Rightarrow \psi \mid \Rightarrow \psi$ by Proposition 15.50 and the Definition of $\supset$. □

Theorem 15.72 (cut-elimination for GRM) If $\vdash_{GRM} G$ then G has in GRM a cut-free proof.

Proof. Immediate from Proposition 15.56 and Theorem 15.68. □

Note 15.73 Theorems 15.68, 15.69, and 15.72 can be generalized to derivations from set of hypersequents. Thus a hypersequent G follows in GRM from a set $\mathcal{S}$ of hypersequents iff every model of $\mathcal{S}$ is a model of G. Moreover, the strong cut-elimination theorem applies to GRM: If $\mathcal{S} \vdash_{GRM} G$ then there is a proof in GRM of G from $\mathcal{S}$ in which all cuts are on formulas from $\bigcup_{H \in \mathcal{S}} \mathcal{F}_H$, where $\mathcal{F}_H$ is the set of formulas (not subformulas!) which appear in some component of H.

Note 15.74 We have proved that $\vdash_{GRM} \varphi$ iff $\vdash_{\mathbf{RM}} \varphi$ and that the cut-elimination theorem obtains for GRM by using semantic methods. However, both facts can also be proved purely syntactically. In the case of the first fact one has to directly prove that $\vdash_{GRM} G$ iff $\vdash_{HRM} \sigma_G$ (see Note 15.54). This is not difficult. In contrast, the only known syntactic proof of the cut-elimination theorem for GRM (in [27]) is extremely complicated, and it is not clear how to generalize it to derivations from sets of hypersequents.

Here are two examples of the use of the cut-elimination theorem for GRM in order to prove properties of **RM**:

Example 15.75 It is easy to prove by induction on length of cut-free proofs of sequents of the form $G \mid \Rightarrow \mid \cdots \mid \Rightarrow$ that such a sequent is provable in GRM iff G is provable in GRM. Now by definition, $G \mid \Rightarrow$ is valid in $\mathcal{M}(\mathbb{Z})$ iff G is valid in $\mathcal{M}(\mathbb{Z}^*)$. It follows that a hypersequent G is valid in $\mathcal{M}(\mathbb{Z})$ iff it is valid $\mathcal{M}(\mathbb{Z}^*)$. Hence the weak completeness of **RM** for $\mathcal{M}(\mathbb{Z})$ implies the weak completeness of **RM** for $\mathcal{M}(\mathbb{Z}^*)$ (Proposition 15.26).

Example 15.76 That the disjunctive syllogism is admissible in **RM** (Corollary 15.28) can easily been proved using *GRM* as follows: Suppose that $\vdash_{RM} \neg\varphi$, and $\vdash_{RM} \varphi \vee \psi$. Then $\vdash_{GRM} \Rightarrow \neg\varphi$, and $\vdash_{GRM} \Rightarrow \varphi \vee \psi$. By Proposition 15.50, this implies that $\vdash_{GRM} \varphi \Rightarrow$ and $\vdash_{GRM} \Rightarrow \varphi \mid\Rightarrow \psi$. Using a cut we infer that $\vdash_{GRM} \Rightarrow\mid\Rightarrow \psi$. It follows by the previous example that $\vdash_{GRM} \Rightarrow \psi$, and so $\vdash_{RM} \psi$.

15.4 RM with Propositional Constants

In this section we briefly describe what happens if we enrich **RM** with the propositional constants considered in Section 14.1.

As in the case of $\mathbf{RMI}_{\neg\rightarrow}$, the addition of F (or T - see Proposition 14.4) is not problematic. In fact, it is rather natural, since an analogue of Theorem 15.15 holds for $\mathbf{RM}^F$. Before presenting it, we need some definitions.

Definition 15.77 (bounded Sugihara matrices)

- A Sugihara chain $\langle \mathcal{V}, \leq, - \rangle$ is *bounded* if $\langle \mathcal{V}, \leq \rangle$ has a minimal element.[2]

- A *bounded Sugihara matrix* for $\mathcal{L}_R^F$ is a Sugihara matrix which is based on a bounded Sugihara chain $\langle \mathcal{V}, \leq, - \rangle$, and in which the interpretation $\tilde{F}$ of F is the minimal element of $\langle \mathcal{V}, \leq \rangle$.

Here is a particularly important example of a bounded Sugihara matrix:

Definition 15.78 $\mathcal{M}^F([0,1])$ is the extension of $\mathcal{M}([0,1])$ to the language $\mathcal{L}_R^F$ which is obtained by letting the interpretation $\tilde{F}$ of F to be 0.

Theorem 15.79

1. $\mathbf{RM}^F$ *is strongly sound and complete for bounded Sugihara matrices.*

2. $\mathbf{RM}^F$ *is strongly sound and complete for* $\mathcal{M}^F([0,1])$.

Proof. A straightforward extension of the proof of Theorem 15.15. □

Corollary 15.80 $\mathbf{RM}^F$ *is a conservative extension of* **RM**.

Things are again more complicated concerning the addition of t (or f) to **RM**. On one hand we do have the following counterparts of the first part of Theorem 15.79 and of Corollary 15.80:

---

[2] Obviously, if a is a minimal element then $-a$ is a maximal one. Hence, being bounded means here having both a minimal element and a maximal one.

Theorem 15.81 $\mathbf{RM^t}$ *is strongly sound and complete for Sugihara matrices in which the set $\mathcal{D}$ of designated elements has a minimal element, and the interpretation $\tilde{t}$ of t is that element.*

Proof. An easy extension of the proof of part 3 of Theorem 15.15. □

Corollary 15.82 $\mathbf{RM^t}$ *is a conservative extension of* $\mathbf{RM}$.

Proof. This easily follows from the soundness part of Theorem 15.81 and part 4 of Theorem 15.15, since the set of designated elements of $\mathcal{M}([0, 1])$ has a minimal element $(1/2)$. □

In contrast, the analogue of the second part of Theorem 15.79 fails:

Proposition 15.83 *Let $\mathcal{M}^t([0, 1])$ be the extension of $\mathcal{M}([0, 1])$ to the language $\mathcal{L}_R^t$ which is obtained by letting the interpretation $\tilde{t}$ of t be $1/2$. Then $\mathbf{RM^t}$ is strongly sound for $\mathcal{M}^t([0, 1])$, but it is not complete for it.*

Proof. The strong soundness part follows from Theorem 15.81. For the failure of completeness it suffices to note that $\vdash_{\mathcal{M}^t([0,1])} t \to \neg t$, but $\not\vdash_{\mathbf{RM^t}} t \to \neg t$. The latter fact can be shown with the help of the matrix $\mathcal{M}^t(\mathbb{Z}^*)$, which is the extension of $\mathcal{M}(\mathbb{Z}^*)$ to $\mathcal{L}_R^t$ that is obtained by letting $\tilde{t} = 1$. (See again Theorem 15.81.) □

Note 15.84 $\vdash_{\mathbf{RM^t}} \varphi \supset \psi \leftrightarrow \varphi \wedge t \to \psi$ (where $\supset$ is the connective defined in Definition 14.46). Hence in $\mathbf{RM^t}$ the implication we have used for $\mathbf{RM}$ and the standard one used in $\mathbf{R^t}$ (item 8 of Section 14.2.3) are identical.

15.5 RM as a Fuzzy Logic

Fuzzy logics are logics which are designed to deal with propositions that involve imprecise concepts, like "tall" or "old". Their semantics is based on the idea of *degrees of truth*, according to which the truth-value assigned to a proposition of this sort might not be one of the two classical values 0 and 1, but any real number between them. Now, in all the standard fuzzy logics investigated in the literature (see [120] for extensive surveys) the consequence relation is based on truth-functional semantics in which what is preserved is absolute truth, i.e., 1 is taken as the only designated value. Hence, Corollary 3.57 implies that none of these logics is paraconsistent. Therefore the obvious way to develop useful paraconsistent fuzzy logics is to use a more comprehensive set of designated values. This is precisely what is done in the semantics of $\mathbf{RM}$ as given in the last item of Theorem 15.15

(i.e. the matrix $\mathcal{M}([0,1])$). Hence, **RM** can serve as an excellent candidate for paraconsistent fuzzy logic. However, to view and use **RM** as a paraconsistent fuzzy logic it would be better to take the implication $\supset$ of **RM** (see Definition 14.46 and Propositions 15.33, 14.47) rather than $\rightarrow$ as a primitive connective. This is possible, since by the next proposition this choice does not affect the expressive power of the language.

Proposition 15.85 *The connective $\rightarrow$ of **RM** is definable in $\mathcal{L}_{CL}$.*

Proof. This follows from Proposition 14.61, using Proposition 15.2. (Note that the equivalence in **RM** of $\varphi \rightarrow \psi$ to $(\varphi \supset \psi) \wedge (\neg \psi \supset \neg \varphi)$ is particularly easy to show by using $\mathcal{M}([0,1])$ and the next Note.)[3] $\square$

Note 15.86 It is easy to see that in any Sugihara matrix we have:

$$a \tilde{\supset} b = \begin{cases} -a & a \leq b \leq -a, \\ b & \text{otherwise.} \end{cases}$$

Next we show that not only is **RM** a fuzzy logic according to the above characterization of this notion, but it (more exactly: its natural conservative extension **RM**^{F}) is in fact a conservative extension of one of the three most basic *standard* fuzzy logics ([120]): Gödel-Dummett logic $\mathbf{G}_\infty$.

Definition 15.87 (Gödel-Dummett logic $\mathbf{G}_\infty$) Let HG_∞ be the extension of HIL by the following linearity axiom:

[Li] $(\varphi \supset \psi) \vee (\psi \supset \varphi)$.

$\mathbf{G}_\infty$ is the logic in $\mathcal{IL}$ which is induced by HG_∞, and $\mathbf{G}_\infty^+$ is its positive (i.e., F-free) fragment.[4]

Theorem 15.88 **RM**^{F} *(see Definition 14.1 and Notation 14.2) is a conservative extension of $\mathbf{G}_\infty$, and **RM** is a conservative extension of $\mathbf{G}_\infty^+$.*

Proof. We show the first part. The proof of the second is almost identical.

We start by noting that since $\supset$ is an implication for **RM** (Proposition 14.47), it follows from Proposition 1.70, Theorem 1.90, Note 1.25, and Theorem 14.18 that HIL is included in **RM**^{F}. It is easy to verify (e.g. by using $\mathcal{M}([0,1])$) that the extra axiom [Li] of HG_∞ is also a theorem of **RM**^{F}. Hence **RM**^{F} is an extension of $\mathbf{G}_\infty$.

---

[3]In [25], it is noted that $\varphi \rightarrow \psi$ is equivalent in **RM** also to $\neg(\varphi \supset \psi) \supset \neg(\psi \supset \varphi)$, so it is definable in terms of just $\neg$ and $\supset$.

[4]Recall that $\mathbf{G}_\infty$ and a hypersequential cut-free Gentzen-type system GLC for it have already been presented in Example 1.144.

15.5 RM as a Fuzzy Logic

To show that $\mathbf{RM^F}$ is a *conservative* extension of $\mathbf{G_\infty}$, assume that $\mathcal{T} \nvdash_{HG_\infty} \psi$, where both $\mathcal{T}$ and ψ are in $\mathcal{L}_{CL}^F$. By applying Theorem 1.97 to $\mathcal{T}, \psi$ and HG_∞ we get an extension $\mathcal{T}^*$ of $\mathcal{T}$ such that:

1. $\mathcal{T}^* \nvdash_{HG_\infty} \psi$.

2. For every φ and τ, $\mathcal{T}^* \vdash_{HG_\infty} \varphi \wedge \tau$ iff both $\mathcal{T}^* \vdash_H \varphi$ and $\mathcal{T}^* \vdash_{HG_\infty} \tau$.

3. For every φ and τ, $\mathcal{T}^* \vdash_{HG_\infty} \varphi \vee \tau$ iff either $\mathcal{T}^* \vdash_{HG_\infty} \varphi$ or $\mathcal{T}^* \vdash_{HG_\infty} \tau$.

Now define that $\psi \equiv \sigma$ iff both $\vdash_{HG_\infty} \psi \supset \sigma$ and $\vdash_{HG_\infty} \sigma \supset \psi$. Since HG_∞ is an (axiomatic simple) extension of HIL, $\equiv$ is an equivalence relation (and actually a congruence relation). Let $\mathcal{V}$ be the set of equivalence classes, and define $\leq$ on $\mathcal{V}$ by letting $[\tau] \leq [\sigma]$ iff $\vdash_{HG_\infty} \tau \supset \sigma$. The fact that HG_∞ is an extension of HIL easily implies this time that $\leq$ is well defined, and is a partial order on $\mathcal{V}$. In addition, the $\vee$-primeness of $\mathcal{T}^*$ (item 3 above) and the special axiom [Li] of HG_∞ entail that $\leq$ is a *linear* order. Obviously, [F] is the minimal element of $\mathcal{V}$ according to this linear order, while axiom $[\supset 1]$ of HIL ensures that $\{\varphi \mid \mathcal{T}^* \vdash_{HG_\infty} \varphi\}$ is its maximal element. Since $\mathcal{V}$ is countable, these facts imply (see the beginning of the proof of Lemma 15.7) that there is a function $e : \mathcal{V} \to [0, 1/2]$ such that e is order preserving, $e([\mathsf{F}]) = 0$, and $e(\{\varphi \mid \mathcal{T}^* \vdash_{HG_\infty} \varphi\}) = 1/2$. Define a valuation ν in $\mathcal{M}([0,1])$ by letting $\nu(p) = e([p])$ for every variable p. We show that the following is true for every formula φ of $\mathcal{L}_{CL}^F$:

(a) If $\mathcal{T}^* \vdash_{HG_\infty} \varphi$ then $\nu(\varphi) \geq 1/2$.

(b) If $\mathcal{T}^* \nvdash_{HG_\infty} \varphi$ then $\nu(\varphi) = e([\varphi])$ (and so $\nu(\varphi) < 1/2$).

Since $\mathcal{T} \subseteq \mathcal{T}^*$ and $\mathcal{T}^* \nvdash_{HG_\infty} \psi$, these two facts imply that ν is a model of $\mathcal{T}$ in $\mathcal{M}([0,1])$ which is not a model of ψ. Hence, Theorem 15.79 entails that $\mathcal{T}^* \nvdash_{\mathbf{RM^F}} \psi$, which is what we want to prove.

We prove (a) and (b) by induction on the complexity of φ.

- The case where φ is an variable or the constant F easily follows from the definition of ν and the properties of e mentioned above.

- Suppose that $\varphi = \tau \supset \sigma$.

  (a) Suppose $\mathcal{T}^* \vdash_{HG_\infty} \varphi$. Then $[\tau] \leq [\sigma]$, and so $e([\tau]) \leq e([\sigma])$. If $\nu(\sigma) \geq 1/2$ then $\nu(\varphi) \geq 1/2$ (see Note 15.86). If not, then by (a) of the induction hypothesis $\mathcal{T}^* \nvdash_{HG_\infty} \sigma$, and so $\mathcal{T}^* \nvdash_{HG_\infty} \tau$. Hence (b) of the induction hypothesis implies that $\nu(\tau) = e([\tau])$ and $\nu(\sigma) = e([\sigma])$. Therefore $\nu(\tau) \leq \nu(\sigma)$, and so $\nu(\varphi) \geq 1/2$.

(b) Suppose that $\mathcal{T}^* \not\vdash_{HG_\infty} \varphi$. Because of Axiom $[\supset 1]$ this implies that also $\mathcal{T}^* \not\vdash_{HG_\infty} \sigma$, and so by (a) $\nu(\sigma) = e([\sigma]) < 1/2$. The assumption also implies that $[\tau] \not\leq [\sigma]$, and so $e([\sigma]) < e([\tau])$. Since by (a) and (b) $\nu(\tau) \geq 1/2$ or $\nu(\tau) = e([\tau])$, it follows that $\nu(\sigma) < \nu(\tau)$, and so (see Note 15.86) $\nu(\varphi) = \nu(\sigma) = e([\sigma])$. It remains to show that $e([\varphi]) = e([\sigma])$ in this case, i.e. that $\varphi \equiv \sigma$. That $\mathcal{T}^* \vdash_{HG_\infty} \sigma \supset \varphi$ is immediate from Axiom $[\supset 1]$. For the converse implication, note that since $\tau \supset (\tau \supset \sigma) \vdash_{HIL} \tau \supset \sigma$ (immediate from the deduction theorem of HIL), our assumption implies that $\mathcal{T}^* \not\vdash_{HG_\infty} \tau \supset (\tau \supset \sigma)$. Hence Axiom $[Li]$ and the $\vee$-primeness of $\mathcal{T}^*$ entail that $\mathcal{T}^* \vdash_{HG_\infty} (\tau \supset \sigma) \supset \tau$. But $\vdash_{HIL} ((\tau \supset \sigma) \supset \tau) \supset ((\tau \supset \sigma) \supset \sigma)$. It follows that $\mathcal{T}^* \vdash_{HG_\infty} (\tau \supset \sigma) \supset \sigma$, i.e., $\mathcal{T}^* \vdash_{HG_\infty} \varphi \supset \sigma$.

- Suppose that $\varphi = \tau \vee \sigma$.

    (a) Suppose that $\mathcal{T}^* \vdash_{HG_\infty} \varphi$. Then the $\vee$-primeness of $\mathcal{T}^*$ implies that either $\mathcal{T}^* \vdash_{HG_\infty} \tau$ or $\mathcal{T}^* \vdash_{HG_\infty} \sigma$. It follows by (a) of the induction hypothesis that either $\nu(\tau) \geq 1/2$ or $\nu(\sigma) \geq 1/2$. In both cases also $\nu(\varphi) \geq 1/2$.

    (b) Suppose that $\mathcal{T}^* \not\vdash_{HG_\infty} \varphi$. By Property 3 of $\mathcal{T}^*$ this means that $\mathcal{T}^* \not\vdash_{HG_\infty} \tau$ and $\mathcal{T}^* \not\vdash_{HG_\infty} \sigma$. It follows by (b) of the induction hypothesis that $\nu(\tau) = e([\tau])$ and $\nu(\sigma) = e([\sigma])$. Assume without loss of generality that $[\sigma] \leq [\tau]$. Then $\mathcal{T}^* \vdash_{HG_\infty} \sigma \supset \tau$, and $e([\sigma]) \leq e([\tau])$. The former fact implies (with the help of the axioms $[\supset \vee]$ and $[\vee \supset]$) that $\varphi \equiv \tau$, and so $e([\varphi]) = e([\tau])$, while the latter — that $\nu(\varphi) = e([\tau])$. Hence $\nu(\varphi) = e([\varphi])$.

- We leave the case where $\varphi = \tau \wedge \sigma$ to the reader.

This ends the proof of (a) and (b), and so of the theorem. □

Note 15.89 The standard semantics of Gödel-Dummett logic $\mathbf{G}_\infty$ as described in the literature on fuzzy logics is provided by the matrix $\langle [0,1], 1, \mathcal{O} \rangle$, where the interpretations in $\mathcal{O}$ of $\vee$, $\wedge$, and F are like in $\mathcal{M}([0,1])$, while $a \widetilde{\supset} b$ is 1 if $a \leq b$ and b otherwise. However, Theorem 15.88 shows that it is not essential to take 1 as the only designated value when we use this logic.

It is also interesting to note that the interpretation of $\neg$ in $\mathcal{M}^\mathsf{F}([0,1])$ (the strongly characteristic matrix for $\mathbf{RM}^\mathsf{F}$) is identical to that used in another one of the most famous fuzzy logics: Łukasiewicz logic.

We end our survey of $\mathbf{RM}$ with Hilbert-type systems in $\mathcal{CL}$ and $\mathcal{CL} \cup \{\mathsf{F}\}$ (respectively) for the full systems $\mathbf{RM}$ and $\mathbf{RM}^\mathsf{F}$. Like the usual systems for fuzzy logics, these systems have [MP] for $\supset$ as their sole rule of inference.

15.5 RM as a Fuzzy Logic

Definition 15.90 ($HRM_\supset$, $HRM_\supset^F$) The systems $HRM_\supset$ and $HRM_\supset^F$ are respectively obtained from the Hilbert-type systems $H_{\mathbf{PAC}}$ and $H_{\mathbf{J_3}}$ (Figure 4.7) by deleting axiom [⊃3] of HCL^+, and replacing [¬⊃⇒1] with:

$$[\neg\supset\Rightarrow 1]_r \quad (\varphi \supset \psi) \supset (\neg(\varphi \supset \psi) \supset \varphi).$$

Theorem 15.91 (strong completeness) *The system $HRM_\supset$ is strongly sound and complete for $\mathbf{RM}$ in $\mathcal{L}_{CL}$, and the system $HRM_\supset^F$ is strongly sound and complete for $\mathbf{RM^F}$ in $\mathcal{L}_{CL} \cup \{F\}$.*

Proof. Checking the validity of the axioms of $HRM_\supset$ in $\mathcal{M}([0,1])$ is easy (with the help of Proposition 14.47 and Note 15.86), and is left to the reader. By Theorem 15.15 this implies the strong soundness of $HRM_\supset$ for $\mathbf{RM}$. The strong soundness of $HRM_\supset^F$ is proved similarly, using Theorem 15.79.

To prove completeness of $HRM_\supset$, assume that $\mathcal{T} \nvdash_{HRM_\supset} \psi$. We show that $\mathcal{T} \nvdash_{\mathbf{RM}} \psi$. For this, apply Theorem 1.97 to get a theory $\mathcal{T}^*$ with the properties listed there (with $H = HRM_\supset$). In particular, $\mathcal{T}^* \nvdash_{HRM_\supset} \psi$, and $\mathcal{T}^*$ is ∨-prime: for every φ and τ, $\mathcal{T}^* \vdash_{HRM_\supset} \varphi \vee \tau$ iff either $\mathcal{T}^* \vdash_{HRM_\supset} \varphi$ or $\mathcal{T}^* \vdash_{HRM_\supset} \tau$. Because of axiom [t], this in turn implies that $\mathcal{T}^*$ is ¬-complete: for every φ either $\mathcal{T}^* \vdash_{HRM_\supset} \varphi$ or $\mathcal{T}^* \vdash_{HRM_\supset} \neg\varphi$. Let $\tau \equiv \sigma$ iff $\tau \supset \sigma$, $\sigma \supset \tau$, $\neg\sigma \supset \neg\tau$, and $\neg\tau \supset \neg\sigma$ are all theorems of $\mathcal{T}^*$. Since $HRM_\supset$ is an extension of $HIL_\supset$, $\equiv$ is easily seen to be an equivalence relation. Let $\mathcal{V}$ be the set of equivalence classes. Define on $\mathcal{V}$ the relation $\leq$ and the operation $-$ by: $[\tau] \leq [\sigma]$ iff $\tau \supset \sigma$ and $\neg\sigma \supset \neg\tau$ are theorems of $\mathcal{T}^*$, and $-[\sigma] = [\neg\sigma]$. Using the double negation axioms ([¬¬⇒] and [⇒¬¬]) and the fact that $HRM_\supset$ is an extension of $HIL_\supset$, it is easy to see that $\leq$ and $-$ are well-defined, that $\leq$ partially orders $\mathcal{V}$, and that $-$ is an involution for $\leq$ on $\mathcal{V}$. (See Definition 15.4.) To show that $S = \langle \mathcal{V}, \leq, - \rangle$ is a Sugihara chain, i.e. $\leq$ is linear, note that $[\neg\supset\Rightarrow 1]_r$ entails (with the help of [⊃1]):

$$(*) \quad \vdash_{HRM_\supset} \neg(\tau \supset \sigma) \supset (\sigma \supset \tau).$$

Now, if $\mathcal{T}^* \vdash_{HRM_\supset} \tau \supset \sigma$ and $\mathcal{T}^* \vdash_{HRM_\supset} \neg\sigma \supset \neg\tau$ then $[\tau] \leq [\sigma]$. If not, then the ¬-completeness of $\mathcal{T}^*$ implies that either $\mathcal{T}^* \vdash_{HRM_\supset} \neg(\tau \supset \sigma)$, or $\mathcal{T}^* \vdash_{HRM_\supset} \neg(\neg\sigma \supset \neg\tau)$. If $\mathcal{T}^* \vdash_{HRM_\supset} \neg(\tau \supset \sigma)$, then by $(*)$ $\mathcal{T}^* \vdash_{HRM_\supset} \sigma \supset \tau$, while by axiom [¬⊃⇒2] $\mathcal{T}^* \vdash_{HRM_\supset} \neg\sigma$, and so $\mathcal{T}^* \vdash_{HRM_\supset} \neg\tau \supset \neg\sigma$. Hence $[\sigma] \leq [\tau]$ in this case. If $\mathcal{T}^* \vdash_{HRM_\supset} \neg(\neg\sigma \supset \neg\tau)$, then by $(*)$ $\mathcal{T}^* \vdash_{HRM_\supset} \neg\tau \supset \neg\sigma$, while by axiom [¬⊃⇒2] $\mathcal{T}^* \vdash_{HRM_\supset} \neg\neg\tau$, and so $\mathcal{T}^* \vdash_{HRM_\supset} \sigma \supset \tau$ (by [¬¬⇒] and [⊃1]). Hence $[\sigma] \leq [\tau]$ in this case too.

$S = \langle \mathcal{V}, \leq, - \rangle$ is therefore a Sugihara chain. Next, we show that $[\tau] \in \mathcal{V}$ is designated in $\mathcal{M}(S)$ iff $\mathcal{T}^* \vdash_{HRM_\supset} \tau$. So, suppose $[\tau]$ is designated. Then by definitions, $-[\tau] \leq [\tau]$, and so $\mathcal{T}^* \vdash_{HRM_\supset} \neg\tau \supset \tau$. Since axiom [t] entails that $\vdash_{HRM_\supset} (\neg\tau \supset \tau) \supset \tau$, it follows that $\mathcal{T}^* \vdash_{HRM_\supset} \tau$. For the converse, assume that $\mathcal{T}^* \vdash_{HRM_\supset} \tau$. Then obviously $\mathcal{T}^* \vdash_{HRM_\supset} \neg\tau \supset \tau$, and by $[\Rightarrow\neg\neg]$ also $\mathcal{T}^* \vdash_{HRM_\supset} \neg\tau \supset \neg\neg\tau$. Hence $[\neg\tau] \leq [\tau]$, i.e. $-[\tau] \leq [\tau]$.

Define now a valuation ν in $\mathcal{M}(S)$ by letting $\nu(p) = [p]$ for every variable p. We prove that $\nu(\varphi) = [\varphi]$ for every formula φ. The proof is by induction on the complexity of φ.

- If φ is atomic then the claims follows from the definition of ν.

- Suppose $\varphi = \neg\tau$.

  Then $\nu(\varphi) = -\nu(\tau)$. It follows by induction hypothesis that $\nu(\varphi) = -[\tau]$, and so $\nu(\varphi) = [\neg\tau] = [\varphi]$ by the definition of $-$.

- Suppose $\varphi = \tau \supset \sigma$.

  Assume first that $\nu(\tau) \leq \nu(\sigma) \leq -\nu(\tau)$, and so $\nu(\varphi) = -\nu(\tau)$. By induction hypothesis this implies that $[\tau] \leq [\sigma] \leq [\neg\tau]$, and $\nu(\varphi) = [\neg\tau]$. It follows that (i) $\mathcal{T}^* \vdash_{HRM_\supset} \tau \supset \sigma$, (ii) $\mathcal{T}^* \vdash_{HRM_\supset} \sigma \supset \neg\tau$, (iii) $\mathcal{T}^* \vdash_{HRM_\supset} \neg\sigma \supset \neg\tau$, and (iv) $\mathcal{T}^* \vdash_{HRM_\supset} \tau \supset \neg\sigma$. By (ii), (iii), and [t], $\mathcal{T}^* \vdash_{HRM_\supset} \neg\tau$, and so (1): $\mathcal{T}^* \vdash_{HRM_\supset} (\tau \supset \sigma) \supset \neg\tau$. In turn, (i) implies (2): $\mathcal{T}^* \vdash_{HRM_\supset} \neg\tau \supset (\tau \supset \sigma)$, while by (iv) and $[\Rightarrow\neg\supset]$, $\mathcal{T}^* \vdash_{HRM_\supset} \tau \supset \neg(\tau \supset \sigma)$, and so (3): $\mathcal{T}^* \vdash_{HRM_\supset} \neg\neg\tau \supset \neg(\tau \supset \sigma)$. Finally, that (4): $\mathcal{T}^* \vdash_{HRM_\supset} \neg(\tau \supset \sigma) \supset \neg\neg\tau$ follows from (i), $[\Rightarrow\neg\neg]$, and $[\neg\supset\Rightarrow 1]_r$. Together, (1)–(4) mean that $\tau \supset \sigma \equiv \neg\tau$, as required.

  Now, assume that $\nu(\tau) \not\leq \nu(\sigma)$ or $\nu(\sigma) \not\leq -\nu(\tau)$. Then, by induction hypothesis, $\nu(\varphi) = \nu(\sigma) = [\sigma]$, and at least one of (i)–(iv) above is false. We show that this implies that $\tau \supset \sigma \equiv \sigma$, and so $[\sigma] = [\varphi]$, implying that in this case too $\nu(\varphi) = [\varphi]$. Since $\sigma \supset (\tau \supset \sigma)$ and $\neg(\tau \supset \sigma) \supset \neg\sigma$ are axioms of $HRM_\supset$, we only need to prove:

  (a) $\mathcal{T}^* \vdash_{HRM_\supset} \neg\sigma \supset \neg(\tau \supset \sigma)$,
  (b) $\mathcal{T}^* \vdash_{HRM_\supset} (\tau \supset \sigma) \supset \sigma$.

  The proofs of (a) and (b) depend on which of (i)–(iv) above is false:

  (i) Suppose that $\mathcal{T}^* \nvdash_{HRM_\supset} \tau \supset \sigma$. Then $\mathcal{T}^* \vdash_{HRM_\supset} \neg(\tau \supset \sigma)$, and (a) follows. By using $[\neg\supset\Rightarrow 1]_r$ and $HIL_\supset$ we get (b) as well.

  (ii) Suppose that $\mathcal{T}^* \nvdash_{HRM_\supset} \sigma \supset \neg\tau$. Then $\mathcal{T}^* \vdash_{HRM_\supset} \neg(\sigma \supset \neg\tau)$. It follows by $[\neg\supset\Rightarrow 2]$ and $[\neg\neg\Rightarrow]$ that $\mathcal{T}^* \vdash_{HRM_\supset} \tau$. Hence, (a) follows from $[\Rightarrow\neg\supset]$, and (b) from the deduction theorem for $\supset$.

  (iii) The proof in the case where $\mathcal{T}^* \nvdash_{HRM_\supset} \neg\sigma \supset \neg\tau$ is similar to that in case (ii), and is left to the reader.

(iv) Suppose that $\mathcal{T}^* \not\vdash_{HRM_\supset} \tau \supset \neg\sigma$. Then $\mathcal{T}^* \vdash_{HRM_\supset} \neg(\tau \supset \neg\sigma)$. It follows, by $[\neg \supset \Rightarrow 2]$ and $[\neg\neg\Rightarrow]$, that $\mathcal{T}^* \vdash_{HRM_\supset} \sigma$, and (b) follows. By $[\neg\supset\Rightarrow 1]_r$ we have that $\mathcal{T}^* \vdash_{HRM_\supset} (\tau \supset \neg\sigma) \supset \tau$ too, and so $\mathcal{T}^* \vdash_{HRM_\supset} \neg\sigma \supset \tau$. Using $[\Rightarrow\neg\supset]$, we get (a).

- Suppose that $\varphi = \tau \vee \sigma$.

  By induction hypothesis, $\nu(\tau) = [\tau]$ and $\nu(\sigma) = [\sigma]$. Since $\leq$ is linear we may assume without loss of generality that $[\tau] \leq [\sigma]$. This implies that $\nu(\varphi) = \nu(\sigma) = [\sigma]$, $\mathcal{T}^* \vdash_{HRM_\supset} \tau \supset \sigma$, and $\mathcal{T}^* \vdash_{HRM_\supset} \neg\sigma \supset \neg\tau$. The second fact implies (using $[\vee \supset]$) that $\mathcal{T}^* \vdash_{HRM_\supset} (\tau \vee \sigma) \supset \sigma$, and the third implies (using axiom $[\Rightarrow\neg\vee]$) that $\mathcal{T}^* \vdash_{HRM_\supset} \neg\sigma \supset \neg(\tau \vee \sigma)$. Together with the axioms $[\supset \vee]$ and $[\neg \vee \Rightarrow 2]$, the last two facts entail that $\tau \vee \sigma \equiv \sigma$. Since $\nu(\varphi) = [\sigma]$, it follows that $\nu(\varphi) = [\tau \vee \sigma] = [\varphi]$.

- The proof in case $\varphi = \tau \wedge \sigma$ is similar to that of the previous case.

Since it was shown above that $[\varphi]$ is designated iff $\not\vdash_{HRM_\supset} \varphi$, and $\mathcal{T} \subseteq \mathcal{T}^*$ while $\mathcal{T}^* \not\vdash_{HRM_\supset} \psi$, the last claim implies that ν is a model of $\mathcal{T}$ in $\mathcal{M}(S)$ which is not a model of ψ. Therefore it follows from Theorem 15.15 that $\mathcal{T} \not\vdash_{\mathbf{RM}} \psi$, as required.

The strong completeness proof of $\mathbf{RM}^\mathsf{F}$ is similar. □

Note 15.92 Let HRM_{ni} be the system in the language $\{\neg, \supset\}$ which is obtained from $HRM_\supset$ in exactly the same way as $H_{\mathbf{PAC}_\supset}$ is obtained from $H_\mathbf{PAC}$ (see Definition 4.98). Then HRM_{ni} is strongly sound and complete for the $\{\neg, \supset\}$-fragment of $\mathbf{RM}$. The proof of this is very similar to the proof of Theorem 15.91 (see [25]).

15.6 Bibliographical Notes and Further Reading

The logic $\mathbf{RM}$ was introduced by Dunn and McCall. (See, e.g., [4, 151] for more information about its history and motivation.) It was later extensively investigated by Meyer and Dunn. As noted in [151], it is "by far the best understood of the Anderson–Belnap style systems."

The notion of a Sugihara Matrix is a generalization (made by Meyer) of a particular such matrix which was introduced in [275].

That $\mathcal{M}_m(\mathbb{Z}_1)$ is weakly characteristic for $\mathbf{L}_{HRM_{\vec{\supset}}}$ and $\mathbf{RM}_{\vec{\supset}}$ was (essentially) shown in [239]. The fact that it is not strongly characteristic for them was (to our best knowledge) first shown in [38].

Theorem 15.15 is essentially due to [147]. However, Dunn has used the countable matrix $\mathcal{M}(\mathbb{Q})$ for strongly characterizing **RM**, rather than the uncountable $\mathcal{M}([0,1])$ used by us here.

The second part of Corollary 15.21, and Propositions 15.22, 15.26 are due to Meyer (see [4, Section 29.3]). Corollary 15.24 and Proposition 15.25 are due to Dunn (see [4, Section 29.4]).

That the disjunctive syllogism is admissible in **RM** (Corollary 15.28), as well as in **R** and in many other systems in Anderson and Belnap's family of logics, is another famous result of Meyer and Dunn. See [214] and Section 25 of [4].

That **RM** satisfies the basic relevance criterion (Proposition 15.35) was first claimed by Meyer (but with a wrong proof) in Section 29.3 of [4]. The decidability of **RM** (Theorem 15.32), and Proposition 15.41 were also first shown by him there.

The use of $\supset$ as an implication for pure **RM** (without propositional constants) was first made in [25].

That **RM** has no finite weakly characteristic *deterministic* matrix (a weaker version Proposition 15.39 in the case of **RM**) was first proved by Dunn in [147]. Proposition 15.39 itself is due to [57].

Theorems 15.43 and 15.44 are due to Dunn ([147], [4] Section 29.4).

The Gentzen-type hypersequential system *GRM* for **RM** was introduced in [27]. Unlike here, the cut-elimination theorem for it was proved there by a complicated syntactic method. The proof given here is taken from [55].

The connection between **RM** and the Gödel-Dummett fuzzy logic $\mathbf{G}_\infty$ was first observed by Dunn and Meyer in [150], where it was proved that $\mathbf{RM}^t$ is a *weakly* conservative extension of the positive fragment of $\mathbf{G}_\infty$. Other, more recent works on substructural fuzzy logics are Slaney's logic **F** ([270]) and Avron's fuzzy extension of **T** ([51]). The approach to developing paraconsistent fuzzy logics described here (which was initiated in [54]) is further developed and extended in [75, 74]. A completely different approach is described in [154]. For more information on fuzzy logics in general we refer the reader to [120, 178, 213].

For an extensive study of fragments of **RM** see [102].

With the exception of Section 15.3, this chapter is mainly based on [54].

Bibliography

[1] J. M Abe. *Paraconsistent Intelligent-Based Systems.* Springer, 2015.

[2] A. Almukdad and D. Nelson. Constructible falsity and inexact predicates. *Journal of Symbolic Logic*, 49:231–333, 1984.

[3] A. R. Anderson and N. D. Belnap. The pure calculus of entailment. *Journal of Symbolic Logic*, 27(19–52), 1962.

[4] A. R. Anderson and N. D. Belnap. *Entailment: The Logic of Relevance and Necessity, Vol.I.* Princeton University Press, 1975.

[5] A. R. Anderson, N. D. Belnap, and M. Dunn. *Entailment: The Logic of Relevance and Necessity, Vol.II.* Princeton University Press, 1992.

[6] O. Arieli. Paraconsistent declarative semantics for extended logic programs. *Annals of Mathematics and Artificial Intelligence*, 36(4):381–417, 2002.

[7] O. Arieli. Paraconsistent reasoning and preferential entailments by signed quantified Boolean formulae. *ACM Transactions on Computational Logic*, 8(3), 2007.

[8] O. Arieli. Conflict-free and conflict-tolerant semantics for constrained argumentation frameworks. *Journal of Applied Logic*, 13(4):582–604, 2015.

[9] O Arieli. On the acceptance of loops in argumentation frameworks. *Journal of Logic and Computation*, 26(4):1203–1234, 2016.

[10] O. Arieli and A. Avron. Logical bilattices and inconsistent data. In *Proc. 9th Annual IEEE Symposium on Logic in Computer Science (LICS'94)*, pages 468–476, 1994.

[11] O. Arieli and A. Avron. Reasoning with logical bilattices. *Journal of Logic, Language, and Information*, 5(1):25–63, 1996.

[12] O. Arieli and A. Avron. The value of the four values. *Artificial Intelligence*, 102(1):97–141, 1998.

[13] O. Arieli and A. Avron. Bilattices and paraconsistency. In D. Batens, C. Mortensen, G. Priest, and J. Van Bendegem, editors, *Frontiers of Paraconsistent Logic*, volume 8 of *Studies in Logic and Computation*, pages 11–27. Research Studies Press, 2000.

[14] O. Arieli and A. Avron. General patterns for nonmonotonic reasoning: from basic entailments to plausible relations. *Logic Journal of the IGPL*, 8(2):119–148, 2000.

[15] O. Arieli and A. Avron. Three-valued paraconsistent propositional logics. In J.-Y. Béziau, M. Chakraborty, and S. Dutta, editors, *New Directions in Paraconsistent Logic*, pages 91–129. Springer, 2015.

[16] O. Arieli and A. Avron. Four-valued paradefinite logics. *Studia Logica*, 105(6):1087–1122, 2017.

[17] O. Arieli, A. Avron, and A. Zamansky. Maximally paraconsistent three-valued logics. In *Proc 12th Int. Conf. on Principles of Knowledge Representation and Reasoning* (KR'10), pages 310–318. AAAI Press, 2010.

[18] O. Arieli, A. Avron, and A. Zamansky. Ideal paraconsistent logics. *Studia Logica*, 99(1–3):31–60, 2011.

[19] O. Arieli, A. Avron, and A. Zamansky. Maximal and premaximal paraconsistency in the framework of three-valued semantics. *Studia Logica*, 97(1):31–60, 2011.

[20] O. Arieli, A. Avron, and A. Zamansky. What is an ideal logic for reasoning with inconsistency? In *Proce. 22nd Int. Joint Conf. on Artificial Intelligence* (IJCAI 2011), pages 706–711, 2011.

[21] O. Arieli and M. Denecker. Reducing preferential paraconsistent reasoning to classical entailment. *Journal of Logic and Computation*, 13(4):557–580, 2003.

[22] O. Arieli and C. Straßer. Deductive argumentation by enhanced sequent calculi and dynamic derivations. *Electronic Notes in Theoretical. Computer Science*, 323:21–37, 2016.

[23] F. G. Asenjo. A calculus of antinomies. *Notre Dame Journal of Formal Logic*, 7:103–106, 1966.

[24] A. Avron. Relevant entailment - semantics and formal systems. *Journal of Symbolic Logic*, 49:334–342, 1984.

[25] A. Avron. On an implication connective of RM. *Notre Dame Journal of Formal Logic*, 27:201–209, 1986.

[26] A. Avron. On purely relevant logics. *Notre Dame Journal of Formal Logic*, 27:180–194, 1986.

[27] A. Avron. A constructive analysis of RM. *Journal of Symbolic Logic*, 52:939–951, 1987.

[28] A. Avron. The semantics and proof theory of linear logic. *Theoretical Computer Science*, 57:161–184, 1988.

[29] A. Avron. Relevance and paraconsistency – A new approach. *Journal of Symbolic Logic*, 55:707–732, 1990.

[30] A. Avron. Relevance and paraconsistency – A new approach. Part II: The formal systems. *Notre Dame Journal of Formal Logic*, 31:169–202, 1990.

[31] A. Avron. Hypersequents, logical consequence and intermediate logics for concurrency. *Annals of Mathematics and Artificial Intelligence*, 4:225–248, 1991.

[32] A. Avron. Natural 3-valued logics: Characterization and proof theory. *Journal of Symbolic Logic*, 56(1):276–294, 1991.

[33] A. Avron. Relevance and paraconsistency – A new approach. Part III: Cut-free Gentzen-type systems. *Notre Dame Journal of Formal Logic*, 32:147–160, 1991.

[34] A. Avron. Simple consequence relations. *Information and Computation*, 92:105–139, 1991.

[35] A. Avron. Whither relevance logic? *Journal of Philosophical Logic*, 21:243–281, 1992.

[36] A. Avron. The method of hypersequents in proof theory of propositional non-classical logics. In W. Hodges, M. Hyland, C. Steinhorn, and J. Truss, editors, *Logic: Foundations to Applications*, pages 1–32. Oxford Science Publications, 1996.

[37] A. Avron. The structure of interlaced bilattices. *Journal of Mathematical Structures in Computer Science*, 6:287–299, 1996.

[38] A. Avron. Multiplicative conjunction as an extensional conjunction. *Logic Journal of the IGPL*, 5:181–208, 1997.

[39] A. Avron. Multiplicative conjunction and an algebraic meaning of contraction and weakening. *Journal of Symbolic Logic*, 63:831–859, 1998.

[40] A. Avron. On the expressive power of three-valued and four-valued languages. *Journal of Logic and Computation*, 9(6):977–994, 1999.

[41] A. Avron. On negation, completeness and consistency. In D. Gabbay and F. Guenther, editors, *Handbook of Philosophical Logic*, volume 9, pages 287–319. Kluwer, 2002.

[42] A. Avron. Classical Gentzen-type methods in propositional many-valued logics. In M. Fitting and E. Orlowska, editors, *Studies in Fuzziness and Soft Computing*, volume 114, pages 117–155. Physica Verlag, 2003.

[43] A. Avron. Combining classical logic, paraconsistency and relevance. *Journal of Applied Logic*, 3:133–160, 2004.

[44] A. Avron. A nondeterministic view on nonclassical negations. *Studia Logica*, 80:159–194, 2005.

[45] A. Avron. Non-deterministic semantics for families of paraconsistent logics. In J. Y. Béziau, W. A. Carnielli, and D. M. Gabbay, editors, *Handbook of Paraconsistency*, volume 9 of *Studies in Logic*, pages 285–320. College Publications, 2007.

[46] A. Avron. Non-deterministic semantics for logics with a consistency operator. *Journal of Approximate Reasoning*, 45:271–287, 2007.

[47] A. Avron. Multi-valued semantics: Why and how. *Studia Logica*, 92(2):163–182, 2009.

[48] A. Avron. A simple proof of completeness and cut-elimination for propositional Gödel logic. *Journal of Logic and Computation*, 21:813–821, 2011.

[49] A. Avron. The classical constraint on relevance. *Logica Universalis*, 8:1–15, 2014.

[50] A. Avron. Paraconsistency, paracompleteness, Gentzen systems, and trivalent semantics. *Journal of Applied Non-Classical Logics*, 24(1-2):12–34, 2014.

[51] A. Avron. Paraconsistent fuzzy logic preserving non-falsity. *Fuzzy Sets and Systems*, 292:75–84, 2014.

[52] A. Avron. What is relevance logic? *Annals of Pure and Applied Logic*, 165:26–48, 2014.

[53] A. Avron. Semi-implication: A chapter in universal logic. In A. Koslow and A. Buchsbaum, editors, *The Road to Universal Logic, Volume I.*, Studies in Universal Logic, pages 59–72. Birkhűser, 2015.

[54] A. Avron. RM and its nice properties. In K. Bimbó, editor, *J. Michael Dunn on Information Based Logics*, volume 8 of *Outstanding Contributions to Logic*, pages 15–43. Springer, 2016.

[55] A. Avron. Cut-elimination in RM proved semantically. *IFCoLog Journal of Logics and their Applications*, 4:605–621, 2017.

[56] A. Avron. Self-extensional three-valued paraconsistent logics. *Logica Universalis*, 11:297–315, 2017.

[57] A. Avron. Paraconsistency and the need for infinite semantic. *Soft Computing*, 2018. To appear.

[58] A. Avron, O. Arieli, and A. Zamansky. On strong maximality of paraconsistent finite-valued logics. In *Proc. 25th Annual IEEE Symposium on Logic in Computer Science* (LICS'10). IEEE Press, 2010.

[59] A. Avron, J. Ben-Naim, and B. Konikowska. Cut-free ordinary sequent calculi for logics having generalized finite-valued semantics. *Logica Universalis*, 1:41–69, 2006.

[60] A. Avron, J. Ben-Naim, and B. Konikowska. Processing information from a set of sources. In *Towards Mathematical Philosophy*, volume 28 of *Trends in Logic*, pages 165–186. Springer, 2009.

[61] A. Avron and J.-Y. Béziau. Self-extensional three-valued paraconsistent logics have no implication. *Logic Journal of the IGPL*, 25:183–194, 2017.

[62] A. Avron and B. Konikowska. Multi-valued calculi for logics based on non-determinism. *Logic Journal of the IGPL*, 13(4):365–387, 2005.

[63] A. Avron and B. Konikowska. Finite-valued logics for information processing. *Fundamenta Informaticae*, 114:1–30, 2012.

[64] A. Avron, B. Konikowska, and A. Zamansky. Cut-free sequent calculi for C-systems with generalized finite-valued semantics. *Journal of Logic and Computation*, 23(3):517–540, 2013.

[65] A. Avron, B. Konikowska, and A. Zamansky. Efficient reasoning with inconsistent information using C-systems. *Information Science*, 296:219–236, 2015.

[66] A. Avron and O. Lahav. A fully maximal paraconsistent extension of S5. 2017. Submitted.

[67] A. Avron and I. Lev. Canonical propositional Gentzen-type systems. In *Proceedings of the 1st International Joint Conference on Automated Reasoning*, volume 2083 of *Lecture Notes in Artificial Intelligence*, pages 529–544. Springer, 2001.

[68] A. Avron and I. Lev. A formula-preferential base for paraconsistent and plausible non-monotonic reasoning. In *Proc. KRR-4, IJCAI'01 Workshop on Inconsistency in Data and Knowledge*, pages 60–70, 2001.

[69] A. Avron and I. Lev. Non-deterministic multi-valued structures. *Journal of Logic and Computation*, 15:241–261, 2005.

[70] A. Avron and A. Zamansky. Paraconsistency, self-extensionality, modality. To appear in a special issue of The Logic Journal of the IGPL.

[71] A. Avron and A. Zamansky. Many-valued non-deterministic semantics for first-order logics of formal (in)consistency. In *Algebraic and Proof-theoretic Aspects of Non-classical Logics*, number 4460 in Lecture Notes in Artificial Intelligence, pages 1–24. Springer, 2007.

[72] A. Avron and A. Zamansky. Non-deterministic semantics for logical systems – A survey. In D. Gabbay and F. Guenther, editors, *Handbook of Philosophical Logic*, volume 16, pages 227–304. Springer, 2011.

[73] A. Avron and A. Zamansky. A paraconsistent view on B and S5. In *Advances in Modal Logic*, volume 11, pages 21–37. College Publications, 2016.

[74] A. Avron and Y. Zohar. Expansions and refinements of non-deterministic matrices: Theory and applications in non-classical logics. 2017. Submitted.

[75] A. Avron and Y. Zohar. Non-deterministic matrices in action: Expansions, refinements, and rexpansions. In *Proc. 17th International Symposium on Multi-valued Logic* (ISMVL'17), 2017.

[76] M. Baaz. Kripke-type semantics for da costa's paraconsistent logic C_ω. *Notre Dame Journal of Formal Logic*, 27:523–527, 1986.

[77] H. Barringer, J. H. Cheng, and C. B. Jones. A logic covering undefinedness in program proofs. *Acta Informatica*, 21:251–269, 1984.

[78] D. Batens. Paraconsistent extensional propositional logics. *Logique et Analyse*, 90–91:195–234, 1980.

[79] D. Batens. Inconsistency-adaptive logics. In E. Orlowska, editor, *Logic at Work*, pages 445–472. Physica Verlag, 1998.

[80] D. Batens. A survey on inconsistency-adaptive logics. In D. Batens, C. Mortensen, G. Priest, and J. Van Bendegem, editors, *Frontiers of Paraconsistent Logic*, volume 8 of *Studies in Logic and Computation*, pages 69–73. Research Studies Press, 2000.

[81] D. Batens. On some remarkable relations between paraconsistent logics, modal logics, and ambiguity logics. In W. A. Carnielli, M. E. Coniglio, and I. D'Ottaviano, editors, *Paraconsistency: The Logical Way to the Inconsistent*, number 228 in Lecture Notes in Pure and Applied Mathematics, pages 275–293. Marcel Dekker, 2002.

[82] D. Batens. A universal logic approach to adaptive logics. *Logica Universalis*, 1:221–242, 2007.

[83] D. Batens, K. de Clercq, and N. Kurtonina. Embedding and interpolation for some paralogics. the propositional case. *Reports on Mathematical Logic*, 33:29–44, 1999.

[84] D. Batens, C. Mortensen, G. Priest, and J. Van Bendegem, editors. *Frontiers of Paraconsistent Logic. Proceedings of the first World Congress on Paraconsistency*, volume 8 of *Studies in Logic and Computation*. Research Studies Press, 2000.

[85] M. J. Beeson. *Foundations of Constructive Mathematics*. Springer, 1985.

[86] N. D. Belnap. How a computer should think. In G. Ryle, editor, *Contemporary Aspects of Philosophy*, pages 30–56. Oriel Press, 1977.

[87] N. D. Belnap. A useful four-valued logic. In J. M. Dunn and G. Epstein, editors, *Modern Uses of Multiple-Valued Logics*, pages 7–37. Reidel Publishing Company, 1977.

[88] J. Y. Béziau. Logiques construites suivant les méthodes de da Costa. *Logique et Analyse*, 33(131–132):259–272, 1990.

[89] J. Y. Béziau. Nouveaux résultats et nouveau regard sur la logique paraconsistent C1. *Logique et Analyse*, 36(141–142):35–58, 1993.

[90] J. Y. Béziau. S5 is a paraconsistent logic and so is first-order classical logic. *Logical Investigations*, 8:301–309, 2002.

[91] J. Y. Béziau. Paraconsistent logic from a modal viewpoint. *Journal of Applied Logic*, 3:6–14, 2005.

[92] J. Y. Béziau. The paraconsistent logic Z. A possible solution to Jaśkowski's problem. *Logic and Logical Philosophy*, 15:99–111, 2006.

[93] J. Y. Béziau. Bivalent semantics for De Morgan logic (The uselessness of four-valuedness). In W. A. Carnielli, M. E. Coniglio, and I. M. L. D'Ottaviano, editors, *The Many Sides of Logic*, pages 391–402. College Publication, 2009.

[94] J. Y. Béziau, W. A. Carnielli, and D. Gabbay, editors. *Handbook of Paraconsistency*. King's College Publications, 2007.

[95] J.-Y. Béziau, M. Chakraborty, and S. Dutta, editors. *New Directions in Paraconsistent Logic*, volume 152 of *Proceedings in Mathematics and Statistics*. Springer, 2015.

[96] J.Y. Béziau. From paraconsistent logic to universal logic. *Sorites*, 12:5–32, 2001.

[97] A. Bialynicki-Birula. Remarks on quasi-boolean algebras. *Bulletin de l'Académie Polonaise des Sciences Classe. III*, V(6):615–619, 1957.

[98] A. Bialynicki-Birula and H. Rasiowa. On the representation of quasi-boolean algebras. *Bulletin de l'Académie Polonaise des Sciences Classe. III*, V(3):259–261, 1957.

[99] K. Bimbó. Relevance logics. In D. Jacquette, editor, *Philosophy of Logic*, pages 723–789. Elsevier, 2006.

[100] K. Bimbó. *Proof Theory: Sequent Calculi and Related Formalisms*. Discrete Mathematics and Its Applications. CRC Press, 2015.

[101] K. Bimbó and M. Dunn. Four-valued logic. *Notre Dame Journal of Formal Logic*, 42(3):171–192, 2001.

[102] W. J. Blok and J. G. Raftery. Fragments of R-mingle. *Studia Logica*, 78:59–106, 2004.

[103] D. A. Bochvar. Ob odnom trechzna čnom iščislenii i ego primenenii k analizu paradoksov klassiceskogo funkcionalńogo iščislenija. *Matematicheskii Sbornik*, 4(46):287–308, 1938. Translated to English by M. Bergmann as "On a Three-valued Logical Calculus and Its Application to the Analysis of the Paradoxes of the Classical Extended Functional Calculus". History and Philosophy of Logic 2, pages 87–112, 1981.

[104] L. Bolc and P. Borowik. *Many-Valued Logics, Volume 1: Theoretical Foundations*. Springer, 1992.

[105] L. Bolc and P. Borowik. *Many-valued logics, Volume 2: Automated Reasoning And Practical Applications*. Springer, 2003.

[106] F. Bou and U. Rivieccio. The logic of distributive bilattices. *Logic Journal of the IGPL*, 19(1):183–216, 2010.

[107] R. T. Brady. The gentzenization and decidability of RW. *Journal of Philosophical Logic*, 19:35–73, 1990.

[108] M. Bremer. *An Introduction to Paraconsistent Logics*. Peter Lang, 2005.

[109] R. A. Bull and K. Segerberg. Basic modal logic. In D. Gabbay and F. Guenther, editors, *Handbook of Philosophical Logic, 2nd edition*, volume 3, pages 1–81. Kluwer, 2001.

[110] D. Busch. Sequent formalization of three-valued logic. In *Partiality, Modality, and Nonmonotonicity*, Studies in Logic, Language and Information, pages 45–75. CLSI Publications, 1996.

[111] W. A. Carnielli and M. E. Coniglio. *Paraconsistent Logic: Consistency, Contradiction and Negation*. Number 40 in Logic, Epistemology, and the Unity of Science. Springer, 2016.

[112] W. A. Carnielli, M. E. Coniglio, and I. D'Ottaviano, editors. *Paraconsistency: The Logical Way to the Inconsistent – Proceedings of the second World Congress on Paraconsistency*. Number 228 in Lecture Notes in Pure and Applied Mathematics. Marcel Dekker, 2001.

[113] W. A. Carnielli, M. E. Coniglio, and J. Marcos. Logics of formal inconsistency. In D. M. Gabbay and F. Guenthner, editors, *Handbook of Philosophical Logic*, volume 14, pages 1–95. Springer, 2007. Second edition.

[114] W. A. Carnielli and M. Lima-Marques. Reasoning under inconsistent knowledge. *Journal of Applied Non-classical Logics*, 2(1):49–79, 1992.

[115] W. A. Carnielli and J. Marcos. Tableau systems for logics of formal inconsistency. In *Proceedings of the 2001 International Conference on Artificial Intelligence*, volume 2, pages 848–852. CSREA Press, 2001.

[116] W. A. Carnielli and J. Marcos. A taxonomy of C-systems. In W. A. Carnielli, M. E. Coniglio, and I. D'Ottaviano, editors, *Paraconsistency: The Logical Way to the Inconsistent*, number 228 in Lecture Notes in Pure and Applied Mathematics, pages 1–94. Marcel Dekker, 2002.

[117] B. F. Chellas. *Modal Logic — An introduction*. Cambridge University Press, 1980.

[118] A. Church. The weak theory of implication. In A. Menne, A. Wilhelmi, and H. Angsil, editors, *Kontroliertes Denken: Untersuchungen zum Logikkalkül und zur Logik der Einzelwissenschften*, pages 22–37. Kommissioas-Verlag Karl Alber, Munich, 1951.

[119] A. Ciabattoni, O. Lahav, L. Spendier, and A. Zamansky. Taming paraconsistent (and other) logics: An algorithmic approach. *ACM Transactions on Computational Logic*, 16(1):1–23, 2014.

[120] P. Cintula, P. Hájek, and C. Noguera. *Handbook of Mathematical Fuzzy Logic*, volume 37–38 of *Studies in Logic*. College Publications, 2011.

[121] D. Ciucci and D. Dubois. A modal theorem-preserving translation of a class of three-valued logics of incomplete information. *Journal of Applied Non-Classical Logics*, 23(4):321–352, 2013.

[122] D. Ciucci and D. Dubois. From possibility theory to paraconsistency. In J.-Y. Béziau, M. Chakraborty, and S. Dutta, editors, *New Directions in Paraconsistent Logic*, pages 229–247. Springer, 2015.

[123] E. F. Codd. *The Relational Model for Database Management*. Addison-Wesley, 1990. Second edition.

[124] C. Cornelis, O. Arieli, G. Deschrijver, and E. Kerre. Uncertainty modeling by bilattice-based squares and triangles. *IEEE Transactions on Fuzzy Systems*, 15(2):161–175, 2007.

[125] J. Crawford and D. Etherington. A non-deterministic semantics for tractable inference. In *Proceedings of the 15th National Conference on Artificial Intelligence and the 10th Conference on Innovative Applications of Artificial Intelligence*, pages 286–291, 1998.

[126] N. C. A. da Costa. Calculs proporitionnels pour les systémes formels inconsistants. *C.R. de l'Académie des Sciences de Paris*, 257:3790–3793, 1963.

[127] N. C. A. da Costa. On the theory of inconsistent formal systems. *Notre Dame Journal of Formal Logic*, 15:497–510, 1974.

[128] N. C. A. da Costa and J. Y. Béziau. Carnot's logic. *Bulletin of the Section of Logic*, 22:98–105, 1993.

[129] N. C. A. da Costa, J.-Y. Béziau, and O. A. S. Bueno. Aspects of paraconsistent logic. *Bulletin of the IGPL*, 3:597–614, 1995.

[130] N. C. A. da Costa and D. Marconi. An overview of paraconsistent logics in the 80's. *The Journal of Non-Classical Logic*, 6(1):5–31, 1989.

[131] C. V. Damásio and L. M. Pereira. A model theory for paraconsistent logic programming. In *Proc. 7th Portuguese Conference on Progress in Artificial Intelligence,*(EPIA'95), volume 990 of *Lecture Notes in Computer Science*, pages 377–386. Springer, 1995.

[132] M. De and H. Omori. Classical negation and expansions of Belnap-Dunn logic. *Studia Logica*, 103(4):825–851, 2015.

[133] M. Denecker, A. Cortés-Calabuig, M. Bruynooghe, and O. Arieli. Towards a logical reconstruction of a theory for locally closed databases. *ACM Transactions on Database Systems*, 35(3), 2010.

[134] M. Denecker and D. De Schreye. Justification semantics: a unifying framework for the semantics of logic programs. In *Proc. of the Logic Programming and Nonmonotonic Reasoning Workshop*, pages 365–379. MIT Press, 1993.

[135] M. Denecker, V. W. Marek, and M. Truszczynski. Fixpoint 3-valued semantics for autoepistemic logic. In *Logical Foundations for Cognitive Agents: Contributions in Honor of Ray Reiter*, pages 113–136. Springer-Verlag, 1999.

[136] M. Denecker, V. W. Marek, and M. Truszczynski. Uniform semantic treatment of default and autoepistemic logics. *Journal of Artificil Intelleligence*, 143(1):79–122, 2003.

[137] M. Denecker and J. Vennekens. The well-founded semantics is the principle of inductive definition, revisited. In *Proc 14th Int. Conf. on Principles of Knowledge Representation and Reasoning* (KR'14). AAAI Press, 2014.

[138] G. Deschrijver, O. Arieli, C. Cornelis, and E. Kerre. A bilattice-based framework for handling graded truth and imprecision. *International Journal of Uncertainty, Fuzziness and Knowledge-Based Systems*, 15(1):13–41, 2007.

[139] A. Dodó and J. Marcos. Negative modalities, consistency and determinedness. *Electronic Notes in Theoretical Computer Science*, 300:21–45, 2014.

[140] P. Doherty, D. Driankov, and A. Tsoukiàs. Partiality, para-consistency and preference modelling, 1992. IDA Research report Lith-IDA-R-92-18, Linköping University.

[141] I. D'Ottaviano. The completeness and compactness of a three-valued first-order logic. *Revista Colombiana de Matematicas*, XIX(1–2):31–42, 1985.

[142] I. D'Ottaviano. On the development of paraconsistent logic and da Costa's work. *Journal of Non-classical Logic*, 7(1–2):89–152, 1990.

[143] I. D'Ottaviano and N. C. da Costa. Sur un problèm de Jakowśki. *C. R. Acad Sc. Paris, Volume 270, Sèrie A*, pages 1349–1353, 1970.

[144] M. Dummett. A propositional calculus with a denumerable matrix. *The Journal of Symbolic Logic*, 24:96–107, 1959.

[145] M. Dummett. *Elements of Intuitionism*. Oxford: Clarendon Press, 1977.

[146] J. M. Dunn. *The Algebra of Intensional Logics*. PhD thesis, University of Pittsburgh, Ann Arbor (UMI), 1966.

[147] J. M. Dunn. Algebraic completeness results for R-mingle and its extensions. *The Journal of Symbolic Logic*, 35:1–13, 1970.

[148] J. M. Dunn. Intuitive semantics for first-degree entailments and 'coupled trees'. *Philosophical Studies*, 29:149–168, 1976.

[149] J. M. Dunn. Partiality and its dual. *Studia Logica*, 66(1):5–40, 2000.

[150] J. M. Dunn and R. K. Meyer. Algebraic completeness results for Dummett's LC and its extensions. *Zeitschrift für mathematische Logik und Grundlagen der Mathematik*, 17:225–230, 1971.

[151] J. M. Dunn and G. Restall. Relevance logic. In D. Gabbay and F. Guenther, editors, *Handbook of Philosophical Logic*, volume 6, pages 1–136. Kluwer, 2002. Second edition.

[152] H. B. Enderton. *A Mathematical Introduction to Logic*. Academic Press, 2001. Second edition.

[153] R. L. Epstein. *The semantic foundation of logic. Vol.I: propositional logics*. Advanced Reasoning Forum, 2012. Third edition.

[154] R. Ertola, F. Esteva, T. Flaminio, L. Godo, and C. Noguera. Paraconsistency properties in degree-preserving fuzzy logics. *Soft Computing*, 19:531–546, 2015.

[155] F. Esteva, P. Garcia, and L. Godo. Enriched interval bilattices and partial many-valued logics: an approach to deal with graded truth and imprecision. *Uncertainty, Fuzziness and Knowledge-based Systems*, 2(1):37–54, 1994.

[156] M. Fitting. A Kripke-Kleene semantics for logic programs. *Journal of Logic Programming*, 2(4):295–312, 1985.

[157] M. Fitting. Bilattices in logic programming. In G. Epstein, editor, *Proc. 20th Int. Symp. on Multiple-Valued Logic*, pages 238–246, 1990.

[158] M. Fitting. Bilattices and the semantics of logic programming. *Journal of Logic Programming*, 11(2):91–116, 1991.

[159] M. Fitting. Kleene's logic, generalized. *Journal of Logic and Computation*, 1:797–810, 1992.

[160] M. Fitting. The family of stable models. *Journal of Logic Programming*, 17:197–225, 1993.

[161] M. Fitting. Kleene's three-valued logics and their children. *Fundamenta Informaticae*, 20:113–131, 1994.

[162] M. Fitting. *First-order Logic and Automated Theorem Proving*. Springer-Verlag, 1996. Second edition.

[163] M. Fitting. Fixpoint semantics for logic programming a survey. *Theoretical Computer Science*, 278(1–2):25–51, 2002.

[164] M. Fitting. Bilattices are nice things. In T. Bolander, V. Hendricks, and S. A. Pedersen, editors, *Self Reference*, volume 178 of *CSLI Lecture Notes*, pages 53–77. CLSI Publications, 2006.

[165] J. M. Font and P. Hájek. On Łukasiewicz's four-valued modal logic. *Studia Logica*, 70(2):157–182, 2002.

[166] J. M. Font and M. Moussavi. Note on a six-valued extension of three-valued logic. *Journal of Applied Non-Classical Logic*, 3(2):173–187, 1993.

[167] D. Gabbay and H. Wansing, editors. *What is Negation?*, volume 13 of *Applied Logic Series*. Springer, 1999.

[168] G. Gargov. Knowledge, uncertainty and ignorance in logic: bilattices and beyond. *Journal of Applied Non-Classical Logics*, 9(2–3):195–283, 1999.

[169] P. Gentilini. Proof theory and mathematical meaning of paraconsistent C-systems. *Journal of Applied Logic*, 9:171–202, 2011.

[170] G. Gentzen. Investigations into logical deduction, 1934. In German. An English translation appears in 'The Collected Works of Gerhard Gentzen', edited by M. E. Szabo, North-Holland, 1969.

[171] G. H. Gessert. Four valued logic for relational database systems. *ACM SIGMOD Record*, 19(1):29–35, 1990.

[172] M. Ginsberg. Multi-valued logics. In M. L. Ginsberg, editor, *Readings in Non-Monotonic Reasoning*, pages 251–258. Morgan Kaufmann, 1988.

[173] M. Ginsberg. Multi-valued logics: A uniform approach to reasoning in AI. *Computer Intelligence*, 4:256–316, 1988.

[174] J.-Y. Girard. Linear logic. *Theoretical Computer Science*, 50:1–102, 1987.

[175] J.-Y. Girard. *Proof theory and logical complexity*, volume 1. Humanities Press, 1987.

[176] S. Gottwald. *A Treatise on Many-Valued Logics, Studies in Logic and Computation*, volume 9. Research Studies Press, Baldock, 2001.

[177] R. Hähnle. Advanced many-valued logic. In D. Gabbay and F. Guenthner, editors, *Handbook of Philosophical Logic*, volume II, pages 297–395. Kluwer, 2001. Second edition.

[178] P. Hájek. *Metamatematics of Fuzzy Logic*. Kluwer, 1998.

[179] S. Halldén. The logic of nonsense, 1949. Uppsala University.

[180] A. Heyting. *Intuitionism. An introduction*. North-Holland Publishing, 1956.

[181] A. Heyting. *L. E. J. Brouwer: Collected Works*, volume 1 of *Philosophy and Foundations of Mathematics*. Elsevier, 1975.

[182] A. Holger and P. Verdée. *Logical Studies of Paraconsistent Reasoning in Science and Mathematics*, volume 45. Springer, 2016.

[183] A. Humberstone. *The Connectives*. MIT Press, 2011.

[184] Y. Ivlev. Quasi-matrix logic as a paraconsistent logic for dubitable information. *Logic and Logical Philosophy*, 2:91–97, 2000.

[185] S. Jaśkowski. On discussive conjunction in the propositional calculus for inconsistent deductive systems. *Logic, Language and Philosophy*, 7:57–59, 1999. Translation of the original paper in *Studia Societatis Scienctiarun Torunesis*, Sectio A, I(8):171–172, 1949.

[186] S. Jaśkowski. A propositional calculus dor inconsistent deductive systems. *Logic, Language and Philosophy*, 7:35–56, 1999. Translation of the original paper in *Studia Societatis Scienctiarun Torunesis*, Sectio A, I(5):57–77, 1948.

[187] I. Johansson. Der minimalkalkul, ein reduzierter intuitionistischer formalismus. *Compositio Mathematica*, 4:119–136, 1937.

[188] J. A. Kalman. Lattices with involution. *Transactions of the American Mathematical Society*, 87:485–491, 1958.

[189] N. Kamide. Gentzen-type methods for bilattice negation. *Studia Logica*, 80:265–289, 2005.

[190] N. Kamide and H. Wansing. Sequent calculi for some trilattice logics. *The Review of Symbolic Logic*, 2(2):374–395, 2009.

[191] N. Kamide and H. Wansing. Proof theory of Nelson's logic: A uniform perspective. *Theoretical Computer Science*, 415, 2012.

[192] N. Kamide and H. Wansing. *Proof Theory of N4-related Paraconsistent Logics*, volume 54 of *Studies in Logic*. College Publications, 2015.

[193] A. S. Karpenko. A maximal paraconsistent logic: The combination of two three-valued isomorphs of classical propositional logic. In D. Batens, C. Mortensen, G. Priest, and J. Van Bendegem, editors, *Frontiers of Paraconsistent Logic*, volume 8 of *Studies in Logic and Computation*, pages 181–187. Research Studies Press, 2000.

[194] S. C. Kleene. *Introduction to Metamathematics*. Van Nostrand, 1950.

[195] S. A. Kripke. The problem of entailment (abstract). *Journal of Symbolic Logic*, 24:324, 1959.

[196] O. Lahav, J. Marcos, and Y. Zohar. It ain't necessarily so: Basic sequent systems for negative modalities. In *Advances in Modal Logic*, pages 449–468, 2016.

[197] C. I. Lewis. *A Survey of Symbolic Logic*. University of California Press, 1918.

[198] C. I. Lewis and C. H. Langford. *Symbolic Logic*. Dover Publications, 1959. First published in 1932.

[199] A. Loparić. A semantical study of some propositional calculi. *The Journal of Non-Classical Logics*, 3:73–95, 1986.

[200] J. Los and R. Suzsko. Remarks on sentential logics. *Indagationes Mathematicae*, 20:177–183, 1958.

[201] J. Łukasiewicz. On 3-valued logic. *Ruch Filosoficzny*, 5:169–171, 1920. English translation: Polish Logic 1920–1939 (S.McCall, editor), Oxford University Press, Oxford, pages 15–18, 1967.

[202] J. Lukasiewicz. *Selected Works*. North Holland, 1970. Edited by L. Borkowski.

[203] J. Łukasiewicz. On the principle of contradiction in aristotle. *Review of Metaphysics*, 24:485–509, 1971. A synopsis in English of the original paper from 1910.

[204] G. Malinowski. *Many-Valued Logics*. Clarendon Press, 1993.

[205] J. Marcos. 8K solutions and semi-solutions to a problem of da Costa. Submitted.

[206] J. Marcos. Modality and paraconsistency. In M. Bilkova and L. Behounek, editors, *The Logica Yearbook 2004*, pages 213–222. Filosofia, 2005.

[207] J. Marcos. Nearly every normal modal logic is paranormal. *Logique et Analyse*, 48(189/192):279–300, 2005.

[208] J. Marcos. On a problem of da Costa. In G Sica, editor, *Essays on the Foundations of Mathematics and Logic*, volume 2, pages 39–55. Polimetrica, 2005.

[209] J. Marcos. On negation: Pure local rules. *Journal of Applied Logic*, 3(1):185–219, 2005.

[210] E. D. Mares. *Relevant Logic*. Cambridge University Press, 2004.

[211] J. McCarthy. A basis for a mathematical theory of computation. *Computer Programming and Formal Systems*, pages 33–70, 1963.

[212] E. Mendelson. *Introduction to Mathematical Logic*. CRC Press, 2009. Fifth edition.

[213] G. Metcalfe, N. Olivetti, and D. Gabbay. *Proof Theory for Fuzzy Logics*. Springer, 2009.

[214] R.K. Meyer and J.M. Dunn. E, R and γ. *Journal of Symbolic Logic*, 34(3):460–474, 1969.

[215] G. E. Mints. Some calculi of modal logic. *Proc. Steklov Inst. of Mathematics*, 98:97–122, 1968.

[216] B. Mobasher, D. Pigozzi, G. Slutzki, and G. Voutsadakis. A duality theory for bilattices. *Algebra Universalis*, 43(2–3):109–125, 2000.

[217] A. Monteiro. Construction des algebres de Lukasiewicz trivalentes dans les algebres de Boole monadiques I. *Mat. Jap.*, 12:1–23, 1967.

[218] C. Mortensen. Every quotient algebra for C_1 is trivial. *Notre Dame Journal of Formal Logic*, 21(4):694–700, 1980.

[219] K. Mruczek-Nasieniewska and M. Nasieniewski. Syntactical and semantical characterization of a class of paraconsistent logics. *Bulletin of the Section of Logic*, 34:229–248, 2005.

[220] K. Mruczek-Nasieniewska and M. Nasieniewski. Paraconsistent logics obtained by J.-Y. Béziau's method by means of some non-normal modal logics. *Bulletin of the Section of Logic*, 37(3/4):185–196, 2008.

[221] K. Mruczek-Nasieniewska and M. Nasieniewski. Béziau's logics obtained by means of quasi-regular logics. *Bulletin of the Section of Logic*, 38:189–203, 2009.

[222] M. Negri, G. Pelagatti, and L. Sbattella. Formal semantics of SQL queries. *ACM Transacxtions on Database Systtems*, 16(3):513–534, 1991.

[223] D. Nelson. Constructible falsity. *Journal of Symbolic Logic*, 14(1):16–26, 1949.

[224] D. Nelson. Negation and separation of concepts in constructive systems. In A. Heyting, editor, *Constructivity in Mathematics*, pages 208–225. North Holland, 1959.

[225] A. Nerode and R. A. Shore. *Logic for Applications*. Springer, 1997.

[226] A. Neto and M. Finger. A KE tableau for a logic for formal inconsistency. In *Proceedings of TABLEAUX'07 position papers and Workshop on Agents, Logic and Theorem Proving*, 2007.

[227] S. P. Odintsov. *Constructive Negations and Paraconsistency*, volume 26 of *Trends in Logic*. Springer, 2008.

[228] S. P. Odintsov. Priestley duality for paraconsistent Nelson's logic. *Studia Logica*, 96(1):65–93, 2010.

[229] S. P. Odintsov and H. Wansing. The logic of generalized truth values and the logic of bilattices. *Studia Logica*, 103(1):91–112, 2015.

[230] N. Olivetti, G. L. Pozzato, and C. Schwind. A sequent calculus and a theorem prover for standard conditional logics. *ACM Transactions on Computational Logic*, 8(4), 2007.

[231] H. Omori and K. Sano. Generalizing functional completeness in belnap-dunn logic. *Studia Logica*, 103(5):883–917, 2015.

[232] H. Omori and H. Wansing. *Special Issue: 40 years of FDE*, volume 106(6) of *Studia Logica*. 2017.

[233] H. Omori and T. Waragai. Negative modalities in the light of paraconsistency. In *The Road to Universal Logic*, pages 539–555. Springer, 2015.

[234] M. Osorio and J. L. Carballido. Brief study of G'_3 logic. *Journal of Applied Non-Classical Logics*, 18(4):475–499, 2008.

[235] M. Osorio, J. L. Carballido, and C. Zepeda. Revisiting ℤ. *Notre Dame Journal of Formal Logic*, 55:129–155, 2014.

[236] M. Osorio, J. L. Carballido, C. Zepeda, and Castellanos J. A. Weakening and extending ℤ. *Logica Universalis*, 9:383–409, 2015.

[237] B. Pahi. On the non-existence of finite characteristic matrices for some implicational calculi (abstract). *The Journal of Symbolic Logic*, 31:682–683, 1966.

[238] F. Paoli. *Substructural Logics: A Primer*. Kluwer, 2002.

[239] R. Parks. A note on R-mingle and Sobociński three-valued logic. *Notre Dame Journal of Formal Logic*, 13:227–228, 1972.

[240] D. Pearce. Default logic and constructive logic. In B. Neumann, editor, *Proc. 10th European Conference on Artificial Intelligence (ECAI'92)*, pages 309–313. John Wiley and Sons, 1992.

[241] P. Poggiolesi. *Gentzen Calculi for Modal Propositional Logic*, volume 32 of *Trends in Logic*. Springer, 2010.

[242] E.L. Post. The two-valued iterative systems of mathematical logic. *Annals of Mathematical Studies*, 5:1–122, 1941.

[243] G. Pottinger. Uniform, cut-free formulations of T, S4 and S5 (abstract). *Journal of Symbolic Logic*, 48:900, 1983.

[244] G. Priest. Logic of paradox. *Journal of Philosophical Logic*, 8:219–241, 1979.

[245] G. Priest. Reasoning about truth. *Artificial Intelligence*, 39:231–244, 1989.

[246] G. Priest. Minimally inconsistent LP. *Studia Logica*, 50:321–331, 1991.

[247] G. Priest. Paraconsistent logic. In D. Gabbay and F. Guenther, editors, *Handbook of Philosophical Logic*, volume 6, pages 287–393. Kluwer, 2002.

[248] Pseudo-Scotus. Questions on Aristotle's prior analytics. In M. Yrjonsuur, editor, *Medieval Formal Logic: Obligations, Insolubles and Consequences*, pages 225–234. Kluwer, 2001.

[249] W. V. O. Quine. *The roots of reference*. Open Court, 1974.

[250] A. R. Raggio. Propositional sequence-calculi for inconsistent systems. *Notre Dame Journal of Formal Logic*, 9:359–366, 1968.

[251] W. Rautenberg. 2-element matrices. *Studia Logica*, 40(4):315–353, 1981.

[252] S. Read. *Relevant Logic*. Basil Blackwell, 1988.

[253] G. Restall. Four-valued semantics for relevant logics (and some of their rivals). *Journal of Philosophical Logic*, 24:139–160, 1995.

[254] G. Restall. *An Introduction to Substructural Logics*. Routledge, 2000.

[255] U. Rivieccio. *An Algebraic Study of Bilattice-Based Logics*. PhD thesis, University of Barcelona, 2010.

[256] J. A. Robinson and A. Voronkov. *Handbook of Automated Reasoning*. Elsevier and MIT Press, 2001. In two volumes.

[257] G. Robles and J. M. Mendéz. A paraconsistent 3-valued logic related to Gödel logic G3. *Logic Journal of the IGPL*, 22(4):515–538, 2014.

[258] J. Rosser and A. R. Turquette. *Many-Valued Logics*. North-Holland Publication, 1952.

[259] R. Routley and R. K. Meyer. Semantics of entailment. In H. Leblanc, editor, *Truth, Syntax, and Modality*, pages 194–243. North Holland, 1973.

[260] L. I. Rozoner. On interpretation of inconsistent theories. *Information Sciences*, 47:243–266, 1989.

[261] K. Schlechta. Some completeness results for stopped and ranked classical preferential models. *Journal of Logic and Computations*, 6(4):599–622, 1996.

[262] K. Schütte. Syntactical and semantical properties of simple type theory. *Journal of Symbolic Logic*, 25(04):305–326, 1960.

[263] D. Scott. Completeness and axiomatization in many-valued logic. In *Proc. of the Tarski Symposium, Proc. of the Symposia in Pure Mathematics*, volume XXV, pages 411–435. AMS, 1974.

[264] S. Scroggs. Extensions of the lewis system S5. *Journal of Symbolic Logic*, 16:112–120, 1951.

[265] A. M. Sette. On propositional calculus P_1. *Mathematica Japonica*, 16:173–180, 1973.

[266] D. J. Shoesmith and T. J. Smiley. Deducibility and many-valuedness. *Journal of Symbolic Logic*, 36:610–622, 1971.

[267] D. J. Shoesmith and T. J. Smiley. *Multiple Conclusion Logic*. Cambridge University Press, 1978.

[268] Y Shramko, J. M. Dunn, and T. Takenaka. The trilattice of constructive truth values. *Journal of Logic and Computation*, 11:761–788, 2001.

[269] Y Shramko and H. Wansing. Some useful 16-valued logics: How a computer network should think. *Journal of Philosophical Logic*, 34:121–153, 2005.

[270] J. Slaney. A logic for vagueness. *Australian Journal of Logic*, 8:100–134, 2010.

[271] J. Słupecki. Der volle dreiwertige aussagenkalkül. *Com. Rend. Soc. Sci. Lett. de Varsovie*, 29:9–11, 1936.

[272] R. Smullyian. *First-Order Logic*. Dover Punlications, 1995. Corrected reprinting of the 1968 original.

[273] B. Sobociński. Axiomatization of a partial system of three-value calculus of propositions. *Journal of Computing Systems*, 1:23–55, 1952.

[274] C. Straßer. *Adaptive logic and defeasible reasoning. Applications in argumentation, normative reasoning and default reasoning*, volume 38 of *Trends in Logic*. Springer, 2014.

[275] T. Sugihara. Strict implication free from implicational paradoxes. *Memoirs of the Faculty of Liberal Arts, Fukui University, Series I*, pages 55–59, 1955.

[276] S. J. Surma. A survey of the methods and results of investigations of the equivalential propositional calculus. In S. J. Surma, editor, *Studies in the History of Mathematical Logic*, pages 33–61. Polish Academy of Science, Institute of Philosophy and Sociology, Wroclaw, 1973.

[277] W. Tait. A non constructive proof of Gentzen's Hauptsatz for second order predicate logic. *Bulletin of the American Mathematical Society*, 72(6):980–983, 1966.

[278] M. Takano. Subformula property as a substitute for cut-elimination in modal propositional logics. *Mathematica Japonica*, 37:1129–1145, 1992.

[279] G. Takeuti. *Proof Theory*. Dover, 2013.

[280] K. Tanaka, F. Berto, E. Mares, and F. Paoli, editors. *Journal of Logic and Logical Philosophy*, volume 19(1). Nicolaus Copernicus University Scientific Publishing, 2010. Special Issue on Paraconsistent Logic.

[281] K. Tanaka, F. Berto, E. Mares, and F. Paoli, editors. *Paraconsistency: Logic and Applications*, volume 26 of *Logic, Epistemology, and the Unity of Science*. Springer, 2013.

[282] A. Tarski. *Introduction to Logic*. Oxford University Press, 1941.

[283] A.S. Troelstra and H. Schwichtenberg. *Basic Proof Theory*. Cambridge University Press, 2000.

[284] A.S. Troelstra and D. van Dalen. *Constructivism in Mathematics, Vol. I and II*. North-Holland, Amsterdam, 1988.

[285] A. Tsoukiàs. A first-order, four valued, weakly paraconsistent logic and its relation to rough sets semantics. *Foundations of Computing and Decision Sciences*, 27:77–96, 2002.

[286] I. Urbas. *On Brazilian paraconsistent logics*. PhD thesis, Australian National University, Canberra, 1987.

[287] I. Urbas. Paraconsistency. *Studies in Soviet Thought*, 39:343–354, 1989.

[288] I. Urbas. Paraconsistency and the C-systems of da Costa. *Notre Dame Journal of Formal Logic*, 30(4):583–597, 1989.

[289] A. Urquhart. The undecidability of entailment and relevant implication. *Journal of Symbolic Logic*, 49:1059–1073, 1984.

[290] A. Urquhart. Many-valued logic. In D. Gabbay and F. Guenthner, editors, *Handbook of Philosophical Logic*, volume II, pages 249–295. Kluwer, 2001. second edition.

[291] D. van Dalen. *Logic and Structure*. Springer, 1994.

[292] A. Van Gelder, K. A. Ross, and J. S. Schlipf. The well-founded semantics for general logic programs. *Journal of the ACM*, 38(3):620–650, 1991.

[293] A. Visser. Four-valued semantics and the liar. *Journal of Philosophical Logic*, 13:181–212, 1984.

[294] F. von Kutschera. Ein verallgemeinerter widerlegungsbegriff fur Gentzenkalkule. *Archiv fur Mathematische Logik und Grundlagenforschung*, 12:104–118, 1969.

[295] G. Wagner. Logic programming with strong negation and inexact predicates. *Journal of Logic and Computation*, 1(6):835–859, 1991.

[296] H. Wansing. Short dialogue between M (Mathematician) and P (Philosopher) on multi-lattices. *Journal of Logic and Computation*, 11:759–760, 2001.

[297] H. Wansing. Sequent systems for modal logics. In D. Gabbay and F. Guenther, editors, *Handbook of Philosophical Logic, 2nd edition*, volume 8, pages 61–145. Kluwer, 2002.

[298] H. Wansing. Connexive modal logic. In R. Schmidt et al, editor, *Advances in Modal Logic*, pages 367–383. King's College Publications, 2005.

[299] H. Wansing. Connexive logic, 2014. Stanford Encyclopedia of Philosophy. https://plato.stanford.edu/entries/logic-connexive/.

[300] H. Wansing and N. Kamide. Intuitionistic trilcattice logics. *Journal of Logic and Computation*, 20(6):1201–1229, 2010.

[301] H. Wansing and S. P. Odintsov. On the methodology of paraconsistent logic. In *Logical Studies of Paraconsistent Reasoning in Science and Mathematics*, volume 45 of *Trends in Logic*. pages 175–204. Springer, 2016.

[302] R. Wójcicki. Some remarks on the consequence operation in sequential logics. *Fundamenta Mathematicae*, 68:269–279, 1970.

[303] R. Wójcicki. *Lectures in Propositional Calculi*. Ossolineum, 1984.

[304] R. Wójcicki. *Theory of Logical Calculi: Basic Theory of Consequence Operations*. Kluwer Academic Publishers, 1988.

[305] Y. Wu, M. Caminada, and D. Gabbay. Complete extensions in argumentation coincide with 3-valued stable models in logic programming. *Studia Logica*, 93(2–3):383–403, 2009.

[306] D. Zaitsev. A few more useful 8-valued logics for reasoning with tetralattice $EIGHT_4$. *Studia Logica*, 92:265–280, 2009.

Index

$+$, 395, 411
$-$, 324
$C(S)$, 207
C_ω, 338, 348
$FOUR$, 140
GB, 244
$GB(S)$, 247
GBk, 244
$GBk(S)$, 247, 253
$GBk^*(S)$, 277
$GCLuN$, 195
GLC, 45
GNB, 296
$GNS5$, 311
$GNS5^h$, 314
$GNS5_2$, 320
GRM, 493, 512
$GRMI$, 474
$GRMI_{\rightarrow}$, 407
$GSRMI_{\rightarrow}$, 439
G^X, 448
$G_{\mathbf{EIP}}$, 227
$G_{\mathbf{PAC}}$, 212
$G_{\mathbf{4All}}$, 158
$G_{\mathbf{J_3}}$, 121
$G_{\mathbf{P_1}}$, 121
$G_{\mathbf{SRM}_{\rightarrow}}$, 121
$GRMI_{min}^{\rightarrow}$, 462
HB, 239
$HB(S)$, 247
HBk, 239
$HBk(S)$, 247, 253
$HBk^\sigma(S)$, 288
HCL, 26
HCL^+, 26

$HCL^+(S)$, 204
HCL^{F}, 28
$HCL^{\mathsf{F}}(S)$, 204
HCL_1, 28
HCL_2, 28
HCL_3, 28
$HCL_{\supset}$, 26
$HCLuN$, 194, 208
HC_1, 289
HIL, 26, 347
$HIL(S)$, 332
HIL^+, 26, 347
$HIL^+(S)$, 332
$HIL_{\supset}$, 26
$HLL_{\rightarrow}$, 372
HNB, 295
HNK, 327
$HNS5$, 310
$HNS5_2$, 319
HNT, 328
HR, 453, 477
HRM, 477
$HRM0_{\rightarrow}$, 373
$HRMI^*$, 466
$HRMI_{-}$, 407, 477
$HRM_{\supset}$, 509
$HRM_{\supset}^{\mathsf{F}}$, 509
HRM_{ni}, 511
$HR_{\rightarrow}$, 331
H^X, 448
$H_{\mathbf{PAC}}$, 212, 349
$H_{\mathbf{4All}}$, 161
$H_{\mathbf{J_3^B}}$, 128
$H_{\mathbf{J_3}}$, 128
$H_{\mathbf{PAC^B}}$, 128

INDEX

$H_{\mathbf{PAC}_\supset}$, 128
$H_{\mathbf{PAC}}$, 128
$HRMI$, 465
$HRMI_{min}$, 460
$HRMI^*_{min}$, 460
$HRMI^{**}_{min}$, 460
IL, 21
LJ, 43, 352
$LJ(S)$, 353
LJ^+, 43, 352
$LJ^+(S)$, 353
$LJ^+(\mathsf{N}_4)$, 353
LJ^+_m, 43, 358
$LJ^+_m(S)$, 358
LJ_m, 43, 358
$LJ_m(S)$, 358
LK, 41
LK^+, 41
$LK^+(S)$, 213
LK^{F}, 42
$LK^{\mathsf{F}}(S)$, 213
$LK^{\mathsf{F}\neg}$, 42
LP, 105
SR, 459
$SRMI_{\rightarrow}$, 435
$SRMI^n_{\rightarrow}$, 437
$TH(\hspace{-1pt}\mid\hspace{-2pt}\sim)$, 372
$Th(\mathcal{M})$, 66
$Th(\mathbf{L})$, 5
Δ, 4
Γ, 4
$\Lambda_\mathcal{M}$, 64, 177
$\Lambda_\mathcal{M}$, 64
$\mathcal{M}^b_4(S)$, 207
$\mathcal{M}^b_{4\mathsf{F}}(S)$, 207
$\mathcal{M}^{\mathsf{t}}$, 505
$\mathcal{M}^{\mathsf{F}}$, 504
$\mathcal{M}_B(S)$, 247
$\mathcal{M}_{Bk}(S)$, 247, 253
$\mathcal{M}^\infty_{Bk}$, 265
$\mathcal{M}^\infty_{Bk}(S)$, 267
$\mathcal{M}_{\mathbf{F}}$, 72
$\mathcal{M}_{\mathsf{8Kb}}$, 201
Π_1, 189
Π_2, 189
$\Rightarrow$, 31
$\Vdash_\mathsf{G}$, 37

$\mathbf{R}_{\rightarrow}$, 449
$\bot$, 84, 136
$\mathcal{IL}$, 395
$\mathcal{SA}$, 465
$\circ$, 91
$\circ$, 112, 233, 234
$\diamond$, 63, 176
f, 447
$\leq_k$, 336
$\leq_\mathsf{k}$, 141, 164
$\leq_\mathsf{t}$, 84, 106, 140, 141, 164, 323
$\mapsto$, 85
$\mathcal{RM}_n$, 484
B, 142
F, 142
$\mathsf{IL}(\mathcal{M})$, 334
N, 142
T, 142
$\models_S$, 14
$\models_\mathcal{M}$, 64
$\models_{\mathcal{EIP}}$, 222
$\neg$, 16, 84, 87, 91, 99, 105, 116, 139, 323
$\odot$, 165
$\oplus$, 141, 164
$\otimes$ (relevance), 382, 395, 405, 411, 432
$\otimes$ (bilattices), 141, 164
$\overline{\mathcal{D}}$, 176
$\rightarrow_L$, 86
$\rightarrow_S$, 103
$\mathcal{R}$, 431
SF, 4
Var, 4
σ_d, 288
σ_l, 288
$\sim$, 323
$\sim_t$, 116
$\supset$ (relevance), 432, 467, 506
$\subset$, 86
$\supset$, 10, 12, 16, 17, 91, 110, 192, 201, 202, 208
$\supset_t$, 71
$\rightarrow$, 371, 374, 375, 506
$\top$, 84, 136
t, 447
$\vdash_{\mathcal{EIP}}$, 222

INDEX

$\vdash_G$, 34
$\vdash_G^m$, 34
$\vdash$, 4, 6
$\vdash_{\mathcal{M}}$, 66, 178
$\vdash_{\mathcal{F}}$, 67, 178
$\vdash_S$, 14
$\vee$, 10, 16, 17, 84, 85, 91, 105, 116, 139, 164, 192, 201, 202, 208, 323
φ_n^R, 487
$\wedge$, 10, 16, 17, 84, 85, 91, 99, 105, 116, 139, 164, 192, 201, 202, 208, 323
$\tilde{\diamond}$, 15, 63, 176
$\mathcal{S}$, 4
$\mathcal{T}$, 4
$\mathrel{\vert\!\sim}^1$, 386
$\mathrel{\vert\!\sim}_{\vec{L}}$, 373–375
f, 14, 63, 84, 136
g_f, 143
g_t, 143
i, 84
t, 14, 63, 84, 136
$\mathcal{L}_C$, 239
$\mathbf{R}_\rightarrow$, xviii, 390
$3_\top^G$, 121
$3_\bot$, 84
$3_\top$, 84
4All, 142
4CL, 156, 208
4Flex, 155
4MCC, 151
4Mon, 150
4Nex, 153
$\mathbf{Bd}(S)$, 271
Bi, 264
Bk, 239
$\mathbf{Bk}(S)$, 247, 253
$\mathbf{Bld}(S)$, 271
Bl, 264
$\mathbf{Bl}(S)$, 271
B, 239, 294, 309
$\mathbf{B}(S)$, 247
$\mathbf{CAR}^\mathbf{I}$, 293
CAR, 293
$\mathbf{CL}^+$, 18, 41, 182

CLuN, xix, 193, 205, 241, 263, 310, 367
CL, 15, 16, 41
$\mathbf{CL}^\mathsf{F}$, 19, 21
$\mathbf{CL}^{\mathsf{F}\neg}$, 19
$\mathbf{C}_{\min}$, 205, 294
$\mathbf{C}_1$, 289
$\mathbf{C}_n$, 233
EIP, 222
E, 458
F, 15
$\mathbf{F}_0$, 392, 399, 415
$\mathbf{F}_{CL}$, 16
$\mathbf{G}_\infty$, 45, 506
$\mathbf{IL}^+$, 20, 331
IL, 43
$\mathbf{IL}(\mathcal{M})$, 333
$\mathbf{IL}^+$, 352
$\mathbf{J}_3^\mathsf{B}$, 110, 112
$\mathbf{J}_3$, 110, 112, 208
KBT, 309
KT45, 314
KT4, 329
KT5, 314
KT, 328
K, 327
LFI1, 112
$\mathbf{LL}_\rightarrow$, 373
$\mathbf{LL}_\rightarrow$, 389
$\mathbf{LL}_\rightarrow$, 388, 389
$\mathbf{LL}_{mm}$, 389
$\mathbf{LP}^\mathsf{B}$, 105
$\mathbf{LP}^\mathsf{F}$, 105
$\mathbf{LP}^{\mathsf{F,B}}$, 105
LP, 105, 456
$\mathbf{L}(\mathrel{\vert\!\sim})$, 370
$\mathbf{L} \upharpoonright CL$, 285
$\mathbf{L}^X$, 448
$\mathbf{L}_{\mathcal{M}}$, 66, 178
$\mathbf{L}_{\mathcal{FOUR}}$, 142
$\mathbf{L}_\mathsf{P}$, 24
$\mathbf{L}_{\mathcal{A}_\omega}$, 432
$\mathbf{L}_{\mathcal{A}_n}$, 432
N3, 349
N4, 349, 352
NB, 295, 309
NK, 327

NS4, 329
NS5, 310
NS5$_2$, 319
NT, 328
N$^-$, see **N4**
N, see **N3**
PAC^B, 110
PAC, 110, 205, 208
PAC^B, 112
PAC$_\to$, 114
PE3, 116
PI, see **CLuN**
P$_1^K$, 99, 101
P$_1^S$, 99, 101
P$_1$, 99, 200
RM0$_\to$, 373
RM0$_\to$, 373
RMI, 458, 465, 477
RMI^{DS}, 486
RMI$_\to$, 407, 408, 449
RMI$_\to^n$, 418
RMI$_{min}$, 460
RM, xx, 367, 373, 477
RM^{DS}, 486
RM$_3$, 110
R$_\to$, xx
R, xx, 393, 451, 453
R$_\to$, 388, 395, 407, 451
R$_\to$, 378
R$_{fde}$, 451
S4, 329
S5, 310, 314
SRMI$_\to^1$, 102
SRM$_\to$, 438
T, 328, 458
mbC, 249
$\mathcal{A}_n^F$, 450
$\mathcal{A}_\omega$, 469
$\mathcal{C}(\mathcal{L})$, 3
$\mathcal{DEFAULT}$, 165
$\mathcal{D}$, 63, 64, 176
$\mathcal{FOUR}$, 138, 140, 141, 165
$\mathcal{L}$, 3
$\mathcal{L}^X$, 448
$\mathcal{L}_{CL}^{F\neg}$, 202
$\mathcal{L}_{All}$, 142
$\mathcal{L}_{CC}$, 151
$\mathcal{L}_{CL}$, 16
$\mathcal{L}_{CL}^+$, 18, 182
$\mathcal{L}_{CL}^F$, 19
$\mathcal{L}_{CL}^{F\neg}$, 19, 155
$\mathcal{L}_{Mon}$, 150
$\mathcal{L}_{Nex}$, 153
$\mathcal{M}_E^4$, 222
$\mathcal{M}_4^B$, 335
$\mathcal{M}_4^b$, 202
$\mathcal{M}_4^{Ib}$, 335
$\mathcal{M}_4^{Ib}(S)$, 337
$\mathcal{M}_2$, 193
$\mathcal{M}_{4F}^b$, 202
$\mathcal{M}_{4F}^{Ib}$, 335
$\mathcal{M}_{4F}^{Ib}(S)$, 337
$\mathcal{M}_{Bk}$, 242
$\mathcal{M}_B$, 242
$\mathcal{M}_{CL}$, 66
$\mathcal{M}_{CL}^+$, 66
$\mathcal{NINE}$, 165
$\mathcal{O}$, 63, 64, 176
$\mathcal{SA}$, 478
$\mathcal{SA}_m$, 424
$\mathcal{THREE}_\top$, 106
$\mathcal{V}$, 63, 64, 176
$\mathcal{W}(\mathcal{L})$, 4
$\mathfrak{T}_\bot^M$, 170
$\mathfrak{T}_\top^M$, 170
$\mathfrak{T}_f^M$, 170
$\mathfrak{T}_t^M$, 170
$\mathcal{A}_\omega$, 411
A$_{PAC}$, 204
B, 87
CL, 51
F, 19, 21, 23, 43, 87, 447
IL, 51
M3$_\bot$, 84
M3$_\top$, 84
M4, 137
N$_3^\top$, 342, 347
N$_4$, 342, 347, 352
PAC, 183
PAC$^+$, 183
T, 447
Var, 4
m, 350
Łos-Suzsko Theorem, 179

INDEX 541

$\mathbf{C}_{min}$, 205
A_{CPAC}, 250
A_C, 250
A_{ld}, 264

A_{PAC}, 331

adaptive logic, xvii, 232
Adjunction, 453, 477
    relevant, 460
algebraic structure, 63
Almukdad, A., 360
analyticity, 65, 175, 178, 303, 313, 322, 341, 342, 345
    $\mathcal{F}$-, 341
Anderson, A., xv, 60, 363, 393, 394, 406, 446, 451, 476
Asenjo, F. G., 105
axiom, 7, 32
    identity, 33
    residuation, 384

Béziau, J. Y., xv, 133, 135, 290, 330
Baaz, M., 338
basic relevance criterion, 52, 68, 364, 365, 434
Batens, D., xv, xvii, 186, 193, 232, 330, 367
Beeson, M. J., 360
Belnap, N., xv, xix, 60, 137, 141, 172, 363, 393, 394, 406, 446, 451, 476
Ben-Naim, J., 97, 186
bifilter, 166
    prime, 166
bilattice, 164
    $\mathcal{DEFAULT}$, 165
    $\mathcal{FIVE}$, 166
    $\mathcal{FOUR}$, 141, 336
    $\mathcal{NINE}$, 165
    Belnap's, 141
    distributive, 164
    interlaced, 164
bivalence, 192
bivaluation, 175, 290
Bochvar, D. A., 84
Bolc, L., 82
Borowik, P., 82

Brouwer, L. E. J., 331, 360

C-system, 232–234
    regular, 235
    strong, 237
Carballido, J. L., 330
Carnielli, W. A., xv, 91, 133, 232, 290
Castellanos J. A., 330
central logic, 458
Church, A., 46, 376, 394
Ciucci, D., 112
classical logic, 14, 66
    positive, 18
classical negation, 388
coherence, 256
    $\mathbf{Bk}$-, 256
compactness, 15, 19, 66, 178
    for $\mathbf{L}_{\mathcal{A}_\omega}$, 438
    for $\mathbf{EIP}$, 231
completeness, 24, 40, 67, 179
    for $\mathbf{N4}$, 349
    for $\mathbf{RMI}$, 468
    for $\mathbf{RM}$, 481
    for $\mathbf{RM}^\mathsf{F}$, 504
    for $\mathbf{RM}^\mathsf{t}$, 505
    for $GCLuN$, 195
    for $GNS5^h$, 316
    for $GSRMI_{\rightarrow}$, 445
    for $H_\mathbf{L}$, 162
    for $H_{\mathbf{J}_3^\mathsf{B}}$, 131
    for $H_{\mathbf{J}_3}$, 131
    for $H_{\mathbf{PAC}^\mathsf{B}}$, 131
    for $H_{\mathbf{PAC}_{\supset}}$, 131
    for $H_{\mathbf{PAC}}$, 131
    for $HCL^+(S)$, 209
    for $HCL^\mathsf{F}(S)$, 209
    for $\tilde{H}IL(S)$, 337, 346
    for $\tilde{H}IL^+(S)$, 337, 346
    for $\tilde{H}RM_\mathsf{F}$, 509
    for $\tilde{H}RM_\supset$, 509
    for LJ, 44
    for $LJ(S)$, 353
    for LJ^+, 44
    for $LJ^+(S)$, 353
    for LJ_m^+, 44
    for $LJ_m^+(S)$, 358

for LJ_m, 44
for $LJ_m(S)$, 358
for LK, 42
for LK^+, 42
for $LK^+(S)$, 215
for $LK^F(S)$, 215
for $LK^{F\neg}$, 42
for LK^F, 42
for $SRMI_{\to}$, 436
for $\mathbf{Bk}(S)$, 258, 277
for $\mathbf{L}_{\mathcal{A}_n}$, 437
for $\mathbf{NB}$, 303
for $\mathbf{NS5}$, 313
for $\mathbf{NS5}_2$, 322
for $\mathbf{RMI}_{\to}$, 413, 425, 430
for $\mathbf{RMI}_{\to}^n$, 418
for 3-val. Gentzen systems, 127
for 4-val. Gentzen systems, 159
for Gentzen systems, 40
for $\mathbf{Bk}$, 244
for $\mathbf{B}$, 244
for $\mathbf{EIP}$, 231
strong, 40
weak, 24, 67, 179
condition
    general refining, 266, 267
    monotonicity, 336, 341
    persistence, 333, 336, 339
cone, 143
conflation, 142
congruence, 9
Coniglio, M. E., xv, 91, 133, 232, 290
conjunction
    associated with $\to$, 383
    for a logic $\mathbf{L}$, 10, 17
    for a matrix $\mathcal{M}$, 69
    for an Nmatrix $\mathcal{M}$, 180
    intensional, *see* internal
    internal, 382
    multiplicative, *see* internal
    relevant, 382
connective, 3, 9
    classically closed, 73
    classically positive, 18
    closed, 69
    conjunction, 10, 17, 380
    defined, 9
    disjunction, 11, 17, 380
    implication, 10, 17, 380
    included in a language, 65
    internal implication, 371
    limited, 69
    non-deterministic, 176
    positive, 13
    primitive, 9
    relevant, 371
    relevant conjunction, 382
    relevant negation, 385
    represented in a matrix, 65
    semi-implication, 10, 12, 17, 378
    strong relevant implication, 376
    weak relevant implication, 371, 374
connexive logic, 48
consequence relation
    induced by a matrix $\mathcal{M}$, 66
    induced by a semantics S, 14
    induced by an Nmatrix $\mathcal{M}$, 178
    Scott-type, *see* scr
    Tarskian, *see* tcr
conservativity, 341
consistency
    operator, 234, 263
    w.r.t. $\neg$, 387
    weak, 368, 371
consistency operator, 234, 263
    regular, 235
    strong, 237
constant
    propositional, 3
constructivity, xx, 293
containment in $\mathbf{CL}$, 16, 48
contraction, 35, 373, 396
    external, 493
contraposition, 396
Crawford, J., 186
cut, 33
cut-elimination, xxi, 38, 39
    (quasi-analytic) for $LJ_m^+(S)$, 359
    (quasi-analytic) for $LJ_m(S)$, 359
    analytic, 38
    for $GBk(S)$, 258
    for $GBk^*(S)$, 277
    for $GCLuN$, 195

INDEX 543

for $GN5^h$, 319
for $GR_{\rightarrow}$, 398
for GRM, 503
for $GRMI$, 475
for $GRMI_{\rightarrow}$, 409
for $GSRMI_{\rightarrow}$, 445
for LJ, 44
for $LJ(S)$, 353
for LJ^+, 44
for $LJ^+(S)$, 353
for LJ_m^+, 44
for LJ_m, 44
for LK, 42
for LK^+, 42
for $LK^+(S)$, 215
for $LK^F(S)$, 215
for 3-valued Gentzen systems, 127
for 4-valued Gentzen systems, 159
for **EIP**, 231
quasi-analytic, 359
strong, 38, 39
strong analytic, 38

D'Ottaviano, I., 86, 110, 290
d-frames, 339
 N4-, 349
da-Costa, N., xix, 50, 53, 60, 86, 110, 135, 233, 290, 338
decidability, 8, 68, 179, 341
 co-semi-, 8
 of $\mathbf{RMI}_{min}$, 462
 of **RM**, 486
 of $HIL(S)$, 347
 of $HIL^+(S)$, 347
 of $\mathbf{IL}^+$, 20
 of **IL**, 22
 of $\mathbf{L}_{HCL^+(S)}$, 210
 of $\mathbf{L}_{HCL^F(S)}$, 210
 of **NB**, 304
 of **NS5**, 313
 of $\mathbf{NS5}_2$, 322
 of **RMI**, 469
 of $\mathbf{RMI}_{\rightarrow}$, 409, 415
 of $\mathbf{R}_{\rightarrow}$, 399
 of $\mathcal{A}_\omega$, 434

of a proof system, 24
of **EIP**, 224
semi-, 8
deduction theorem
 classical, 10, 29
 for $SRMI_{\rightarrow}$, 435
 relevant, 10, 12, 376
deduction theorems for $\mathbf{RMI}_{\rightarrow}$, 409
denotational semantics, 14, 40
 Kripke-, 333
derivation, see proof
deterministic $\mathcal{M}$-frames, 334
disjunction
 for a logic **L**, 11, 17
 for a matrix $\mathcal{M}$, 69
 for an Nmatrix $\mathcal{M}$, 180
Dodó, A., 330
double Kripke frames, see dframes339
double negation, 396
Dubois, D., 112
Dunn, M., xix, 137, 172, 360, 382, 406, 511

entailment
 first degree, 226, 451
 pre, 368
 relevant, 368
 s-, 369
 Scott s-, 371
 Scott type, 370
equivalence, 9
Etherington, D., 186
ex contradictione sequitur quodlibet, xv, 47
excluded middle, 49, 331, 358
expansion, 35
exploding
 controllably, 51
explosion, 51
 principle of-, xv
explosive
 partially, 51
extension
 by definitions, 9
 by rules, 7
 of matrices, 64, 66

of Nmatrices, 176, 178
extension (of a Hilbert-type system)
    axiomatic, 25
extension (of a logic), 6, 8, 10
    axiomatic, 6, 415, 418
    conservative, 6, 10
    proper, 6
    simple, 6, 399, 415, 418
    strongly proper, 6
    weakly conservative, 6

family
    $N_3^\top$, 347
    N_4, 347
finitariness, 5
first-degree entailment, 226, 451
Fitting, M., 46, 141, 150, 164, 172
formula, 3
    atomic, 3
    bottom element, 238
    characterization by, 142
    complex, 3
    congruence, 9
    equivalence, 9
    fully relevance, 408
    isolated, 191
    principal, 32
    side, 32
fragment (of a logic), 6
function
    flexible, 154
    monotonic, 143
functional completeness, 17, 65

Gödel-Dummett logic, 45, 506, 508, 512
Gabbay, D., xv, xxiii
general refining conditions, 266, 267
Gentzen, G., 24, 31, 46, 353
Gentzen-type systems, 34, 121, 158, 352, 474, 493
    semantics for-, 39
Ginsberg, M., 164, 165
Girard, J. Y., 186, 382, 394, 458
Gottwald, S., 82
guard connective, 150

Halldén S., 84

Heyting, A., 331, 360
Hilbert, D., 24, 25
Hilbert-type systems, 128, 161
hypersequent, 44, 493
    RM-valid, 495
    maximal, 496
    translation of-, 474
    valid, 440
Hähnle, R., 82

ideal paraconsistent logic, 59, 60
identity axioms, 33, 372, 396
implication
    Łukasiewicz, 86
    for a logic **L**, 10, 17
    for a matrix $\mathcal{M}$, 69
    for an Nmatrix $\mathcal{M}$, 180
    internal, 371
    material, 85
    relevant, 376–378
    semi-, 378
    Sobociński, 103
    weak relevant, 371, 374
inference rule, 7
interpretation, 15, *see* valuation, *see* valuation
    bivalent, 15, 48
    induced by a matrix, 73
intuitionistic logic, 20, 331
invertibility, 33, 43
isolation condition
    strong, 420
Ivlev, Y., 186

Jaśkowski, S., 60, 330
Johansson, I., 23, 29, 44, 51, 52

Kamide, N., xv, 172, 360
Kleene, S. C., 84, 87, 105
Konikowska, B., 97, 186
Kripke frames, 20
    S-d-, 340
    $\mathcal{M}$-, 333
    NB-, 299
    NS5-, 312
    NS5$_2$-, 321
    d-, 339
Kripke, S., 20, 333, 339, 406

INDEX 545

Lahav, O., 330
language, 3
  sub-, 4
LEM, 49, 331
Lev, I., 186
Lewis argument, xvi, 47
LFI, xix, 52, 233, 305, 365
Lindenbaum algebra, 413, 482
Linear logic, 373, 388
logic, xvii, 3, 5
  ¬-coherent with classical logic, 48
  ¬-contained in classical logic, 48
  ¬-paracomplete, 135
  ¬-paraconsistent, xvii, 50
  ¬-paradefinite, 135
  F-complete, 54
  F-maximal relative to classical logic, 54
  L-contained in classical logic, 16
  Łukasiewicz, 508
  F, 512
  T, 512
  3-valued, 83
  4-valued, 135
  8Kb, 91
  induced by $\mid\sim$, 370
  adaptive, 232
  Asenjo-Priest, 105
  Bochvar, 84
  classical, 14, 66
  classical ¬-paraconsistent, 53
  complete for $\mathcal{M}$, 67
  connexive, 48
  constructive, xx, 331
  decidable, xviii, 8
  Dunn-Belnap, xix
  effective, xviii
  exploding
    non-, 52
  explosive, 51
  extension by definitions, 9
  extension of, 6
  finitary, 5
  fragment of, 6
  fully maximal, 57, 59
  fuzzy, 505
  Gödel-Dummett, 45, 506, 508, 512
  Halldén, 84
  induced by a matrix $\mathcal{M}$, 66
  induced by an Nmatrix $\mathcal{M}$, 178
  intuitionistic, xx, 20, 21, 331
  linear, 394
  maximal relative to classical logic, 54
  maximally paraconsistent, 53
  McCarthy, 84
  minimal, 23, 29, 44, 51, 52
  modal, xx, 293
  multi-valued, 63
  Nelson's 3-valued, 349
  Nelson's 4-valued, 349
  non-monotonic, 5
  normal, 13
  paraconsistent, 49
    boldly, 52
  positive, 13, 18
  positive classical, 18
  positive intuitionistic, 20
  pre-¬-paraconsistent, xvii, 49
  relevant, xx, 363, 404
  replacement property of-, 9
  self-extensional, 9
  semi-normal, 13, 59
  semi-relevant, 367
  sequential Kleene, 84
  sound for $\mathcal{M}$, 67
  sparse, 191
  strong Kleene, 84, 105
  strongly ¬-paraconsistent, xvii, 50, 59
  strongly maximal, 53
  strongly non-exploding, 52
  strongly relevant, 391, 402, 404, 416
  substitution property of-, 9
  substructural, 394
  uniform, 67
  valid formula of, 5
  weak Kleene, 84
Loparić, A., 338
Los-Suszko Theorem, 68
Łukasiewicz logic, 508

Lukasiewicz, J., 58, 60, 86, 133, 172

Maehara, S., 43
Malinowski, G., 82
Marcos, J., xv, xxiii, 91, 133, 232, 290, 330
material implication, 85
    paradox, 366, 379
matrix, xix, 63
    $\mathcal{SA}$, 465
    $\mathcal{RM}_n$, 484
    $\mathcal{SA}$, 473
    ¬-contained in classical logic, 72
    f-, 70
    RMI, 465, 473
    RMI$_\to$, 429
    extension, 64
    reduction, 64
    submatrix, 64
    bilattice-valued, 166, 167
    bounded Sugihara, 504
    characteristic, 67, 365, 424
    classically-closed, 73
    Dunn-Belnap, 138
    effective, 64
    extension, 66
    finite, 64
    fully maximal, 77
    non-deterministic, 176
    paraconsistent, 75
    set of-, 67
    subclassical, 72, 74
    submatrix, 66
    Sugihara, 478, 479, 511
    suitable, 334
    weakly characteristic, 67, 366, 410
maximality, 53
    full, 57
    of **CL**, 19
    of **RM**, 490
    of **CL**$^+$, 19
    of **L**$_\mathcal{M}$, 185
    of **NS5**$_2$, 326
    of **RMI**$_\to$, 415, 417
    paraconsistency, 53
    pre-, 185, 186, 196
    relative to classical logic, 54
McCall, S., 511
McCarthy, J., 84
Meyer, R. K., 373, 393, 458, 511
mingle, 373, 407, 477, 493
    relevant, 408, 462
minimal logic, 23, 44, 51, 52
minimality, 202
    of **CLuN**, 195
    of **IL**$^+$, 21
    of **R**$_\to$, 399, 404
    of **C**$_{min}$, 205
Mints, G., 46
modality, xx, 293
model
    of a formula, 14, 64, 177
    of a hypersequent, 440
    of a sequent, 440
modularity, 187, 202
Modus Ponens, 27, 372, 396, 453
monotonicity
    $\leq_k$-, 107
    (M), 4, 6
Mortensen, C., xv
Mruczek-Nasieniewska, K., 330

Nasieniewski, M., 330
negation, xvii, 19, 47, 48
    associated with $\to$, 387
    classical, 388
    complete, 49, 59
    contrapositive, 49, 294
    Dunn-Belnap, 137
    explosive, 49
    in a matrix, 72
    internal, 385, 386
    intuitionistic, 21
    involutive, 49, 59
    Kleene, 87
    left involutive, 137
    modal operator, 293
    relevant, 385
    right involutive, 137
    Sette's, 87
    Sette's 3-valued, 87
Nelson, D., xx, 60, 349, 360
Nmatrix, xix, 175, 176

INDEX 547

¬-coherence with classical logic, 184
¬-contained in classical logic, 184
¬-fundamental, 187, 188
f-, 181
$\mathcal{F}$-analytic, 341
characteristic, 179
deterministic, 176
determinization of-, 182
effective, 176
expansion of-, 181
extension, 176, 178
finite, 176
paraconsistent, 184, 192
proper, 176
reduction, 176
refinement of-, 182
set of-, 178
simple refinement of-, 182
strictly proper, 176
two-valued, 192
weakly characteristic, 179, 402, 416
non-determinism, xix, 175, 187
non-triviality, 5
normality, 13
   of **CLuN**, 193
   of **RM**, 486
   of **IL**$^+$, 20
   of **IL**, 22
   of **NB**, 304
   of **NS5**, 313

Odinstov, S., 172, 360
Omori, H., 172, 330
operator
   consistency, 234, 263
   strong consistency, 237
Osorio, M., 330

Pahi, B., 406
Paoli, F., 476
paraconsistency, xv, 47, 50
   bold, 52, 416
   classical, 53
   maximal, 53

maximal relative to classical logic, 54
pre-, xvii, 49
strong, 50
strongly maximal, 53
Parks, Z., 446
permutation, 35, 372, 396
Pierce's axiom, 349
positive classical logic, 41
possible worlds, xix
Pottinger, G., 46
Priest, G., xv, 60, 105, 111
proof, 24
   analytic, 37
   derivation tree, 42
   in Gentzen-type systems, 32
   in Hilbert-type systems, 25
proof system, xxi, 23
   GB, 244
   $GB(S)$, 247
   GBk, 244
   $GBk(S)$, 247, 253
   $GBk^*(S)$, 277
   $GC\bar{L}uN$, 195
   GLC, 45
   GNB, 296
   $GNS5$, 311
   $GNS5^h$, 314
   $GNS5_2$, 320
   GRM, 493
   $GRMI$, 474
   $GRMI_{\rightarrow}$, 407
   $GR_{\rightarrow}$, 390
   $GR_{\rightarrow}$, 396
   $GSRMI_{\rightarrow}$, 439
   $G_{\mathbf{EIP}}$, 227
   $G_{\mathbf{4All}}$, 158
   $G_{\mathbf{4CL}}$, 452
   $G_{\mathbf{J_3}}$, 121
   $G_{\mathbf{PE3}}$, 121
   $G_{\mathbf{P_1}}$, 121
   $G_{\mathbf{SRM}_{\rightarrow}}$, 121
   $GRMI_{min}$, 462
   HB, 239
   $HB(S)$, 247
   HBk, 239
   $HBk(S)$, 247, 253

HCL, 26
HCL^+, 26
$HCL^+(S)$, 204
$HCL^F(S)$, 204
$HCL_\supset$, 26
$HCLuN$, 194
HC_1, 289
HIL, 26
$HIL(S)$, 332
HIL^+, 26
$HIL^+(S)$, 332
$HIL_\supset$, 26
$HLL_\to$, 384
$HLL_{\to}$, 388
$HLL_\to$, 372
$HM^F_\supset$, 509
HNB, 295
HNK, 327
$HNS5$, 310
$HNS5_2$, 319
HNT, 328
HR, 453, 477
HRM, 477
$HRM0_\to$, 384
$HRMI^*$, 466
$HRMI_{\to}$, 477
$HRM_\supset$, 509
$HR_{\to}$, 388
$HR_\to$, 378, 395
H_{4All}, 161
$H_{J_3^B}$, 128
H_{J_3}, 128
H_{PAC^B}, 128
$H_{PAC_\supset}$, 128
H_{PAC}, 128
$HRMI$, 465
$HRMI_{min}$, 460
$HRMI^*_{min}$, 460
$HRMI^{**}_{min}$, 460
LJ, 43, 352
$LJ(S)$, 353
LJ^+, 43, 352
$LJ^+(S)$, 353
LJ^+_m, 43, 358
LJ_m, 43, 358
$LJ_m(S)$, 358
$LJ^+_m(S)$, 358

LK, 41
LK^+, 41
$LK^+(S)$, 213
$LK^F(S)$, 213
$HBk^\sigma(S)$, 288
canonical, 57
complete for **L**, 24
cut-free, 38
effectiveness, 24
finitariness, 24
for 3-valued logics, 121
for 4-valued logics, 158
Gentzen-style, xxi, 31, 121, 158, 352, 474, 493
Hilbert-type, xxi, 25, 128, 161
hypersequential, 44, 314, 493
monotonicity, 23
non-triviality, 24
quasi canonical, 58, 97, 160
reflexivity, 23
sound for **L**, 24
standard, 36, 59
structurality, 23
theorem, 24
transitivity, 23
weakly complete for **L**, 24
weakly sound for **L**, 24
well-axiomatized, 28
propositional logic, *see* logic

Quine, W, V. O., 186

Raggio, A. R., 290
RDP, *see* relevant deduction property, 379, 380, 394
reduction
    of matrices, 64
    of Nmatrices, 176
refinement, 182
    simple, 182
    weakest, 207
refining conditions, 206
reflexivity, 370
    (R), 4, 6, 368
relevance, xx, 363
    basic criterion, 52, 364, 365, 367, 434
    connectives, 371

INDEX 549

logic (strong), 402, 404, 416
minimal semantic criterion, 366, 367, 404, 416
strong criterion, 368, 377
relevance criterion
    basic, 52, 68, 364, 365, 434
    minimal semantic, 366, 404, 416
    strong, 368, 377
relevant adjunction, 460
relevant deduction property, 12, 376, 401
relevant entailment, 368
relevant logic, 363, 390, 404
    maximal, 407
    maximal normal, 432
    maximal strong, 407, 415
    minimal, 395
    strong, 391, 402, 404, 416
relevant mingle, 462
replacement property, xx, 9, 293, 368, 377
    for $\mathbf{R}$, 454
    for $\to$, 375
residuation axioms, 384
Restall, G., 406, 476
rexpansion, 183
    preserving, 183
    strongly preserving, 183
Rivieccio, U., 172
Robinson, J. A., 46
Rosser, J., 82
Routley, R., 393, 458
rule, 7
    additive, 36
    admissible, 7, 33
    application of-, 7, 32
    canonical, 57
    conclusion of-, 32
    extension by-, 7
    Gentzen type, 32
    instance of-, 32
    invertible, 33, 43
    logical, 36
    multiplicative, 36
    of proof, 294
    premise of-, 32
    quasi-canonical, 58

structural, 36

S-d-frames, 340
s-entailment, 369
satisfiability, 64, 177
satisfiability relation, 14
Schütte, K., 186
Scott consequence relation, see scr
Scott entailment, 370
Scott, D., 5, 370
scr, 6, 35
Scroggs' property, 319, 417, 489
    for $\mathbf{RMI}_{\to}$, 417
    for $\mathbf{RM}$, 489
self-extensionality, 9, 293
    of $\mathbf{IL}^+$, 20
    of $\mathbf{IL}$, 22
    of $\mathbf{NB}$, 305
semantics, xxi, 23
    bivaluation, 175, 290
    classical, 15
    denotational, 14, 40, 59
    logical, 14
    multi-valued, 63
    possible translation, 290
    Routley and Meyer, 458
semi-implication
    for a logic $\mathbf{L}$, 10, 17, 378
seminframe, 341
sequent, 31
    antecedent of-, 31
    hyper, 44
    multiple conclusion, 31
    semantics of-, 39
    single conclusion, 31
    succedent of-, 31
sequent calculus
    logical, 34
Sette, A. M., 87, 99, 200
Shoesmith, D. J., 5
Shramko, Y., 172, 360
Slaney, J., 512
Smiley, T. J., 5
Smullyan, R., 46
Sobociński, B., 102, 103, 419
soundness, 24, 40, 67, 179
    strong, 40

weak, 24, 67, 179
source valuation, 220
sparse, 191
splitting, 493
Straßer, C., xvii
strong relevance criterion, 368, 377
structural rules, 35
structurality, 5, 368, 370
subformula property, xxi, 37–39
submatrix, 64, 66
substitution, 4
substitution property, 9
substitution-closure, 65, 178
Sugihara chain, 478
    bounded, 504
Sugihara matrix, 478, 479, 504, 511
Sugihara, T., 478, 511
suitability, 334
symmetry condition, 420
Słupecki. J., 86

Tait, W., 186
Takano, M., 330
Takeuti, G., 46
Tarski, A., 4
Tarskian consequence relation, *see* tcr
tautology, 5, 54, 64, 177
tcr, 4, 5, 14, 35
    consistent, *see* non-trivial
    finitary, 5
    non-trivial, 5, 14
    structural, 5, 14
theory, 4
    ¬-consistent, 387
    $\vdash_L$-non trivial, 67
    classically (in)consistent, 19
    satisfiable, 64, 177
    trivial, 64, 177
transitivity, 370, 396
    (Cut), 4, 6, 368, 372
tree, 429, 430

Troelstra, A.S., 360
truth functionality, xix, 14, 65, 175, 178
truth values, xix, 14, 63
    designated ($\mathcal{D}$), 63
    similar, 170
    types of-, 170
Turquette, A. R., 82
Tweety dilemma, 110

uniformity, 67, 364
Urquhart, A., 82, 458

valuation, 64, 440
    k-adequate, 497
    k-completion, 497
    k-semi-, 497
    EIP, 220
    partial, 64, 177
    similar, 170
    source, 220
Van Bendegem, J., xv
van Dalen, D., 360
variable
    propositional, 3
variable sharing property, 364, 375, 415
Vasiliev, N., 60
Voronkov, A., 46
VSP, *see* variable sharing property, 376, 379, 380, 383

Wansing, H., xv, 48, 172, 330, 360
Waragai, T., 330
weakening, 36
    external, 493
    relevant, 408
Wójcicki, R., 82

Zaitsev, D., 172
Zepeda, C., 330
Zohar, Y., 186, 330